THE METHODOLOGY OF PLANT GENETIC MANIPULATION: CRITERIA FOR DECISION MAKING

Developments in Plant Breeding

VOLUME 3

The titles published in this series are listed at the end of this volume.

The Methodology of Plant Genetic Manipulation: Criteria for Decision Making

Proceedings of the Eucarpia Plant Genetic Manipulation Section Meeting held at Cork, Ireland from September 11 to September 14, 1994

Edited by

ALAN C. CASSELLS and PETER W. JONES
Department of Plant Science, University College, Cork, Ireland

Reprinted from *Euphytica* Volume 85: 1–3, 1995

Springer-Science+Business Media, B.V.

Library of Congress Cataloging-in-Publication Data

The Methodology of plant genetic manipulation criteria for decision making / edited by Alan C. Cassells and Peter W. Jones.
p. cm. -- (Developments in plant breeding , v. 3)
Includes index.
ISBN 978-0-7923-3687-7 ISBN 978-94-011-0357-2 (eBook)
DOI 10.1007/978-94-011-0357-2
1. Plant genetic engineering. 2. Crops--Genetic engineering. I. Cassells, A. C. II. Jones, Peter W. III. Series.
SB123.57.M47 1995
631.5'23--dc20 95-34327

ISBN 978-0-7923-3687-7

Printed on acid-free paper

Contents

Special Lecture

Comparison of genetic transformation techniques

Target traits for genetic manipulation

Methods for the production of wide crosses

Exploitation of somaclonal variation and mutagenesis

Selection and evaluation of genetically manipulated plants

Rationalism of plant improvement techniques

Overview

Euphytica **85:** IX, 1995.

Guest Editorial

Advances in molecular biology and in plant cell and tissue culture are providing the plant breeder with novel techniques and systems for plant genetic manipulation and for the selection of improved genotypes. In recent years, advances in plant physiology have enabled us to clone plants *in vitro via* axillary buds and nodes, to regenerate plants adventitiously from explants and from cell and protoplasts *via* callus, and to fuse protoplasts. Genetic variability ('somaclonal variation') has also been detected, particularly in adventitious regenerants from the cells of polysomatic tissues. These advances are seen by tissue culturists as having a role in plant improvement.

In parallel, molecular biologists have developed techniques to identify, isolate and clone genes, and to transform plants using tissue culture systems. Genetic engineering in particular, has caught the public imagination and, more importantly, has attracted high levels of funding for model studies and specific tasks.

Plant breeders have kept a watching brief on these developments and have progressively integrated the new methods, as they were proven, into their programmes. Breeders recognised early the benefits of *in vitro* mutagenesis systems for vegetatively-propagated plants; they have also incorporated techniques such as anther culture, embryo rescue and micropropagation into their breeding programmes.

Currently we have plant breeders and two other groups (tissue culturists and genetic engineers) who are developing methods for plant improvement. Modern plant breeders in general, have formal training in plant genetics, plant physiology, agronomy, plant pathology, biometry and cognate areas. The training inputs for the tissue culturist are more variable but usually include plant physiology, biochemistry and possibly plant pathology. The training of molecular biologists is weighted towards plant cell biochemistry and molecular biology. It is important that these groups collaborate if the methods of plant genetic manipulation are to be integrated efficiently into plant breeding. The primary purpose of meetings of this section of EUCARPIA (European Association for Research on Plant Breeding) is to provide a forum where current developments in plant genetic manipulation are presented. This publication incorporates the proceedings of the meeting of 1994 of the EUCARPIA Section: Genetic Manipulation in Plant Breeding, held in University College, Cork, Ireland.

The editors would like to thank Ann Egan and Majella O'Sullivan who provided secretarial assistance, and Judy Cassells who provided editorial help.

Alan C. Cassells
Peter W. Jones
Guest-Editors

Euphytica **85:** 1–12, 1995.

Genetic manipulation in plant breeding – prospects and limitations

C.N. Law
John Innes Centre, Norwich, NR4 7UH, U.K.

Key words: genetic engineering, gene targets, mapping, markers, transformation, QTLs

Introduction

At the beginning of a meeting on genetic manipulation in plant breeding – it would seem appropriate that we should reflect upon the problems of feeding the predicted population expansion of the next century.

For many European countries, the immediate problems are not those of increasing population but of over-production and the need to reduce costs by lowering inputs. On the world's stage, the long-term requirements for agricultural production are not clear. Although all authorities agree that the world's population will continue to expand, there appears to be some disagreement about the capacity of agriculture to feed the expansion. The FAO believes that food production will continue to increase for the next 20 to 30 years at the same rate as the past 30 years and that this will be sufficient to meet demand. The Worldwatch Institute has, however, just reported that this rate is not sustainable. Moreover, they believe that science and technology can no longer ensure that the onward march of higher and higher yields can continue.

We are also subjected to an almost daily litany of doom and despair: global warming, depletion of the ozone layer, oil running out, water running out and so on. The next century is certainly packed with trouble – if the meteor doesn't get us first, then we will gradually either be roasted, frozen or desiccated, and to make doubly sure, we shall certainly be starved as well!

It seems that if we are to stand any chance whatsoever in the next century, our capacity to adapt the living resources that we depend upon will be extremely important for survival. Because of changes in the environment and depleting energy resources, this capacity to adapt will be every bit as important to those areas of the world which are blessed with over-production like Europe today. Accelerated adaptation through genetic change is, of course, the means by which plant breeders have achieved their aims in the past. Today the new techniques of genetic engineering may offer to the plant breeder the chance of speeding up adaptation to an extent hitherto considered impossible. This new power will undoubtedly bring risks but it could also be the only chance we have of escaping the dangers ahead.

Genetic manipulation

All progress in plant breeding depends upon genetic manipulation; the subject is vast.

To make it more manageable, I shall confine myself to discussing the prospective role in plant breeding of the new cellular and molecular techniques which allow the detection of DNA variation on the one hand, and the isolation and transfer of 'naked' plant, bacterial and viral genes into plants on the other. This, of course, leaves out some important areas, one of which is the public acceptance or rejection of the use of genetic engineering in plant breeding – a possible Achilles heel for the future exploitation of these new technologies.

1. Genetic markers

The use of DNA probes, labelled radio-actively or by fluorescent dyes, is a powerful means of establishing the position of base sequence differences or restriction fragment length polymorphisms (RFLPs) along a chromosome. The use of the polymerase chain reaction to amplify DNA sequences to such an extent that they can be made visible by gel electrophoresis, has provided the basis for a large number of methods for recognising DNA polymorphisms between individuals, e.g., RAPD, SCAR, SPLAT, AFLP – to mention just a few! Such differences can be used as 'markers' and have led to the rapid development of saturated genetic maps for many important crop plants. In 1987, for example, 116 genes had been assigned to positions on the 42 chromosomes of bread wheat (Worland et al., 1987); today this has risen to over 1100 (Gale pers. communication). Similar dramatic extensions of genetic maps have occurred in maize, tomato, cotton, rice, tobacco, millet, barley, oil seed rape and sunflower, as well as in other crops.

Aside from the increased understanding that these maps provide about the physical organisation of genes along chromosomes and the location of recombinational hotspots, the availability of many markers offers great possibilities for exploitation in plant breeding.

(i) The use of markers to tag genes of agronomic importance

Examples of this are becoming more and more frequent. One of the best, although not exploiting a molecular marker, is the tagging of the gene for eyespot resistance in bread wheat (McMillan et al., 1986; Worland et al., 1988). Resistance to this disease is quantitative and of low heritability. The established method of screening is laborious and involves the inoculation of extensively replicated seedlings grown under controlled conditions for several months before the degree of penetration of the fungus through each leaf sheath is scored. The extent of penetration is the criterion used to select for resistance. In 1983, a gene *Pch1* for eyespot resistance introduced previously from *Aegilops ventricosa* was identified using cytogenetic methods and employing established screening procedures. About the same time a unique allele of the endopeptidase gene was found to be very tightly linked to *Pch1*. This allele could therefore be used as diagnostic for the eyespot gene. The need for the laborious and costly screening procedure was therefore removed and it was possible to select for the presence of the gene with 100% certainty in half a day and, moreover, to do the screening non-destructively on a portion of the seed itself.

Although the list of gene tagging examples is still short – Weeden (1991) for example gives only twelve – more and more gene tags will appear in the future. Other than the need to identify the useful gene in the first place, this approach is limited by the need to find a marker closely linked to the useful gene or pairs of markers straddling it. For some crops e.g., wheat, where using currently available technologies, polymorphisms appear not to be frequent, this is an important limitation and other approaches such as actually inserting appropriate markers may have to be contemplated.

Gene tagging is finding its greatest use in breeding for disease resistance. This partly reflects its importance as well as the fact that the genetics of this character is better understood than for many others. As I shall mention later, once genes affecting other important characters have been identified, then gene tagging will begin to have a higher profile for the breeding of these characters also.

Being able to tag disease resistance genes may have great value in accumulating two or more genes for resistance to a particular pathogen in one variety. At present this so-called pyramiding of genes for resistance may be difficult because it is impossible to distinguish a plant carrying two genes for resistance from a plant carrying just one or other of the genes. The pathogen races necessary to separate the three host genotypes may, for instance, be unavailable. With appropriate gene tags this difficulty can be overcome.

Similar benefits from gene tagging would arise whenever there is a need to distinguish between single, duplicate or triplicate gene doses such as occur frequently in polyploids. In bread wheat, for example, it is not easy to distinguish phenotypically between a genotype *Ppd1Ppd1 Ppd2Ppd2* for day-length insensitivity from *Ppd1Ppd1 ppd2ppd2* or *ppd1ppd1 Ppd2Ppd2*, even though there is almost certainly a difference in the degree of insensitivity.

(ii) Marker assisted selection

Given a large number of marker differences between parental lines, it is attractive to think that these can be used to improve the efficiency of selection particularly for characters of low heritability. This must be the case if the markers are already known to tag genes of agronomic importance. At the moment this type of detailed information is not available. If a proportion of the markers is closely linked with the genes of agronomic importance, however, then screening within the early generations of a breeding programme should establish those markers which are associated with improving the characters in the desired direction. An alternative way of establishing this is to screen just the tails of the distribution of individuals or families for differences in the frequency of the markers. Selection for the critical markers revealed by this preliminary screening should improve the response to selection compared with phenotypic selection for the characters only.

The efficiency of this type of selection depends upon the number of markers and their closeness to the desired genes (Lande & Thompson, 1990; Zhang & Smith, 1993). When the numbers are high and the linkages close, then a high proportion of the variation under selection can be accounted for by the markers. Selection can then be very effective. In practice this is never the case; the number of markers is usually few and the proportion of the variation accountable is correspondingly low. Moreover, the cost of screening individual plants for many markers is likely to be prohibitive.

Markers are proving to be of value in the breeding of hybrids. In maize, RFLPs have been found to distinguish fairly readily between different groups of inbred lines. Furthermore, these measurements of dissimilarity could be used as good predictors of the performance of the inbreds in producing high yielding hybrids; the more dissimilar, the higher the hybrid yield (Walton & Helentjaris, 1987).

(iii) Markers and the location of QTLs

The use of markers is likely to have its greatest impact in unravelling the genetical basis of quantitative characters. Over the last few years many developments have occurred in the methodology of identifying and locating within the genome, so-called quantitative trait loci or QTLs by the presence of one or more linked markers. All of the statistically based methods claim to accurately locate and map a single QTL given markers that straddle the locus (Lander & Botstein, 1989; Martinez & Curnow, 1992; Luo & Kearsey, 1991). Often this location can be identified using a rather limited number of F2 or backcross plants.

The trouble arises when there is more than one QTL situated on a region of a chromosome. When this occurs, it is impossible, with any of the current methodologies, to be absolutely certain that there is one, two or more QTLs involved. The location of a single QTL to a region of a chromosome may therefore be correct or then again it may be just a 'ghost effect' brought about by two or more linked QTLs – there is no unequivocal way of distinguishing between these possibilities.

The idea of locating polygenes (an earlier definition of a QTL!) was first mooted by Thoday (1961). He used markers in *Drosophila* to identify the chromosome responsible for a particular effect on a quantitative character such as sternopleural chaetae. He was then at pains to minimise the potentially obscuring genetic variation contributed by background chromosomes as well as environmental variation. The next step was to use a range of markers on the critical chromosome to define the region responsible for the effect on the character. Finally, by progeny testing, he assayed the recombinants within the critical region to see whether or not they could be classified into two or more groups. This segregational test was the unique feature of the Thoday analyses, and enabled him to deal with his located polygene as if it were a classical Mendelian gene with tests for linkage, calculation of recombination frequencies and so on (Thoday, 1979).

A very similar approach was adopted by myself and colleagues in wheat, using inter-varietal chromosome substitution lines as a starting point (Law et al., 1981). These are lines where each chromosome of a wheat variety is replaced in turn by its homologue from another variety. Since wheat has twenty-one pairs of chromosomes, then twenty-one different substitution lines are possible, each differing from the donor variety by just one chromosome, all the other chromosomes being identical. This is ideal plant material for locating QTLs because by crossing a substitution line to the donor variety, you are dealing with just the gene differences on one chromosome – all the others being constant. Moreover, by further crossing such a single chromosome heterozygote to the donor monosomic for the same chromosome, homozygous recombinant lines can be obtained. This enables extensive replication, thereby reducing the effects of environmental variation.

This approach has now been used on several occasions and in many of the cases the phenotypes of the derived, single chromsome recombinant lines have fallen into two distinct classes, indicating the location of single QTLs (Law, 1966, 1967; Law et al., 1976; Worland & Law, 1986).

Both these approaches are time-consuming and involve a great deal of effort in producing suitable experimental material. Current approaches to locating QTLs have taken on board the need to reduce environmental variation by developing what are called recombinant inbreds i.e., random inbred lines or doubled haploid lines. Those engaged in this research, however, are seemingly much less attracted to trying to control or minimise background variation, an essential step for segregational genetics to succeed. In 1965, Wherhahn and Allard used a technique which addressed the problem of background variation and which is worth considering in the light of today's marker-driven analyses. They produced many lines by first backcrossing for two or three generations to a recipient wheat variety and then selfing each of the lines for several generations. The majority of the resulting lines for any character was similar, if not identical, to the recipient but some lines differed, the number of lines differing being related to the number of QTLs. It would seem to me that this approach is worth combining with markers. Beckmann and Soller (1989) have indeed come up with a similar recommendation and have further developed some of the theory, but as far as I am aware, no-one has taken up this advice.

An alternative is, of course, to develop chromosome substitution lines just as in wheat. This was possible in wheat because of the availability of aneuploid stocks which effectively provided cytological markers for maintaining chromosomes intact during transfer by backcrossing. In most crop plants such stocks are not available – however, molecular markers are. It should be possible therefore to use markers to select for a non-recombinant chromosome and to substitute this whilst re-constituting the recipient background by backcrossing. The major difficulty here would be the low frequency of non-recombinants which would make selection rather difficult. A way around this is to substitute a region of a chromosome such as an arm, so that frequencies would be manageable. It is interesting to note that at least one group of researchers in oil-seed rape is using this approach (see Lydiate, this volume).

(iv) The need to locate QTLs

Is it really necessary to locate a QTL precisely? Even if it is possible to use a segregational test, the limitation on numbers that can be screened inevitably means that the QTL is at least 1–2 cM in length or between 100,000 to 500,000 base pairs long and therefore of sufficient size to cover 100–200 coding genes. Given that this must be so, then why bother? For plant breeding purposes all that is required is the knowledge that when segregating, this marker or pair of markers can account for most of the effect determined by a region of a chromosome.

There are three reasons for wanting to locate a QTL with some degree of precision. The first of these is the need to be sure that the located QTL is a realistic entity and is not just a statistical artefact owing its origins to two or more linked (even loosely linked) genes. The assigned QTL needs to have some stability if its role within the breeding context is to be understood. The second is the hope that such QTLs may be useful in leading to the cloning of the genes directly concerned with plant performance. Undoubtedly, some QTLs will be complex genetically, several of the component genes contributing to the effects associated with the QTL. The more frequent occurrence, though, is surely going to be where the effect is due to just one of the genes in the region defined by the QTL. The task is to target that one gene. This in itself may be exceedingly difficult and may have to depend on the identification and subsequent isolation of related genes in other plant species such as *Arabidopsis* and rice, where the techniques of genetic analysis and gene isolation may be easier. Other procedures for gene isolation may be used, but whatever the approach it will be necessary to know that there is a QTL and where it is located in the crop plant. The third reason stems directly from this requirement – the need to have confidence in a QTL location so that it can be used in comparative genetics with other crop species.

The analysis of quantitative characters through markers is therefore a very attractive area of research activity which should contribute to improving the efficiency of plant breeding. The most productive scenario for genetic analysis and the identification of QTLs will be to screen (using markers) for chromosomal effects in the first place, and then to follow this up with more detailed analyses using further markers and reducing both background and environmental variation.

(v) Limitations on the use of markers

There is, however, one important constraint on the use of markers in plant breeding and indeed in research for that matter. This is the cost and time required to undertake such analyses. Improvements in screening techniques which will reduce costs can be expected in the future but it is likely, even given such reductions, that the large-scale application of marker selection to plant breeding will be restricted to large breeding organisations. Although some gene tagging can be expected to occur in most breeding programmes, greater emphasis is likely to be given to characterising potential parents rather than undertaking the more costly marker-assisted selection. This demand may encourage the diagnostics industry to provide a service in gene tagging for a limited number of important genes.

A potentially more serious limitation is the lack of detectable polymorphisms in certain crops at the present time. Although saturated maps may exist for all crops, many of these have been put together by using very wide crosses as the base mapping material. When breeding populations are considered, the extent of polymorphism is much reduced. This is a problem for inbreeding crops rather than outbreeders. The one crop where this appears to be a severe limitation at the moment is wheat. Further research is required to try and overcome this weakness.

(vi) Comparative gene mapping

As mentioned above, a very useful spin-off from gene mapping using molecular probes is the discovery that many probes are effective in mapping related genes and sequences across groups of distantly-related crop species. It is thus possible to look at the cereals as a group and relate the maps of wheat with those of barley, rice, millet, sorghum, rye and maize. Another group is tomato, potato and pepper. Yet another includes *Brassica* spp. which also shows some relationship with *Arabidopsis* so that it may be possible to relate the genes from this 'plant model' directly with those occurring in crop plants. A new era of comparative plant genetics has therefore dawned which could have important consequences for plant genetics as a whole as well as for plant breeding.

For some years it has been possible to identify homoeologous relationships between the chromosomes of wheat, rye and barley through the cytogenetic transfer and substitution of chromosomes. Comparative mapping allows this to be done in great detail and already homologies between genes across these species as well as with maize and rice are now being recognised (Moore et al., 1993). Such comparisons will inevitably prompt the search for related genes, recognised and mapped in one species but as yet, not known in another. An excellent example occurs in barley and wheat where the location of the genes for day-length response and vernalization requirement match very closely. One exception is the vernalization gene located on chromosome 4H of barley. No such gene has been located on the group 4 chromosomes of wheat. This may mean that the genes on the group 4 chromosomes have been lost or inactivated but it is more likely that it is due to the failure of wheat geneticists to find them.

It should therefore be possible to build up lists of genes that have been conserved across a range of different crops. Such lists will be of considerable interest in plant breeding. Genes which have been conserved over long periods of evolutionary divergence must obviously be of importance to all crops. Perhaps of greater interest, though, will be those genes that have not been conserved. These may have potential today in the crops lacking them since their loss may have occurred long before the ancestral species had been exposed to the needs of agriculture.

As already mentioned, the opportunities offered by comparative mapping emphasises strongly the need for precise identification and location of QTLs. The identification of QTLs is obviously based upon a detailed analysis of an agronomic character which is far removed from the gene product. The ability to compare related QTLs across species will almost certainly add new attributes to the gene and may even lead to a re-definition of the QTL's role in terms of its influence on biochemical and physiological characters. This would obviously increase the understanding of the gene's function, helping its exploitation in breeding programmes, and improving the chances of isolating the gene through reverse genetics.

(vii) The use of markers in alien gene transfer

Plant breeders have had a long love affair with wide crossing and the introduction of genes from related species. In very many cases the affair has led to a successful marriage and useful genes have been introduced into new commercial varieties. This has even occurred when it can been shown that the introduced chromosome segment containing the gene has deleterious side effects, e.g., the majority of 'alien' genes for resistance in wheat, have been associated with a yield penalty (The et al., 1988). Plant breeders have nevertheless accepted such transfers into their programmes, preferring to compensate for the weaknesses by selecting appropriate backgrounds whilst accepting the benefits. To-day RFLPs, RAPDs and even *in situ* hybridisation techniques are available to make the process of transfer more precise so that just small segments containing the desired gene can be selected. The presence of adverse linked genes should therefore be less of a problem.

Of greater value is the very high probability that the introduced segments, although much reduced in size by earlier standards, will still carry polymorphic markers which will prove to be diagnostic for the desired gene that has been transferred. 'Alien' gene transfers will thus have in-built tags already available for exploitation. In many crops such as wheat, where gene tagging through molecular markers is proving to be difficult, 'alien' gene transfers are likely to be very attractive.

In this volume, there is a section devoted to new methods of wide crossing. These offer the opportunity to extend the range of inter-specific crosses and

Table 1. Annual approvals world-wide of genetically-engineered plants classified by character and year (OECD Report, 1993)

Characters	Numbers granted each year							Total
	1986	1987	1988	1989	1990	1991	1992	
Herbicide tolerance		6	13	36	74	97	263	489
Disease resistance		2	1	3	13	16	35	
Virus resistance		1	4	6	24	34	46	115
Insect resistance			6	7	15	26	35	89
Use of markers	1		9	16	44	91	221	382
Quality traits			2	4	11	21	34	72
Flower colour					1	3	1	5
Research studies					2	7	9	18
Male sterility				1	7	8	23	39
Stress resistance				1	5		3	9
Metal tolerance				1	2		3	
Other							1	1
Total approvals	1	7	36	73	188	300	652	1257

include ways of limiting the extent of gene transfer from one of the parents through asymmetric hybridisation. Past experience suggests that, even given limited gene transfer, the resultant derivatives will still require a great deal of screening and assessment before a worthwhile product having breeding potential will emerge. In this complex undertaking, markers will surely have an important role to play in selecting the most desirable transfers.

Indeed, there are very few activities being contemplated by both researchers and practical plant breeders that are unlikely to benefit from the use of markers in one way or another. Even somaclonal variants and the results of induced mutagenesis need analysis to determine what part of the genome has changed, and whenever analysis is mentioned, then the use of markers must be a worthwhile option.

2. Direct gene transfers

We now turn to genetic engineering proper. This is the direct introduction into a plant of an isolated or modified single gene using transformation techniques. I do not wish to discuss the variety of techniques which are currently being used because this will be dealt with later. It is sufficient to say that transformation is now possible for almost all of the plants cultivated by man. Where this has not happened, it is more to do with the insignificance of the plant in agriculture or forestry rather than a reflection of a stubborn resistance to all attempts to transform it. A year or two ago, cereals were regarded as being recalcitrant species. This has now changed and transformation has been reported as being successful for nearly all cereal species. Transformation is no longer a problem. This is a marvellous achievement.

The process of plant breeding has therefore been transformed overnight, although there is much still to be done. The list of successful transfers, however, is getting longer by the day, and examples of commercially released plants are just beginning to appear. This increasing activity is illustrated in Table 1 where the number and types of field release applications throughout the world are listed. There are also more different crops involved in these applications, the most frequently represented being oil seed rape (see Table 2). Almost certainly, now that transformation is available more widely, including the more prevalently-grown cereals, the numbers of applications will increase. It is also of interest to note that most of the applications refer to resistance to viruses, insects and herbicides. This is perhaps not unexpected, knowing the striking successes that have been achieved through the introduction of virus coat protein genes to give viral resistance (Powell et al., 1986), the exploitation of insect toxins by transferring the *Bt* gene from *Bacillus thuringiensis* (Barton et al., 1987) or the trypsin inhibitor gene from cowpea (Hilder et al., 1987) to give insect resistance and the insertion of genes for resistance to different herbicides

Table 2. The number of approvals granted per crop for the years 1986 to 1992 (OECD Report, 1993)

Crop	Approvals
Lucerne	21
Maize	65
Cotton	37
Flax	49
Oil seed rape	290
Potato	133
Soybean	40
Sugarbeet	28
Tobacco	72
Tomato	72

(De Block et al., 1987; Jordan & Hughen, 1988). Other gene transfer successes are being added to this list, an important example being the use of anti-sense genes to suppress the activity of the ripening gene, *PG*, coding for the enzyme polygalacturonase in tomato (Smith et al., 1990). More extreme transfers have also been reported and include the gene for an anti-freeze protein from the fish, winter flounder (*Pseudopleuronectes americanus*), in tomato and tobacco (Hightower et al., 1991), and the gene for human serum albumin in potato (Sijmons et al., 1990). It is indeed an impressive record of achievement and it is, of course, only the start. We can expect to see an ever-increasing list of transfers in the future.

Over and above this deserved acclamation, there are still a few loose ends that I, as a geneticist interested in plant breeding, should like to see cleared up. The first of these concerns the mechanics of the transfer process. The means by which genes are introduced using *Agrobacterium* exploit an already perfected and adapted process of gene transfer but this cannot be said for the particle gun. How is a piece of naked DNA, shot violently into a cell, integrated into the chromosome? We appear to know nothing about the factors influencing the process. Transformation commonly results in the transfer of many transgenes which is a useful property if one wishes to study the effect of gene dosage. Amongst independently-derived, transformed plants, however, the expression of the introduced gene can vary and the transmission and behaviour of the gene in subsequent crosses can be non-Mendelian and in extreme cases, the gene may not be transmitted at all. Pleiotropic effects of the transgene have rarely ever been reported, and in many cases the impression is that such effects have not even been considered. One of the major difficulties in trying to find answers to these concerns is that most transgenes have been manipulated within the Company sector, and they are loth to release such sensitive and possibly adverse information likely to affect the prospects for commercialisation of any of their material.

The recommended way of coping with the difficulties of transmission and expression is to produce several (> 100) transformants each carrying a single copy of the transgene and then select for stability and desired expression amongst them (Dale et al., 1993). Although this follows the well-tried custom of 'suck it and see' practised successfully by all plant breeders, it is nevertheless (without knowing something of the basis of the differences in stability and expression) not a very satisfactory position to be in. This is particularly the case where the transgene is a vital ingredient of the new varieties' commercial attractiveness – a position which is likely to be the case for most genetically-engineered plants. Plant breeders have long been aware of the difficulties in achieving uniformity and the loss of just one transgene in 500 plants could cause problems in meeting the requirements for varietal purity in some crop plants. There is therefore a need to research these aspects of transgenic stability.

3. Targets for genetic engineering

Which characters and genes to target has always been a fundamental question for all plant breeders. It is even more relevant to genetic engineering. Table 3 shows a list of target characters which most plant breeders, genetic engineers and plant biotechnologists would nominate, if asked to compose a shopping list.

Those characters where genetic engineering is making an impact are highlighted in the list as are those where an impact might be made. The majority of these are to be found in the categories of altering inputs, hybrid production and products. This is because for these characters, either the genes are known and have been or can be isolated, or the enzymes affecting them are well-characterised and the prospects for gene isolation are considered to be good.

For some of the characters, like photosynthetic efficiency, for example, a great deal of background research has been going on. Many of the genes involved have been described and some isolated, but there is little indication that such a complex character can

Table 3. Target characters: those being genetically-engineered or those which are likely to be so in the near future, are given in italics

Category	Target characters
Hybrid production	Self-incompatibility
	Male-sterility
Plant growth & development	Structure (height, branching, leaves, roots)
	Flowers (structure, colour, timing)
Altering inputs	*Herbicide resistance*
	Pest resistance
	Bacterial, *viral* and fungal resistance
	Nitrogen fixation in non-legumes
	Improved nutrient uptake
	Improved photosynthetic efficiency
Products	*Sugar and starch (different composition* and/or higher content)
	Oils (different composition and/or higher content)
	Storage proteins (different composition and/or higher content)
	Flavours and fragrances (in food or as extracts)
	pharmaceuticals
	Fibres (textiles)
	Fruit (ripening and quality)
Environment	Drought, salinity and heat tolerance
	Resistance to water logging
	Cold adaptation
	Metal uptake and accumulation

be improved by genetic engineering. This is a pity because, arguably, of all the characters mentioned, a more efficient photosynthetic process would have the most profound effect on global biological productivity. However, this is surely a case where the genetic engineer is competing with millions of years of perfecting photosynthetic efficiency by natural selection, so workers can be excused if they feel that the dice is loaded very much against them!

Some of the most attractive characters for the genetic engineer occur in the products category. Several glutenin and gliadin genes in wheat, hordein genes in barley and zein genes in maize have been cloned, and at least one wheat glutenin gene has been transferred into tobacco. What is now required is the transfer of these genes into more interesting and novel backgrounds such as sorghum for instance, or to increase the gene dosage to see whether it is possible to increase the amounts of a particular protein. There is also much activity altering the composition of oil, mostly in oilseed rape. The genes for different enzymes controlling additional steps in the biosynthetic pathway of rape oil have been introduced and have been shown to produce a new oil (lauric acid) of value in producing detergents (Voelker et al., 1992). Other genes introducing new enzymes from coriander and castor bean and affecting the pathway in different ways are being explored with the hope that oils valuable in making plasticisers, cosmetics and pharmaceuticals will be obtained (Murphy, this volume).

The other categories not designated as likely areas for genetic engineering at the moment, have one thing in common and that is that little is known about their detailed genetics, or about the biochemical and physiological processes underpinning their expression. Many of them would be regarded as the traditional characters of plant breeding: morphological features such as height, leaf width and length, grain numbers and weight, and yield itself. They would also include most of the characters in the environment category. In most cases they are likely to involve multi-step pathways controlled by many genes. In some cases they will be genes for transcription factors whose function will be to regulate whole batteries of genes.

These are the types of character where the marker analyses referred to earlier could be employed most usefully. The location and full evaluation of QTLs affecting these characters would be the first step down the road to their exploitation, followed by their subsequent directed manipulation, either through gene tagging or, having isolated the genes responsible, by gene transfer. To pursue a genetic engineering approach with such characters is a formidable undertaking because apart from the difficulties of identifying and then isolating the gene, it is very likely that the gene involved is only one of many affecting the character concerned. Before embarking on this course it is essential to know the importance of the gene within the overall breeding programme, hence the need for evaluation, not just of the gene itself but of other genes influencing the character. Almost certainly these are the genes which are directly involved in plant productivity and are therefore the genes of greatest importance to plant breeding in the long term when yield requirements from agriculture are likely to be high on the list.

4. An example: the genetic engineering of flowering time

As an illustration of the difficulties of genetically engineering genes within one of these categories, I shall describe some current work on the genetics of flowering and the prospects of engineering such genes in plant breeding.

The control of flowering time has always been an important breeding objective. Because the character is usually of high heritability, it is not regarded as being particularly difficult to manipulate by selection. This does, of course, depend upon sufficient variation being available within the breeding population to make progress. When this is not the case, problems have arisen in the past adapting a crop to a new environment. The same problems are likely to arise in the future.

Such a problem occurred in the course of the spread of wheat around the globe during the last century. To establish wheat in parts of South America, Southern Africa and Australia, it was essential to grow it during the winter and to commence flowering as soon as day-lengths began to lengthen in the early Spring. In this way it was possible to avoid the yield-reducing, high temperatures and desiccating conditions normally present in these countries in the late Spring and early Summer. This was achieved by the introduction through wide-varietal crossing of genes for photoperiodic insensitivity, *Ppd1*, *Ppd2* and *Ppd3*. Indeed, without the introgression of these genes, it is doubtful whether the Australian wheat industry in particular would ever have existed.

Even today, the fine tuning of allelic variation at one or other of these genes can have an important influence on the yield of UK wheats, and almost certainly the best combination of these genes has yet to be established in many wheat-breeding programmes throughout the world. To be able to manipulate these genes quickly and in a desired way by means of genetic engineering would obviously be attractive to a breeder. Before we can do this, however, we need to be able to understand why it is that certain genes are more appropriate than others in achieving higher yield stabilities.

Perhaps an even more striking example of a need for a new gene for flowering control comes from sugar-beet. Beet is usually grown from a spring sowing so that its inherent vernalization requirement is never satisfied and the majority of the plants never run up to flower, ensuring that most of their growth goes into their roots. If beet were to be grown from an autumn sowing, then the potential for growth in the following season would be that much greater. With such a growing strategy, the vernalization requirement of the plant would be satisfied and the beet would flower or 'bolt' in the summer rather than putting growth into increasing the root yield. If it were possible to prevent this, either by enhancing the vernalization requirement or inhibiting the initiation of flowering in another way, then the yields of beet could be increased at one step by about 15%.

Unfortunately, genetic variation allowing this to happen does not appear to be available in sugar beet or any of its relatives. The transfer of genes from other plant species or the introduction of anti-sense constructs of genes known to be involved in vernalization might therefore offer a solution and provide a means of achieving this objective. Professor Tudor Thomas and his colleagues at Broom's Barn Experimental Station in the UK are using this strategy and are encouraged in this endeavour by current successes in isolating genes for flowering control in *Arabidopsis*.

There is, however, a long way to go yet. The understanding of the genetic control of flowering in our crop plants is still rather rudimentary (Law et al., 1993). In the wheats, the control seems to be based on a series of genes sensitive either to vernalization or day-length, acting as inhibitors or suppressors of a probable multi-step progression to flower initiation. In *Arabidopsis* the number of genes affecting flowering stands at 16, and there are almost certainly many more. At least one and possibly two of these genes have been cloned and I suspect, if it has not happened already, it will not be long before these will have been used in transformation experiments. An important next stage so far as plant breeding is concerned, will be to match these genes to related genes in crop plants with the hope that this will lead to their subsequent isolation and characterisation. As the understanding of the flowering process advances in *Arabidopsis*, then the mechanisms revealed will need to be related to those occurring in crop plants. In this way it may be possible to intervene in a directed and predictable way in the flowering process.

There is thus a great deal of uncertainty about the consequences of using transgenes for flowering time in crop plants, mainly because we still know very little about the genetics and absolutely nothing about the actual mechanisms involved. If this is the case for flowering time, then the position is much less promising for most of the other characters listed under plant development and growth in Table 3, because flowering time is probably better understood genetically than most of these other characters.

To achieve successful transformation and to exploit this in the breeding of many characters, a great deal of prior genetic analysis needs to be undertaken. For the most part, this has yet to be done.

5. Gene isolation or cloning

Up to now, I have been emphasising the need to identify genes for subsequent genetic engineering. This is not, of course, the whole story. These genes need to be isolated before they can be used. Many of the transgenes referred to earlier were cloned because the proteins they encode and the tissues in which they were most active, were known. For many genes, given the state of understanding of their function, this luxury is unavailable. There is a range of techniques, though, which seek to get around this problem. One is to find a closely-linked RFLP marker, enabling the region to be cloned and the desired gene to be isolated by chromosome walking. Another is to use transposon mutagenesis. Yet another is the creation of small deletions encompassing the desired gene and then the cloning of the gene by differential hybridisation. In all these cases the task of isolating just a single gene in a crop plant is not easy. There is, therefore, much hope that the use of a model plant such as *Arabidopsis*, where genome size, small generation time and many other aspects are more favourable to the genetic engineer, will lead to the cloning of related genes to those identified in crop plants. An alternative model plant which may be of more relevance to monocots is rice, which has many of the advantageous attributes of *Arabidopsis*. Again the hope is that such cloned genes may be used either as probes to pull out the genes from the crop plant or as a means of recognising the processes they control in *Arabidopsis*, the proteins they produce and so on, and that these may be a useful guide to identifying similar gene products in crop plants and thereby leading to their isolation indirectly.

One cannot help feeling, though, that the shear size of effort to clone even just a few genes by this route, will be beyond the resources and indeed the patience of most plant breeders. I therefore am not surprised to find that when I have questioned plant breeders about their preferences among the new technologies, they nearly always favour the use of 'markers' rather than 'transgenes'. Are plant breeders wrong, however?

6. The plant breeder has been there before

Finally, I would like to draw this paper to a close by highlighting the rather irritating fact, especially for a research worker in plant breeding, that breeders, through empirical means or by just following a hunch, often seems to have got there without any of the trimmings of scientific endeavour to help them.

An example of this comes from my own work and that of my colleagues working with the semi-dwarf character in wheat. As everybody knows, this character was the basis of the so-called Green Revolution wheats of the 1960s which changed the agriculture of many developing countries (Borlaug, 1978). The genetical story underlying this success is rather complex, however. It transpires that there have really been two semi-dwarf Green Revolutions. The first of these was in the 1930s when the Italian breeder Strampelli introduced semi-dwarfism from the Japanese variety, Akakomugi, into Italian and Mediterranean wheats (Strampelli, 1932). The second was in the 1960s when Borlaug followed the same route, again using a Japanese source of dwarfism, but in this case from the variety Norin 10. The genetical basis of these two sources of dwarfism are different, however. The Strampelli semi-dwarfs depend on the genes *Rht8* and *Ppd1* (Law et al., 1981;

Worland et al., 1988), the Borlaug semi-dwarfs on the genes *Rht1* and *Rht2* (Gale & Youssefian, 1984).

Further work has revealed even further complexity. *Rht1* and *Rht2* are, in fact, just part of a multiple allelic series occurring at each locus. Some alleles are more potent in their effects on height than *Rht1* or *Rht2*, others less so. The degree of reduction of height is correlated with the level of insensitivity to exogenous GA but more importantly with the numbers of grain set within an ear; the shorter the plant the greater the numbers of grain. It is this property that is responsible for the marked increases in yield obtained from these semi-dwarfs rather than the often-cited reduction in height and increased lodging resistance.

Yet it is not just seed numbers which are important. If this were so, then the shortest plants would be the highest yielders. The plant has to have the capacity to fill out the grain. This means that very short plants have insufficient biomass to produce the assimilate required to meet the demands of an enlarged 'sink'. This is why the semi-dwarf phenotype is preferred.

The environment is also important; it appears that breeders are not happy with the use of *Rht1* and *Rht2* in Italy, Austria, Germany, Hungary, the former Yugoslavia and other Balkan countries, and prefer to use either the Strampelli semi-dwarf or a less potent allele of *Rht1* derived from the Japanese variety Saitama 27 (Worland, 1986). Investigations using near isogenic lines for these genes have shown that in these countries the yields are indeed lower for the isogenics with the *Rht1* or *Rht2* genes, and higher for *Rht8* plus *Ppd1* or when *Rht1S* from Saitama 27 is used. This is the reverse of the situation if these same isogenics are grown further north or to the west. The exact nature of these genotype/environment interactions is unknown but the reduced yields almost certainly involve higher temperatures and possibly reduced water availability at the time of ear development and later.

Two important points emerge from this example. The first is the complex nature of achieving high wheat yields in different environments. The second, and of most interest, is that the plant breeder had managed to achieve this 'fine tuning' without knowing anything at all about the details of the genetics. In my view this is a most salutary illustration of the difficulties that will be faced by the genetic engineer or indeed anyone trying to make a directed genetical change in breeding in the future. To be able to predict the consequences of a particular gene transfer across different environments and into different genetic backgrounds, is not going to be easy.

7. Conclusion

In Berlin in 1988, I was asked to give a similar talk to the then Eucarpia Genetic Manipulation meeting (Law, 1988). I chose to entitle my talk 'The need for a multi-disciplinary approach to genetic manipulation in plant breeding'. I see no need to change my views on this score. Directed genetic manipulation will have to take account of many aspects of plant growth and behaviour if it is to succeed in plant breeding, and the traditional ways of breeding for close adaptation will still be required.

I ended my talk then by urging the plant breeder to have the wider vision that genetic engineering can offer, whilst at the same time goading the molecular biologist by saying that

> 'There are more things in heaven and earth . . .
> Than are dreamt of in your philosophy'.

I think that these views are still in part true today. The molecular biologist in plant breeding has come a long way since 1988 but is still only at the stage of 'dipping a big toe in the ocean'. The plant breeder on the other hand, while being much more involved in the technology, is still inhibited in its use, partly because of the cost and partly because of the lack of knowledge about the genes themselves. It is up to researchers in many disciplines to overcome these deficiencies, giving breeders the chance to use their imagination to the full in order to achieve the breakthroughs needed in the future.

References

Barton, K.A., H.R. Whiteley & N.-S. Yang, 1987. *Bacillus thuringienesis*-endotoxin expressed in *Nicotiana tabacum* provides resistance to lepidopteran insects. Plant Physiology 85: 1103–1109.

Beckmann, J.S. & M. Soller, 1989. Backcross inbred lines for mapping and cloning of loci of interest. In: B. Burr, T. Helentjaris & S. Tanksley (Eds). Development and Application of Molecular Markers to Problems in Plant Genetics. pp. 117–122. Cold Spring Harbor, New York.

Borlaug, N.E., 1968. Wheat breeding and its impact on world food supply. Proc. 3rd Int. Wheat Genet. Symp. Australian Academy of Sciences, Canberra, pp. 1–36.

Dale, P.J., J.A. Irwin & J.A. Scheffler, 1993. The experimental and commercial release of transgenic crop plants. Plant Breeding 111: 1–22.

Gale, M.D. & S. Youssefian, 1984. Dwarfing genes in wheat. In: G.E. Russell (Ed). Progress in Plant Breeding 1. pp. 1–35. Butterworths, London.

Hightower, R., C. Baden, E. Penzes, P. Lund & P. Dunsmuir, 1991. Expression of antifreeze proteins in transgenic plants. Plant. Mol. Biol. 17: 1013–1021.

Hilder, V.A., A.M.R. Gatehouse, S.E. Sheerman, R.F. Barker & D. Boulter, 1987. A novel mechanism of insect resistance engineered into tobacco. Nature 330: 160–163.

Lande, R. & R. Thompson, 1990. Efficiency of marker-assisted selection in the improvement of quantitative traits. Genetics 124: 743–756.

Lander, E.S. & D. Botstein, 1989. Mapping Mendelian factors underlying quantitative traits using RFLP linkage maps. Genetics 121: 185–199.

Law, C.N., 1966. The location of genetic factors affecting a quantitative character in wheat. Genetics 53: 478–498.

Law, C.N., 1967. The location of genetic factors controlling a number of quantitative characters in wheat. Genetics 56: 445–461.

Law, C.N., 1988. The need for a multi-disciplinary approach to genetic manipulation in plant breeding. In: W. Horn, C.J. Jensen, W. Odenbach & O. Schieder (Eds). Genetic Manipulation in Plant Breeding. pp. 867–882. Walter de Gruyter, Berlin, New York.

Law, C.N., A.J. Worland & B. Giorgi, 1976. The genetic control of ear-emergence time by chromosomes 5A and 5D in wheat. Heredity 36: 49–58.

Law, C.N., J.W. Snape & A.J. Worland, 1981. Intraspecific chromosome manipulation. Phil. Trans. R. Soc., London B292: 509–518.

Law, C.N., C. Dean & G. Coupland, 1993. Genes controlling flowering and strategies for their isolation and characterization. In: B.R. Jordan (Ed). The Molecular Biology of Flowering. pp. 47–68. CAB International.

Luo, Z.W. & M.J. Kearsey, 1989. Maximum likelihood estimation of linkage between a marker gene and a quantitative locus. II. Application to backcross and doubled haploid populations. Heredity 66: 117–124.

Martinez, O. & R.N. Curnow, 1992. Estimating the locations and the sizes of the effects of quantitative trait loci using flanking markers. Theor. Appl. Genet. 80: 480–488.

McMillan, D.E., R.E. Allen & D.E. Roberts, 1986. Association of an isozyme locus and strawbreaker foot rot resistance derived from *Aegilops ventricosa* in wheat. Theor. Appl. Genet. 72: 743–747.

Moore, G., M.D. Gale, N. Kurata & R.B. Flavell, 1993. Molecular analysis of small grain cereal genomes: current status and prospects. Biotechnol. 11: 584–589.

Powell-Abell, P., R.S. Nelson, B. De, N. Hoffman, S.G. Rogers, R.T. Fraley & R.N. Beachy, 1986. Delay of disease development in transgenic plants that express the tobacco mosaic virus coat protein gene. Science 232: 738–743.

Sijmons, P.C., B.M.M. Dekker, B. Schrammeijer, T.C. Verwoerd, P.J.M. Van Den Elsen & A. Hoekema, 1990. Production of correctly processed human serum albumin in transgenic plants. Biotechnol. 8: 217–221.

Smith, C.J.S., W. Schuch & D. Grierson, 1990. Inheritance and effect on ripening of antisense polygalacturonase genes in transgenic tomatoes. Plant Mol. Biol. 14: 369–379.

Strampelli, N., 1932. Origini, soiluppi, lavori i rigultati. Roma.

The, T.T., B.D.H. Latter, R.A., McIntosh, F.W. Ellison, P.S. Brennan, J. Fisher, G.J. Hollamby, A.J. Rathjen & R.E. Wilson, 1988. Grain yields of near-isogenic lines with added genes for stem rust resistance. In: T.E. Miller & R.M.D. Koebner (Eds). Proc. 7th Intl. Wheat Gen. Symp. pp. 901–906.

Thoday, J.M., 1961. The location of polygenes. Nature 191: 368–370.

Thoday, J.M., 1979. Polygene mapping: uses and limitations. In: J.N. Thompson Jr & J.M. Thoday (Eds). Quantitative Genetic Variation. pp. 219–233. Academic Press.

Voelker, T.A., A.C. Worrell, L. Anderson, J. Bleibaum, C. Fan, D.J. Hawkins, S.E. Radke & H.M. Davies, 1992. Fatty acid biosynthesis redirected to medium chains in transgenic oilseed plants. Science 257: 72–73.

Walton, M. & T. Helentjaris, 1987. Application of restriction fragment polymorphism (RFLP) technology to maize breeding. In: D. Wilkinson (Ed). Proc. 42nd Annual Corn and Sorghum Research Conference. ASTA. pp. 48–75.

Weeden, N.F., 1991. Chromosomal organisation and gene mapping. In: D.R. Murray (Ed). Advanced Methods in Plant Breeding and Biotechnology. pp. 23–49. CAB International.

Wherhahn, C. & R.M. Allard, 1965. The detection and measurement of the effects of individual genes involved in the inheritance of a quantitative character in wheat. Genetics 51: 109–119.

Worland, A.J., 1986. Gibberellic acid insensitive dwarfing genes in southern European wheats. Euphytica 35: 857–866.

Worland, A.J., C.N. Law, T.W. Hollins, R.M.D. Koebner & A. Giura, 1988. Location of a gene for resistance to eyespot (*Pseudocercosporella herpotrichoides*) on chromosome 2D of bread wheat. Plant Breeding 101: 43–51.

Worland, A.J. & C.N. Law, 1986. Genetic analysis of chromosome 2D of wheat. 1. The location of genes affecting height, daylength insensitivity, hybrid dwarfism and yellow-rust resistance. Z. Pflanzenzüchtg. 96: 331–345.

Worland, A.J., M.D. Gale & C.N. Law, 1987. Wheat genetics. In: F.G.H. Lupton (Ed). Wheat Breeding – Its Scientific Base. pp. 129–171. Chapman & Hall.

Worland, A.J., C.N. Law & S. Petrovic, 1988. Pleiotropic effects of the chromosome 2D genes *Ppd1*, *Rht8* and *Yrl6*. In: T.E. Miller & R.M.D. Koebner (Eds). Proc. 7th Int. Wheat Symp. pp. 669–674.

Zhang, W. & C. Smith, 1993. Simulation of marker-assisted selection utilizing linkage disequilibrium: the effects of several additional factors. Theor. Appl. Genet. 86: 492–496.

Euphytica **85:** 13–27, 1995.

Strategies for variety-independent genetic transformation of important cereals, legumes and woody species utilizing particle bombardment

Paul Christou
John Innes Centre, Norwich Research Park, Norwich NR4 7UH, U.K.

Key words: gene transfer, crop species, particle bombardment, transgenic plants, cereals, legumes, woody plants

Summary

The limiting component in the creation of transgenic crops has been the lack of effective means to introduce foreign genes into elite germplasm. However, the development of novel direct DNA transfer methodology, by-passing limitations imposed by *Agrobacterium*-host specificity and cell culture constraints, has allowed the engineering of almost all major crops, including formerly recalcitrant cereals, legumes and woody species. The creation of transgenic rice, wheat, maize, barley, oat, soybean, phaseolus, peanut, poplar, spruce, cotton and others, in an efficient and in some cases, variety-independent fashion, is a significant step towards the routine application of recombinant DNA methodology to the improvement of most important agronomic crops. In this review we will focus on key elements and advantages of particle bombardment technology in order to evaluate its impact on the accelerated commercialization of products based on agricultural biotechnology and its utility in studying basic plant developmental processes and function through transgenesis. Fundamental differences between conventional gene transfer methods, utilizing *Agrobacterium* vectors or protoplast/suspension cultures, and particle bombardment will be discussed in depth.

Introduction

Agricultural biotechnology promises to improve crop productivity, complementing traditional breeding by decreasing our dependence on harmful chemicals, pesticides, fertilizers and antibiotics, etc. Contemporary social and environmental trends emphasize improved safety and quality of agricultural products. Genetic engineering is now available to breeders and can supplement conventional practices in improving crop performance. It is important for the breeder to identify goals and objectives for the genetic modification of crops, which are both scientifically feasible and economically viable. In the past, agriculture has been an energy- and labour-intensive industry. Biotechnology offers the opportunity to reduce both these costs. In order to achieve this objective it is important to identify the genetic need for crop improvement as defined by plant breeders. Factors including basic physiology and genetics of pest resistance, the many years and locations needed to evaluate and identify stress tolerance, and the length of time (in generations) required to break up undesirable genetic linkages or to assemble desirable traits, need to be examined very carefully (Cullis, 1987). Even in cases involving successful separation of the introduced gene from linked deleterious genes, the gene's inheritance and expression may be altered unpredictably in the new genetic background.

In order to apply genetic engineering successfully to crop improvement, one should be able to identify and isolate agronomically-useful genes, modify them according to strict specifications, and transfer them between species; ultimately, of course, these genes need to be recovered in mature plants and used in a breeding programme. Currently-available methods are designed only to transfer single or few genes. Many traits connected with final plant productivity are the result of multigene families that are only recognized in the context of their ability to be manipulated in a breeding programme, but with little or no knowledge of the biochemical basis of their action (Goodman et al., 1987). In contrast, success in plant genetic engineering will rely, to a great degree, on a thorough understanding of the molecular, genetic and metabolic characteristics and properties of traits to be transferred (Barton & Brill, 1983). In the years ahead, applications of new biotechnology techniques to agriculture are likely to enable the high growth rate of crop productivity to continue by following three general strategies: increase yields by using more productive plant strains, cut losses from pests and disease, and improve storage.

The available gene pool with traditional breeding methods is restricted by the sexual incompatibility of many interspecific and intergeneric crosses (Nisbet & Webb, 1990). Genetic manipulation and *in vitro* culture provide a means for the gene pool to be substantially broadened by allowing transfer of specific genes controlling well-defined traits from one organism to another, thus improving crops in a less haphazard way. By avoiding back-crossing programmes, which may take years in some cases, considerable time and financial resources can be conserved.

With conventional transformation methods, application of biotechnology to crop improvement has been limited to species amenable to such techniques. Too much emphasis has been bestowed on model plant systems that were shown to be easily amenable to genetic and cellular manipulation. As a result, we developed a good understanding of the molecular biology of tobacco, potato, tomato and petunia with very little effort directed towards the important agronomic and industrial crops. In any biotechnology programme it is important to work directly with elite cultivars. As Norman Borlaug observed, variety development is dynamic and subject to constant change due to the way pathogens mutate. Agronomic advances also dictate the constant evolution of improved cultivars. Working with model systems amenable to laboratory technologies, may appear to be attractive in the early stages of any research programme. However, short-term gains will be offset by the need to transfer all the technology from the model system to elite cultivars, a proposition that may not always be feasible (Borlaug, 1984).

Elaboration of methods for gene transfer into a wide range of plant species is a direct result of the development of *in vitro* techniques for the culture and propagation of cells and tissues. Genes can be accessed from many diverse sources, e.g. plants, animals, bacteria, fungi and even viruses. It is theoretically possible to design a gene to be expressed in a tissue/organ-specific manner at particular developmental stages in the plant's life cycle.

Targets for genetic transformation include: resistance to broad-spectrum, environmentally-safer herbicides which cannot normally be used with susceptible crops; isolation, characterization and cloning of disease- and pest-resistance genes from other plant species or bacteria such as *Bacillus thuringiensis*; elimination of bloating in animals grazing on forage crops such as clover by introducing tannin genes; improving protein content of grain and forage legumes for increased nutritional quality; oil quality improvements; understanding the mechanism of nitrogen fixation by *Rhizobium*-legume symbiotic relationships in order to improve nitrogen fixation in legumes and extend it to non-leguminous crops.

Comparative evaluation of gene transfer methods and criteria for stable transformation

Recognition of the ability of the soil bacterium *Agrobacterium tumefaciens* to transfer a portion of its DNA to plants was perhaps one of the most important milestones in plant biotechnology (Barton et al., 1983; Caplan et al., 1983; Herrera-Estrella et al., 1983): Major advances contributing to the popularity of *Agrobacterium*-based transformation systems include the development of disarmed strains where the oncogenes, which result in tumorigenesis and thus difficulties in the regeneration capabilities of transformed tissues, are deleted from the plasmid. The development of binary vectors, in which the T-DNA borders are located on a small but wide-host range plasmid and the virulence genes of the Ti-plasmid are located on an independent plasmid and act *in trans* to effect the excision of the T-DNA from the vector plasmid, was also crucial in expanding the use of *Agrobacterium*-based vectors in plant transformation. Any DNA fragment can be transferred from the Ti plasmid of *Agrobacterium tumefaciens* (or Ri plasmid from *A. rhizogenes*) as long as it is located in between the T-DNA borders flanking the wild type Ti/Ri plasmid T-DNA. These small plasmids can be easily manipulated in *E. coli* and *Agrobacterium* (Binns, 1988). Chimeric genes containing selectable resistance gene markers, such as the antibiotics kanamycin (Becker et al., 1982) or hygromycin (van den Elzen et al., 1985), methotrexate (Pua et al., 1987), herbicides such as Basta (De Block et al., 1987) and glyphosate (Amrhein et al., 1983), or screenable markers such as Lux (Ow et al., 1986) or GUS (Jefferson et al., 1987), can be introduced between the T-DNA borders. These genes are driven by bacterial (e.g. nopaline synthase; Depicker et al., 1982) viral (e.g. Cauliflower Mosaic Virus; Gardner et al., 1981) or plant promoters (e.g. Rubisco; Moses & Chua, 1988). As a result, such disarmed strains carrying Ti plasmids can be used to produce transgenic plants that differ from the wild type only by the presence of the foreign gene(s). Despite the many advantages of *Agrobacterium*-based transformation systems, major problems remain. The limited host range of *Agrobacterium* needs to be expanded if this method is to be used for the introduction of foreign DNA into the major crop species. Monocotyledonous crops, and cereals in particular, appear to be very recalcitrant to *Agrobacterium* infection; this may be attributed to either a lack of a wound response or a lack of effective DNA transfer from the bacterium to the plant. Even for dicotyledonous crops, severe host-specificities restrict the use of *Agrobacterium* as a transformation vector to specific cultivars that are of limited commercial importance. Regeneration from tissue culture is a prerequisite, and somaclonal variation resulting from tissue culture-induced mutations needs to be considered in any *Agrobacterium*-based transformation system. The long time-frames required to regenerate plants from transformed cells is another disadvantage of the method.

Direct DNA transfer involving electroporation or chemical fusagens such as polyethylene glycol with calcium phosphate, require the use of protoplasts. Unfortunately most crops cannot be regenerated from single cells. Even in cases where regeneration from single cells has been reported, genotype specificities severely hinder the transformation and subsequent regeneration of elite cultivars in these crops. The mechanical introduction of plasmid DNA into cellular organelles using microscopic needles, i.e. microinjection (Morikawa & Yamada, 1985) is not subject to host range limitations, but regeneration from protoplasts is a requirement. One report was published describing recovery of transgenic plants by injecting organized tissues such as microspore-derived proembryos of *Brassica* (Neuhaus et al., 1987).

Particle bombardment employs high velocity metal particles to deliver biologically active DNA into plant cells. The concept has been described in detail by Sanford (1988). The ability to deliver foreign DNA into regenerable cells, tissues or organs, appears to provide the best method for achieving truly genotype-independent transformation in agronomic crops, by-passing *Agrobacterium* host-specificity and tissue culture-related regeneration difficulties. Due to the physical nature of the process, there is no biological constraint to the actual DNA delivery; thus genotype is not a limiting factor. Combining the relative ease of DNA introduction into plant cells with an efficient regeneration protocol (avoiding protoplast or callus culture) and somatic embryogenesis, we appear to have the best system in place for transformation of crops that are refractory to alternative gene tranfer methods. Important advancements and refinements in the process described subsequently, using soybean and rice as model systems for dicotyledons and monocotyledons respectively, demonstrated effectively the power of the technique (Christou et al., 1990, 1992).

Dekeyser et al. (1990) demonstrated transient gene expression in intact and organized rice tissues using electroporation of leaf bases. Subsequently, the recovery of transgenic maize plants via electroporation of partially-digested immature zygotic embryos was reported (D'Halluin et al., 1992).

Development of genetic engineering processes to improve plants relies heavily on the establishment of firm criteria for integrative transformation. It is important to realize that neither phenotypic nor physical data alone are acceptable. A checklist developed by Potrykus (190) to establish conclusively integrative transformation in plants is very useful.

Particle bombardment

Klein et al. (1987) described a procedure in which high velocity microprojectiles were utilized to deliver nucleic acids into living cells. In those experiments, transient expression of exogenous RNA or DNA was demonstrated in epidermal cells of onion (*Allium cepa*). Following these experiments, the technique of particle bombardment (otherwise known as biolistics, microprojectile bombardment, particle acceleration, etc.) was shown to be the most versatile and effective way for the creation of transgenic organisms, incuding microorganisms, mammalian cells and a large number of plant species. Table 1 provides a comprehensive listing of plant species successfully engineered using particle bombardment technology. An estimated two hundred papers have been published on various aspects of the technique, including a number of comprehensive reviews (Birch & Franks, 1991; Christou, 1992; Klein et al., 1992).

Several advantages make microprojectile bombardment the method of choice for engineering crop species: (a) transformation of organized tis-

Table 1. Transgenic major species recovered through particle bombardment

Plant species	Common name	Explant	Reference
Glycine max	Soybean	Meristems	Christou et al., 1990
Arachis hypogaea	Peanut	Meristems	Brar et al., 1992
Phaseolus vulgaris	Common bean	Meristems	Russell et al., 1993
Zea mays	Corn	Susp. culture	Gordon-Kamm et al., 1990
		Susp. culture	Fromm et al., 1990
		Imm. embryos	Koziel et al., 1993
Oryza sativa	Rice	Immature embryos	Christou et al., 1991
Triticum aestivum	Wheat	Embr. calllus/Imm. embryos	Vasil et al., 1992
		Imm. embryos	Weeks et al., 1993
Hordeum vulgare	Barley	Immat. embryos/Embr. callus	Wan & Lemaux, 1994
Avena sativa	Oat	Embryogenic callus	Sommers et al., 1992
Saccharum officinarum	Sugarcane	Embryogenic callus	Bower & Birch, 1992
Gossypium hirsutum	Cotton	Meristems	McCabe & Martinell, 1993
Vaccinium macrocarpon	Cranberry	Stem sections	Serres et al., 1992
Carica papaya	Papaya	Embryos; hypocotyls	Fitch et al., 1990
Populus spp.	Poplar	Embryogenic suspension	Dayton et al., 1992
Picea glauca	Spruce	Embryogenic callus	Ellis et al., 1993
Helianthus annuus	Sunflower	Apical meristems	Bidney et al., 1992
Nicotiana tabacum	Tobacco	Leaves, suspension cultures	Tomes et al., 1990

sue; the ability to engineer organized and potentially-regenerable tissue permits introduction of foreign genes into elite germplasm; (b) universal delivery system; transient gene expression has been demonstrated in numerous tissues representing many different species. In particular cases in which recovery of transgenic plants has not been reported, this deficit is due more to the lack of a favourable regeneration response than to the DNA delivery method; (c) transformation of recalcitrant species; engineering of important agronomic crops such as soybean, cotton, maize, rice, etc., is restricted to a few non-commercial varieties when conventional methods are used. Particle bombardment technology allows recovery of transgenic plants from many commercial cultivars; (d) study of basic plant development; by utilizing chromogenic markers it is possible to study developmental processes and also to clarify the origin of germline in regenerated plants.

A number of parameters have been identified and need to be considered carefully in experiments involving transformation using particle bombardment. Particles should be of sufficiently high mass in order to possess adequate momentum to penetrate the appropriate tissue. Suitable metal particles include gold, tungsten, palladium, rhodium, platinum, iridium and possibly other second and third row transition metals. Metals should be chemically inert to prevent adverse reactions with the DNA or cell components. Additional desirable properties for the metal include size and shape, as well as agglomeration and dispersion properties. The nature, form and concentration of the DNA need also to be considered. In the process of coating metal particles with DNA, certain additives such as spermidine and calcium chloride appear to be useful. The nature of the DNA, e.g. single- versus double-stranded may also be important under some conditions, even though this was shown not to be a significant variable in specific cases. It is very important to target the appropriate cells that are competent for both transformation and regeneration. It is apparent that different tissues have different requirements; extensive histology needs to be performed in order to ascertain the origin of regenerating tissue in a particular transformation study. Depth of penetration thus becomes one of the most important variables and the ability to tune a system to achieve particle delivery to specific cell layers may be the difference between success and failure in recovering transgenic plants from a given tissue. In cases in which the Biolistic® device has been used, particularly with the original version of the instrument, cells near the centre of the target are injured and cannot proliferate. This injury was attributed to physical trauma of the cells from the gas blast and acoustic shock generated by the device. The use of baffles or mesh screens reduced cell death and increased transformation frequency significantly (Gordon-Kamm et al., 1990; Russell et al., 1992a).

The temperature, photoperiod and humidity under which donor plants, explants and bombarded tissues are maintained are also important. These parameters have a direct effect on the physiology of tissues and this can be an important variable. Such factors will influence receptiveness of target tissue to foreign DNA delivery and also affect its susceptibility to damage and injury that may adversely affect the outcome of the transformation process. Some explants may require a 'healing' period after bombardment under specific regimens of light, temperature and humidity. Choice and nature of explants and pre- and post-bombardment culture conditions are factors that may determine whether experiments utilizing particle bombardment are successful. In addition, explants derived from plants that are under stress, e.g. infected with bacteria or fungi, over- or under-watered, etc., will provide inferior material for bombardment experiments. Considerable evidence has been accumulating to indicate that in order to achieve high transformation frequencies, metal particles need to be directed to the nucleus (Yamashita et al., 1991). Osmotic pretreatment of target tissues has also been shown to be of importance (Russell et al., 1992b; Vain et al., 1993). In addition, experiments performed with synchronized cultured cells indicate that transformation frequencies may also be influenced by cell cycle stage (Iida et al., 1991). However, such physical treatments appear to be more important in experiments in which de-differentiated tissues are used as targets for bombardment experiments. Physical trauma and tungsten toxicity were found to reduce efficiency of transformation in experiments performed with tobacco cell suspension cultures (Russell et al., 1992a).

Many investigators have over-stressed the significance of transient expression. Transient expression studies should only be used as a guide in the development of systems for the stable transformation of a given species. In some cases exhaustive experiments were performed using transient expression data in an attempt to achieve complete protocol optimization for the recovery of stable transformants. This, however, may be unwise as optimization or maximization of transient activity does not necessarily result in optimal or even *any* stable transformation. Therefore, studies involving numbers of transiently expressing cells and foci per unit mass or volume of recipient cells may be meaningless and in a lot of cases irrelevant to the final outcome, particularly when the object is recovery of transgenic plants. It is important to utilize data from stable transformation experiments to draw conclusions pertaining to stable transformation. Of course, if no transient activity is observed following a bombardment experiment, it is unlikely that any stable transformants will be recovered.

A number of different instruments based on various accelerating mechanisms are currently in use. These include the original gunpowder device (Sanford et al., 1987), an apparatus based on electric discharge (McCabe & Christou, 1993), a microtargeting apparatus (Sautter et al., 1991), a pneumatic instrument (Iida et al., 1990), an instrument based on flowing helium (Finer et al., 1992; Takeuchi et al., 1992), and an improved version of the original gunpowder device utilizing compressed helium (Sanford et al., 1991). Hand-held versions of both the original Biolistics® device and the Accell® device are also in use. The most widely-used instrument is the one currently marketed by Bio-Rad, Inc. (Biolistics®) but Accell®-based methodology has been particularly useful in developing variety-independent gene transfer methods for the more recalcitrant cereals and legumes. Detailed descriptions of the various acceleration devices, principles of operation and other details may be found in the primary references.

Until recently, the key barrier to achieving effective transformation of agronomically-important species was the DNA delivery method. Microprojectile bombardment has had a tremendous impact on this limitation. The challenge now is shifting back to the biology of the explant used in bombardment experiments. It is apparent that the conversion frequency of transient to stable transformation events is low. This does not mean, however, that transgenic plants from most of the crops that have been engineered, cannot be obtained at high enough frequencies to make the process commercially useful and economical. More attention needs to be paid to the biology of explants prior to and following bombardment. We need to identify how more cells can be induced to become competent for stable DNA uptake and regeneration. Optimization of biological interactions between physical parameters and target tissue needs to be better understood. Not much is known about the fate of DNA from the time particles are introduced into plant cells. Recipient tissue variation and variability due to bombardment conditions complicate the picture even further. Additional issues such as irregular particle size and improvements in hardware design need to be resolved.

Transgenic cereals

Engineering of cereal crops became possible as a result of the development of particle bombardment-based methodology. Monocotyledonous species exhibit natural resistance to infection by *Agrobacterium.* Even in situations in which such resistance was overcome, problems inherent to low efficiencies, host specificities, and tissue culture-induced mutations, including infertility of recovered plants, prevented the wider adoption of *Agrobacterium*-based vectors. A recent report of the utilization of *Agrobacterium* for the transformation of rice illustrates some of these problems (Chan et al., 1993). Similar difficulties were encountered with protoplast systems, the major problems in this case being limited success of regeneration of plants from protoplasts, and sterility.

Rapid progress in the recovery of transgenic cereals has been the result of the utilization of organized and regenerable tissues such as immature embryos, in combination with particle bombardment technology. Embryogenic suspension and callus cultures were utilized initially, particularly with maize and wheat, but the severe constraints imposed by genotype and sterility in regenerated plants limited the utility of these systems.

*Corn (*Zea mays*)*

Recovery of maize plants from electroporated transformed protoplasts resulted in a major disappointment when all transgenic plants obtained in the early experiments were shown to be sterile (Rhodes et al., 1988). *Agrobacterium* was shown to be effective in introducing viral DNA into maize, but was incapable of achieving germline transformation (Grimsley et al., 1987). Two groups utilized embryogenic maize suspension cultures for the recovery of transgenic, fertile plants (Fromm et al., 1990; Gordon-Kamm et al., 1990). A specific cross was utilized by both groups to develop a regenerable embryogenic suspension culture amenable to transformation. Particle bombardment of such a suspension culture followed by selection using the herbicide bialaphos, resulted in transformed embryogenic calli in independent experiments. Fertile transgenic plants were regenerated and stable inheritance and expression of *bar*, and functional activity of the enzyme phosphinothricin acetyltransferase, were observed in a number of subsequent generations. The transformation process and the presence of foreign gene(s) did not detrimentally influence either plant vigour or fertility (Gordon-Kamm et al., 1991). Similar results were obtained in experiments in which the herbicide chlorsulfuron was used (Fromm et al., 1990).

Transformation of regenerable maize tissues by electroporation of wounded immature zygotic embryos or embryogenic type I callus provides an alternative to particle bombardment (D'Halluin et al., 1992). Tissues were wounded either mechanically or enzymatically and were subsequently electroporated with a neomycin phosphotransferase II gene. Transformed embryogenic calli were selected from electroporated tissues on kanamycin-containing medium and fertile transgenic plants were regenerated. Absence of chimerism and Mendelian segregation in R1 progeny showed that transgenic plants were derived from a single cell or a small number of cells.

Recovery of transgenic maize, expressing a number of genes of agronomic interest, including insect resistance, resulted in successful field trials. Koziel et al. (1993) introduced a synthetic gene encoding a truncated version of the CryIA(b) protein derived from *Bacillus thuringiensis* into immature embryos of an elite line of maize via particle bombardment. High levels of protein expression were obtained. Hybrid maize plants resulting from crosses of transgenic elite inbreds with commercial lines were evaluated for resistance to the European corn borer under field conditions. Plants expressing high levels of the insecticidal protein exhibited excellent resistance to repeated heavy infestations of the insect.

*Rice (*Oryza sativa*)*

Recovery of transgenic rice was reported less than five years ago. These experiments utilized direct DNA transfer methods such as electroporation (e.g. Toriyama et al., 1988) or PEG-mediated gene transfer (e.g. Zhang et al., 1988) into protoplasts. Subsequent reports established these two techniques as the methods of choice for gene transfer into rice protoplasts and recovery of fertile transgenic plants. A severe limitation of these methods, however, is the requirement to establish systems to regenerate plants from single cells. Even though such systems do exist in some of the japonica varieties, e.g. Taipei 309, most elite japonicas as well as the vast majority of indica varieties are very difficult to regenerate from protoplasts. In experiments in which regeneration of indica varieties from protoplasts was accomplished, transgenic plant recovery has not been reported. Notable exceptions are the reports by Datta and co-workers, which described recovery of fertile transgenic plants from the variety Chinsurah Boro II (indica type; Datta et al., 1990), and more recently, primary transformants from the elite variety IR72, which were infertile (Datta et al., 1992). In both cases embryogenic suspension cultures were established from immature pollen grains and these cultures were used to isolate protoplasts for subsequent transformation experiments. These constraints demonstrated the strong genotype- and culture-dependent component of rice *in vitro* culture. Another serious problem of protoplast-based systems was identified as reduced fertility or complete infertility of primary transformants.

Efficient systems for the regeneration of rice plants from the scutellum of immature embryos through somatic embryogenesis have been described (Hartke & Lorz, 1989). These studies demonstrated that essentially all rice cultivars could be regenerated from immature embryos even though regeneration efficiencies varied substantially. Exposure of immature rice embryos to the auxin 2,4-D results in the development of embryogenic callus which can be converted to plants upon transfer to media supplemented with indole-3-acetic acid (IAA) and zeatin. Following particle bombardment as described elsewhere, Christou et al. (1991) were able to recover transformed embryogenic callus and plants either with or without selection. Transformation frequencies in experiments in which selection was applied were very high, in the range of 2 to 15%, calculated in terms of independently-derived transgenic plants per unit number of bombarded explants. The antibiotic hygromycin B was used to select transformed embryogenic callus. Transgenic callus and plants were also recovered from experiments in which selection was not used, illustrating the high level of efficiency of the transformation process. All recovered plants were clonal in nature, demonstrating their derivation from single or a small number of cells (Christou et al., 1992). Field trials in which resistance to the herbicide Basta® was demonstrated, were carried out in 1993 (Fig. 1). Aspects of these procedures were recently reproduced by another group utilizing a Biolistics® instrument; however, transformation frequencies were significantly lower, the range of genotypes transformed was limited, and fertility of transgenic plants was very low with some important varieties exhibiting total sterility (Li et al., 1993).

*Wheat (*Triticum aestivum*)*

Particle bombardment technology again provided the necessary technological breakthrough for engineering wheat. A plasmid encoding *bar* was introduced into type C long-term regenerable embryogenic callus (Vasil et al., 1992). Following selection in bialaphos-containing medium, plants from two lines were regenerated. The transgenic nature of these plants was confirmed by molecular analysis and out-crossing to wild-type plants. Transformed progeny was obtained by crossing transgenic primary transformants with wild-type pollen.

Weeks et al. (1993) utilized callus derived from immature wheat embryos in bombardment experiments to recover transgenic fertile plants. The cultivar Bobwhite was chosen for use in transformation experiments because of its high frequency of regeneration from tissue culture. Immature embryos (0.5–1 mm) were bombarded with a plasmid containing the *bar* and *gus* genes driven by the maize ubiquitin promoter. Bombardment was performed 5 days after embryo excision, just after initiation of callus proliferation. Nine independent lines of fertile transgenic plants were obtained in the experiments described. The transformation frequency was approximately 0.1–0.2%. Some of the lines had normal fertility and seed set; however, most exhibited reduced seed set compared with wild type Bobwhite plants regenerated from tissue culture; one line was completely sterile. Amongst the various phenotypic criteria used to assess transformation, the ability of regenerated plantlets to develop a normal root system in the presence of bialaphos, was the most reliable indication of stable integration of the *bar* gene. Each plant analyzed which exhibited resistance at the rooting stage was shown to contain DNA fragments homologous to the *bar* coding region. As with reports from other crops, the level of PAT activity was not correlated to the DNA copy number.

*Barley (*Hordeum vulgare*)*

Until very recently, barley was exceptionally recalcitrant to *in vitro* manipulation and genetic engineering. However, a recent report describing recovery of transgenic barley plants via particle bombardment appears to have removed that constraint (Wan & Lemaux, 1994). Immature zygotic embryos, young callus, and microspore-derived embryos, were bombarded with plasmids containing the *bar* and *gus* genes, as well as a gene for the barley yellow dwarf virus coat protein (BYDVcp). A total of ninety one indepen-

Fig 1. Rice field trial, performed by Drs S. Linscombe and M. Braveman, Louisiana State University, Rice Research Experiment Station, Crawley, Louisiana (Summer 1993). A number of independently-derived lines representing diverse indica and japonica cultigens, resistant to the herbicide Basta® were evaluated in the field following successful greenhouse studies. All lines exhibited complete resistance to high levels of herbicide application

dent bialaphos-resistant callus lines were recovered. A large number of fertile transgenic plants were obtained from 36 of these lines; forty one lines gave albino plants. Transmission of introduced genes to progeny was confirmed by germinating immature embryos on bialaphos-containing medium, and also by spraying seedlings with the herbicide.

*Oat (*Avena sativa*)*

Transgenic, friable, embryogenic oat cultures were selected for phosphinothricin resistance following particle bombardment (Somers et al., 1992). From three bombardment experiments 111 PPT-resistant callus lines were derived. One to more than 20 copies of the transgene(s) were integrated into the genome of the cultures. Co-expression of the *gus* gene with the *bar* gene was 75%. Plants were regenerated from 38 of the transgenic lines. Regenerated plants generally exhibited male sterility, although one transgenic line regenerated fully-fertile plants.

*Sugarcane (*Saccharum officinarum*)*

Bower & Birch (1992) reported recovery of mature transgenic sugarcane (*Saccharum* spp.) plants. Optimal bombardment conditions for embryogenic callus required microprojectile velocities higher than those previously found effective for sugarcane suspension culture cells (Franks & Birch, 1991). Stable transformants were obtained following bombardment with the *nptII* gene under the control of the *Emu* promoter. NPT II levels in transformed plants were 20–50 times the background levels in control plants in ELISA assays. DNA analysis showed that transgenic plants had 1–3 copies of the introduced gene integrated into the sugarcane genome. The procedure afforded 1–3 transgenic plants per treated plate within 16 weeks of bombardment.

Transgenic legumes

Creation of transgenic grain legumes has been accomplished using a combination of particle bombardment technology and *de novo* regeneration systems. Suspension cultures have also been utilized, although success has been very limited. A series of reports claiming soybean transformation using protoplasts and electroporation were recently retracted by the senior author (Widholm, 1993a, b, c). For important species such as soybean, peanut and phaseolus, elite varieties can be engineered relatively easily. Frequencies vary depending on the particular cultivar used; however, at least in the case of soybean, all 30 cultivars attempted resulted in the recovery of transgenic plants (Christou et al., 1990). It is likely that additional legumes such as pea, and various under-exploited species such as *Vigna, Vicia, Lens* and *Cicer* will be amenable to this technology, provided adequate resources are devoted to such projects. It is unlikely that *Agrobacterium*, for example will replace particle bombardment for the engineering of large-seeded legumes. This is because legumes exhibit severe host specificity to infection by various *Agrobacterium* strains, as exemplified by a number of comprehensive studies in which large numbers of strains were shown to be ineffective in infecting most soybean varieties (Lianzheng et al., 1984; Owens & Cress, 1985). In addition, it is very difficult to regenerate intact plants from de-differentiated tissues in most legumes, thus alternative DNA transfer methods based on protoplast technology are not practical.

*Soybean (*Glycine max*)*

Early attempts at soybean engineering focused on regeneration from protoplasts and embryogenic suspension cultures. However, despite some initial successes, progress was very slow and the recovery of transgenic soybean plants remained a distant and elusive objective. Soybean genetic engineering became a reality following the invention and optimization of particle bombardment. In fact soybean was used as a model system to develop the technology for a large number of crop species that were shown to be extremely recalcitrant to genetic manipulation. Following the initial observation by Klein et al. (1987), that macromolecules could be introduced and expressed transiently into epidermal cells of onion, Christou et al. (1987) demonstrated that DNA-coated metal particles could deliver biologically-active DNA into soybean organized tissue, with subsequent recovery of stable transformants in the form of callus lines and transgenic roots. Soon after that report, the same group reported recovery of the first genetically-engineered staple crop, soybean (McCabe et al., 1988). Embryonic axes from mature and immature soybean seeds were found to be the optimum explant. Following mechanical isolation of the axes, tissue was bombarded and cultured on MS medium supplemented with high levels of cytokinin. After a two-week culture period, bombarded tissues initiated multiple shoots. Six to ten weeks later, shoots elongated sufficiently to allow their transfer to the greenhouse. Depending on the particular soybean genotype used, 5–20 shoots could be recovered per bombarded explant.

By combining a simple genotype-independent regeneration protocol, based on the proliferation of multiple shoots from the general area of the meristem of soybean embryonic axes, and electric discharge particle acceleration for the delivery of foreign DNA, a commercial process for the introduction of any gene into any soybean variety was developed (Christou et al., 1987). Many hundreds of independently-derived soybean plants were obtained and analyzed, representing a wide phenotypic spectrum of transformation events (Christou, 1990). The pattern of transformed phenotypes formed the basis for optimizing the bombardment protocol and permitted focusing particle delivery primarily to the appropriate cells (L2 cell layer) from which the germline of the plant originates. Parameters such as bombardment intensity, DNA and gold bead loading rates, number and timing of bombardment, were adjusted in order to recover the maximum number of what was predicted, and later confirmed to be phenotypes associated with germline transformation events. Expression patterns of marker genes in specific tissue types in R0 plants and correlation with transmission of the introduced gene(s) to progeny (Christou & McCabe, 1992) resulted in the development of an efficient procedure for the genetic improvement of this important staple crop. This process is the only method, at present, that is in commercial use for the engineering of soybean in a variety-independent fashion. Any gene can be introduced into any soybean variety, and hundreds of transgenic plants can be recovered following a simple screening procedure. Field trials of herbicide-resistant soybeans have been ongoing for the last 3 years (Fig. 2) with expected commercialization in the very near future. Additional targets for soybean

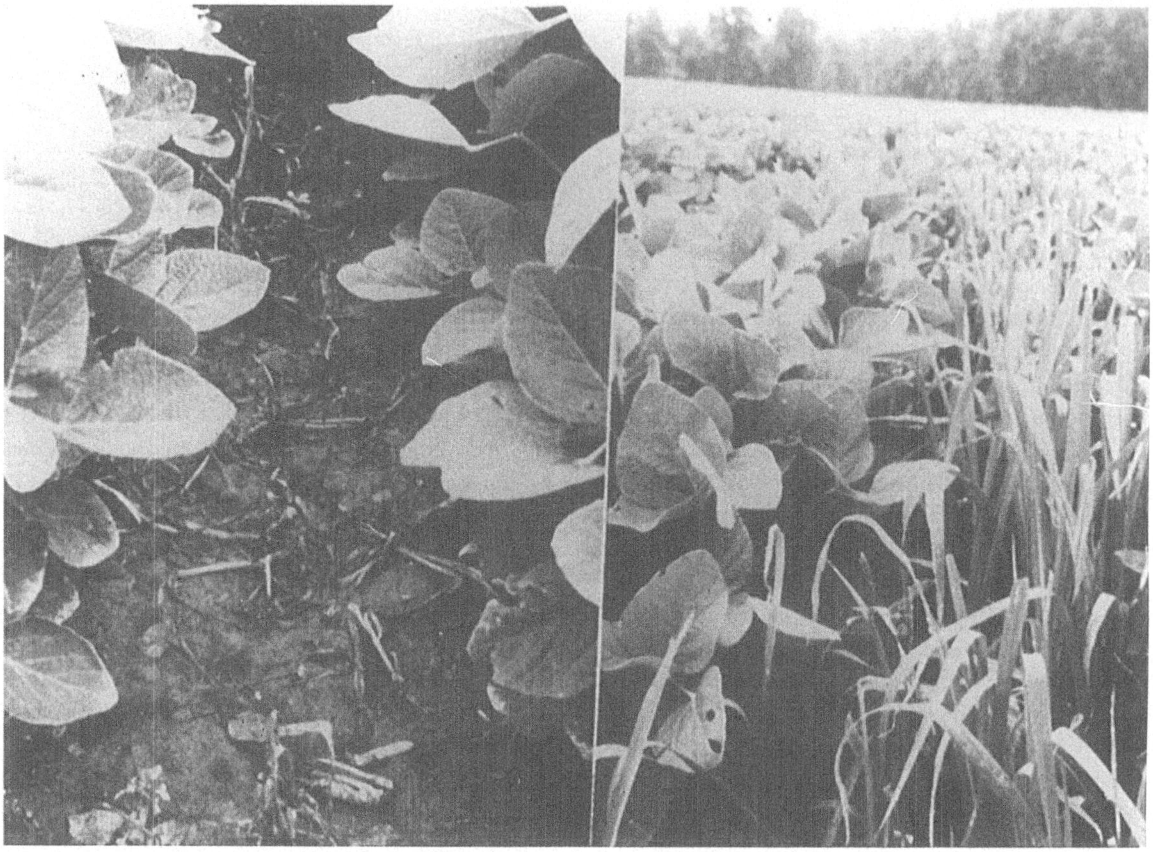

Fig. 2. Herbicide-resistant soybean field trial. Resistance to the herbicides Basta® and Roundup® has been demonstrated in the field over a three year period.

improvement include modification of protein and oil composition as well as resistance to viruses and other diseases. The generation and field testing of herbicide-resistant/tolerant soybean plants represents a milestone in the commercial application of plant genetic engineering to food crops.

*Common bean (*Phaseolus vulgaris*)*

Genetic transformation studies in *Phaseolus* have until recently met with limited success. There are no reports of protoplast transformation and recovery of transgenic callus as a result of electroporation, PEG-mediated transformation or other direct or indirect DNA transfer methods. Two general approaches have been employed in attempts to engineer *Phaseolus*, namely *Agrobacterium* infection and particle bombardment. Only the latter method resulted in the recovery of transgenic plants. Soon after the development of this procedure for the development of transgenic soybeans, *Phaseolus* species were shown to be amenable to the same methodology. Transgenic beans expressing marker genes such as *gus, bar* as well as the coat protein gene from the bean golden mosaic virus were recovered (Russell et al., 1993). Transgenic plants were characterized over five generations of self-fertilization with no loss of introduced genes or expression.

*Peanut (*Arachis hypogaea*)*

Experience with other legumes suggests that introduction of foreign genes into peanut using *Agrobacterium* will not be a straightforward task. Brar and co-workers used particle bombardment to engineer two elite peanut varieties (Brar et al., 1992, 1994). Shoot meristems of mature embryonic axes were bombarded with foreign genes, and transgenic plants expressing GUS and resistance to the herbicide Basta® were recovered. The tomato spotted wilt virus nucleocapsid protein was introduced into peanut and resistance to infection by the virus is currently being investigated. The transgenic nature of these plants was confirmed by molecular analyses, herbicide spraying and inheritance of introduced genes in subsequent generations.

Transgenic woody species

Incorporation of foreign DNA into trees and other woody species, with subsequent regeneration of transformed plants has been reported for a very limited number of genera primarily utilizing *Agrobacterium*. As with other crops, however, complex host-bacterium interactions and other factors significantly limited the scope of such vectors in woody species engineering.

*Cotton (*Gossypium hirsutum*)*
The technique originally developed to engineer cotton utilized *Agrobacterium tumefaciens* vectors and was applicable to only a specific variety, Coker 312 (Umbeck et al., 1987, Firoozabady et al., 1987). Finer & McMullen (1990) reported recovery of transgenic cotton plants from the same cultivar utilizing bombardment of embryogenic suspension cultures. Most varieties of commercial interest proved difficult or impossible to regenerate into plants from the obligatory callus phase. Traits introduced into regenerable varieties could be bred into other lines but the process was lengthy and prone to somaclonal variation. McCabe & Martinell (1993) developed a procedure, based on ACCELL^{TM} technology, to deliver foreign genes directly into the meristematic tissue of excised embryonic axes of cotton, that resulted in the recovery of transgenic plants in a variety-independent fashion. R1 seeds recovered from pollen-positive R0 plants were germinated and grown in the greenhouse. Leaf punches were taken from each plant and analyzed for GUS activity to determine segregation patterns. Subsequently pollen staining for GUS allowed selection of homozygous individuals for the production of R2 plants for further analysis. Southern blot and enzyme analyses of R2 and R3 plants confirmed inheritance of foreign genes by successive generations in a Mendelian fashion.

*Cranberry (*Vaccinium macrocarpon*)*
Genetic transformation of the American cranberry (*Vaccinium macrocarpon* Ait.) was accomplished via electric discharge particle bombardment (Serres et al., 1992). Stem sections derived from *in vitro* cultures were bombarded with *gus, npt II* and *bt* genes. Following bombardment, stem sections were cultured on solid bud-inducing medium containing kanamycin. A thin overlay of water containing kanamycin was added to inhibit growth of non-transformed cells. Within seven weeks, green shoots emerged most of which were shown to contain all three genes (by PCR and Southern blot analyses).

*Poplar (*Populus *sp.)*
Particle bombardment technology was utilized to extend the range of transgenic *Populus* species that could not be regenerated from protoplasts. McCown et al. (1991) utilized three different target tissues including protoplast-derived callus, nodules and stems to generate transgenic plants through electric discharge particle bombardment. Pretreatment of explants, fine-tuning of bombardment parameters, and the use of a selection technique employing flooding of the target tissue with media containing the selective agent, were found to be important for reliable recovery of transformed plants.

*Yellow poplar (*Liriodendron tulipifera*)*
Single cells and small cell clusters isolated from embryogenic suspension cultures of yellow poplar (*Liriodendron tulipifera* L.) were bombarded with genes encoding for GUS and NPT II (Dayton-Wilde et al., 1992). Between 3 and 30 copies of the intact *gus* gene were detected in independently-transformed callus lines. Somatic embryos induced from transformed cell cultures were found to be uniformly positive for GUS expression. Transgenic plants expressed both genes in roots and leaves.

*White spruce (*Picea glauca*)*
Gymnosperms have been particularly difficult to engineer. The only report describing recovery of transgenic plants from conifers involved transgenic white spruce (*Picea glauca*) following electric discharge particle acceleration of plasmid DNA into regenerable embryogenic callus (Ellis et al., 1993). Successful transformation was dependent on three factors: ability to form embryogenic callus, ability of the putatively-transgenic callus to survive selection, and competence of transformed tissue to express foreign DNA.

*Papaya (*Carica papaya*)*
Following the original report describing recovery of transgenic papaya plants via particle bombardment of

embryogenic tissue (Fitch et al., 1990), Fitch et al. (1992) characterized papaya plants engineered with the coat protein gene of papaya ringspot virus (PRV). By utilizing 2,4-D-treated immature zygotic embryos as targets, and following regeneration of transgenic plants on kanamycin-containing media, they obtained plants that exhibited varying degrees of resistance to PRV. One line appeared to be completely resistant when challenged with the severe Hawaiian PRV HA. Tests to recover PRV from the inoculated resistant plants by means of transferring leaf extracts to a local lesion host were negative, indicating complete resistance presumably due to inhibition of PRV replication. A number of virus-resistant lines are currently being tested in the field.

Comparison of conventional DNA transfer methods with particle bombardment

In this section we will summarize conclusions from examples involving generations of transgenic crop plants utilizing *Agrobacterium* infection, electroporation, chemical fusagens and particle bombardment. Success in terms of recovering fertile transgenic plants in reasonable frequencies from elite varieties should be a key criterion for evaluating a specific gene transfer method. Cereals are not susceptible to infection by *Agrobacterium* so direct DNA transfer into protoplasts was the most widely-used method for transformation prior to the development of particle bombardment. Early experiments with protoplast-derived callus indicated that partial or complete sterility and albinism in regenerated transgenic plants posed a serious problem. For leguminous species, host specificities restricted the use of *Agrobacterium*-based vectors to those few cultivars susceptible to infection. Even in situations in which *Agrobacterium* was shown to work well, for example in the case of alfalfa (*Medicago sativa*), limitations due to the culture system and regeneration again restricted the effectiveness of *Agrobacterium*-based vectors. In a number of cases it was possible to transform tissue which unfortunately was shown not to be regenerable (Baldes et al., 1987; Lin et al., 1987).

If we were to identify a key factor that has made particle bombardment the most widely-used gene transfer method for engineering recalcitrant species, this would be the identification of intact explants as targets. By minimizing or eliminating completely callus or suspension cultures as targets, the regeneration constraint for most elite cultivars is eliminated. Consequently, breeders may concentrate on the engineering of their most advanced breeding lines and this will result in significant time and labour savings in ongoing plant improvement programmes.

Despite the tremendous success and versatility of particle bombardment-based gene transfer methods, the technique is not without its problems. Aspects pertaining to underlying mechanisms of the gene transfer process, which is likely to have an impact on the creation and nature of transgenic plants in terms of DNA content and integration patterns, as well as the limited availability of instruments, are issues requiring urgent attention if the technique is to realize its full potential. Recent results indicate strongly that given adequate resources for the engineering of a specific plant species, particle bombardment still represents the best method available for introducing foreign genes into a given species. The problem is thus reduced to one of technology transfer between sophisticated laboratories having access to the tools needed for the successful engineering of a given species, and a large number of laboratories, particularly in developing countries which do not have access to such technology. Issues pertaining to intellectual property rights and commercialization of the products of plant genetic engineering need to be balanced very carefully with the need to improve crop species, especially in developing countries, and also the need to advance scientific frontiers using recombinant DNA technology.

Certainly alternative gene transfer technologies will continue to be developed. Such technologies may offer alternatives and distinct advantages over existing transformation methods. It is likely that no single gene transfer method will prove to be the method of choice for the engineering of all plant species.

Conclusions and future prospects

Tremendous advances in gene transfer methodology have occurred during the past few years. We are now in a position to embark on meaningful improvement programmes for a number of species utilizing the tools of molecular and cell biology. In specific cases such as soybean, rice, maize, barley, etc., gene transfer methods have been developed that do not appear to be restricted by variety or genotype. A number of 'orphan' crops, including most of the ones cultivated for human consumption in developing countries have received very little attention. It is reasonable to expect that given enough resources, practical gene transfer methods for all the major plant species will be developed. This is of extreme importance particularly in developing countries, the population of which derives most of its nutritional calories from grain legumes and cereals.

Plant genetic engineering is now at a crucial crossroad. The gene transfer constraint appears to have been removed from a number of important crops. Technical problems still remain but they are not insurmountable. The attention of the scientific community is gradually shifting to other areas such as identification and cloning of genes responsible for multigenic traits. One area which should not be neglected encompasses issues of public perception and environmental risk assessment of products derived from recombinant DNA technology. Emphasis on the improvement of under-exploited crops, crucial for the survival of people in developing countries, should receive more attention.

The first transgenic plants were recovered in 1983. It is indeed remarkable that in just one decade the tools of recombinant DNA technology and cell biology are now at the disposal of plant breeders. Important practical issues can now be addressed and increased agricultural productivity should be the direct beneficiary of advances in this field. Alternative uses for surplus crops resulting from recombinant DNA technology, have the potential to provide new resources for industry and the consumer, thus expanding the economic basis in industrialized countries.

Note added in proof

While this manuscript was in print, a report describing recovery of transgenic rice plants utilizing *Agrobacterium* was published (Hiei et al. 1994. The Plant J. 6: 271–282).

References

Amrhein, N., D. Johanning, J. Schab & A. Schulz, 1983. Biochemical basis for glyphosate tolerance in a bacterium and a plant tissue culture. FEBS Letts. 157: 191–196.

Baldes, R., M. Moos & K. Geider, 1987. Transformation of soybean protoplasts from permanent suspension cultures by cocultivation with cells of *Agrobacterium tumefaciens*. Plant Mol. Biol. 9: 135–141.

Barton, K.A., A.N. Binns, A.J.M. Matzke & M.D. Chilton, 1983. Regeneration of intact tobacco plants containing full length copies of genetically engineered T-DNA and transmission of T-DNA to R1 progeny. Cell 32: 1033–1043.

Barton, K.A. & W.J. Brill, 1983. Prospects in Plant Genetic Engineering. Science 219: 671–676.

Bidney, D., C. Scelonge, J. Martich, M. Burrus, L. Sims & G. Huffman, 1992. Microprojectile bombardment of plant tissues increases transformation frequency by *Agrobacterium tumefaciens*. Plant Mol. Biol. 18: 301–303.

Beck, E., E.A. Ludwig, B. Reiss & H. Schaller, 1982. Nucleotide sequence and exact localization of the neomycin phosphotransferase gene from transposon Tn5. Gene 19: 327–336.

Binns, A.N., 1988. Cell biology of *Agrobacterium* infection and transformation of plants. Ann. Rev. Microbiol. 42: 575–606.

Birch, R.G. & T. Franks, 1991. Development and optimization of microprojectile systems for plant genetic transformation. Aust. J. Plant Physiol. 18: 453–469.

Bower, R. & R. Birch, 1992. Transgenic sugarcane plants via microprojectile bombardment. Plant J. 2: 409–416.

Borlaug, N.E., 1984. Plant breeding goals and strategies for the 80s. In: D.A. Evans, W.R. Sharp & P. Ammirato (Eds). Handbook of Plant Cell Culture. Vol. 4, pp. 3–11. MacMilland Publishing Co., New York.

Brar, G.S., B.A. Cohen & C.L. Vick, 1992. Germline transformation of peanut (*Arachis hypogaea* L.) utilizing electric discharge particle acceleration (ACCELL^{TM}) technology. Proceedings of the American Peanut Research and Education Soc., Inc. Norfolk, Virginia. Vol. 24, p. 21.

Brar, G.S., B.A. Cohen, C.L. Vick & G.W. Johnson, 1994. Recovery of transgenic peanut (*Arachis hypogaea* L.) plant from elite cultivars utilizing Accell technology. Plant J. 5: 745–753.

Caplan, A., L. Herrera-Estrella, D. Inze, E. van Haute, M. van Montagu, J. Schell & P. Zambryski, 1983. Introduction of genetic material into plant cells. Science 220: 815–821.

Chan, M.T., H.H. Chang, S.L. Ho, W.F. Tong & S.M. Yu, 1993. *Agrobacterium*-mediated production of transgenic rice plants expressing a chimeric a-amylase promoter/β-glucuronidase gene. Plant Mol. Biol 22: 491–506.

Christou, P., J.E. Murphy & W.F. Swain, 1987. Stable transformation of soybean by electroporation and root formation from transformed callus. Proc. Natl. Acad. Sci. USA 84: 3962–3966.

Christou, P., 1990. Morphological description of transgenic soybean chimeras created by the delivery, integration and expression of foreign DNA using electric discharge particle acceleration. Ann. Bot. 66: 379–386.

Christou, P., D.E. McCabe, B.J. Martinell & W.F. Swain, 1990. Soybean genetic engineering – Commercial production of transgenic plants. Trends Biotech. 8: 145–151.

Christou, P., T. Ford & M. Kofron, 1991. Production of transgenic rice (*Oryza sativa* L.) plants from agronomically important indica and japonica varieties via electric discharge particle acceleration of exogenous DNA into immature zygotic embryos. Bio/Technol. 9: 957–962.

Christou, P., 1992. Genetic transformation of crop plants using microprojectile bombardment. Plant J. 2: 275–281.

Christou, P., T.L. Ford & M. Kofron, 1992. The development of a variety-independent gene-transfer method for rice. Trends Biotech. 10: 239–246.

Christou, P., & D.E. McCabe 1992. Prediction of germline transformation events in chimeric R0 transgenic soybean plantlets using tissue specific expression patterns. Plant J. 2: 283–290.

Cullis, C.A., 1987. Biotechnology and plant productivity. Ohio J. Sci. 87: 143–147.

Datta, S.K., A. Peterhans, K. Datta & I. Potrykus, 1990. Genetically engineered fertile Indica-rice recovered from protoplasts. Bio/Technol. 8: 736–740.

Datta, S.K., K. Datta, N. Soltanifar, G. Donn & I. Potrykus, 1992. Herbicide-resistant Indica rice plants from IRRI breeding line

IR72 after PEG-mediated transformation of protoplasts. Plant Mol. Biol. 20: 619–629.

Dayton-Wilde, H., R.B. Meagher & S.A. Merkle, 1992. Expression of foreign genes in transgenic yellow-poplar plants. Plant Physiol. 98: 114–120.

De Block, M., J. Botterman, M. Vandewiele, J. Dockx, C. Thoen, V. Gossele, N. Rao Movva, C. Thompson, M. Van Montagu & J. Leemans, 1987. Engineering herbicide resistance in plants by expression of a detoxifying enzyme. EMBO 6: 2513–2518.

Dekeyser, R.A., B. Claes, R.M.U. De Rycke, M.E. Habets, M. Van Montagu & A.B. Caplan, 1990. Transient gene expression in intact and organized rice tissues. Plant Cell 2: 591–602.

Depicker, A., S. Stachel, P. Dhaese, P. Zambryski & H.M. Goodman, 1982. Nopaline synthase transcript mapping and DNA sequence. J. Mol. Appl. Gen. 1: 561–568.

D'Halluin, K., E. Bonne, M. Bossut, M. De Beuckeleer & J. Leemans, 1992. Transgenic maize plants by tissue electroporation. Plant Cell 4: 1495–1505.

Ellis, D.D., D.E. McCabe, S. McInnis, R. Ramachandran, D.R. Russell, K.M. Wallace, B.J. Martinell, D.R. Roberts, K.F. Raffa & B.H. McCown, 1993. Stable transformation of *Picea glauca* by particle acceleration – A model system for conifer transformation. Bio/Technol. 11: 84–89.

Finer, J.J. & M.D. McMullen, 1990. Transformation of cotton (*Gossypium hirsutum* L.) via particle bombardment. Plant Cell Rep. 8: 586–589.

Finer, J.J., P. Vain, M.W. Jones & M.D. McMullen, 1992. Development of particle inflow gun for DNA delivery to plant cells. Plant Cell Rep. 11: 323–328.

Firoozabady, E., D.L. DeBoer, D.J. Merlo, E.L. Halk, L. Amerson, K.E. Rashka & E.E. Murray, 1987. Transformation of cotton (*Gossypium hirsutum* L.) by *Agrobacterium tumefaciens* and regeneration of transgenic plants. Plant Mol. Biol. 10: 105–116.

Fitch, M.M.M., R.M. Manshardt, D. Gonsalves, J.L. Slightom & J.C. Sanford, 1990. Stable transformation of papaya via microprojectile bombardment. Plant Cell Rep. 9: 189–194.

Fitch, M.M.M., R.M. Manshardt, D. Gonsalves, J.L. Slightom & J.C. Sanford, 1992. Virus resistant papaya plants derived from tissues bombarded with coat protein gene of papaya ringspot virus. Bio/Technol. 10: 1466–1472.

Franks, T. & R. Birch, 1991. Gene transfer into intact sugarcane cells using microprojectile bombardment. Austr. J. Plant Physiol. 18: 471–480.

Fromm, M.E., F. Morrish, C. Armstrong, R. Williams, J. Thomas & T.M. Klein, 1990. Inheritance and expression of chimeric genes in the progeny of transgenic maize plants. Bio/Technol. 8: 833–844.

Gardner, R., A. Howarth, P. Hahn, M. Brown-Luedi, R. Shepherd & J. Messing, 1981. The complete nucleotide sequence of an infectious clone of Cauliflower mosaic virus by M13mp7 shotgun sequencing. Nucl. Acids Res. 9: 2871–2882.

Goodman, R.M., H. Hauptly, A. Crossway & V.C. Knauf, 1987. Gene transfer in crop improvement. Science 236: 48–54.

Gordon-Kamm, W.J., T.M. Spencer, M.L. Mangano, T.R. Adams, R.J. Daines, W.G. Start, J.V. O'Brien, S.A. Chambers, W.R. Adams, N.G. Willetts, T.B. Rice, C.J. Mackey, R.W. Krueger, A.P. Kausch & P.G. Lemaux, 1990. Transformation of maize cells and regeneration of fertile transgenic plants. Plant Cell 2: 603–618.

Gordon-Kamm, W.J., T.M. Spencer, M.L. Mangano, T.R. Adams, R.J. Daines, W.G. Start, J.V. O'Brien, S.A. Chambers, A.P. Adams, N.G. Willetts, C.J. Mackey, R.W. Krueger, S.J. Zachwieja, A.P. Kausch & P.G. Lemaux, 1991. Transformation of maize using microprojectile bombardment: an update and perspective. In Vitro Cell. Dev. Biol. 27: 21–27.

Grimsley, N., T. Hohn, J.W. Davies & B. Hohn, 1987. *Agrobacterium*-mediated delivery of infectious maize streak virus into maize plants. Nature 325: 177–179.

Hartke, S. & H.J. Lorz, 1989. Somatic embryogenesis and plant regeneration from various indica rice (*Oryza sativa* L.) genotypes. J. Genet. Breed. 43: 205–214.

Herrera-Estrella, L., A. Depicker, M. Van Montague & J. Schell, 1983. Expression of chimeric genes transferred into plant cells using a Ti-plasmid-derived vector. Nature 303: 209–213.

Iida, A., T. Yamashida, Y. Yamada & H. Morikawa, 1991. Efficiency of particle bombardment-mediated transformation is influenced by cell cycle stage in synchronized cultured cells of tobacco. Plant Physiol. 97: 1585–1587.

Iida, A., M. Seki, M. Kamada, Y. Yamada & H. Morikawa, 1990. Gene delivery into cultured plant cells by DNA-coated gold particles accelerated by a pneumatic particle gun. Theor. Appl. Genet. 80: 813–816.

Jefferson, R.A.,, T.A. Kavanagh & M.W. Bevan, 1987. GUS fusions: β-glucuronidase as a sensitive and versatile gene fusion marker in higher plants. EMBO 6: 3901–3907.

Klein, T.M., E.D. Wolf, R. Wu & J.C. Sanford, 1987. High-velocity microprojectiles for delivering nucleic acids into living cells. Nature 327: 70–73.

Klein, T.M., R. Arentzen, P.A. Lewis & S. Fitzpatrick-McElligott, 1992. Transformation of microbes, plants and animals by particle bombardment. Bio/Technol. 10: 286–291.

Koziel, M.G., G.L. Beland, C. Bowman, N.B. Carozzi, R. Crenshaw, L. Crossland, J. Dawson, N. Desai, M. Hill, S. Kadwell, K. Launis, K. Lewis, D. Maddox, K. McPherson, M.R. Meghji, E. Merlin, R. Rhodes, G.W. Warren, M. Wright & S.V. Evola, 1993. Field performance of elite transgenic maize plants expressing an insecticidal protein derived from *Bacillus thuringiensis*. Bio/Technol. 11: 194–200.

Lianzheng, W., Y. Guangchu, L. Jiaofen, L. Bojun, W. Jian, Y. Zhenchun, L. Xiulan, S. Qiquan, J. Xingcun & Z. Zeqi, 1984. A study on tumor formation of soybean and gene transfer. Scientia Sinica B. 27: 391–397.

Li, L., R. Qu, A. de Kochko, C. Fauquet & R.N. Beachy, 1993. An improved rice transformation system using the biolistic method. Plant Cell Rep. 12: 250–255.

Lin, W., J.T. Odell & R.M. Schreiner, 1987. Soybean protoplast culture and direct gene uptake and expression by cultured soybean protoplasts. Plant Physiol. 84: 856–861.

McCabe, D.E., W.F. Swain, B.J. Martinell & P. Christou, 1988. Stable transformation of soybean (*Glycine max*) by particle acceleration. Bio/Technol. 6: 923–926.

McCabe, D.E. & P. Christou, 1993. Direct DNA transfer using electric discharge particle acceleration (ACCELL technology). Plant Cell Tissue Organ Cult. 33: 227–236.

McCabe, D.E. & B.J. Martinell, 1993. Transformation of elite cotton cultivars via particle bombardment of meristems. Bio/Technol. 11: 596–598.

McCown, B., D. McCabe, D. Russell, D. Robinson, K. Barton & K. Raffa, 1991. Stable transformation of Populus and incorporation of pest resistance by electric discharge particle acceleration. Plant Cell Rep. 9: 590–594.

Morikawa, H. & Y. Yamada, 1985. Capillary microinjection into protoplasts and intranuclear localization of injected materials. Plant Cell Physiol. 26: 229–236.

Moses, P.B. & N.H. Chua, 1988. Light switches for plant gene. Sci. Amer. 1988: 88–118.

Neuhaus, G., G. Spangenberg, O.M. Scheid & H.G. Schweiger, 1987. Transgenic rapeseed plants obtained by the microinjection of DNA into microspore-derived embryoids. Theor. Appl. Genet. 70: 30–36.

Nisbet, G.S. & K.J. Webb, 1990. Transformation in legumes. In: Y.P.S. Bajaj (Ed). Biotechnology in Agriculture and Forestry. Vol. 10. Legumes and Oilseed Crops I. pp. 38–48. Springer-Verlag, Berlin, Heidelberg.

Ow, D.W., K.V. Wood, M. Deluca, J.R. De Wet, D.R. Helinski & S.H. Howell, 1986. Transient and stable expression of the firefly luciferase gene inplant cells and transgenic plants. Science 234: 856–859.

Owens, L.D. & D.E. Cress, 1985. Genotypic variability of soybean response to *Agrobacterium* strains harboring the Ti or Ri plasmids. Plant Physiol. 77: 87–94.

Potrykus, I., 1990. Gene transfer to plants: assessment and perspectives. Physiol. Plant. 79: 125–134.

Pua, E.C., A. Mehra-Palta, F. Nagy & N.H. Chua, 1987. Transgenic plants of *Brassica napus* L. Biotechnol. 5: 815–817.

Rhodes, C.A., D.A. Pierce, I.J. Mettler, D. Mascarenthas & J.J. Detmer, 1988. Genetically transformed maize plants from protoplasts. Science 240: 204–207.

Russell, J.A., M.K. Roy & J.C. Sanford, 1992a. Physical trauma and tungsten toxicity reduce the efficiency of biolistic transformation. Plant Physiol. 98: 1050–1056.

Russell, J.A., M.K. Roy & J.C. Sanford, 1992b. Major improvement in biolistic transformation of suspension-cultured tobacco cells. *In Vitro* Cell. Dev. Biol. 28: 97–105.

Russell, D.R., K. Wallace, J. Bathe, B.J. Martinell & D.E. McCabe, 1993. Stable transformation of *Phaseolus vulgaris* via electric-discharge mediated particle acceleration. Plant Cell Rep. 12: 165–169.

Sanford, J.C., 1988. The biolistic process. Trends Biotechnol. 6: 299–302.

Sanford, J.C., T.M. Klein, E.D. Wolf & N.J. Allen, 1987. Delivery of substances into cells and tissues using a particle bombardment process. J. Part. Sci. Technol. 6: 559–563.

Sanford, J.C., M.J. Devit, J.A. Russell, F.D. Smith, P.R. Harpending, M.K. Roy & S.A. Johnston, 1991. An improved, helium driven biolistic device. Technique 3: 3–16.

Sautter, C., H. Waldner, G. Neuhaus-Url, A. Galli, G. Neuhaus & I. Potrykus, 1991. Micro-targeting: High efficiency gene transfer using a novel approach for the acceleration of micro-projectiles. Bio/Technol. 9: 1080–1085.

Serres, R., E. Stang, D.E. McCabe, D. Russell, D. Mahr & B. McCown, 1992. Gene transfer using electric discharge particle bombardment and recovery of transformed cranberry plants. J. Amer. Soc. Hort. Sci. 117: 174–180.

Somers, D.A., H.W. Rines, W. Gu, H.F. Kaeppler & W.R. Bushnell, 1992. Fertile, transgenic oat plants. Bio/Technol. 10: 1589–1594.

Takeuchi, Y., M. Dotson & N.T. Keen, 1992. Plant transformation: a simple particle bombardment device based on flowing helium. Plant Mol. Biol. 18: 835–839.

Tomes, D.T., A.K. Weissinger, M. Ross, R. Higgins, B.J. Drummond, S. Schaaf, J. Malone-Schoneberg, M. Staebell, P. Flynn, J. Anderson & J. Howard, 1990. Transgenic tobacco plants and their progeny derived from microprojectile bombardment of tobacco leaves. Plant Mol. Biol. 14: 261–268.

Toriyama, K., Y. Arimoto, H. Uchimiya & K. Hinata, 1988. Transgenic rice plants after direct gene transfer into protoplasts. Bio/Technol. 6: 1072–1074.

Umbeck, P., G. Johnson, K. Barton & W. Swain, 1987. Genetically transformed cotton (*Gossypium hirsutum* L.) plants. Bio/Technol. 5: 263–266.

Vain, P., M.D. McMullen & J.J. Finer, 1993. Osmotic treatment enhances particle bombardment-mediated transient and stable transformation of maize. Plant Cell Rep. 12: 84–88.

Van den Elzen, M., J. Townsend, K. Lee & J. Bedbrook, 1985. A chimaeric hygromycin resistance gene as a selectable marker in plant cells. Plant Mol. Biol. 5: 299–302.

Vasil, V., A.M. Castillo, M.E. Fromm & I. Vasil, 1992. Herbicide resistant fertile transgenic wheat plants obtained by microprojectile bombardment of regenerable embryogenic callus. Bio/Technol. 10: 667–674.

Wan, Y. & P.G. Lemaux, 1994. Generation of large numbers of independently-transformed fertile barley plants. Plant Physiol. 104: 37–48.

Weeks, T.J., O.D. Anderson & A.E. Blechl, 1993. Rapid production of multiple independent lines of fertile transgenic wheat (*Triticum aestivum*). Plant Physiol. 102: 1077–1084.

Widholm, J.M., 1993a. Retraction. Plant Physiol. 102: 331.

Widholm, J.M., 1993b. Retraction, Physiol. Plant. 87: 199.

Widholm, J.M., 1993c. Retraction. Plant Cell Rep. 12: 478.

Yamashita, T., A. Iida & H. Morikawa, 1991. Evidence that more than 90% of b-glucuronidase-expressing cells after particle bombardment directly receive the foreign gene in their nucleus. Plant Physiol. 97: 829–831.

Zhang, W. & R. Wu, 1988. Efficient regeneration of transgenic plants from rice protoplasts and correctly regulated expression of the foreign gene in the plants. Theor. Appl. Genet. 75: 835–840.

Euphytica **85**: 29–34, 1995.

Transforming the plastome: genetic markers and DNA delivery systems

P.J. Dix & T.A. Kavanagh[1]
Biology Dept., St. Patrick's College, Maynooth, Co. Kildare, Ireland; [1] *Genetics Dept., Trinity College, Dublin, Ireland*

Key words: antibiotic resistance, biolistics, chloroplast mutations, *Nicotiana*, 16S ribosomal RNA genes

Summary

Stable chloroplast transformants were first obtained following particle bombardment of tobacco leaves, and later by PEG-mediated uptake of DNA by protoplasts. The transforming DNA in these studies was itself of plastid origin and carried double (streptomycin, spectinomycin) antibiotic resistance which was used to select transformants. Integration was by homologous recombination, and both donor and recipient were *Nicotiana* species. Recent characterisation of plastid mutants of *Solanum nigrum* has allowed the extension of this gene replacement approach to include *Nicotiana:Solanum* combinations.

The introduction of functional heterologous genes into the plastome is an alternative approach based on the use of constructs in which a bacterial resistance gene is flanked by sequences homologous to a region of the recipient plastome. Thus homologous recombination in the flanking sequences allows introduction of a foreign gene. A large number of putative transformants can be generated by the method, but this apparent attraction is partly offset by the need for repeated cycles of re-selection to obtain homoplasmic plants. In contrast, homoplasmy can be accomplished in a single selection step using plastome-encoded antibiotic resistance markers.

The plastome is an attractive target for the introduction of useful genes into crop plants, as maternal inheritance acts as an insurance against unwanted spread of the foreign gene, and the large plastome copy number ensures immediate gene amplification and may influence levels of expression. Specific characters encoded on the plastid DNA, including components of photosynthesis and other aspects of metabolism, will also become open to manipulation as a consequence of developments in plastid transformation.

Introduction

Plant cells contain a number of nonphotosynthetic plastid types. In addition to those capable of developing into chloroplasts (proplastids and etioplasts), mature plastids such as amyloplasts and chromplasts are usually found in specific tissues and are associated with the accumulation and storage of particular products (e.g. starch, carotenoids). Collectively, plastids have a number of important metabolic and developmental functions. In view of the fundamental role of chloroplasts, it is inevitable that their photosynthetic efficiency should be a major target for contemporary plant breeders. Consequently an understanding of the subtle interaction between a number of nuclear and plastid genes is important, and since both genomes act in concert in producing functional organelles, both may be identified as sites where productive alterations might be achieved through genetic manipulation.

A great deal of information about the plastid genome (plastome) has accrued over the past two decades and there have been thorough reviews on both the organisation and the coding capacity of the genome (Dyer, 1985; Mullet, 1988). The chloroplast genome generally contains 120–170 kbp and codes for around 70 proteins as well as tRNAs and rRNAs. In most higher plants the plastome contains a large (10–75 kbp) inverted repeat, separating the remainder of the genome into large and small unique regions (Palmer, 1985). While the majority of proteins associated with structure or functioning of chloroplasts, either directly or indirectly (e.g. those concerned with the biosynthesis of photopigments), are encoded on nuclear genes, a number of vital ones are of plastome origin. Probably the best known of these are *rbc*L which codes for the large subunit of ribulose-1,5-biphosphate carboxylase, and *psb*A which codes for the 32 kDa Q_B protein, a key component of the electron transport chain on the reducing side of Photosystem II. In addition, there are a number of other genes for thylakoid proteins, including the cytochromes and ATP synthase as well as both Photosystem complexes, many of the chloroplast ribosomal proteins, and several envelope proteins (Dyer, 1985). Some detailed genetic maps are now available, and these have been enhanced by the sequencing of the entire genome in the liverwort *Marchantia polymorpha* (Ohyama et al., 1986) and tobacco (Shinozaki et al., 1986). In the most recent update of the tobacco plastome genetic map (Maliga, 1993), however, there remain 11 open reading frames (ORFs) of unknown function.

Genetic studies on the plastome have always been complicated by the uniparental-maternal pattern of inheritance of chloroplasts in most plants, and the large and variable number of plastome copies per plastid, and plastids per cell, which complicates the sorting out of mutant or recombinant types. To some extent these difficulties have been overcome by the use of protoplast fusion, allowing complete cytoplasmic mixing (Medgyesy, 1990), and an increasing number of selectable plastome markers (Fluhr et al., 1985; McCabe et al., 1989). Consequently, there have been significant advances, notably the demonstration of genetic recombination between plastomes (Medgyesy et al., 1985; Thanh & Medgyesy, 1989), and of occasional paternal transmission of plastids in species previously thought to exhibit strictly maternal inheritance (Medgyesy et al., 1986; McCabe et al., 1989). Taken together, these observations suggest that the possibility may exist for breeders to recombine plastid genes through conventional crossing in crop plants. The obstacles remain substantial, however, as both events (contribution of paternal plastids to the zygote, and recombination of plastomes in a mixed cytoplasm) are rare, and detection of the desired recombinants is dependent on suitable selectable markers. For these reasons, as well as the greater diversity of genes which could ultimately be introduced, plastome engineering by genetic transformation is a more attractive option. Here also there are difficulties however, residing in the bilayer membrane surrounding the organelle. Despite one early report (De Block et al., 1985), there is little evidence to support *Agrobacterium*-mediated transformation as a route to stable integration of foreign genes into the plastome. Success in this area has therefore been dependent on the development of direct DNA delivery methods and it is very recent. The first reports of stable transformation of chloroplasts only came in 1988 for a green alga, *Chlamydomonas* (Boynton et al., 1988), and in 1990 for a higher plant, tobacco (Svab et al., 1990). Since then there have been enough refinements to suggest the procedure could become more generally applicable, and of wide utility for the introduction of foreign genes into the plastome. Key factors in the transformation protocols are the choices of appropriate DNA delivery methods and selectable markers. The purpose of this short review is to evaluate these and to survey the possible breeding goals which may be tackled through genetic manipulation of the plastome.

Genetic markers

Chloroplast mutations

On the basis that integration of transforming DNA into the plastome will ordinarily proceed through homologous recombination, plastome-encoded antibiotic-resistant mutants are an attractive source for the donor DNA. In recent years a number of such mutants have been isolated and characterised, particularly in *Nicotiana* species. These include several alterations in the 16S rRNA, conferring resistance to streptomycin or spectinomycin (summarised by Svab & Maliga, 1991). A concentration of spectinomycin resistance sites is found in a conserved stem in the secondary structure of the RNA molecule. These include mutants spe^{R}40, altered at nucleotide 1012, and spe^{R}4 (nucleotide 1140), described by Fromm et al. (1987), and SPC1 (nucleotide 1138) and SPC2 (nucleotide 1139), reported by Svab & Maliga (1991). A recent spectinomycin plus streptomycin resistant double mutant (StSp1) of *Solanum nigrum* (Dix et al., 1990), also has an alteration at nucleotide 1140 (Kavanagh et al., 1993). An additional site for spectinomycin resistance in tobacco has been identified (nucleotide 1333) close to the tRNA binding region of the molecule (Svab & Maliga, 1991). Streptomycin resistance mutations have been found at nucleotides 860 (Etzold et al., 1987; Fromm et al., 1989) and 472 (Fromm et al., 1989). The latter is in the conserved '530 loop' believed to be associated with ribosomal proofreading and frame monitoring.

Streptomycin resistance mutations have also been found in the gene encoding a ribosomal protein. The 3′ part of this split gene (3′ *rps*12) is located in the inverted repeat, close to the 16s rRNA gene. A streptomycin-resistant mutant of tobacco was found to have an alteration in codon 90, converting a proline to serine (Galili et al., 1989), while the *S. nigrum* mutant, StSp1, has an A → C change (converting a lysine to glutamine) three codons upstream in codon 87 (Kavanagh et al., 1993).

A number of plastome-encoded lincomycin-resistant mutants have been isolated in *Nicotiana plumbaginifolia* (Cséplö & Maliga, 1984), and several have been shown to result from point mutations in the 23S rRNA gene. Three sites have been identified, at nucleotides 2032, 2058 and 2059, all in domain V of the secondary structure of the RNA (Cséplö et al., 1988). One of these sites (*E. coli* co-ordinate 2058) has also been identified in a newly reported mutant of *S. nigrum* (Kavanagh et al., 1993).

These binding type antibiotic resistance mutations provide highly effective markers for stable chloroplast transformation in higher plants, since they ensure an efficient selection pressure for the homoplasmic state. This is because only plastids homozygous for the mutations are resistant to the antibiotics, due to the recessive nature of these resistance traits. Furthermore, resistant plastids afford no cross protection in a mixed population. Therefore sensitive plastids are efficiently counterselected in the presence of the antibiotics. The mutations can be most effectively used in pairs as this eliminates the complications caused by forward mutation to antibiotic resistance in the recipient cell population. Svab & Maliga (1990) used a pUC-based plasmid into which a double mutant (streptomycin plus spectinomycin-resistant) tobacco 16S rRNA gene was cloned. To facilitate confirmation of stable transformation, a novel *Pst* I restriction site was also introduced close to the selectable markers. Recipient leaves were tobacco, with their own or *N. plumbaginifolia* cytoplasm (an alloplasmic substitution line), transformation was by microprojectile bombardment (to be discussed later), and initial selection was for spectinomycin resistance. Subsequent screening revealed that a high proportion of the putative transformants also carried the flanking markers (streptomycin resistance and the *Pst* I site). A subsequent study (Staub & Maliga, 1992) suggested that transformation resulted in integration of long regions of homologous DNA. More recently, other reports have confirmed the effectiveness of these mutations in the 16S rRNA for the selection of plastome transformants in *Nicotiana* (Golds et al., 1993; O'Neill et al., 1993). The latter authors highlight the difficulty of detecting the integration borders within this highly conserved region of the plastome, when both donor and recipient DNA are *Nicotiana* species.

In previously unpublished experiments we have (in collaboration with P. Medgyesy) now used DNA from a different streptomycin plus spectinomycin, resistant double mutant (StSp1), this time of *Solanum nigrum*, for transformation. In this case the mutations are in different genes (*rps* 12 and 16S rRNA respectively) on the inverted repeat (Kavanagh et al., 1993), and there are a number of other diagnostic sequence differences between *Nicotiana* and *Solanum* in the region. Wild type *N. plumbaginifolia* was the recipient and transformation of protoplasts was carried out as described previously (O'Neill et al., 1993). Selection was for spectinomycin resistance followed by screening for streptomycin resistance, or simultaneously for both antibiotics, and preliminary characterisation of putative transformant plants has confirmed the presence of *S. nigrum* plastid DNA. *Nicotiana* and *Solanum* have previously been shown to be mutually incompatible from the viewpoint of chloroplast transfer (Thanh et al., 1988), and the present experiment is the first to demonstrate plastid gene transfer between taxonomically distant species, incompatible both somatically and sexually.

An additional plastic marker which may be of use for plastome transformation is triazine herbicide resistance located on the *psb*A gene (Goloubinoff et al., 1984). This may be an insufficiently powerful marker for primary selection of transformants, particularly since expression is dependent on photosynthetic activity (Cséplö & Medgyesy, 1986), but the gene's location, proximal to the right hand repeat means it might be possible to use it in conjunction with one of the antibiotic resistance markers. A lincomycin-resistant mutant of *S. nigrum* altered in the 23S rRNA gene (Kavanagh et al., 1993) also has this triazine resistance mutation.

Heterologous genes

This strategy is a gene insertion approach, as compared to gene substitution in the case of homologous integration of an uninterrupted stretch of plastome, and has proven effective for transforming the *Chlamydomonas* chloroplast (Goldschmidt-Clermont, 1991). It relies on the use of plasmids containing a dominant resistance marker, with plastid regulatory elements, and flanked by suitable plastome sequences to permit homologous recombination. Recently it has also been used for higher plant plastome transformation. As in the case of *Chlamydomonas* the gene *Aad*A, encoding aminoglycoside adenine transferase and conferring resistance to aminoglycoside antibiotics, was used. Svab & Maliga (1993) generated plasmid pZS197 in which a chimeric *Aad*A gene (with a modified plastid ribosomal RNA promoter, and the 3′ region of the *psb*A gene) is situated between two chloroplast genes, rbcL and ORF512 and used this to transform tobacco leaves by microprojectile bombardment. They report 100 fold higher transformation frequencies than was earlier achieved with selection based on 16S rRNA mutations (Svab et al., 1990) and have since described similar results using a kanamycin resistance gene (Carrer et al., 1993). However, while the dominant marker clearly gives a high incidence of initial putative transformant shoots, it does not effectively select against retention of sensitive plastids, but instead confers resistance on the whole cell, a problem which does not arise with *Chlamydomonas* with its single plastid. This means that primary transformants are typically heteroplastidic, necessitating repeated cycles of re-selection (shoot regeneration on selective medium) to obtain the homoplasmic state achieved in a single step with binding type antibiotic resistance markers. This protracted process with its possible consequences (chromosomal abnormalities, loss of regeneration capacity) could be a drawback if the emphasis of research shifts away from 'model' species, and towards commercially important crops.

These considerations, in the opinion of the present authors, favour the use of binding type antibiotic resistance mutations, rather than dominant markers, for most applications of chloroplast transformation. Ultimately however, this may prove a question of personal preference, as both strategies have been used successfully and both will doubtless be subject to further refinements.

Delivery systems

The biolistic approach

A direct approach to getting DNA into the cells and across the double plastid membrane, is the procedure of bombardment of plant tissue with microprojectiles (tungsten or gold) coated with the DNA, using protocols based on that developed by Klein et al. (1988), for stable nuclear transformation. The first report on algal plastid transformation (Boynton et al., 1988) used this method as did the series of reports on higher plant plastid transformation from the laboratory of P. Maliga (Svab et al., 1990; Svab & Maliga, 1993; Staub & Maliga, 1992, 1993; Carrer et al., 1993). The big attraction of this method is that it is technically simple. Whole plant organs (usually leaves) can be bombarded, and transformants regenerated from them under selection, without the need for efficient procedures for isolation and culture of protoplasts. Thus it should be possible to adapt the method for a wide range of crop species for which adventitious shoots can be initiated on explants. Initial work used an explosive (gun powder) charge to project the particles, although more recently a high pressure pulse of helium gas is favoured. Both types of biolistic gun, produced by Du Pont, have been used effectively for plastome transformation by Maliga and co-workers (Maliga, 1993). A perceived drawback of the approach is the high cost associated with lease or purchase of the equipment. However, detailed instructions for making a particle inflow gun (costing only a few hundred dollars), using flowing helium, have now been published (Takeuchi et al., 1992; Finer et al., 1992) and should make the approach more widely accessible.

Direct delivery of DNA to protoplasts

DNA delivery to protoplasts can be effected by electroporation (Shillito et al., 1985) or by polyethylene glycol (PEG) treatment (Negrutiu et al., 1987). Two recent reports describe stable plastome transformation in *Nicotiana* species using the latter approach (O'Neill et al., 1993; Golds et al., 1993). This method requires no specialised equipment and is therefore relatively inexpensive. It is, however, more tedious than the biolistic approach and requires an efficient protoplast isolation and culture procedure, which clearly limits the range of species to which it can be applied.

It is difficult to make direct comparisons of the efficiencies of the biolistic and PEG-mediated methods for plastome transformation, due to the very different nature of the recipient plant material (whole leaves v. protoplasts) and the difficulty in defining the regenerable target cell population in the former. In a number of recent experiments using the PEG method and plasmids with double spectinomycin plus streptomycin resistance, of *N. tabacum* or *S. nigrum* origin, we find frequencies of about 1 transformant per 10^4 viable plated protoplasts. This seems quite respectable and, with the added attraction of a single-cell cloning step reducing the danger of generating chimeras, suggests that PEG treatment is a very attractive method for species fulfilling the protoplast culture requirements.

Applications

Fundamental investigations into interactions between nuclear and plastid genes, and the roles and regulation of the latter, will certainly benefit from our capacity to transform the plastome. Strategies will include examining the consequences of alteration of regulatory elements, gene replacement using mutant forms of plastid genes, and the use of antisense constructs to study the role of open reading frames. Insertion mutagenesis using selectable foreign sequences is less likely to be valuable in the small, fully-sequenced plastome, made up mostly of cloned or easily clonable genes, and with efficient homologous recombination. In most cases these studies can be effected through the gene replacement approach, utilising two binding type antibiotic resistance mutations linked to, or co-transformed with, modified plastid sequences. Where possible this should be the method of choice because of the rapid and reliable attainment of homoplasmy in transformed cells. The obvious target of most of these investigations will be photosynthetic activity, and the insights gained might suggest means of improving its efficiency by selective modification of the plastome.

A number of the traits being engineered into crop plants, e.g. resistance to agrochemicals, insect predation and viral diseases, are dependent on high levels of expression of a particular protein. The plastome could be an attractive site for the genes concerned, providing gene amplification up to 10,000 times (Maliga, 1993) and high levels of gene product in the leaves (Staub & Maliga, 1993). For an objective like the introduction of *Bacillus thuringiensis* insecticidal protein genes, the chloroplast may be a suitable location for the product. In some other cases it may be necessary to tackle the problem of export from the chloroplast. A further attraction of this approach is the exclusion of paternal plastids from the zygote in most crop plants, which would help ensure containment of the foreign gene in the modified crop variety.

Amplification of the foreign gene, when incorporated into the plastome, may cause re-appraisal of transgenic plants as an alternative to transgenic animals for the production of human gene products, such as anti-hemophiliac factor IX (Clark et al., 1989) as well as that of recombinant proteins by micro-organisms. There may be major barriers to realization of some of these ideas and it would be easy to overstate the significance of gene amplification for product accumulation. Additionally, the transformation protocols still need to be adapted for species other than *Nicotiana*. However, the availability of procedures for transforming plastids, appears set to put some new strategies and targets on the agenda for genetic manipulation of crop plants.

Acknowledgements

The authors would like to thank P. Medgyesy for critical reading of the manuscript.

References

Boynton, J.E., N.W. Gillham, E.H. Harris, P. Hosler, A.M. Johnson, A.R. Jones, B.L. Randolph-Anderson, D. Robertson, T.M. Klein, K.B. Shark & J.C. Sanford, 1988. Chloroplast transformation in *Chlamydomonas* with high velocity microprojectiles. Science 240: 1534–1538.

Carrer, H., T.N. Hockenberry, Z. Svab & P. Maliga, 1993. Kanamycin resistance as a selectable marker for plastid transformation in tobacco. Mol. Gen. Genet. 241: 49–56.

Clark, A.J., H. Bessos, J.O. Bishop, P. Brown, S. Harris, R. Lathe, M. McClenaghan, C. Prowse, J.P. Simons, C.B.A. Whitelaw & I. Wilmut, 1989. Expression of human anti-hemophiliac factor IX in the milk of transgenic sheep. Bio/Technology 7: 487–492.

Cséplö, A. & P. Maliga, 1984. Large scale isolation of maternally inherited lincomycin resistant mutations, in diploid *Nicotiana plumbaginifolia* protoplast cultures. Mol. Gen. Genet. 196: 407–412.

Cséplö, A. & P. Medgyesy, 1986. Characteristic symptoms of photosynthesis inhibition by herbicides are expressed in photomixotrophic tissue cultures of *Nicotiana*. Planta 168: 24–28.

Cséplö, A., T. Etzold, J. Schell & P. Schreier, 1988. Point mutations in the 23S rRNA genes of four lincomycin resistant *Nicotiana plumbaginifolia* mutants could provide new selectable markers for chloroplast transformations. Mol. Gen. Genet. 214: 295–299.

De Block, M., J. Schell & M. Van Montagu, 1985. Chloroplast ciencetransformation by *Agrobacterium tumefaciens*. EMBO J. 4: 1367–1372.

Dix, P.J., C.P. McKinley & P.F. McCabe, 1990. Antibiotic resistant mutants in *Solanum nigrum*. In: H.J.J. Nijkamp, L.H.W. Van der Plas & J. Van Aartrijk (Eds). Progress in Plant Cell and Molecular Biology, pp. 169–174. Kluwer Academic Publishers, Dordrecht.

Dyer, T.A., 1985. The chloroplast genome and its products. Oxford Surveys of Plant Molecular and Cell Biology 2: 147–177.

Etzold, T., C.C. Fritz, J. Schell & P.H. Schreier, 1987. A point mutation in the chloroplast 16S rRNA gene of a streptomycin resistant *Nicotiana tabacum*. FEBS Lett. 219: 343–346.

Finer, J.J., P. Vain, M.W. Jones & M.D. McMullen, 1992. Development of the particle inflow gun for DNA delivery to plant cells. Plant Cell Rep. 11: 323–328.

Fluhr, R., D. Aviv, E. Galun & M. Edelman, 1985. Efficient induction and selection of chloroplast encoded antibiotic resistant mutants in *Nicotiana*. Proc. Natl. Acad. Sci. USA 82: 1485–1489.

Fromm, H., M. Edelman, D. Aviv & E. Galun, 1987. The molecular basis for rRNA-dependent spectinomycin resistance in *Nicotiana* chloroplasts. EMBO J. 6: 3233–3237.

Fromm, H., E. Galun & M. Edelman, 1989. A novel site for streptomycin resistance in the '530 loop' of chloroplast 16S ribosomal RNA. Plant Mol. Biol. 12: 499–505.

Galili, S., H. Fromm, D. Aviv, M. Edelman & E. Galun, 1989. Ribosomal protein S12 as a site for streptomycin resistance in *Nicotiana* chloroplasts. Mol. Gen. Genet. 218: 289–292.

Golds, T., P. Maliga & H.-U. Koop, 1993. Stable plastid transformation in PEG-treated protoplasts of *Nicotiana tabacum*. Bio/Technology 11: 95–97.

Goldschmidt-Clermont, M., 1991. Transgenic expression of aminoglycoside adenine transferase in the chloroplast: A selectable marker for site-directed transformation of *Chlamydomonas*. Nucleic Acids Res. 19: 4083–4089.

Goloubinoff, P., M. Edelman & R.B. Hallick, 1984. Chloroplast coded atrazine resistance in *Solanum nigrum*: psbA loci from susceptible and resistant biotypes are isogenic except for a single codon change. Nucleic Acids Res. 12: 9489–9496.

Kavanagh, T.A., K.M. O'Driscoll, P.F. McCabe & P.J. Dix, 1993. Mutations conferring lincomycin, spectinomycin, and streptomycin resistance in *Solanum nigrum* are located in three different chloroplast genes. Mol. Gen. Genet., in press.

Klein, T.M., T. Gradziel, M.E. Fromm & J.C. Sanford, 1988. Factors influencing gene delivery into *Zea mays* cells by high velocity microprojectiles. Bio/Technology 6: 559–563.

Maliga, P., 1993. Towards plastid transformation in flowering plants. TIBTECH 11: 101–107.

McCabe, P.F., A.M. Timmons & P.J. Dix, 1989. A simple procedure for the isolation of streptomycin resistant plants in *Solanaceae*. Mol. Gen. Genet. 216: 132–137.

Medgyesy, P., 1990. Selection and analysis of cytoplasmic hybrids. In: P.J. Dix (Ed). Plant Cell Line Selection, pp. 287–316. VCH Publishers, Weinheim.

Medgyesy, P., E. Fejes & P. Maliga, 1985. Interspecific chloroplast recombination in a *Nicotiana* somatic hybrid. Proc. Natl. Acad. Sci. USA 82: 6960–6964.

Medgyesy, P., A. Páy & L. Márton, 1986. Transmission of paternal chloroplasts in *Nicotiana*. Mol. Gen. Genet. 204: 195–198.

Mullet, J.E., 1988. Chloroplast development and gene expression. Ann. Rev. Plant Physiol. Plant Mol. Biol. 39: 475–502.

Negrutiu, I., R.D. Shillito, I. Potrykus, G. Biasini & F. Sala, 1987. Hybrid genes in the analysis of transformation conditions, I. Setting up a simple method for direct gene transfer in plant protoplasts. Plant Mol. Biol. 8: 363–373.

Ohyama, K., H. Fukuzawa, T. Kohchi, H. Shirai, T. Sano, S. Sano, K. Umesono, Y. Shiki, M. Takeuchi, Z. Chang, S.-I. Aota, H. Inokuchi & H. Ozeki, 1986. Chloroplast gene organization deduced from complete sequence of liverwort *Marchantia polymorpha* chloroplast DNA. Nature 322: 572–574.

O'Neill, C., G.V. Horváth, E. Horváth, P.J. Dix & P. Medgyesy, 1993. Chloroplast transformation in plants: polyethylene glycol (PEG) treatment of protoplasts is an alternative to biolistic delivery systems. Plant J. 3: 729–738.

Palmer, J.D., 1985. Comparative organization of chloroplast genomes. Ann. Rev. Genet. 19: 325–354.

Shillito, R.D., M.W. Saul, J. Paszkowski, M. Müller & I. Potrykus, 1985. High efficiency direct gene transfer to plants. Bio/Technology 3: 1099–1103.

Shinozaki, K., M. Ohme, M. Tanaka, T. Wakasugi, M. Hayashida, T. Matsubayashi, N. Zaita, J. Chunwongse, J. Obokata, K. Yamaguchi-Shinozaki, C. Ohto, K. Torozawa, B.Y. Meng, M. Sugita, H. Deno, T. Kamogashira, K. Yamada, J. Kusuda, F. Takaiwa, A. Kato, N. Tohdo, H. Shimada & M. Sugiura, 1986. The complete nucleotide sequence of the tobacco chloroplast genome: its gene organization and expression. EMBO J. 5: 2043–2049.

Staub, J.M. & P. Maliga, 1992. Long regions of homologous DNA are incorporated into the plastid genome by transformation. Plant Cell 4: 39–45.

Staub, J.M. & P. Maliga, 1993. Accumulation of D1 polypeptide in tobacco plastids is regulated *via* the untranslated region of psbA mRNA. EMBO J. 12: 601–606.

Svab, Z. & P. Maliga, 1991. A mutation proximal to the tRNA binding region of the *Nicotiana* plastid 16S rRNA confers resistance to spectinomycin. Mol. Gen. Genet. 228: 316–319.

Svab, Z. & P. Maliga, 1993. High-frequency plastid transformation in tobacco by selection for a chimeric *aad*A gene. Proc. Natl. Acad. Sci. USA 90: 913–917.

Svab, Z., P. Hajdukiewitz & P. Maliga, 1990. Stable chloroplast transformation in higher plants. Proc. Natl. Acad. Sci. USA 87: 8526–8530.

Takeuchi, Y., M. Dotson & N.T. Keen, 1992. Plant transformation: a simple particle bombardment device based on flowing helium. Plant Mol. Biol. 18: 835–839.

Thanh, N.D. & P. Medgyesy, 1989. Limited chloroplast gene transfer via recombination overcomes plastome-genome incompatibility between *Nicotiana tabacum* and *Solanum tuberosum*. Plant Mol. Biol. 12: 87–93.

Thanh, N.D., A. Páy, M.A. Smith, P. Medgyesy & L. Márton, 1988. Intertribal chloroplast transfer by protoplast fusion between *Nicotiana tabacum* and *Salpiglossis sinuata*. Mol. Gen. Genet. 213: 186–190.

Euphytica **85:** 35–44, 1995.

Genetic engineering of cereal crop plants: a review

A. Jähne, D. Becker & H. Lörz
Institut für Allgemeine Botanik, Zentrum für angewandte Molekularbiologie der Pflanzen, AMP II, Ohnhorststr. 18, D-22609 Hamburg, Germany

Key words: cereals, protoplast transformation, tissue electroporation, particle bombardment

Summary

Many aspects of basic and applied problems in plant biology can be investigated by transformation techniques. In dicotyledonous species, the ability to generate transgenic plants provides the tools for an understanding of plant gene function and regulation as well as for the directed transfer of genes of agronomic interest.

For many dicotyledonous plants *Agrobacterium tumefaciens* can be routinely used to introduce foreign DNA into their genome. However, cereals seem to be recalcitrant to *Agrobacterium*-mediated transformation.

In cereals, many efforts have been made in recent years to establish reliable transformation techniques. Several transformation techniques have been developed but to date only three methods have been found to be suitable for obtaining transgenic cereals: transformation of totipotent protoplasts, particle bombardment of regenerable tissues and, more recently, tissue electroporation. The current state of transformation methods used for cereals will be reviewed.

Introduction

The transfer of defined genes is theoretically the most straightforward method for improvement of crop plants. Methods for crop plant transformation have only been developed in recent years. Generally the method of choice for the delivery of genes to dicotyledonous species is the use of *Agrobacterium tumefaciens*. Transformed plants are being obtained for an increasing number of species including agronomically-important crops (Hooykaas & Schilperoort, 1992). The *Agrobacterium* vector system is not only used extensively for the transfer of various traits to crop plants, but also for the study of gene function in plants. Applications include the transfer of genes affecting such widely diverse traits as resistance to pests, diseases or herbicides and tolerance to environmental stress. The transfer of genes in order to modify metabolic pathways to change the quality of plant products for industrial purposes is also an important goal. Some of the transgenic crops produced are now ready for marketing, and thus transformation techniques will supplement classical breeding methods.

Cereals, as a major group of crop plants, are important targets for the application of genetic manipulation techniques. Unfortunately, most monocotyledons are not among the natural hosts of *Agrobacteria*. Only members of the orders Liliales and Arales have proved to be susceptible; all members of the Poales tested have shown to be nonsusceptible (De Cleene, 1985). Consequently, the prospects for successful genetic engineering of cereals utilising *Agrobacterium* would not seem to be very promising. Nevertheless, there is evidence that under certain conditions an *Agrobacterium*-mediated gene transfer to some monocotyledonous species is possible (Bytebier et al., 1987; Raineri et al., 1990; Gould et al., 1991; Mooney et al., 1991; Li et al., 1992). However, the regeneration of stably-transformed plants and the inheritance of the transferred gene has been discussed more controversially by Langridge et al. (1992).

Furthermore, several groups have reported the phenomenon of 'agroinfection' where viral genomic sequences have been transferred to cereal meristematic cells resulting in systemic viral infection in the recovered plants (Hohn et al., 1987; Grimsley et al.,

Table 1. Methods investigated for gene transfer to cereals (for review see Potrykus, 1990)

- Chemically induced DNA uptake into protoplasts
- Electrically induced DNA uptake into protoplasts
- Bombardment of cells and tissues with DNA-coated particles
- Electroporation of tissues with DNA
- Macroinjection of DNA into floral tillers
- Microinjection of DNA into microspores, microspore-derived cells and tissues with DNA-coated particles
- Electroporation of tissues with DNA
- Macroinjection of DNA into floral tillers
- Microinjection of DNA into microspores, microspore-derived pro-embryos or zygotic pro-embryos
- DNA-uptake by germinating pollen
- Imbibition of embryos with DNA

1987; Dale et al., 1989). As cereals cannot be readily transformed by *Agrobacterium* research activities have focussed on the development of alternative gene transfer methods. Various techniques have been tested (Table 1) but presently only three methods have proved to be suitable for obtaining transgenic cereals: tissue electroporation, transformation of protoplasts and particle bombardment of regenerable tissue cultures.

Tissue electroporation

One successful method is the delivery of DNA to regenerable tissues by electroporation. Tissue electroporation has been used to transfer DNA to enzymatically- or mechanically-wounded zygotic or somatic maize embryos. Transgenic plants have been obtained reproducibly and the transferred neo-gene segregated according to Mendelian rules (D'Halluin et al., 1992). Although this is a very recent development, the future prospects of this technique are very promising since there has been a further report showing successful gene transfer to scutellum cells of wheat by transient expression experiments (Klöti et al., 1993).

Transformation of cereal protoplasts

Another technique which has reproducibly yielded transgenic cereals is DNA uptake by protoplasts, stimulated either by PEG treatment or induced by electroporation. For a long time, cereals seemed to be recalcitrant in tissue and especially in protoplast culture. However, considerable progress has been made in establishing reliable and efficient *in vitro* culture systems. In many dicotyledonous species, plants can be regenerated from mesophyll protoplasts, but in cereals there is only scant evidence that protoplasts isolated from leaves are capable of sustained divisions (Hahne et al., 1990). There is, however, a recent report of plant regeneration from mesophyll protoplasts of rice (Gupta & Pattanayak, 1993).

So far, embryogenic suspension cultures are the main source of totipotent cereal protoplasts. Embryogenic suspensions originating either from immature embryos or from microspores, have been generated for nearly all important cereals (Vasil & Vasil, 1992). The establishment of embryogenic suspensions suitable for the release of protoplasts has been an important prerequisite for the progress achieved in cereal protoplast research. Nevertheless, it is very difficult and labour-intensive to initiate and maintain these suspensions. Furthermore, regeneration capacity has been observed to decline gradually during cultivation in cereal suspensions (Jähne et al., 1991a). Therefore the long-term availability of embryogenic suspensions is a limiting factor in protoplast research. A solution to this problem could be the cryopreservation of suspension cells. Cryopreserved maize suspensions have been shown to provide a long-term and reliable source of totipotent cells (Shillito et al., 1989). However, efficient freezing protocols are not available for all cereal cell suspension cultures and considerable time and effort must be spent for the establishment of novel suspension cultures (Fretz et al., 1992).

Embryogenic suspension cultures have been used for the isolation of totipotent protoplasts and regeneration of plants from these single cells is possible for most important cereals such as rice, maize, wheat and barley (Table 2). However, plant regeneration from cereal protoplasts remains a difficult and often unreliable process, depending on many parameters not under experimental control (Potrykus, 1989). Protoplast regeneration is currently an efficient and routinely-used method only in rice and in specific genotypes of maize.

The use of protoplasts in genetic engineering has very significant applications not only for stable transformation. For several types of experiment such as the analysis of promotor function and gene expression, it is possible to use protoplasts for transient expression studies.

Protoplasts can be induced to take up DNA either by PEG or by electric pulses. Both methods have proved

Table 2. Regeneration and transformation of protoplasts of cereal crops

	Regeneration of fertile plants	Transgenic plants
Rice	Abdullah et al., 1986	Toriyama et al., 1988
	Toriyama et al., 1986	Zhang et al., 1992
	Kyozuka et al., 1987	Zhang & Wu, 1988
	Kyozuka et al., 1988	Shimamoto et al., 1989
	Datta et al., 1990	Datta et al., 1990
	Li & Murai, 1990	Tada et al., 1990
	Datta et al., 1992	Terada et al., 1993
		Rathore et al., 1993
Maize	Prioli & Söndahl, 1989	Rhodes et al., 1988
	Shillito et al., 1989	Donn et al., 1992
	Morocz et al., 1990	Golovkin et al., 1993
	Donn et al., 1992	Omirulleh et al., 1993
Barley	Jähne et al., 1991b	
	Funatsuki et al., 1992	
	Golds et al., 1993	
Wheat	Ahmed & Sagi, 1993	

Table 3. Biolistic transformation of cereals using different target tissues

Embryogenic cell suspension cultures	
Rice	Cao et al., 1992
Maize	Gordon-Kamm et al., 1990
	Fromm et al., 1990
Oat	Somers et al., 1992
Embryogenic callus cultures	
Maize	Genovesi et al., 1992
	Walters et al., 1992
Wheat	Vasil et al., 1992
Barley	Wan & Lemaux, 1993
Sugarcane	Bower & Birch, 1992
Primary explants	
Rice	Christou et al., 1991
Maize	Koziel et al., 1993
Wheat	Weeks et al., 1993
	Becker et al., 1994
Barley	Wan & Lemaux, 1994
	Ritala et al., 1994
	Jähne et al., 1994
Tritordeum	Barcelo et al., 1994
Triticale	Zimny et al., 1995

to be suitable for stable transformation of cereal protoplasts and transformed cell lines could be obtained (Fromm et al., 1986; Rhodes et al., 1988; Lazzeri et al., 1991). Although direct DNA-uptake is a successful and routinely-used method, the regeneration of transgenic plants remains difficult. Until now, it has only been possible to obtain transgenic plants by protoplast transformation in rice and maize, from which protoplast-derived fertile plants can be obtained reproducibly (Table 2). This underlines that the regeneration of protoplast-derived plants still remains a significant limiting factor in obtaining transgenic cereals.

In both maize and rice it has now been shown that the transferred genes are inherited by the progeny and are as stable as original plant genes. Furthermore, direct DNA-uptake into protoplasts provides the possibility of co-transformation and thus the recovery of transgenic plants without selectable marker genes can be achieved by subsequent conventional breeding methods.

Particle bombardment

Embryogenic cell suspensions represent not only a source of totipotent protoplasts, but can also be used as target cells for an alternative successful transformation technique: particle bombardment. The particle bombardment process is a method for the delivery of genes into intact cells and tissues through the use of DNA-coated microprojectiles (tungsten or gold). It was developed by Sanford & co-workers (1987) and has become the second most widely-used method for plant genetic transformation after *Agrobacterium*-mediated gene transfer (Gray & Finer, 1993). Several laboratories have demonstrated that microprojectiles are suitable for the transfer of genes to a wide range of plant tissues and species. Apparently, there is no difference in the efficiency of biolistic transformation of monocotyledonous and dicotyledonous species (Sanford, 1990).

For the transformation of cereals, the choice of appropriate target tissue is of major importance as there are only a few tissues capable of plant regeneration. Regenerable tissues of cereals have been tested by transient expression assays and have proved to be suitable for biolistic transformation experiments (Table 3). The most common tissues used for this purpose are embryogenic suspension cells and embryogenic callus cultures. For the stable transformation and regeneration of maize (Gordon-Kamm et al., 1990; Fromm et al., 1990), rice (Cao et al., 1992) and oat (Somers et al., 1992) suspension cells have been used as target tissue. However, the morphogenetic competence of cells is significantly reduced during maintenance and the phenomenon of somaclonal variation limits the suitability of this tissue. Accordingly, embryogenic callus has been considered as a target tissue because the time needed for establishment of cultures and plant regeneration is shorter for callus cultures than for suspension cultures. Using callus cultures, it was possible to regenerate transgenic sugarcane (Bower & Birch, 1992), wheat (Vasil et al., 1992) maize (Genovesi et al., 1992; Walters et al., 1992) and barley (Wan & Lemaux, 1993).

Although these cultures could be successfully used for the production of transgenic cereals, it would be more desirable to deliver DNA directly into primary explants with a high regeneration capacity. The time necessary for the preparation of the target tissue is comparatively low and the risk of somaclonal variation is reduced as the period in culture is shortened to a few weeks. In cereals, scutellar tissues of rice (Christou et al., 1991), maize (Koziel et al., 1993) and wheat (Weeks et al., 1993) have been used for the regeneration of stably-transformed plants. In the following section, the progress made in our laboratory towards biolistic transformation of cereals using various explants is summarized.

Strategies for the biolistic transformation of primary explants

A prerequisite for the production of transgenic cereals has been the development of efficient *in vitro* culture systems from which fertile plants can be regenerated at a high frequency. In our experiments the explants of choice have been the scutellar tissue of immature embryos of wheat and triticale, immature inflorescences of Tritordeum (Barcelo et al., 1989) and barley microspores. Currently, the culture of barley microspores is considered superior to anther culture as the regeneration frequency can be significantly increased (Olsen, 1991; Hoekstra et al., 1993; Mordhorst & Lörz, 1993). In barley, a spontaneous autoendoreduplication of the genome during the first division of the microspore leads to homozygous, diploid regenerants. Therefore, this haploid target tissue makes the regeneration of homozygous R_0-plants possible.

Each target tissue has been the subject of individual optimization experiments to improve conditions for particle bombardment. The most convenient method to measure the efficiency of DNA delivery into intact cells, is the determination of the number of cells which transiently express the *uid*A-gene (β-glucuronidase). Because of the relatively low sensitivity of the histochemical GUS-assay, the use of a strong promoter which enhances the expression of the marker gene is important. A construct harbouring the *uid*A-gene driven by an Act-1-D-promoter (McElroy et al., 1990) provided best results in all tissues.

For optimization of the bombardment process and for selection of stably-transformed plants, a plasmid (pDB1; Fig. 1) containing the visualizable marker gene *uid*A under the control of the Actin 1 promoter from rice, and the selectable marker gene *bar* under the control of the CaMV 35S promoter has been used.

For Tritordeum the method of co-transformation using the plasmids pAct-1-D/GUS and pCalneo has been successfully demonstrated (S. Lütticke, unpublished).

The aim of these optimization studies is to achieve a high frequency of transiently expressing cells. However, it is also very important that the target tissue does not suffer significantly from the bombardment. The degree of tissue damage depends on the type of explant, the particle density and the acceleration pressure.

Immature embryos of wheat and triticale have proved to be the most sensitive tissue. In wheat, high particle densities (116 μg gold particles average size 0,4–1,2 μm) per bombardment caused severe tissue damage, whereas Tritordeum inflorescences were not reduced in their morphogenetic competence under the same conditions. The viability of barley microspores has not been influenced by the amount of particles used for bombardment in our studies.

The results from our optimization and selection experiments demonstrate that the optimal particle density has to be determined carefully for each type of explant, whereas the acceleration pressure can be relatively wide-ranging without having a negative effect on tissue viability and competence.

Strategies for the improvement of the biolistic method not only concern the parameters of the particle bombardment process, but also the culture conditions for explants. In spring varieties of wheat and in triticale, a clear reduction of tissue damage has been achieved when the immature embryos were precultured *in vitro* for two to seven days. In Tritordeum, a preculture of one day led to an increase in the number of transformed plants, a finding which reinforces the importance of the determination of optimal preculture

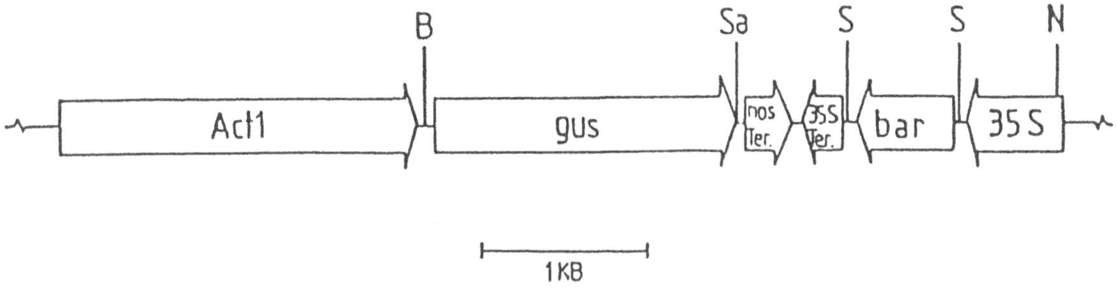

Fig. 1. Schematic representation of plasmid pDB1 used in transformation experiments: S: Sal I; Sa: SacI; B: BamHI; N: NcoI.

conditions (Barcelo et al., 1994). In barley, no effect on transient expression of the β-glucuronidase gene or on viability has been observed when freshly-isolated or one- to three-day old microspores were bombarded. Under optimal culture and bombardment conditions, an average number of 100 transient GUS-signals per embryo has been counted in wheat and triticale (Fig. 2A). In barley, about 1% of the bombarded mirospores transiently expressed the *uid*A-gene (Fig. 2B).

The establishment of an efficient transformation system requires careful choice of an appropriate selectable marker gene. For grasses, antibiotics such as hygromycin, kanamycin of G-418 have been used successfully. In comparison, the major advantage of herbicide selection is that plants can be selected in late stages of development by a simple spray test.

For wheat, triticale and barley, phosphinothricin (PPT) resistance conferred by the *bar* gene has proved to be a useful selectable marker for the regeneration of transformed plants. The successful use of PPT selection depends on the cell type and the developmental stage. Therefore, the selection conditions were optimized individually. The optimal selection conditions varied; in wheat a PPT-concentration of 0.5–2 mg/l was most suitable, whereas in barley 3–5 mg/l PPT was optimal. Two to three weeks after bombardment the calli were transferred to selection medium. The selection pressure was applied both during callus induction and the plant regeneration phase (Fig. 2C). Furthermore, the regenerants were sprayed in a later stage of development with a solution containing 150–200 mg/l PPT in order to verify their resistance (Fig. 2D). This selection system represents a fast and highly-efficient method for identifying transformed plants.

Using the pDB1-construct (Fig. 1) it is also possible to screen non-selected regenerants by spraying with PPT or by histochemically assaying the expression of the *uid*A gene (Fig. 2G). By this method several transgenic triticale (J. Zimny, personal communication) and one transformed barley plant have been identified.

The transformation efficiency depended on the quality of the explant material and varied from experiment to experiment. In wheat, an average frequency of one transgenic plant per 83 bombarded embryos could be achieved (Becker et al., 1994). This frequency is substantially higher than the 1–2 plants per 1000 bombarded embryos reported by Weeks et al. (1993). In barley, independent transformation events led to the average recovery of one plant per 1×10^7 bombarded microspores (Jähne et al., 1994). Southern analysis of selected regenerants demonstrated that in most cases plants contained intact copies of both marker genes (Fig. 3). Single copy as well as multi-copy integration of one or both marker genes (*uid*A/*bar* or *uid*A/*neo*) was detectable. Furthermore larger or smaller fragments were observed, suggesting that deletions, rearrangements and/or methylation at restriction sites had occurred.

In our experiments, no phenotypic abnormalities or reduced fertility (Fig. 2H) was observed. This has been reported for transgenic maize (Gordon-Kamm et al., 1990), wheat (Vasil et al., 1992) and oat plants (Somers et al., 1992) obtained from microprojectile bombardment of embryogenic suspension or callus cultures.

The segregation of the introduced marker gene *uid*A was visualized histochemically in pollen grains of R0-plants (Fig. 2E and F) and in leaves of the progeny. The functional activity of the *bar* gene was tested by spraying the progeny with an aqueous solution of the herbicide Basta.

A 1 : 1 segregation as well as segregation of the introduced marker gene in a non-Mendelian fashion was observed in wheat, triticale and Tritordeum. In barley, the introduced marker genes were inherited by

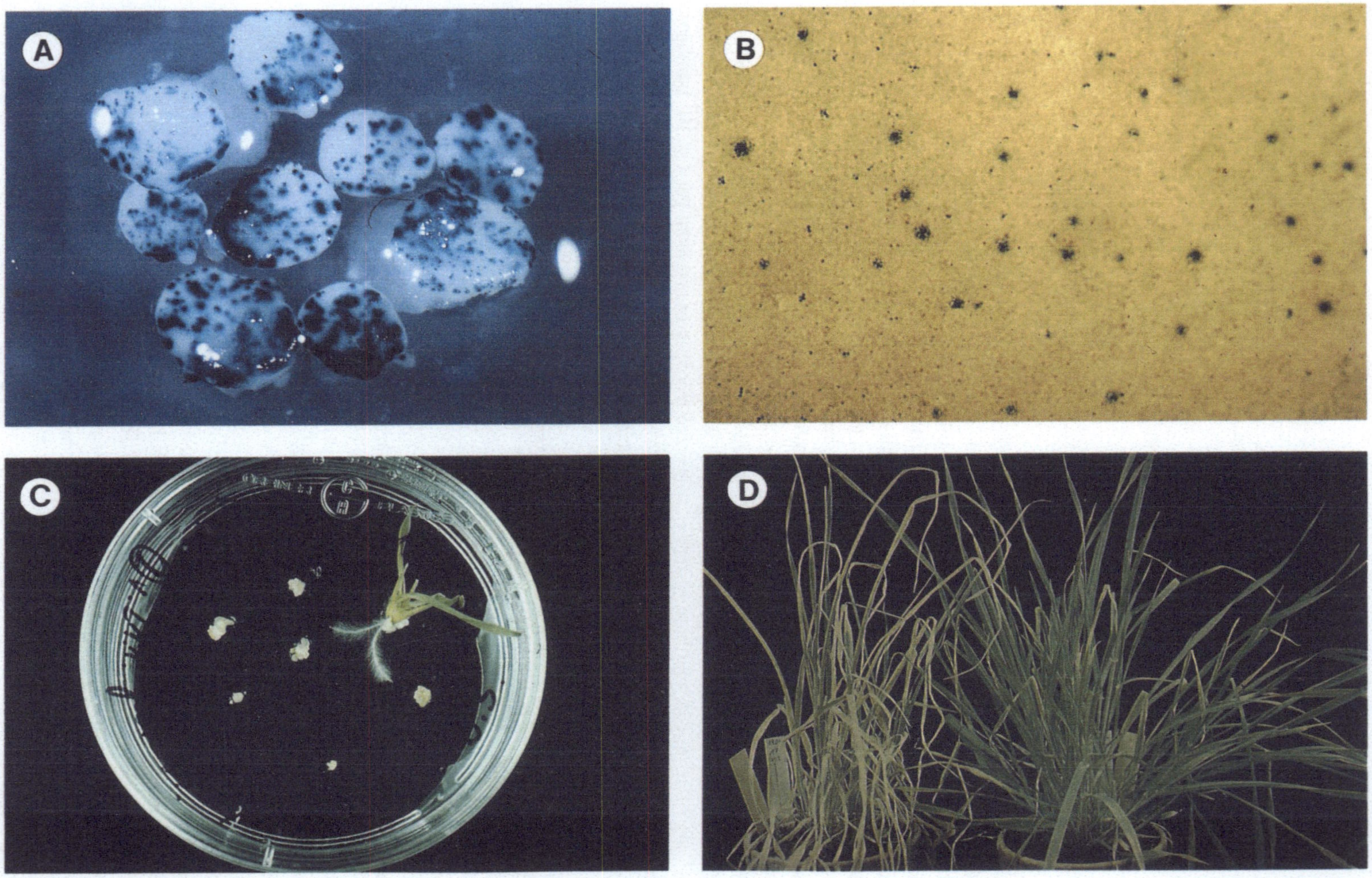

Fig 2 *2A + B* Transient GUS activity in scutellar tissue of wheat (A) and in barley microspores (B) 48 h after bombardment *C* Regeneration of a barley plant under selection conditions using 5 mg/l PPT *D* Selection of a transformed barley plant by spraying an aqueous solution of 150 mg/l PPT, left sensitive control plant, right resistant transgenic plant

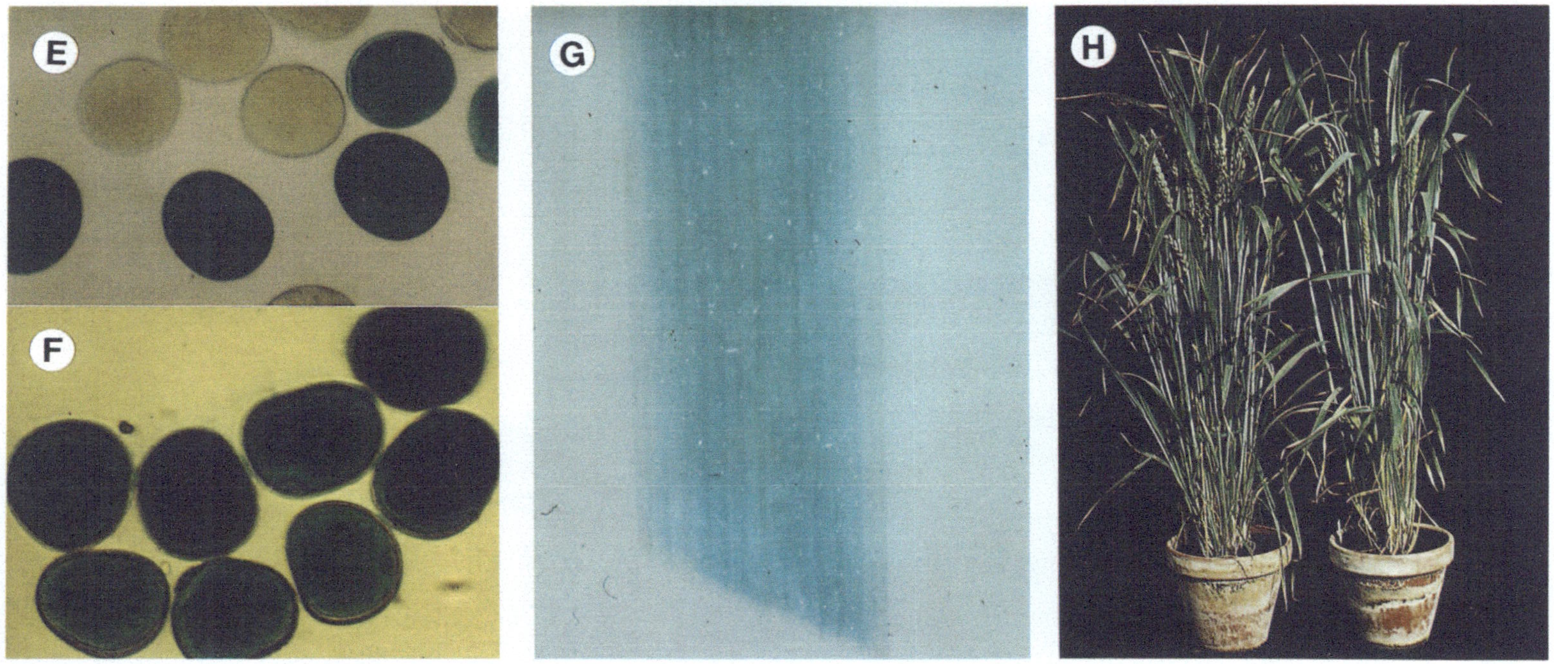

Fig 2 *2E + F* GUS activity in pollen of transgenic R0 plants A segregation of 1 1 in a wheat plant (E) and pollen of a homozygous barley plant (F) are shown *G* GUS activity in leaf tissue of transgenic wheat *H* Mature transgenic R0 plant (left) and a seed derived wheat plant (right)

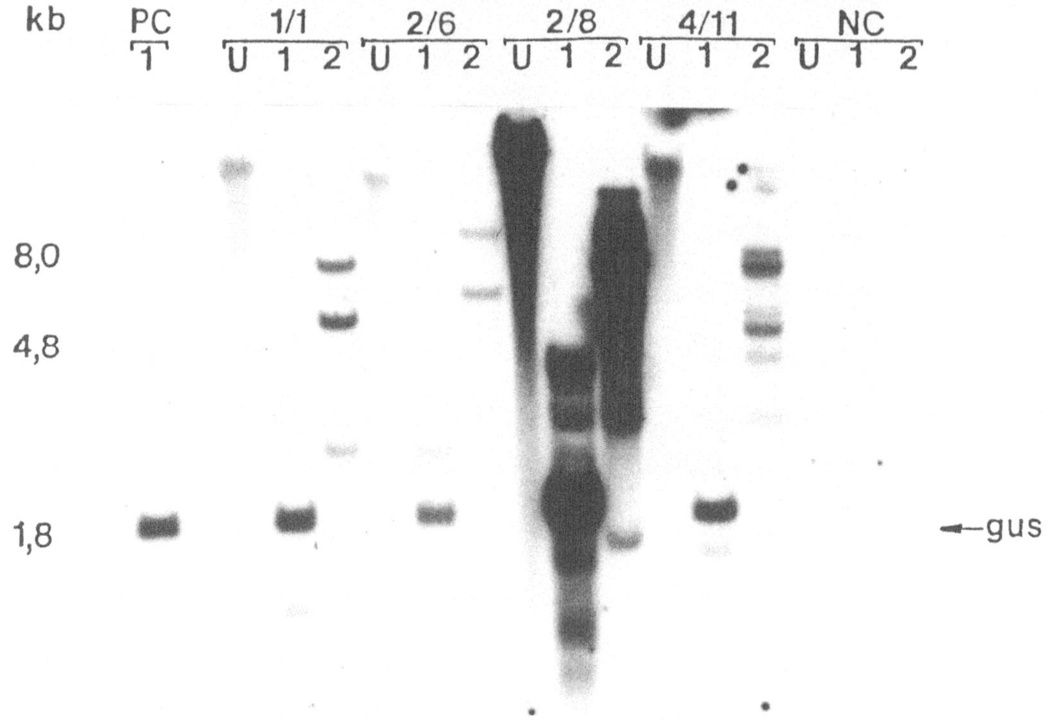

Fig. 3. Integration of the *gus* gene in four GUS positive R_0-plants of wheat (1/1, 2/6, 2/8 and 4/11). Southern blots of genomic DNA (25 μg/lane). Hybridization was carried out using a DIG-11-dUTP-labelled *gus* probe. NC: Negative wheat control, PC: Positive control; plasmid pDB1 digested with BamHI and SacI. U: Undigested. 1: Digested with BamHI/SacI to cut out the *gus* gene. 2: Digested with NcoI to determine the number of integration sites per genome.

all the progeny, indicating the homozygous genotype of the transformed plants.

Progeny have been further analysed by Southern hybridization. Wheat plants showing a 1 : 1 segregation in pollen grains and a 3 : 1 inheritance of the marker genes in the progeny, had the same integration pattern as the parental line. This indicates a close linkage of the introduced marker genes and their inheritance as a genetic unit. However, individuals containing multi-copy insertions did not always inherit the genes in a Mendelian fashion.

Conclusion

Although the establishment of protoplast regeneration and transformation systems has improved significantly, a routine combination of both systems is not yet possible.

On the other hand, the development of biolistic transformation systems has allowed rapid progress towards the recovery of transgenic cereals. Embryogenic suspension and callus cultures were used initially but recent results show that primary explants seem to be more advantageous for the routine production of fertile transgenic cereals. The recently-developed method of tissue electroporation promises to be another very attractive approach for the transformation of primary explants. In summary, the tools of genetic manipulation of cereals have been significantly improved, providing increased opportunities to transfer agronomically-interesting genes.

Note added in proof

In the meantime several new reports on the transformation of cereals have been published. Most of these reports present transgenic cereal plants obtained either by particle bombardment or by tissue electroporation giving further evidence for the suitability of these methods.

However, an unexpected publication was the one of Hiei et al. 1994. The Plant J. 6: 271–282, who reported the regeneration of transgenic Japonica rice plants from scutellar tissues co-cultivated with *Agrobacterium tumefaciens*.

References

Abdullah, B., E.C. Cocking & J.A. Thompson, 1986. Efficient plant regeneration from rice protoplasts through somatic embryogenesis. Bio/Technology 4: 1087–1090.

Ahmed, K.Z. & F. Sagi, 1993. Culture of and fertile plant regeneration from regenerable embryogenic suspension cell-derived protoplasts of wheat (*Triticum aestivum* L.). Plant Cell Rep. 12: 175–179.

Barcelo, P., A. Vazquez & A. Martin, 1989. Somatic embryogenesis and plant regeneration from Tritordeum. Plant Breeding 103: 235–240.

Barcelo, P., C. Hagel, D. Becker, A. Martin & H. Lörz, 1994. Transgenic cereal (Tritordeum) plants obtained at high efficiency by microprojectile bombardment of inflorescence tissue. Plant J. 5(4): 583–592.

Becker, D., R. Brettschneider & H. Lörz, 1994. Fertile transgenic wheat from microprojectile bombardment of scutellar tissue. Plant J. 5(2): 299–307.

Bower, R. & R.G. Brich, 1992. Transgenic sugarcane plants via microprojectile bombardment. Plant J. 2(3): 409–416.

Bytebier, B., F. Deboeck, H. De Greve, M. Van Montagu & J.-P. Hernalsteens, 1987. T-DNA organization in tumour cultures and transgenic plants of the monocotyledon *Asparagus officinalis*. Proc. Natl. Acad. Sci. USA 84: 5345–5349.

Cao, J., X. Duan, D. McElroy & R. Wu, 1992. Regeneration of herbicide resistant transgenic rice plants following microprojectile-mediated transformation of suspension culture cells. Plant Cell Rep. 11: 586–591.

Christou, P., T.L. Ford & M. Kofron, 1991. Production of transgenic rice (*Oryza sativa* L.) plants from agronomically important Indica and Japonica varieties via electric discharge acceleration of exogenous DNA into immature zygotic embryos. Bio/Technology 9: 957–962.

Dale, P.J., M.S. Marks, M.M. Brown, C.J. Woolston, D.F. Chen, D.M. Gilmour & R.B. Flavell, 1989. Agroinfection of wheat: Inoculation of *in vitro* seedlings and embryos. Plant Sci. 63: 237–245.

Datta, S.K., K. Datta & I. Potrykus, 1990. Fertile Indica rice plants regenerated from protoplasts isolated from microspore-derived cell suspensions. Plant Cell Rep. 9: 253–256.

Datta, S.K., A. Peterhans, K. Datta & I. Potrykus, 1990. Genetically engineered fertile Indica rice recovered from protoplasts. Bio/Technology 8: 736–740.

Datta, K., I. Potrykus & S.K. Datta, 1992. Efficient fertile plant regeneration from protoplasts of the Indica rice breeding line IR72 (*Oryza sativa* L.). Plant Cell Rep. 11: 229–233.

De Cleene, M., 1985. The susceptibility of monocotyledons to *Agrobacterium tumefaciens*. Z. Phytopathol. 113: 81–89.

D'Halluin, K., E. Bonne, M. Bossut, M. De Beuckeleer & J. Leemans, 1992. Transgenic maize plants by tissue electroporation. Plant Cell 4: 1495–1505.

Donn, G., P. Eckes & H. Müller, 1992. Genübertragung auf Nutzpflanzen. BioEngineering 8: 40–46.

Fretz, A., A. Jähne & H. Lörz, 1992. Cryopreservation of embryogenic suspension cultures of barley (*Hordeum vulgare* L.). Botanica Acta 105: 140–145.

Fromm, M., L.P. Taylor & V. Walbot, 1986. Stable transformation of maize after gene transfer by electroporation. Nature 319: 791–793.

Fromm, M.E., F. Morrish, A. Armstrong, R. Williams, J. Thomas & T.M. Klein, 1990. Inheritance and expression of chimeric genes in the progeny of transgenic maize plants. Bio/Technology 8: 833–839.

Funatsuki, H., H. Lörz & P.A. Lazzeri, 1992. Use of feeder cells to improve barley protoplast culture and regeneration. Plant Sci. 85: 179–187.

Genovesi, D., N. Willetts, S. Zachwieja, M. Mann, T. Spencer, C. Flick & W. Gordon-Kamm, 1992. Transformation of an elite maize inbred through microprojectile bombardment of regenerable embryogenic callus. *In Vitro* Cell Dev. Biol. 18: 189–200.

Golovkin, M.V., M. Abraham, S. Morocz, S. Bottka, A. Feder & D. Dudits, 1993. Production of transgenic maize plants by direct DNA uptake into embryogenic protoplasts. Plant Sci. 90: 41–52.

Golds, T.J., J. Babczinsky, A.P. Mordhorst & H.-U. Koop, 1993. Protoplast preparation without centrifugation: plant regeneration from barley (*Hordeum vulgare* L.). Plant Cell Rep. 13: 188–192.

Gordon-Kamm, W.J., T.J. Spencer, M.L. Mangano, T.R. Adams, R.J. Daines, W.G. Start, J.V. O'Brien, S.A. Chambers, W.R. Adams, N.G. Willetts, T.B. Rice, C.J. Mackey, R.W. Krueger, A.P. Kausch & P.G. Lemaux, 1990. Transformation of maize cells and regeneration of fertile transgenic plants. Plant Cell 2: 603–618.

Gould, J., M. Devey, O. Hasegawa, E.C. Ulian, G. Peterson & R.H. Smith, 1991. Transformation of *Zea mays* L. using *Agrobacterium tumefaciens* and the shoot apex. Plant Physiol. 95: 426–434.

Gray, D.J. & J.J. Finer, 1993. Development and operation of five particle guns for introduction of DNA into plant cells. Plant Cell Tiss. & Org. Cult. 33: 219.

Grimsley, N., T. Hohn, J.W. Davies & B. Hohn, 1987. *Agrobacterium*-mediated delivery of infectious maize streak virus into maize plants. Nature 325: 177–179.

Gupta, H.S. & A. Pattanayak, 1993. Plant regeneration from mesophyll protoplasts of rice (*Oryza sativa* L.). Bio/Technology 11: 90–94.

Hahne, B., H. Lörz & G. Hahne, 1990. Oat mesophyll protoplasts: their response to various feeder cultures. Plant Cell Rep. 8: 590–593.

Hoekstra, S., M.H. van Zijderveld, F. Heidekamp & F. van der Mark, 1993. Microspore culture of *Hordeum vulgare* L.: the influence of density and osmolarity. Plant Cell Rep. 12: 661–665.

Hohn, B., T. Hohn, M.I. Boulton, J.W. Davies & N. Grimsley, 1987. Agroinfection of *Zea mays* with maize streak virus DNA. p. 459–468. In: N.-H. Chu (Ed). Plant Molecular Biology. Plenum Publishing Corporation, New York.

Hooykaas, P.J.J. & R.A. Schilperoort, 1992. *Agrobacterium* and plant genetic engineering. Plant Mol. Biol. 19: 15–38.

Jähne, A., P.A. Lazzeri, M. Jäger-Gussen & H. Lörz, 1991a. Plant regeneration of embryogenic cell suspensions of barley (*Hordeum vulgare* L.). Theor. Appl. Genet. 82: 74–80.

Jähne, A., P.A. Lazzeri & H. Lörz, 1991b. Regeneration of fertile plants from protoplasts derived from embryogenic cell suspensions of barley (*Hordeum vulgare* L.). Plant Cell Rep. 10: 1–6.

Jähne, A., D. Becker, R. Brettschneider & H. Lörz, 1994. Regeneration of transgenic, microspore-derived, fertile barley. Theor. Appl. Genet. 89: 525–533.

Klöti, A., V.A. Iglesias, J. Wünn, P.K. Burkhardt, S.K. Datta & I. Potrykus, 1993. Gene transfer by electroporation into intact scutellum cells of wheat embryos. Plant Cell Rep. 12: 671–675.

Koziel, M.G., G.L. Beland, C. Bowman, N.B. Carozzi, R. Crenshaw, L. Crossland, J. Dawson, N. Desai, M. Hill, S. Kadwell, K. Launis, K. Lewis, D. Maddox, K. McPherson, M.R. Meghji, E. Merlin, R. Rhodes, G.W. Warren, M. Wright & S.E. Evola, 1993. Field performance of elite transgenic maize plants expressing an insecticidal protein derived from *Bacillus thuringiensis*. Bio/Technology 11: 194–200.

Kyozuka, J , Y Hayashi & K Shimamoto, 1987 High frequency plant regeneration from rice protoplasts by novel nurse culture methods Mol Gen Genet 206 408–413

Kyozuka, J , E Otoo & K Shimamoto, 1988 Plant regeneration from protoplasts of Indica rice genotypic differences in culture response Theor Appl Genet 76 887–890

Langridge, P, R Brettschneider, P Lazzeri & H Lorz, 1992 Transformation of cereals via *Agrobacterium* and the pollen pathway a critical assessment Plant J 2 631–638

Lazzeri, PA , R Brettschneider, R Luhrs & H Lorz, 1991 Stable transformation of barley via PEG-induced direct DNA uptake into protoplasts Theor Appl Genet 81 437–444

Li, X Q , C -N Liu, S W Ritchie, J -Y Peng, S B Gelvin & T K Hodges, 1992 Factors influencing *Agrobacterium* mediated transient expression of *gus*A in rice Plant Mol Biol 20 1037–1048

Li, Z & N Murai, 1990 Efficient plant regeneration from rice protoplasts in general medium Plant Cell Rep 9 216–220

McElroy, D , W Zhang, J Cao & R Wu, 1990 Isolation of an efficient Actin promotor for use in rice transformation The Plant Cell 2 163–171

Mooney, PA , PB Goodwin, E S Dennis & D J Llewllyn, 1991 *Agrobacterium tumefaciens*-gene transfer into wheat tissues Plant, Cell Tiss & Org Cult 25 209–218

Mordhorst, A P & H Lörz, 1993 Embryogenesis and development of isolated barley (*Hordeum vulgare* L) microspores are influenced by the amount and composition of nitrogen sources in culture media J Plant Physiol 142 485–492

Morocz, S , G Donn, J Nemeth & D Dudits, 1990 An improved system to obtain fertile regenerants via maize protoplasts isolated from a highly embryogenic suspension culture Theor Appl Genet 80 721–726

Olsen, FL , 1991 Isolation and cultivation of embryogenic microspores from barley (*Hordeum vulgare* L) Hereditas 115 255–266

Omirulleh, S , M Abraham, M Golovkin, I Stefanov, M K Karabaev, M Mustardy, S Morocz & D Dudits, 1993 Activity of a chimeric promoter with the doubled CaMV 35S enhancer element in protoplast-derived cells and transgenic plants in maize Plant Mol Biol 21 415–428

Petersen, WL , S Sulc & C L Armstrong, 1992 Effect of nurse cultures on the production of macro calli and fertile plants from maize embryogenic suspension culture protoplasts Plant Cell Rep 10 591–594

Potrykus, I , 1989 Gene transfer to cereals an assessment Tibtech 7 269–273

Potrykus, I , 1990 Gene transfer to plants assessment and perspectives Physiol Plant 79 125–134

Prioli, L M & M R Sondahl, 1989 Plant regeneration and recovery of fertile plants from protoplasts of maize (*Zea mays* L) Bio/Technology 7 589–594

Raineri, D M , P Bottino, M P Gordon & E W Nester, 1990 *Agrobacterium*-mediated transformation of rice (*Oryza sativa* L) Bio/Technology 8 33–38

Rathore, K S , V K Chowdhury & T K Hodges, 1993 Use of *bar* as selectable marker gene and for the production of herbicide-resistance in rice plants from protoplasts Plant Mol Biol 21 871–884

Rhodes, C A , K S Lowe & K L Ruby, 1988 Plant regeneration from protoplasts isolated from embryogenic maize cell cultures Bio/Technology 6 56–60

Ritala, A , K Aspegren, U Kurten, M Salmenkallio-Marttila, L Mannonen, R Hannus, V Kauppinen, T H Teeri & T M Enari, 1994 Fertile transgenic barley by particle bombardment of immature embryos Plant Mol Biol 24 317–325

Sanford, J C , T M Klein, E D Wolf & N Allen, 1987 Delivery of substances into cells and tissues using a particle bombardment process J Part Sci Tech 5 27–37

Sanford, J C , 1990 Biolistic plant transformation Physiol Plant 79 206–209

Shillito, R D , G K Carswell, C M Jonsons, J J DiMaio & C T Harms, 1989 Regeneration of fertile plants from protoplasts of elite inbred maize Bio/Technology 7 581–587

Shimamoto, K , R Terada, T Izawa & H Fujimoto, 1989 Transgenic rice plants regenerated from transformed protoplasts Nature 338 274–276

Somers, D A , H W Rines, W Gu, H F Kaeppler & W R Bushnell, 1992 Fertile transgenic oat plants Bio/Technology 10 1589–1594

Tada, Y, M Sakamoto & T Fujimura, 1990 Efficient gene introduction into rice by electroporation and analysis of transgenic plants use of electroporation buffer lacking chloride ions Theor Appl Genet 80 475–480

Terada, R , T Nakayama, M Iwabuchi & K Shimamoto, 1993 A wheat histone H3 promotor confers cell division dependent and -independent expression of the *gus*A gene in transgenic rice plants Plant J 3 241–252

Toriyama, K , K Hinata & T Sasaki, 1986 Haploid and diploid plant regeneration from protoplasts of anther callus in rice Theor Appl Genet 73 16–19

Toriyama, K , Y Arimoto, H Uchimiya & K Hinata, 1988 Transgenic rice plants after direct gene transfer into protoplasts Bio/Technology 6 1072–1074

Vasil, I K & V Vasil, 1992 Advances in cereal protoplast research Physiol Plant 85 279–283

Vasil, V , A M Castillo, M E Fromm & I K Vasil, 1992 Herbicide resistant fertile transgenic wheat plants obtained by microprojectile bombardment of regenerable embryogenic callus Bio/Technology 10 667–674

Walters, D A , C S Vetsch, D E Potts & R C Lundquist, 1992 Transformation and inheritance of a hygromycin phosphotransferase gene in maize plants Plant Mol Biol 18 189–200

Wan, Y , & PG Lemaux, 1994 Generation of large numbers of independently transformed fertile barley plants Plant Physiol 104 37–48

Weeks, J T , O D Anderson & A E Blechl, 1993 Rapid production of multiple independent lines of fertile transgenic wheat (*Triticum aestivum*) Plant Physiol 102 1077–1084

Zhang, W & R , Wu, 1988 Efficient regeneration of transgenic rice plants from rice protoplasts and correctly regulated expression of foreign genes in the plants Theor Appl Genet 76 835–840

Zhang, H M , H Yang, E L Rech, T J Golds, A S Davis, B J Mulligan, E C Cocking & M R Davey, 1988 Transgenic rice plants produced by electroporation-mediated plasmid uptake into protoplasts Plant Cell Rep 7 379–384

Zimny, J , D Becker, R Brettschneider & H Lorz, 1995 Fertile, transgenic *Triticale* (× *Triticosecale* Wittmack) Molecular Breeding 1 155–164

Euphytica **85:** 45–51, 1995.

Shoot apical meristems as a target for gene transfer by microballistics

C. Sautter, N. Leduc[1], R. Bilang[2], V.A. Iglesias, A. Gisel, X. Wen & I. Potrykus
Department of Plant Sciences, Federal Institute of Technology, Zürich, Switzerland; [1] *present address: R.C.A.P. lab, ENS-46 Allée d'Italie, Lyon, France;* [2] *present address: Department of Cellular and Developmental Biology, Harvard University, Cambridge, MA 02138, USA*

Key words: meristem, shoot apex, ballistic microtargeting, gene transfer, wheat

Summary

The classical approach of gene transfer to a given plant species delivers the foreign gene to transformable cells and then puts the effort into generating plants. This approach is very difficult in many important crop plants, including cereals, and the results of regeneration are very genotype-dependent. In contrast, we use regenerable cells and try to transform them. Shoot apical meristems provide a tissue which regenerates *in situ* a fertile plant for most given genotypes or species. Transformation of meristem cells may lead to transgenic sectors in chimeras. These sectors may contribute to the gametes and, thus, to transgenic offspring, which then should be homohistonts and not sectorial chimeras like their parents. Our model plant for these studies is wheat. Microtargeting is a ballistic approach which is particularly suitable for the controlled delivery of microprojectiles to meristem cells *in situ* (Sautter et al., 1991). We summarize in this paper our experience with ballistic microtargeting of transgenes to wheat shoot apical meristem cells *in situ*.

Abbreviations: GUS – β-glucuronidase

Introduction

The shoot apical meristem of plants generates the whole green part of the plant body, including the flowers, thus it is a very important tissue in the developmental biology of plant. In particular, the meristem is a tissue with a very high regeneration potential. Gene transfer to meristem cells could, therefore, overcome many of the regeneration problems due to difficult tissue culture systems in many important crop species (Potrykus, 1990).

However, little is known about the biology of this tissue (as reviewed by Medford, 1992). The reason for this gap in our knowledge is that the meristem is a tiny structure which makes biochemical or molecular biological studies very difficult. It is impossible to collect enough material from meristems for such studies without contamination by other tissues. Thus, the meristem is often confused with the complete shoot apex, which also contains the leaf primordia and the young leaves. Furthermore, the meristem as a tissue may represent a complicated pattern of cells. Each of these cells may differ physiologically due to its unique position in the meristem.

One possible approach to the study of meristem biology would be a combination of cytochemical or histochemical methods together with cell fate studies. This approach would provide biochemical or molecular biological data for individual cells at a microscopic level, and on the other hand would make the position of any cell clear within the meristem. It is possible to obtain such data by the delivery of biologically active material, e.g. marker genes, to meristem cells with microprojectiles.

Ballistic microtargeting (Sautter et al., 1991) is the method of choice for delivering biologically-active material to meristem cells *in situ*. It allows us to aim at small tissues, control the penetration of the microprojectiles, and adjust the particle density at the target site. Microtargeting is highly efficient in model systems if used for gene transfer (Sautter et al., 1991) and it offers a way of transporting not only genes but any biologically-active material (Sautter, 1993).

In the present paper, we summarize our experience with microtargeting of marker genes to wheat meristems. This includes vegetative apical shoot meristems of seedlings ca. 10 days after germination, shoot meristems of immature proembryos ca. 7 days after anthesis, and floral meristems ca. 20 days after germination.

Materials and methods

Plant material

Seeds of spring wheat (*Triticum aestivum* cv. Sonora) were grown under greenhouse conditions as described by Bilang et al. (1993). From these plants we took either immature embryos or seeds. For the production of floral meristems, seeds were sown in soil under long days (18 h photoperiod at 18°/14° C) and grown for 20 days according to Leduc et al. (1994). For the production of vegetative shoot apical meristems, seeds were sterilized by soaking in 70% ethanol for 2 min, followed by sodium hypochloride and four water rinses (Bilang et al., 1993). The sterile seeds were sown in glass tubes on MS medium (Murashige & Skoog, 1962) supplemented with 100 mg l^{-1} Cefotaxime (Serva, Heidelberg, Germany), 2% sucrose, and 0.8% Difco agar. Shoot apical meristems from 6- to 10-day-old plantlets were exposed by removal of the coleoptiles and the first three to five leaves (Bilang et al., 1993). Roots were trimmed to about 5 mm. The explants were then placed on MS-basal medium supplemented with different sucrose concentrations (optimum: 10%), and 0.8% agarose. After particle delivery, the explants were transferred to MS-agarose, and for prolonged culture onto MS-agar 5 days following bombardment and cultured as described above (Bilang et al., 1993).

Immature embryos were prepared according to Iglesias et al. (1994a). Briefly, ears, ca. 7 days post anthesis, were surface-sterilized with 70% ethanol and air dried. The seed coat of the caryopsis was cut with a scalpel and the embryo was rescued. It was then placed with the scutellum downwards and the embryo axis upwards on MS-medium supplemented with 200 mg l^{-1} casein hydrolysate and 100 mg l^{-1} inositol. For bombardment, the embryos were mounted on medium as above supplemented with different sucrose concentrations, the optimum of which was 15%. For further details of handling and culture of immature wheat embryos, see Iglesias et al. (1994a, 1994b).

Inflorescences were collected about 20 days after sowing and sterilized by immersion in 3.5 (w/v) filtered calcium hypochloride solution for 20 min (Leduc et al., 1994) and then rinsed three times in water. After dissection the meristems were placed on MS-medium containing 0.06 M sucrose, 4.6 μM kinetin, and 0.53 μM 1-natphthylacetic acid, solidified with 8 g/l Difco agar according to Leduc et al. (1994). The osmolarity of this medium was adjusted to 0.52 M sucrose prior to bombardment. After bombardment, inflorescences were placed on fresh medium, liquid or solidified, depending on the culture period followed (Leduc et al., 1994).

Microtargeting

Treatment with ballistic microprojectiles accelerated by the microtargeting instrument was performed with 5×10^6 particles per μl and between 1.0 and 2.5 mg/ml plasmid DNA in the gold particle suspension. Particles were prepared according to Sautter et al. (1991), and plasmid DNA was isolated using resin columns (Quiagen Inc., USA). The optimum physical parameters for bombardment with the microtargeting instrument varied depending on the type of meristem. In all cases, a restriction of 140 μm in diameter, ca. 5 mm working distance, and between 100 and 130 bar nitrogen pressure were used to create the acceleration force (Sautter et al., 1991; Bilang et al., 1993; Iglesias et al., 1994a; Leduc et al., 1994).

Histochemistry

For transient expression studies the plant material was incubated for two days after bombardment. For observation of larger sectors, the plant material was grown for several weeks under conditions similar to those applied before bombardment. Subsequently, the tissue was stained for GUS activity (Jefferson et al., 1987) as modified according to Iglesias et al. (1994a).

Results and discussion

Marker genes

Visual marker genes have been used in our studies to follow gene delivery after shooting into the target tissue. In most of our experiments we used bacterial

β-glucuronidase (GUS) as a visual marker (Jefferson et al., 1987). The *gus* gene was controlled in all our experiments by the rice actin promoter in the plasmid pAct1-D, kindly provided by R. Wu (McElroy et al., 1990). Although the conventional histochemical assay is destructive, GUS probably does not influence differentiation in meristem cells. In contrast, we can not exclude the possibility that the genes *Bperu* (Goff et al., 1990) and *C1* (Cone et al., 1986), coding for regulatory proteins which stimulate anthocyanin production (Ludwig et al., 1990), might interfere with cell development in general. A construct (pBC17) containing both these genes under the control of the CaMV 35S promoter-Adh intron and nos 3′ was kindly provided by T. Klein (DuPont Agricultural Products, Wilmington, Delaware, USA). We observed expression with both markers in meristems, but sector formation in meristem cells was more frequent with GUS (Bilang et al., 1993; Iglesias et al., 1994a).

Particle penetration

In general the penetration of microprojectiles depends on the target tissue and the kinetic energy of the projectile. Since we use only round gold particles, the particle shape and material were unchanged throughout the experiments. According to Newton, the kinetic energy of the particles is proportional to their mass and the square of their velocity ($mv^2/2$). As the heavy mass of the projectiles is proportional to their radius cubed, the size of the particles is considered to be the most important parameter for their penetration.

We are able to produce round gold particles of any desired diameter, and the standard deviation of the particle size is very low (Sautter et al., 1991). Therefore, we can control this parameter very precisely. With increasing particle size (Fig. 1B), the number of hits in the second cell layer (LII) increases in shoot meristems (Bilang et al., 1993). In contrast, the impact of the acceleration force (Fig. 1A) which determines the particle velocity, has comparably little effect (Bilang et al., 1993). In conclusion, the effect of the velocity is less dramatic than that of the size of the projectiles (Bilang et al., 1993). This is in accord with the theory, and applies also to floral meristems (Leduc et al., 1994).

Using a small particle size, we can prevent particles from penetrating into deeper cell layers and confine their transport to the first cell layer (LI), from which the epidermal cell layer develops. An example is shown in Fig. 2a. These microtargeting parameters could be used to enrich periclinal chimeras which have the transgene only in the epidermal cell layer. Such a periclinal chimera should prevent the offspring from inheriting a foreign gene, because the gametes develop exclusively from LII cells and the cell layering is maintained during apical and axillary bud proliferation (Winkler, 1907; Satina et al., 1940; Tilney-Bassett, 1987). Periclinal chimeras could be interesting in many vegetatively-propagated crops (like apple or grapevine) as a biosafety containment mechanism. Higher kinetic energy leads to expression of the microtargeted transgene in the second cell layer (LII, Fig. 2b). Simultaneously, particles ending up in the LI can obviously have a detrimental effect on their target cells due to the high kinetic energy of the projectiles (Fig. 2b, arrow).

Cells do not usually survive histochemistry for transient GUS expression. Therefore, one can not conclude from a positive GUS expression test two days after bombardment that such a cell would have developed further. However, GUS histochemistry 3 weeks after bombardment, showed us that at least some of these cells have divided and given rise to small sectors (Fig. 2c), although many more cells were expressing GUS 2 days after bombardment (Fig. 2d).

Osmoticum

The impact of the projectile on the cells can be reduced by pre-incubation of the tissue in media with high osmolarity. This is in accord with other ballistic systems (Perl et al., 1992; Vain et al., 1993). It is not clear why osmotica improve gene expression and, probably, the survival of cells which received a projectile. It can be speculated that membranes of partially-plasmolysed cells better withstand the pressure pulse in cells at the moment of projectile penetration. Flower meristems need up to 20% sucrose in the medium (Leduc et al., 1994). Although sucrose as an osmoticum is no problem for transient expression, non-metabolizable sugars such as mannitol or sorbitol improve the regeneration of the tissue after ballistic treatment (Iglesias et al., 1994a).

Meristem models

In order to cover all the developmental stages of a plant, we studied vegetative shoot apical meristems of immature embryos and of seedlings and, additionally, flower meristems. Immature embryos represent a very young developmental stage of the plant but they nevertheless can germinate normally (Iglesias et al.,

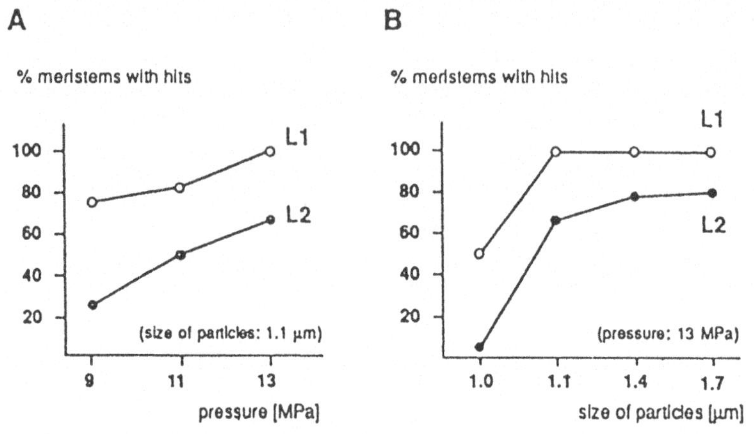

Fig. 1. Parameters affecting particle penetration into wheat meristems. Effects of accelerating pressure (a) and particle diameter (b) on frequency of particle penetration into wheat meristems. After bombardment, meristems were cleared and the position of particles determined microscopically as described (Bilang et al., 1993). Each treatment included 15 meristems.

1994b). Transgenic cells in the meristem of an immature embryo have, therefore, a chance to grow into large sectors in the plant. The larger the sectors, the better the chance that they will contribute to the development of the gametes. Furthermore, the meristem of young immature embryos is already exposed and therefore need not be dissected. Dissection always involves the removal of leaves, which are unavailable afterwards for photosynthesis. This is a disadvantage in the use of meristems of seedlings, since they have to be dissected to remove the coleoptile and the primary leaves. The advantage is that their meristems can be transformed immediately, if seeds are available. This shortens the time schedule by one vegetative cycle. Flower meristems could lead to transgenic offspring very soon after ballistic treatment, with only small sectors being required.

Examples of flower meristems are shown in Fig. 2. The flower meristem must be removed from the plant and cultivated further *in vitro*, which is difficult in wheat (Leduc et al., 1994), in contrast to maize (Dupuis & Pace, 1993a, 1993b). Experiments are in progress to bombard the flower meristem *in situ*, and to stimulate further normal development in isolated flower meristems. The expression frequency is high in floral meristems. We have observed an average of 6 sectors per inflorescence in 86% of the explants 12 days after ballistic treatment (Leduc et al., 1994).

Immature wheat embryos ca. 7 days after anthesis have an exposed shoot meristem which is not yet covered by the primary leaf and the coleoptile (Iglesias et al., 1994a). An immature embryo in this developmental stage is shown on the scanning electron micrograph (Fig. 3a). After microtargeting these meristems, cells at the very tip of the meristem can express GUS transiently (Fig. 3b, arrow 'm'). Since the targeting area of the instrument is not sharply bordered (Iglesias et al., 1994a), other particles hit the area around the meristem, which will generate the coleoptile, and lead to expression there (Fig. 3b, arrow 'c').

Similar observations have been made with vegetative meristems of wheat seedlings ca. 10 days after the beginning of germination (Bilang et al., 1993). Bombardment using microtargeting leads to transient GUS expression in cells of the meristem *sensu stricto* (Fig. 3c) and some other GUS-positive cells are found in the leaf primordia (Fig. 3c). In contrast with earlier attempts to transform shoot apical meristem cells of wheat with conventional guns, using a macroprojectile (Oard et al., 1990), after microtargeting the meristems show a high number of transiently-expressing cells, the tissue looks healthy, and they are able to grow further.

The meristems of seedlings and of immature embryos grow after microtargeting with a certain time delay. It takes about two weeks for them to restore their normal growth rate (Bilang et al., 1993; Iglesias et al., 1994a, 1994b). Meristems from immature embryos then grow to form normally fertile plants. In contrast, plants grown from seedling meristems show a reduced seet set, possibly as a consequence of meristem dissection though they are fertile.

Besides the *uidA* gene, which encodes for the β-glucuronidase, we also used the genes *Bperu* and *C1*

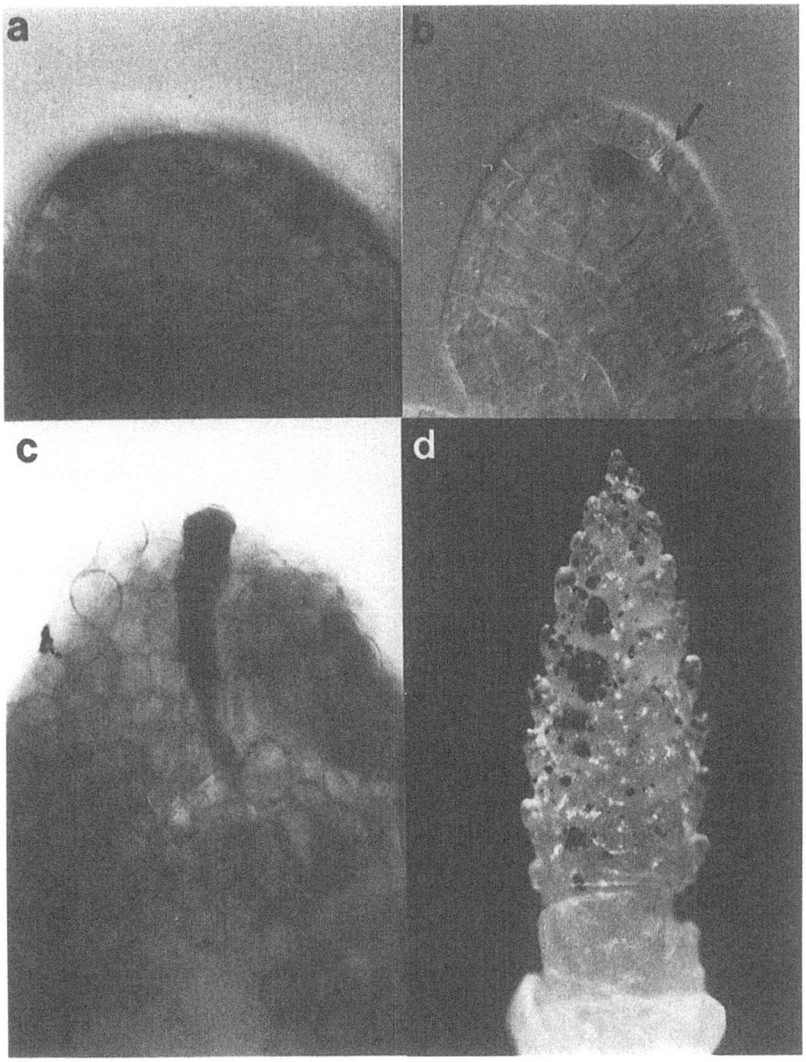

Fig. 2. Expression of GUS after microprojectile treatment of wheat floral meristems. (a), (b), and (d) represent transient expression after 48 hours of incubation for expression; (c) was stained for GUS 3 weeks after microtargeting. (a): GUS expressing cells in a spikelet primordium confined to the first cell layer (× 400). (b): Example of a transiently-expressing cell in the second cell layer of a spikelet primordium. Arrow. damaged cell in the first cell layer containing a particle (× 400). (c): Small sector in an ovary primordium stained for GUS 3 weeks after bombardment (× 400). (d): Overview of a flower meristem with many GUS-expressing cells (× 50).

encoding for regulatory proteins inducing the synthesis of anthocyanin. These represent a marker which is visible via an endogenous reaction and, therefore, needs no external interference with the cells. Using this marker, we observed significantly less transient expression in the cells of the meristems, whereas the expression outside of the meristems was comparable with both markers (Bilang et al., 1993; Iglesias et al., 1994). Looking for sectors instead of transient expression in single cells, we observed the opposite effect. In tissues outside the meristem, e.g. in the area from which the coleoptile will arise, we found many more sectors with the anthocyanin marker after two to three weeks of incubation following bombardment. In particular, the coleoptile exhibited many anthocyanin sectors of different size, some of them longer than 1 cm and containing more than 50 cells. Figure 3d shows a part of such a sector, which indicates stable integration, although molecular data have not yet been obtained.

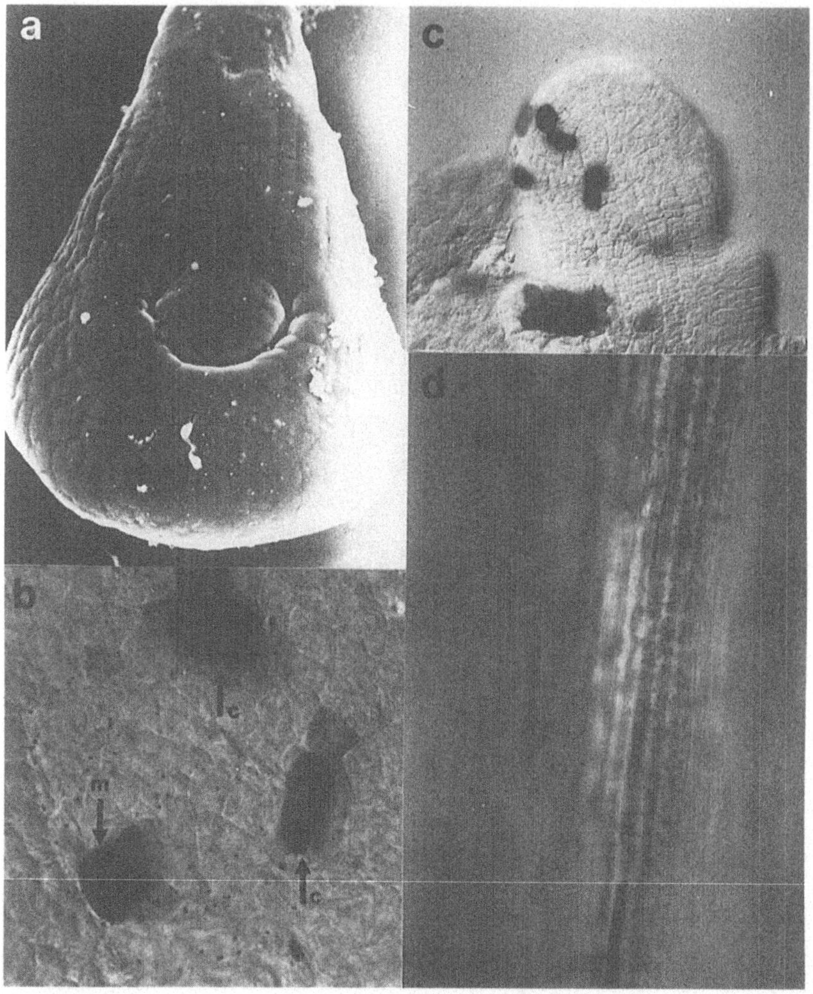

Fig. 3. Ballistic microtargeting to vegetative apical meristems. (a): Scanning electron micrograph of an immature wheat embryo ca 7 days post anthesis (× 190). (b): Transient GUS expression after 48 hours of incubation in the shoot apical meristem of an immature embryo shot 7 days post anthesis. Arrow 'm': cell at the tip of the meristem; arrows 'c': cells in the area from which the coleoptile arises (× 400). (c): Transient GUS expression after 48 hours of incubation following bombardment of the vegetative shoot apical meristem of a 10 days old seedling (× 150). (d): Part of a longer sector in a coleoptile three weeks after ballistic treatment of the meristem of an immature embryo similar to that in Fig. 3a and 3b, but with the anthocyanin inducing genes *Bperu* and *C1* (× 150).

Conclusions

In all meristem models, transient expression has been obtained after microtargeting with high frequency. We found sectors of different size in all three meristem types examined. Up to now, large sectors have been restricted to vegetative organs and only when using anthocyanin as a marker. Thus, the ballistic microtargeting technique is ready to be used in meristem studies. This could include cell fate studies or promoter studies of any apex-specific genes (Medford, 1992). Microtargeting and transient expression of marker genes provide rapid localization with precise resolution, which allows us to distinguish between the meristem, its cell layers and the leaf primordia.

Acknowledgement

We are indebted to Ted Klein (DuPont Agric. Products, Wilmington, Del., USA) and Ray Wu (Section of Biochemistry, Cornell University, Ithaca, N.Y., USA) for providing the plasmids pBC17 and pAct1-D, respectively. We gratefully acknowledge the permission of

Blackwell Scientific Publ. Ltd. (Oxford, UK) to reproduce figures from the Plant Journal (Bilang et al., 1993) and to Springer Verl. (Heidelberg, Germany) to reproduce figures from Planta (Iglesias et al., 1994) and Sexual Plant Reproduction (Leduc et al., 1994). We are indebted to Dr Andrew Fleming for help with the English.

References

Bilang, R., S. Zhang, N. Leduc, V.A. Iglesias, A. Gisel, J. Simmonds, I. Potrykus & C. Sautter, 1993. Transient expression in vegetative shoot apical meristems of wheat after ballistic microtargeting. The Plant Journal 4: 735–744.

Cone, K., F. Burr & B. Burr, 1986. Molecular analysis of the maize anthocyanin regulatory locus C1. Proc. Natl. Acad. Sci. USA 83: 9631–9635.

Dupuis, I. & G.M. Pace, 1993a. Factors affecting *in vitro* maturation of isolated maize microspores. Plant Cell Rep. 12: 564–568.

Dupuis, I. & G.M. Pace, 1993b. Gene transfer to maize male reproductive structure by particle bombardment of tassel primordia. Plant Cell Rep. 12: 607–611.

Iglesias, V.A., A. Gisel, R. Bilang, N. Leduc, I. Potrykus & C. Sautter, 1994a. Transient expression of different marker genes in meristem cells of wheat immature embryos after ballistic microtargeting. Planta 192: 84–91.

Iglesias, V.A., A. Gisel, I. Potrykus & C. Sautter, 1994b. *In vitro* germination of wheat proembryos to fertile plants. Plant Cell Rep. 13: 377–380.

Jefferson, R.A., T.A. Kavanagh & M.W. Bevan, 1987. GUS-fusions: β-glucuronidase as a sensitive and versatile gene fusion marker in higher plants. EMBO J. 6: 13901–13907.

Goff, S.A., T.M. Klein, B.A. Roth, M.E. Fromm, K.C. Cone, J.P. Radicella & V.L. Chandler, 1990. Transactivation of anthocyanin biosynthetic genes following transfer of B regulatory genes into maize tissues. EMBO J. 9: 2517–2522.

Leduc, N., V.A. Iglesias, R. Bilang, A. Gisel, I. Potrykus & C. Sautter, 1994. Gene transfer to inflorescence and flower meristem using ballistic micro-targeting. Sexual Plant Reproduction 7: 135–143.

Ludwig, S.R., B. Bowen, L. Beach & S.R. Wessler, 1990. A regulatory gene as a novel visible marker for maize transformation. Science 247: 449–450.

Medford, J., 1992. Vegetative apical meristems. Plant Cell 4: 1029–1039.

McElroy, D., W. Zhang, J. Cao & R. Wu, 1990. Isolation of an efficient actin promoter for use in rice transformation. The Plant Cell 2: 163–171.

Murashige, T. & F. Skoog, 1962. A revised medium for rapid growth and bioassay with tobacco tissue cultures. Physiol. Plant. 15: 473–497.

Oard, J.H., D.F. Paige, J.A. Simmonds & T.M. Gradziel, 1990. Transient gene expression in maize, rice, and wheat cells using an airgun apparatus. Plant Physiol. 92: 334–339.

Perl, A., H. Kless, A. Blumenthal, G. Galili & E. Galun, 1992. Improvement of plant regeneration and GUS expression in scutellar wheat calli by optimization of culture conditions and DNA-microprojectile delivery procedures. Mol. Gen. Genet. 235: 279–284.

Potrykus, I., 1990. Gene transfer to cereals: an assessment. Biotechnology 6: 535–542.

Satina, S., A.F. Blakeslee & A.G. Avery, 1940. Demonstration of the three germ layers in the shoot apex of *Datura* by means of induced polyploidy in periclinal chimeras. Am. J. Bot. 27: 895–905.

Sautter, C., H. Waldner, G. Neuhaus-Url, A. Galli, G. Neuhaus & I. Potrykus, 1991. Micro-targeting: High efficiency gene transfer using a novel approach for the acceleration of micro-projectiles. Biotechnology 9: 1080–1085.

Sautter, C., 1993. Development of a microtargeting device for particle bombardment of plant meristems. Plant Cell, Tissue and Organ Culture 33: 251–257.

Tilney-Bassett, R.A.E., 1987. Plant chimeras. Edward Arnold Ltd., Baltimore.

Vain, P., M.D. McMullen & J.J. Finer, 1993. Osmotic treatment enhances particle bombardment-mediated transient and stable transformation of maize. Plant Cell Rep. 12: 84–88.

Winkler, H., 1907. Über Pfropfbastarde und pflanzliche Chimären. Ber. dt. Bot. Gesell. 23: 568–576.

Euphytica **85:** 53–61, 1995.

Studies of the mechanism of transgene integration into plant protoplasts: improvement of the transformation rate

Indridi Benediktsson[1], Claudia P. Spampinato[2] & Otto Schieder
Institut für Angewandte Genetik, Albrecht-Thaer-Weg 6, 14195 Berlin, Germany; [1] Present address: Institute for Experimental Pathology, Keldur, 112 Reykjavik, Iceland; [2] CEFOBI, Suipacha 531, 2000 Rosario, Argentina

Key words: bleomycin, direct gene transfer, expression, irradiation, petunia, protoplasts

Summary

The production of transgenic plants by means of direct gene transfer to protoplasts is now a widely-used technique. The biological mechanisms underlying the transformation are still poorly understood, but many investigations have attempted to shed light on some components of this process. Varying the experimental conditions has in some cases led to better transformation rates, but further improvements of the protocols are possible. Such improvements will require a better understanding of how the alien DNA enters the cells, becomes integrated into the chromosomes and is treated as a part of the plant genome. Irradiation with sublethal doses of X-rays or UV-light has been shown to increase the transformation frequency, while certain drugs have been shown to act in a similar manner. The effects of these and other factors are discussed.

Abbreviations: Aph – aphidicolin, ATF – absolute transformation frequency, BLM – bleomycin, CaMV – cauliflower mosaic virus, CAT – chloramphenicol acetyl transferase, CHO – Chinese hamster ovary cells, EF – enhancement factor, Nos – nopaline synthase, NPTII – neomycin phosphotransferase II, Ocs – octopine synthase, PEG – polyethyleneglycol, RTF – relative transformation frequency

Introduction

Transformation of plants with DNA has received considerable attention in recent years. The natural transformation system of *Agrobacterium tumefaciens* has proved useful in many dicotyledonous species and is the method of choice for producing transgenic plants of those species at high efficiency. Protoplast transformation, i.e. direct gene transfer (PEG/Mg^{++}; electroporation; sonication; microinjection) or the co-culture of protoplasts and *Agrobacterium tumefaciens*, is a different type of strategy that requires a successful plant cell culture and regeneration system. Although the number of species that fulfil this criterion is growing steadily (Puite, 1992) and recently included some of the recalcitrant graminaceous species (Shimamoto et al., 1989), this remains a major obstacle in protoplast transformation. It seems that any plant that can be manipulated by cell culture and which can regenerate shoots, can be transformed with alien DNA.

In this paper we want to concentrate on direct gene transfer and look at the circumstances that allow the introduction of alien genes into the protoplast, the nucleus and finally the genome. For a general review on plant transformation methods, see Potrykus (1991).

Numerical calculations

There are two common ways of expressing transformation rates in the literature: Absolute transformation frequency, ATF, and relative transformation frequency, RTF (in %).

$$RTF = \frac{TG/ml}{NC/ml} \times 100 \qquad ATF = \frac{TG/ml}{NP/ml}$$

TG: colonies expressing the transgene (e.g. resistance)
NC: total number of growing colonies
NP: initial number of protoplasts
(all volumes refer to the original protoplast suspension at the time of plating)

The influence of irradiation or chemical treatment on the transformation rate is expressed as the enhancement factor, EF

$$EF = \frac{RTF_{treated}}{RTF_{not\ treated}}$$

Background

As soon as aseptic culture of protoplasts and regeneration of plants became possible, researchers started to consider the potential for genetically transforming plants through naked DNA transfer to protoplasts. In 1972, Hoffmann & Hess reported the uptake of radioactively-labelled *Petunia* genomic DNA into nuclei of *Petunia* protoplasts. Davey et al. (1980) succeeded in transforming protoplasts to hormone autotrophy with isolated Ti plasmid from *Agrobacterium tumefaciens*, showing that transformation does not require the bacterium itself. By analogy with protoplast fusion and animal cell transformation, the protoplasts were treated with divalent cations (Ca^{++} or Mg^{++}) and polyethyleneglycol (PEG) together with the DNA.

The transformation frequencies in these early attempts were in general very low and the work suffered from difficulties in the screening of transformants. The latter was overcome with the construction of chimeric antibiotic-resistance genes that could be expressed in plants. Herrera-Estrella et al. (1983) and Fraley et al. (1983) designed such chimeric genes and transferred them into tobacco and *Petunia* cells via *Agrobacterium*.

Paszkowski et al. (1984) showed that direct gene transfer to protoplasts does not require any special sequences from the Ti plasmid. They transformed *Nicotiana tabacum* cells with a bacterial plasmid carrying a hybrid gene for expression of the kanamycin resistance gene *nptII* in plants under the control of the CaMV 35S promoter and terminator. The transformation rate (ATF) was in the range of 10^{-6}. Hain et al. (1985) reported the successful transformation of *N. tabacum* with another hybrid gene, *nptII* under control of the Nos promoter and the Ocs terminator, and harvested transgenic colonies at a rate of 10^{-5} to 10^{-4} (ATF). Shillito et al. (1985) introduced the use of electroporation in plant protoplast transformation and obtained transformation rates as high as 1.6×10^{-3} (ATF) (equal to RTF of 2%). Although electroporation sometimes has a positive effect on the transformation rate, the chemical method is much simpler and higher numbers of protoplasts can be treated leading to higher absolute numbers of transgenic clones. Negrutiu et al. (1987) tested various parameters for transformation in *Nicotiana* and optimized PEG and Mg^{++} concentrations, resulting in transformation rates of up to 4×10^{-3} (ATF). In that study electroporation was less efficient than the chemical method. Krüger-Lebus & Potrykus (1987) presented a simple protocol for transformation of *Petunia*. The last two protocols have been used (with minor variations) for transformation of protoplasts of different species in many laboratories.

The efficiency of transformation varies significantly between species and even between cultivars within the same species. Transformation studies of two cultivars of moth bean, *Vigna aconitifolia*, revealed relative transformation rates that differed by one order of magnitude (1.0% and 0.1%) (Köhler et al., 1987). The same paper reported similar differences between two genotypes of *N. tabacum* (SR1 and W38) when treated simultaneously with identical methods.

One has to be very cautious when comparing transformation rates in the literature. Even transformation of the same material by one person and with identical methods on different days can lead to different transformation rates. Parameters such as initial protoplast number can be interpreted quite differently by different persons. For this reason it is more reliable to compare the relative transformation frequency. Using RTF for comparison eliminates the fluctuations caused by the variable plating efficiency. For exact comparison of the influence of a given parameter on the transformation rate, it is essential to run both situations (with and without the new parameter) simultaneously and not change any other conditions. Of course, this can be difficult e.g. when testing the effect of cell synchronization (Meyer et al., 1985) where some protoplasts have to be treated with chemicals for many days and the control (non-synchronized) is freshly isolated.

Apart from testing various gene constructs, which lead to different transformation rates as the result of the regulatory sequences or the ease of selection, researchers have tested different configurations of the DNA offered to the protoplasts. Notable among these studies is the comparison of linear vs circular plasmids.

Negrutiu et al. (1987) achieved better results with the linear than with the circular form of plasmid pABD1 when transforming *Nicotiana*. Ballas et al. (1988) studied the transforming efficiency of three topological forms (supercoiled/relaxed/linear) of a plasmid carrying the *cat* gene in transient expression analysis. They found a significantly higher expression when using linear rather than circular forms of the plasmid. Bates et al. (1990) however, did not find the same correlation when using a different plasmid that carried the CAT gene. They also observed that the linearized plasmids could become recircularized inside the protoplasts. This shows that, as in vertebrate cells (Miller & Temin, 1983), foreign DNA undergoes many rearrangements after uptake into the plant cell, which is also reflected by the appearance of gene-concatamers that can become integrated at a single site in the genome (Miller & Temin, 1983; Czernilowsky et al., 1986). This can be utilized when intending to incorporate non-screenable genes or other sequences together with a screenable gene in co-transformation. The seond gene is integrated at a surprisingly high frequency into the genome of the selected clones (Schocher et al., 1986). We co-transformed *Petunia* protoplasts with two different antibiotic-resistance genes and selected transformants first for one of the traits and then for the other in reciprocal tests (Benediktsson et al., 1991). While the RTF for the transgenic clones with regard to the first gene was in the range of 0.1–1%, the other marker was expressed in 50% of these clones no matter which was selected for first. When testing the integration of silent copies of one type of gene in clones that only expressed the other type, we saw that all tested clones (n = 25) had incorporated both gene types. When looking for integrations in cells that had been treated with DNA but did not express any marker, inactive copies of the gene were found in a high proportion of the clones.

Where is the bottle neck?

There are many barriers that the alien DNA has to pass on its way from the outside of the protoplast into the genome. The most obvious barrier of normal plant cells, the cell wall, has been removed in protoplasts. In a typical transformation assay the number of plasmid molecules is in great excess to the number of protoplasts (some million copies per cell). This DNA is brought in close vicinity to the surface of the cell membrane by PEG/Mg^{++}-DNA co-precipitation. The surrounding medium can influence the uptake of the DNA into the cytosol, but probably does not greatly affect any of the subsequent steps. Once inside the cell, the DNA has to be transported through the cytosol, where it can encounter many dangers like nucleases, must pass the nuclear membrane, traverse the karyoplasm, be linearized and integrated into the genomic DNA, which itself has to be cut and ligated, and eventually the gene has to be expressed. It is not clear which of these tasks is the limiting step for successful transformation.

Meyer et al. (1985) found that synchronized tobacco protoplasts were more efficiently transformed in the S- and M-phases of the cell cycle than in other phases. This led to the assumption that the nuclear membrane was a substantial barrier to the import of alien DNA. Tyagi et al. (1989) however, did not find this correlation between cell cycle phases and transformation efficiency.

X-ray irradiation

When protoplasts that had gone through the PEG-transformation procedure were irradiated with low doses of X-rays and then cultured and screened, it turned out that the irradiation highly-stimulated the transformation rate compared to the identical, non-irradiated control (Köhler et al., 1989). This was demonstrated for all the plants that were tested: *Nicotiana tabacum* SR1 and W38, *Petunia* Mitchell, *Vigna aconitifolia* and *Brassica nigra.* In all cases the irradiation had a positive effect on the transformation rate leading to a 2- to 10-fold increase in transformants (Fig. 1). Doses of 10 or 15 Gy, which gave at least a 2-fold increase, only slightly affected the plating efficiency, if at all, and did not alter the phenotype of regenerated shoots.

Further investigation of the effect on transformation (Köhler et al., 1990) revealed that the irradiation of protoplasts one hour before the PEG/DNA treatment also increased the rate of successful transformation. As the cells but not the plasmid DNA were irradiated, it would be unlikely that the primary irradiation effect, reflected in the higher transformation rates, was due to some conformational or other changes in the plasmid DNA. In our system the irradiation effect was not based on increased nicking or linearization of the plasmids by the X-rays.

When irradiated before DNA transfer, the protoplasts were at much higher densities (1×10^6 ml^{-1}) than when irradiated in the culture medium after DNA transfer (3×10^4 ml^{-1}). Despite this, irradiation of protoplasts before or after the addition of plasmid DNA

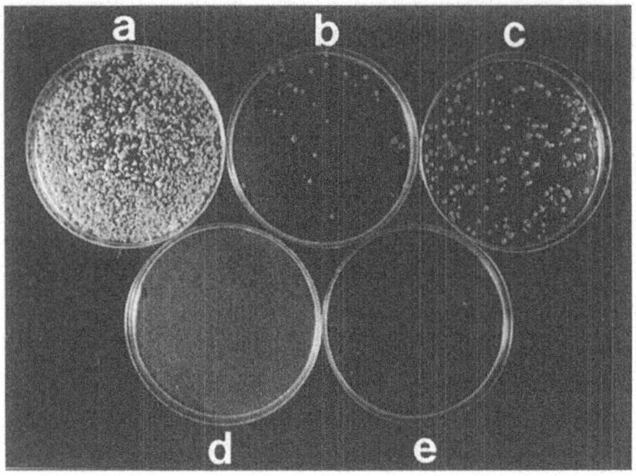

Fig. 1. Effect of X-ray irradiation on the transformation rate. *Petunia* protoplasts were isolated and treated with plasmid DNA (pHP23/kan^R) as described (Köhler et al., 1990). After ten days the growing calli were subjected to selection pressure (kanamycin, 50 mg l^{-1}). The picture shows four-week old colonies: a) growth control, no selection pressure; b) kanamycin^R calli, not irradiated; c) kanamycin^R calli, irradiated with 10Gy; d) control treated with carrier DNA only, irradiated with 10 Gy and subjected to selection (no resistant colonies); e) control irradiated with 120 Gy (lethal dose).

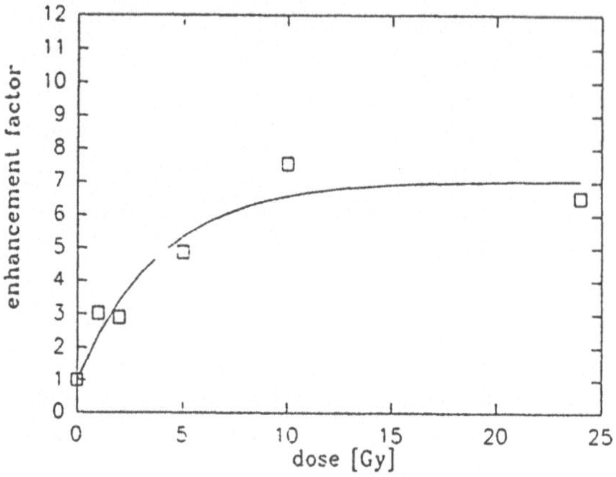

Fig. 2. Enhancement of transformation rates at different doses of X-rays. *Petunia* cells were treated as described in the legend to Fig. 1, and the enhancement ratio for different irradiation doses calculated. The relative transformation frequency, RTF, at each dose was divided by the RTF without irradiation, giving the enhancement factor, EF. Each value is the average of three independent experiments.

resulted in similar RTF values for the same doses. This shows that the effect that leads to improved transformation rates is to be found inside the cell and not in the surrounding medium (the protoplasts were also washed after irradiation and before DNA transfer). Irradiation could have produced some active compounds in the medium, e.g. peroxides, that could act upon the membrane or the plasmid molecules before they entered the cell. This is rejected by our data.

The effects of increasing X-ray dosage were tested on *Petunia* (1–24 Gy) and *Brassica* (2–62.5 Gy) protoplasts. When the enhancement factor, EF, was compared for different doses of irradiation, the effect increased with X-ray dosage at the beginning, but reached a maximum at 10 Gy for *Petunia* (Fig. 2). In *Brassica* the highest dose tested (62.5 Gy) gave a considerably higher response than the second highest (25 Gy) but the dose-effect curve did not show linearity; the EF value increased more rapidly at the lower doses, with a maximum value between 25 and 62.5 Gy.

The effects of X-rays on the RTF in both *Petunia* and *Brassica* were highest when applied shortly before (1 h) or after (1 h–4 h) the treatment with DNA. At 8 h thereafter the EF had dropped by 50% and after 24 h only a slight stimulation could be detected. At 48 h no stimulation was seen. This decline probably reflects the degradation rate of biologically active plasmids in the plant protoplasts.

UV irradiation

When using UV light for irradiation of protoplasts, a similar rise in RTF is found as with X-rays (Benediktsson et al., 1991; Gharti-Chhetri et al., 1990). Spivak et al. (1984) and van Duin et al. (1985) tested the effect of UV-irradiation of plasmid molecules upon the transformation frequency of human cells. The irradiated plasmids were shown to increase the transformation efficiency up to 11-fold over the unirradiated control plasmids. Irradiated plasmids that had been treated (repaired) with photoreactivating enzymes failed to stimulate the transfection. Treatment with X-rays or DNase (nicking/linearization) proved ineffective or negative in this regard. In contrast with our results using protoplasts (Benediktsson et al., 1991), UV-irradiation of the target human cells alone only slightly increased the transformation frequency (van Duin et al., 1985). In CHO cells, UV-irradiation increased transformation frequency when applied to the cells, but not when applied to the plasmid DNA (Nairn et al., 1991), indicating that foreign gene integration in those cells follows a different type of mechanism than that in human cells.

Gharti-Chhetri et al. (1990) tested UV-irradiation of plasmid molecules before transformation of *Nicotiana plumbaginifolia*, but did not observe any increase in the transformation rates, whereas irradiation of protoplasts resulted in a stimulation. Dzelkalus & Bogorad (1985) successfully transformed cyanobacterial cells with foreign DNA after UV-irradiation of the cells. Unirradiated cells could not be transformed with this nonhomologous, non-autonomously replicating plasmid DNA. They suggested that the reason for this induction might rely on an effect occurring at a step after uptake of DNA into the cells, as these can be transformed without UV treatment by homologous DNA. The authors also suggested that the observed UV induction was based on an effect similar to the SOS-response in *E. coli*, either by preventing DNA restriction or by inducing synthesis of recombination enzymes, leading to illegitimate incorporation into the host DNA.

Treatment with specific drugs

Treatment of tobacco protoplasts with 3-aminobenzamide (3-BA), a compound that interferes with the DNA replication, has been shown to increase the transformation rate in *Nicotiana plumbaginifolia* (Gharti-Chhetri et al., 1990). Mytomycin C is another drug that stimulates RTF in the same manner in plants (Paszkowski et al., 1992). In an animal cell system, 5-fluorodeoxyuridine has been shown to enhance transformation (Postel, 1985).

To test whether the DNA-breaking activity of irradiation might be responsible for its effect on the transformation rates, we performed experiments with ultra-low doses of bleomycin (BLM), an antibiotic that has been shown to act by inducing breaks in the DNA strands (Umezawa, 1971; Burger et al., 1981; Bianchi & López-Larraza, 1991). The toxicity of BLM on haploid *Petunia* protoplasts was tested. BLM was added at various concentrations after the freshly-isolated mesophyll protoplasts had been suspended in liquid medium. The protoplasts were extremely sensitive to the drug (Fig. 3). For further studies a final concentration of 0.05 $\mu g\ ml^{-1}$ was chosen (approx. 80% survival).

In order to determine the effect of BLM upon the transformation rates after direct gene transfer, protoplasts were treated with plasmid DNA carrying the *nptII* gene for kanamycin resistance. After completion of the transformation procedure, the protoplasts were exposed to BLM or X-rays (10 Gy) or both. The results

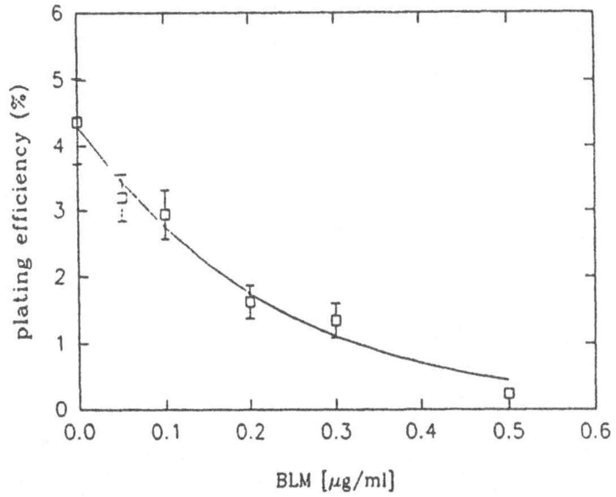

Fig. 3. Toxicity of bleomycin to *Petunia* protoplasts. Mesophyll protoplasts of *Petunia* Mitchell were isolated as previously described (Köhler et al., 1989) and treated with pHP23 (Paszkowski et al., 1988). One hour after the DNA transfer the protoplasts were plated as described (Benediktsson et al., 1991) and different concentrations of BLM (Sigma Chemical Co., St. Louis, USA) were added. The BLM was not removed. After five weeks of culture the plating efficiency was evaluated (percentage of protoplasts developing callus). The bars show the standard deviation.

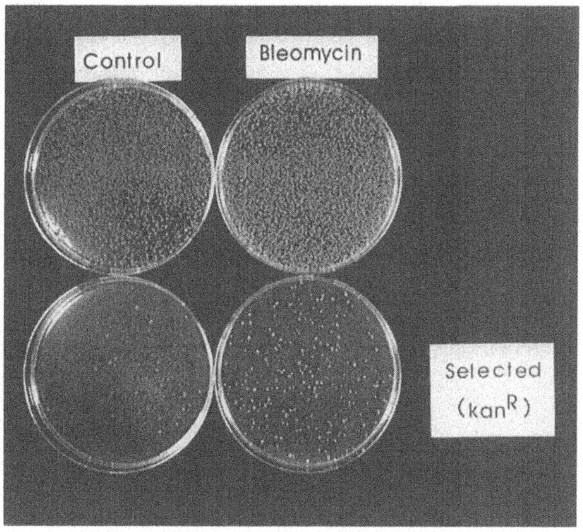

Fig. 4. Effect of bleomycin on the transformation rate. *Petunia* protoplasts were transformed and cultured as described by Benediktsson et al. (1991). BLM (0.05 μg/ml) was added to half of the protoplast suspension directly after transformation (right row). The other half was not treated ('control'; left row). After plating on Petri-dishes (day 10) one half of the protoplasts, both treated and not, was exposed to selection pressure (50 mg l^{-1} kanamycin; lower plates). The other half was cultivated without kanamycin (upper plates). The photograph was taken four weeks after the transformation.

Table 1. Influence of BLM and X-ray on the transformation rates

Exp.[a]	Type[b]	Protoplasts ($\times 10^6$)	PE[c] (%)	Kan^R calli	RTF[d] (%)	EF[e]
I	contr.	1.88	5.5	80	0.078	1
I	BLM	2.81	4.8	1015	0.752	9.6
II	contr.	1.96	7.6	89	0.060	1
II	BLM	1.96	6.0	516	0.436	7.3
III	contr.	0.75	7.6	182	0.318	1
III	BLM	1.50	7.8	1637	1.406	4.4
III	X-ray	1.50	5.7	2022	2.368	7.5
III	both	1.50	6.2	2279	2.435	7.7

[a] Data from three independent experiments.
[b] 'BLM' was treated with 0.05 μg/ml bleomycin; 'X-ray' with 10 Gy of X-rays; 'both' with BLM and X-ray.
[c] Plating efficiency (percentage of protoplasts making callus).
[d] Relative transformation frequency.
[e] Enhancement factor.

are shown in Table 1 and Fig. 4. It is clear from these observations that BLM is capable of inducing a similar increase in transformation rates as is irradiation. These results support the assumption that DNA repair mechanisms are responsible for the elevation of transformation rates. As mentioned before, the inducible effect of X-rays reaches a maximum at about 10 Gy for *Petunia*. Here, we show that the use of BLM and X-rays simultaneously does not result in a higher transformation frequency than with only one of these agents. This is comparable with using a higher dose of X-rays and indicates that in both cases the increased transformation frequency is probably based on the same phenomenon. If not, we would expect an additive effect when using both stimulating agents.

Transient expression

In contrast to the so-called 'stable' or 'integrative' expression dealt with above, transient expression is not detectably affected by either UV- or X-ray-irradiation (Benediktsson et al., 1991). It is assumed that non-integrated DNA is responsible for transient expression and that the copy number of a gene within the cell nucleus is the limiting factor for the strength of the signals (Pröls et al., 1988). As irradiation does not increase transient expression, we suggested that the principal effect of irradiation must act after the plasmid molecules have entered the nucleus, i.e. at the integration or thereafter. It must be regarded as unlikely that irradiation eases the access of DNA into the nucleus, otherwise we would expect increased transient expression after that treatment.

Measurement of the activity of DNA polymerases in intact protoplasts (Benediktsson et al., 1994) showed that adding external template DNA increased the activity of these nuclear enzymes. This effect of added DNA was further increased in the presence of PEG. This suggests that the DNA, or at least small fragments of it, is relatively easily taken up into the protoplast and the nucleus. Another result supporting this hypothesis comes from experiments with co-transformation (Benediktsson et al., 1991). Irradiation did not alter the balance of cells effectively transformed by two different genes. *Petunia* cells that had been treated with two antibiotic-resistance genes showed expression of both genes in approximately 50% of the cells positive for one trait, regardless of irradiation. If the number of plasmid molecules in the nucleus had been increased by irradiation, we would have expected the co-transformation frequency to rise. Further, this observation argues against the hypothesis that the principal effect of irradiation lies in increased integration into each successfully transformed cell, as in this case we would also expect higher co-transformation rates.

It seems that, as suggested by Gould & Dains (1985), not all protoplasts in a transformation assay are potentially capable of incorporating alien genes effectively. The existence of a fixed population of competent cells does not, however, seem likely. It rather appears that some cells were in a 'competent state', in a metabolic sense, which can be induced by e.g., irradiation. Following this hypothesis, cells in a competent state can be transformed by one gene or co-transformed by two or more, without necessarily increasing the copy number of the foreign genes in each cell.

The question as to whether the copy number of integrated genes in transgenic clones is increased by irradiation has not yet been fully answered. The problem is that after direct gene transfer, the copy number varies considerably between different clones. This means that a very high number of both irradiated and non-irradiated clones would have to be analysed for workers to come to a definite conclusion. Examination of seven irradiated and seven non-irradiated hygromycin-resistant *Brassica nigra* clones revealed an approximately 80% higher copy number in the irradiated cells, while the transformation rate was sevenfold higher than in the unirradiated control (Köhler et al., 1990). Gharti-Chhetri et al. (1990) reported a slightly higher number of integration sites of the *nptII* gene after UV-irradiation of *Nicotiana plumbaginifolia* protoplasts. In *Petunia* we found no difference in the average gene copy number among forty tested clones, half of which had been irradiated (Benediktsson et al., 1991). Using a human cell system, van Duin et al. (1985) did not detect any difference in copy number of UV-irradiated plasmid DNA integrated into fibroblasts compaired to unirradiated plasmids. They used batches of 300–600 independent clones for this examination. In this connection, it is important to keep in mind that raising the copy number of a transgene in a clone does not necessarily improve the expression conditions. The genes can in some cases be negatively influenced by homologous copies elsewhere in the genome (Hobbs et al., 1993).

Conclusion

Van Duin et al. (1985) suggested that the likely reason for the observed stimulation of transformation of mammalian cells by UV is to be found in the lesions induced in the plasmids by UV light, especially pyrimidine-dimers, the principal lesion caused by UV. They did not, however, exclude the influence of other DNA lesions. According to our experiments, we must look for another explanation for the observed stimulation of protoplast transformation, as X-rays and BLM also show the same influence as UV. A common effect of all agents, including 3-BA and other chemicals that have been shown to act in this manner, is the breaking of DNA strands. For BLM this is reported to be the main activity whereas X-rays have a much wider influence, although DNA breakage is biologically the most effective. Treatment with 3-BA can lead to strand breaks as it interferes with the ligation of DNA ends during replication (Morgan & Cleaver, 1983). Also, UV-irradiation can lead to strand breaks after removal of the pyrimidine dimers by excision repair enzymes. In fact, Hall et al. (1992) reported substantial fragmentation of genomic DNA molecules after UV-irradiation of sugarbeet protoplasts. This breakage of DNA strands of the nuclear DNA seems to be a crucial step in direct gene transfer.

Cell phase synchronization of protoplasts has in some cases led to increased transformation rates (Meyer et al., 1985). The synchronization was achieved through incubation with aphidicholin (APH) which inhibits the DNA polymerase α. The enhanced transformation was detected shortly after removal of the drug from the incubation medium but not 15 hours later. Possibly the strand breaks produced by APH were the reason for the observed effect. Irradiation with X-rays or UV-light is known to delay the cell cycle progression in many cells but Chatterjee & Raman (1988) found that exposure of lymphocytes to BLM did not result in such a delay and suggested that BLM was only causing DNA strand breaks and no other cellular damage. Therefore it is unlikely that the cell cycle delay after irradiation is responsible for improved transformation rates.

The exposure of protoplasts to irradiation or BLM during direct gene transfer can lead to increased mutations in the cells, above the level caused by the extra DNA itself. As the dose of irradiation of BLM we applied was extremely low, this mutagenicity should be minimal, e.g. albino mutants, that can be caused by a number of different defects, have not been observed in the thousands of transgenic calli or shoots we have examined so far.

If it turns out as a general phenomenon that alien DNA is integrated at a high frequency into all protoplast genomes but the rate limiting factor in transformation is the expression of those integrated genes, one could suggest that the bottle-neck for transformation would not lie in the integration *per se* but rather in the quality of the integration. As the exposure to external influences (irradiation, certain chemicals) near the time of DNA treatment can improve the transformation frequency but not when applied later, we conclude that the importance lies in the activation or inactivation of the genes shortly after integration or the accessibility of the genes, i.e. integration near to active genes or enhancers.

Two further considerations also fit well with this hypothesis. First, irradiation and BLM can cause more damage to open structures like euchormatin or active genes than to tightly-packed DNA (Kuo, 1981). Secondly, there exist many types of DNA repair mechanisms that have a higher affinity for active genes (Bohr et al., 1985; Mellon et al., 1986) or even active strands of genes (Terleth et al., 1991) than for the rest of the DNA. Maybe this is also the case for mechanisms that are important in integration of genes into plant protoplasts and can be induced by DNA strand breaks. This situation would cause increased incorporation of transgenes into expression-susceptible loci after irradiation. We are currently running experiments to test this hypothesis. If it proves true, the use of BLM or irradiation in routine transformations, and where a high number of clones is not required, would be recommended.

Acknowledgements

The work was supported by a grant from the Commission of the European Communities No. CI 1*/0402-D(AM).

References

Ballas, N., N. Zakai, D. Friedberg & A. Loyter, 1988. Linear forms of plasmid are superior to supercoiled structures as active templates for gene expression in plant protoplasts. Plant Molec. Biol. 11: 517–527.

Bates, G.W., S.A. Carle & W.C. Piastuch, 1990. Linear DNA introduced into carrot protoplasts by electroporation undergoes ligation and recircularization. Plant Molec. Biol. 14: 899–908.

Benediktsson, I., F. Köhler & O. Schieder, 1991. Transient and stable expression of marker genes in cotransformed *Petunia* protoplasts in relation to X-ray and UV-irradiation. Transgenic Res. 1: 38–44.

Benediktsson, I., C.P. Spampinato, C.S. Andreo & O. Schieder, 1994. Analysis of DNA polymerase activity in *Petunia* protoplasts treated with clastogenic agents. Physiol. Plant 90: 445–450.

Bianchi, N.O. & D.M. López-Larraza, 1991. DNA damage and repair induced by bleomycin in mammalian and insect cells. Environm. Molec. Mutagen. 17: 63–68.

Bohr, V.A., C.A. Smith, D.S. Okumoto & P.C. Hanawalt, 1985. DNA repair in an active gene: removal of pyrimidine dimers from the DHFR gene of CHO cells is much more different than in the genome overall. Cell 40: 359–369.

Burger, R.M., J. Peisach & S.B. Horwitz, 1981. Mechanism of bleomycin action: *in vitro* studies. Life Sci. 28: 715–727.

Chatterjee, A. & M.J. Raman, 1988. A comparison of aberation distribution and cell-cycle progression in cells treated with bleomycin with those exposed to X-rays. Mutat. Res. 202: 51–57.

Czernilowski, A.P., R. Hain, L. Herrera-Estrella, H. Lörz, E. Goyvaerts, B.J. Baker & J. Schell, 1986. Fate of selectable marker DNA integrated into the genome of *Nicotiana tabacum*. DNA 5: 101–113.

Davey, M.R., E.C. Cocking, J. Freeman, N. Pearce & I. Tudor, 1980. Transformation of *Petunia* protoplasts by isolated *Agrobacterium* plasmids. Plant Sci. Lett. 18: 307–313.

Dzelzkalns, V.A. & L. Bogorad, 1985. Stable transformation of the cyanobacterium *Synechocystis* sp. PCC 6803 induced by irradiation. J. Bacteriol. 165 (3): 964–971.

Fraley, R.T., S.G. Rogers, R.B. Horsch, P.R. Sanders, J.S. Flick, S.P. Adams, M.L. Bittner, L.A. Brand, C.L. Fink, J.S. Fry, G.R. Galluppi, S.B. Goldberg, N.L. Hoffmann & S.L. Woo, 1983. Expression of bacterial genes in plant cells. Proc. Natl. Acad. Sci. USA 80: 4803–4807.

Gharti-Chhetri, G.B., W. Cherdshewasart, J. Dewulf, J. Paszkowski, M. Jacobs & I. Negrutiu, 1990. Hybrid genes in the analysis of transformation conditions. 3. Temporal/spatial fate of NTPII gene integration, its inheritance and factors affecting these processes in *Nicotiana plumbaginifolia*. Plant Molec. Biol. 14: 687–696.

Gould, A.R. & R.J. Daines, 1985. Plant protoplasts and the cell cycle. In: L.C. Fowke & F. Constable (Eds). Plant Protoplasts, pp. 67–76. CRC-Press, Boca Raton, Florida.

Hain, R., P. Stabel, A.P. Czernilofsky, H.H. Steinbiß, L. Herrera-Estrella & J. Schell, 1985. Uptake, integration, expression and genetic transmission of a selectable chimaeric gene by plant protoplasts. Mol. Gen. Genet. 199: 161–168.

Hall, R.D., F.A. Krens & J.A. Rouwendal, 1992. DNA radiation damage and asymmetric somatic hybridization: Is UV a potential substitute or supplement to ionizing radiation in fusion experiments? Physiol. Plant 85: 319–324.

Herrera-Estrella, L., M. De Block, E. Messens, J.-P. Hernalsteens, M. Van Montagu & J. Schell, 1983. Chimeric genes as dominant selectable markers in plant cells. EMBO J. 2; 6: 987–995.

Hobbs, S.L.A., T.D. Warkentin & C.M.O. DeLong, 1993. Transgene copy number can be positively or negatively associated with transgene expression. Plant Molec. Biol. 21: 17–26.

Hoffmann, F. & D. Hess, 1972. Die Aufnahme radioaktiv markierter DNS in isolierte Protoplasten von *Petunia hybrida*. Z. Pflanzenphysiol. 69: 81–83.

Köhler, F., C. Golz, S. Eapen & O. Schieder, 1987. Influence of plant cultivar and plasmid-DNA on transformation rates in tobacco and moth bean. Plant Sci. 53: 87–91.

Köhler, F., G. Cardon, M. Pöhlmann, R. Gill & O. Schieder, 1989. Enhancement of transformation rates in higher plants by low-dose irradiation: are DNA repair systems involved in the incorporation of exogenous DNA into the plant genome? Plant Molec. Biol. 12: 189–199.

Köhler, F., I. Benediktsson, G. Cardon, C.S. Andreo & O. Schieder, 1990. Effect of various irradiation treatments of plant protoplasts on the transformation rates after direct gene transfer. Theor. Appl. Genet. 79: 679–685.

Krüger-Lebus, S. & I. Potrykus, 1987. A simple and efficient method for direct gene transfer to *Petunia hybrida* without electroporation. Plant Molec. Biol. Reporter 5 (2): 289–294.

Kuo, T.M., 1981. Preferential damage of active chromatin by bleomycin. Cancer Res. 41: 2439–2443.

Mellon, I., A.B. Vilhelm, C.A. Smith & P.C. Hanawalt, 1986. Preferential DNA repair of an active gene in human cells. Proc. Natl. Acad. Sci. USA 83: 8878–8882.

Meyer, P., E. Walgenbach, K, Bussmann, G. Hombrecher & H. Saedler, 1985. Synchronized tobacco protoplasts are efficiently transformed by DNA. Mol. Gen. Genet. 201 (3): 513–518.

Miller, C.K. & H.M. Temin, 1983. High efficiency ligation and recombination of DNA fragments by vertebrate cells. Science 220: 606–609.

Morgan, W.F. & J.E. Cleaver, 1983. Effect of 3-aminobenzamide on the rate of ligation during the repair of alkylated DNA in human fibroblasts. Cancer Res. 43: 3104–3107.

Nairn, R.S., G.M. Adair, C.B. Christmann & R.M. Humphrey, 1991. Ultraviolet stimulation of intermolecular homologous recombination in Chinese hamster ovary cells. Mol. Carcinogen. 4: 519–526.

Negrutiu, I., R. Shillito, I. Potrykus, G. Biasini & F. Sala, 1987. Hybrid genes in the analysis of transformation conditions. I. Setting up a simple method for direct gene transfer in plant protoplasts. Plant Molec. Biol. 8: 363–373.

Paszkowski, J., R.D. Shillito, M. Saul, V. Mandák, T. Hohn, B. Hohn & I. Potrykus, 1984. Direct gene transfer to plants. EMBO J. 3: 2717–2722.

Paszkowski, J., M. Baur, A. Bogucki & I. Potrykus, 1988. Gene targeting in plants. EMBO J. 7: 4021–4026.

Paszkowski, J., A. Peterhans, H. Schlüpmann, C. Basse, E.G. Lebel & J. Masson, 1992. Protoplasts as tools for plant genome modifications. Physiol. Plant 85: 352–356.

Postel, E.H., 1985. Enhancement of genetic transformation frequencies of mammalian cell cultures by damage to the cell DNA. Mol. Gen. Genet. 201: 136–139.

Potrykus, I., 1991. Gene transfer to plants: Assessment of published approaches and results. Ann. Rev. Plant Physiol. Plant Mol. Biol. 42: 205–225.

Pröls, M., R. Töpfer, J. Schell & H.H. Steinbiß, 1988. Transient gene expression in tobacco protoplasts: I. Time course of CAT appearance. Plant Cell Rep. 7: 221–224.

Puite, K.J., 1992. Progress in plant protoplast research. Physiol. Plant 85: 403–410.

Schocher, R.J., R.D. Shillito, M.W. Saul, J. Pazkowski & I. Potrykus, 1986. Co-transformation of unlinked foreign genes into plants by direct gene transfer. Bio/Technology 4: 1093–1096.

Shillito, R.D., M.W. Saul, J. Paszkowski, M. Müller & I. Potrykus, 1985. High efficiency direct gene transfer to plants. Bio/Technology 3: 1099–1103.

Shimamoto, M., R. Terada, T. Izawa & H. Fujimoto, 1989. Fertile transgenic rice plants regenerated from transformed protoplasts. Nature 338: 274–276.

Spivak, G., A.K. Ganesan & P.C. Hanawalt, 1984. Enhanced transformation of human cells by UV irradiated pSV2 plasmids. Mol. Cell Biol. 4: 1169–1171.

Terleth, C., P. van de Putte & J. Brouver, 1991. New insights in DNA repair: preferential repair of transcriptionally active DNA. Mutagenesis 6: 103–111.

Tyagi, S., B. Spörlein, A.K. Tyagi, R.G. Herrmann & H.U. Koop, 1989. PEG- and electroporation-induced transformation in *Nicotiana tabacum*: influence of genotype on transformation frequency. Theor. Appl. Genet. 78: 287–292.

Umezawa, H., 1971. Natural and artificial bleomycins: chemistry and antitumor activities. Pure Appl. Chem. 28: 665–680.

Van Duin, M., A. Westerveld & J.H.J. Hoeijmakers, 1985. UV stimulation of DNA-mediated transformation of human cells. Mol. Cell Biol. 5 (4): 734–741.

Euphytica **85:** 63–74, 1995.

Expression of foreign genes in sunflower (*Helianthus annuus* L.) – evaluation of three gene transfer methods

Hélène Laparra, Monique Burrus[1], Reiner Hunold, Brigitte Damm[2], Ana-Maria Bravo-Angel[3], Roberte Bronner & Günther Hahne
Institut de Biologie Moléculaire des Plantes, Centre National de la Recherche Scientifique et Université Louis Pasteur, 12, Rue du Général Zimmer, 67084 Strasbourg Cedex, France; [1] *present address: Laboratoire de Biologie et Physiologie Végétales, UFR Sciences, Université de Reims, BP 374, 51062 Reims, France;* [2] *present address: MOGEN Intl., Einsteinweg 97, 2333 LB Leiden, The Netherlands;* [3] *present address: Friedrich-Miescher-Institut, Postfach 273, 4002 Basel, Switzerland*

Key words: Agrobacterium tumefaciens, electroporation, particle gun, polyethylene glycol, regeneration, transformation

Summary

Suitable sunflower tissues and cells were transformed either by direct gene transfer into protoplasts, particle bombardment, or *Agrobacterium* co-culture. While all techniques allowed efficient short-term or transient expression of the introduced gene(s) in the respective tissues, stable transformation was only observed after transformation with *Agrobacterium*. The latter technique was suitable for the production of transgenic callus from seedling cotyledons and occasional shoots with chimaeric expression of the transgene. Detailed analysis of the interaction of *Agrobacterium* with this explant showed that infection efficiency was critically dependent on the co-culture conditions, and that the preferentially-transformed cells were not the ones competent for regeneration.

Abbreviations: BAP – benzyl adenine, CAT – chloramphenicol acetyl transferase, 2,4-D – 2,4-dichloro phenoxy acetic acid, GUS – β-D-glucuronidase, MS – medium according to Murashige & Skoog (1962), NAA – naphthalene acetic acid, NPTII – neomycin phospho transferase II, PEG – polyethylene glycol, PIG – particle inflow gun, SH – medium according to Schenk & Hildebrandt (1972)

Introduction

Despite the fact that the production of transgenic sunflower (*Helianthus annuus* L.) plants has been described in several instances (Everett et al., 1987; Schrammeijer et al., 1990; Malone-Schoneberg et al., 1991; Espinasse-Gellner, 1992), a reliable, efficient and universally applicable protocol has not yet been published. It is obvious that some key factors implicated in the complex process of regenerating a fertile plant from a transgenic cell are difficult to master in this species. It is not known which factors precisely are at the origin of the difficulties, but genotypic and physiological effects certainly figure on the list.

A successful protocol for plant transformation must involve an efficient gene delivery method addressing cells for which a regeneration system exists. For such a protocol to be practically useful, the combination must show a sufficiently high overall efficiency.

For sunflower, a number of regeneration protocols have been described which allow regeneration of plants from explants such as hypocotyl, cotyledons, or immature zygotic embryos, as well as from protoplasts obtained from most of these tissues. Most regeneration systems appear to be direct, i.e., without an intervening callus phase, and even protoplast-derived calli loose their regeneration potential rapidly. Prolonged culture in the non-differentiated state appears to be incompatible with plant regeneration, at least for the currently available protocols and genotypes.

In this study, we have evaluated three methods for introducing foreign DNA into sunflower cells, namely, direct gene transfer into protoplasts, particle bombardment, and gene transfer using *Agrobacterium tumefaciens* as vector. We have studied some parameters important for the gene transfer efficiency for each system. Expression of marker genes was measured after short culture periods and thus must be considered transient in most cases. The respective potential of the three approaches for the production of transgenic plants is discussed with particular references to the localisation of regenerating vs transformed cells.

Materials and methods

Plant material and plasmids

All experiments were performed with a public inbred sunflower line, HA 300B. Plasmids used for direct gene transfer were pCG35S (Lepetit et al., 1991) containing the *uid*A, *cat*, and *npt*II genes all under control of the CaMV 35S promoter, or exceptionally, pVT GUS containing the *uid*A gene under control of a modified 35S-promoter (see text). *Agrobacterium*-mediated transformation was obtained with GV 2260 (pGUS-Int), a binary vector containing the *uid*A (35S promoter) and *npt*II (*nos* promoter) genes (Vancanneyt et al., 1990).

Culture and transformation techniques

Seeds were grown in the greenhouse to produce immature embryos as described by Jeannin & Hahne (1991), or sterilized and allowed to germinate *in vitro* (Knittel et al., 1991). Protoplast isolation was according to the protocol described by Fischer et al. (1992). Conditions for PEG-mediated gene transfer have been described by Lepetit et al. (1991), and for electroporation by Jung et al., (1992) and Veidt et al. (1992). Estimation of GUS- and CAT-activities were made as described by Lepetit et al. (1991). Preparation of particles and their coating with plasmid DNA has been described elsewhere (Hunold et al., 1993), but followed standard procedures. Immature embryos to be transformed were precultured on the respective media for three days, treated by bombardment or with *Agrobacterium* , and cultured further on the same medium until analysis. If selection pressure was applied, this medium was supplemented with either kanamycin (50 mg/l) or paromomycin (25 mg/l). Usual culture media for immature embryos were modified MS media ((Murashige & Skoog, 1962), supplemented with 0.1 mg/l NAA and 0.5 mg/l BAP). The presence of blue spots indicating GUS expression after histological staining (Atanassova et al. 1992) was evaluated with the help of a dissecting microscope, or, where necessary, by microscopical observation of free-hand sections. Shoot regeneration from cotyledons of seedlings was performed according to Knittel et al. (1991), except that explants were not sectioned lengthwise because this operation provoked necrosis after co-cultivation with *Agrobacterium.* If not indicated otherwise, explants were dissected, dipped into *Agrobacterium* suspension (O.D. = 1.0) for 10 min, blotted dry, and co-cultured for 2–3 days. According to the experiments, *A. tumefaciens* was grown in LB or YEB medium (Maniatis et al., 1982). The explants were then washed with MS medium containing 0.5 g/l cefotaxim, and further cultured on the appropriate media. Media employed in this study were as follows: MS: hormone-free medium according to Murashige & Skoog (1962); MSBAP: MS medium supplemented with BAP (2 mg/l, if not indicated otherwise); MSC: MS medium supplemented with BAP (1 mg/l) and NAA (0.5 mg/l); PER: medium according to Paterson & Everett (1985); PERC: PER medium modified to contain 0.5 mg/l NAA and 1.0 mg/l BAP; SH: hormone-free medium after Schenk & Hildebrandt (1972); 1/10 SH: SH-medium diluted tenfold; SHC: SH medium modified to contain 0.5 mg/l NAA and 1.0 mg/l BAP.

Results and discussion

We introduced similar genetic constructs, containing visible, quantifiable or selectable marker genes (coding for β-D-glucuronidase, chloramphenicol acetyl transferase, and neomycin phosphotransferase, respectively) into sunflower cells either by direct gene transfer into protoplasts, particle bombardment, or *Agrobacterium tumefaciens* infection. This parallel study allows us to compare these different approaches for their performance with respect to production of transgenic cells and, eventually, plants. All experiments were conducted on cell or tissue types suitable for the particular technique, and for which regeneration protocols were established in our laboratory. Since long-term clonal propagation by cuttings is not possible

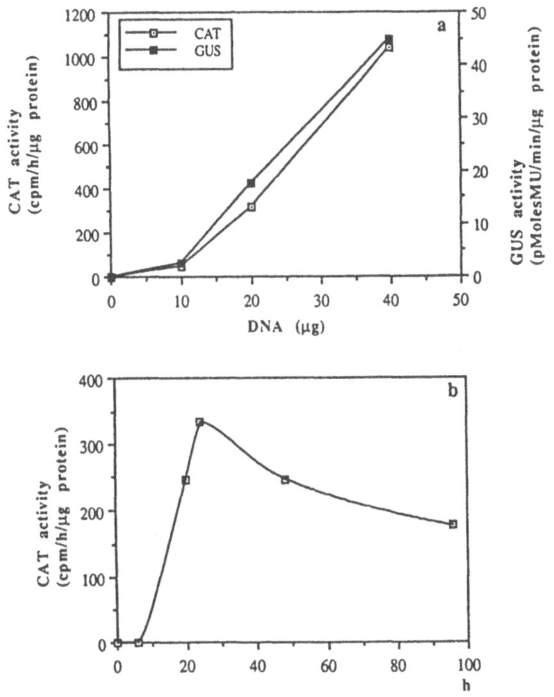

Fig. 1. PEG-mediated direct gene transfer into 10^6 protoplasts. *a.* Expression of two marker genes as a function of plasmid DNA (pCG35S) concentration. Carrier DNA has been added at twice the concentration of the plasmid DNA. Expression time 24 h. *b.* Kinetics of CAT expression after transformation with 20 μg plasmid DNA and 40 μg carrier DNA.

with sunflower, seedlings or immature embryos from an inbred line were used throughout the study.

Direct gene transfer to protoplasts

Transient expression studies require large numbers of viable protoplasts. For sunflower, this condition is fulfilled by protoplasts isolated from either hypocotyl or cotyledons of young seedlings. Both protoplast types can be used equally well, but since hypocotyl protoplasts are more easily obtained in large numbers, most experiments have been performed with this material.

Two principal methods are available for the introduction of DNA into freshly-prepared protoplasts, i.e., electroporation and polyethylene (PEG)-mediated uptake. For both techniques, preliminary experiments showed that the presence of a two-fold excess of carrier DNA increased the enzyme activity by a factor of two, for both GUS and CAT. Carrier DNA was therefore incorporated in all experimental procedures.

For PEG-mediated DNA transfer, best results were obtained with PEG 6000. Under the experimental conditions used, 40% of the protoplast preparation remained viable after the PEG treatment. The expression level increased with the amount of plasmid DNA, for both GUS and CAT (Fig. 1a). For practical reasons, we used 20 μg of plasmid DNA and 40 μg carrier DNA as standard concentrations in most experiments. A kinetic study showed that the transgenic enzyme activity was already detectable after 6 h, and was remarkably stable in sunflower protoplasts. More than 50% of the maximal CAT activity (measured after 24 h) was still present 4 days after the transformation experiment (Fig. 1b). The use of linearized plasmid DNA was without effect compared with circular plasmids, and in contrast to tobacco protoplasts (Negrutiu et al., 1987), a heat shock of 45 °C did not increase transgene expression in sunflower protoplasts.

For electroporation, the electrical parameters must be adjusted to obtain the optimal compromise between transfer efficiency and protoplast lysis. For our protoplast preparations (diameter 20–80 μm), the minimal field strength necessary for any detectable expression level of the introduced *uid*A gene (encoding the GUS protein) was 375 V/cm. The expression increased up to an optimal level (700 V/cm) and then decreased (Fig. 2a). Increasing field strength caused protoplast damage. This is probably the reason for the decreasing expression level beyond a certain field strength. We measured protoplast damage by quantifying the protein content in recoverable intact protoplasts (Fig. 2b). The optimal electroporation conditions (700 V/cm) corresponded to a survival rate of approx. 50%. The expression level after 24 h was comparable to that obtained with tobacco mesophyll protoplasts, but lower than the expression levels observed with sunflower protoplasts after PEG-mediated gene transfer, even though we used 40 μg DNA and 80 μm carrier DNA in all electroporation treatments.

We conclude from our transient expression studies that PEG-mediated gene transfer is more efficient than electroporation for sunflower protoplasts. High transient expression levels can be consistently obtained with a simple routine protocol. The protoplasts surviving the transformation treatment can be cultured and proceed through the first divisions like untreated cultures. As in control cultures, transformed protoplasts developed in a different fashion according to the physical culture conditions (liquid medium vs agarose- or alginate-solidified medium; Fischer & Hahne, 1992). Stable integration of foreign DNA introduced into sun-

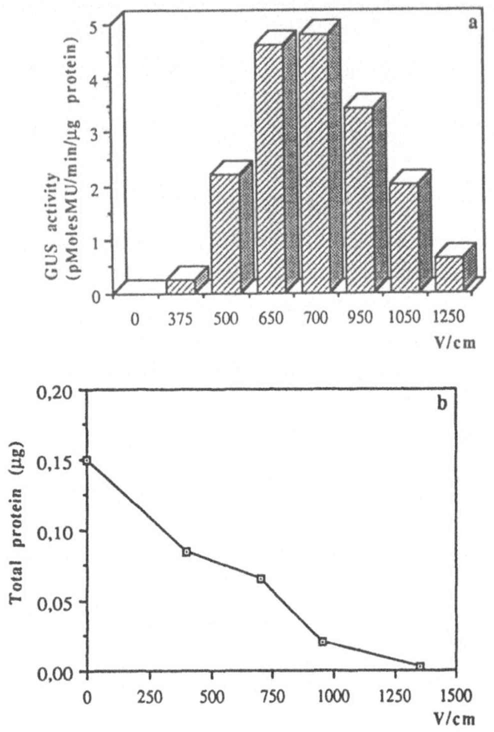

Fig. 2. Direct gene transfer into protoplasts by electroporation (8.10^5 protoplasts/point, 40 μg plasmid DNA (pCG35S), 80 μg carrier DNA, 125 μF capacity). *a*. GUS activity as a function of electric field strength. *b*. Total protein content of the pellet of intact protoplasts 24 h after electroporation, as a function of electric field strenght. This value measures the fraction of intact protoplasts having survived the treatment.

flower protoplasts occurs at frequencies of approx. 10^{-4} (Moyne et al., 1989). Kanamycin (50 mg/l) is a suitable selective agent to enrich these cultures for transgenic cell lines. However, since the overall regeneration efficiency of plants transferable to the greenhouse at present is also low (10^{-4}; Fischer et al., 1992), we did not try to obtain transgenic shoots from protoplasts. We are now working to increase the efficiency of plant regeneration to a level which allows the recovery of rare events such as stable integration of introduced foreign DNA.

Particle bombardment

Several methods are available which allow us to introduce DNA into intact plant cells, but of those not involving a biological vector, only particle bombardment is widely used. Indeed, this technique has been successfully used for the production of transgenic plants from a number of species which cannot be transformed otherwise, such as maize (Gordon-Kamm et al., 1990), sugarcane (Bower & Birch, 1992), and white spruce (Ellis et al., 1993). Furthermore, particle bombardment may be used as an efficient means to induce transient expression in tissues not amenable to protoplast isolation (Knudsen & Müller, 1991; Twell et al., 1989), and to injure tissues in order to prepare them for *Agrobacterium* infection. Under certain conditions, this latter approach may increase the infection efficiency considerably (Bidney et al., 1992a). We have used two different particle gun designs with various sunflower tissues, and compared the efficiencies for particle bombardment and *Agrobacterium* treatment alone and in combination.

One of our guns is the gunpowder-driven device described by Zumbrunn et al. (1989), the other the helium particle inflow gun (PIG) described by Finer et al. (1992). Although their mechanism of action is quite different, and optimal bombardment conditions must be determined separately for each case, the results are comparable between the two systems. For everyday use, the gunpowder-driven model is more cumbersome and requires more DNA-loaded particles than the PIG for a comparable efficiency. We therefore only use the latter model for routine experiments, but most of the results described have been verified with both models.

The cell type attained by the particles is a critical factor for the success of experiments aimed at the regeneration of transgenic plants. Penetration depth of the particles, and thus the targeted cell type, depend to a large degree on the physical bombardment conditions, some of which can be chosen at will (e.g., chamber vacuum, distance, acceleration pressure). The particle type (material, size) may also be very important for reaching the intended cell type. In our experience, the denser gold particles (Alpha Products, 1.5–3 μm) can penetrate into deeper cell layers than the less dense tungsten particles (BioRad M17) which, in addition, have a more pronounced tendency towards aggregation.

With both gun models and both particle types, it was impossible to penetrate the cotyledons of young seedlings any further than the cuticle (tungsten) or the epidermis (gold). Transformation events were rare in this material (on average, 1–8 'blue spots' in max. 27% of the explants) even after intense efforts to optimise the system on the technical (gun) and the biological side (preculture conditions, age). Although a good regeneration system existed for this material (see below), we abandoned this approach and concentrated

Table 1. GUS expression 3, 14, and 28 days after bombardment by the particle inflow gun with plasmid-coated microprojectiles, and after *Agrobacterium* co-cultivation of immature zygotic embryos with or without prior bombardment with naked particles

Technique/culture time before GUS assay (days)	Treated embryos (no.)	GUS-positive explants (%)	Blue sectors or spots per embryo (no.)
Co-culture			
28	206	13.6 ± 9.0	3.7 ± 4.1
Bombardment + Co-culture			
28	229	8.5 ± 3.7	2.1 ± 1.8
Bombardment			
3	171	90.2 ± 9.7	35.0 ± 4.9
14	254	52.3 ± 11.2	9.0 ± 3.5
28	259	22.3 ± 21.0	9.6 ± 5.9

Data were obtained in 3 independent experiments. Immature embryos were precultured for 3 days in the dark on the same medium as used for bombardment and co-culture.

our bombardment experiments on immature embryos which proved to be a more suitable material.

Immature zygotic embryos of sunflower were most responsive to particle bombardment when harvested 8–9 days post fertilization (approx. 2 mm in size), and precultured for 3 days on MS medium supplemented with NAA (0.1 mg/l), BAP (0.5 mg/l), and sucrose (3%). Such immature embryos were used as standard material for all subsequent bombardment experiments. Sunflower immature embryos were found to be sensitive to kanamycin (50 mg/l) and paromomycin (25 mg/l) which are both useful selective agents for this material.

Immature embryos were either bombarded with plasmid-coated tungsten particles, with uncoated tungsten particles with or without subsequent *Agrobacterium* infection, or cut in half and infected with *Agrobacterium* alone. The samples were analyzed after 3 days, and two and four weeks after treatment. Particle bombardment gave high levels of transient expression which diminished with prolonged culture time, even on selective medium (Table 1). Most of the 'blue spots' persisting after longer culture had not developed into larger sectors, indicating that no active cell division had occurred at these sites. However, in some cases, expanding GUS-expressing areas were observed. Samples bombardment with naked particles never gave rise to any blue spot which can therefore be considered as a specific indication of GUS expression. In embryos treated with *Agrobacterium*, GUS expression was analyzed 4 weeks after selection on kanamycin-containing medium. The number of embryos expressing the transgene was lower after this treatment than after bombardment with plasmid-coated particles (Table 1). Finally, a combination of bombardment with uncoated particles and *Agrobacterium* infection could not increase the transformation efficiency above the values obtained for *Agrobacterium* alone (Table 1). This finding is different from the results of Bidney et al. (1992a) who observed a marked increase in transformation efficiency after bombardment. This contradiction is probably related to the specific conditions under which the respective experiments have been conducted. Immature embryo halves expose a very large wound which seems to allow good access to *Agrobacterium*, while the parts of the explants reached by the particles, mainly belonging to the cotyledon, do not develop under our experimental conditions.

Bombardment with particles, DNA-coated or not, did not influence the further development of the embryos when these were cultured on non-selective medium. However, to date we could not recover any plantlets from embryos submitted to any one of the three treatments after selection on kanamycin-containing medium. This failure to regenerate plants can be explained by either the low frequency of stable integration events, or the severe reduction of the morphogenic potential of sunflower tissues under the influence of kanamycin which has been previously described by other authors (e.g., Everett et al., 1987).

Although particle bombardment so far did not prove to be a useful technique for stable transformation of sunflower tissues under our conditions, it is still a pow-

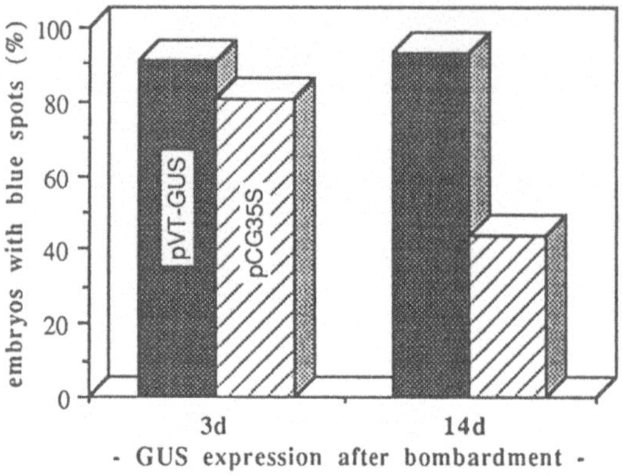

Fig. 3. GUS expression in immature embryos after bombardment with plasmid-coated microprojectiles. Expression of two different constructs (see text for explanation) was measured after 3 and 14 days. Embryos were counted as positive if at least one blue spot, indicating a single transformation event, was visible under the dissection microscope.

erful technique for transient expression studies in cell types which do not readily yield protoplasts, such as immature embryos, anthers, or pollen grains. As an illustration of this fact, we have compared the expression of different chimaeric constructs where the GUS reporter gene was coupled to promoters of different strength (Fig. 3). Briefly, several sequences reputed to enhance expression were added to the CaMV 35S promoter, such as the Kozak consensus sequence (Kozak, 1986), the Ω' sequence (Sleat et al., 1988), and a doubled enhancer sequence of the 35S promoter (Kay et al., 1987). Indeed, these modifications, incorporated in a plasmid called pVT-GUS, rendered the promoter more efficient than the unmodified 35S promoter in the plasmid pCG35S (Lepetit et al., 1991), both at 3 days and two weeks after the bombardment experiment (Fig. 3).

Transformation of cotyledons by Agrobacterium

A multitude of shoots can be induced on cotyledons of young seedlings (Fig. 4a; Knittel et al., 1991), and plant regeneration is easily possible from this material. For this reason, cotyledons were chosen to study *Agrobacterium*-mediated transformation of sunflower.

Unlike other sunflower tissues (Escandón & Hahne, 1991), seedling cotyledons were not easily transformed by *Agrobacterium*. Under standard culture conditions optimized for plant regeneration (Knittel et al., 1991), only a few transformation events could be detected after one week or one month of culture on selective or non-selective medium (Fig. 4c). The transformation efficiency could be increased (Fig. 4b) by optimizing a number of factors influencing the physiology of the explant (cotyledon age, germination conditions), of the agrobacteria (growth medium, induction by phenolic compounds), and also their interaction (co-culture conditions). The physiological age of the cotyledon influences strongly the morphogenic response (Knittel et al., 1991), but was of noticeable influence on the transformation frequency only beyond day 5. Similarly, germination of the donor plants on media containing different combinations of NAA and BAP, considerably influenced viability of the explants taken from these plants, negatively for the most part. Transformation efficiency, however, remained unaffected by this factor, and blue sectors appeared as early as 72 h after the beginning of co-culture under all conditions. Transformation events were almost exclusively limited to a rather small area of the cotyledon (Figs 4d, f) which curiously corresponded to the zone to which regeneration was limited. For all further experiments, we maintained the conditions most favourable for regeneration as defined by Knittel et al. (1991), i.e., cotyledons isolated from plantlets were allowed to germinate for 3–5 days on 1/4 concentrated mineral salts of MS medium without added hormones.

Any modification of the growth conditions of the agrobacteria prior to the co-culture, such as different growth media (LB, YEB, liquid MS, or preculture in $MgSO_4$) or addition of acetosyringone (20 μM) were without effect on the transformation efficiency.

In contrast, the co-culture conditions proved to be very important, both for bacterial growth and transformation efficiency (Table 2). Factors influencing the percentage of transformed explants (i.e., showing at least one transformation event after staining for GUS activity), included the mineral composition of the medium, and, more importantly, its hormonal composition. Best results were obtained with MS- or SH-based media and in the absence of hormones. However, under these conditions shoot induction was completely abolished and only non-morphogenic callus could be recovered from such cultures. When a layer of feeder cells was included between the co-culture medium and the explants, the hormonal influence on the plant/bacterial interaction was considerably attenuated (Table 3). Two different sunflower suspensions (root- and cotyledon-derived) and a tobacco suspension were

Table 2. Influence of co-culture medium on bacterial growth and transformation efficiency

Co-culture condition[a]	Bacterial growth[b]	Transformation efficiency (%)
MS	+++	38
MSC	++	36
SHC	++	23
1/10 SHC	+	15
PER	++	29
MSBAP	-	27
MS + aa	+++++	23
MS + ye	+++	60
MS fl	-	78
MSBAP + fl	-	46
1/10SH	+	23
1/10SH + aa	+++++	24
1/10SH + ye	+++	13
1/10SH + fl	-	49

[a] Explants were placed on indicated media (see Materials and methods for composition) during the co-culture period (3 d), then transferred to PERC medium supplemented with 0.5 g/l Claforan, cultured for further 12 days, and submitted to a histological GUS assay. aa = medium was supplemented with asparagine and glutamine (400 mg/l each); ye = medium was supplemented with 10 g/l yeast extract; fl = co-culture was in the presence of feeder cells.
[b] Bacterial growth was evaluated by visual inspection of the culture at the time of subculture.

Table 3. Influence of feeder cells on the modulation of co-culture efficiency by the medium

Co-culture medium[a]				Blue spots/explant
2,4-D (mg/l)	NAA (mg/l)	BAP (mg/l)	Feeder layer	(max.)
-	-	-	-	5
-	-	-	+	40
-	0.5	1.0	-	5
-	0.5	1.0	+	40
1.0	-	0.02	+	50

[a] Explants were placed on indicated media (MS supplemented with the indicated hormones) during the co-culture period (3 d), then transferred to PERC medium supplemented with 0.25 g/l Claforan, 0.25 g/l carbenicillin, and 50 mg/l paromomycin. They were cultured for further 12 days, and submitted to a histological GUS assay.

Table 4. Shoot formation after co-culture on different media. All of these media allowed shoot formation from untransformed explants in the presence or absence of a feeder layer, but were unsuitable for transformation in the absence of a feeder layer

Co-culture medium[a]			Blue spots/explant	Shoot
NAA (mg/l)	BAP (mg/l)	Feeder layer	(max.)	formation
-	2	+	5	+
-	5	+	25	+++
-	10	+	35	+++
-	20	+	35	-
-	40	+	20	-
0.5	10	+	50	-

[a] Explants were placed on indicated media (MS supplemented with the indicated hormones) during the co-culture period (3 days), then transferred to PERC medium supplemented with 0.25 g/l Claforan, 0.25 g/l carbenicillin, and 50 mg/l paromomycin. They were cultured for further 12 days, and submitted to a histological GUS assay.

all equally effective. Such a setup thus allowed us to employ compromise conditions suitable for efficient transformation and shoot induction (Fig. 4g; Table 4), i.e., co-culture on a medium containing BAP (5–10 mg/l). To a certain extent, the action of the feeder culture could be mimicked by $AgNO_3$ (1 mg/l). Both approaches also enhanced the intensity of GUS expression.

Explants thus treated could then be subcultured to a medium suitable for shoot development, supplemented with cytostatic compounds to limit bacterial growth and antibiotics for selection of transformed cells expressing the corresponding marker gene. The most frequently used selective agent, kanamycin, is equivocal for sunflower explants. It is completely inefficient for hypocotyl (Escandón & Hahne, 1991) and cotyledonary explants (this study) of several genotypes, but is useful for the selection of protoplasts derived from these tissues (Moyne et al., 1989), for cultured embryo axes (Burrus et al., unpublished), and for immature embryos (this study). We therefore tested a whole range of antibiotics and herbicides for their toxicity towards cotyledonary explants (Table 5). Paromomycin at 50–100 mg/l appeared a good choice in conjunction with the *npt*II gene.

These experiments resulted in an optimized protocol. Cotyledons were taken from 3-day old plantlets, and their proximal halves co-cultured with *Agrobacterium* on a modified MS medium containing 5 mg/l BAP and 1 mg/l $AgNO_3$. As an alternative to silver nitrate, feeder cultures could be used. After 3 days of

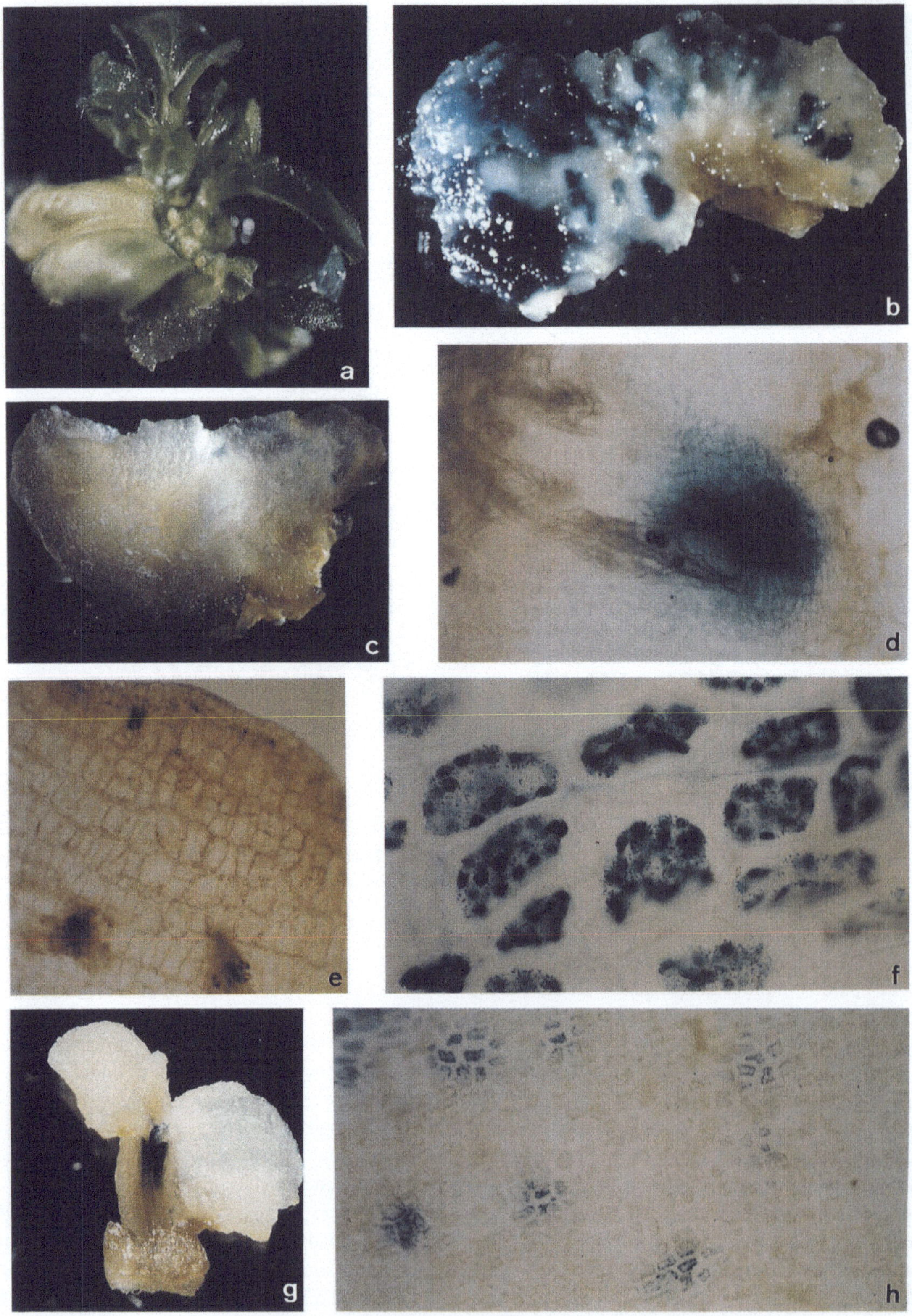
a
b
c
d
e
f
g
h

Fig. 4. Localization of transformation events in *Agrobacterium*-infected sunflower cotyledons. *a*. Regeneration from the petiole-cotyledon transition zone (7x). *b, c*. Cotyledons after *Agrobacterium* infection under conditions optimized for transformation (*b*, 14x) and for regeneration (*c*, 14x). *d, e*. Close view of a cotyledon transformed under conditions allowing both regeneration and transformation. The GUS expressing site (*d*, 36x) was located in the lower part of the explant, close to the vascular bundles. The region developing towards regeneration (*e*, 20x) showed no sign of transgene expression. *f*. Detail of *d*, showing that every cell is transformed in this region. This can be taken as an indication of cell proliferation (180x). *g*. In rare cases, regenerating shoots were chimaeric for the introduced *u i d A* gene. This example, transformed in the apical meristem, might have given rise to transformed offspring (7x). *h*. Isolated transformation events in the distal part of the cotyledon. No indication of cell proliferation is visible (36x).

Table 5. Sensitivity of sunflower seedling cotyledons to selective agents

mg/l	Resistant explants[a] (%)						
	kanamycin	G 418	paromomycin	hygromycin	bromoxynil	phleomycin	phosphinotricin
0	100	100	100	100	100	100	100
5	-	77	-	46	50	17	-
10	-	45	-	11	29	8	10
20	-	0	-	0	0	-	10
25	-	0	-	0	0	8	-
50	100	-	36	-	-	-	0
75	-	-	0	-	-	0	-
100	95	-	0	-	-	-	-
200	84	-	-	-	-	-	-
400	100	-	-	-	-	-	-

[a] Explants were cultured in the presence of the respective selective agent and considered as resistant if tissue proliferation (callus or shoots) was detectable after 1 month of culture on PERC. Each point corresponds to 2 replications with 20 explants each.

co-culture, the explants were subcultured to the same medium, without silver nitrate, but supplemented with 100 mg/l proline, 5 g/l KNO_3, 0.25 g/l claforan, 0.25 g/l carbenicillin, and 50 mg/l paromomycin. Under these conditions, shoots could be regularly obtained which showed GUS activity at least in certain sectors (Fig. 4g). However, the overall frequency of such chimaeric shoots was low, much below one shoot per 100 explants, and many of these GUS expressing structures did not develop any further. This observation, valid for all five tested genotypes, prompted us to study in detail the origin and structure of these small shoots, as well as to localize the tissue preferentially transformed by *Agrobacterium*.

First divisions were rapidly initiated after the cotyledon sections were brought into culture; cell clusters (Fig. 4e) became visible from outside during the next 5 days, and developed into organized structures within two weeks (Fig. 4a). All shoots observed to date developed from a multicellular complex and originated from the subepidermal cell layers. A close inspection of the initiated structures (corresponding to the 'slow shoots' described by Knittel et al., (1991)) showed that some were complete shoots with a functional apical meristem, while the vast majority resembled a bundle of leaf-like structures without meristematic organization (Burrus et al., 1993).

When cotyledons were sampled at various times after co-culture with *Agrobacterium*, stained for GUS activity, and sectioned for histological analysis, cells containing the indigo dye were found, almost exclusively, close to the conductive tissue in the most basal part of the cotyledon (Fig. 4d), but palisade cells (Fig. 4e) were only occasionally observed to be GUS positive in this region. Thus, transformation occurs in the region, but not the cell type, competent for shoot regeneration. In contrast, in the distal part of the explant, single cells weakly expressing the transgene were most often found near the upper epidermis (Fig. 4h). They thus contained the correct cell type, but not a favourable region, since shoots have never been observed from there. The interaction between *Agrobacterium* and the cotyledonary tissue appeared to be highly influenced by the culture conditions, and particularly the hormonal content of the medium. It is therefore conceivable that different cell types might correspond under different culture conditions. However, when we analyzed the localization of GUS positive cells in cotyledons co-cultured under different conditions (no hormones, excess of auxin (NAA, 2 mg/l) or cytokinin (BAP, 5 mg/l), or a hormonal balance suitable for regeneration (BAP, 1 mg/l; NAA 0.5 mg/l)) it was always the same tissue which contained the majority of the indigo-stained cells. We are convinced that these results reflect the true situation rather than being a staining artifact, because (i) GUS positive cells are observed in other parts of the cotyledons after particle bombardment using the same staining technique, (ii) these observations explain the low frequencies observed for transformation events in shoots or shoot-like structures, and (iii) uniformly-transformed plant tissues from other species, used as controls, did stain uniformly under our conditions.

Our experiments concerning the *Agrobacterium*-mediated transformation of sunflower cotyledons illustrate the complexity of transformation in a system where regeneration is direct, limited to a certain cell type, and has relatively stringent requirements concerning the integrity of the explant. In the described case, the regenerating cell type is not the one preferentially transformed by *Agrobacterium.* An indirect regeneration system, where the culture can be enriched for transformed cells which maintain their morphogenic potential and can be induced to regenerate at a later moment, would not suffer from the same problems.

We now have at our disposal a reliable regeneration system, although not all of the induced shoot-like structures have functional meristems. As in other species, this may possibly be overcome with the identification of a more suitable genotype. The conditions for efficient infection and transformation by *Agrobacterium* have also been established, but as yet the two events only coincide in the same cells at low frequency. Since the frequency of (chimaeric) transgenic shoots is unpredictable but low, we did not even try to bring plants to the flowering stage. Even if we had succeeded in obtaining transformed plants, we still would not have had a routine protocol. In view of the fact that the areas containing either the regenerating or the transformed cells, retained their localization within the cotyledon under all experimental conditions, in spite of a systematic study of several factors, we must conclude that it will be difficult to increase the incidence of transformed shoots beyond the chance level observed today.

Conclusion

None of the three different approaches to transformation of sunflower cells described here has allowed regeneration of transgenic plants. Although each approach has its proper merits and difficulties, we can nevertheless identify some common problems which could explain why it is so difficult to transform sunflower. Even in systems where regeneration is efficient from sunflower tissues, this regeneration is almost exclusively direct and thus only affects a relatively small proportion of the cells in the explant. It is these particular cells that must be transformed, in contrast to indirect systems which can be enriched for transformed cells which are then induced to regenerate at a later time. In the rare cases where regeneration from sunflower callus has been described, the selection procedure may reduce the regeneration efficiency below a useful level. The two publications existing to date, describing transgenic sunflower plants, have each suffered from one of the described problems. In the case described by Everett et al. (1987), the major problem was regeneration after selection, while the study published by Schrammeijer et al. (1990) suffered from low efficiency: only one sterile plant could be recovered from 1500 treated meristems.

We can consider the conditions for gene transfer to sunflower to be established for all three methods used in this study. Direct transfer into protoplasts as well as particle bombardment represent powerful means for transient expression studies, each technique being suitable for addressing different problems. Stable integration of the introduced DNA is possible with all three described techniques, but because of the problems discussed above, *Agrobacterium*-mediated transformation appears to be the most promising technique for the recovery of transformed plants, provided that a suitable combination of bacteria and tissue can be identified. A modification of the technique described by Schrammeijer et al. (1990) apparently has allowed the production of sufficient transgenic sunflower plants to warrant a field test (USDA, 1993), but information is only available in the form of conference abstracts (Malone-Schoneberg et al., 1991; Bidney et al., 1992b) and detailed technical information is not available. We are now evaluating meristems and other tissues, and have obtained encouraging results with this material.

Acknowledgements

This work was supported by a joint fellowship of the French 'Ministère de l'Enseignement Supérieur et de la Recherche' and Sanofi Elf Bio Recherches to B.D. and M.B., and by a fellowship of the same Ministry and the 'Deutsche Forschungs-Gemeinschaft' to R.H.

Note

Since the submission of this manuscript, we have obtained transgenic plants using a modification of the approach described by Schrammeijer et al. (1990) and Bidney et al. (Malone-Schoneberg et al., 1991; Bidney et al. 1992b), using *Agrobacterium tumefaciens* and embryonic axes (Knittel et al., 1994).

References

Atanassova, R., N. Chaubet & C. Gigot, 1992. A 126 bp fragment of a plant histone gene promoter confers preferential expression in meristems of transgenic *Arabidopsis*. Plant J. 2: 291–300.

Bidney, D., C. Scelonge, J. Martich, M. Burrus, L. Sims & G. Huffman, 1992a. Microprojectile bombardment of plant tissues increases transformation frequency by *Agrobacterium tumefaciens*. Plant Mol. Biol. 18: 301–313.

Bidney, D.L., C.J. Scelonge & J.B. Malone-Schoneberg, 1992b. Transformed progeny can be recovered from chimearic plants regenerated from *Agrobacterium tumefaciens* treated embryonic axes of sunflower. p. 1408–1412. In: Proc. 13th Int. Sunflower Conf. Vol. II. International Sunflower Soc., Pisa, Italy.

Bower, R. & R.G. Birch, 1992. Transgenic sugarcane plants via microprojectile bombardment. Plant J. 2: 409–416.

Burrus, M., R. Bronner & G. Hahne, 1993. Shoot regeneration from sunflower: a histological study. Biotechnol. Biotechnol. Equip., 7: 126–128.

Ellis, D.D., D.E. McCabe, S. McInnis, R. Ramachandran, D.R. Russell, K.M. Wallace, B.J. Martinell, D.R. Roberts, K.F. Raffa & B.H. McCown, 1993. Stable transformation of *Picea glauca* by particle acceleration. Bio/Technol. 11: 84–89.

Escandón, A. & G. Hahne, 1991. Genotype and composition of culture medium are factors important in the selection for transformed sunflower (*Helianthus annuus* L.) callus. Physiol. Plant. 81: 367–376.

Espinasse-Gellner, A., 1992. A simple and direct technique of transformation in sunflower. p. 50. In: Sunflower Research Workshop. National Sunflower Association, Fargo.

Everett, N.P., K.E.P. Robinson & D. Mascarenhas, 1987. Genetic engineering of sunflower (*Helianthus annuus* L.). Bio/Technol. 5: 1201–1204.

Finer, J.J., P. Vain, M.W. Jones & M.D. McMullen, 1992. Development of the particle inflow gun for DNA delivery to plant cells. Plant Cell Rep. 11: 323–328.

Fischer, C. & G. Hahne, 1992. Structural analysis of colonies derived from sunflower (*Helianthus annuus* L.) protoplasts cultured in liquid and semisolid media. Protoplasma 169: 130–138.

Fischer, C., P. Klethi & G. Hahne, 1992. Protoplasts from cotyledon and hypocotyl of sunflower (*Helianthus annuus* L.): shoot regeneration and seed production. Plant Cell Rep. 11: 632–636.

Gordon-Kamm, W.J., T.M. Spencer, M.L. Mangano, T.R. Adams, R.J. Daines, W. Start, J.V. O'Brien, S.A. Chambers, W.R. Adams Jr., N.G. Willets, T.B. Rice, C.J. Mackey, R.W. Krueger, A.P. Kausch & P.G. Lemaux, 1990. Transformation of maize cells and regeneration of fertile transgenic plants. Plant Cell 2: 603–618.

Hunold, R., R. Bronner & G. Hahne, 1993. GUS expression in sunflower following microprojectile bombardment. Biotechnol. Biotechnol. Equip., 7: 91–95.

Jeannin, G. & G. Hahne, 1991. Donor plant growth conditions and regeneration of fertile plants from somatic embryos induced on immature zygotic embryos of sunflower (*Helianthus annuus* L.). Plant Breed. 107: 280–287.

Jung, J.L., S. Bouzoubaa, D. Gilmer & G. Hahne, 1992. Visualization of transgene expression at the single protoplast level. Plant Cell Rep. 11: 346–350.

Kay, R., A. Chan, M. Daly & J.C. McPherson, 1987. Duplication of CaMV 35S promoter sequences creates a strong enhancer for plant genes. Science 236: 1299–1302.

Knittel, N., A.S. Escandón & G. Hahne, 1991. Plant regeneration at high frequency from mature sunflower cotyledons. Plant Sci. 73: 219–226.

Knittel, N., V. Gruber, G. Hahne & P. Lénée, 1994. Transformation of sunflower (*Helianthus annuus* L.): a reliable protocol. Plant Cell Rep., 14: 81–86.

Knudsen, S. & M. Müller, 1991. Transformation of the developing barley endosperm by particle bombardment. Planta 185: 330–336.

Kozak, M., 1986. Point mutations define a sequence flanking the AUG initiator codon that modulates translation by eukaryotic ribosomes. Cell 44: 283–292.

Lepetit, M., M. Ehling, C. Gigot & G. Hahne, 1991. An internal standard improves the reliability of transient expression studies in plant protoplasts. Plant Cell Rep. 10: 401–405.

Malone-Schoneberg, J.B., D. Bidney, C. Scelonge, M. Burrus & J. Martich, 1991. Recovery of stable transformants from *Agrobacterium tumefaciens* treated split shoot axes. In: 1991 World Congress on Cell and Tissue Culture, Anaheim, USA.

Maniatis, T., E.F. Fritsch & J. Sambrook, 1982. Molecular cloning: a laboratory manual. Cold Spring Harbour Laboratory, Cold Spring Harbour, NY (USA).

Moyne, A.L., D. Tagu, C. Bergounioux & R. Gadal, 1989. Transformed calli obtained by direct gene transfer into sunflower protoplasts. Plant Cell Rep. 8: 97–100.

Murashige, T. & F. Skoog, 1962. A revised medium for rapid growth and bioassays with tobacco tissue cultures. Physiol. Plant. 15: 473–497.

Negrutiu, I., R. Shillito, I. Potrykus, G. Biasini & F. Sala, 1987. Hybrid genes in the analysis of transformation conditions. I. Setting up a simple method for direct gene transfer in plant protoplasts. Plant Mol. Biol. 8: 363–373.

Paterson, K.E. & N.P. Everett, 1985. Regeneration of *Helianthus annuus* inbred plants from callus. Plant Sci. 42: 125–132.

Schenk, R.U. & A.C. Hildebrandt, 1972. Medium and techniques for induction and growth of monocotyledoneous and dicotyledoneous plant cell cultures. Can. J. Bot. 50: 199–204.

Schrammeijer, B., P.C. Sijmons, P.J.M. van den Elzen & A. Hoekema, 1990. Meristem transformation of sunflower via *Agrobacterium*. Plant Cell Rep. 9: 55–60.

Sleat, D., R. Hull, P.C. Turner & T.M.A. Wilson, 1988. Studies on the mechanism of translational enhancement by the 5' leader

sequence of tobacco mosaic virus RNA. Eur. J. Biochem. 175: 75–86.
Twell, D., T.M. Klein, M.E. Fromm & S. McCormick, 1989. Transient expression of chimaeric genes delivered into pollen by microprojectile bombardment. Plant Physiol. 91: 1270–1274.
USDA, 1993. Record of environmental release permits to February 1993. US Department of Agriculture Animal and Plant Health Inspection Service, Biotechnology, Biologies and Environmental Protection, Biotechnology Permit Unit, 6506 Belcrest Road, Hyattsville, Maryland, USA.
Vancanneyt, G., R. Schmidt, A. O'Connor-Sanchez, L. Willmitzer & M. Rocha-Sosa, 1990. Construction of an intron-containing marker gene: splicing of the intron in transgenic plants and its use in monitoring early events in *Agrobacterium*-mediated plant transformation. Mol. Gen. Genet. 220: 245–250.
Veidt, I., S.E. Bouzoubaa, R.M. Leiser, V. Ziegler-Graff, H. Guilley, K. Richards & G. Jonard, 1992. Synthesis of full-length transcripts of beet western yellows virus RNA: messenger properties and biological activity in protoplasts. Virology 186: 192–200.
Zumbrunn, G., M. Schneider & J.D. Rochaix, 1989. A simple particle gun for DNA-mediated cell transformation. Technique 1: 204–216.

Euphytica **85:** 75–80, 1995.

Maize transformation utilizing silicon carbide whiskers: a review

J.A. Thompson[1], P.R. Drayton[1], B.R. Frame[2], Kan Wang[2] & J.M. Dunwell[1]
[1] *Plant Biotechnology, ZENECA Seeds, Jealott's Hill, Bracknell, Berkshire RG12 6EY, U.K.;* [2] *ICI Seeds, Box 500, Slater, IA 50244, U.S.A*

Key words: transformation, silicon carbide, whiskers, maize

Summary

We review here the most recently developed technique for maize transformation which involves the vortexing of silicon carbide whiskers with maize cells in the presence of plasmid DNA. Fertile transgenic plants have been regenerated following whisker-mediated transformation which is compared with the alternatives described to date, namely protoplast uptake, particle bombardment and electroporation of intact tissue.

Introduction

Almost thirty years have passed since the first attempt was made to introduce exogenous DNA into maize through physical intervention (Coe & Sarkar, 1966). This work, although unsuccessful, identified two principal problems – that the cell wall was an effective barrier to large structures and that it is not possible to confirm transformation if the number of cells penetrated is small. In recent years these two issues have been addressed through the development of a range of physical and chemical methods which enable efficient DNA delivery to recipient cells, and through effective selection systems which permit only the growth of transformed cells.

Despite occasional optimistic claims (Kivilaan & Blaydes, 1974) and circumstantial evidence (Korohoda & Strzalka, 1979), maize transformation, defined as the production of fertile transgenic plants which produce transgenic seed, was reported for the first time only four years ago (Fromm et al., 1990; Gordon-Kamm et al., 1990). Since then small-scale field trials have been conducted of transgenic maize carrying single gene traits, with 142 maize trials notified to the USDA in the twelve-month period to April 1994 (USDA, APHIS Permits Unit), demonstrating the speed with which progress is being made.

In this paper the relative merits of the four transformation approaches which to date have been proved to generate transgenic maize plants will be discussed. These include particle bombardment, direct gene transfer to protoplasts and tissue electroporation, with particular emphasis on the method developed most recently, silicon carbide whisker-mediated transformation.

Proven transformation methods for maize

Once plant cells have been transformed they must be capable of sustained division (to enable their selection) whilst remaining totipotent in order that plants can subsequently be regenerated. This has proved to be an elusive formula in maize, made more difficult until recently by the lack of effective DNA delivery approaches. Embryogenic cell cultures or tissue such as immature zygotic embryos which can produce an embryogenic response, have been identified as the most suitable targets for maize (and other cereal) transformation. Table 1 summarises reports of transgenic plant production with the four methods used to transform such tissue (Wilson et al., 1994).

Silicon carbide whisker-mediated transformation

Whisker transformation is technically very simple and does not require expensive equipment or consumables. Whiskers, cells and plasmid DNA are combined in,

Table 1. Reports of fertile transgenic maize plants produced with different transformation methods (after Wilson et al., 1994)

Transformation method	Target tissue	Reference
Silicon carbide whiskers	Embryogenic suspension	Frame et al. (1994)
Particle bombardment	Embryogenic callus	Fromm et al. (1990), Fromm (1994)
	Embryogenic suspension	Gordon-Kamm et al. (1990)
	Immature zygotic embryos	Koziel et al. (1993)
Protoplasts/PEG	Embryogenic suspension	Golovkin et al. (1993)
Tissue electroporation	Immature zygotic embryos	D'Halluin et al. (1992)
	Embryogenic suspension	Laursen et al. (1994)

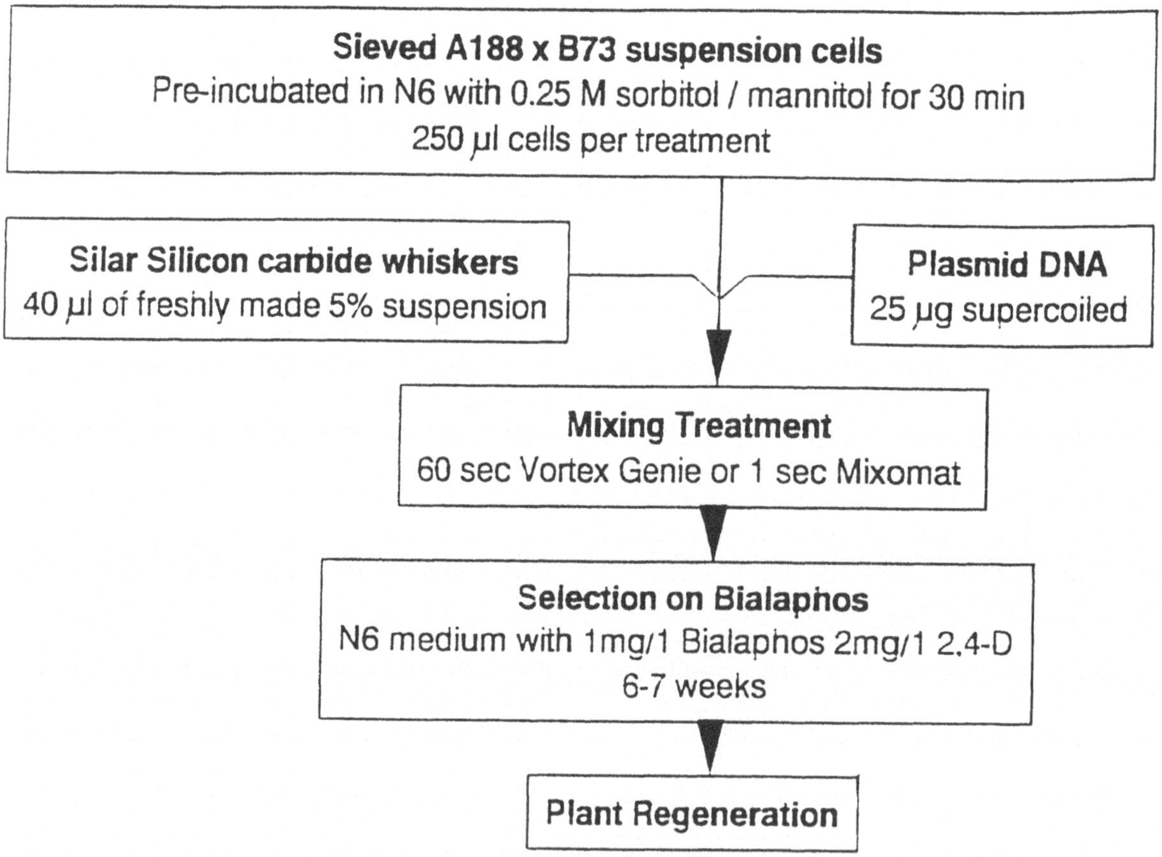

Fig. 1. Whisker transformation method.

for example, an Eppendorf tube and mixed on a vortex or oscillating mixer (Fig. 1, for full protocol details see Wang et al., 1994). Silicon carbide whiskers have great intrinsic hardness and fracture readily to give sharp cutting edges (Greenwood & Earnshaw, 1984). They are obtained by the thermal reduction of silica in a reducing atmosphere, one source of silica being rice husks (Mutsuddy, 1990). Industrially, silicon carbide whiskers are used as abrasives in the manufacture of cutting tools and in the production of composite materials.

When whiskers, typically with a mean diameter of less than 1 μm, are mixed with plant cells and plasmid DNA, cell penetration appears to occur (Kaeppler et al., 1990). Whilst the exact mechanism for whisker transformation is unclear, it seems likely that the whiskers function as numerous needles, facilitating DNA entry into cells during the mixing process. The method described by Kaeppler et al. (1990) involved pre-mixing whiskers and DNA prior to the addition of cells, however this step does not enhance DNA delivery in our experience. It is important that whiskers are not stored in aqueous suspension for extended periods, since we have noted that DNA delivery is reduced as a consequence (unpublished results), a finding confirmed by Kaeppler et al. (1992). It is possible that whisker-DNA interaction is reduced under these circumstances.

Most whisker preparations are highly heterogenous, comprising whiskers from 5–500 μm in length. We have observed significant differences in DNA delivery between batches of silicon carbide whiskers from different manufacturers, with the highest efficiency to date obtained using the whiskers produced by Advanced Composite Materials Corporation (Greer, South Carolina, USA; Wang et al., 1994). There is clear scope for the production of a whisker type with greater uniformity and, for example, altered surface characteristics, which may give more efficient DNA delivery to plant cells.

DNA delivery with whiskers, as judged by transient gene expression, has been reported in a number of plant species (maize, tobacco, Kaeppler et al., 1990; *Agrostis alba*, Asano et al., 1991).

Stable whisker-mediated transformation of non-regenerable Black Mexican Sweet corn (BMS) and tobacco cells has been described (Kaeppler et al., 1992; Kaeppler & Somers, 1994). The transformation efficiency of BMS was rather low (an average of 3.4 transformed clones were obtained per treatment following Basta selection) compared with that achieved using particle bombardment (Kaeppler et al., 1992). Co-expression of linked genes *bar* and *gus*) was reported with a frequency (65%) similar to that reported in maize cells transformed by microprojectile bombardment (Spencer et al., 1990).

We have observed co-expression of linked *bar* and *gus* genes at a frequency of 90% in whisker-transformed BMS cells and co-expression of unlinked genes (*pat* together with the anthocyanin genes *c1* and *b-peru* introduced on 3 separate plasmids) at a frequency of 60%.

Recently, stable transformation of *Chlamydomonas reinhardtii* has also been achieved with the whisker approach (Dunahay, 1993) with a transformation efficiency comparable to other methods and high levels of cell survival.

Our group has recently produced fertile transgenic maize plants carring the *bar* and *gus* genes using whisker transformation of embryogenic suspension cultures. This represents the first example of transgenic plant production in any higher plant species with this transformation method (Frame et al., 1994). The current efficiency of whisker transformation is one tenth to one fifth that achieved in our labs with particle bombardment of A188 × B73 suspension cultures, based on treated cell volume. In 9 representative whisker transformation experiments, 40 bialaphos-resistant stably-transformed clones were produced from 75 ml of treated suspension cells using a construct carrying the *bar* gene. A total of 311 plants were regenerated from 22 of these clones and the *bar* gene was inherited in a Mendelian fashion in those families subsequently analyzed (Frame et al., 1994).

The simplicity of the whisker approach and the ability to treat 30–50 tubes of cells at one time, result in there being scope to alter a large number of variables associated with the components of the system i.e. the target cells, whiskers and mixing method. Various parameters can be altered and assessed using transient gene expression to ensure that efficient DNA delivery is achieved without compromising cell survival. For example, the speed and duration of mixing can be easily controlled and has a clear influence on transformation efficiency (Frame et al., 1994). The use of a combined sorbitol/mannitol pre-treatment of tissue prior to whisker mixing also enhances transient expression 3–5 fold, as reported with particle bombardment (Vain et al., 1993).

Unlike particle bombardment and PEG treatment of protoplasts, whiskers deliver plasmid DNA in a non-precipitated form, making it possible to control the quantity of DNA available for transformation. We have achieved stable transformation of BMS cells with as little as 0.1 μg of input plasmid DNA, the transformation efficiency being 50% that achieved using 25 μg. Direct gene transfer tends to introduce multiple gene copies into cells and these gene copies often display complex integration patterns (e.g. Spencer et al., 1992). It remains to be shown definitively whether these integration characteristics have any functional significance but the ability to control input DNA quantity with whisker transformation may make it feasible to influence copy number.

We have recently obtained stably-transformed callus lines from whisker treatment of embryogenic callus of an elite inbred stiff stalk maize variety, indicating the potential for DNA delivery to more organised tissue.

Particle bombardment

The use of high velocity microprojectiles to carry DNA into maize was first reported in 1988 (Klein et al.). Stable transformation of cultured maize cells (Klein et al., 1989) and regeneration of fertile transgenic maize plants from transformed, cultured callus and suspension cells and zygotic embryos (Fromm et al., 1990; Gordon-Kamm et al., 1990; Koziel et al., 1993 respectively) have been described. Particle bombardment is currently used in a number of commercial laboratories to generate transformed maize plants and there has been a significant quantity of work published on various components of this transformation system (see Fromm, 1994).

DNA-coated particles of tungsten, or preferably gold, can be introduced into virtually any plant cell accessible to bombardment. However the great majority of cells in the target tissue do not receive DNA during bombardment and a large proportion of those that do (~ 98%) do not survive (Hunold et al., 1994). Recently, the use of bacterial cells as microprojectiles has been reported (Rasmussen et al., 1994).

There are two areas where limitations in efficiency are encountered with bombardment of embryogenic cell cultures. First, the number of independent transgenic events from which plants can be regenerated is relatively low and, second, transgenic regenerants often show poor fertility. The causes of poor fertility and phenotypical abnormality in transgenic regenerants are unknown. These problems are encountered to varying degrees with regenerated corn plants in general and are thought to be related to the culture process. The phenomenon appears more marked in plants recovered from selection *in vitro*. Keeping the periods spent in culture and under selection to a minimum may be the most practical short-term solution to overcoming these problems. The work of Koziel et al. (1993) is of note in this respect; this group used zygotic embryos (of an elite inbred of Lancaster parentage) as targets for bombardment, introducing a synthetic gene coding for an insecticidal protein. Insect resistance was demonstrated in the resulting plants.

Direct gene transfer to protoplasts

Procedures for direct gene transfer to protoplasts involving polyethylene glycol (PEG) treatment or electroporation are well-established for many species and have been used with a range of cereals including maize. There has been speculation that the process of enzymatic protoplast isolation may constitute the trigger in cereals for something akin to a 'wound response', thus increasing cell competence for transformation and regeneration (Potrykus, 1990).

Although protoplasts are attractive as a transformation target because a large number of treated cells receive DNA, maize protoplast culture remains technically difficult. Whilst embryogenic maize cultures offer the best possibility for the recovery of plants from protoplasts, the development of embryogenic suspension cultures capable of yielding division-competent cells is a time-consuming and poorly-understood process.

Several attempts have been made to develop an efficient system for the production of fertile plants from immature embryo-derived, embryogenic maize suspension protoplasts which would be useful for transformation studies (Shillito et al., 1989; Prioli & Sondahl, 1989; Petersen et al., 1992). However, in all these reports, problems with plant fertility were encountered.

In contrast, Mórocz et al. (1990) found relatively low frequencies of developmental abnormalities in plants regenerated from embryogenic suspension protoplasts of the complex synthetic maize genotype He/89.

This efficient but unique regeneration system was successfully used to achieve stable transformation following PEG-mediated DNA uptake and subsequent selection on phosphinothricin, with the majority of plants regenerated from the resulting transformed callus being fertile (Golovkin et al., 1993). This therefore represents the first report of fertile transgenic plant production from maize protoplasts. At present, genotype constraints and the reduced vigour and fertility of plants regenerated from protoplasts as a consequence of their extended culture history, probably outweigh the benefits of protoplasts as recipients for the integration of foreign DNA.

Tissue electroporation

D'Halluin et al. (1992) described an electroporation method for the production of transgenic maize plants. This method utilises as target tissue immature zygotic embryos of a size from which callus can be readily initiated or finely chopped, Type 1 callus. The authors of this method claim that wounding and/or degrading of intact tissue is an important prerequisite for success with DNA uptake. This wounding may be achieved by physical means, e.g. cutting, or by a chemical means such as a one to two minute enzyme treatment.

The transgenic plants produced by the method of D'Halluin et al. (1992) contained the barnase gene targeted to the tapetal tissue and were male sterile, but otherwise phenotypically normal. Progeny from these plants showed a 1:1 segregation for the *npt II* gene. The *npt II* negative progeny were male fertile, while the *npt II* positive progeny were male sterile. The barnase gene was detected by Southern analysis in all of the male sterile plants.

More recently, Laursen et al. (1994) reported the recovery of transgenic plants by electroporation of suspension culture cells treated with a pectin-degrading enzyme. Two different suspensions were used, one derived from A188 × B73 and one derived from a B73-related inbred, and plants were regenerated from thirteen different transformed callus lines selected on bialaphos.

Other approaches to transformation

Various other forms of DNA delivery to maize have also been reported. These methods have ranged from the soaking of seed with DNA solution (Korohoda & Strzalka, 1979), the treatment of pollen (Ohta, 1986; Booy et al., 1989), various injection techniques (Bennetzen et al., 1988) and *Agrobacterium* co-cultivation (Graves & Goldman, 1986; Gould et al., 1991). However, in none of these cases has unequivocal evidence of stable transformation been presented. In particular, the possibility of transformation of cryptic endophytes known to occur in maize (Konstantinov et al., 1991) as elsewhere, cannot be excluded.

Despite this lack of success to date, there seems little doubt that progress is being made with methods such as microinjection (Gaillard et al., 1992) and transgenic plants will be produced at some stage.

Concluding remarks

Progress in producing transgenic maize has been rapid over the last five years, moving from occasional reports of rare transformation events to success with several methods. The simplest proven method reported to date, and the most recent, is that involving the vortexing of silicon carbide whiskers with cells and plasmid DNA (Frame et al., 1994). It is to be hoped that such simplicity will be combined with an increasing frequency of regeneration and thereby the extension of transgenic capability to a wider and wider range of maize laboratories.

References

Asano, Y., Y. Otsuki & M. Ugaki, 1991. Electroporation-mediated and silicon carbide whisker-mediated DNA delivery in *Agrostis alba* L. (Redtop). Plant Science 9: 247–252.

Bennetzen, J.L., C. Lin, S. McCormick & B.J. Staskawicz, 1988. Transformation of *Adh* null pollen to *Adh+* by 'macroinjection'. Maize Genetics Cooperative Newsletter 62: 113–114.

Booy, G., F.A. Krens & H.J. Huizing, 1989. Attempted pollen-mediated transformation of maize. J. Plant Physiol. 135: 319–324.

Coe, E.H. & K.R. Sarkar, 1966. Preparation of nucleic acids and a genetic transformation attempt in maize. Crop Science 6: 432–435.

D'Halluin, K., E. Bonne, M. Bossut, M. De Beuckeleer & J. Leemans, 1992. Transgenic maize plants by tissue electroporation. The Plant Cell 4: 1495–1505.

Dunahay, T.G., 1993. Transformation of *Chlamydomonas reinhardtii* with silicon carbide whiskers. BioTechniques 15: 452–460.

Frame, B.R., P.R. Drayton, S.V. Bagnall, C.J. Lewnau, W.P. Bullock, H.M. Wilson, J.M. Dunwell, J.A. Thompson & K. Wang, 1994. Production of fertile transgenic maize plants by silicon carbide whisker-mediated transformation. The Plant Journal, 6: 941–948.

Fromm, M.E., F. Morrish, C.L. Armstrong, R. Williams, J. Thomas & T.M. Klein, 1990. Inheritance and expression of chimeric genes in the progeny of transgenic maize plants. Bio/Technology 8: 833–839.

Fromm, M.E., 1994. Production of transgenic maize plants by microprojectile-mediated gene transfer. In: M. Freeling & V. Walbot (Eds). The Maize Handbook, pp. 677–684. Springer-Verlag, New York.

Gaillard, A., E. Matthys-Rochon & C. Dumas, 1992. Selection of microspore derived embryogenic structures in maize related to transformation potential by microinjection. Bot. Acta 105: 313–318.

Golovkin, M.V., M. Abraham, S. Mórocz, S. Bottka, A. Fehér & D. Dudits, 1993. Production of transgenic maize plants by direct DNA uptake into protoplasts. Plant Science 90: 41–52.

Gordon-Kamm, W.J., T.M. Spencer, M.L. Mangano, T.R. Adams, R.J. Daines, W.G. Start, J.V. O'Brien, S.A. Chambers, W.R. Adams, N.G. Willetts, T.B. Rice, C.J. Mackay, R.W. Krueger, A.P. Kausch & P.G. Lemaux, 1990. Transformation of maize cells and regeneration of fertile transgenic plants. The Plant Cell 2: 603–618.

Gould, J., M. Devey, O. Hasegawa, E.C. Ulian, G. Peterson & R.H. Smith, 1991. Transformation of *Zea mays* L. using *Agrobacterium tumefaciens* and the shoot apex. Plant Physiol. 95: 426–434.

Graves, A.C. & S.L. Goldman, 1986. The transformation of *Zea mays* seedlings with *Agrobacterium tumefaciens*. Plant Mol. Biol. 7: 43–50.

Greenwood, N.N. & A. Earnshaw, 1984. Silicon carbide. In: Chemistry of the Elements, pp. 386–394. Pergamon Press, Oxford.

Hunold, R., R. Bronner & G. Hahne, 1994. Eearly events in microprojectile bombardment: cell viability and particle location. The Plant Journal 5: 593–604.

Kaeppler, H.F., W. Gu, D.A. Somers, H.W. Rines & A.F. Cockburn, 1990. Silicon carbide fiber-mediated DNA delivery into plant cells. Plant Cell Rep. 9: 415–418.

Kaeppler, H.F., D.A. Somers, H.W. Rines & A.F. Cockburn, 1992. Silicon carbide fiber-mediated stable transformation of plant cells. Theor. Appl. Genet. 84: 560–566.

Kaeppler, H.F. & D.A. Somers, 1994. DNA delivery to maize cell cultures using silicon carbide fibers. In: M. Freeling & V. Walbot (Eds). The Maize Handbook, pp. 610–613. Springer-Verlag, New York.

Kivilaan, A. & D.F. Blaydes, 1974. Attempts to achieve genetic transformation in plants. Michigan State University Research Report 246: 2–5.

Klein, T.M., M. Fromm, A. Weissinger, D. Tomes, S. Schaaf, M. Sletten & J.C. Sanford, 1988. Transfer of foreign genes into intact maize cells using high velocity microprojectiles. Proc. Natl. Acad. Sci. USA 85: 4305–4309.

Klein, T.M., L. Kornstein, M.E. Fromm & J.C. Sanford, 1989. Genetic transformation of maize cells by particle bombardment. Plant Physiol. 91: 440–444.

Konstantinov, K., S. Mladenovic & M. Denic, 1991. Recombinant DNA technology in maize breeding. V. Plant and indigenous bacterial strains transformed by the gene controlling kanamycin resistance. Genetika (Beograd) 23: 121–135.

Korohoda, J. & K. Strzalka, 1979. High efficiency genetic transformation in maize induced by exogenous DNA. Z. Pflanzenphysiol. 94: 95–99.

Koziel, M.G., G.L. Beland, C. Bowman, N.B. Carozzi, R. Crenshaw, L. Crossland, J. Dawson, N. Desai, M. Hill, S. Kadwell, K. Launis, K. Lewis, D. Maddox, K. McPherson, M.R. Meghji, E. Merlin, R. Rhodes, G.W. Warren, M. Wright & S.V. Evola, 1993. Field performance of elite transgenic maize plants expressing an insecticidal protein derived from *Bacillus thuringiensis*. Bio/Technology 11: 194–200.

Laursen, C.M., R.A. Kryzek, C.E. Flick, P.C. Anderson & T.M. Spencer, 1994. Production of fertile transgenic maize by electroporation of suspension culture cells. Plant Mol. Biol. 24: 51–61.

Mutsuddy, B.C., 1990. Electrokinetic behaviour of aqueous silicon carbide whisker suspensions. J. Am. Ceram. Soc. 9: 2747–2749.

Mórocz, S., G. Donn, J. Nemeth & D. Dudits, 1990. An improved system to obtain fertile regenerants via maize protoplasts isolated from a highly embryogenic suspension culture. Theor. Appl. Genet. 80: 721–726.

Ohta, Y., 1986. High-efficiency genetic transformation of maize by a mixture of pollen and exogenous DNA. Proc. Natl. Acad. Sci. USA 83: 715–719.

Petersen, W.L., S. Sulc & C.L. Armstrong, 1992. Effect of nurse cultures on the production of macro-calli and fertile plants from maize embryogenic suspension culture protoplasts. Plant Cell Rep. 10: 591–594.

Potrykus, I., 1990. Gene transfer to plants: assessment and perspectives. Physiologia Plantarum 79: 125–134.

Prioli, L.M. & M.R. Sondahl, 1989. Plant regeneration and recovery of fertile plants protoplasts of maize (*Zea mays* L.). Bio/Technology 7: 589–594.

Rasmussen, J.L., J.R. Kikkert, M.K. Roy & J.C. Sanford, 1994. Biolistic transformation of tobacco and maize suspension cells using bacterial cells as microprojectiles. Plant Cell Rep. 13: 212–217.

Shillito, R.D., G.K. Carswell, C.M. Johnson, J.M. DiMaio & C.T. Harms, 1989. Regeneration of fertile plants from protoplasts of elite inbred maize. Bio/Technology 7: 581–587.

Spencer, T.M., W.J. Gordon-Kamm, R.J. Daines, W.G. Start & P. Lemaux, 1990. Bialaphos selection of stable transformants from maize cell cultures. Theor. Appl. Genet. 79: 625–631.

Spencer, T.M., J.V. O'Brien, W.G. Start, T.R. Adams, W.J. Gordon-Kamm & P.G. Lemaux, 1992. Segregation of transgenes in maize. Plant Mol. Biol. 18: 201–210.

Vain, P., M.D. McMullen & J.J. Finer, 1993. Osmotic treatment enhances particle bombardment-mediated transient and stable transformation of maize. Plant Cell Rep. 12: 84–88.

Wang, K., B.R. Frame, P.R. Drayton & J.A. Thompson, 1994. Silicon carbide whisker-mediated transformation: Regeneration of transgenic maize plants. In: I. Potrykus (Ed). Gene Transfer to Plants. Springer-Verlag, in press.

Wang, K., P.R. Drayton, B.R. Frame, J.M. Dunwell & J.A. Thompson, 1994. Whisker-mediated plant transformation: An alternative technology. In Vitro Cell. Devel. Biol., in press.

Wilson, H.M., B.P. Bullock, J.M. Dunwell, J.R. Ellis, B.R. Frame, J.R. Register & J.A. Thompson, 1994. Maize transformation. In: K. Wang, A. Herrera-Estrella & M. Van Montagu (Eds). Transformation of Plants and Soil Microorganism, pp.65–80. Cambridge University Press, Cambridge.

Euphytica **85:** 81–88, 1995.

Transgenic barley by particle bombardment. Inheritance of the transferred gene and characteristics of transgenic barley plants

Anneli Ritala[1,2], Reino Aikasalo[3], Kristian Aspegren[2], Marjatta Salmenkallio-Marttila[1], Satu Åkerman[1], Leena Mannonen[1], Ulrika Kurtén[1], Riitta Puupponen-Pimiä[1], Teemu H. Teeri[2] & Veli Kauppinen[1]

[1] *VTT, Biotechnology and Food Research, P.O. Box 1505, FIN-02044 VTT, Espoo, Finland;* [2] *Institute of Biotechnology, P.O. Box 45, FIN-00014 University of Helsinki, Finland;* [3] *Boreal Plant Breeding, Myllytie 8, FIN-31600, Jokioinen, Finland*

Key words: gene transfer, *Hordeum vulgare*, neomycin phosphotransferase II, particle bombardment, transgenic barley

Summary

Transgenic barley plants (*Hordeum vulgare* L. cv. Kymppi) were obtained by particle bombardment of various tissues. Immature embryos and microspore-derived cultures were bombarded with gold particles coated with plasmid DNA carrying the gene coding for neomycin phosphotransferase II (NPTII), together with plasmid DNA containing the gene for β-glucuronidase (GUS).

Bombarded immature embryos were grown to plants without selection and NPTII activity was screened in small plantlets. One plant proved to be transgenic (T_0). This chimeric plant passed the transferred *nptII* gene to its T_1 progeny. The presence of the *nptII* gene was demonstrated by the PCR technique and enzyme activity was analyzed by an NPTII gel assay. Four T_0 spikes and 15 T_1 offspring were transgenic. The integration and inheritance was confirmed by Southern blot hybridization. Transgenic T_2 and T_3 plants were produced by isolating embryos from green grains of transgenic T_1 and T_2 plants, respectively and growing them to plants. After selfing, the ratio of transgenic to non-transgenic T_2 offspring was shown to follow the rule of Mendelian inheritance. The general performance of transgenic plants was normal and no reduction in fertility was observed.

Microspore-derived cultures were bombarded one and four weeks after microspore isolation. After bombardment, cultures were grown either with or without antibiotic selection (geneticin^R or kanamycin). When cultures were grown without selection and regenerated plants were transferred to kanamycin selection in rooting phase, one out of a total of about 1500 plants survived. This plant both carried and expressed the transferred *nptII* gene. The integration was confirmed by Southern blot hybridization. This plant was not fertile.

Introduction

Gene transfer has become an important tool in plant breeding. Even the recalcitrant monocotyledonous crop plants have come within the range of genetic engineering. Important landmarks since the invention of particle bombardment (Sanford et al., 1987; Klein et al., 1987) have been the production of transgenic maize (Gordon-Kamm et al., 1990), rice (Christou et al., 1991), wheat (Vasil et al., 1992) and oat plants (Somers et al., 1992). Finally, reports of the production of transgenic barley plants have also been published (Ritala et al., 1994; Wan & Lemaux, 1994).

The key problems in crop plant transformation are still the establishment of suitable cell and tissue culture systems for target materials, low transformation frequencies and problems in the regeneration of transformed tissues. The marker gene and selection systems used, also require further development. This report examines problems associated with kanamycin and

Table 1. Results from PCR- and enzyme activity analyses of T_2 plants

Parent tiller (T_0)	Number of T_1 parents of T_2 plants	Number of T_2 plants analyzed	Number of T_3 plants carrying the *nptII* gene (PCR)	Number of *nptII*-positive T_2 plants showing NPTII activity
I	5	103	82	65
II	1	17	12	10
III	4	48	40	34
IV	1	9	5	5
Total	11	177	139	114

geneticin^R selection systems in producing transgenic barley plants. It also confirms that barley can be stably modified with foreign DNA without affecting the fertility or general performance of the plants.

Materials and methods

Plant material

Immature embryos of barley (*Hordeum vulgare* L. cv. Kymppi) were dissected from surface-sterilized seeds about twelve days after anthesis. They were kept overnight in liquid L1 medium (Lazzeri et al., 1991) with 2.0 mg/l 2,4-D, at 4° C in the dark. Isolation of barley (*H. vulgare* L. cv. Kymppi) microspores after 21 to 28 days of cold pretreatment was carried out according to Olsen (1991). Isolated microspores were cultured in modified N6 medium (Chu et al., 1975) for four weeks, after which they were regenerated according to Olsen's protocol (1991).

Plasmids

The pasmids pAT13 and pHTT303 were described in Ritala et al. (1993). Plasmid DNA was propagated in the *E. coli* strain DH5α, isolated on a large scale by the alkaline lysis method of Birnboim & Doly (1979) and dissolved in 10 mM Tris-HCl, 1 mM EDTA, pH 8.

Coating particles

Gold particles (Bio-Rad, 1.0 μm in diameter; 60 mg in 1 ml sterile water) were coated with a 1:1 mixture of the plasmids pHTT303 and pAT13, or for transient expression studies with pAT13 alone as described in Ritala et al. (1993).

Fig. 1. Mature seed set of a transgenic T_1 plant.

Particle bombardment

The helium modification of the Biolistic^R PDS-1000 (Bio-Rad) was used and the bombardment conditions were the same as described by Ritala et al. (1994), except that for one week old microspore cultures, 6.2 MPa rupture disks in addition to 9.0 MPa

Table 2. Characteristics of transgenic T_2 plants (averages of 23 individuals) grown in greenhouse conditions

	Plant height/cm	Ears/plant	Grains/ ears	1000 s.w./ g	Grain yield/ear mg	Grain yield/plant g	Partially sterile ears	
							per plant	sterile flowers per ear %
Average	83.2	21.4	23.3	39.5	919	19.5	0.9	17.8
Range	73–93	12–31	15–27	28.4–54.3	600–1410	7.8–24.9	0–4	0–80
SD	5.5	5.0	3.2	6.3	196	4.8	1.2	26.3

SD = Standard deviation.
1000 s.w. = weight of thousand seeds.

A)

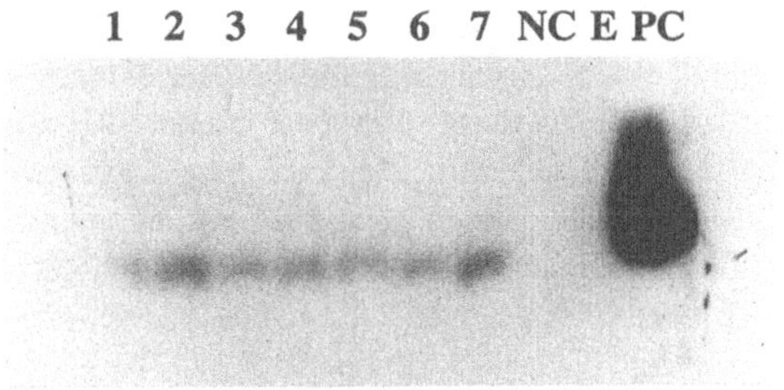

B)

Fig. 2. A) NPTII gel assay of leaf extracts from transgenic T_3 barley plants (lanes 1 to 7). Lane PC represents a positive control from tobacco transgenic 35S-*nptII* and lane NC a negative control from a non-transformed barley plant. Lane E is empty. B) Transgenic T_3 offspring in potting soil.

disks were used. For particle bombardment, about 20 embryos were placed scutellum down on an area of 1–1.5 cm^2 on MS3018 medium (Murashige & Skoog, 1962; Foroughi-Wehr et al., 1976; Hunter, 1987). Microspore-derived cultures were bombarded either one or four weeks after isolation. Approximately 10 to 50 mg or 100 to 350 mg fresh weight, respectively, were plated on nylon on MS3017 medium for bombardment (Murashige & Skoog, 1962; Foroughi-Wehr et al., 1976). After bombardment the petri dishes were sealed with Nescofilm (Bando Chemical Ind. Ltd.) and samples were kept at 23° C in the dark.

Plant recovery

Immature embryos were transferred to light (50 μmol m^{-2} s^{-1}) one day after bombardment. After three weeks green shoots were transferred to MS3019 rooting medium (Murashige & Skoog, 1962; Foroughi-Wehr et al., 1976). Rooted plants were potted in soil mix (Vermiculite^R:peat:soil, 2:1:1) and grown in growth room conditions (40–50% relative humidity, 16 hours light (170 μmol m^{-2} s^{-1}) at 18° C, 8 hours darkness at 14° C). Bombarded microspore-derived cultures were grown according to Olsen (1991). Plants were regenerated without selection or by using different antibiotics and other strategies. Selection protocols were: a) 175 ppm geneticin^R on MS3017 for two weeks and subculturing to MS3018 containing 175 ppm geneticin^R, from which small plantlets were transferred to non-selective medium, b) two weeks on MS3017 with 200 ppm kanamycin, transfer to MS3018 with 400 ppm kanamycin, from which small plantlets were transferred to non-selective medium, and c) cultures were grown without selection and green plantlets were transferred in the rooting phase to MS3019 with 100 ppm kanamycin. Continuous selection with 175 ppm geneticin^R or 200 ppm kanamycin was also tested.

The T_1 generation was obtained from the T_0 plant, the T_2 from T_1 and the T_3 from T_2 by dissecting embryos from green grains and growing them to plants on MS3019 rooting medium. Rooted plants were further potted in soil mix and grown in growth room conditions. For the follow-up of the general growth performance of the transgenic T_2 plants, they were grown in normal greenhouse conditions at a density of one plant per pot.

Assays for GUS and NPTII activities

Transient expression of the bombarded *uidA* gene was analyzed according to Jefferson's (1987) histochemical method, with the modifications of Ritala et al. (1993). Neomycin phosphotransferase II was assayed as described in Ritala et al. (1994), using the dot blot method according to McDonnell et al. (1987) and by the gel assay of Van den Broeck et al. (1985) and Reiss et al. (1984).

DNA isolation

Total DNA from leaves and from the remnants of the spikes after grain removal was isolated according to Dellaporta et al. (1983).

Screening of plants by the PCR technique

The polymerase chain reaction (PCR) was carried out in a Techne PHC-2 thermocycler as described in Ritala et al. (1994). Primers were designed to amplify a 5′ sequence of the chimeric *nptII* gene in pHTT303.

Southern blot hybridization

For Southern blots, isolated total DNA was restricted with the enzymes *Bam*HI, *Eco*RI, *Hind*III, *Dpn*I and/or *Dpn*II in the buffers recommended by the enzyme manufacturers. The protocols used follow Southern (1975) and Sambrook et al. (1989) and are described in detail in Ritala et al. (1994).

Results

Bombarded immature embryos were grown to plants without antibiotic selection. One out of 227 plants expressed the transferred *nptII* gene (T_0). Embryos from green grains of the transgenic T_0 plant were isolated for further cultivation to T_1 progeny and the presence of the transferred *nptII* gene in the spike remnants after grain removal was screened by the polymerase chain reaction (PCR). Of the 116 spikes produced to date, the *nptII* gene was found by PCR in fifteen. Offspring of these fifteen spikes were further analyzed by PCR and enzyme activity tests. The presence of the *nptII* gene was demonstrated in four sets of these siblings, so that fifteen out of 25 offspring plants carried the transferred *nptII* gene and fourteen of them expressed it. The integration and inheritance were confirmed by Southern blot hybridization. Detailed data are presented in Ritala et al. (1994).

The T_2 and T_3 generations were obtained by growing isolated embryos to plants from green grains of transgenic T_1 and T_2 plants, respectively. The yield of 79% transgenic plants observed in the T_2 generation (Table 1) was consistent with the expected Mendelian ratio of 83%. Of these transgenic plants 82% also expressed the transferred gene, indicating that in some of the plants the foreign gene had been eliminated. In order to estimate the germination of the transgenic barley, seeds were allowed to mature on the spikes of transgenic T_1 and T_2 plants (Fig. 1). The germination frequency of the seeds was determined according to the International Rules for Seed Testing (1993). Germination from T_2 seeds was 89% and from T_3 seeds 99%. The performance of transgenic T_2 plants was monitored and some of the key characteristics are shown in Table 2. T_2 plants were grown in pots in a greenhouse (one plant per pot, distance between pots 40 cm). The number of partially sterile ears, which is a normal phenomenon in greenhouse conditions, was slightly increased in the transgenic plants. The mean numbers of partially sterile ears per transgenic and non-transformed plant were 0.9 and 0.4, respectively. Although no exact comparison was carried out, the overall appearance of the transgenic plants was normal.

In the T_3 generation 26 plants were assayed for NPTII activity and the enzyme was detected in all of them (Fig. 2A and B). The *nptII* gene was found by PCR in 22 of these T_3 plants. This indicated an error frequency of the PCR of 15%, which was in agreement with our existing estimate of general PCR error frequency.

Approximate estimations of transformation frequencies in the bombardment of microspore-derived cultures were made by analyzing the transient expression of the bombarded *uidA* gene. In one week old microspore cultures, about 0.17% of the viable microspores expressed the transferred gene (Fig. 3A).

A)

B)

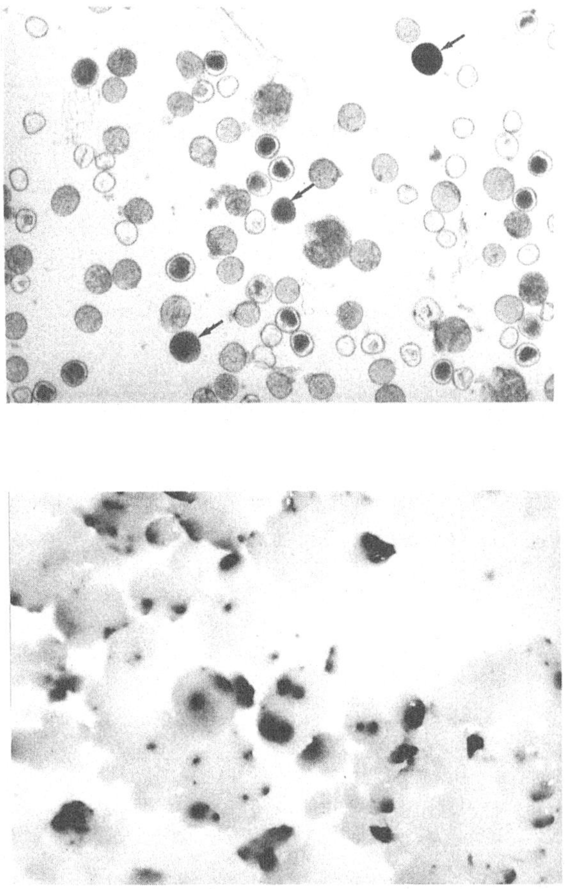

Fig. 3. Transient expression of bombarded *uidA* gene in one week old (A) and four week old (B) barley microspore-derived cultures.

In four week old microspore-derived cultures, clusters of many cells had formed and it was not possible to determine exact cell counts. There was approximately one blue spot per microspore-derived aggregate (Fig. 3B).

Microspore-derived cultures were bombarded one and four weeks after microspore isolation. After bombardment, cultures were grown either with or without antibiotic selection (geneticin^R or kanamycin) using different selection strategies. When bombarded cultures were grown without selection, about 2000 green plants were obtained and three plants showed NPTII activity in dot blot assay. One of these three plants was fertile and PCR screening of the spike remnants after grain removal indicated that 19 spikes were transgenic. PCR and Southern blot hybridization revealed that the transferred gene was not passed to the T_1 progeny. Continuous cultivation in the presence of 175 ppm geneticin^R resulted in only albino plants. However, if the selection pressure was released by transferring the small plantlets to non-selective medium (selection protocol a), one green among about 300 albinos was produced. The green plant showed no NPTII activity, but 23% of the analyzed albinos were transgenic in NPTII activity assay. Continuous cultivation in kanamycin (200 ppm) resulted in several thousand albino plants. When the selection pressure was released by transferring the small plantlets to non-selective medium (selection protocol b), 40 green plants among 2000 albinos were produced. One of the green plants and 21% of the analyzed albinos showed NPTII activity. The NPTII-positive green plant was fertile and PCR screening of the remaining spikes after grain removal indicated that three spikes were transgenic, but PCR and Southern blot hybridization showed that the transferred *nptII* gene was not passed to the T_1 progeny. When cultures were grown without selection and regenerated green plants were transferred to kanamycin selection in the

Table 3. Bombardment of microspore-derived barley cultures

Bombardment Selection	Four weeks after microspore isolation a)	b)	c)	without selection	One week after microspore isolation c)	without selection
Number of samples bombarded	80	76	23	61	17	8*
Total number of plants regenerated	317	2001	1162	6144	2596	486
Number of albino plants (% of albinism)	316 (99.7%)	1961 (98.0%)	629 (54.1%)	4175 (68.0%)	1566 (60.3%)	361 (74.3%)
Number of plants assayed for NPTII/ **Number of positive signals**	1 green/**0** 53 albinos/ **12 (23%)**	40 green ones/**1** 115 albinos/ **24 (21%)**	0/**0**	1869 green ones/**3**	1 green/**1**	108 green ones/**0**
Number of green plants selected in rooting phase/ **Number of survivals**	ND	ND	533/0	ND	1030/1	ND

Selection codes:
a) 175 ppm geneticin^R on MS3017 and MS3018, from which small plantlets were transferrred to non-selective medium
b) 200 ppm kanamycin on MS3017 for two weeks, then to MS3018 with 400 ppm kanamycin, from which small plantlets were transferred to non-selective medium
c) Cultures were grown without selection, green plantlets were rooted in MS3019 with 100 ppm kanamycin
* When samples were bombarded, 6.2 MPa rupture disks were used instead of 9.0 MPa disks.
ND = Not done

rooting phase, one out of about 1500 plants survived. This plant both carried and expressed the transferred *nptII* gene (Fig. 4A). The integration was confirmed by Southern blot hybridization (Fig. 4B). This plant was however not fertile and thus no offspring were produced. Results from the bombardment of microspore-derived barley cultures are summarized in Table 3.

Discussion

We have obtained transgenic barley plants by particle bombardment of immature embryos and microspore-derived cultures. The bombardment of immature embryos and direct NPTII activity screening without selection steps resulted in one transgenic barley plant. The transferred *nptII* gene was stably integrated in the genome and passed to T_1 and further to T_2 and T_3 progeny. In T_2 progeny the transgene was found to segregate according to Mendelian ratios and the gene was functional in 82% of the cases. The general appearance, fertility, seed set and germination of seeds of transgenic barley plants were normal.

Some antibiotic selection systems (Kanamycin, Geneticin^R) increase albinism to such an extent that selection is no longer practical in producing transgenic barley plants, even though 20% of the albinos are transgenic. Chimerism also affects the capacity of plants to survive under intense selection pressure. On the other hand, screening of plants by PCR or by enzyme activity analyses is laborious, time-consuming and expensive. Therefore more attention should be paid to the choice of marker gene and selection systems. In the case of maize (Gordon-Kamm et al., 1990), rice (Christou et al., 1991), wheat (Vasil et al., 1992) and oat (Somers et al., 1992), the herbicide resistance conferred by the *bar* gene was successfully used. Of the selection strategies tested in this work, selection with kanamycin in the rooting phase appears at the moment to be the most promising. Bombardment of a one week

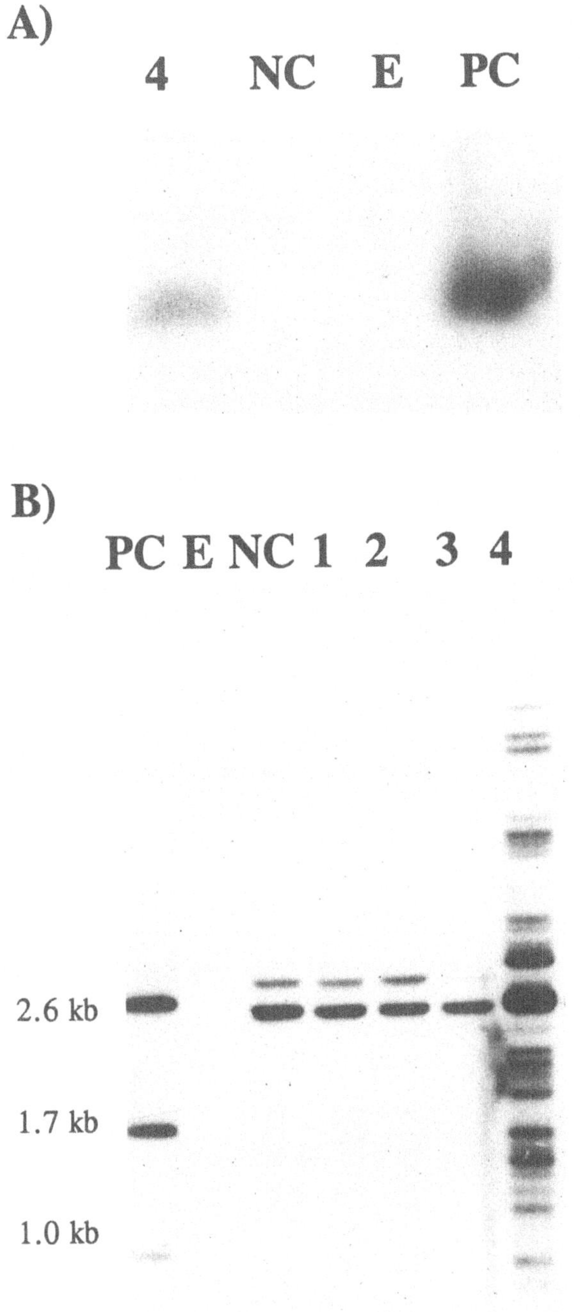

Fig. 4. NPTII activity (A) and Southern blot analysis (B) of a transgenic microspore-derived barley plant (lane 4). NPTII gel assay was carried out from leaf extract. Lane PC represents a positive control from tobacco transgenic for 35S-*nptII* and lane NC a negative control from a non-transformed barley plant. Lane E is empty. For Southern blot hybridization total DNA samples were digested with enzymes *Bam*HI, *Eco*RI and *Hind*III and probed with plasmid pHTT303. The positive control (PC) is *Bam*HI, *Eco*RI and *Hind*III digested pHTT303 (5 pg, representing approximately one copy per haploid genome), mixed with 5 μg *Bam*HI, *Eco*RI and *Hind*III digested DNA from a non-transformed barley. This digestion releases vector (2.6 kb), *nptII* gene (1.7 kb) and 35S promoter (1.0 kb) fragments. The NC lane represents a negative control containing DNA isolated from leaves of a non-transformed barley plant. Samples 1 to 3 are untransformed barley plants and sample 4 is a transgenic microspore-derived barley plant.

old microspore culture combined with selection in the rooting phase resulted in the recovery of one transgenic plant. This plant turned out to be sterile and no offspring were produced. Nevertheless, this result demonstrates that the bombardment of microspore-derived cultures can also produce transgenic barley plants.

Acknowledgements

Michael Bailey is thanked for critically reading this manuscript. The plasmid pAT13 was a kind gift from the Carlsberg Research Center. The excellent technical assistance of Taina Ala-Hakuni, Tuuli Teikari, Jaana Juvonen, Leena Toivonen, Anne Heikkinen, Marja Huovila, Anu Immonen and Seija Rissanen is greatly appreciated. This work was supported by the Foundation for Biotechnical and Industrial Fermentation Research, the Finnish Ministery of Agriculture and Forestry, the Technology Development Centre (TEKES), the Emil Aaltonen Foundation and the Finnish brewing and malting industry.

References

Birnboim, H.C. & J. Doly, 1979. A rapid alkaline extraction procedure for screening recombinant plasmid DNA. Nucl. Acid Res. 7: 1513–1523.

Christou, P., T.L. Ford & M. Kofron, 1991. Production of transgenic rice (*Oryza sativa* L.) plants from agronomically important indica and japonica varieties via electric discharge particle acceleration of exogenous DNA into immature zygotic embryos. Bio/Technology 9: 957–962.

Chu, C.C., C.C. Wang, C.S. Sun, C. Hsū, K.C. Yin, C.Y. Chu & F.Y. Bi, 1975. Establishment of an efficient medium for anther culture of rice through comparative experiments on the nitrogen sources. Sci. Sin. 18: 659–668.

Dellaporta, S.L., J. Wood & J.B. Hicks, 1983. A plant DNA minipreparation: Version II. Plant Mol. Biol. Rep. 1: 19–21.

Foroughi-Wehr, B., G. Mix, H. Gaul & H.M. Wilson, 1976. Plant production from cultured anthers of *Hordeum vulgare* L. Z. Pflanzenzüchtg. 77: 198–204.

Gordon-Kamm, W.J., T.M. Spencer, M.L. Mangano, T.R. Adams, R.J. Daines, W.G. Start, J.V. O'Brien, S.A. Chambers, W.R. Adams Jr., N.G. Willets, T.B. Rice, C.J. Mackey, R.W. Krueger, A.P. Kausch & P.G. Lemaux, 1990. Transformation of maize cells and regeneration of fertile transgenic plants. Plant Cell. 2: 603–618.

Hunter, C.P., 1987. Plant generation method. European patent application 0 245 898 A2: 1–8.

International Rules for Seed Testing, 1993. International Seed Testing Association, Seed Sci. & Technol. 21, Supplement.

Jefferson, R.A., 1987. Assaying chimeric genes in plants: The GUS fusion system. Plant Mol. Biol. Rep. 5: 387–405.

Klein, T.M., E.D. Wolf, R. Wu & J.C. Sanford, 1987. High-velocity microprojectiles for delivering nucleic acids into living cells. Nature 327: 70–73.

Lazzeri, P.A., R. Brettschneider, R. Lührs & H. Lörz, 1991. Stable transformation of barley via PEG-induced direct DNA uptake into protoplasts. Theor. Appl. Genet. 81: 437–444.

McDonnell, R.E., R.D. Clark, W.A. Smith & M.A. Hinchee, 1987. A simplified method for the detection of neomycin phosphotransferase II activity in transformed plant tissues. Plant Mol. Biol. Rep. 5: 380–386.

Murashige, T. & F. Skoog, 1962. A revised medium for rapid growth and bio assays with tobacco tissue cultures. Physiol. Plant 15: 473–497.

Olsen, F.L., 1991. Isolation and cultivation of embryogenic microspores from barley (*Hordeum vulgare* L.). Hereditas 115: 255–266.

Reiss, B., R. Sprengel, H. Will & H. Schaller, 1984. A new sensitive method for qualitative and quantitative assay of neomycin phosphotransferase in crude cell extracts. Gene 30: 211–218.

Ritala, A., L. Mannonen, K. Aspegren, M. Salmenkallio-Marttila, U. Kurtén, R. Hannus, J. Mendez Lozano, T.H. Teeri & V. Kauppinen, 1993. Stable transformation of barley tissue culture by particle bombardment. Plant Cell Rep. 12: 435–440.

Ritala, A., K. Aspegren, U. Kurtén, M. Salmenkallio-Marttila, L. Mannonen, R. Hannus, V. Kauppinen, T.H. Teeri & T.M. Enari, 1994. Fertile transgenic barley by particle bombardment of immature embryos. Plant Mol. Biol. 24: 317–325.

Sambrook, J., E.F. Fritsch & T. Maniatis, 1989. Molecular Cloning: A Laboratory Manual, 2nd Ed. Cold Spring Harbor Laboratory Press, Cold Spring Harbor, NY.

Sanford, J.C., T.M. Klein, E.D. Wolf & N. Allen, 1987. Delivery of substances into cells and tissues using a particle bombardment process. Particulate Sci. Technol. 5: 27–37.

Somers, D.A., H.W. Rines, W. Gu, H.F. Kaeppler & W.R. Bushnell, 1992. Fertile, transgenic oat plants. Bio/Technology 10: 1589–1594.

Southern, E.M., 1975. Detection of specific sequences among DNA fragments separated by gel electrophoresis. J. Mol. Biol. 98: 503–517.

Van den Broeck, G., M.P. Timko, A.P. Kausch, A.R. Cashmore, M. Van Montagu & L. Herrera-Estrella, 1985. Targeting of a foreign protein to chloroplasts by fusion to the transit peptide from the small subunit of ribulose 1,5-bisphosphate carboxylase. Nature 313: 358–363.

Vasi, V., A.M. Castillo, M.E. Fromm & I.K. Vasil, 1992. Herbicide resistant fertile transgenic wheat plants obtained by microprojectile bombardment of regenerable embryogenic callus. Bio/Technology 10: 667–674.

Wan, Y. & P.G. Lemaux, 1994. Generation of large numbers of independently transformed fertile barley plants. Plant Physiol. 104: 37–48.

Euphytica **85**: 89–95, 1995.

Transient gene expression in transformed banana (*Musa* cv. Bluggoe) protoplasts and embryogenic cell suspensions

László Sági, Serge Remy, Bert Verelst, Bart Panis, Bruno P.A. Cammue[1], Guido Volckaert[2] & Rony Swennen
Laboratory of Tropical Crop Husbandry, [1] F.A. Janssens Laboratory of Genetics, and [2] Laboratory of Gene Technology, Catholic University of Leuven, Belgium

Key words: banana, biolistic transformation, electroporation, embryogenic cell suspension, *Musa* spp., protoplast

Summary

In order to introduce currently-available genes with agronomical value into banana, two genetic transformation protocols have been optimized.

Firstly, regenerable protoplasts isolated from embryogenic cell suspensions of the cultivar Bluggoe have been used for the introduction of several chimaeric *uidA* gene constructs by electroporation. With the inclusion of polyethylene glycol and heat shock, the frequency of transiently expressing protoplasts reached 1.8% as shown by an *in situ* β-glucuronidase assay. A duplicated 35S promoter with an alfalfa mosaic virus leader sequence (pBI-426) induced the highest expression rate among the constructs tested.

Embryogenic cell suspensions of cv. Bluggoe have also been bombarded with accelerated particles coated with a high expression *uidA* gene construct (pEmuGN) using a biolistic gun. After a partial optimization of the procedure, transient GUS assays reproducibly demonstrated the presence of 400 blue foci in 30 μl of settled cell volume (approximately 25 mg cells). Selection and characterization of antibiotic-resistant transformed cultures is in progress.

Abbreviations: AMV – alfalfa mosaic virus, GUS – β-glucuronidase, TGE – transient GUS expression, *uidA* – gene for β-glucuronidase

Introduction

Banana (including plantain) is the world's largest fruit crop with an annual production of 74 million tons (FAO, 1993) and is the staple food for nearly 400 million people. The main factors threatening this level of production are fungal and viral diseases. The sigatoka disease complex (caused by *Mycophaerella* spp.) may cause yield losses of up to 30–50% (Mobambo et al., 1993), while plants infected with the banana bunchy top virus may become completely unproductive.

Since most cultivated bananas are triploid and highly sterile, application of classical breeding methods for disease resistance has resulted in rather limited success (Vuylsteke et al., 1993). By a one-step introduction of well-characterized genes conferring disease resistance to banana, genetic engineering may provide a powerful tool for breeding programmes. To exploit this potential, (i) highly performing *in vitro* regeneration systems and (ii) efficient genetic transformation procedures coupled with high gene expression are needed for banana.

Very recently, significant progress has been achieved in the *in vitro* manipulation of banana. Regenerable embryogenic cell suspensions have been established in the authors' laboratory from proliferating meristems (Dhed'a et al., 1991) and this technique has been successfully applied to several genetically distinct cultivars (Dhed'a, 1992). Protoplasts can be isolated from these embryogenic cell suspensions and plants have been regenerated therefrom through somatic embryogenesis at a high frequency (Panis et al.,

1993b). Furthermore, the established embryogenic cell suspensions can be stored by cryopreservation without loss of regenerating ability (Panis et al., 1992). Other laboratories have also succeeded in regenerating plants from cell suspensions (Novak et al., 1989) as well as from protoplast cultures (Megia et al., 1993).

Direct DNA transfer by electroporation (Fromm et al., 1985) into viable and highly-regenerative protoplasts provides an opportunity for efficient genetic transformation of banana. The technique is effective for transformation of a range of dicotyledonous (Riggs & Bates, 1986; Lindsey & Jones, 1989; Chupeau et al., 1989) and monocotyledonous species (Toriyama et al., 1988; Huang & Dennis, 1989). However, numerous factors affect the efficiency of gene transfer, including capacitance and field strength, duration and shape of electrical pulses, buffer composition and temperature, type of chimaeric gene constructs, and the concentration and form of DNA. In a recent report, we have demonstrated the introduction of foreign DNA into banana protoplasts by electroporation at a high frequency as assessed by a transient *in situ* GUS assay (Sagi et al., 1994).

Alternatively, banana cell suspensions may be directly transformed by particle bombardment obviating the need for protoplast isolation and culture techniques. An inexpensive but efficient biolistic gun model accelerates particles directly in a stream of helium without macrocarriers (Takeuchi et al., 1992; Finer et al., 1992). This technique has produced stable transformation of soybean and maize (Finer et al., 1992; Vain et al., 1993). According to our initial observations, this method resulted in high transient GUS expression in banana embryogenic cell suspensions (Panis et al., 1993a).

In the present report, data are provided on transient expression of the *uidA* gene of *E. coli* in banana protoplasts after transformation by electroporation. Transient GUS expression is also demonstrated after particle bombardment of suspension cells using a flowing helium gun.

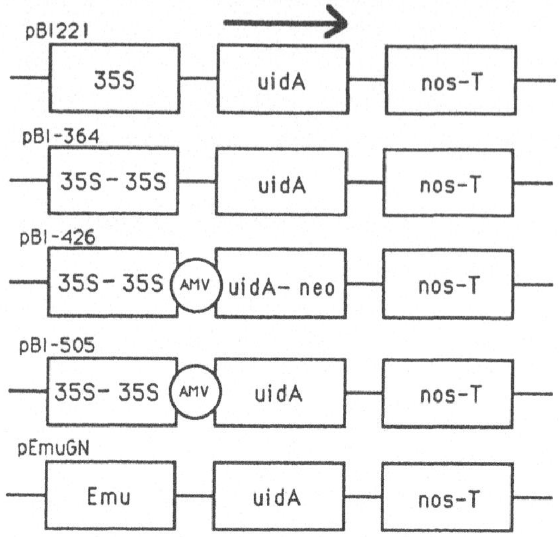

Fig. 1. Schematic representation of chimaeric GUS constructs used for transformation experiments. 35S = CaMV 35S promoter, 35S-35S = the tandem repeat CaMV 35S promoter, AMV = alfalfa mosaic virus leader sequence, uidA = GUS reporter gene, uidA-neo = GUS-NPTII fusion gene, nos-T = NOS terminator, Emu = Emu promoter (Last et al., 1991).

Materials and methods

Cell suspensions and plasmids. Embryogenic cell suspension lines of cv. Bluggoe (*Musa* spp., ABB group) described by Dhed'a et al. (1991) were maintained and subcultured weekly in MS medium (Murashige & Skoog, 1962) supplemented with 5 μM 2,4-D and 1 μM zeatin (ZZ medium). The cell suspensions were cultured for at least 1 year prior to use in the experiments and consisted of small clusters of isodiametric, cytoplasm-rich cells.

The plasmid pBI221 was obtained from Clontech Laboratories, Inc. This 5.7 kb vector consists of pUC19 and a 3.0 kb *Hind*III-*Eco*RI fragment carrying the CaMV 35S promoter, the GUS gene from *E. coli* and the NOS polyA site from *Agrobacterium tumefaciens*. The plasmids pBI-364 pBI-426 pBI-505 and pEmuGN were kindly provided by William Crosby, Plant Biotechnology Institute, Saskatoon, Sask., Canada and David Last, CSIRO, Canberra, Australia, respectively. The structure of chimaeric *uidA* gene constructs located on these plasmids is shown in Fig. 1. The plasmids were propagated in *E. coli* DH5α strain in the presence of 50 mg l^{-1} ampicillin. Plasmid DNA prepared by alkaline lysis was purified on Qiagen columns (Diagen GmbH). DNA concentrations were determined by absorption measurements at 260 nm and by gel electrophoresis.

Protoplast isolation. Isolation and purification of protoplasts from embryogenic cell suspensions was done according to Panis et al. (1993). The purified protoplasts were resuspended in electroporation buffer at a protoplast density of 1.25×10^6 ml^{-1}. The electroporation buffers are described in Table 1. Protoplasts were counted using a modified Neubauer haemocytometer. Complete removal of the cell wall was confirmed by Calcofluor white staining while the viability of freshly-isolated or electroporated protoplasts was assessed by staining either with fluorescein diacetate or Evans' blue.

Transformation procedures. Protoplast transformation by electroporation was carried out as described by Sagi et al. (1994). Briefly, a 800 μl aliquot containing 10^6 protoplasts in electroporation buffer was placed into cuvettes with a 0.4 cm gap. After addition of plasmid DNA to a concentration of 60 $\mu g\ ml^{-1}$, cuvettes were stored on ice for 10 min, then electroporated with a 960 μF capacitor of a Bio-Rad Gene Pulser^{TM} transfection apparatus at a field strength of 800 V cm^{-1}. After electroporation, the cuvettes were placed on ice for 10 min and then for 10 min at room temperature. Protoplasts were diluted with ZZ medium supplemented with 0.55 M mannitol to a density of $10^5\ ml^{-1}$ and incubated in the dark at 24° C. The following controls were used: (1) samples electroporated with pUC19 DNA, (2) samples electroporated without plasmid DNA, (3) non-electroporated samples incubated with plasmid DNA.

For particle bombardment of suspension cells, a flowing helium gun has been constructed based on the description of Takeuchi et al. (1992) and the procedure will be described in detail elsewhere (Sagi et al., in press). Briefly, for one bombardment 1 μg DNA was precipitated according to Sanford et al. (1993) onto 300 μg gold (Bio-Rad) or M-17 tungsten particles (Sylvania) and applied into a Swinney syringe filter unit. Particles were prepared and stored at 4° C as suspensions in 50% glycerol or 40% polyethylene glycol 6000 (PEG). Suspension cells were collected 4 to 6 days after subculture and 30 μl settled cell volume (approximately 25 mg cells) was used for bombardment. Particles were accelerated at a pressure of 3 to 6 bars. Cells were then cultured in ZZ medium for 1 or 2 days and assayed for transient GUS expression.

Transient β-glucuronidase expression assay. For the histochemical *in situ* assays, protoplasts were collected 48 hours after electroporation, resuspended in 50 mM sodium phosphate buffer, pH 7.0 and incubated for periods ranging from overnight up to 10 days at 37° C in the presence of 1 mM X-Gluc (5-bromo-4-chloro-3-indolyl-β-D-glucuronide) as described by Jefferson (1987). Protoplast transformation frequency was assessed by counting blue-stained protoplasts and relating this to the total number of protoplasts. Two to four internal repeats were counted and averaged for each treatment in each experiment. On average 10^5 protoplasts were observed for each repeat. The total number of treated protoplasts was determined independently for each repeat. For bombarded cells, transient GUS expression frequencies were expressed as number of blue foci per shot averaged over three to six replicates per treatment. Cultures of *E. coli* were used in parallel as positive controls for the GUS assay.

Statistical evaluation was carried out after *arcsin* transformation ($y' = 2\ arcsin\ y^{1/2}$) of data from electroporation experiments. The ANOVA and Duncan's multiple range tests were performed using the statistical software package SAS (SAS Institute, Inc., Cary, North Carolina, USA). Where significant differences in the treatment means were found at the 5% probability level of an F-test, means were compared using Duncan's multiple range test at 5% level of significance.

Results and discussion

Transient GUS expression (TGE) in electroporated protoplasts

Viability of freshly-isolated protoplasts was always over 90% with an average of 93%, as assessed by fluorescein diacetate staining or by the dye exclusion test with Evans' blue. After 24 hours, viability of the control cultures (no electroporation) in ASP-buffer (Tada et al., 1990) or in Cl-buffer (Fromm et al., 1985) was reduced from 93% to 75% and 67%, respectively.

Representative results in Table 1 illustrate that ASP buffer was superior to Cl-buffer, and that heat shock (45° C, 5 min) in ASP-buffer further increased TGE. Maximum TGE reached 1.8% of total electroporated protoplasts (Fig. 2a). Similar frequencies were observed by other groups in different species (Zhang & Wu, 1988; Dhir et al., 1991; Diaz & Carbonero, 1992).

Comparison of different plasmid constructs

In the absence of PEG, the rather low frequency of transformation was associated only with limited differences between the plasmid constructs tested (Table 2). However, when 5% PEG was added and transformation frequencies increased in accordance with our previous findings (Sagi et al., 1994), significant differ-

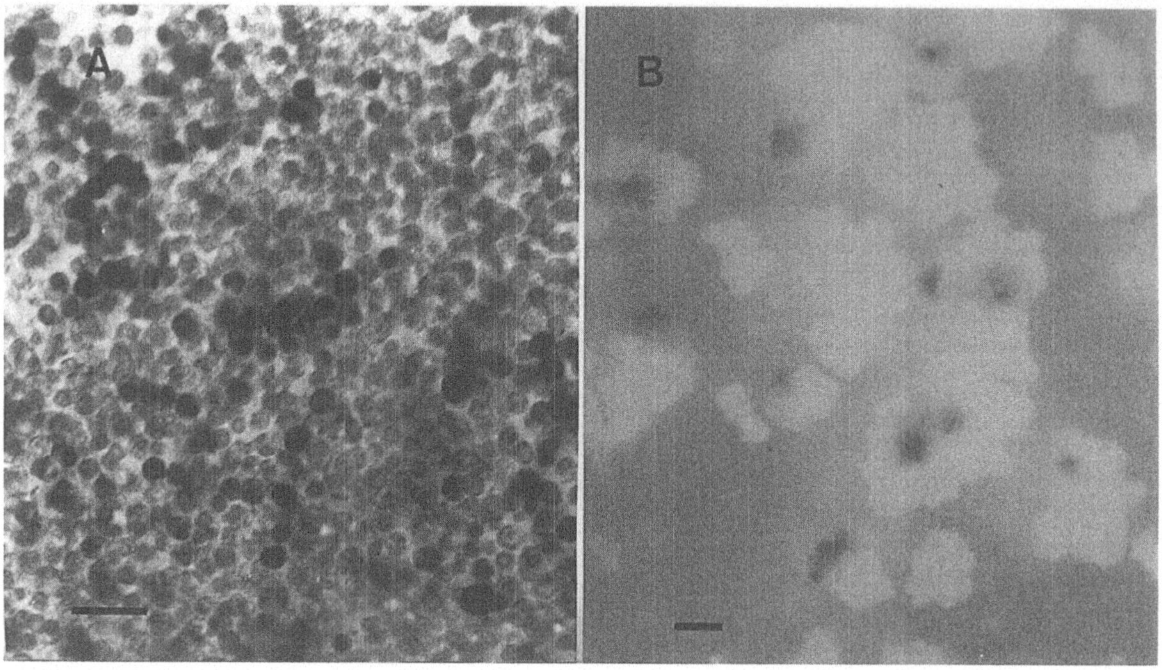

Fig. 2. Transient GUS expression in a) cv. Bluggoe banana protoplasts electroporated with plasmid pBI-426 in 5% PEG, b) cv. Bluggoe embryogenic cell suspensions after particle bombardment with plasmid pEmuGN coated gold particles. Bar = 100 μm.

Table 1. Effect of electroporation buffers and heat shock (45° C, 5 min) on transient GUS expression in banana protoplasts

Electroporation buffer	Treatment[a] Control	Treatment[a] Shocked	Mean[a]
ASP[1]	1.082 b	1.868 a	1.468 a
Cl[2]	0.734 b	0.758 b	0.746 b

TGE frequencies are expressed as percentages based on the number of blue protoplasts in two replicates of 10^5 protoplasts for each treatment electroporated with plasmid pBI-426 in 5% PEG and assessed 48 hours after electroporation.

[a] Entries within these headings followed by the same letter are not significntly ($P \leq 0.05$) different by Duncan's test after *arcsin* transformation. Standard deviation did not exceed ± 20% of the mean.

[1] 70 mM K-aspartate, 5 mM Ca-gluconate, 5 mM MES, 0.55 M mannitol, pH 5.8 (Tada et al., 1990).

[2] 150 mM NaCl, 4 mM $CaCl_2$, 10 mM HEPES, 0.55 M mannitol, pH 7.2 (Fromm et al., 1985).

Table 2. Effect of plasmid constructs on transient GUS expression in electroporated banana protoplasts

Plasmid	Construct	PEG concentration (%) 0	PEG concentration (%) 5
pBI221	35S	0.048 b	0.198 c
pBI-364	35S-35S	0.092 ab	0.205 c
pBI-426	**35S-35S-AMV**	0.073 ab	**0.434 a**
pBI-505	**35S-35S-AMV**	0.104 a	**0.329 b**

TGE frequencies are expressed as percentages based on the number of blue protoplasts in at least two replicates of 10^5 protoplasts for each treatment and assessed 48 hours after electroporation. Entries within columns followed by the same letter are not significantly ($P \leq 0.05$) different by Duncan's test after *arcsin* transformation. Standard deviation did not exceed ± 20% of the mean.

ences were found between the constructs in the following order: pBI-426 > pBI-505 > PBI-364 ≈ pBI221. Obviously, constructs with the AMV leader sequence gave significantly higher TGE than those without this sequence. The tandem repeat 35S promoter alone was not sufficient to increase transient gene expression significantly.

In accordance with the results obtained by Charest et al. (1993) in conifer species and Hobbs et al. (1990) in pea protoplasts, this observation supports the conclusion of Jobling & Gehrke (1987) that the AMV untrans-

Table 3. Effect of age suspension cells on transient GUS expression in electroporated banana protoplasts

PEG concentration (%)	Age of suspension cells (week) 1	2
0	0.053 e	0.057 e
3	0.485 c	0.607 bc
5	**0.765 ab**	0.276 d
8	0.815 a	0.007 f

TGE frequencies are expressed as percentages based on the number of blue protoplasts in two replicates of 10^5 protoplasts for each treatment electroporated with plasmid pBI-426 and assessed 48 hours after electroporation. Entries followed by the same letter are not significantly ($P \leq 0.05$) different by Duncan's test after *arcsin* transformation. Standard deviation did not exceed ± 10% of the mean.

Table 4. Transient GUS expression in banana suspension cells after particle bombardment

Particle Type	Bombarded in	Storage solution 50% glycerol	40% PEG
Tungsten	water	9.3 ± 7.1	49.0 ± 3.7
	ethanol	188.3 ± 56.1	166.3 ± 49.0
Gold	water	56.3 ± 31.4	74.3 ± 39.6
	ethanol	242.0 ± 14.4	324.3 ± 89.0

TGE frequencies are expressed as the number of blue foci per shot (± S.E.) with plasmid pEmuGN coated tungsten or gold particles and assayed 2 days after bombardment.

lated leader sequence increases translational efficiency of chimaeric mRNAs.

Effect of age of suspension cells

The reaction of protoplasts isolated from 1- or 2-week old suspension cells to various PEG levels was compared (Table 3). It was found that PEG exposure was not damaging to protoplasts isolated from 1-week old cell suspension. In contrast, protoplasts from 2-week old suspensions showed a declining TGE to increasing PEG concentration. Though 8% PEG treatment resulted in the highest TGE in 1-week old protoplasts, at this concentration results were highly unreproducible and aggregation of protoplasts was observed. One-week old cell suspensions were clearly superior to 2-week old cells for protoplast transformation. Moreover, in agreement with our previous results (Sagi et al., 1994), inclusion of 5% PEG during electroporation had the best influence on TGE.

Transient GUS expression in bombarded suspension cells

Effect of particles and precipitation factors

Table 4 illustrates TGE frequencies in bombarded banana suspension cells in relation to particle type, storage solution of particles and suspending agent prior to bombardment. When gold particles were used for coating, TGE frequencies were clearly higher for all treatments than with tungsten particles. It may be that gold particles bind DNA more efficiently and that they are more inert to plant cells than tungsten. Russell et al. (1992) have indeed observed that tungsten particles were toxic to tobacco suspension cells while gold particles were not toxic.

Using 40% PEG to prepare particle stock solutions usually resulted in higher TGE than 50% glycerol. This difference was more marked when particles were bombarded as an aqueous suspension, and was probably caused by more complete precipitation of DNA in the presence of PEG (Lis, 1980). Moreover, particles were more easily resuspended in PEG and could be preserved for coating for as long a period of time as in glycerol. Finally, in our laboratory, resuspension of coated particles in ethanol prior to bombardment was clearly more efficient than the removal of most of the aqueous solution from precipitation reactions and the use of the rest for bombardment. Double bombardment of cells further improved TGE frequency and resulted in an average of 494 ± 60.3 blue foci per sample.

Effect of age of suspension cells

There were only slight differences between suspensions of various ages when TGE was determined 2 days after transformation. Five- and six-day old suspensions produced more blue foci per shot than 4-day old suspension cells. However, 1 day after bombardment, 4-day old suspension cells showed higher TGE than cells at 5 days after subculture (Table 5). When comparing gold and tungsten particles, gold proved to be superior, as already shown in Table 4.

As the result of partial optimization of bombardment conditions, we have reproducibly observed at least 400 blue foci per shot (Fig. 2b) where for each shot approximately 25 mg cells are prepared. This frequency is comparable to what has been previously reported in other monocotyledons (Wang et al., 1988; Chibbar et al., 1993; Ritala et al., 1993) and it appears to be high enough to allow the production of stable transformants. Recently, we have been able to select regenerable cul-

Table 5. Effect of age of banana suspension cells on transient GUS expression after particle bombardment

Particle bombardment	GUS-assay days after	Age of suspension cells (days after subculture) 4	5	6
Gold	1 day	508.3 ± 178.3	387.3 ± 112.9	ND
	2 days	341.3 ± 57.8	468.3 ± 31.5	423.0 ± 48.2
Tungsten	2 days	ND	229.3 ± 15.2	300.3 ± 43.4

TGE frequencies are expressed as the number of blue foci per shot (± S.E.) with plasmid pEmuGN coated gold or tungsten particles and assessed 1 or 2 days after bombardment. ND – not determined.

tures after bombarding suspension cells with plasmids carrying the gene for hygromycin resistance. Molecular and histochemical analysis of these cultures is now in progress (Sagi et al., in press).

Based on these results, our goal is to introduce agronomically-important genes into banana. Our first targets are genes coding for new types of antifungal proteins (AFPs). These AFPs are stable, cysteine-rich small peptides which are isolated from seeds of different plant species (Broekaert et al., 1992; Cammue et al., 1992; Terras et al., 1992). More uniquely, they have a broad antifungal effect, and showed high antifungal activity *in vitro* to *Mycosphaerella fijiensis* and *Fusarium oxysporum*, the main fungal pathogens in banana, while they exert no toxicity for human and plant (including banana) cells (Cammue et al., 1993). In addition to fungal resistance, engineering virus resistance in banana should also be considered. Banana bunchy top virus has recently been isolated and the virus genome is now being sequenced (J. Dale, personal communication) with the aim of isolating the essential genes for replication of the virus.

Acknowledgements

The authors are grateful to Dr. William Crosby (Plant Biotechnology Institute, Saskatoon, Canada) and Dr. David Last (CSIRO, Canberra, Australia) for providing the plasmids pBI-364 pBI-426 pBI-505 and pEmuGN, respectively. We thank Dr. Djailo Dhed'a for the embryogenic cell suspensions and Ines Van den Houwe, Ann Janssen, and Willem Dillemans for valuable technical assistance. This work was financed by contract ETC-007 of the 'Vlaamse Aktieprogramma Biotechnologie' of the Flemish Ministry of Economy and by the Belgian Administration for Cooperation in Development.

References

Broekaert, W.F., W. Marien, F.R.G. Terras, M.F.C. De Bolle, P. Proost, J. Van Damme, L. Dillen, M. Claeys, S.B. Rees, J. Vanderleyden & P.B.A. Cammue, 1992. Antimicrobial peptides from *Amaranthus caudatus* seeds with sequence homology to the cysteine/glycine-rich domain of chitin-binding proteins. Biochemistry 31: 4308–4314.

Cammue, B.P.A., M.F.C. De Bolle, F.R.G. Terras, P. Proost, J. Van Damme, S.B. Rees, J. Vanderleyden & W.F. Broekaert, 1992. Isolation and characterization of a novel class of plant antimicrobial peptides from *Mirabilis jalapa* L. seeds. J. Biol. Chem. 267: 2228–2233.

Cammue, B.P.A., M.F.C. De Bolle, F.R.G. Terras & W.F. Broekaert, 1993. Fungal disease control in *Musa*: application of new antifungal proteins. In: J. Ganry (Ed). Proc. Int. Symp. on Genetic Improvement of Bananas for Resistance to Diseases and Pests. Montpellier, France, September 7–9 1992, pp. 221–225. CIRAD/INIBAP/CTA.

Charest, P.J., N. Caléro, D. Lachance, R.S.S. Datla, L.C. Duchesne & E.W.T. Tsang, 1993. Microprojectile-DNA delivery in conifer species: factors affecting assessment of transient gene expression using the β-glucuronidase reporter gene. Plant Cell Rep. 12: 189–193.

Chibbar, R.N., K.K. Kartha, R.S.S. Datla, N. Leung, K. Caswell, C.S. Mallard & L. Steinhauer, 1993. The effect of different promoter-sequences on transient expression of *gus* reporter gene in cultured barley (*Hordeum vulgare* L.) cells. Plant Cell Rep. 12: 506–509.

Chupeau, M.-C., C. Bellini, P. Guerche, B. Maisonneuve, G. Vastra & Y. Chupeau, 1989. Transgenic plants of lettuce (*Lactuca sativa*) obtained through electroporation of protoplasts. Bio/Technology 7: 503–508.

Dhed'a, D., 1992. Culture de suspensions cellulaires embryogéniques et régénération en plantules par embryogénèse somatique chez le bananier et le bananier plantain (*Musa* spp.). Ph.D. thesis No. 1743. Faculty of Agricultural Sciences, Catholic University of Leuven, Belgium, 167 pp.

Dhed'a, D., F. Dumortier, B. Panis, D. Vuylsteke & E. De Langhe, 1991. Plant regeneration in cell suspension cultures of the cooking banana cv. Bluggoe (*Musa* spp. ABB group). Fruits 46: 125–135.

Dhir, S.K., S. Dhir, A. Hepburn & J.M. Widholm, 1991. Factors affecting transient gene expression in electroporated *Glycine max* protoplasts. Plant Cell Rep. 10: 106–110.

Diaz, I. & P. Carbonero, 1992. Isolation of protoplasts from developing barley endosperm: a tool for transient expression studies. Plant Cell Rep. 10: 595–598.

FAO, 1993. Production yearbook 1992. Food and Agriculture Organisation of the United Nations, Rome, Italy, 265 pp.

Finer, J.J., P. Vain, M.W. Jones & M.D. McMullen, 1992. Development of the particle inflow gun for DNA delivery to plant cells. Plant Cell Rep. 11: 323–328.

Fromm, M., L.P. Taylor & V. Walbot, 1985. Expression of genes transferred into monocot and dicot cells by electroporation. Proc. Natl. Acad. Sci. USA 82: 5824–5828.

Hobbs, S.L.A., J.A. Jackson, D.S. Baliski, C.M.O. DeLong & J.D. Mahon, 1990. Genotype- and promoter-induced variability in transient β-glucuronidase expression in pea protoplasts. Plant Cell Rep. 9: 17–20.

Huang, Y.-W. & E.S. Dennis, 1989. Factors influencing stable transformation of maize protoplasts by electroporation. Plant Cell Tissue Organ Culture 18: 281–296.

Jefferson, R.A., 1987. Assaying chimeric genes in plants: the GUS fusion system. Plant Molec. Biol. Rep. 5: 387–405.

Jobling, S.A. & L. Gehrke, 1987. Enhanced translation of chimaeric messenger RNAs containing a plant viral untranslated leader sequence. Nature 325: 622–625.

Last, D.I., R.I.S. Brettell, D.A. Chamberlain, A.M. Chaudhury, P.J. Larkin, E.L. Marsh, W.J. Peacock & E.S. Dennis, 1991. pEmu: an improved promoter for gene expression in cereal cells. Theor. Appl. Genet. 81: 581–588.

Lindsey, K. & M.G.K. Jones, 1989. Stable transformation of sugarbeet protoplasts by electroporation. Plant Cell Rep. 8: 71–74.

Lis, J.T., 1980. Fractionation of DNA fragments by polyethylene glycol induced precipitation. Meth. Enzymol. 65: 347–353.

Megia, R., R. Haicour, S. Tizroutine, V. Bui Trang, L. Rossignol, D. Sihachakr & J. Schwendiman, 1993. Plant regeneration from cultured protoplasts of the cooking banana cv. Bluggoe (*Musa* spp., ABB group). Plant Cell Rep. 13: 41–44.

Mobambo, K., F. Gauhl, D. Vuylsteke, R. Ortiz, C. Pasberg-Gauhl & R. Swennen, 1993. Yield losses in plantain from black sigatoka leaf spot and field performance of resistant hybrids. Field Crops Res. 35: 35–42.

Murashige, T. & F. Skoog, 1962. A revised medium for rapid growth and bioassays with tobacco tissue cultures. Physiol. Plant 15: 473–497.

Novak, F.J., R. Afza, M. Van Duren, M. Perea-Dallos, B.V. Conger & T. Xiaolang, 1989. Somatic embryogenesis and plant regeneration in suspension cultures of dessert (AA and AAA) and cooking (ABB) bananas (*Musa* spp.). Bio/Technology 7: 154–159.

Panis, B., D. Dehd'a & R. Swennen, 1992. Freeze-preservation of embryogenic *Musa* suspension cultures. In: R.P. Adams & J.E. Adams (Eds). Conservation of Plant Genes: DNA Banking and *In Vitro* Biotechnology, pp. 183–195. Academic Press, New York.

Panis, B, D. Dhed'a, K. De Smet, L. Sagi, B. Cammue & R. Swennen, 1993a. Cell suspensions from somatic tissue in *Musa*: applications and prospects. In: J. Ganry (Ed). Proc. Int. Symp. on Genetic Improvement of Bananas for Resistance to Disease and Pests. Montpellier, France, September 7–9 1992, pp. 317–325. CIRAD/INIBAP/CTA.

Panis, B., A. Van Wauwe & R. Swennen, 1993b. Plant regeneration through direct somatic embryogenesis from protoplasts of banana (*Musa* spp.). Plant Cell Rep. 12: 403–407.

Riggs, C.D. & G.W. Bates, 1986. Stable transformation of tobacco by electroporation: Evidence for plasmid concatenation. Proc. Natl. Acad. Sci. USA 83: 5602–5606.

Ritala, A., L. Mannonen, K. Aspegren, M. Salmenkallio-Marttila, U. Kurtén, R. Hannus, J. Mendez Lozano, T.H. Teeri & V. Kauppinen, 1993. Stable transformation of barley tissue culture by particle bombardment. Plant Cell Rep. 12: 435–440.

Russell, J.A., M.K. Roy & J.C. Sanford, 1992. Physical trauma and tungsten toxicity reduce the efficiency of biolistic transformation. Plant Physiol. 98: 1050–1056.

Sagi, L., S. Remy, B. Panis, R. Swennen & G. Volckaert, 1994. Transient gene expression in electroporated banana (*Musa* spp., cv. Bluggoe, ABB group) protoplasts isolated from regenerable embryogenic cell suspensions. Plant Cell Rep. 13: 262–266.

Sagi, L., B. Panis, S. Remy, H. Schoofs, K. De Smet, R. Swennen & B. Cammue, 1995. Genetic transformation of banana and plantain (*Musa* spp.) via practicle bombardment. Bio. Technology. in press.

Sanford, J.C., F.D. Smith & J.A. Russell, 1993. Optimizing the biolistic process for different biological applications. Meth. Enzymol. 217: 483–509.

Tada, Y., M. Sakamoto & T. Fujimura, 1990. Efficient gene introduction into rice by electroporation and analysis of transgenic plants: use of electroporation buffer lacking chloride ions. Theor. Appl. Genet. 80: 475–480.

Takeuchi, Y., M. Dotson & T. Keen, 1992. Plant transformation: a simple particle bombardment device based on flowing helium. Plant Molec. Biol. 18: 835–839.

Terras, F.R.G., H.M.E. Schoofs, M.F.C. De Bolle, F. Van Leuven, S.B. Rees, J. Vanderleyden, B.P.A. Cammue & W.F. Broekaert, 1992. Analysis of two novel classes of plant antifungal proteins from radish (*Raphanus sativus* L.) seeds. J. Biol. Chem. 267: 15301–15309.

Toriyama, K., Y. Arimoto, H. Uchimiya & K. Hinata, 1988. Transgenic rice plants after direct gene transfer into protoplasts. Bio/Technology 6: 1072–1074.

Vain, P., M.D.M. McMullen & J.J. Finer, 1993. Osmotic treatment enhances particle bombardment-mediated transient and stable transformation of maize. Plant Cell Rep. 12: 84–88.

Vuylsteke, D., R. Swennen & R. Ortiz , 1993. Development and early evaluation of black sigatoka-resistant tetraploid hybrids of plantain (*Musa* spp., ABB group). Euphytica 65: 33–42.

Wang, Y.-C., T.M. Klein, M. Fromm, J. Cao, J.C. Sanford & R. Wu, 1988. Transient expression of foreign genes in rice, wheat and soybean cells following particle bombardment. Plant Molec. Biol. 11: 433–439.

Zhang, W. & R. Wu, 1988. Efficient regeneration of transgenic plants from rice protoplasts and correctly regulated expression of the foreign gene in the plants. Theor. Appl. Genet. 76: 835–840.

Euphytica **85**: 97–100, 1995.

Transformation of lily by *Agrobacterium*

Simon A. Langeveld, Merel M. Gerrits, Anton F.L.M. Derks, Piet M. Boonekamp & John F. Bol[1]
Bulb Research Centre, Vennestraat 22, 2160 AB Lisse, The Netherlands; [1] *Institute for Plant Molecular Sciences, Gorlaeus Laboratory, State University Leiden, Einsteinweg 55, 2300 RA Leiden, The Netherlands*

Key words: Agrobacterium, transformation, lily, β-glucuronidase

Summary

Lily cv. Harmony was inoculated with several *Agrobacterium* strains to study its susceptibility to *Agrobacterium* infection and transformation. Tumorous tissue formation on inoculated stem internodes of sterile-grown plantlets, as well as expression of a β-glucuronidase marker gene interrupted by an intron in cells of inoculated stem nodes, indicate that the monocotyledon *Lilium* is a host for *Agrobacterium*.

Introduction

Lilium is a monocotyledonous ornamental crop belonging to the Liliaceae, some members of which can be transformed and infected by *Agrobacterium. Chlorophyticum capense* was the first species in the Liliaceae reported to be susceptible to transformation by *Agrobacterium tumefaciens* (Hooykaas-Van Slogteren et al., 1984). Moreover, *Asparagus officinalis* and *Allium cepa*, also members of the Liliaceae, were shown to be hosts for *Agrobacterium* (Hernalsteens et al., 1984; Dommisse et al., 1990). To improve the quality of lily, an economically-important ornamental crop, it would desirable to have a transformation procedure that allows the introduction of viral coat protein genes into the lily genome in order to establish coat protein-mediated protection against pathogenic lily viruses. Thus lily was investigated as a potential host for *Agrobacterium*. Because of its highly regenerative capacity, lily can easily be micropropagated by adventitious regeneration of bulblets on bulb scales, callus, leaves or stem nodes. This adventitious regeneration on many different tissues of lily greatly increases the chance of finding a tissue type that is both regenerative and susceptible to transformation by *Agrobacterium*. In general, the first step in the detection of *Agrobacterium*-susceptibility is to study tumorous tissue formation after inoculation with virulent strains. This is followed by detection of opines present in the tumour tissue as proof of the Ti-DNA transfer into the plant's cells. On most dicotyledons, wild-type *Agrobacterium* strains induce crown galls at the sites of infection. However, monocotyledonous species susceptible to *Agrobacterium* may develop less severe symptoms, making it more difficult to analyze the presence of opines in infected tissue (De Cleene, 1985). Sterile plantlets of the lily cv. Harmony, an Asiatic hybrid, were used for inoculation with virulent and non-virulent strains of *Agrobacterium* and analyzed for tumour induction. As an alternative to the detection of opines to prove transformation, the transformed state of individual plant cells was more accurately determined by the use of a marker gene. The β-glucuronidase (GUS) encoding marker gene has been recently (Vancanneyt et al., 1990) provided with an intron to prevent bacterial expression. This makes it possible to analyse the *Agrobacterium*-infected tissue shortly after infection for expression of the *gus* gene by histochemical staining. In this study a binary vector carrying the GUS-intron gene between its Ti-DNA borders was introduced into different *Agrobacterium* strains to test the capability of these strains to transform lily tissue.

Materials and methods

Agrobacterium *strains*

Tissue culture

Sterilized bulb scales of lily cv. Harmony were micropropagated as described by Van Aartrijk & Blom-Barnhoorn (1983); scales freshly cut from sterilized bulblets were incubated on regeneration medium for 3 months at 20° C without light to let new bulblets form. Sterile plantlets of 5 cm length were obtained from these bulblets by prolonged incubation in daylight at 25° C for a period of 2 weeks on the same medium. These plantlets were used for inoculation with *Agrobacterium*. Three to 4-week old stems from soil grown bulbs were surface-sterilized with 10% bleach for 5 min and used for inoculation with *Agrobacterium* after removal of the leaves. Stem segments of 3 cm containing 3 to 4 stem nodes were incubated on regeneration medium after cutting and inoculation.

Tobacco plants (*Nicotiana tabacum*) were grown from seed and cultured under sterile conditions on Murashige & Skoog medium (Murashige & Skoog, 1962).

Agrobacterium *inoculation*

The binary vector p35SGUSINT, a pBIN19 derivative containing the β-glucuronidase marker gene provided with an intron in the coding frame to prevent bacterial expression (Vancanneyt et al., 1990), was introduced into the *Agrobacterium* strains by electroporation (Dower et al., 1988). Transformed bacteria were selected and grown in YMB medium (Sambrook et al., 1989) with 25 mg l^{-1} kanamycin. For plant infection 100 ml of medium was inoculated with a single colony and grown to a density of 10^6–10^7 cells ml^{-1}. Acetosyringone was added to the medium at a concentration of 100 μM and incubation was prolonged for 4 hours. Cells were centrifuged, washed with medium without kanamycin and resuspended at a density of 10^8 cells ml^{-1}. This suspension was used as inoculum. With a syringe needle dipped into the inoculum, sterile-grown tobacco plants and sterile regenerated lily plantlets were punctured at stem internode sections over a length of 3 mm. Surface-sterilized stem segments without leaves were scratched with a syringe needle dipped into the inoculum at both stem internode sections and stem nodes.

Scales freshly cut from sterile bulblets were inoculated by dipping into the inoculum and puncturing at the wound site with a needle.

Histochemical analysis

For detection of GUS expression the plant tissue was incubated in a solution of 1 mM X-gluc (5-bromo-4-chloro-3-indolyl-β-D-glucuronic acid cyclohexylammonium) at pH 7.5 after vacuum infiltration as described by Jefferson (1987).

Results

Induction of callus

Ten sterile grown plantlets of the lily cultivar Harmony were inoculated at internode sections of the stems with the *Agrobacterium* C58 wild-type strain. Six plantlets of the same cultivar were inoculated with the *Agrobacterium* strain C58-C9, a C58 strain cured of the pTiC58 plasmid. After 6 weeks the sites of inoculation were scored for development of tumorous tissue. On 6 plantlets infected with the C58 strain, swelling of the wound sites and callus-like outgrowths approximately 3 mm^2 in size were observed. On none of the plantlets infected with the C58-C9 strain were callus outgrowths observed.

β-glucuronidase expression

Ten surface-sterilized stem segments of cv. Harmony with 3 to 4 nodes per stem were inoculated with each strain of *Agrobacterium* provided with the p35SGUSINT binary vector (Table 1). After 3 weeks the tissues were analyzed by histochemical staining for GUS activity. The strain C58 appeared to be the most efficient in transfecting stem node tissue of cv. Harmony. Four stems showed 15 clusters of blue cells spread over 11 nodes. The disarmed strain C58-C1 gave only one single cluster of transformed cells. All clusters of intensely-stained blue cells were observed at the sites of wounding and no callus formation was observed. The control inoculation with the strain C58-C9 without the binary vector was negative for X-gluc staining. The *A. rhizogenes* strain 1855 and ATCC 15834 did not induce wound swellings and callus-like outgrowths on lily plantlets but induced normal hairy roots on the internode sections of sterile tobacco plants.

Table 1. Strains of *Agrobacterium* and their efficiency in GUS activity (blue spots)

Strain[a]	Ri/Ti plasmid	Vector	Number of inoculated stem nodes	Total number of blue spots
C58-C9	–		36	0
C58	pTiC58	p35SGUSINT	32	15
C58-C1	pMP90	p35SGUSINT	24	1
1855	pRi1855	p35SGUSINT	30	0
ATCC 15834	pRi15834	p35SGUSINT	33	0

[a] C58 is an *A. tumefaciens* strain, 1855 and ATCC 15834 are *A. rhizogenes* strains. C58-C1 is a disarmed strain harbouring pMP90 (Koncz & Schell, 1986).

After X-gluc staining these sections showed clusters of intensely-stained blue cells.

Discussion

From the three different *Agrobacterium* strains tested in this study only the nopaline strain C58 was able to transform certain tissues of lily cv. Harmony. By now there have been several reports of tumour induction on monocotyledons by the C58 strain (Hooykaas-Van Slogteren et al., 1984; Hernalsteens et al., 1984; Domisse et al., 1990). On lily this strain induces tumorous swellings with callus-like outgrowths at the sites of inoculation. These tumorous reactions could be easily discerned from the general wound responses found in the plants inoculated with the C58-C9 strain that lacks the Ti plasmid. Although it is known that some monocotyledons produce a compound upon wounding which induces the *vir* genes of *Agrobacterium* (Usami et al., 1988), acetosyringone was used in all experiments because it is unknown yet whether lily has a wound reaction that induces *vir* genes. Detection of nopaline was not considered because of the small volume of the tumorous tissues. Inoculation of larger, surface-sterilized stem segments with *Agrobacterium* strains harboring a binary vector with the GUS-intron marker gene, allowed a definitive conclusion about transformation events and the type of cells that were transformed. The stem segments were inoculated both at the internodes and at the nodes and, in contrast to the experiment with sterile plantlets, only the node tissue proved to be susceptible to *Agrobacterium* infection and transformation. Possibly the developmental state of the stem influences its susceptibility to *Agrobacterium*. The frequency of transformed cell clusters obtained with the virulent C58 strain was significantly higher than the number obtained with the disarmed strain C58-C1. To find an optimum for transformation, the inoculation was repeated after growth of the C58 strain at other pH values showing a similar number of blue cell clusters at pH 5.5 and pH 6.0 but no transformed cell clusters at pH 6.5 or higher values. In the case of the virulent strain, co- transformation of the Ti-DNA may occur and plant cells harboring the Ti-DNA may proliferate predominantly due to expression of the hormone genes encoded by the Ti-DNA. As no antibiotic selection was applied to the inoculated tissue this may explain the difference in the numbers of transformed cells found with the virulent and disarmed strain. The cells transformed in the node tissue were all located at the wounding sites and in tissue that is regenerative on MS-medium with hormones. The low frequency of transformation found with the disarmed strain however, may put several constraints upon the selection of transformed regenerants in the future.

Acknowledgements

This research has been part of the 'Bloembollenurgentie programma' funded by the 'Produktschap voor Siergewassen' and the Ministry of Agriculture of the Netherlands.

References

De Cleene, M., 1985. The susceptibility of monocotyledons to *Agrobacterium tumefaciens*. Phytopath. Z. 113: 81–89.

Dommisse, E.M., D.W.M. Leung, M.L. Shaw & A.J. Conner, 1990. Onion is a monocotyledonous host for *Agrobacterium*. Plant Sci. 69: 249–257.

Dower, W.J., J.F. Miller & C.W. Ragsdale, 1988. High efficiency transformation of *E. coli* by high voltage electroporation. Nuc. Acids Res. 16: 6127–6145.

Hernalsteens, J-P., L. Thia-Toong, J. Schell & M. Van Montagu, 1984. An *Agrobacterium* transformed cell culture from the monocot *Asparagus officinalis*. EMBO J. 3: 3039–3041.

Hooykaas-Van Slogteren, G.M.S., P.J.J. Hooykaas & R.A. Schilperoort, 1984. Expression of Ti plasmid genes in monocotyledonous plants infected with *Agrobacterium*. Nature 311: 763–764.

Jefferson, R.A., 1987. Assaying chimeric genes in plants: The GUS gene fusion system. Plant Mol. Biol. Rep. 5: 387–405.

Koncz, C. & J. Schell, 1986. The promoter of T_l-DNA gene 5 controls the tissue-specific expression of chimaeric genes carried by a novel type of *Agrobacterium* binary vector. Mol. Gen. Genet. 204: 383–396.

Murashige, T. & F. Skoog, 1962. A revised medium for rapid growth and bioassays with tobacco tissue cultures. Physiol. Plant 15: 473–497.

Sambrook, J., E.F. Fritsch & T. Maniatis, 1989. Molecular Cloning, A Laboratory Manual. Cold Spring Harbor Laboratory Press.

Van Aartrijk, J. & G.J. Blom-Barnhoorn, 1983. Adventitious bud formation from bulb scale explants of *Lilium speciosum* Thunb. *in vitro*. Effects of wounding, TIBA and temperature. Z. Pflanzenphysiol. 110: 355–363.

Vancanneyt, G., R. Schmidt, A. O'Connor-Sanchez, L. Willmitzer & M. Rocha-Rosa, 1990. Construction of an intron-containing marker gene: Splicing of the intron in transgenic plants and its use in monitoring early events in *Agrobacterium* mediated plant transformation. Mol. Gen. Genet. 220: 245–250.

Usami, S., S. Okamoto, I. Takebe & Y. Machida, 1988. Factor inducing *Agrobacterium tumefaciens vir* gene expression is present in monocotyledonous plants. Proc. Natl. Acad. Sci. USA 85: 3748–3752.

Euphytica **85**: 101–108, 1995.

An assessment of morphogenic and transformation efficiency in a range of varieties of potato (*Solanum tuberosum* L.)

Philip J. Dale & Kaija K. Hampson
John Innes Centre, Colney Lane, Norwich NR4 7UJ, U.K.

Key words: Agrobacterium, plant regeneration, potato, *Solanum tuberosum*, tissue culture, transformation

Summary

To provide a truly genotype-independent transformation system, it is necessary to be able to transform a wide range of potato genotypes. The ability to regenerate shoots *in vitro* was determined for 34 potato varieties using tuber disc explants. Following a culture regime used extensively in previous studies with the variety Desiree, half of the varieties could be regenerated from tuber discs and half could not. From a sample of varieties that could be regenerated from tuber discs, all but one variety gave transgenic plants. Twelve varieties were evaluated for the capacity to regenerate shoots from leaf and internode explants excised from *in vitro* grown plants. All of the varieties tested regenerated adventitious shoots. Leaf and internode explants from 5 varieties were subsequently used for transformation, and transgenic plants were produced from two potato varieties that did not give transgenic plants from tuber disc explants. Some varieties could not be transformed by either method, and will require modification of the *in vitro* regeneration and transformation system to be successful.

Abbreviations: 2,4-D – 2,4-dichlorophenoxyacetic acid, GA_3 – gibberellic acid, IAA – indole-3-acetic acid

Introduction

Advances in transformation and recombinant DNA methods have made it possible to introduce genes from many different organisms into crop plants. As a consequence there are now opportunities to modify agricultural and horticultural crops in ways unimaginable by conventional breeding methods (Dale et al., 1993). For transformation to be of practical use in plant breeding, it is desirable to be able to transform particular plant genotypes. This is especially so for crop species that rely on vegetative reproduction for the maintenance and propagation of superior heterozygous gene combinations. In self-pollinating, seed-propagated crop species, it is often possible to use a genotype that can be transformed easily and to transfer the introduced gene into other genetic backgrounds by back-crossing (e.g. oilseed rape). This approach is laborious in potato (*Solanum tuberosum* L.), and not an attractive option when a transgene needs to be inserted into advanced breeding lines or into established and proven crop varieties.

There are now several reports of transformation in potato, and *Agrobacterium* methods are in routine use in many laboratories for certain plant genotypes (Sheerman & Bevan, 1988; De Block, 1988; Stiekema et al., 1988; Visser et al., 1989; Wenzler et al., 1989; Newell et al., 1991). Although there are reports of genotype-independent methods (De Block, 1988), the number of genotypes that have been tested is limited. Differences in transformation efficiency between potato varieties are reported in the literature (Sheerman & Bevan, 1988; Wenzler et al., 1989), and an analysis of 4 varieties and 5 breeding lines using the *gus* reporter gene (Higgins et al., 1992) has revealed varietal differences in the early events of the transformation process.

The objective in this study was to compare a wide range of potato cultivars, first to determine their morphogenic potential in culture, and second in a smaller study, their efficiency in producing transgenic shoots and plants. Explants were taken from stored tubers, and from *in vitro* grown plants.

Materials and methods

Agrobacterium *and vector plasmid construct*

All the transformation experiments were conducted using *Agrobacterium tumefaciens* strain LBA 4404 carrying the binary vector pBin 19 (Bevan, 1984). The plasmid contained the neomycin phosphotransferase gene (*nptII*) regulated by the nopaline synthase promoter, and the β-glucuronidase reporter gene (*gus*) (Jefferson, 1987) regulated by the 35S cauliflower mosaic virus promoter.

Agrobacterium was prepared by inoculating 10 ml liquid cultures of Luria broth containing 50 mg/l kanamycin monosulphate (Sigma), and grown overnight at 27° C in an orbital shaker at 100 rpm. The cultures were centrifuged at 2000 rpm, and after discarding the supernatant, *Agrobacterium* was resuspended in 10 ml of MS basal liquid medium (Murashige & Skoog, 1962) with 30 g/l sucrose. Inoculum prepared in this way was used for co-cultivation with tuber discs and stem and leaf sections as described below.

Tuber disc explants

Tubers were obtained from 34 varieties of virus-tested potato (*Solanum tuberosum* L.) plants, grown under standard agricultural field conditions in experimental plots at Cambridge (see Fig. 1 for list of varieties). Tubers were harvested in September, stored in darkness at 5° C and used for transformation experiments between the November and January immediately following harvest. Extensive experience with the variety Desiree has shown that the rate of plant regeneration declines with time in storage (Sheerman & Bevan, 1988; Dale & McPartlan, 1992). At least 6 tubers were selected from each variety for each experiment.

Tubers were peeled and surface-sterilised in 10% sodium hypochlorite for 15 min and rinsed three times in sterile distilled water. Using a 10 mm diameter sterilized cork borer, plugs of tuber tissue were transferred to a 9 cm plastic Petri dish and cut into 1–2 mm thick discs with a sterile scalpel. Approximately 30 discs per 9 cm Petri dish were flooded with 10 ml of the *Agrobacterium* inoculum, and left for 30 min. Discs were transferred without blotting to 9 cm plastic Petri dishes (5 discs per dish), on a co-cultivation medium. Tuber discs were distributed at random among the replicates.

The co-cultivation medium contained MS salts, 30 mg/l sucrose, 1 mg/l thiamine HCl, 0.5 mg/l nicotinic acid and 0.5 mg/l pyridoxine HCl (Sheerman & Bevan, 1988). The pH was adjusted to 5.8 before adding 8 g/l Difco Bacto agar and autoclaving at 121° C for 15 min. After autoclaving, the following were added before pouring into 9 cm plastic Petri dishes: 1.8 mg/l zeatin riboside (Sigma), 0.87 mg/l IAA-aspartic acid (Research Organics) and 20 mg/l acetosyringone (Aldrich).

Tuber discs were co-cultivated for 2 d at 25° C with continuous cool white fluorescent light at 60 μE/m^2/s. Petri dishes were sealed with Parafilm^{TM}. After co-cultivation, discs were transferred to a medium containing the same components as the co- cultivation medium, but without acetosyringone and with the addition of 500 mg/l carbenicillin (Sigma) and 100 mg/l kanamycin monosulphate.

Cultured discs were maintained under culture room conditions (as above) for two weeks. Explants were then transferred to new Petri dishes containing the same medium as above but with the carbenicillin concentration reduced to 200 mg/l. Plant regeneration was scored after a further two weeks (4 weeks from the end of co-cultivation).

In vitro *shoot explants*

In vitro shoot cultures were initiated from nodes excised from tuber shoots, and grown in Magenta GA7 vessels (Sigma) containing MS medium (including the MS organic components) with 30 g/l sucrose, pH 5.8 and solidified with 6 g/l agar (Sigma, A7002). Shoot cultures were maintained by transferring cuttings containing 1 or 2 nodes to fresh medium every 6–8 weeks. Plantlets used for transformation had been subcultured to new culture medium about 6 weeks prior to use, and grown at 25° C in continuous light at an intensity of 60 μE/m^2/s. Each vessel contained approximately 6 plantlets.

Before co-cultivation with *Agrobacterium*, the *in vitro*-grown plants were prepared by cutting internodes into 5–8 mm lengths (excluding node tissue). Leaves were prepared for co-cultivation by making a cut (at right-angles to the mid-rib) across the petiole end of the leaf, and at the leaf tip end, to give a 5 mm wide section of leaf.

Leaf and internode explants were incubated in 8 ml of *Agrobacterium* suspension in 60 ml screw cap containers (Sterilin) for 10 min and swirled gently. Leaf and internode segments were removed and placed without blotting onto 9 cm Petri dishes containing culture medium. The dishes were sealed with Parafilm^{TM} and incubated for 2 d at 25° C with continuous light at 60 $\mu E/m^2/s$.

The culture medium contained MS basal medium, 30 g/l sucrose, pH 5.8, 6 g/l agar (Sigma) and autoclaved for 15 min. Following autoclaving the following were added after filter sterilization: 0.5 mg/l 2,4-D (Sigma), 0.5 mg/l zeatin riboside, 5 mg/l GA_3 (Calbiochem).

Following co-incubation, the explants were transferred to Petri dishes containing the same medium but with the addition of 500 mg/l carbenicillin and 100 mg/l kanamycin monosulphate. After a further two weeks, explants were transferred to the same medium but with 200 mg/l carbenicillin and without 2,4-D. Plant regeneration was scored after a further 2 weeks (4 weeks from the end of co-cultivation).

The conditions used to determine the morphogenic potential of the three explants from the different potato varieties were the same as those used for transformation, except that explants were not incubated in *Agrobacterium*, and antibiotics were omitted from the culture media. In controls where antibiotics were used, but without *Agrobacterium* treatment of the explants, normal shoot regeneration was not observed.

The plants regenerated following *Agrobacterium* incubation were screened for expression of the *gus* gene using the histochemical method described by Jefferson (1987). The *gus* reporter gene has been used extensively in other potato studies (Jefferson, 1990; Dale & McPartlan, 1992), but Southern blotting (Sambrook et al., 1989) analysis was used on a sample of plants to confirm that plants scored as positive by the histochemical analysis also carried *gus* DNA sequences.

Results

Regeneration from tuber discs

For transformation to be effective, it is necessary for the chosen explant to have the potential to regenerate shoots in culture. Tuber discs from 34 varieties of potatoes were tested to determine morphogenic potential in a range of cultivars. For each variety, 100 tuber discs were cultured in two replicates. In regeneration and transformation studies, morphogenic potential is usually expressed as a number of shoots regenerated, as a percentage of the total number of explants cultured. Morphogenic potential is expressed in Fig. 1 as the number of shoots regenerated from 100 tuber discs, so the values presented in the histogram also correspond to the percentage regeneration rates. Because of the importance in later transformation studies of being able to distinguish between independently-transformed shoots, care was taken to count only shoots of independent origin from the tuber disc. Where several shoots emerged from the same position on the disc, these were counted as one regeneration event. Where shoots were produced, discs mostly had 1 to 5 regeneration events, so some discs did not regenerate any shoots.

Seventeen varieties regenerated shoots from tuber discs and seventeen varieties did not. Shoot regeneration rates ranged from 3 ('Cara') to 133 ('Pentland Squire') shoots from 100 tuber discs. An analysis of variance (ANOVA) showed that the differences between varieties were highly significant ($P = < 0.1\%$). There was no significant difference in regeneration rate between the two replicates.

Transformation using tuber discs

Tuber discs were used for transformation from twelve varieties that displayed a range of shoot regeneration responses, and included two varieties that gave no regeneration ('Drayton' and 'Maris Piper'). One hundred and fifty discs were cultured per variety, in three replicates. The number of shoots regenerating in the presence of kanamycin selection ranged from zero to 125 shoots per 100 discs cultured (Fig. 2).

There were statistically significant differences between varieties in the number of shoots produced (ANOVA; $P = < 5\%$). All the regenerated shoots were tested for expression of the *gus* reporter gene and 87% were positive. There was no indication of differences between varieties in the proportion of the regenerated plants that were GUS positive (range 71–100%). GUS negative plants grew normally in the presence of kanamycin.

There were two unexpected observations. 'Pentland Dell', which displayed the potential to regenerate shoots from tubers (19 shoots per 100 tuber discs; Fig. 1) gave no transgenic shoots, and 'Drayton', which did not give shoots in the regeneration test, regenerated transgenic shoots (49 shoots per 100 discs; Fig.

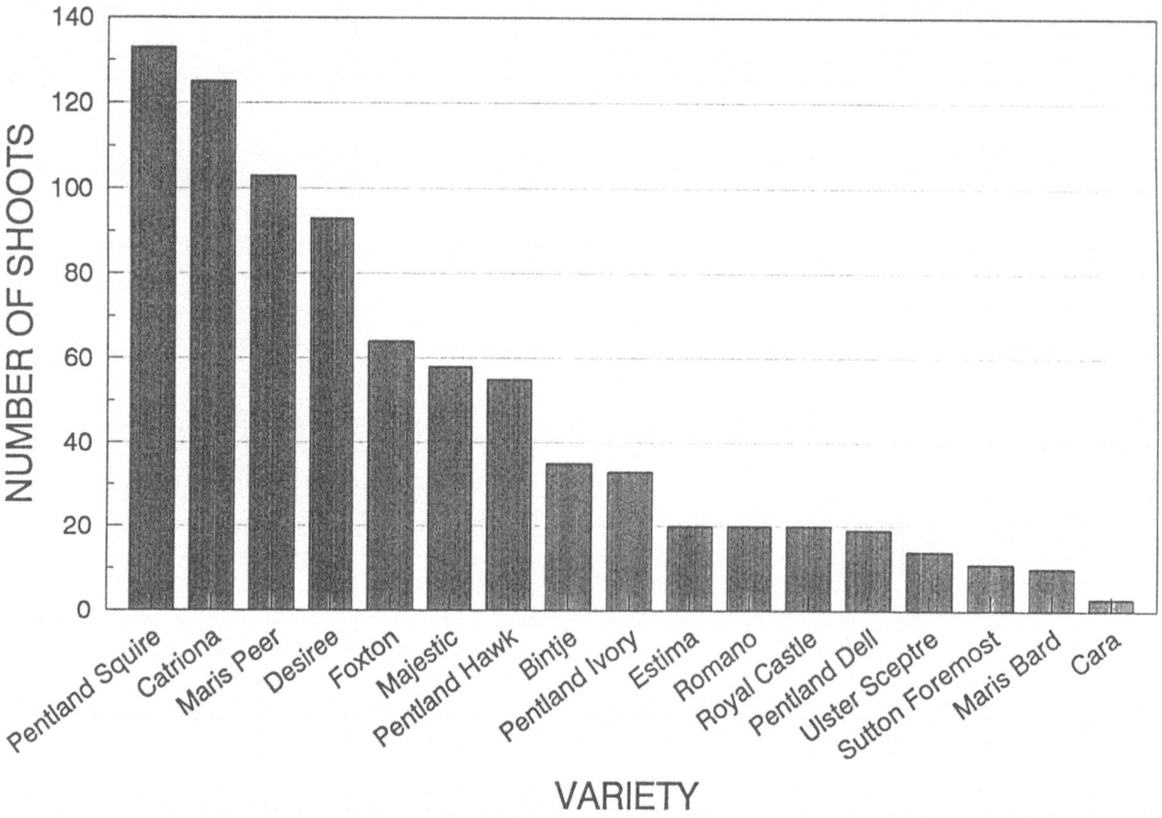

Fig. 1. Number of shoots of independent origin regenerating from 100 cultured tuber discs of 34 varieties of potatoes during a 4–5 week period. The following varieties did not regenerate shoots: Arran Banner, Arran pilot, Baille, Corsair, Cromwell, Drayton, Edzel Blue, Home Guard, King Edward, Kingston, Maris Anchor, Maris Piper, Pentland Crown, Pentland Javelin, Record, Stormont Enterprise, Wilja.

2). This observation will be referred to again later (see Discussion).

Using the varieties presented in Fig. 2, a correlation analysis between the number of shoots regenerating from a variety, with and without transformation (data in Fig. 2 compared with data in Fig. 1), gave a correlation coefficient of $r_{10} = 0.59$. This shows a significant ($P = < 0.5\%$) positive association between the two values. It is, of course, reasonable to expect a variety to have the potential for transformation, only if it is able to regenerate plants from the chosen explant. However, these data indicate that, among the varieties selected for the transformation experiment, only 35% ($r^2 = 0.348$) of the variation in transformation efficiency between varieties, is attributable to variation in the capacity of a variety to regenerate shoots from tuber discs.

Regeneration from in vitro *grown plants*

In vitro-grown plants from twelve varieties were also tested for their ability to regenerate adventitious shoots from two different explants: leaf sections and stem internodes. A plant regeneration protocol from these explants was established in another study, using *in vitro*-grown virus tested plant stocks (Dale, unpublished). Varieties shown to regenerate from tuber discs were included in the sample, along with varieties that did not regenerate from tuber discs ('Arran Banner', 'Drayton', 'King Edward', 'Kingston', 'Pentland Crown' and 'Record'). Thirty five to 190 (average 108) explants were cultured per variety, in two replicates. Regeneration is expressed as the number of shoots regenerated from 100 explants, so as before, the values also correspond to the percentage regeneration rates. The number of regenerating shoots of independent origin was more difficult to estimate for the very small leaf and internode explants, than for tuber discs.

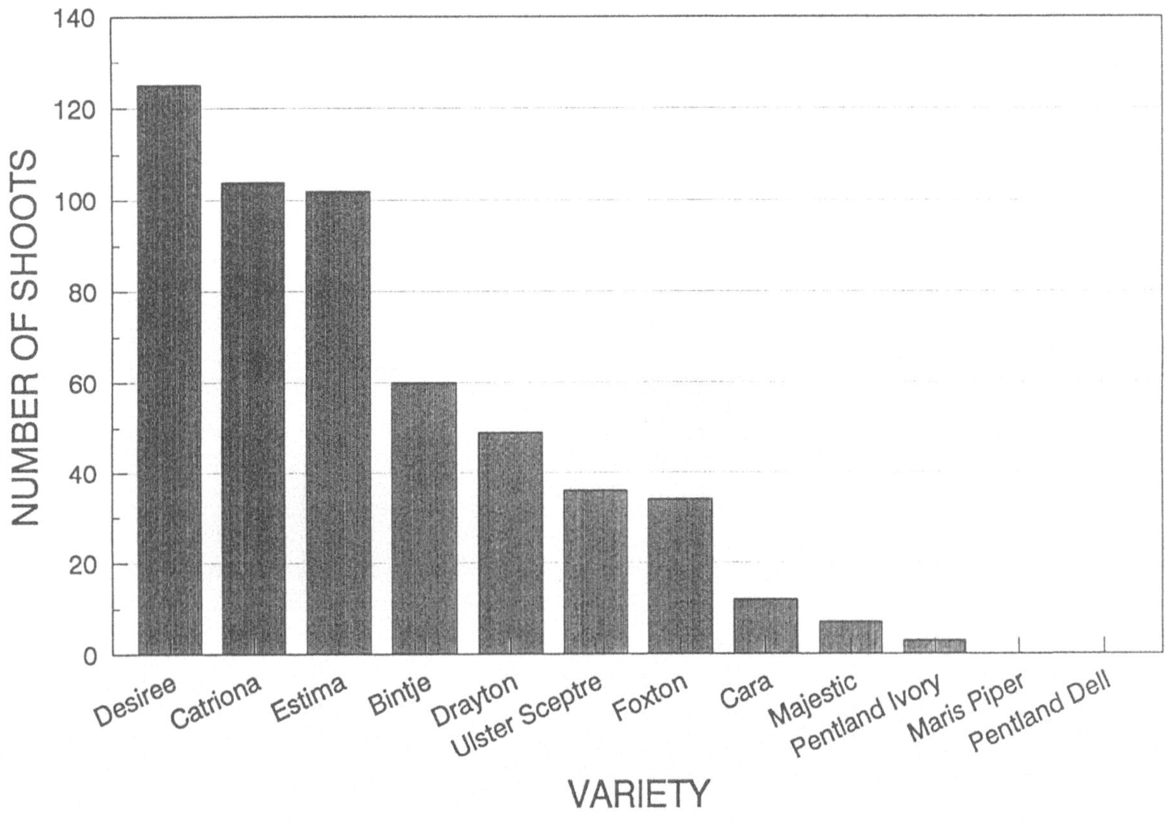

Fig. 2. Number of shoots of independent origin regenerating from *Agrobacterium*-treated discs cultured on shoot regeneration medium containing kanamycin. Regeneration rate is expressed as the number of shoots produced per 100 tuber discs cultured.

Explants mostly had 1–5 shoots of independent origin, so some explants did not regenerate any shoots.

Adventitious shoots regenerated from both explants in all twelve varieties tested, and there were significant differences (ANOVA) between varieties for regeneration from both leaf explants (Fig. 3a; P = < 1%) and internode explants (Fig. 3b; P = < 5%).

A correlation analysis of the shoot regeneration data between leaf and internode explants in the 12 varieties gave a highly significant positive correlation (r_{10} = 0.86; P = < 0.1%). This indicates that the two explants have a similar genetic potential to regenerate; however, both explants were from the same sample of *in vitro*-grown plants and would be subject to similar *in vitro* environmental effects.

Transformation using leaves and stem internodes from in vitro-*grown plants*

Five varieties were selected that represented a range of response in culture. All except 'Maris Piper' (not tested) had previously shown the potential to regenerate shoots from leaf and internode explants, but had displayed very different regeneration responses from cultured tuber discs. In this investigation the variety Desiree displayed a high capacity for regeneration and for transformation, and has in other studies been used extensively for potato transformation and the field evaluation of transgenes (Dale & McPartlan, 1992). 'Pentland Dell' could be regenerated from tuber discs but failed to produce transgenic plants. 'Pentland Crown', 'Record' and 'Maris Piper' failed to regenerate shoots from tuber discs. Between 48 and 107 leaf explants and internode explants, in two replicates, were incubated in *Agrobacterium*, and shoots were regenerated.

The frequencies of shoots regenerated in the presence of kanamycin selection (expressed as shoots per 100 explants) were: 3.3/9.0 (leaf/stem explants) for 'Desiree', 22/43 for 'Pentland Dell', 3.5/0 for 'Pentland Crown', 0/0 for 'Record' and 'Maris Piper'. Usually only one transgenic shoot was produced per explant. All regenerated plants were tested for expres-

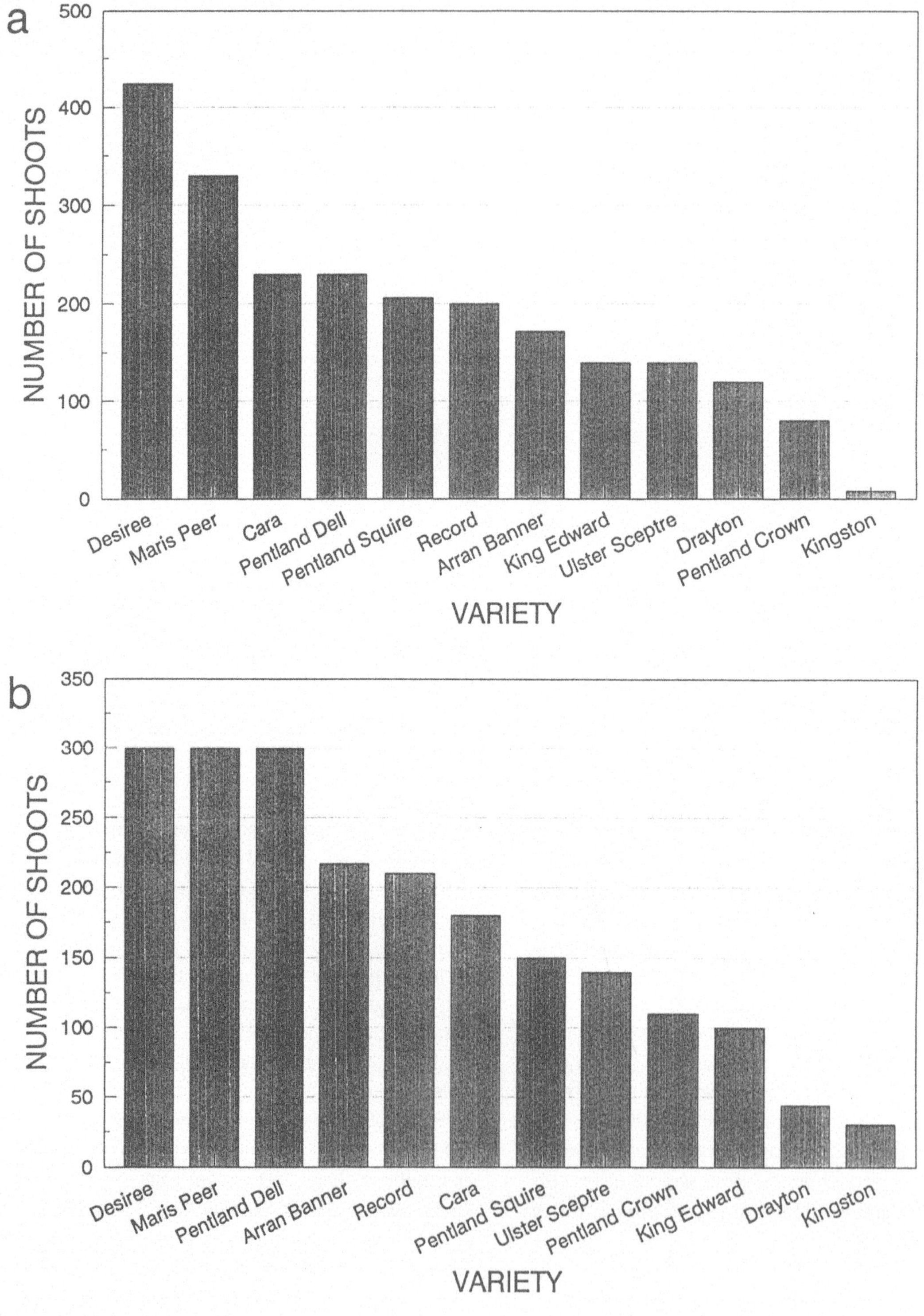

Fig. 3. Number of shoots judged to be of independent origin from (a) leaf and (b) internode explants derived from *in vitro* grown plants of twelve potato varieties. Regeneration rate is presented as the number of shoots regenerating from 100 cultured explants.

sion of the *gus* reporter gene and 68% were positive. As was observed for tuber discs, there was no evidence that the proportion of GUS positive plants was different between varieties (range 65–100%).

Discussion

Using a regeneration medium used successfully for many experiments with the variety Desiree, half of the 34 varieties regenerated shoots and half did not. Care was taken to use healthy tubers that were grown, harvested and stored together under the same conditions. In all experiments tubers were used only if they were less than 5 months from the date of harvest. Of the 12 varieties that were used for transformation using tuber discs, ten of them gave transgenic shoots.

Several of the varieties gave a higher frequency of plant regeneration following transformation and selection than in the initial regeneration experiment. The reason for this was partly a consequence of estimating the number of shoots of independent origin. In the regeneration experiment, clusters of shoots would frequently be counted as one regeneration event, whereas in the transformation experiment, regenerated shoots were often physically isolated from one another on the disc surface. It is feasible also that there was a degree of shoot dominance within a tuber disc in the regeneration experiment. Without the presence of transformation and antibiotic selection, shoots often grew quickly and vigorously. It is possible that the more rapidly growing shoots inhibited the regeneration of further shoots.

It was noted that 'Pentland Dell', which displayed the potential to regenerate shoots from tubers gave no transgenic shoots, and that 'Drayton', which did not give shoots in the regeneration test, gave transgenic shoots in the transformation test. The behaviour of 'Pentland Dell' is not surprising because factors other than regeneration potential can affect the efficiency of transformation. The behaviour of 'Drayton' in these experiments is more difficult to interpret. The tuber storage times and procedures used for all the varieties were based principally on experience over several years of transforming the variety Desiree using tuber discs. In this variety, the use of a sample of 6 tubers with the discs randomly-distributed over the replicates and treatments, has been found to be sufficient to even out any tuber effects. It is possible that the change in regeneration potential with storage, or the variation between tubers, is influenced by variety, and that this might account for the observed variability for 'Drayton'.

It is worthy of note that the correlation analysis for the regeneration rates of tuber discs of the varieties with and without transformation, was significant. This analysis indicates that 35% of the observed variation in regeneration response was attributable to variety. It also indicates, however, that 65% of the variation in response is attributable to physiological and other sources of variation.

The leaves and internode stem sections from *in vitro*-grown plants regenerated adventitious shoots from all twelve varieties tested. The variety Desiree, which consistently regenerates well from tuber discs, also regenerated shoots at a high frequency from these explants, and gave transgenic plants. The varieties Pentland Dell and Pentland Crown, which did not produce transgenic plants from tuber discs, did give transgenic plants from the *in vitro* grown explants; but the varieties Record and Maris Piper did not produce transgenic plants with any of the explants used. Leaf and internode stem sections of 'Maris Piper' tended to produce more callus than the other varieties tested, so it is possible that the auxin level needs to be reduced for this variety, and would be an important variation to test the variety Record also.

Despite claims that transformation methods can be genotype-independent (De Block, 1988) it is unlikely that any one method is efficient for all varieties. From the data obtained in this study, it is likely that the method of choice will depend to some extent on the variety to be transformed. There are many advantages of the tuber disc transformation method: a large number of aseptic explants can be obtained quickly and easily, there is no requirement to maintain plants *in vitro* and the transformation rates can be very high. The disadvantages are that some varieties do not respond to the transformation system used in this study, and tubers need to be used within about 5 months from harvest. The tuber age limitation can be overcome by carrying out transformation work immediately after harvest. An alternative is to grow plants in a glasshouse to produce tubers out of season. 'Desiree' tubers have been produced for transformation studies in this way at various times of the year.

One of the potential complications of transformation is that the transgenic plants can display somaclonal variation (Karp, 1991) resulting from the transformation process. Variation of this kind was observed in a field experiment in which the performance of 70 independently-transformed plants was compared (Dale & McPartlan, 1992). The transgenic plants in that experiment were obtained using the tuber disc transformation system, and small but statistically-significant somaclonal variation was observed. The transformation system used for *in vitro*-grown explants in this study, displayed significantly more callus than is observed in the tuber disc transformation system. As somaclonal variation is usually associated with callus formation, it is possible that more variation of this kind would be present among transgenic plants from the leaf and internode explant method.

The variation in the proportion of shoots expressing the *gus* gene ranged from 65–100% for the different varieties. There was no indication of a difference in this GUS expression between varieties or between the two transformation methods. All the shoots scored as regenerating in the transformation experiments, were able to grow healthily in the presence of kanamycin selection. A sample containing GUS-expressing and non-expressing shoots were all positive when probed with the *gus* gene by Southern blotting. Variation in expression of the *gus* gene has been reported for potato in other studies (Stiekema et al., 1988; Jefferson, 1990).

In conclusion, the tuber disc transformation system used in this study can give very high transformation rates that are significantly better than other frequencies reported (0.3–8.1% Visser et al., 1989; 2–13% Newell et al., 1991; c 17%, Higgins et al., 1992). In some instances high transformation frequencies have been obtained (c. 100%, De Block, 1988) but this involved passing through a distinct callus phase. As indicated above, with the tuber disc system the shoots are often observed to emerge directly from the tuber disc with little evidence of associated callus. Our view is that where tuber disc transformation works, it is preferable to the use of other explants. Where transformation is not observed from tuber discs, the *in vitro*-derived leaf and internode explants should be used. In the less responsive varieties it will be necessary to modify the regeneration and transformation conditions, and the auxin level in the regeneration medium would be an important variable to test.

Acknowledgements

We thank David Lander for his help with the GUS histochemical analysis and Southern blot analysis, Judith Irwin for comments during preparation of the manuscript and AFRC for financial support. We thank the Scottish Crops Research Institute for providing the original source of virus-tested tubers.

References

Bevan, M., 1984. Binary *Agrobacterium* vectors for plant transformation. Nuc. Acids Res. 12: 8711–8721.

Dale, P.J., J.A. Irwin & J.A. Scheffler, 1993. The experimental and commercial resease of transgenic crop plants. Plant Breeding 111: 1–22.

Dale, P.J. & M.C. McPartlan, 1992. Field performance of transgenic potato plants compared with controls regenerated from tuber discs and shoot cuttings. Theor. Appl. Genet. 84: 585–591.

De Block, M., 1988. Genotype-independent leaf disc transformation of potato (*Solanum tuberosum*) using *Agrobacterium tumefaciens*. Theor. Appl. Genet. 76: 767–774.

Higgins, E.S., J.S. Hulme & R. Shields, 1992. Early events in transformation of potato by *Agrobacterium tumefaciens*. Plant Sci. 82: 109–118.

Jefferson, R.A., 1987. Assaying chimeric genes in plants: the GUS gene fusion system. Plant Mol. Biol. Rep. 5: 387–405.

Jefferson, R.A., 1990. New approaches for agricultural molecular biology: from single cells to field analysis. In: J.P. Gustafson (Ed). Gene manipulation in plant improvement II, pp. 365–400. Plenum Press, New York.

Karp, A., 1991. On the current understanding of somaclonal variation. In: B.J. Miflin (Ed). Oxford Survey of Plant Molecular and Cell Biology, pp. 1–58. Oxford University Press, Oxford.

Murashige, T. & F. Skoog, 1962. A revised medium for rapid growth and bio assays with tobacco tissue cultures. Physiol. Plant. 15: 473–497.

Newell, C.A., R. Rozman, M.A. Hinchee, E.C. Lawson, L. Haley, P. Sanders, W. Kaniewski, N.E. Tumer, R.B. Horsch & R.T. Fraley, 1991. *Agrobacterium*-mediated transformation of *Solanum tuberosum* L. cv. Russet Burbank. Plant Cell Rep. 10: 30–34.

Sambrook, J., E.F. Fritisch & T. Maniatis, 1989. Molecular Cloning. Cold Spring Harbor Laboratory Press, Cold Spring Harbor, New York.

Sheerman, S. & M.W. Bevan, 1988. A rapid transformation method for *Solanum tuberosum* using binary *Agrobacterium tumefaciens* vectors. Plant Cell Rep. 7: 13–16.

Stiekema, W.J., F. Heidekamp, J.D. Louwerse, H.A. Verhoeven & P. Dijkhuis, 1988. Introduction of foreign genes into potato cultivars Bintje and Desiree using an *Agrobacterium tumefaciens* binary vector. Plant Cell Rep. 7: 47–50.

Visser, R.G.F., E. Jacobsen, A. Hasseling-Meinders, M.J. Schans, B. Witholt & W.J. Feenstra, 1989. Transformation of homozygous diploid potato with an *Agrobacterium tumefaciens* binary vector system by adventitious shoot regeneration on leaf and stem segments. Plant Mol. Biol. 12: 329–337.

Wenzler, H., G. Mignery, G. May & W. Park, 1989. A rapid and efficient transformation method for the production of large numbers of transgenic potato plants. Plant Sci. 63: 79–85.

Euphytica **85**: 109–112, 1995.

Transgenic apples display stable gene expression in the fruit and Mendelian segregation of the transgenes in the R1 progeny

D.J. James, A.J. Passey & S.A. Baker
Horticulture Research International, East Malling, West Malling, ME19 6BJ, U.K.

Key words: *Malus pumila* Mill, apple fruit, inheritance, stable gene expression, transgenes, trees

Summary

The transfer of genes via *Agrobacterium* to a perennial tree crop such as apple, requires monitoring of the stability of the genes in the target tissues such as the fruit and leaves. If the same genes are required for introgression into a conventional breeding programme, their expression also needs to be stable and their inheritance should follow a normal Mendelian pattern. In the following report we show, for the first time, the stable expression and Mendelian segregation of transgenes in a tree species. We have evidence for a 1:1 segregation of the *nos* and *nptII* genes among R1 progeny from a transgenic apple parent. In addition, we present evidence for stable gene expression of both *nos* and the co-transferred gene *nptII* in the flesh of apple fruit 7 years after the initial transformation.

Abbreviations: *nos* – nopaline synthase, *nptII* – neomycin phosphotransferase, ACC – 1-aminocyclopropane carboxylic acid, PVP – polyvinylpyrrolidone, PCR – polymerase chain reaction, EFE – ethylene forming enzyme, B.t. – *Bacillus thuringiensis*

Introduction

Transgenic apple clones (cv. Greensleeves) were first produced in 1986 (James et al., 1989) using the disarmed Ti-binary vector pBIN6 (Bevan, 1984) in an *Agrobacterium*-mediated leaf disc transformation procedure (Horsch et al., 1985). The plants were transgenic for a reporter gene nopaline synthase (*nos*) and a selectable marker neomycin phosphotransferase (*nptII*) conferring resistance to kanamycin.

Procedure

In 1987 six transgenic clones of *in vitro*-rooted plants of pBIN6 apple, clones A, B, D, E, H and J, were transferred to soil and grown under glass for 2 years before being grafted onto M27 dwarfing rootstock to reduce size and encourage early flowering. Annual tests, since 1987, for expression of *nos* in the leaves of all clones have always been positive. All clones appeared phenotypically normal except clone B, which showed a tendency for premature leaf fall and senescence. This clone had previously shown reduced rooting ability *in vitro* (James et al., 1989). In 1992 two clones, and in 1993 four others, flowered under growth room conditions.

In 1992 controlled pollinations were performed on the first two clones, B and E, using pollen of the apple cv. Baskatong, which carried a dominant homozygous gene for anthocyanin pigmentation (R) (Church & Williams, 1978) and which permitted the segregation of hybrid zygotic embryos from any selfed or apomictic embryos. Transgenic flowers were emasculated before opening and then hand-pollinated. Fruit set was normal on all pollinated trees. To expedite segregation analyses, more than 70 immature embryos were excised aseptically from clone E fruit and some 20 from clone B trees 30–40 days post-anthesis. All were germinated *in vitro* and micropropagated by standard procedures (James & Dandekar, 1991) to provide early leaf material for biochemical and molecular analyses.

Table 1. ELISA assays for the NPTII protein performed on transgenic and control apple tissues

Clone/Seedling	ng NPTII/mg protein	
	Callus from fruit 'flesh'	Leaf extract
Untransformed tissue Replicate 1	0	0
Untransformed tisssue Replicate 2	0	0
Transformed tissue Clone A Replicate 1	Not available	1.4
Transformed tissue Clone A Replicate 2	Not available	1.0
Transformed tissue Clone E Replicate 1	7.8	Not done
Transformed tissue Clone E Replicate 2	11.2	Not done
Mean and S.E. of seedlings containing *nptII* gene fragment	–	4.8 ± 0.9 (n = 23)
Mean and S.E. of seedlings not containing *nptII* gene fragment	–	0.31 ± 0.07 (n = 20)

Gene expression assays of apple fruit

*Nopaline synthase (*nos*)*

Gene expression assays for *nos* were conducted using paper electrophoresis (Otten & Schilperoort, 1978). Samples of apple 'flesh' were excised directly from mature fruit raised under greenhouse conditions. Freshly excised 'flesh' tissue always gave a positive response for *nos* from several different apple fruits. Callus tissues derived from flesh tissue after 5 weeks growth on a callus-inducing medium containing 50 μg/μl of the antibiotic kanamycin also tested positive.

*Neomycin phosphotransferase (*nptII*)*

A double antibody sandwich ELISA method for the NPTII protein was used for biochemical assays (Nagel et al., 1992) with the addition of 10% (w/v) insoluble PVP to the extraction buffer. NPTII ELISA assays were performed on different tissues of transformed and non-transformed 'Greensleeves' R0 apple plants. Transgenic fruit flesh was able to form callus on callus-induction media containing either 0 or 50 μg/ml of kanamycin during a 5 week growth period, whereas control flesh from untransformed fruit of the same cultivar and grown under the same conditions failed to grow in the presence of the antibiotic. The NPTII protein was detectable in both fruit 'flesh' callus (8–11 ng/mg protein) and leaf material (1–1.5 ng/mg protein) from transgenic plants, but not in equivalent material from control untransformed plants (Table 1).

Inheritance data

When micropropagated, rooted seedlings were grown to the glasshouse stage of development, a striking variation in size and habit of many of the seedlings was apparent. Rosette, prostrate and upright phenotypes with a range of leaf sizes and shapes were represented amongst the population.

Expression assays

*Nopaline synthase (*nos*)*

Micropropagated, immature embryos derived from two independent transgenic apple clones, B and E, were assessed for segregation and expression of the *nos* gene using paper electrophoresis. Data from these assays showed that the gene segregated 1:1 according to a normal Mendelian ratio. Clone E showed 35 'seedlings' positive for expression of nopaline synthase and 39 negative from a population of 74 (chi-square = 0.108, ns.) whilst clone B showed 7 positive and 5 negative from a population of 12 'seedlings' (chi-square = 0.33, n.s.).

*Neomycin phosphotransferase (*nptII*)*

A total of 53 seedlings from clone E were examined for the presence of the NPTII protein in leaf extracts. In those seedlings segregating for the presence of the gene, the amount of protein varied from 0.03 to 30 ng/mg protein, with most of the values falling within the category 1–10 ng/mg protein (mean value of 4.8 ± 0.9, Table 1). Among 20 seedlings that did not show the presence of the *nptII* fragment, 5 gave values of 0 NPTII protein and the remaining values were all less than 1 ng NPTII/mg protein with a mean of 0.31 ± 0.07.

Molecular genetic analysis of R1 seedling population

Tissue (mainly leaf) samples from micropropagated seedlings were frozen in liquid nitrogen, vacuum dried for at least two days, and then stored at - 80° C for processing. The method of Dellaporta et al. (1983) was used to isolate genomic DNA with yields averaging 5–10 μg/g fresh weight. The polymerase chain reaction (PCR) was used to detect diagnostic fragments from both the *nos* (0.5 kb) and *nptII* (0.7 kb) genes in a randomly-selected sample of 30 R1 seedlings. The primers used for *nos* and *nptII* were those used by Hamill et al. (1991). The reagents and protocol used were those described for the kit supplied by Perkin Elmer. The reactions were performed on an Omni-Gene Thermocycler (Hybaid Ltd.) using 25 or 50 μl reaction volumes with the following programme: initial denaturing, 6 min at 94° C followed by 30 cycles of 1 min denaturing at 94° C, 1 min annealing at 65° C, 1 min extension at 72° C. Extension at 72° C was allowed to continue for 6 min at the end of the programme. Approximately 50–100 ng DNA per sample was loaded onto 1.5% agarose gels for resolution of the diagnostic fragments.

Among the 30 seedling selected from clone E, 15 showed the presence of both gene fragments of *nos* and *nptII*, 13 failed to display the presence of either (chi-square = 0.07 for 1 d.f.) and two seedlings, E2 and E41, consistently showed only the presence of the *nptII* gene fragment. The presence or absence of nopaline synthase activity always correlated with PCR data. E2 and E41 failed to show nopaline synthase expression as expected. However, both showed no reactivity on the ELISA test although PCR showed the presence of the diagnostic *nptII* fragment. In the same 15 seedlings, all displayed the presence of NPTII protein in the ELISA although some had very low levels no greater than others lacking the gene.

Among the 13 seedlings that segregated for absence of both genes, none of them showed any nopaline synthase activity after paper electrophoresis. However two of them, E65 and E70, indicated presence of the NPTII protein in the ELISA assay although only low levels (< 1 ng/mg protein) were measured. This could be due to cross-reaction with endogenous non-specific kinases that would vary in amounts in different seedlings (Draper et al., 1988).

Discussion

The data presented here suggests that long-lived stable transgene expression in apple fruit tissues many years after gene insertion, should in the future permit the introduction of novel genes that control the developmental physiology and biochemistry of apple fruit development. Already two genes that code for key enzymes involved in the ripening of climacteric fruit by modulating endogenous levels of ACC and thereby ethylene biosynthesis, have been isolated i.e. ACC synthase and ACC oxidase or the ethylene forming enzyme (EFE) (see review by Theologis, 1992). By inverting these 'ripening genes' with respect to their promoter i.e. making the 'anti-sense form' of the genes, ethylene biosynthesis can be reduced markedly and tomato ripening can be altered for improved fruit quality (Picton et al., 1993). Since the corresponding genes have also been isolated, cloned and sequenced from ripening apple fruit (Dong et al., 1991, 1992), the same anti-sense strategy could possibly be applied to control apple ripening. At the same time, fruit-specific genes and their corresponding promoters can be identified for the targeting of fruit gene expression, governing improvements in both fruit quality and enhanced pest and disease resistance (Edwards & Corruzzi, 1990). The *nos* promoter drives the expression of both transgenes in the pBIN6 vector (Bevan, 1984). This may help future workers when considering vector design for transgene expression in apple fruit.

Expression assays and the PCR molecular evidence for the segregating R1 progeny confirms the inheritance of the transgenes in a Mendelian fashion. Only two seedlings, E2 and E41, failed to show segregation of both genes. The loss of the *nos* gene in these cases may have different causes. It could have been eliminated during embryo sac differentiation; it could have been lost during chiasmata formation at meiosis or sister chromatid exchange could have occurred within the 500 bp region of the 'primed' sequence of the gene. Of these the last is presumably the least likely. Either of the two former explanations is more likely since the *nos* gene in the construct pBIN6 has already been shown to be unstable and has been shown to be absent from clone A although multiple copies of the *nptII* gene are present (James et al., 1989).

These data suggest that transgenes can now be used in conventional apple tree breeding programmes where economically important genes such as B.t. (Dandekar et al., 1992), that encode insect endotoxins, are available. However the variation in transgene expression observed here for the NPTII protein would dictate a need for screening of the transgenic progeny for the desired introduced gene prior to selection and propagation.

References

Bevan, M.W., 1984. Binary *Agrobacterium* vectors for plant transformation. Nuc. Acids Res. 12: 8711–8721.

Church, R. & R.R. Williams, 1978. Pollination of pome fruits. Report of the Long Ashton Research Station for 1977. pp. 21–22.

Dandekar, A.M., G.H. McGranahan, S.L. Uratsu, C. Leslie, P.V. Vail, S. Tebbets, D. Hoffman, J. Driver, P. Viss & D.J. James, 1992. Engineering for apple and walnut resistance to codling moth. In: Brighton Crop Protection Conference – Pests and Diseases, Vol. 2. pp. 741–747.

Dellaporta, S.L., J. Wood & J.B. Hicks, 1983. A plant DNA minipreparation: version II. Plant Mol. Biol. Rep. 1: 19–21.

Dong, J.G., W.T. Kim, W.K. Yip, G.A. Thompson, L. Li, A.B. Bennett & S.F. Yang, 1991. Cloning of a cDNA encoding 1-aminocyclopropane-1-carboxylate synthase and expression of its mRNA in ripening apple fruit. Planta 185: 38–45.

Dong, J.G., D. Olson, A. Silverstone & S.F. Yang, 1992. Sequence for a 1-aminocyclopropane-1-carboxylate oxidase from apple fruit. Plant Physiol. 98: 1530–1531.

Draper, J., R. Scott & R. Walden, 1988. Plant Genetic Transformation and Gene Expression – A Laboratory Manual. Blackwell Scientific Publications, Oxford. pp. 355.

Edwards, J.W. & G.M. Corruzzi, 1990. Cell-specific gene expression in plants. Ann. Rev. Genet. 24: 275–303.

Hamill, J.D., S. Rounsley, A. Spencer, G. Todd & M.J.C. Rhodes, 1991. The use of the polymerase chain reaction in plant transformation studies. Plant Cell Rep. 10: 221–229.

Horsch, R.B., J.E. Fry, N.L. Hoffman, D. Eicholtz, S.G. Rogers & R.T. Fraley, 1985. A simple and generalised method of transferring genes into plants. Science 227: 1229–1231.

James, D.J., A.J. Passey, D.J. Barbara & M.W. Bevan, 1989. Genetic transformation of apple (*Malus pumila* Mill.) using a disarmed Ti-binary vector. Plant Cell Rep. 7: 658–661.

James, D.J. & A.M. Dandekar, 1991. Regeneration and transformation of apple (*Malus pumila* Mill.). Plant Tissue Culture Manual B8: 1–18.

Nagel, R.J., J.M. Manners & R.G. Birch, 1992. Evaluation of an ELISA assay for rapid detection and quantitation of neomycin phosphotransferase II in transgenic plants. Plant Molec. Biol. Rep. 10: 263–272.

Otten, L.A.B.M. & R.A. Schilperoort, 1978. A rapid microscale method for the detection of lysopine and nopaline dehydrogenase activities. Biochim. Biophys. Acta 527: 494–500.

Picton, S., S.L. Barton, M. Bouzayen, A.J. Hamilton & D. Grierson, 1993. Altered fruit ripening and leaf senescence in tomatoes expressing an antisense ethylene-forming enzyme transgene. The Plant J. 3: 469–482.

Theologis, A., 1992. One rotten apple spoils the whole bushel: The role of ethylene in fruit ripening. Cell 70: 181–184.

Euphytica **85**: 113–118, 1995.

Transformation studies in *Hordeum vulgare* using a highly regenerable microspore system

W.A. Harwood, S.J. Bean, D.-F. Chen, P.M. Mullineaux & J.W. Snape
John Innes Centre, Colney, Norwich NR4 7UH, U.K.

Key words: Hordeum vulgare, isolated microspores, particle bombardment, transformation

Summary

A highly regenerable, isolated microspore system for barley, *Hordeum vulgare* L. cv. Igri, has been developed which is amenable to transformation studies using particle bombardment. The system allows DNA to be delivered to microspores at the single cell stage and both transient and stable transformation events have been demonstrated. The potential advantages of using isolated microspores as the target tissue in routine transformation systems are discussed.

Introduction

Recent advances in particle gun technology have led to considerable progress towards the routine transformation of cereals. Particle bombardment is not limited by the type of target tissue unlike the direct DNA uptake methods (PEG-mediated uptake and electroporation) which in general depend on the availability of protoplasts to plant systems. By far the most popular target tissue for particle bombardment is immature embryos or cultures derived from them. All of the major cereals have now been transformed using this target tissue, rice (Christou et al., 1991; Li et al., 1993), maize (Gordon-Kamm et al., 1990; Fromm et al., 1990), wheat (Vasil et al., 1992; Weeks et al., 1993; Becker et al., 1994; Nehra et al., 1994), oats (Somers et al., 1992), and barley (Wan & Lemaux, 1994). We have examined an alternative target tissue, isolated microspores, for transformation studies of barley. Barley has a long history of anther culture used to produce doubled haploids and now a number of procedures have been published for the culture of isolated microspores (Kohler & Wenzel, 1985; Ziauddin et al., 1990; Olsen, 1991; Hoekstra et al., 1992; Ziauddin et al., 1992; Hoekstra et al., 1993; Kao, 1993). The use of barley microspores as a target for DNA uptake induced by PEG was examined by Kuhlmann et al. (1991) and transient expression of the *gus* gene was indicated following the treatment. Bolik & Koop (1991) and Olsen (1991) used isolated microspores of barley in a study examining the feasibility of introducing DNA by microinjection, however no DNA delivery was demonstrated. Wan & Lemaux (1994) carried out provisional transformation studies with barley microspores using particle bombardment and observed transient GUS expression but did not pursue this approach as other target tissues, immature embryos or microspore-derived embryos, enabled them to obtain transformed plants. Here we describe a method for the delivery of DNA to isolated microspores at the single cell stage, using the PDS1000 He particle delivery system, which allows microspores to retain their high regeneration potential following particle bombardment. The *gus* gene was used to demonstrate both transient and stable transformation events.

Materials and methods

Microspore culture

Donor plants for microspore culture were sown in a 2:2:1 mixture of Levington M3 compost:Perlite:Grit and grown in a controlled environment room at 12° C, 16 h day, 80% relative humidity with light levels of 600 μmol·m^{-2}·s^{-1}. At the 2–3 leaf stage seedlings were vernalised for 10 weeks at 4° C, 16 h day and 160 μmol·m^{-2}·s^{-1}. Plants were then returned to the original growth conditions, transferred to 11 cm pots after 1 week and fed with 100 ml of a 2:1 mixture of Chempak No 1:Chempak No 2 twice weekly.

When microspores were at the mid to late uninucleate stage, spikes were harvested, surface sterilised with 70% ethanol and the anthers removed and placed in 0.3 M mannitol pH 5.8, 100 anthers in 5 ml mannitol solution. Anthers at the top and bottom of the spike contained microspores at different developmental stages and therefore were not used. Dishes were sealed and kept in the dark at 25° C for 3–4 days. After this time a small proportion of the microspores had been released into the mannitol. These were collected into a 15 ml centrifuge tube and the microspores remaining within the anther walls were released by gentle squashing with a flattened glass rod. Microspores released in this way were collected by washing twice with 3 ml mannitol solution and added to the centrifuge tube. Cells were gently pelleted by spinning at 100 G for 4 mins and resuspended in FW medium, a modification of the medium used by Foroughi-Wehr et al. (1976) containing 0.175 M maltose as the carbon source (Hunter, 1987), 1 mg l^{-1} phenylacetic acid (PAA) and 0.2 mg l^{-1} kinetin. The medium was adjusted to pH 5.8 and filter sterilised.

Initial microspore culture density was 1×10^5 ml^{-1} in 5 cm Petri dishes, 3–4 ml per dish. Dishes were shaken at 32 rpm and kept at 25° C in the dark. After 2 weeks in liquid medium, microspore-derived calli were transferred to hydrophobic-edged cellulose nitrate filters (0.45 μm pore size, Sartorius) over FW medium solidified with 0.8% SeaPlaque agarose (FMC Bioproducts) in 6 cm dishes. Dishes were kept at 25° C under low light, 16 h day. After a further 2 weeks the microspore-derived material was removed from the filters and plated directly onto FW medium in 9 cm plates. Green shoots were removed to 150 ml jars (Sterilin) containing FW medium and once they were well established they were vernalised for 10 weeks. After a recovery period at 25° C for 2 weeks they were transferred to the compost mixture described above and kept in an incubator (Vindon Scientific Ltd.) at 15° C under individual propagators for 2 weeks. Plants were then grown to maturity under the same conditions as the donor plants.

Particle bombardment

Particle bombardment was carried out using the PDS1000 He gun (BioRad). Microspores were bombarded with 1 μm gold particles coated with plasmid pEmuGN (kindly provided by D.I. Last, CSIRO, Canberra) which contains the gus gene under control of the Emu promoter. 60 mg gold particles were washed three times with 1 ml 100% ethanol followed by 3 washes in sterile distilled water. Particles were resuspended in 1 ml sterile distilled water and stored at 4° C for up to 2 weeks. Gold particles were coated with DNA by adding 5 μl DNA (1 μl/μl), 50 μl 2.5 M $CaCl_2$ and 20 μl 0.1 M spermidine to a 50 μl aliquot of suspended gold particles. Particles were continually mixed by vortexing during the additions and then allowed to continue mixing for 3 minutes after which they were pelleted and the supernatant removed. After washing in 250 μl 100% ethanol, particles were resuspended in 60 μl 100% ethanol.

Microspores for particle bombardment were transferred to cell culture inserts (3 μm pore size, Falcon) 1–4 days after isolation. Inserts each containing 1–1.5 × 10^5 microspores were held in 6-well plates (Falcon) with each well containing a total volume of 1.5 ml FW medium. 0.5 ml of medium was added to the base of each well before the microspores were added to the insert. This allowed the medium from the insert to be drawn through, leaving the microspores distributed across the membrane of the insert but still in contact with the culture medium beneath it. For bombardment, each insert was removed in turn to a deep 5 cm dish (Sterilin). Immediately after bombardment the insert was returned to the same well of the 6-well plate so that the microspores were only out of contact with the culture medium for a very short time.

The optimised set of bombardment conditions for microspores included 450 psi rupture discs, a distance between the rupture disc retaining cap and the macrocarrier cover lid (measured with PDS-1000 He gap tools) of 6 mm, stopping screen support in the bottom position in the fixed nest giving a macrocarrier flight distance of 13 mm, the target tissue placed 6 cm from the stopping screen, a gold particle size of 1 μ and a chamber vacuum of 26 ins Hg.

Histochemical GUS assay

Microspore-derived tissue was incubated overnight at 37° C in a 5-Bromo-4-chloro-3-indolyl-β-D-glucuronic acid (X-gluc) solution prepared by dissolv-

Table 1. Effect of rupture disc strength on transient GUS expression

Experiment number	Rupture disc-breakage point (psi)	Number of times bombarded	Number of individual blue cells
1	900	1	451
2	900	2	182
3	900	2	428
4	450	1	3934
5	450	2	5150
6	450	2	1435

ing 25 mg of X-gluc in 250 μl Dimethyl sulphoxide then adding 40 ml 0.1 M phosphate buffer, pH 7 and 10 ml methanol. Alternative preparations were as above but without methanol, or without methanol but including 0.5 mM potassium ferricyanide and potassium ferrocyanide. X-gluc stocks were filter sterilised and stored at - 20° C

Results

Frequencies of microspore isolation and regeneration

The microspore isolation procedure described allowed the recovery of 70% of the total microspore population compared with only 15% from a shed isolation procedure, when only those shed into the mannitol solution without additional disruption of the anthers were collected and 34% when a mechanical isolation was used. Another advantage of the isolation procedure was that it yielded a very clean microspore population with little debris present. Freshly isolated microspores are shown in Fig. 1a. After 4 days a sub-population of enlarged microspores was observed which had undergone nuclear division, these represented 10% of the total population and usually underwent cell division and eventually formed embryos or callus (Fig. 1b).

An average of 400 green plants were regenerated from each 100 anthers used for microspore culture but the potential exists to considerably increase this number, and an example of a regeneration plate is shown in Fig. 1c. One of the advantages of the microspore system is that 95% of the regenerants were green and only 5% albino. From anther culture the percentage of albinos is 40%, the marked improvement from the microspore system may be due to the rapid regeneration time, as little as 6 weeks from isolation to plantlet; this does not allow the accumulation of genetic instabilities which can occur during long tissue culture phases.

Optimisation of particle bombardment

The PDS1000 He particle gun has a number of variable parameters which were optimised for DNA delivery to microspores. This was done by bombarding microspores with the plasmid pEmuGN which contains the *gus* gene, encoding β-glucuronidase, under the control of the Emu promoter which is based on a truncated maize *Adh1* promoter with multiple copies of the Anaerobic Responsive Element from the maize *Adh1* gene and ocs-elements from the octopine synthase gene of *Agrobacterium tumefaciens* (Last et al., 1991). Microspores to be used for the assay of GUS activity 2 or 3 days after bombardment were left in the cell culture inserts and during the assay, the culture medium in the base of the well was simply replaced with X-gluc solution which allowed the distribution of transient expression events to be examined. The single variable which had the most dramatic effect on transient expression levels was the rupture disc. Levels of transient expression increased as the pressure at which the disc ruptured decreased. Table 1 shows the effect of reducing the rupture disc from 900 psi to 450 psi, which is the lowest disc available. Although there is variation between shots, 450 psi rupture discs consistently gave the highest transient expression. Figure 1d shows cell culture inserts containing microspores bombarded using 900 psi and 450 psi discs and controls bombarded without DNA. The visible dark areas contain large numbers of GUS expressing cells whereas towards the edge of the membrane there are many individual blue microspores. Near the centre of the membrane there was some leakage of the blue GUS reaction product from the microspores. This occurred with or without methanol in the X-gluc solution. When potassium ferricyanide and potassium ferrocyanide were added there was no leakage from the microspores, however only those strongly expressing GUS were easily visualised. Figure 1e shows transient expression in microspores 3 days after bombardment. Care was taken to discount any microspores which could have been stained by leakage from neighbouring cells. Controls bombarded without DNA showed no staining. Other variable parameters in the gun had a smaller effect, with the position of the target tissue in the chamber being

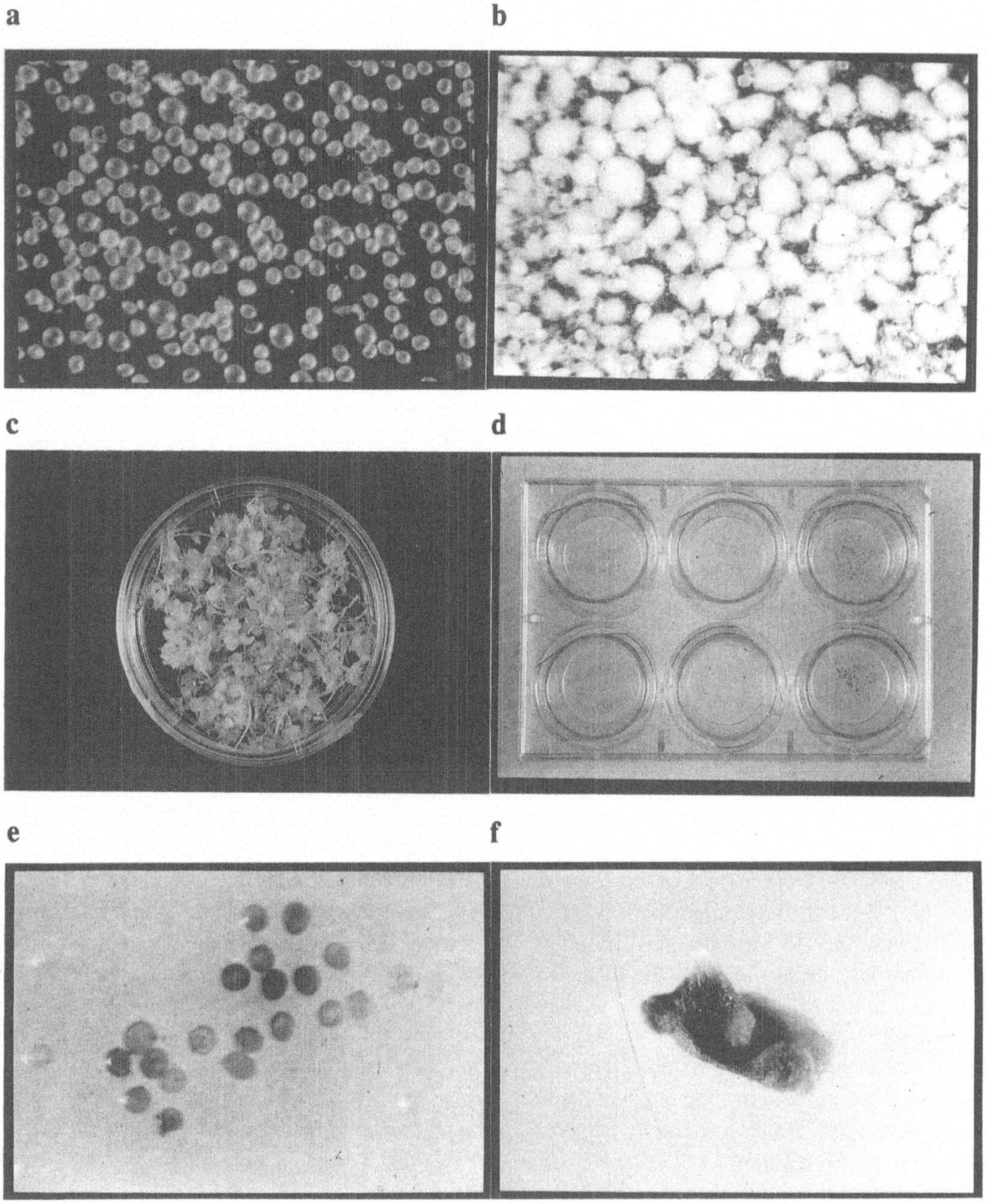

Fig 1 (a) Freshly isolated microspores of barley cv Igri (b) Microspore derived embryos/callus after 2 weeks in culture (c) Regeneration from microspore derived material (d) Six cell culture inserts following bombardment and staining for GUS activity, the left pair were bombarded without DNA, the middle pair bombarded with plasmid pEmuGN using 900 psi rupture discs and the right pair of inserts were bombarded with plasmid pEmuGN using 450 psi rupture discs (e) Bombarded microspores showing transient GUS expression (f) A microspore-derived embryo expressing GUS

the next most important (data not shown). A detailed description of the PDS1000 He gun is given by Kikkert (1993).

Using the optimised set of conditions described, transient expression over time after bombardment was examined by counting the number of GUS positive microspores/embryos from 3 to 21 days after bombard-

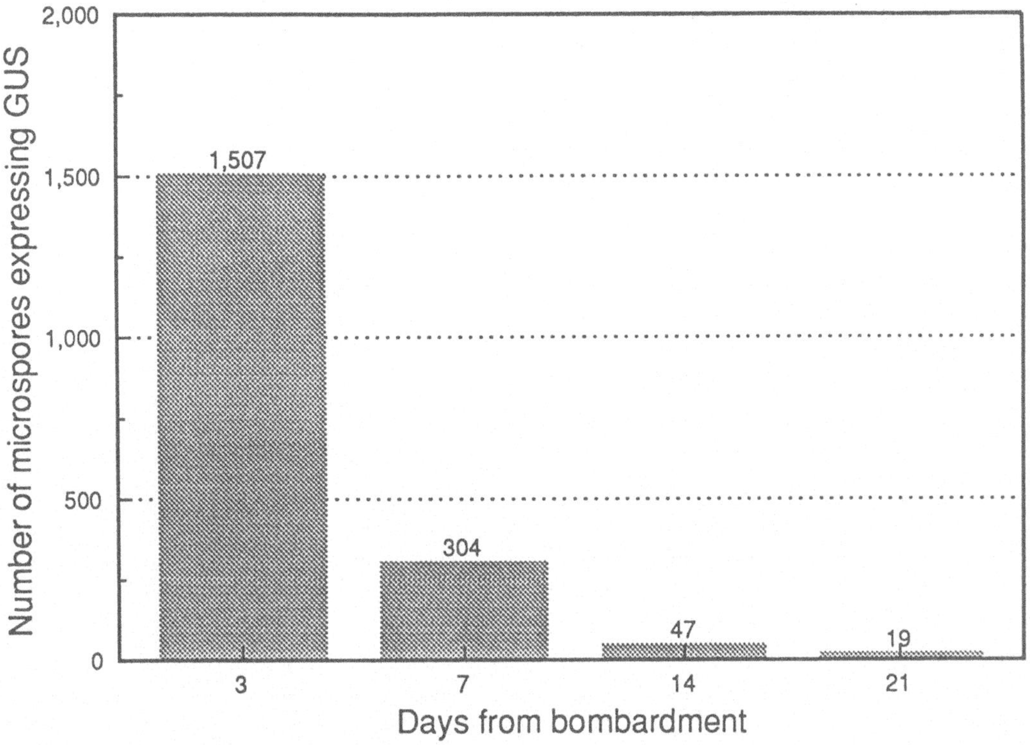

Fig. 2. GUS expression in microspores from 3 to 21 days after bombardment. Each bar represents the average from 4 individual bombardments.

ment (Fig. 2). As late as 21 days following bombardment, single GUS expression microspores could still be seen along with microspore-derived embryos/calli which strongly expressed GUS (Fig. 1f). These large GUS positive structures were assumed to represent stable transformation events and were observed at a frequency of one per bombardment. The time between microspore isolation and bombardment was shown to have an important effect on transient expression levels. In most cases best results were obtained by bombarding microspores two days after isolation. However, there was some variation between populations, presumably due to the exact stage which the microspores had reached and more detailed examination of the population may allow the best time for bombardment to be accurately determined.

Discussion

The potential of the barley isolated microspore system in providing a suitable target tissue for transformation has been demonstrated. The advantages of this culture system include the highly regenerable nature of microspore cultures. By paying particular attention to density of the microspore cultures and the osmolarity of the medium, the regeneration frequencies can be increased even more (Hoekstra et al., 1993). The fact that the majority of the regenerants are green is another important advantage. Wan & Lemaux (1994) found that of 91 transgenic barley callus lines generated from immature embryos or microspore derived embryos, 41 were able to regenerate only albino plants. The time which these target tissues spend in tissue culture prior to regeneration is thought to influence the level of albinism and would explain why, with the very rapid regeneration from microspores, albinos are not a problem. Also the system would allow homozygous transgenic plants to be produced in only one generation which would offer advantages to breeders utilising transgenic material. Culture of microspores is also possible in other barley varieties although the efficiency is lower than with cv. Igri.

Barley anther cultures have previously been used as the target for particle bombardment and DNA delivery (Creissen et al., 1990). The ability to introduce

DNA to single microspores soon after isolation, however, means that the material has not been adversely affected by tissue culture conditions prior to transformation and regenerated transformants should not be chimeric. There are some problems with the selection of transformed material from microspore cultures, as we have found that the regeneration potential of the microspores is adversely affected by phosphinothricin (PPT) commonly used as a selective agent. Use of alternative compounds introduced at the correct stage should overcome this problem.

The conditions necessary for the introduction of DNA to microspores using the PDS1000 He particle gun and to obtain transient and stable transformation events have been demonstrated in this preliminary study. Although immature embryos of barley have been shown to provide good target tissue for particle bombardment and have led to the production of transgenic plants (Wan & Lemaux, 1994) this alternative system may be more suitable for utilisation by breeders in some cases because of its unique advantages. Work is now in progress to isolate stably-transformed plants from material derived from bombarded microspores.

Acknowledgements

This work was supported by an AFRC-DTI Link programme involving: Advanced Technologies (Cambridge) Ltd., Agricultural Genetics Co., Ceiba Geigy R & D Seeds and Unilever Research.

References

Becker, D., R. Brettschneider & H. Lorz, 1994. Fertile transgenic wheat from microprojectile bombardment of scutellar tissue. The Plant Journal 5: 299–307.

Bolik, M. & H.U. Koop, 1991. Identification of embryogenic microspores of barley (*Hordeum vulgare* L.) by individual selection and culture and their potential for transformation by microinjection. Protoplasma 162: 61–68.

Christou, P., T.L. Ford & M. Kofron, 1991. Production of transgenic rice (*Oryza sativa* L.) plants from agronomically important Indica and Japonica varieties via electric discharge particle acceleration of exogenous DNA into immature embryos. Biotechnology 9: 957–962.

Creissen, G., C. Smith, R. Francis, H. Reynolds & P. Mullineaux, 1990. *Agrobacterium*- and microprojectile-mediated viral DNA delivery into barley microspore-derived cultures. Plant Cell Rep. 8: 680–683.

Foroughi-Wehr, B., G. Mix, H. Gaul & H.M. Wilson, 1976. Plant production from cultured anthers of *Hordeum vulgare* L. Z. Pflanzenzüchtg. 77: 198–204.

Fromm, M.E., F. Morrish, C. Armstrong, R. Williams, J. Thomas & T.M. Klein, 1990. Inheritance and expression of chimeric genes in the progeny of transgenic maize plants. Biotechnology 8: 833–839.

Gordon-Kamm, W.J., T.M. Spencer, M.L. Mangano, T.R. Adams, R.J. Daines, W.G. Start, J.V. O'Brian, S.A. Chambers, W.R. Adams, N.G. Willetts, T.B. Rice, C.J. Mackey, R.W. Krueger, A.P. Kausch & P.G. Lemaux, 1990. Transformation of maize cells and regeneration of fertile transgenic plants. The Plant Cell 2: 603–618.

Hoekstra, S., M.H. van Zijderveld, J.D. Louwerse, F. Heidekamp & F. van der Mark, 1992. Anther and microspore culture of *Hordeum vulgare* L. cv. Igri. Plant Sci. 86: 89–96.

Hoekstra, S., M.H. van Zijderveld, F. Heidekamp & F. van der Mark, 1993. Microspore culture of *Hordeum vulgare* L.: the influence of density and osmolarity. Plant Cell Rep. 12: 661–665.

Hunter, C.P., 1987. European Patent Application No. 0 245 898 A2.

Kao, K.N., 1993. Viability, cell division and microcallus formation of barley microspores in culture. Plant Cell Rep. 12: 366–369.

Kikkert, J.R., 1993. The biolistic PDS-1000/He device. Plant Cell, Tissue and Organ Culture 33: 221–226.

Kohler, F. & G. Wenzel, 1985. Regeneration of isolated barley microspores in conditioned media and trials to characterize the responsible factor. J. Plant Physiol. 121: 181–191.

Kuhlmann, U., B. Foroughi-Wehr, A. Graner & G. Wenzel, 1991. Improved culture system for microspores of barley to become a target for DNA uptake. Plant Breeding 107: 165–168.

Last, D.I., R.I.S. Brettell, D.A. Chamberlain, A.M. Chaudhury, P.J. Larkin, E.L. Marsh, W.J. Peacock & E.S. Dennis, 1991. pEmu: an improved promoter for gene expression in cereal cells. Theor. Appl. Genet. 81: 581–588.

Li, L., R. Qu, A. de Kochko, C. Fauquet & R.N. Beachy, 1993. An improved rice transformation system using the biolistic method. Plant Cell Rep. 12: 250–255.

Nehra, N.S., R.N. Chibbar, N. Leung, K. Caswell, C. Mallard, L. Steinhauer, M. Baga & K.K. Kartha, 1994. Self-fertile transgenic wheat plants regenerated from scutellar tissues following microprojectile bombardment with two distinct gene constructs. The Plant Journal 5: 285–297.

Olsen, F.L., 1991. Isolation and cultivation of embryogenic microspores from barley (*Hordeum vulgare* L.). Hereditas 115: 255–266.

Somers, D.A., H.W. Rines, W. Gu, H.F. Kaeppler & W.R. Bushnell, 1992. Fertile, transgenic oat plants. Biotechnology 10: 1589–1594.

Vasil, V., A.M. Castillo, M.E. Fromm & I.K. Vasil, 1992. Herbicide resistant fertile transgenic wheat plants obtained by microprojectile bombardment of regenerable embryogenic callus. Biotechnology 10: 667–674.

Wan, Y. & P.G. Lemaux, 1994. Generation of large numbers of independently transformed fertile barley plants. Plant Physiol. 104: 37–48.

Weeks, J.T., O.D. Anderson & A.E. Blechl, 1993. Rapid production of multiple independent lines of fertile transgenic wheat (*Triticum aestivum*). Plant Physiol. 102: 1077–1084.

Ziauddin, A., E. Simion & K.J. Kasha, 1990. Improved plant regeneration from shed microspore culture in barley (*Hordeum vulgare* L.) cv. Igri. Plant Cell Rep. 9: 69–72.

Ziauddin, A., A. Marsolais, E. Simion & K.J. Kasha, 1992. Improved plant regeneration from wheat anther and barley microspore culture using phenylacetic acid (PAA). Plant Cell Rep. 11: 489–498.

Euphytica **85**: 119–123, 1995.

Biolistic introduction of a synthetic *Bt* gene into elite maize

M. Hill, K. Launis, C. Bowman, K. McPherson, J. Dawson, J. Watkins, M. Koziel & M.S. Wright
Ciba Biotechnology, P.O. Box 12257, Research Triangle Park, NC 27709-2257, U.S.A

Key words: Bacillus thuringiensis, maize, microprojectile bombardment, transformation

Summary

A synthetic *Bt* gene encoding a truncated version of the CryIA(b) protein derived from *Bacillus thuringiensis* was successfully introduced into elite maize using microprojectile bombardment of immature embryos. The method used to initiate and identify transformation events is described. We describe the detailed parameters used for the Biolistics device as well as the plasmids used for the transformations. The plasmids contained the synthetic *Bt* gene driven by either the 35S CaMV promoter or a combination of two tissue-specific promoters, leaf and pollen, derived from maize. Specific conditions for the culture of Type I callus from immature embryos, the phosphinothricin (PPT) selection protocol, and the regeneration of plants are discussed. T0 and T1 plants were initially identified using the pH-dependent chlorophenol red test and/or the histochemical β-glucuronidase (GUS) assay. PCR and Southern data confirm the presence of the 35S CaMV promoter and the synthetic *Bt* gene.

Introduction

Cereal transformation via biolistic particle bombardment, and maize transformation in particular, remained elusive until recently (Potrykus, 1990). This report details the successful transformation of maize via microprojectile bombardment of immature embryos of an elite inbred. Transformants are recovered from Type I callus (Tomes, 1985) initiated after bombardment of the embryos. Other successful reports of maize transformation utilized suspension cultures (Fromm et al., 1990; Gordon-Kamm et al., 1990) or callus cultures of A188 × B73 crosses (Genovesi et al., 1992; Walters et al., 1992; Songstad et al., 1992).

Materials and methods

Source tissue

In 11 experiments, more than 2000 immature embryos of CG00526, a Lancaster-type maize inbred, were aseptically excised 14–15 days after pollination from surface-sterilized ears of greenhouse-grown plants. In each experiment, embryos, ranging in size from 1.5 to 2.5 mm in length, were cultured with the scutellum uppermost, on callus initiation medium, 2DG4 + 5 mg l^{-1} chloramben (Schweizerhal Inc., South Plainfield, NJ). 2DG4 medium is Duncan's (Duncan et al., 1985) 'D' medium modified to contain 20 mg l^{-1} glucose.

Preparation of DNA-coated microcarrier

Plasmid DNA was precipitated onto 1 μm gold microcarriers as described by the DuPont Biolistic manual (DuPont, 1990).

Two plasmids contained the synthetic *Bt* gene encoding for the CryIA(b) protein (Fig. 1) (Koziel et al., 1993). Plants from two transformation events were chosen for further study and for field evaluation. The two events, 171 and 176 were obtained using the synthetic *Bt* plasmids pCIB4431 or pCIB4418 (Fig. 1). The selectable marker for both events was supplied on pCIB3064 which carries a phosphinothricin acetyl transferase (*bar*) gene driven by the 35S promoter (Koziel et al., 1993). The DNA/gold mixture was prepared so as to deliver approximately 1 μg of

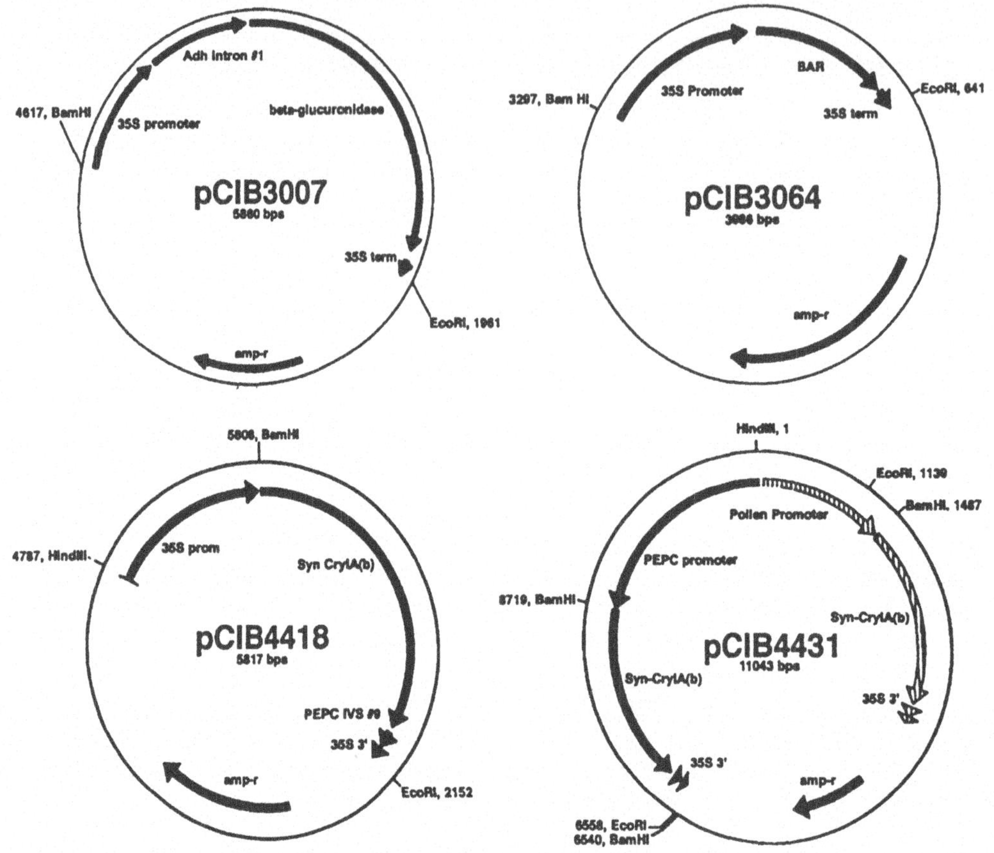

Fig. 1. Plasmid maps.

DNA per bombardment. The 171 event also contained pCIB3007 to encode the *gus* gene (Fig. 1).

Bombardment

Thirty-six embryos on a plate were bombarded using the Biolistics® PDS-1000HE device (DuPont, Wilmington, DE) according to the manufacturer's protocol. The tissue was placed on the shelf 8 cm below the stopping screen shelf. A 10 × 10 μm stainless steel screen hand-punched at Ciba or a 24 × 24 μm standard screen supplied by the Biolistic manufacturer was used with rupture discs of 1550 psi value for the bombardments. Following bombardment, the embryos were cultured in the dark at 25° C.

Callus formation

One to 14 days after bombardment, embryos were transferred to callus initiation medium containing 1–3 mg l^{-1} phosphinothricin (PPT) and incubated in the dark at 25° C. Embryos were scored for Type I embryogenic callus formation at 2 and 3 weeks after bombardment. Embryogenic tissue was transferred to callus maintenance medium, 2DG4 + 0.5 mg l^{-1} 2,4-dichlorophenoxyacetic acid (2,4-D) containing 1–3 mg l^{-1} PPT. Type I callus was subcultured every two weeks for a total of 12 weeks to fresh maintenance medium containing 3–10 mg l^{-1} of the selection agent.

Regeneration

Regeneration from the selected callus was initiated after 12 weeks on PPT. Type I callus was subcultured onto a modified Murashige & Skoog medium (MS) (Murashige & Skoog, 1962) containing 3% sucrose, 0.25 mg l^{-1} 2,4-D and 5 mg l^{-1} benzylaminopurine and cultured under 16 hours of light (50 μE $m^{-2}s^{-1}$), 8 hours dark, at 25° C. Two weeks later, the tissue was transferred to MS medium containing 3% sucrose without phytohormones. Plants regenerated during the following 4–10 weeks. Regenerated plants were cul-

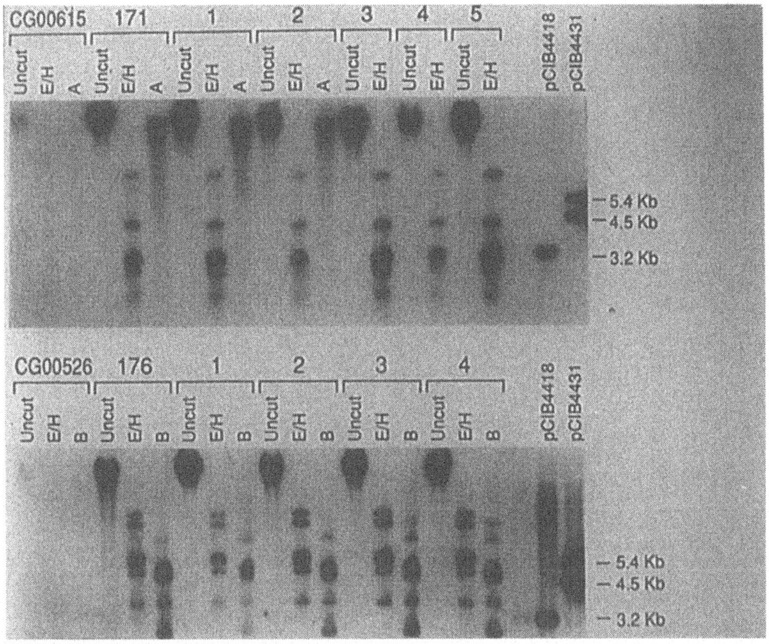

Fig. 2. Southern analysis of 176 event and progeny.

tured on MS medium modified to contain half the concentration of salts and 3% sucrose.

Assays for verification of transformation

Multiple leaf pieces from all plants were first tested histochemically for GUS expression (Jefferson, 1989), if applicable, and expression of the *bar* selectable marker gene using the chlorophenol red test (CR) (Kramer et al., 1993). In this assay, tissue that is resistant to PPT can actively grow on medium containing the compound and can acidify the pH of the medium which is indicative of healthy tissue. Within 4–7 days, resistant tissue changes the medium colour from deep red to bright yellow. PCR analysis (Perkin Elmer Cetus, 1991) followed the GUS and CR assays, and plants positive in these assays were transferred to the greenhouse for insect testing (Koziel et al., 1993) and maturation. Several plants were also tested by ELISA (Clark et al., 1986) for the synthetic *Bt* protein. Southern blot analysis was also done using standard methods (Sambrook et al, 1989), see Fig. 2.

Embryo rescue

To meet a field planting deadline, traditional harvesting methods could not be used. Embryos from T1 progeny transformed plants were 'embryo rescued' (Weymann et al., 1993). Fourteen to 16 days after pollination, the ear tip with 25–50 kernels was cut from the ear with a coping saw. The excised ear piece was surface-sterilized and individual embryos were excised and plated with the scutellum uppermost on B5 medium (Gamborg, 1968) containing 2% sucrose. Rescued seedlings were tested for the presence of the inserted genes in the same manner as the transformed parent.

Results

Table 1 lists the eleven experiments that yielded maize transformants when immature embryos of CG00526 were bombarded. Thirty-two percent of the total 2000 treated embryos initiated Type I callus, e.g., for the 171 event, 18% of the embryos bombarded produced Type I callus; for the 176 event, the callus initiation frequen-

Table 1. Experiments and bombardment conditions from which transformation events were recovered

Expt.	Embryos plated	Responding embryos	Events recovered	Selection conditions[1]	Plants regenerated
1	235	65	7	3/14	1355
2	129	64	1	1/5→3/35→10/63	251
3	102	59	1	1/3	171
4	36	19	1	1/3→3/33→10/60	465
5	140	80	4	1/3→3/33→10/60	2322
6	137	86	1	3/1	1676
7[2]	247	60	2	3/1	693
8[3]	249	44	2	3/1	515
9	251	37	1	3/1	727
10	162	46	1	3/3	679
11	318	86	2	3/3	6460
Total	2006	646	23		15314

[1] mg l^{-1} PPT day^{-1} applied after bombardment.
[2] Event 171 recovered.
[3] Event 176 recovered.

cy was 15%. Every embryo which produced callus was cultured separately thus representing an individual transformation event. Selection was carried out using 1–3 mg l^{-1} PPT at 1–14 days after bombardment. Since these were early experiments, the optimal selection protocol had not been determined (Table 1). A total of 15,314 plants were regenerated from the 11 experiments. All of the plants recovered were assayed by the CR test for the presence of the *bar* gene and by histochemical analysis for *gus* gene expression, if applicable. Positive plants identified by these assays were analysed by PCR for the presence of the 35S promoter and/or the *Bt* gene. Plants that assayed positive in all these tests were subsequently assayed by ELISA for the *Bt* protein and subsequently fed to insects.

Two events, 171 and 176, were chosen from the positive events for further development. Thirty-three plants were recovered from the 171 event and 38 from the 176 event. Plants that assayed positive in the CR and GUS (if applicable) assays were in turn assayed by PCR analysis for the presence of the 35S promoter on the selectable marker and the synthetic *Bt* gene. Plants that were positive by PCR assay were moved to the greenhouse for insect bioassay and maturation. For the 171 event, 25 plants were positive and 8 negative by the CR assay. All positive plants expressed the *gus* gene. For the 176 event, 8 plants were positive and 30 were negative. In eleven experiments, a total of twenty-three independent events were recovered or an average of two independent events per experiment.

All transformants were fertile, producing abundant pollen and full ears. Plants were backcrossed and outcrossed to several Ciba Seeds elite inbreds.

In some cases, embryo rescue was employed to produce the T1 hybrid seedlings for planting in the field. Leaf samples were taken from the seedlings for analysis by GUS histochemical assay, PCR analysis, CR assay, ELISA for CryIA(b) protein and insect bioassay with European corn borer (ECB) larvae. After analysis, almost 1,000 plants were shipped to the field.

Discussion

The results presented in this paper demonstrate that maize transformation can be routinely accomplished using immature embryos as the source tissue. In this study, independent transformants were recovered from 11 experiments, multiple transformation events were recovered from single experiments, and multiple plants were recovered from each event.

Interestingly, a range of culture parameters can be employed to achieve maize transformation via microprojectile bombardment of immature embryos. The selection agent (PPT) can be applied from 1 to 3 mg l^{-1} from 1 to 14 days after bombardment. Transformants can be recovered from 'step-up' as well as constant-level selection pressure.

It is possible that a range of bombardment conditions such as rupture disk pressure and distance to the target may also be applied. The success of this maize transformation procedure is likely to be due, in part, to the initial age and general good health of the immature embryos as well as the subsequent culture maintenance of the lines under selection pressure. Using a genotype with a proven history of high callus initiation frequency, simplifies the procedure.

Further, the use of embryo rescue techniques enabled us to meet deadlines for the first field testing of elite maize engineered for insect resistance. This procedure saved 30 days in recovering T0 progeny (T1 plants) for field evaluation and allowed for testing during the field season in which the transgenics were produced.

References

Clark, M.F., R.M. Lister & M. Bar-Joseph, 1986. Elisa Techniques. Methods Enzymol. 118: 742–766.

Duncan, D.R., M.E. Williams, B.E. Zehr & J.M. Widholm, 1985. The production of callus capable of plant regeneration from immature embryos of numerous *Zea mays* genotypes. Planta 165: 322–332.

E.I. du Pont de Nemours & Co., 1990. Biolistic PDS-1000 Particle Delivery System. Wilmington, DE.

Fromm, M.E., F. Morrish, C. Armstrong, R. Williams, J. Thomas & T.M. Klein, 1990. Inheritance and expression of chimeric genes in the progeny of transgenic maize plants. Bio/Technology 8: 833–839.

Gamborg, O., R. Miller & K. Ojima, 1968. Nutrient requirements of suspension cultures of soybean root cells. Exp. Cell Res. 50: 151–158.

Genovesi, D., N. Willetts, S. Zachwieja, M. Mann, T. Spencer, C. Flick & W. Gordon-Kamm, 1992. Transformation of an elite maize inbred through microprojectile bombardment of regenerable embryogenic callus. *In Vitro* Cell Dev. Bio. 28: 124A.

Gordon-Kamm, W.J., T. Spencer, M. Mangano, T. Adams, R. Daines, W. Start, J. O'Brien, S. Chambers, W. Adams, N. Willetts, T. Rice, C. Mackey, R. Krueger, A. Kausch & P. Lemaux, 1990. Transformation of maize cells and regeneration of fertile transgenic plants. The Plant Cell 2: 603–618.

Jefferson, R.A., 1989. The *GUS* gene fusion system. Nature 342: 837–838.

Koziel, M., G. Beland, C. Bowman, N. Carozzi, B. Crenshaw, L. Crossland, J. Dawson, N. Desai, M. Hill, S. Kadwell, K. Launis, K. Lewis, D. Maddox, K. McPherson, M. Meghji, E. Merlin, R. Rhodes, G. Warren, M. Wright & S. Evola, 1993. Field performance of elite transgenic maize plants expressing an insecticidal protein derived from *Bacillus thuringiensis*. Bio/Technology 11: 194–200.

Kramer, C., J. DiMaio, G. Carswell & R. Shillito, 1993. Selection of transformed protoplast-derived *Zea mays* colonies with phophinothricin and a novel assay using the pH indicator chlorophenol red. Planta 190: 454–458.

Murashige, T. & F. Skoog, 1962. A revised medium for rapid growth and bioassays with tobacco tissue cultures. Physiol. Plant 15: 473–479.

Perkin Elmer Cetus, 1991. Protocol for DNA amplification. Perkin Elmer Cetus Instruments, Norwalk, CT.

Potrykus, I., 1990. Gene transfer to cereals: an assessment. Bio/Technology 8: 535–542.

Sambrook, J., E.F. Fritsch & T. Maniatis, 1989. Molecular Cloning: A Laboratory Manual, Second edition. Cold Spring Harbor Laboratory Press.

Songstad, D., K. Lowe, S. Betz & J. Cabrera-Ponce, 1992. Callus cultures as alternative target tissues in microprojectile bombardment. Agronomy Abstracts, p. 198.

Tomes, D.T., 1985. Cell culture, somatic embryogenesis and plant regeneration in maize, rice, sorghum and millet. In: S.W.J. Bright & M.G.K. Jones (Eds). Cereal Tissue and Cell Culture, pp. 175–203. Martinus Nijhoff/Dr. W. Junk Publishers, Dordrecht.

Walters, D.A., C.S. Vetsch, D.E. Potts & R.C. Lundquist, 1992. Transformation and inheritance of a hygromycin phosphotransferase gene in maize plants. Plant Molec. Bio. 18: 189–200.

Weymann, K., K. Urban, D. Ellis, R. Novitzky, E. Dunder, S. Jayne, D. Murray, G. Jen & G. Pace, 1993. Isolation of transgenic progeny of maize by embryo rescue under selective conditons. *In Vitro* Cell Dev. Biol. 29: 33–37.

Euphytica **85:** 125–130, 1995.

Genetic transformation of *Eustoma grandiflorum* Griseb. by microprojectile bombardment

Laura Semeria[1], Anna Maria Vaira[2], Gian Paolo Accotto[2] & Andrea Allavena[1]
[1] *Istituto Sperimentale per la Floricoltura, C.so Inglesi, 508, 18038 Sanremo (IM), Italy;* [2] *Istituto Fitovirologia Applicata, CNR, Strada Delle Cacce, 73, 10135 Torino, Italy*

Key words: particle gun, *gus*, lisianthus, *nptII*, PCR, prairie gentian

Summary

Foreign DNA was introduced into cell suspension cultures and leaf tissue of *Eustoma grandiflorum* Griseb. (lisianthus) by microprojectile bombardment. For this purpose a low-cost bombardment device that uses a helium flux to accelerate microprojectiles was built. When cell suspensions were used, an average of 4.1 Kan resistant calli were recovered per shot after 4 months' cultivation on selective medium. Most of the Kan resistant plants regenerated from calli were positive to GUS assay. Both the *nptII* and *gus* genes were successfully amplified from alkali-treated leaves of putative transgenic plants by PCR analysis. Transgenic plants were not recovered from bombarded leaves. Considering the host range specificity of *Agrobacterium*, and the response of the species to plant regeneration from suspension culture, microprojectile bombardment is, at present, the most efficient procedure for genetic transformation of lisianthus.

Abbreviations: BA – 6-benzyladenine, Cx – cefotaxime, 2,4 D – (2,4-dichlorophenoxy) acetic acid, FDA – fluorescein diacetate, *gus* – β-glucuronidase, IAA – indole-3-acetic acid, IBA – indole-3-butyric acid, 2iP – (2-isopentenyl) adenine, Kan – kanamycin, *nptII* – neomycin phosphotransferase II, PCR – polymerase chain reaction

Introduction

The integration of foreign genes into plants has been achieved by several methods: *Agrobacterium* vectors, microprojectile bombardment, microinjection, electroporation and direct DNA uptake into protoplasts, cells or intact tissue.

The major obstacles to *Agrobacterium*-mediated gene transfer have been attributed to the host range specificity. The success of most of the other approaches is hampered, in crop species, by severe difficulties with multiple step plant regeneration procedures from protoplasts, single cells or calli. These problems can be solved by microprojectile bombardment. In fact many plant tissues, including cell suspensions, calli, meristems, embryos, pollen, somatic tissues of *in vitro* and even *in vivo* grown plants may be targeted by microprojectiles and some targets do not require *in vitro* re-differentiation steps for plant regeneration. Virtually any species might be transformed. Nuclear DNA together with mitochondrial and chloroplast DNA was successfully transformed. For all these distinguishing traits, microprojectile bombardment is considered a broadly-applicable system for gene delivery.

Eustoma grandiflorum Griseb. (Gentianaceae) commonly known as lisianthus or prairie gentian, is a flower crop introduced into cultivation as a cut flower and pot plant (Halevy & Kofranek, 1984). The species is amenable to micropropagation and for plant regeneration from somatic tissues (Ruffoni et al., 1993). Preliminary results on genetic transformation of lisianthus were reported by Handa (1992) and by Semeria & Allavena (1993). Both papers report abundant hairy root induction following co-cultivation with wild-type *A. rhizogenes* strains. Hairy roots spontaneously regenerated into full plants. *A. tumefaciens* showed a narrow host range specificity and a low frequency of transformation. In fact only three transgenic plants were regenerated, following co-cultivation of 300 explants of a specific lisianthus clone, with the strain A281 pKIWI105. The other vector/clone combinations tested were unable to produce transgenic plants (Semeria & Allavena, 1993).

In this paper we describe an efficient method for gene transfer into lisianthus, based on bombardment of cell suspension cultures with DNA-coated microprojectiles. An inexpensive device was built for this purpose, simplifying the model proposed by Finer et al. (1992).

Materials and methods

Plant materials

Cell suspensions and leaf explants, derived from a selected plant of the cv. Jodel Blue (Royal Sluis) were used. Cell suspensions were maintained on a rotary shaker (90 rpm) and subcultured weekly on CSM (Table 1). Leaf explants were excised from *in vitro* plants propagated on LPM.

Bombardment device

The home-made microprojectile accelerator employed was similar to the particle inflow gun described by Finer et al. (1992). The vacuum chamber was substituted in our device with a 25 cm diameter plastic desiccator (Kartell, Italy, cat. no. 250). We also used a manual electric switch instead of a timer relay. Other main components were: solenoid valve (Sirai, Italy, type Z323D); swinnex (Millipore, cat. no. SX0001300); vacuum gauge.

Microprojectile preparation and bombardment

Plasmid pKIWI105 (Janssen & Gardner, 1989), containing the *gus* and the *nptII* genes, was amplified in the *E. coli* strain DH5α and purified using a Plasmid Maxi Kit (Qiagen, Germany, cat. no. 12162).

Tungsten (Fluka cat. no. 95390: analysis no. 511945 and no. 295295-1) and gold (Aldrich cat. no. 32.658-5) microprojectiles were coprecipitated with plasmid DNA according to Morikawa et al. (1989). Prior to bombardment, 2 μl of the microprojectile suspension was loaded in the center of the swinnex screen. To discharge microprojectiles on the target tissue by the helium flux, the solenoid valve of the device was opened for 200–300 ms using the electric switch. The target tissue was placed on the bottom of the desiccator at a distance of 18 cm from the outlet of the swinnex. Six bombardments per treatment were performed (one bombardment/target).

Growth conditions and media composition

All plant material was grown *in vitro* at 23° C to 25° C under 16 h photoperiod using 40 W Osram cool white fluorescent tubes, providing a quantum flux density of 28 $\mu E \cdot m^{-2} \cdot s^{-1}$. MS basal medium (Murashige & Skoog, 1962) was used in all experiments. Additions to basal medium are reported in Table 1. Agar (1%) was added to solidify media when required. The pH of the media was adjusted to 5.7 before autoclaving. Antibiotics were filter-sterilized and added to autoclaved media.

Tissue culture procedure

Aliquots of 3 ml of 4–7 days old cell suspension cultures were dispensed on 50 mm filter disks (Whatman no. 4) and vacuum dried on a Buchner funnel for 3 to 5 s. The filters were transferred into Petri dishes without substrate for bombardment. Filters were then moved to agarized selective CIM (Table 1). Nine transfers were performed over a period of 4 months. The concentration of Kan in CIM was raised from the initial level of 50 $mg \cdot l^{-1}$ to 100 $mg \cdot l^{-1}$ (first subculture) and to 300 $mg \cdot l^{-1}$ (5th subculture). Putative transgenic calli, with a diameter of 3–5 mm, were transferred on solidified CRM plus Kan (300 $mg \cdot l^{-1}$). Putative transgenic regenerated plants were micropropagated and rooted on appropriate media (Table 1).

When leaves from lisianthus plantlets were used for microprojectile bombardment, 10–12 explants were layered in the centre of Petri dishes, shooted and transferred to agarized LRM for direct regeneration via organogenesis. Explants were subcultured every two weeks for three months.

Cell viability test

The effect of the transformation treatments (plating, vacuum drying and bombardment) on lisianthus cell viability was measured by FDA test (Widholm, 1972).

Table 1. Composition of the media used for transformation of lisianthus. MS basic ingredients were included in all the media. Concentrations are expressed in $mg \cdot l^{-1}$ if not otherwise indicated

Medium code	Tissue culture steps	Additional ingredients	References
CSM	Maintenance of suspension cultures	2,4D (2) BA (0.5)	B. Ruffoni, personal communication
CIM	Callus induction from bombarded cells	Sorbitol (0.125 M) Mannitol (0.125 M) 2,4D (2) BA (0.5) Km (300) Cx (100)	Russell et al., 1992
CRM	Regeneration from transgenic calli	2iP (3) Cx (100) Km (50, 100, 300)	This work
LRM	Direct regeneration from leaves	2iP (10) IBA (0.1) Km (300) Cx (100)	Ruffoni et al., 1990
LPM	Micropropagation	BA (0.3) Km (300) Cx (100)	Damiano et al., 1986
RM	Plantlet rooting	IAA (1) Km (100) Cx (100)	Damiano et al., 1986

PCR analysis

Pieces of leaves were prepared for PCR analysis (Saiki et al., 1988) using the alkali treatment described in Klimyuk et al. (1993). *gus* and *nptII* specific primers were as described (Dong & McHughen, 1993). The expected size of amplified fragments was 1180 bp for *gus* and 679 for *nptII*. Each reaction was performed in a 50 μl volume, using the GeneAmp PCR Core Reagents (Perkin Elmer Cetus) with a final concentration of 0.5 μM for each primer and of 2.5 mM for $MgCl_2$. Forty cycles were performed, with denaturation at 95° C for 40 s, annealing at 60° C for 2 minutes and extension at 72° C for 2 minutes.

Statistical analysis

Data regarding the effect of different microprojectile types and DNA/microprojectile concentration on transient gene expression were analyzed with ANOVA; mean comparisons were made using Student-Newman-Keuls test.

Table 2. Effect of treatment requested for transformation (plating, vacuum drying and bombardment) on lisianthus cell viability measured by FDA test (Widholm, 1972). Cells were dispersed on Whatman # 4 paper disks placed in Petri dishes containing MS basal medium added with osmotic compound (0.125 M mannitol plus 0.125 M sorbitol) and growth factors (2 $mg \cdot l^{-1}$ 2,4D plus 0.5 $mg \cdot l^{-1}$ BA) in all combinations. Data are expressed as percentage of fluorescent (viable) cells (1: $\leq$ 20%; 2: 21–40%; 3: 41–60%; 4: 61–80%; 5: 81–100%)

Media composition		Applied treatment		
Growth factors	Osmotic components	Plating	Plating; vacuum drying	Plating; vacuum drying; bombardm.
-	-	3	2	1
+	-	3	3	2
-	+	4	3	3
+	+	5	4	4

Results and discussion

Laborious set-up procedures were not required to optimize the gun parameters since preliminary experiments revealed the high performances of the home-made helium gun. The device was easy to use and inexpensive.

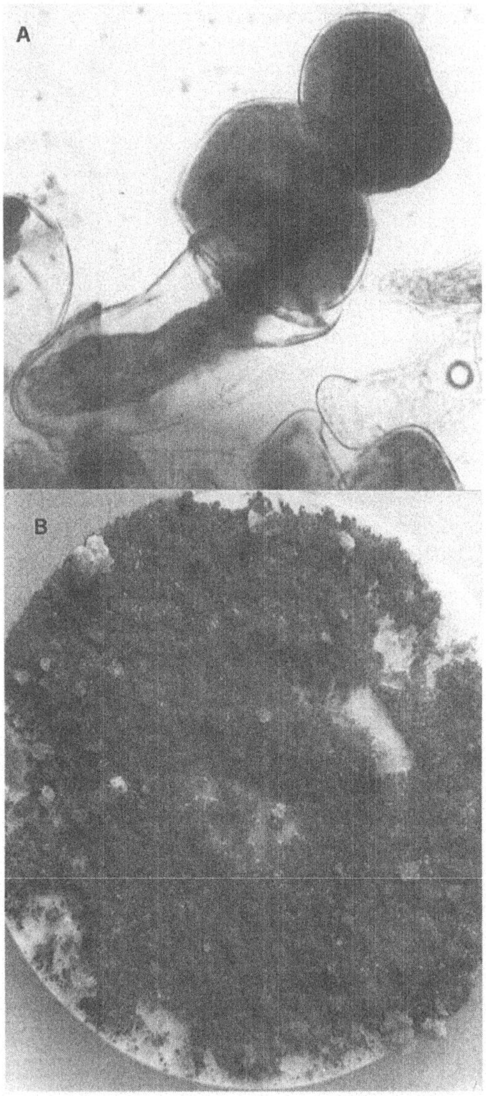

Fig. 1. Fate of lisianthus cells bombarded with pKIWI105 and grown on selective CIM. A: early division of cells expressing the *gus* gene (histochemical assay); B: Km resistant calli developed after 4 months of culture; most of them were also GUS positive.

Micro-organism contamination was not found on target plant materials as a consequence of bombardment. Minimum time elapsing between bombardments was 2–3 minutes, the limiting factor being the time needed to evacuate the desiccator (60 s) and to pressurize it (30 s).

Plant regeneration in lisianthus occurs from most of the somatic tissues (Ruffoni et al., 1993). Leaf explants and morphogenic cell suspension cultures were used in our experiments. Transient GUS expression was observed by histochemical assay (Jefferson et al., 1987) in both types of tissue two days after bombardment.

Table 3. Effect of two types of tungsten, with distinct analysis no, and DNA concentration on transient GUS expression in lisianthus cell culture after microprojectile bombardment. The average number of blue foci represent the mean of 6 bombardments (one bombardment per target)

Exper.	Tungsten (Fluka Cat. no. 95390)	DNA/Tungsten ($\mu g \cdot mg^{-1}$)	Transient GUS expression (number blue foci/shot)
1	An. no. 511945	2	1675 a*
	An. no. 295295-1	2	1195 b
2	An. no. 295295-1	2	1195 a
	An. no. 295295-1	4	983 b

* Within each experiment means followed by different letter are significantly different at $p \leq 0.05$ by Student-Newman-Keuls test.

Tissue manipulation and composition of media affected cell viability (Table 2). Notably, cell viability was improved by an osmotic treatment after bombardment in agreement with recent reports (Finer et al., 1992; Russell et al., 1992). When growth factors were included in the medium together with osmoticum, cell viability was maintained sufficiently high (60–80%). The CIM medium was therefore chosen for further experiments.

Tungsten microprojectiles were the most suitable to transform lisianthus cells. DNA-coated gold particles adhered to the swinnex screen and only a small portion of them was discharged on the target by the helium flux so that transformation efficiency was very low (5–10 blue spots/shot). A comparison between two types of microprojectiles (Table 3) revealed that those with the wider range of particle diameter and bigger average size (Fluka analysis no. 511945) were significantly more efficient in inducing transient gene espression. Like gold, small tungsten particles stick to the screen of the swinnex and are no longer available. The DNA concentration is related, at least in part, to the adhesion of the particles to the screen and to the formation of clusters. The efficiency of transformation was significantly improved, using the smaller microprojectiles, by co-precipitation at the lower DNA concentration (Exp. 2 of Table 3). As a consequence, a relatively large plasmid, like pKIWI105 may be not the best choice for plant transformation with the bombardment device. A

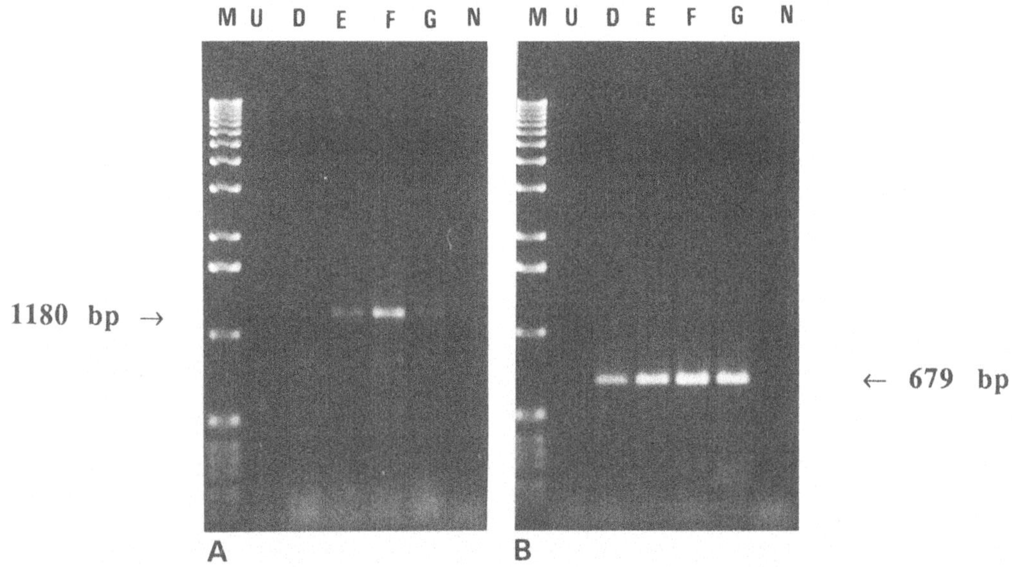

Fig. 2. PCR analysis of alkali-treated lisianthus leaf tissue from plants D, E, F and G. 10 μl aliquots of the reaction volume were loaded on a 1% agarose gel. M: size marker (1 Kb DNA ladder, BRL). U: untransformed lisianthus. N: reaction without template. Panel A: amplification of *gus* gene. Panel B: amplification of *nptII* gene. In all plants the fragments of the expected size were amplified but the GUS fragment from plant D in panel A was very faint and can hardly be seen on the printed photograph. Arrows indicate the expected size of amplified fragments.

smaller plasmid could provide more copies of the *gus* (or any other) gene per unit of total plasmid DNA, thus allowing the use of lower DNA/microprojectile ratio and the achievement of higher transformation efficiency.

ANOVA and mean comparison also revealed significant differences in transient transformation efficiency among shoots. The variability of the amount of microprojectiles loaded on the swinnex screen and the imperfect control of the solenoid valve activation by the manual electric switch, may have partially affected shot reliability. The average efficiency of the home-made helium gun was in any case satisfactory in comparison with more sophisticated devices.

When lisianthus cell suspensions were used, an average of 4.1 calli resistant to Kan were recovered per bombarded filter after 4 months' cultivation on selective medium (average of 15 bombarded filters) (Fig. 1). Plants were regenerated from calli following 1–3 months of growth on CRM and screened by the histochemical GUS assay). Most of the regenerated plants were GUS positive.

Three Kan resistant, GUS positive plants (plants D, E and G) together with one Kan resistant GUS negative plant (F) were screened for the presence of the *nptII* and *gus* genes by PCR (Fig. 2). In all plants the fragments of the expected size were amplified, demonstrating the presence of the two genes in the plant genome. However the GUS specific fragment was amplified with reduced efficiency, probably because of its size. Fragments larger than 1 kb are not always efficiently amplified from alkali-treated samples (Klimyuk et al., 1993). Plant F, although negative in the GUS histochemical assay, contained the *gus* gene according to PCR analysis. The foreign DNA could have been integrated in an untranscribed region of a chromosome, or it could have been mutated, losing, for example, transcription signals. Our results show that the alkali treatment of leaf tissues previously described for tomato (Klimyuk et al., 1993) can be successfully used for lisianthus. It was found that 25 mg$\cdot$l^{-1} Kan definitely inhibits the growth of normal lisianthus cells (data not shown) in reported experimental conditions and this, together with the PCR results confirms the transgenic nature of the resistant plants. PCR can be routinely used for rapid screening of particle bombardment-derived calli and plants because tissues are not contaminated with the bacterial DNA vector.

When leaf material was used, 50–150 blue foci per shot were observed by the histochemical assay. As a common feature one intensely blue stained cell was surrounded by 5–10 less coloured cells. In comparison stained foci on cell suspension cultures were of only 2–3 cells. Considering that in our suspension culture experiments, the rate of conversion of transiently expressing foci into stable integrations was 0.24% on average (4.1 resistant calli/1675 foci), we should expect 2–3 transgenic plants every 10 shots (1000 foci). No transgenic plants were recovered, however. It is possible that microprojectile-targeted cells were not, or no longer, totipotent.

Recent experiments demonstrated that lisianthus plantlets can be regenerated in one month from suspension culture via somatic embryogenesis (B. Ruffoni, personal communication). The utilization of the improved tissue culture protocol should accelerate the transformation process. Vain et al. (1993) found a 2.6-fold enhancement in transient *gus* gene expression and a 6.8-fold increase in stable transformation frequency by osmotic treatment of embryogenic suspension culture cells of maize before bombardment. Such treatment should be tested to verify the effect on lisianthus cell cultures.

Conclusions

The use of the home-made helium gun has provided a satisfactory method for transformation of lisianthus cells and recovery of stable transformed plants. The efficiency of the method can be further improved by small modifications to the device, using appropriate plasmids and choosing a tissue culture procedure that speeds up the plant regeneration steps. Efforts are in progress to introduce into lisianthus, agronomically important genes.

Acknowledgements

Plasmid pKIWI105 was a kind gift from R. Gardner and B. Janssen, Aukland, New Zealand. We thank our colleague B. Ruffoni, for providing lisianthus cell suspension.

References

Damiano, C., P. Curir, P. Esposito & B. Ruffoni, 1986. La propagazione *in vitro* di *Lisianthus russelianus* Hook. Ann. Ist. Sper. Flor. Sanremo Vol. XVII-1.

Dong, J.Z. & A. McHughen, 1993. An improved procedure for production of transgenic flax plants using *Agrobacterium tumefaciens*. Plant Sci. 88: 61–71.

Finer, J.J., P. Vain, M.W. Jones & M.D. McMullen, 1992. Development of the particle inflow gun for DNA delivery to plant cells. Plant Cell Rep. 11: 323–328.

Halevy, A.H. & A.M. Kofranek, 1984. Evaluation of Lisianthus as a new flower crop. HortScience 19: 845–847.

Handa, T., 1992. Regeneration and characterization of prairie gentian (*Eustoma grandiflorum*) plant transformed by *Agrobacterium rhizogenes*. Plant Tissue Culture Letters 9: 10–14.

Janssen, B. & R.C. Gardner, 1989. Localized transient expression of GUS in leaf discs following cocultivation with *Agrobacterium*. Plant Mol. Biol. 14: 61–72.

Jefferson, R.A., T.A. Kavanagh & M.W. Bevan, 1987. GUS fusion: – glucuronidase as a sensitive and versatile gene fusion marker in higher plants. EMBO J. 6: 3901–3907.

Klimyuk, V.I., B.J. Carrol, C.M. Thomas & J.D.G. Jones, 1993. Alkali treatment for rapid preparation of plant material for reliable PCR analysis. The Plant Journal 3: 493–494.

Morikawa, H., A. Iida & Y. Yamada, 1989. Transient expression of foreign genes in plant cells and tissues obtained by a simple biolistic device (particle gun). Appl. Microbiol. Biotechnol. 31: 320–322.

Murashige, T. & F. Skoog, 1962. A revised medium for rapid growth and bioassay with tobacco tissue cultures. Physiol. Plant. 15: 473–497.

Ruffoni, B., C. Damiano, F. Massabó & P. Esposito, 1990. Organogenesis and embryogenesis in *Lisianthus russelianus* Hook. Acta Horticulture 280: 83–88.

Ruffoni, B., A. Giovannini, L. Semeria, F. Massabó, C. Costantino & A. Allavena, 1993. Tissue culture in lisianthus (*Eustoma grandiflorum*). In: T. Schiva & A. Mercuri (Eds). Creating Genetic Variation in Ornamentals. Proc. XVII Eucarpia Symp. Sanremo, Italy, 1–5 March 1993.

Russell, J.A., K.R. Mihir & J.C. Sanford, 1992. Major improvement in biolistic transformation of suspension-cultured tobacco cells. *In Vitro* Cell Dev. Biol. 28: 97–105.

Saiki, R.K., D.H. Gelfand, S. Stoffel, S.J. Scharf, R. Higuchi, G.T. Hom, K.B. Mullis & H.A. Erlich, 1988. Primer directed amplification of DNA with thermostable DNA polymerase. Science 239: 487–491.

Semeria, L. & A. Allavena, 1993. Genetic transformation of *Eustoma grandiflorum* Griseb. In: T. Schiva & A. Mercuri (Eds). Creating Genetic Variation in Ornamentals. Proc. XVII Eucarpia Symp. Sanremo, Italy, 1–5 March 1993.

Widholm, J.M., 1972. The use of fluorescein diacetate and phenosafranine for determining viability of cultured plant cells. Stain Tech. 47: 189–194.

Vain, P., M.D. McMullen & J.J. Finer, 1993. Osmotic treatment enhances particle bombardment-mediated transient and stable transformation of maize. Plant Cell Rep. 12: 84–88.

Euphytica **85:** 131–134, 1995.

Some methodological aspects of apple transformation by *Agrobacterium*

J.G. Schaart, K.J. Puite, L. Kolova[1] & N. Pogrebnyak[2]
DLO-Centre for Plant Breeding and Reproduction Research (CPRO-DLO), Department of Cell Biology, P.O. Box 16, NL-6700 AA Wageningen, The Netherlands; [1] *Fruitgrowing Research Institute, Plovdiv, Bulgaria;* [2] *Institute of Cell Biology and Genetic Engineering, Kiev, The Ukraine*

Key words: apple, transformation, *Agrobacterium*, preculture, azacytidine

Summary

Leaf explants of apple cvs Gala and Golden Delicious were infected with the *Agrobacterium tumefaciens* strain AGL0(pMOG410). The effects of a 2 d preculture of the explants before infection and the addition of 5-azacytidine to the selection medium were studied. The percentages of GUS-positive explants after 5 w did not significantly alter due to these treatments. One of the 'Gala' shoots, which was removed from a leaf explant cultured for 8 w on selection medium, proved to be GUS-positive and will be analyzed further. In general, however, it should be concluded that regeneration of transgenic shoots directly from leaf tissue was not very effective.

Abbreviations: MS – Murashige and Skoog medium, GD – 'Golden Delicious'

Introduction

Transformation of apple has been successful for a number of cultivars, such as Greensleeves (James et al., 1989), M26 (Maheswaran et al., 1992), and Red Delicious (Goodwin et al., 1993). Use was made of leaf explants and *Agrobacterium tumefaciens*-mediated gene transfer with the *nptII* gene as selection marker. Transgenic shoots appeared after selection on kanamycin-containing medium either on callus formed ('Red Delicious') or directly from the leaf tissue (M26). This last system, if successful, should be preferred in order to avoid culture-induced genetic variation.

In this report results are given of *A. tumefaciens*-mediated transformation of the apple cultivars Gala and Golden Delicious, using leaf explants. The experiments were aimed at testing whether transgenic shoots can be obtained directly on leaf tissue without an intermediate callus phase.

Two other factors of interest were studied: the use of a preculture of the explants before *Agrobacterium* infection to enhance the number of de-differentiating cells along the cut edge at the time of infection, and the use after co-cultivation with *Agrobacterium* of 5-azacytidine, being a cytosine methylation inhibitor. Preculture of explants was found to be effective in the transformation of other species (flax, McHughen et al., 1989; *Arabidopsis thaliana*, Sangwan et al., 1992). A dramatic increase in the number of transformants of *A. thaliana* was found using 5-azacytidine (Mandal et al., 1993).

Materials

In vitro shoot cultures of cvs Gala and Golden Delicious were initiated from 1–2 cm shoot tips, after sterilization with a 0.7% solution of sodium hypochloride for 10 min, followed by thorough washing with sterile water. Proliferation with monthly subculture took place on a shoot propagation medium (SPM) of MS with 3% sucrose, 1 mg/l BAP, 0.1 mg/l IBA and 0.9% agar, at 22–24° C and 9 W/m^2 (3000 lux).

Leaf segments (3–4 mm) from the first four unfolded leaves of 4–6 w old shoot cultures were cut transversely to the midrib and placed with the adaxial side on shoot regeneration medium (SRM). This medium consisted of MS with 3% sorbitol, 2.2 mg/l thidiazuron and 1 mg/l NAA, solidified with 0.3% gelrite.

For transformation, the *A. tumefaciens* strain AGL0 (pMOG410) was used. AGL0 is a supervirulent strain derived from A218, and the plasmid pMOG410 contains the selection marker *nptii* with *nos* promoter and the reporter gene *gus*-intron with CaMV35S promoter between the left and right border of the T-DNA. Control experiments were carried out using the empty AGL0 strain, while regeneration controls were also included.

A. tumefaciens infection of the leaf explants was performed after 2 d preculture on SRM, or without preculture. The period of 2 d was chosen based on shoot regeneration experiments with leaf explants 0–4 d on MS with 3% sucrose, 0.8% agar and only NAA (0.2 mg/l) before transfer to a similar medium, but with BAP (5 mg/l) instead of NAA (see Results). Explants were immersed in bacterium suspension, diluted with MS containing 3% sucrose (pH 5.2) to an OD of 0.3–0.5 at 600 nm for 4–6 min. After 2 d co-cultivation in the dark at 25° C, the explants were transferred to a selection medium of SRM with 50 mg/l kanamycin + 200 mg/l cefotaxim. Explants were cultured for 2 w in the dark before being exposed to normal light conditions.

In a number of experiments 3 mg/l 5-azacytidine was added over 3 d to the selection medium starting 3 d after the end of the co-cultivation period. All these explants routinely had a 2 d preculture before *A. tumefaciens* infection.

After 4 w explants were transferred to fresh selection medium containing 3% sucrose and 0.8% agar instead of sorbitol and gelrite, respectively. In order to determine the transformation frequency, a GUS assay (Jefferson et al., 1987) was performed 5 w after *A. tumefaciens* infection. Shoots appeared on the leaf explants near the cut edge and were transferred 6–8 w after *A. tumefaciens* infection to SPM + 200 mg/l cefotaxim with or without 25 mg/l kanamycin.

Table 1. Effect of a preculture of the leaf explants before *Agrobacterium* infection on transformation efficiency, i.e. the percentage of GUS-positive explants, the number of GUS spots per GUS-positive explant (n = 12 explants for 'Gala' and n = 24 explants for 'Golden Delicious'), and the number of putative transgenic shoots (n = 170 explants)

	Percentage of GUS$^+$ explants (GUS spots/GUS+ explant)				Shoots	
	0 d	2 d	2 d/0 d		0 d	2 d
Gala	96 (4.3)	100 (5.4)	1.0 (1.3)		15	24
GD	61 (3.0)	50 (3.4)	0.8 (1.1)		6	1
	56 (8.2)	43 (4.9)	0.8 (0.6)		3	1
	30 (2.3)	30 (2.3)	1.0 (1.0)		0	1
	54 (2.2)	75 (1.7)	1.4 (0.8)		1	0
		Mean	1.0 (1.0)	Total	25	27

Results and discussion

Preculture

Figure 1 gives the percentage of shoot regenerating leaf explants of cv. Golden Delicious when these were cultured on a medium with only NAA for 0–4 d prior to transfer to a second medium with BAP as the only hormone present. Shoot induction is highest for the explants at least after 1 or 2 d on the NAA-containing medium. The 1–2 d period may reflect a period of developmental changes of the tissue at the cut edge, such as cell de-differentiation, needed to reach competence for shoot induction. Accordingly, as competence for transformational events may also require de-differentiation of cells (Sangwan et al., 1992), a preculture period of 2 d was chosen for the transformation experiments.

The results of 2 d preculture of the explants before transformation for the percentage of GUS-positive explants and GUS spots/GUS-positive explant are given in Table 1. The third column represents the ratio of the effects of 2 d and 0 d preculture. The number of putative transgenic shoots which could be removed from the leaf explants after 6 w are given in the last two columns of Table 1. There is no clear evidence for an increase in transformation efficiency using a 2 d preculture period.

The shoots which were transferred to SPM + 25 mg/l kanamycin + 200 mg/l cefotaxim did not survive. Those which were first cultured on SPM without kanamycin, and subsequently on SPM + 50 mg/l kanamycin + 200 mg/l cefotaxim, also died. These shoots proved to be GUS-negative and can be considered as escapes. It should be mentioned here that with the empty AGL0 strain no shoot regeneration on selection medium took place, except in the last experiment with 'Golden Delicious' where 3 shoot primordia were observed.

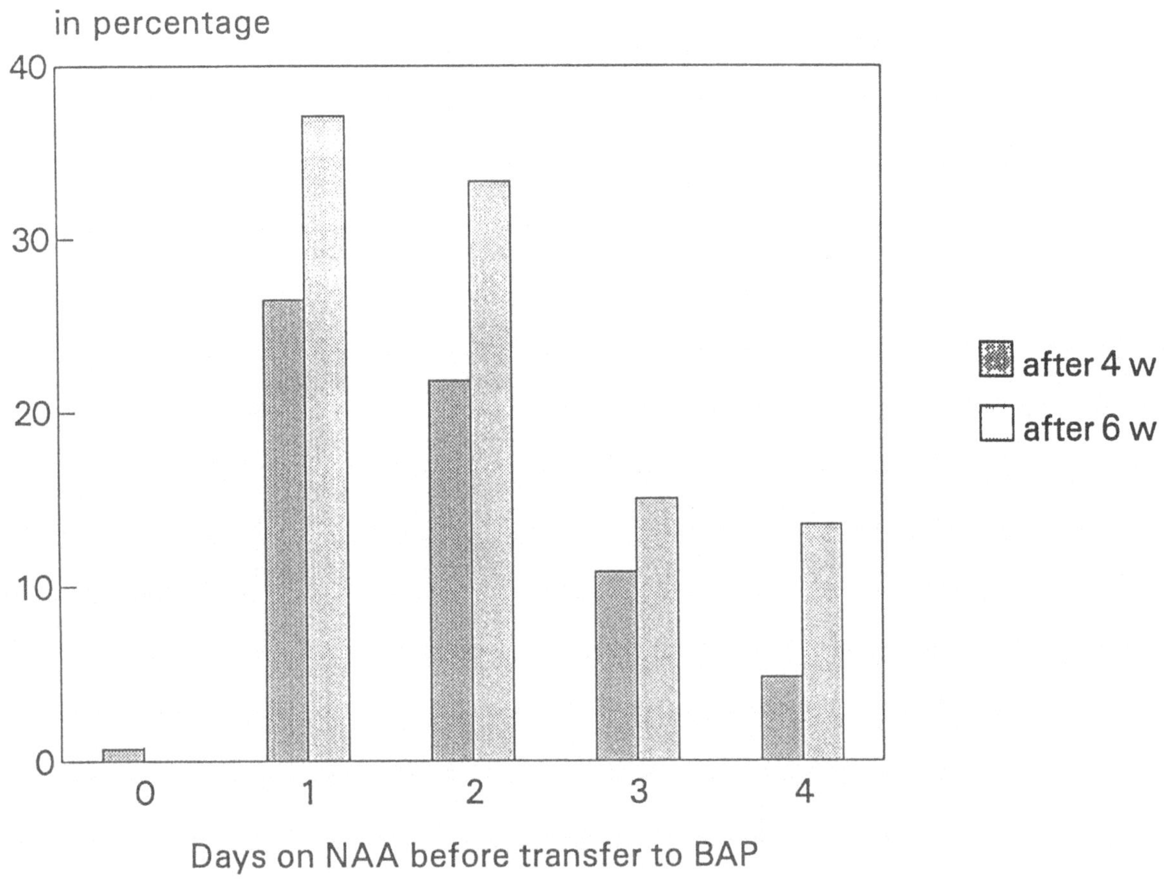

Fig. 1. The percentage of shoot regenerating explants of cv. Golden Delicious at 4 and 6 w. The explants were cultured 0–4 d on MS with 0.2 mg/l NAA before they were transferred to MS with 5 mg/l BAP (n = 120 per treatment).

Table 2. The effect of 5-azacytidine added to the selection medium on transformation efficiency (n = 12 explants for GUS assays, n = 120 explants for putative transgenic shoots)

	Percentage of GUS$^+$ explants (GUS spots/GUS$^+$ explant)				Shoots	
	+ aza	- aza	+ aza/- aza		+ aza	- aza
Gala	94 (15.5)	73 (15.8)	1.3 (1.0)		7	1
	90 (9.5)	100 (7.2)	0.9(1.3)		12	7
GD	77 (12.8)	93 (9.0)	0.8 (1.4)		2	12
		Mean	1.0 (1.2)	Total	21	20

Azacytidine

The percentages of GUS-positive explants and GUS spots/GUS-positive explant, when explants were cultured on a medium with or without 5-azacytidine after transformation, is shown in Table 2. There is no indication of azacytidine inhibition. The same conclusion holds for the number of shoots obtained with or without azacytidine (Table 2). These shoots have been tested for their transgenic nature. One of the 'Gala' shoots, excised from a leaf explant cultured for 8 w on selection medium without 5-azacytidine, proved to be GUS-positive and will be further analyzed. It should be noted that azacytidine was added at a certain stage after co-cultivation, based on the procedure used for *Arabidopsis* (Mandal et al., 1993). This may not necessarily be the optimal choice for apple.

Conclusion

A preculture of the leaf explants before *A. tumefaciens* infection or the addition of 5-azacytidine to the selection medium after co-cultivation does not seem to influence the frequency of transformation, as indicated by the percentage of GUS-positive explants after 5 w. The number of shoots appearing on the explants after 6 w on selection medium is likewise not clearly influenced by these treatments.

One GUS-positive 'Gala' shoot was obtained so far. The non-transgenic nature of the other shoots suggests that regeneration of transgenic shoots directly from leaf tissue will be difficult and that in future experiments, shoot regeneration via a callus phase should be considered.

References

Goodwin, P.B., S. Sriskandarajah & J. Speirs, 1993. Genetic transformation of apple. In: T. Hayashi et al. (Eds). Techniques on Gene Diagnosis and Breeding in Fruit Trees, pp. 178–183. FTRS, Japan.

James, D.J., A.J. Passey, D.J. Barbara & M. Bevan, 1989. Genetic transformation of apple (*Malus pumila* Mill.) using a disarmed Ti-binary vector. Plant Cell Rep. 7: 658–661.

Jefferson, R.A., T.A. Kavanagh & M.W. Bevan, 1987. GUS fusions: β-glucuronidase as a sensitive and versatile gene fusion marker in higher plants. EMBO J. 6: 3901–3907.

Mandal, A., V. Lang, W. Orczyk & E.T. Palva, 1993. Improved efficiency for T-DNA-mediated transformation and plasmid rescue in *Arabidopsis thaliana*. Theor. Appl. Genet. 86: 621–628.

McHughen, A., M. Jordan & G. Feist, 1989. A preculture period prior to *Agrobacterium* inoculation increases production of transgenic plants. J. Plant Physiol. 135: 245–248.

Maheswaran, G., M. Welander, J.F. Hutchinson, M.W. Graham & D. Richards, 1992. Transformation of apple rootstock M26 with *Agrobacterium tumefaciens*. J. Plant Physiol. 139: 560–568.

Sangwan, R.S., Y. Bourgeois, S. Brown, G. Vasseur & B. Sangwan-Norreel, 1992. Characterization of competent cells and early events of *Agrobacterium*-mediated genetic transformation in *Arabidopsis thaliana*. Planta 188: 439–456.

Euphytica **85**: 135–147, 1995.

Physiological complexity and plant genetic manipulation

Marcello Buiatti & Patrizia Bogani
Department of Animal Biology and Genetics, Via Romana, 17, 50125 Firenze, Italy

Key words: genetic manipulation, physiological networks, homeostasis, metabolism, breeding

Summary

One of the limiting factors in the development of new cultivars in a reasonable time using recombinant DNA techniques, is an inability to predict the interaction between the introgressed gene(s) and the host genome and metabolism.

This review presents a survey of the literature on the constraints determining the coherence between alien sequences and their products, and the organization of the receiving genome and its physiological equilibrium. An hypothesis supported by preliminary experimental data is put forward that such constraints derive from co-adaptation during the evolution of gene complexes driven by external selection pressure, and by changes in genes coding for key factors of plant metabolism. Conclusions are finally drawn on a series of possible methods to be used in genetic engineering, in relation to breeding practice, compatible with the rules governing the organization of physiological networks.

Introduction

Until the end of the eighties the general approach to the use of recombinant DNA techniques for genetic engineering of crop plants was founded on what Mayr (1963) called 'beanbag genetics', i.e. the classical Mendelian methodology 'of studying each gene locus separately and independently' therefore pragmatically excluding the study of interactions and pleiotropic effects. Such an approach, extremely useful for its high resolving power, avoids the fact that adaptation derives from the harmonic and dynamic interaction of genes in a complex network (the genome) whose fitness is not simply the sum of components attributed to single genes.

As reported by Dobzhansky (1970), Darwin himself introduced this concept stating that 'the whole organism is so tied together during its growth and development, that when slight variations in any one part occur, and are accumulated through natural selection, other parts become modified'. This sentence introduces the concept of evolution through co-adaptation of gene complexes a consequence of which is the possible pleiotropic effect of single gene modifications. It should be stressed that plant breeders have always been aware of the need for long-term remodelling of plant genotypes following a change in a key character for the development of a new, competitive cultivar. Moreover, a great deal of information on the genetics and the physiology of key genes with pleiotropic effects has been obtained by geneticists studying the introduction of single gene mutations into known and productive genetic backgrounds using refined genetic methods (interspecific hybridization and backcrossing, induced mutations, selection of tissue and cell culture derived mutants, etc.) available before the widespread use of recombinant DNA techniques.

A classical example of this kind of work (we will report on relevant experiments using somaclonal mutants later in this review) are the thorough studies carried out mainly by the group at the Plant Breeding Institute in Cambridge on the effect of introgression of dwarf mutant alleles from a Japanese cultivar (Norin 10) into cultivated bread wheat. In this case it has been shown that insensitivity to gibberellin in mutants not only induced dwarfism but also increased tillering, change in flowering time etc., i.e. modified a series of characters which together led the new cultivars to be an important part of the Green Revolution (Gale & Law, 1977; Flintham & Gale, 1983; Gale & Youssefian, 1985). Although these and other similar data are widely known, present knowledge of the regulation of the dynamic interacting networks of life is still not sufficient to predict the limits within which single components can be changed and yet allow the general metabolic architecture of the network to be maintained. This is obviously of great relevance to genetic manipulation i.e. the recently developed techniques of genetic engineering, *in vitro* selection and protoplast fusion. The need for a thorough discussion on these points is suggested by the fact that both selection *in vitro* and genetic engineering have given unpredictable results, often hampering the utilization in the market of modified and/or selected plants.

Several groups have been building models which allow simulation experiments on the dynamics of self-organizing biological networks of interacting elements where the effects of perturbations can be studied. According to Stuart Kauffman (1991, 1992a, 1992b), living systems can be simulated through the use of Boolean networks of interacting elements and be shown to be continuously at the 'edge of chaos' (Langton, 1991), in the sense that, although several of the components may be 'frozen' in stable combinations (states), the others may change following small or large perturbations. In these simulations, every perturbation elicited by a change in a component of the system, leads to a cascade of events whose entity depends on the number and strength of the connections of this specific component with the rest of the network.

This introduces the biological concept of hierarchy of genes, well known to physiological geneticists, according to which some genes have a larger pleiotropic effect than others. Connections between elements (genes) are quantitatively weighed and, in many cases, the input in one component may derive from a non-additive equilibrium between the outputs of two or more connected elements. The variable consequences of the amount of change and its 'diffusion' throughout the network on the weights of outputs of the changing component, introduces the biological concept of a hierarchical functional organization of the genome, some genes being more critical than others for metabolism and development. The non-additive character of connective weights, on the other hand, is in line with the fact that the control of metabolism and/or development, often involves the establishment of specific equilibria between two or more genes and gene products, with the presence of threshold values: in general, a series of processes led by non-linear interactions between effectors. Examples of this kind of behaviour range from the control of the switch between the lytic and lysogenic cycle of phage lambda to the relevance of auxin/cytokinin ratios and thresholds for cell proliferation and differentiation (see later sections). If all this is true, we may draw four working hypotheses relevant to the evaluation of side effects resulting from genetic manipulation of plants.

Firstly, the pleiotropic effect of a gene should depend on the number of strength of its connections with the other genes and gene products of the organism. Secondly, one should expect the host genotype to exert a regulatory pressure by modifying and controlling the expression of homologous and heterologous 'alien' genes. Thirdly, the amount of change induced by integration of an alien sequence should depend on the genetic background in which it is inserted in terms of interactions between gene products. Fourthly, the effect being quantitative and regulated in a non-linear way, constitutive regulation of expression of the inserted sequence may modify its effects.

In the following sections of this review, we will try to verify these hypotheses with data both from our laboratory and from the published literature.

Genomic homeostatic processes

Recent evidence shows that low expression of an engineered gene in plants may derive from a lack of harmony between DNA composition in the host and donor organisms, or from the presence of homeostatic mechanisms (Lerner, 1954) aimed at the maintenance of evolutionarily fixed levels of transcription. In this context, harmony with host genome GC/AT ratios and codon usage and the number of copies of the inserted homologous or heterologous genes coding for the same product, seem to be two relevant factors. The first phenomenon has been shown by Perlak et al. (1991, 1993) who were able to improve dramatically the expression of three *Bacillus thuringiensis* genes (*cry* I A (b) *cry* I A (c), *cry* III A) by increasing their G + C content (lower in *B. thuringiensis* than in host plants) and changing A + T-rich sequences potentially responsible for mRNA destabilization or polyadenylation or similar to introns. A different approach has been followed by Koziel et al. (1993) who designed a *cry* I A (b) gene from *B. thuringiensis* to resemble a maize gene in terms of codon usage and also found increased expression in transgenic plants. These data are consistent with the computational demonstration, based on a model developed by Lió & Bagnoli (unpublished), of a clear positive correlation between expression levels and coherence between codon usage in the genes and relative frequencies of synonymous tRNAs.

The influence of copy number of inserted genes on their expression was discovered in *Agrobacterium tumefaciens*-induced tumour lines where a negative effect of multiple copies was observed (Gelvin et al., 1983). Since this finding, a large amount of evidence has been collected pointing to the existence of a series of regulatory mechanisms directed at maintaining the overall metabolic equilibrium of the plant within a 'sustainable' range. This is particularly evident in the case of insertion of sequences homologous to genes already present in the host plant. This process (for an early review see, Jorgensen, 1990) may involve the suppression of expression only of host sequences (Elkind et al., 1990), or only of transgenes when more than one copy is inserted (De Carvalho et al., 1992), or of both (co-suppression) (Napoli et al., 1990). Multiple copy inhibition may also occur when sequences inserted are of bacterial origin, particularly when multiple transformations are sequential (Matzke et al., 1989). In all cases, the process is strictly dependent on the structural sequence inserted, even if in a truncated form (Smith et al., 1990), and may be limited to a particular organization of the transgenes. This is the case in the inhibition of the expression of a T-DNA carrying *uid* A and *GUS* in tobacco, occurring only when the sequences are inserted as inverted or tandem repeats (Hobbs et al., 1993). In some cases, but not always, the relative locations of transgenes may also play a relevant role (Matzke & Matzke, 1990), probably due to upstream sequences. Inhibition may occur both at the transcriptional and post-transcriptional level (De Carvalho et al., 1992) and has been attributed to methylation (see Finnegan et al., 1993 for a recent review), changes in RNA stability, synthesis of antisense RNA (Grierson et al., 1991), or other unknown causes probably dependent on the activation of different feedback homeostatic regulators. For instance, mechanisms through which silencing of host and inserted sequences may be developmentally regulated require investigation, inhibition being generally more intense in older than in younger cells, and influenced by the environment (Hart et al., 1992).

Genetic network dynamics

Apart from direct genome-to-gene homeostatic mechanisms, an introduced alien gene has to cope with a series of interactions with the existing plant metabolic architecture. For the sake of simplicity these can be divided into two large groups. In the first, we may classify the effects of the introduction of genes which, although having rather loose or no interactions between themselves, deeply affect a number of 'downstream' genes and functions, often switching on and off cascades of pleiotropic metabolic processes. In other words these genes receive weak (loose) inputs from the network while giving strong outputs to all sequences connected at lower levels of a hierarchical order in the metabolic organization. The second group of interactions is typical of 'circular' metabolic networks such as those related to the protein, carbohydrate and lipid pathways, with more homogeneous connection 'weights' between components.

The hierarchy of developmental processes

Development in eukaryotes derives from the interaction between cell division, expansion and differentiation and may be thought of as a series of bifurcations (canalization events) leading to diversification and organization of the cellular architecture of the plant. Developmental processes, in other words, are controlled by a hierarchy of genes coding for a vast array of substances which act as the inducers of a series of sequential bifurcations (Buiatti, 1988).

So-called phytoregulators are situated at the top of the hierarchy, acting, unlike animal hormones, as fairly non- specific inducers of a large range of events, as shown in the last few years with the aid of mutants and genetic engineering for sense or antisense sequences involved in their synthesis and/or in the reception of signals. Some examples of the effects of plant regulatory molecules are reported in Table 1.

There are a number of indications that phytohormones influence each other's synthesis. For instance, gibberellic acid affects IAA levels (Jindal & Hemberg, 1976; Low & Hamilton, 1984), auxin triggers ethylene formation (Kende, 1993; Yeng & Hoffmann, 1984), auxin and cytokinin syntheses are reciprocally regulated, etc. Moreover, many mutants selected for 'resistance' to high concentrations of one hormone are 'cross-resistant' to others. Phytohormones, as suggested by Bonner (1949), probably act through interactions with rather non-specific receptors which may be

Table 1. Some examples of pleiotropic effects by major phytoregulators

Phytoregulators	Effects	References
Phytochrome	Plant height, hypocotyl, stems, petioles elongation, leaf chlorophyll content, apical dominance, leaf senescence, chloroplast development, flowering, number of leaves, etc.	Furuya, 1993
Ethylene	Germination, senescence, abscision, sex determination, response to stress, stem elongation, geotropism, hypocotyl growth, chlorophyll loss, etc.	Kieber & Ecker, 1993
Abscisic acid	Response to stress, germination, plant height, number of leaves, geotropism, senescence, etc.	King, 1988; Taylor, 1992
Gibberellins	Cell elongation and division, rooting, germination, etc.	Davies, 1988
Cytokinins	Senescence, shoot and root production, plant height, leaf area, chloroplast functions, auxin synthesis, cell division and growth, etc.	Kaminek, 1992
Auxins	Cell wall extension, cell growth and division, cytokinin, ethylene, gibberellin activity, apical dominance, vascular development, anther and flower development, etc.	Jacobs & Ray, 1976; Shininger, 1979

present in a limited number, thus determining competition between different effectors. Therefore, developmental choices are determined by specific optimal equilibria between such effectors and in many cases will show very clear threshold effects. Hence the effect of introducing alien genes involved in the synthesis and/or reception of phytohormones will necessarily be pleiotropic, and, being largely dependent on the equilibria in the regulatory network of host plants, may be inconsistent, depending on host genotype. Some examples of pleiotropy are reported in Table 2.

Variable effects of the modification of phytohormone equilibria depending on the endogenous synthetic capacity of the host genotype have also been observed as expected. Increasing the cytokinin levels of *Nicotiana glutinosa* through insertion of the *ipt* gene from *A. tumefaciens*, determined auxin autotrophy (Binns et al., 1987), while an enhancement of cytokinin utilization in *N. langsdorffii* and in the non-tumorous mutant of the amphidiploid hybrid *N. glauca* × *N. langsdorffii*, produced via *rol* C integration in our laboratory, resulted in both auxin and cytokinin habituation (Bogani et al., unpublished). Integration of *A. tumefaciens* hormone synthetic elements (*iaa*M, *iaa*H, *ipt*) unbalanced in cytokinin or auxin production through inactivation of one gene at a time, in a range of species of *Nicotiana* (Nacmias et al., 1987), led to very different results in terms of shoot and root formation and cell proliferation depending on host genotype. Notably, tumour formation was observed in most cases with the exception of one species, which was completely resistant to all *Agrobacterium* strains; the other genotypes either needed the combined action of the *iaa*M, *iaa*H and *ipt* oncogenes, or were able to complement with only a single oncogene. Data obtained for root and shoot regeneration suggested either an inhibitory effect on root formation of the insertion mutations in genes *iaa*M or *iaa*H, or a positive one by mutations in the *ipt* gene, in genotypes capable *per se* of root differentiation. Moreover, the *ipt* integration particularly promoted shoot formation in species with probably a higher endogenous auxin content. These data could be at least partially explained by the differentiation, during the evolution of this genus, of species with varying auxin/cytokinin equilibria (Buiatti & Bennici, 1970; Bogani et al., 1985; Bogani et al., unpublished). This suggests the hypothesis that genes coding for key metabolic products for example, enzymes needed for the synthesis or utilization of phytohormones, may 'lead' to the co-adaptation of harmonious gene complexes during evolution.

Table 2. Some cases of pleiotropic effects exerted by hormone-related genes in morphogenetic mutants and in transgenic plants

Genetic modifications	Plant effects	Pleiotropic	References
ethylene-resistant mutants	*Arabidopsis*	Inhibition of hypocotyl elongation, promotion of germination, loss of chlorophyll, increase in peroxidase activity.	Bleecker et al., 1988
auxin-resistant mutants	*Nicotiana*	Cross resistance to ABA and paclobutrazol, reduction in seed dormancy, increased tendency to wilt, promotion of root formation	Bitoun et al., 1990; Rousselin et al., 1992
ABA-deficient mutants	*Arabidopsis*; Tomato	Inhibition of carotenoid biosynthesis, wiltness, abnormal stomatal behaviour.	Tal & Nevo, 1973; Rock & Zeevaart, 1991
integration of *ipt* gene from *Agrobacterium tumefaciens*	*Arabidopsis*; *Nicotiana* shoots, reduction in	Delay in senescence, increase in adventitious Medford et al., 1989 plant height and leaf area, root inhibition, increase in the expression of chloroplast gene rbc L.	Smart et al., 1991; Yusibov et al., 1991
integration of *rol* B gene from *Agrobacterium rhizogenes*	*Nicotiana*	Modification of anther and flower development, increase in IAA levels and decrease in gibberellin activity.	Spena et al., 1992
integration of *iaa* L gene from *Pseudomonas savastanoi*	*Nicotiana*	Reduction of apical dominance inhibition of vascular development.	Medford et al., 1989; Smigocki & Owens, 1989

To test this hypothesis we carried out in our laboratory two series of experiments. Firstly, we analyzed the *in vitro* behaviour of explants from a high number of *Nicotiana* species to ascertain whether competence for the formation of roots (high auxin), shoots (high cytokinin) or for habituation have a random distribution in the genus, or suggest the existence of specific phytohormone-controlled branching points during evolution. Secondly, we used a highly mutable system (*in vitro* cultured tomato cells grown on different hormone combinations) to test the possible selection of 'concerted' genome changes depending on phytohormone levels. As shown in Table 3, root formation was indeed found to be concentrated in the *Paniculatae*, shoots being often differentiated in the *Alatae*, and habituation being more frequently observed in the *Repandae, Bigelovianae* and *Suaveolentes* (Bogani et al., unpublished). Moreover, the calculation of correlation values between the presence of different RAPD bands in tomato cell clones, carried out with a mutual information algorithm (Simoni et al., 1993; Lió,

unpublished), showed patterns which varied according to the hormone content of culture media (Fig. 1a, b, c).

The fixation throughout evolution of co-adapted gene complexes possibly partially controlled by genes involved in highly pleiotropic processes, has a dual effect on breeding practices and genetic manipulation using recombinant DNA techniques. While changes in one component of a co-adapted network may lead to unforeseen problems, these can be sometimes balanced by corresponding modifications in the regulation complex. Moreover, conscious manipulation of genes affecting the physiological background may bring profitable results as in the case of dwarf mutants in wheat already mentioned. Relevant examples of this may be taken from experiments on the effects of different combinations on resistance to biotic stresses.

Thus, tomato cell lines not competent for the induction of active defence mechanisms against *Fusarium oxysporum*, acquired this competence when transformed with the *ipt* gene from *A. tumefaciens* (Storti et

Table 3. Habituation and differentiation in calluses of different species of *Nicotiana* grown on hormone-free medium

Sections	*Nicotiana* species	Habituation	Root formation	Shoot formation
Paniculatae	*N. glauca*	±	-	-
	N. paniculata	-	+	-
	N. knightiana	±	++	-
	N. benavidesii	-	++	-
	N. raimondii	-	++	-
Rusticae	*N. rustica*	±	-	-
Tomentosae	*N. otophora*	+	-	-
	N. tabacum	-	-	-
Alatae	*N. langsdorffii*	-	-	+
	N. alata	-	-	+
	N. longiflora	+	-	+
	N. plumbaginifolia	+	-	+
Repandae	*N. repanda*	+	-	++
Acuminatae	*N. acuminata*	-	-	-
	N. pauciflora	+	-	-
Bigelovianae	*N. bigelovii*	+	-	-
	N. clevelandii	+	-	-
Sauveolentes	*N. debneyi*	+	-	-
	N. gossei	±	-	-

al., 1994). Resistance to fungal and bacterial pathogens has also been improved in potato through the introduction of *rol* C, the resistance being correlated with a reduced level of glucose and with low productivity (Fladung & Gieffers, 1993). In this case, however, the results of the integration of hormone-related genes may be partially controlled by modulation of transgene expression, for instance, through promoter manipulation. Fladung et al. (1993) showed that the negative pleiotropic effects resulting from the integration of *rol* C into potato plants could be almost eliminated when a light-inducible promoter, active only in leaves, was used in place of a constitutive 35 S CaMV promoter used in the first experiments.

A possible connection between carbohydrate metabolism, resistance and hormonal background, is suggested by the observation that tomato cells, selected for high polysaccharide content, also showed acquired competence for active defence and, when regenerated into plants, displayed a mutated phenotype resembling those obtained in other plants by increasing the endogenous cytokinin levels (Guardiola et al., 1994; Storti et al., 1994). This may imply that the selected cells had an altered physiological equilibrium, and would give a partial explanation of the pleiotropic changes observed by Cassells et al. (1992) in potato plants regenerated from *in vitro* cultures and selected for disease resistance. In other words, given the known effects on resistance of altering levels of cytokinins (Fladung & Gieffers, 1993), or other phytohormones like ethylene (Broglie et al., 1989), it may be suggested that selection for high polysaccharide levels or directly for resistance from *in vitro* cultures does not necessarily lead to the isolation of single resistance genes but, rather, of physiological backgrounds favouring the induction of active defence-related genes. Such backgrounds may also enhance the 'alien-recognition' capacity of the cells as suggested by the fact that the physiological phenotype of the above-mentioned tomato cells selected for high polysaccharide content, seem to be extremely similar to those of cell mutants directly selected for increased *Fusarium* elicitor recognition (Buiatti et al., 1987) and, finally, of resistant tomato cultivars (Storti et al., 1992).

At lower levels in the hierarchy of developmental control, other genes code for components of signal transduction chains and/or for substances needed for more specific bifurcations. Signals of this kind include,

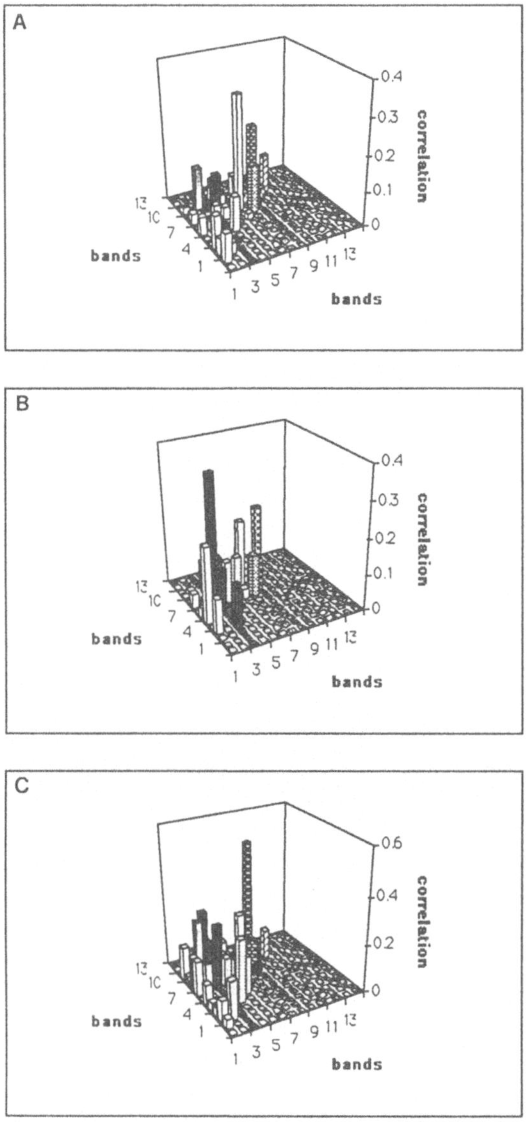

Fig. 1. Correlation analysis between random polymorphic amplification products (RAPD bands) measured for each somaclone grown on LS medium with 0.4 mg/l 2,4-D and 1 mg/l kinetin (A), or 1 mg/l 2,4-D and 1 mg/l kinetin (B) or 0.8 mg/l 2,4-D and 0.2 mg/l kinetin (C).

for instance, messengers of hormone action like the phytochrome-related G-proteins of peas (Yoshida et al., 1993), the ethylene negative regulator (Kieber et al., 1993), probably the protein kinase in the 'tousled' mutant of *Arabidopsis* which influences both leaf and flower development (Roe et al., 1993), and compounds like the lipo-oligosaccharides which influence carrot embryogenesis (De Jong et al., 1993). These and other signal transduction components are related to processes as diverse as light reception, self-incompatibility, cell division, resistance to pathogens, regulation of carbon metabolism, maintenance of K^+ flux across the plant and membrane changes in response to osmotic shocks. Although of lower general relevance, they still influence a series of processes and may have gross pleiotropic effects when the corresponding genes are mutated or integrated in a plant genome (Alderson, 1993; Kieber et al., 1993; Martin et al., 1993). A mutation in a negative regulator of ethylene response (CTR1), for instance, induced in *Arabidopsis* smaller plants, low fertility, reduction of rooting, smaller inflorescences with elongated gynoecium and reduction in size of epidermal cells (Kieber et al., 1993). Similarly *cdc* 25, a mitotic inducer gene from fission yeast inserted in tobacco, altered leaf morphology, cell dimension and reduced flowering time, increasing the number of flowers, which, in turn, had an altered morphology (Bell et al., 1993).

At even lower levels in the hierarchy, genes have been discovered which seem to act in a very specialized manner on specific, determined bifurcations very much like similar sequences in *Drosophila* or other animals. These genes include homeobox-containing sequences like kn, a regulator of leaf development in maize, OSH 1 from rice, the leucine zipper proteins in *Arabidopsis*, or homeotic genes like the flower-controlling sequences found in *Arabidopsis* and *Antirrhinum* (see Dennis & Bowman, 1993 for a recent review). Homeobox genes of *Drosophila* are believed to activate batteries of genes acting as transcription factors; the plant homologues probably act in a similar way. Mutations in kn 1 (Hake, 1992) affect the ligules, lateral veins and sheaths of maize leaves and induce the formation of 'knots' of dividing cells in already expanding leaves similar to those induced in tobacco by *cdc* 25 (Bell et al., 1993). The gene OSH 1 isolated from rice by Matsuoka et al. (1993), when introduced into rice, induced leaf morphology changes similar to those observed with knotted mutants. It should be stressed, however, that the situation changed when the same gene was inserted into the genome of a dicotyledonous plant (*Arabidopsis*). In this case, flower morphology was also modified and the plants formed a large number of adventitious shoots leading to a bushy appearance similar to that induced by the gene coding for a rice GTP-binding protein (Kamada et al., 1992) and by the genes for cytokinin synthesis and/or utilization (see before). This led Matsuoka et al. (1993) to suggest that OSH 1, as with the other rice gene, may act by stimulating limited synthesis of cytokinins with a decrease of auxin, in transgenic shoot apices.

In this case, as in those of cytokinin and auxin genes or of resistance-related sequences, a different genetic background (*Arabidopsis* vs rice) induced a differential expression of the integrated gene leading to different modifications in the phenotype.

The homeotic genes discovered so far in plants are strictly concerned with the determination of the identity of floral parts. They all seem to share a conserved, putative DNA-binding domain, the MADS box, found in several transcription factors of animal and yeast origin, like products of oncogenes and other regulatory proteins generally involved in cell differentiation or in transduction of environmental signals (Schwarz-Sommer et al., 1992). Homeotic genes control the choice between anther and carpels or petals or sepals in developing flowers and have similar mechanisms of action in *Arabidopsis* and *Antirrhinum.* In all cases mutation effects seem to be essentially limited to the flowers and not to other plant parts. Therefore changes induced in plants transgenic for these sequences may be almost exactly predicted once the rather simple interactions between components of the system are known (Mandel et al., 1992; Mizukami & Ma, 1992; Tröbner et al., 1992; Bradley et al., 1993).

Metabolic networks

As mentioned above, the effects of change in one component of a network result from the interaction of that component with one or more metabolic pathways. This leads to the 'spreading' of modifications in the affected system and, potentially, to others connected to it, due to a series of feedbacks and to the inherent homeostatic defences of the general co-adapted metabolic complex. A good example of this behaviour is demonstrated by the effects of increased (through insertion of functional alien sequences) or decreased (through antisense) synthesis of enzymes and other proteins connected with carbohydrate and oil metabolism. For instance, the insertion of a yeast-derived invertase into tobacco (Sonnewald et al., 1991) resulted in an unexpected increase in leaf sucrose and glucose with a disproportionate increase in fructose, and the inhibition of photosynthesis, probably due to a negative feedback from glucose (Sheen, 1990). Moreover, surprisingly, several developmental parameters were also affected. Transgenic plants showed reduced height, leaf and root growth, reduced seed setting and late flowering, were longer-lived and had curled leaves with evenly-distributed light green sectors. It is worth noting that the authors attributed part of the 'stunting' to osmotic effects due to higher fructose concentrations. This is consistent with the observation (Muleo et al., unpublished) that *Actinidia* and tomato plants regenerated from shoots selected *in vitro* for resistance to high concentrations of methyl-glucose, showed a similar phenotype. Growth was also inhibited in potato and tobacco plants with increased pyrophosphatase activity due to the insertion of the corresponding *E. coli* gene, which induced accumulation of sucrose, starch and hexoses (Jelitto et al., 1992). Reduction in photosynthesis and growth retardation were observed in potato plants where reduced triose phosphate translocator-exporting activity has been achieved through antisense technology (Riesmejer et al., 1993). Similarly, the diversion of the acetylacyl CoA intermediates from the cytosolic mevalonic acid pathway into polyhydroxy-butyrate (PHB), a polymer of industrial interest, resulted in stunting (Fentem, P.A., personal communication). Interference into lipid biosynthetic pathways, although less severe (Murphy D.J., personal communication), may also induce unforeseen modifications of development. Thus, an alteration in fatty acid composition induced in *Arabidopsis* by the introduction of an *E. coli* glycerol-3-phosphate-acyl-transferase, was associated with an increased susceptibility to chilling, slow growth and a low yield of viable seeds. Moreover, some of the transgenic plants were smaller, flowered earlier and formed a multitude of adventitious shoots (Wolter et al., 1992). Changes in amino-acid metabolism have also recently been shown to bring unexpected results. *Arabidopsis* tryptophan biosynthesic mutants showed a wide range of grossly-modified phenotypes ranging from small, bushy, infertile auxotrophs to leaky mutants with nearly normal appearance (Raikhel & Last, 1993). Moreover, transformation with *E. coli* feedback-insensitive forms of aspartate kinase (the rate-limiting enzyme of the threonine and lysine pathways) and dihydrodipicolinate synthase (the first enzyme of lysine biosynthesis) led to a dramatic increase in threonine but, unexpectedly, not in lysine due probably to an enhanced rate of lysine catabolism (Karchi et al., 1993).

All these data clearly show that, on one hand, metabolic pathways have inter-network connections and, on the other, may affect in some instances developmental control systems (see preceding section). In both cases a change induced in one step of one of the pathways may spread its effects to other pathways and to development. Spreading is due both to the modification of metabolic chains and to the homeostatic mechanisms of the plant which try to restore the normal equilibrium by counteracting induced change, or by organizing a new balance of metabolism. The adjustments then can be large or limited to the bypass of the modification through the exploitation of the genetic and phenotypic plasticity of the plant. An example of such a counter-effect is the physiological compensation observed in *Nicotiana sylvestris* and *N. tabacum* by Beffa et al. (1993), following transformation with an antisense class I glucan endo 1,3-β-glucosidase. In this case, transformation did not result in lower β-1,3-glucanase overall activity, because a serologically-distinct glucanase activity was induced in antisense transformants through some unknown damage upon recognition mechanism. As commented upon by the authors, this is a limited variant of the known physiological processes of alternative metabolic pathways activated in the presence of stress.

Concluding remarks

Practical applications of the use of modern techniques of genetic manipulation in plants seem to be on the verge of a critical change. The number of available genes and sequences of potential use for improving the productivity and quality of cultivated plants is steadily and rapidly increasing. They range from genes for resistance to insects, fungi or bacteria (of great relevance in this field is the recent isolation of two plant genes for resistance to fungi and bacteria by, respectively, Johal & Briggs, 1992 and Martin et al., 1993), to sequences for male sterility, control of fruit maturation, cut flower vase life, flower colour etc., to quote some of the most interesting examples (Raikhel & Last, 1993). New avenues of applied research are also being opened up by the growing interest in non-food products from agriculture, often based on specific molecules whose production may be improved via genetic manipulation from both the quantitative and qualitative points of view.

At the same time, research, initially centred on the introduction of single genes, often of bacterial, fungal or animal origin, is now being directed to the isolation of sequences concerned with important metabolic and developmental plant processes, as discussed in this review. This new direction may lead to a real breakthrough in the field and allow full use of the powerful tools of molecular biology for unravelling the complex network of plant physiological processes, but on the other hand is bringing to light problems similar to those which have delayed the applications of medical and veterinary knowledge gained in the animal and human field in the last decade. Failure to apply the vast amount of data obtained on the control of cell proliferation and the origin of tumours to their cure and prevention, and the fact that genetic engineering is now known not to be sufficient *per se* for the production of new marketable plant cultivars, are examples of such difficulties. Plant breeders and geneticists have known for a long time, that lengthy genotype remodelling is needed in all cases it seems, after the insertion of a useful alien gene. All failures in fact, can be attributed to the discovery of a series of side (pleiotropic) effects of genetic transformations.

It should be stressed that plants, due to their open developmental systems and to the high frequency of somatic mutations and epigenetic changes like sequence amplification and methylation/hypomethylation processes (Buiatti, 1989), have a higher homeostatic capacity and can therefore tolerate drastic changes in genome organization better than animals. This is proven by the survival of plants carrying gross chromosomal anomalies like aneuploidy and polyploidy, the behaviour of interspecific hybrids which compensate for the problems created by the union of two distant genotypes through the creation of new equilibria, and the relative success in plants, in comparison with animals, of mutation breeding. This is the reason for the limited pleiotropic effects found when the gene introduced or the mutation selected, influences small areas of plant metabolism with little or no interaction (as in the case of some genes of alien origin) with the rest of the network. Conversely, however, homeostasis is also the cause of low or null expression of introduced genes obtained by direct inactivation or by the induction of compensating alternative pathways, (functional redundancy). Limits to the number of copies inserted in terms of overproduction of the coded protein are imposed by the buffering power of the host genome. Negative effects connected with the recognition of alien sequences, different in codon usage from host plants, or localized in chromosomal regions endowed with low transcription capacity have also been observed. Besides these self-regulated mechanisms which tend to induce the recovery of a stable equilibrium (the optimal 'attractor' of the network) acquired during evolution, other unpredicted changes may occur due to the plastic nature of the system and create gross anomalies leading to a re-organization of the genome. Similar 'hopeful monsters' (Goldschmidt, 1940) probably also occur spontaneously in nature following mutations in key metabolic steps and may have contributed to the 'sudden' formation and isolation of new species (Gould, 1993). It should be stressed, however, that only a few hopeful monsters successfully survive in nature, as generally changes in key components of the genome cause a decrease in fitness which often means a reduction in fertility and/or productivity. This is often due to 'amplification' of the change, inherent, as suggested in the Introduction, in the nature of complex networks, where connections rapidly propagate waves of change. Thus, the initial modification may be a small quantitative alteration in the level of key substances already approaching a threshold, a change in the time of expression of key genes or simply a slightly variant state of the system which brings it near to a phase transition threshold and thereby makes it more sensitive to specific or non-specific environmental stimuli. An example of such process may be, for instance, the response to an environmental stimulus determined by a small change in a promoter, or even a transcribed, sequence which makes it sensitive to environmentally-induced changes in transcription or degradation rates (Green, 1993). All this does not imply by any means that genetic engineering is a hopeless practice. As we have described, negative effects can be avoided by manipulating the regulatory framework, inserting genes with a limited number of connections, etc. Moreover plant plasticity and homeostatic capacity, once their regulation is better known, can be exploited for the improvement of crop plants in the direction of better and new products, and higher productivity when desired, as has been shown by single gene introgression in the past and, now, by the reported examples of improvement in resistance to stress. This necessarily implies, however, a better knowledge of the processes which leads from the translation of one-dimensional information strands of DNA into the three-dimensional phenotype and of its interactions with the environment. Within this framework an interdisciplinary approach is becoming increasingly useful, leading to modelling of sequences according, not only to their meaning in terms of translated products, but also to the 'internal rules' which modulate their expression in absolute terms and/or spatially and temporally during development. This means studying specific (reiterated or not) regulatory sequences, periodical motifs with possible meanings in terms of the physico-chemical structure and plasticity of DNA and the corresponding RNAs, and rules of coherence for codon usage or average composition of host genotype used for transformation etc. This kind of knowledge may be very important for the control of the expression of alien genes for a better integration into the metabolism of the host plant, given the realization that there are indeed limits to the amount of change which can be tolerated within a positive cost-benefit balance. The choices, of course, will have to depend on the level of the hierarchy of genes into which putative inserted sequences may be placed and/or on the weights of interconnections existing with the rest of the genome and its products. Specific interdisciplinary modelling may then be used with the aim of simulating the fate of transformed plants using mathemathical methods similar to those currently employed by students of non-linear dynamics (see for instance Casdagli & Ecbank, 1992).

All this points to the fact, also crucial, that mere knowledge of the sequence of a gene and of its primary function is not sufficient, when its action interferes with wide areas of plant metabolism, for the prediction of the phenotype of a plant transformed with that gene or carrying a mutated copy of it. The necessary know-how may be gained at least partially (living systems as stated clearly by Waddington, 1968, are in a continuous state of 'homeorhesis') if the problem is tackled in an interdisciplinary way, uniting knowledge of genetics, physiology, molecular biology, mathematics, physics, and exploiting reductionist methods but within a non-reductionist frame of mind. We may conclude by agreeing with Yates (1993) that 'the physical basis of life involves at least the six attributes of nonlinearity, broken symmetry, dissipation of free energy, complexity, orderly disorder and marginal dynamic stability', characters so fundamental to the understanding of life as to make Yates say 'perhaps these attributes collectively constitute a sufficiently unique set to justify considering the living system as a fourth state of matter'.

Acknowledgements

This work was supported by the National Research Council of Italy, Special Project RAISA and the ECC Biotechnoology Programme.

References

Alderson, A., P.A. Sabelli, J.R. Dickinson, D. Cole, M. Richardson, M. Kreis, P.R. Shewry, N.G. Halford, 1993. Complementation of snf1, a mutation affecting global regulation of carbon metabolism in yeast, by a plant protein kinase cDNA. Proc. Natl. Acad. Sci. USA 88: 8602–8605.

Beffa, R.S., J.M. Neuhaus & F. Meins, 1993. Physiological compensation in antisense transformants: specific induction of an 'ersatz' glucan endo-1,3-β-glucosidase in plants infected with necrotizing viruses. Proc. Natl. Acad. Sci. USA 90: 8792–8796.

Bell, M.H, N.G. Halford, J.C. Ormrod & D. Francis, 1993. Tobacco plants transformed with *cdc* 25, a mitotic inducer gene from fission yeast. Plant Mol. Biol. 23: 445–451.

Binns, A.N., J. Labriola & R.C. Black, 1987. Initiation of auxin autonomy in *Nicotiana glutinosa* cells by the cytokinin biosynthesis gene from *Agrobacterium tumefaciens*. Planta 171: 539–548.

Bitoun, R., P. Rousselin & M. Caboche, 1990. A pleiotropic mutation results in cross-resistance to auxin, abscisic acid and paclobutrazol. Mol. Gen. Genet. 220: 234–239.

Bleecker, A.B., M.A. Estelle, C. Somerville & H. Kende, 1988. Insensitivity to ethylene conferred by a dominant mutation in *Arabidopsis thaliana*. Science 241: 1086–1089.

Bogani, P., M. Buiatti, S. Tegli, M.G. Pellegrini, P. Bettini & A. Scala, 1985. Interspecific differences in differentiation and dedifferentiation patterns in the genus *Nicotiana*. Plant Systematic and Evolution 151: 19–29.

Bonner, J., 1949. Relation of respiration and growth in the *Avena* coleoptile. Am. J. F. Bot. 36: 429–436.

Bradley, D., R. Carpenter, H. Sommer, N. Hartley & E. Coen, 1993. Complementary floral homeotic phenotypes result from opposite orientation of a transposon at the locus of *Antirrhinum*. Cell 72: 85–95.

Broglie, K.E., Ph. Biddle, R. Cressuran & R. Broglie, 1989. Functional analysis of DNA sequences responsible for ethylene regulation of a bean chitinase gene in transgenic tobacco. The Plant Cell 1: 599–607.

Buiatti, M., 1988. Information flux and constraints in development and evolution: a critical view. p. 331–349. In: S. Ciliberto, R. Livi, S. Ruffo & M. Buiatti (Eds). Chaos and Complexity. World Scientific Singapore.

Buiatti, M., 1989. The use of cell and tissue culture for mutation breeding. p. 179–201. In: P. Parey (Ed). Science of Plant Breeding, Eucarpia, Gottingen.

Buiatti, M. & A. Bennici, 1970. Callus formation and habituation in *Nicotiana* species in relation to the specific ability for dedifferentiation. Rendiconti XLVIII 2: 261–269.

Buiatti, M., C. Simeti, S. Vannini, G. Marcheschi, A. Scala, P. Bettini, P. Bogani & M.G. Pellegrini, 1987. Isolation of tomato cell lines with altered response to *Fusarium* cell wall components. Theor. Appl. Genet. 75: 37–40.

Casdagli, M. & S. Ecbank, 1992. Non linear modelling and forecasting. Addison Wesley, Publishing Company Reading Mass. (USA).

Cassells, A.C., M.L. Deadman, C.A. Brown & E. Griffin, 1991. Field resistance to late blight (*Phytophthora infestans* (Mont) de Bary) in potato (*Solanum tuberosum* (L.) somaclones associated with instability and pleiotropic effects. Euphytica 56: 75–80.

Davies, P.J., 1988. The plant hormones: their nature, occurrence, and functions. p. 1–11. In: P.J. Davies (Ed). Plant Hormones and Their Role in Plant Growth and Development. Kluwer Academic Publ., Boston.

De Carvalho, F., G. Gheysen, S. Kushinir, M. Van Montagu, D. Inzé & C. Castresana, 1992. Suppression of β-1,3-glucanase transgene expression in homozygous plants. EMBO J. 11: 2595–2602.

De Jong, A.J., R. Heldstra, H.P. Spaink, M.V. Hartog, E.A. Meijer, Th. Hendriks, F. Lo Schiavo, M. Terzi, T. Bisseling, A. Van Kammen & S.C. De Vries, 1993. *Rhizobium* lipooligosaccharides rescue a carrot somatic embryo mutant. The Plant Cell 5: 615–620.

Dennis, E. & J.L. Bowman, 1993. Manipulating floral organ identity. Current Biology 3: 90–93.

Dobzhansky, Th., 1970. Genetics of the evolutionary process. Columbia University Press, NY.

Elkind, Y., R. Edwards, R. Mavandad, S. Hedrick, O. Ribak, R.A. Dixon & C.J. Lamb, 1990. Abnormal plant development and down regulation of phenyl-propanoid biosynthesis in transgenic tobacco containing a heterologous phenylalanine ammonia lyase gene. Proc. Natl. Acad. Sci. USA 87: 9075–9061.

Finnegan, E.J., R.I.S. Brettel & E.S. Dennis, 1993. The role of DNA methylation in the regulation of plant gene expression in DNA methylation. p. 218–261. In: P.J. Jost & H.P. Saluz (Eds). Molecular Biology and Biological Significance. Birkhauser Verlag, Basel, Switzerland.

Fladung, M., A. Ballvora & Th. Schmiling, 1993. Constitutive or light-regulated expression of the *rol* C gene in transgenic potato plants has different effects on yield attributes and tuber carbohydrate composition. Plant Mol. Biol. 23: 749–757.

Fladung, M. & W. Gieffers, 1993. Resistance reaction of leaves and tubers of *rol* C transgenic tetraploid potato to bacterial and fungal pathogens. Physiol. Mol. Plant Pathol. 42: 123–132.

Flintham, J.E. & M.D. Gale, 1983. The Tom Thumb dwarfing gene, Rht3 in wheat. II. Effects on height, yield and grain quality. Theor. Appl. Genet. 66: 30–45.

Furuya, M., 1993. Phytochromes: their molecular species, gene families, and functions. Ann. Rev. Plant. Physiol. Mol. Biol. 44: 617–645.

Gale, M.D. & S. Youssefian, 1985. Dwarfing genes in wheat. p. 1–35. In: J.E. Russell (Ed). Progress in Plant Breeding. Butterworths, London.

Gale, M.D. & C.N. Law, 1977. The identification and exploitation of Norin 10 semidwarfing genes. Pl. Br. Inst. Camb. Annu. Rep., 1976: pp. 21–33.

Gelvin, S., S.J, Karcher & V.J. Di Rita, 1983. Methylation of T-DNA in *Agrobacterium tumefaciens* and in several crown gall tumors. Nucl. Acids Res. 11: 159–174.

Goldschmidt, R.A., 1940. The material basis of evolution. Yale University Press, New Haven.

Gould, S.J., 1933. Evolution of organisms. p. 15–42. In: C.A.R. Boyd & O. Noble (Eds). The Logic of Life. Oxford University Press, Oxford.

Green, P.J., 1993. Control of mRNA stability in higher plants. Plant Physiol. 102: 1065–1070.

Grierson, D., R.G. Fray, A.J. Hamilton, C.J.S. Smith & C.F. Watson, 1991. Does cosuppression of sense genes in transgenic plants involve antisense RNA? Trends Biotech. 9: 122–123.

Guardiola, M.L., P. Bettini, P. Bogani, M.G. Pellegrini, E. Storti, P. Bittini & M. Buiatti, 1994. Modification of competence for *in vitro* response to *Fusarium oxysporum* in tomato cells: I. Selection from a susceptible cultivar for high and low polysaccharide content. Theor. Appl. Genet. 87: 988–995.

Hake, S., 1992. Unraveling the knots in plant development. Trends Genet. 8: 109–114.

Hart, C.M., B. Fisher, J.M. Nevhans Jr. & F. Meins, 1992. Regulated inactivation of homologous gene expression in transgenic *Nicotiana sylvestris* plants containing a defense-related tobacco chitinase gene. Mol. Gen. Genet. 235: 179–188.

Hobbs, S.L.A., Th.D. Narkentin & C.M.O. De Long, 1993. Transgene copy number can be positively or negatively associated with transgene expression. Plant Mol. Biol. 21: 17–26.

Jacobs, M. & P. Roy, 1976. Rapid auxin-induced decrease in free space pH and its relationship to auxin-induced growth in maize and pea. Plant Physiol. 58: 203–209.

Jelitto, T., V. Sonnewald, L. Willmitzer, M. Hajirezai & M. Stitt, 1992. Inorganic pyrophosphate content and metabolites in potato and tobacco plants expressing *E. coli* pyrophosphatase in their cytosol. Planta 188: 238–244.

Jindal, K.K. & T. Hemberg, 1976. Influence of gibberellic acid on growth and endogenous auxin levels in epicotyl and hypocotyl tissue of normal and dwarf bean plants. Physiol. Plant. 38: 78–82.

Johal, D.S. & S.P. Briggs, 1992. Reductase activity encoded by the HM1 disease resistance gene of maize. Science 258: 985–987.

Jorgensen, R., 1990. Altered gene expression in plants due to *trans* interactions between homologous genes. Trends Biotechnol. 8: 340–344.

Kamada, Y., S. Yamauchi, S. Youssefian & H. Sano, 1992. Transgenic tobacco plants expressing *rgp* 1 encoding a ras-related GTP binding protein from rice, show distinct morphological characteristics. Plant J. 2: 799–807.

Kaminek, M., 1992. Progress in cytokinin research. Trends Biotech. 10: 159–164.

Karchi, H., O. Shaul & G. Galili, 1993. Seed specific expression of a bacterial desensitized aspartate kinase increases the production of seed threonine and methionine in transgenic tobacco. Plant J. 3: 721–727.

Kauffman, S.A., 1991. Antichaos and adaptation. Sci. Amer. p. 78–84.

Kauffman, S.A., 1992a. Origins of order: Self organization and selection in evolution. Oxford University Press, Oxford.

Kauffman, S.A., 1992b. The sciences and complexity and origins of order. p. 303–319. In: J.E. Mittenthal & A.B. Baskin (Eds). Principles of Organization in Organisms. Addison Wesley Publ. Company Reading, N.Y.

Kende, H., 1993. Ethylene biosynthesis. Ann. Rev. Plant Physiol. Plant. Mol. Biol. 44: 283–307.

Kieber, J.J. & J.R. Ecker, 1993. Ethylene gas: it's not just for ripening any more. Trends Genet. 9: 356–361.

Kieber, J.J., M. Rothenberg, G. Roman, K.A. Feldmann & J.R. Ecker, 1993. CTR 1, a negative regulator of the ethylene response pathway in *Arabidopsis* encodes a member of the *raf* family of protein kinases. Cell 72: 427–441.

King, P.J., 1988. Plant hormone mutants. Trends Genet. 4: 157–162.

Koziel, M.G., L.G. Beland, C. Bowman, B. Carozzi, R. Crenshaw, L. Crossland, J. Dawson, N. Desai, M. Hill, S. Kadwell, K. Lannis, K. Lewis, D. Maddox, K. McPherson, M.R. Meghji, E. Merlin, R. Rhodes, G. Warren, M. Wright & S.V. Evola, 1993. Field performance of elite transgenic maize plants expressing an insecticidal protein derived from *Bacillus thuringiensis*. Biotechnology 11: 194–200.

Langton, C., 1991. Life at the edge of chaos. p. 41–91. In: C. Langton, C. Taylor, J.D. Farmer & S. Rassmussen (Eds). Artificial Life II. Addison Wesley Publ. Company Reading, N.Y.

Lerner, I.M., 1954. Genetic homeostasis. Oliver and Boyd, Edinburgh, London.

Low, D.M. & R.H. Hamilton, 1984. Effects of gibberellic acid on endogenous indole-3-acetic acid levels in dwarf pea. Plant Physiol. 75: 255–256.

Mandel, A., J.L. Bowman, S.A. Kempin, H. Ma, E.M. Meyerovitz & M.F. Yanofsky, 1992. Manipulation of flower structure in transgenic tobacco. Cell 71: 133–143.

Martin, B., S.H. Brommonschenkel, C. Julapark, S. Frary, M.N. Ganal, R. Spivey, T. Wu, E.P. Earle & S.D. Tanksley, 1993. Map based cloning of a protein kinase gene conferring disease resistance in tomato. Science 262: 1432–1435.

Matsuoka, M., H. Ichikawa, A. Saito, Y. Tada, T. Fujimora & Y. Kano-Murakami, 1993. Expression of rice homoeobox gene causes altered morphology of transgenic plants. The Plant Cell 5: 1039–1048.

Matzke, M.A. & A.J.M. Matzke, 1990. Gene interactions and epigenetic variation in transgenic plants. Devel. Genet. 11: 214–223. Matzke, M.A., M. Primig, J. Trnovskj & A.J.M. Matzke, 1989. Reversible methylation and inactivation of marker genes in sequentially transformed tobacco plants. EMBO J. 8: 643–649.

Mayr, E., 1963. Animal species and evolution. Belknap, Cambridge.

Medford, J.I., R. Horgan, Z. El-Sawi & H.J. Klee, 1989. Alterations of endogenous cytokinins in transgenic plants using a chimeric isopentenyl transferase gene. The Plant Cell 1: 403–413.

Mizukami, Y. & J. Ma, 1992. Ectopic expression of the floral homeotic gene AGAMOUS in transgenic *Arabidopsis* plants alters floral organ identity. Cell 71: 119–131.

Nacmias, B., S. Ugolini, M.D. Ricci, M.G. Pellegrini, P. Bogani, P. Bettini, D. Inzé & M. Buiatti, 1987. Tumor formation and morphogenesis on different *Nicotiana* species and hybrids induced by *Agrobacterium tumefaciens* T-DNA mutants. Devel. Genet. 8: 61–71.

Napoli, C., C. Lemieux & R. Jorgensen, 1990. Introduction of a chimeric chalcone synthase gene into petunia results in reversible cosuppression of homologous genes *in* trans. Plant Cell 2: 279–289.

Perlak, F.J., R.L. Fuchs, D.A. Dean, S.L. McPherson & D.A. Fischoff, 1991. Modification of the coding sequences enhances plant expression of insect control protein genes. Proc. Natl. Acad. Sci. USA 88: 3324–3328.

Perlak, F.J., T.B. Stone, Y.M. Muskopf, L.S. Petersen, G.B. Parker, S.A. McPherson, J. Nyman, S. Love, G. Reed, D. Bierer & D.A. Fishoff, 1993. Genetically improved potatoes: protection from damage by Colorado potatoes beetles. Plant Mol. Biol. 2: 313–321.

Raikhel, N.V. & R.L. Last, 1993. The wide world of plant molecular genetics. The Plant Cell 5: 821–830.

Riesmejer, J.N., V.I. Fluegge, B. Schulz, D. Heineke, H.W. Heldt, L. Willmitzer & W.B. Frommer, 1993. Antisense repression of the chloroplast triose phosphate translocator affects carbon partitioning in transgenic potato plants. Proc. Natl. Acad. Sci. USA 90: 6160–6164.

Rock, C.D. Jr. & A.D. Zeevart, 1991. The *aba* mutant of *Arabidopsis thaliana* is impaired in epoxy-carotenoid biosynthesis. Proc. Natl. Acad. Sci. USA 88: 7496–7499.

Roe, J.L., C.J. Rivin, A. Sessions, K.A. Fieldmann & P.C. Zambryski, 1993. The tousled gene in *A. thaliana* encodes a protein kinase homolog that is required for leaf and flower development. Cell 75: 939–950.

Rousselin, P., Y. Kraepiel, R. Maldiney, E. Miginiac & M. Caboche, 1992. Characterization of three hormone mutants of *Nicotiana plumbaginifolia*: evidence for a common ABA deficiency. Theor. Appl. Genet. 85: 213–221.

Schwarz-Sommer, Z., I. Hue, P. Huijser, P.J. Flor, R. Hunsen, F. Tetens, L. Wolf-Ekkerard, H. Saedler & H. Sommer, 1992. Characterization of the *Antirrhinum* floral homoeotic MADS-box gene deficiens: evidence for DNA binding and autoregulation of its persistent expression throughout flower development. EMBO J. 11: 251–263.

Sheen, J., 1990. Metabolic repression of transcription in higher plants. The Plant Cell 2: 1027–1038.

Shininger, T.L., 1979. The control of vascular development. Annu. Rev. Plant. Physiol. 30: 313–317.

Simoni, A., P. Bogani, P. Bettini, M.G. Pellegrini & M. Buiatti, 1993. Analisi del polimorfismo del DNA in somacloni di pomodoro mediante RAPD. Atti XXXVII Convegno Annuale SIGA p. 120.

Smart, C.M., S.R. Scofielf, M.W. Bevan & T.A. Dyer, 1991. Delayed leaf senescence in tobacco plants transformed with *tmr*, a gene for cytokinin production in *Agrobacterium*. Plant Cell 3: 647–656.

Smigocki, A.C. & L.D. Owens, 1989. Cytokinin to auxin ratios and morphology of shoots and tissue transformed by a chimeric isopentenyl transferase gene. Plant Physiol. 91: 808–811.

Smith, C.J.S., C.F. Watson, C.R. Bird, J. Ray, W. Schuch & D. Grierson, 1990. Expression of a truncated tomato polygalacturonase gene inhibits expression of the endogenous gene in transgenic plants. Mol. Gen. Genet. 224: 477–481.

Sonnewald, V., M. Braver, A. Von Schaewer, M. Stitt & L. Willmitzer, 1991. Transgenic tobacco plants expressing yeast derived invertase in either cytosol, vacuole or apoplast: a powerful tool for studying sucrose metabolism and sink/source interactions. The Plant J. 1: 95–106.

Spena, A., J.J. Estruch, E. Prinsen, W. Nacken, H. Van Onckelen & H. Sommer, 1992. Anther specific expression of the *rol* B gene of *Agrobacterium rhizogenes* increases IAA content in anthers and alters anther development and whole flower growth. Theor. Appl. Genet. 84: 520–527.

Storti, E., C. Latil, S. Salti, P. Bettini, P. Bogani, M.G. Pellegrini, C. Simeti, A. Molnar & M. Buatti, 1992. The *in vitro* physiological phenotype of tomato resistance to *Fusarium oxysporum* f.sp. *lycopersici*. Theor. Appl. Genet. 84: 123–128.

Storti, E., P. Bogani, P. Bettini, P. Bittini, M.L. Guardiola, M.G. Pellegrini, D. Inzé & M. Buiatti, 1994. Modification of competence for *in vitro* response to *Fusarium oxysporum* in tomato cells: II. Effect of the integration of *Agrobacterium tumefaciens* genes for auxin and cytokinin synthesis. Theor. Appl. Genet. 88: 89–96.

Tal, M. & Y. Nevo, 1973. Abnormal stomatal behaviour and root resistance, and hormonal imbalance in three wilty mutants of tomato. Biochem. Genet. 8: 291–300.

Taylor, I.B., 1992. Genetics of ABA synthesis. p. 23–37. In: W.D. Davies & M.G. Jones (Eds). Abscisic Acid. Bios Scientific Publ.

Tröbner, W., L. Ramirez, P. Motte, I. Hue, P. Huijser, L. Wolf Ekkehard, H. Saedler, H. Sommer & Z. Schwarz-Sommer, 1992. GLOBOSA: a homeotic gene which interacts with DEFICIENS in the control of *Antirrhinum* floral organogenesis. EMBO J. 11: 4693–4704.

Waddington, C.H., 1968. The basic ideas of biology. p. 12–18. In: C.H. Waddington (Ed). Towards a Theoretical Biology 1: Prolegomenon. Aldine, Chicago.

Wolter, F.B., R. Schmidt & E. Heniz, 1992. Chilling sensitivity of *Arabidopsis thaliana* with genetically engineered membrane lipids. EMBO J. 11: 4685–4692.

Yates, F.E., 1993. Self organizing systems. p. 189–218. In: C.A.R. Boyd & D. Noble (Eds). The Logic of Life. Oxford University Press, Oxford.

Yeng, S.F. & N.E. Hoffmann, 1984. Ethylene biosynthesis and its regulation in higher plants. Annu. Rev. Plant Physiol. 35: 155–189.

Yoshida, K., Y. Nagano, N. Murai & Y. Sasaki, 1993. Phytochrome-regulated expression of genes encoding the small GTP-binding protein in peas. Proc. Natl. Acad. Sci. USA 90: 6636–6640.

Yusibov, V.M., P. Chun, V.M. Andrianov & E.S. Piruzian, 1991. Phenotypically normal transgenic T-*cyt* tobacco plants as a model for the investigation of plant gene expression in response to phytohormonal stress. Plant Mol. Biol. 17: 825–836.

Euphytica **85**: 149–158, 1995.

Strategies for engineering virus resistance in transgenic plants

T.A. Kavanagh & C. Spillane
Department of Genetics, Trinity College, Dublin 2, Ireland

Key words: plant genetic engineering, virus-resistant transgenic plants

Summary

Transgenic virus-resistant plants were first produced in 1986 by genetically engineering tobacco plants to express the coat protein of tobacco mosaic virus. The introduction of coat protein transgenes has since proved to be an extremely effective and generally applicable approach to engineering virus resistance in crop plants. Extensive field trials with transgenic, virus-resistant tobacco, tomato, potato and cucumber lines have confirmed not only the durability of the resistance under natural conditions but the ease with which virus-resistant lines retaining the original cultivar traits can be recovered.

A number of alternative anti-viral strategies based on transgenes from a surprisingly wide variety of sources have also been developed. These include the use of viral genes coding for proteins involved in the replication cycle and in systemic transport of viruses within the plant, the use of interfering viral RNA sequences, and the use of transgenes derived from plant and animal sources. In the latter category, the use of mammalian antibodies to confer disease resistance in plants is a particularly exciting new development. Considerable progress has also been made towards the molecular cloning of natural anti-viral resistance genes in plants.

Introduction

Most crop plant species are susceptible to a number of different pathogens, some of which may cause severe systemic infection resulting in significant crop losses. Hence a major preoccupation of both breeders and growers alike has been the development of strategies that protect against infection. From the grower's point of view, strategies that limit crop exposure to potential pathogens and that control pathogen concentration in the environment (for example by controlling insect vector populations and planting only certified virus-free stock), are a vital part of crop husbandry. By contrast the breeder's objective is to maintain an effective level of resistance to viral pathogens through a programme of continuous cultivar improvement.

For most crop plants, evolutionary forces in the pathogen population act in time to subvert the effectiveness and durability of resistance. Consequently, the identification of new sources of resistance is a major challenge. Most new resistance genes are identified amongst the wild relatives of crop species but many cannot be incorporated into conventional breeding programmes because of barriers to sexual or somatic hybridization. Fortunately, over the past decade, advances in molecular genetics and in particular, the development of plant transformation technologies, have made it possible to engineer high-level resistance to viral pathogens in plants using transgenes from a wide variety of sources.

Crucially, because the transgenic approach for generating virus-resistant cultivars involves the single-step introduction of individual highly characterized transgenes, the recovery of plants that retain the original cultivar traits is relatively straightforward (Jongedijk et al., 1992). This is in marked contrast to conventional resistance-breeding strategies which often involve wide crosses followed by extensive (and expensive) backcrossing to achieve the same objective. Consequently, from a commercial point of view, the transgenic approach is very attractive.

Genetic engineering for virus resistance is one of the major success stories in plant biotechnology. New sources of antiviral transgenes that confer effective resistance in plants continue to be discovered. In this paper, we detail the variety of transgenes that has been used to engineer virus resistance, and comment on future directions.

Pathogen-derived transgenes

Paradoxically, the first virus-resistant, transgenic plants were generated using a transgene derived from a viral pathogen. The idea that a pathogen's genome might itself be the source of resistance genes was formally proposed by Sandford & Johnson (1985). Anticipating rapid progress in transgenic plant technology, they proposed that the inappropriate overexpression of wild-type or mutant viral genes in a host plant would disrupt the life-cycle of an incoming virus and so confer resistance-by-default on the host. A working precedent for such a scenario already existed in the phenomenon of viral cross protection, in which plants infected with one strain of a virus are found to be immune to superinfection by another strain of the same virus. This natural phenomenon has been exploited for many years by horticulturists in the form of deliberate infection of plants with a mild strain of a virus in order to confer protection against more severe strains (Fulton, 1986). With the development of efficient techniques for generating transgenic plants, it became possible to investigate which viral determinants, if any, were capable of conferring a cross-protection-like effect.

Coat protein-mediated resistance

A gene encoding the viral coat or capsid protein of tobacco mosaic virus (TMV) was the first pathogen-derived transgene to be tested for its cross protection potential in transgenic plants (Powell-Abel et al., 1986). The resulting plants were found to be resistant to TMV infection, with those expressing the highest levels of coat protein showing the highest levels of resistance. Furthermore it was clear that the resistance was not due to somaclonal variation and was stably inherited. The general applicability of this strategy for engineering virus resistance has since been widely documented for members of at least 20 different RNA viruses (reviewed in Hull & Davies, 1992). This includes tomato spotted wilt virus (TSWV), a member of the Tospoviruses (lipid-enveloped viruses) where transgenic plants expressing the nucleocapsid protein gene were found to be resistant to infection (Gielen et al., 1991). More recently, resistance to a single-stranded DNA virus, the geminivirus tomato yellow leaf curl virus, has been demonstrated in transgenic tomato plants expressing the viral coat protein gene (Kunik et al., 1994).

Typically, coat protein-mediated resistance is manifested as a reduction in the number of lesions on inoculated leaves, a reduced rate of systemic disease development and very low levels of virus accumulation in transgenic compared with control plants following deliberate inoculation with the parent virus. In many instances, the strength of coat protein-mediated resistance can approach near immunity to infection (exemption from infection) even with high concentration inocula.

Several lines of evidence suggest that the underlying mechanism responsible for the resistant phenotype differs depending on the particular viral group or viral transgene being studied. In the case of TMV (Powell Abel et al., 1986), alfalfa mosaic virus (AIMV) (Loesch-Fries et al., 1987) and PVX (Hemenway, 1988) the strength of resistance correlated positively with the levels of coat protein in transgenic plants. Indeed plants that accumulated only coat protein transcripts of TMV or AIMV and not the coat protein itself were not resistant. However, in the case of potato virus Y (PVY) (Farinelli & Malnoe, 1993) and potato leafroll virus (PLRV) (Barker et al., 1993) resistance correlated with the levels of coat protein transcripts and not with levels of the coat protein. Moreover, in both these studies some virus resistant transgenic lines were identified in which the coat protein was undetectable. TSWV complicates the picture still further; virus-resistant plants were obtained when an intact nucleocapsid protein transgene was used but also when the transgene was rendered untranslatable through removal of the initiating ATG codon (deHaan et al., 1992).

The specificity of coat protein-mediated resistance has been extensively studied in several systems. Transgenic tobacco plants expressing the coat protein gene of soybean mosaic virus (SMV), for which tobacco is a nonhost species, were found to be resistant to two serologically unrelated potyviruses, PVY and tobacco etch virus (TEV) (Stark & Beachy, 1989). The coat proteins of SMV, TEV and PYV share approximately 60% amino acid sequence homology. Similarily, transgenic plants expressing the coat protein of TMV also showed significant levels of resistance to infection with viruses whose coat proteins shared 60% or greater amino acid homology with the TMV coat protein (Nejidat & Beachy, 1990). Significant levels of heterologous resistance have also been reported in other virus groups.

Transgenic tobacco plants expressing the coat protein gene of cucumber mosaic virus (CMV) were found to be protected against chrysanthemum mild mottle virus (Nakajima et al., 1993) while tobacco plants expressing the coat protein gene of lettuce mosaic virus (LMV), a potyvirus, were found to be resistant not only to LMV but also to PVY (Dinant et al., 1993).

Field testing of virus-resistant transgenic plants is vital in order to test the durability of the resistance under natural conditions and to determine whether important agronomic characteristics of the original cultivar have been retained following the transformation procedure. Because the coat protein-mediated resistance strategy has been widely applied since the mid-1980's, many large scale field trials of transgenic virus-resistant potato (Kaniewski et al., 1990; Jongedijk et al., 1992), tomato (Sanders et al., 1992), and cucumber (Gonsalves et al., 1992) lines have now been completed after several years of evaluation. These studies confirm the durability of the resistance under field conditions and further show that commercially valuable, highly resistant, true-to-type lines can be obtained.

Non-coat protein transgenes

The possibility of engineering resistance using viral genes other than the coat protein gene has also been extensively investigated. These include genes coding for functions such as (i) the RNA-dependant RNA polymerase or replicase protein involved in virus replication (ii) viral proteins involved in cell-to-cell movement and (iii) viral proteases involved in processing polyprotein gene products. In some of the following examples, it has not yet been definitively established whether the RNA transcript or the protein product of the transgene plays the active role in resistance.

Replicase-mediated resistance

The initial discovery that expression of a viral replicase transgene could confer resistance to infection, came from experiments designed to test whether the hypothetical 54 kD subunit of the replicase of TMV was involved in viral replication (Golemboski et al., 1990). Surprisingly, transgenic tobacco plants containing a cDNA copy of this portion of the replicase gene were found to be highly resistant to TMV infection even though the 54 kD protein product could not be detected. There are now several reports of similar 'replicase-mediated' resistance to different viruses: PVX (Braun & Hemenway, 1992; Longstaff et al., 1993), PVY (Audy et al., 1994), pea early browning virus (Macfarlane & Davies, 1992), CMV (Anderson et al., 1992) and cymbidium ringspot virus (CyRSV) (Rubino et al., 1993). These include transgenic plants containing either the entire wild-type viral replicase gene or a truncated or mutated version of the replicase gene in which particular sequences have been altered or deleted.

Replicase-mediated resistance can confer virtual immunity to infection. In general the resistance tends to be very strain-specific although Donson et al. (1993) report broad spectrum resistance to tobamoviruses in transgenic plants expressing a modified TMV replicase gene. The mechanism of replicase-mediated resistance is unclear. Depending on the particular system under investigation, there can either be an inverse relationship between the degree of resistance and the quantity of the expressed viral gene product or no relationship at all. In many cases the translational product of the transgene cannot be detected in plants that show high levels of resistance. This suggests that resistance may involve RNA transcripts of the transgene rather than the replicase protein itself.

Movement protein-mediated resistance

Systemic viral infections can only be established if progeny viruses can move from cell to cell away from their initial site of entry. Without movement, the infection will be effectively terminated and the plant will appear to be resistant to infection. Studies with TMV have shown that movement is dependent on a specific viral protein (the 30 kD 'movement protein') which alters the diameter of the plasmodesmatal pores connecting adjacent cells and so facilitates virus transport (Deom et al., 1992). Because systemic movement is a virus-encoded function, it was a logical candidate for attempts to generate virus-resistant plants. Transgenic plants expressing the wild-type natural gene were not found to be resistant to TMV whereas plants expressing a defective, mutated form of the 'movement protein' were resistant (Lapidot et al., 1993). Furthermore, unlike the resistance resulting from transgenic expression of the replicase function, the resistance was not strongly strain-specific. This suggests that movement proteins might be exploited to engineer broad spectrum resistance to viruses.

Polyprotein protease-mediated resistance

In viruses of the PVY group (potyviruses), the viral genome codes for a single large protein, the polyprotein, rather than several discrete protein products. Following its translation, the polyprotein is proteolytically processed to produce the mature functional gene products. Recent studies in transgenic plants of factors controlling the proteolytic processing of the polyprotein, led to the unexpected finding that plants engineered to express the viral protease domain of either PVY (Vardi et al., 1993) or tobacco vein mottling virus (Maiti et al., 1993), exhibited a high degree of resistance to the respective viruses. The resistance was very strain-specific and is presumably due to interference with normal processing of the primary polyprotein product resulting in disruption of the viral life-cycle.

RNA-based strategies for engineering symptom attenuation or resistance

Several lines of evidence suggest that expression of viral RNA sequences that may or may not contain open reading frames can also be used to confer virus resistance in transgenic plants. Disruption of the virus replication cycle would be achieved not by the overproduction of viral proteins, but by overproduction of sense (+ strand) or antisense (- strand) RNA copies of specific regions of the viral genome. In theory, if a plant cell contained a sufficiently high concentration of these molecules, they might act either (i) to sequester the minute quantities of essential gene products required for virus replication or (ii) to disrupt virus replication by promoting inappropriate RNA:RNA interactions.

Satellite RNAs

Satellite RNAs are small viral RNAs (approximately 300 nucleotides) that cannot infect or replicate by themselves and are essentially molecular parasites of replication-competent viruses. In this capacity, they can act either to increase or decrease the severity of symptoms produced by the helper virus (for a review see Palukaitis et al., 1992). The first indication that viral RNA sequences could be used to engineer resistance came from studies of transgenic plants that expressed cDNA copies of the symptom-attenuating satellite RNAs of CMV (Harrison et al., 1987) and tobacco ringspot virus (Gerlach et al., 1987). In both cases inoculation of transgenic plants with the parent virus led to amplification of the satellite RNA transcripts to very high levels. This in turn conferred protection against the otherwise severe effects of infection with the parent virus alone. Protection is almost certainly based on RNA-RNA interactions, since satellite RNAs do not contain open reading frames, but the precise mechanism is not known.

Antisense and sense RNAs

An alternative RNA-based strategy that has been successfully employed to down-regulate the expression of numerous plant nuclear genes, is that based on antisense RNA. So far, there have been few reports of effective resistance mediated by transgenes coding for viral antisense (- strand) RNAs. This is puzzling since antisense RNAs have been shown to be effective inhibitors of viral RNA translation *in vitro* but it may reflect the fact that only specific regions of the genome are appropriate targets for antisense inhibition. Antisense-mediated resistance has however been reported for the geminivirus tomato golden mosaic virus (TGMV) and for TMV. Transgenic tobacco plants expressing antisense transcripts of the AL1 gene of TGMV showed significantly reduced symptom development when challenged with the virus, the reduction showing a broad correlation with the level of antisense transcripts (Day et al., 1991). More recently, antisense transgenes directed against the 5′ end of the viral genome have been successfully employed to engineer resistance to TMV (Nelson et al., (1993).

The ability of viral sense RNA transcripts (apart from those derived from coat protein or replicase transgenes) to confer resistance in transgenic plants has also been investigated. Zaccomer et al. (1993) reported that transgenic plants expressing the 3′ untranslated region of the genome of turnip yellow mosaic virus (TYMV) were partially protected against TYMV while Kollar et al. 1993) reported that *Nicotiana benthamiana* plants expressing transgenes encoding defective interfering RNAs of CyRSV, were resistant to CyRSV infection.

Plant-derived transgenes

Genes encoding pathogenesis-related proteins

The most common manifestation of resistance to viral pathogens in plants is the so-called hypersensitive response (HR). Typically, hypersensitivity is triggered by recognition of the pathogen by the host and results in programmed cell death around the initial site of infection. The recognition event itself has been shown, in some instances, to depend on a single host resistance gene (Dixon & Lamb, 1990). Recognition in turn leads to the induction of a large number of defence-related genes. Some of these act locally at the site of infection to cause HR-associated cell death but many are expressed systemically and are involved in the development of a higher than normal resistance state throughout the plant, termed systemic acquired resistance (SAR) (Ward et al., 1991). Chief among this latter class of genes whose expression correlates with the development of SAR, are those coding for the so-called pathogenesis-related (PR) proteins. A large number of PR proteins, first detected in tobacco leaves that were reacting hypersensitively to TMV, have now been characterized. They include at least 10 major acidic proteins which are located predominantly in the intercellular spaces, and a set of basic proteins having a vacuolar location (Hooft van Huijsduinen, 1986a). Several PR proteins have been characterized biochemically and shown to possess glucanase (Kauffman et al., 1987) or chitinase activity (Legrand et al., 1987) or to function as permatins (Roberts & Selitrennikoff, 1990).

In tobacco, SAR confers a high degree of protection not only to further challenges with TMV but also to other viral, fungal and bacterial pathogens. Because the development of SAR and the synthesis of PR proteins occur simultaneously, it has been suggested that PR proteins play an active part in defense. Experimental evidence supporting this conclusion derived initially from the observation that salicylic acid, when sprayed onto tobacco leaves, induces a subset of PR proteins whose appearance parallels the development of SAR (Hooft van Huijsduijnen et al., 1986b). Salicylic acid is a natural plant metabolite and has recently been shown to play a critical role in the development of SAR; plants genetically engineered to degrade endogenous salicylic acid are unable to establish SAR (Gaffney et al., 1993).

The expression of genes encoding PR proteins in transgenic plants would provide the most direct way of assessing whether their appearance merely correlates with the development of SAR or whether they play an active role in resistance. Should the latter be the case, then the range of available strategies for engineering broad spectrum resistance to plant pathogens would be greatly increased. Unfortunately however, evidence supporting an active role for PR proteins in the resistance response to viral pathogens has not been forthcoming. Linthorst et al. (1989) generated transgenic tobacco plants that constitutively expressed one of three different PR proteins (PR-1b, PR-S and GRP) but found no evidence of increased resistance to viral infection when compared with non-transgenic control plants. However, a more recent study of transgenic tobacco plants expressing high levels of PR-1a, a protein of unknown function (Alexander et al., 1993), found the plants exhibited increased tolerance to two fungal pathogens. Neither of these studies suggests a critical role for PR proteins in SAR. It remains to be seen, however, whether this is generally true for PR proteins.

Plant genes encoding anti-viral proteins

A class of polypeptides variously called anti-viral or ribosome-inactivating proteins (RIPs) have been identified in a number of plant species (Stirpe et al., 1992) of which the best known source is pokeweed, *Phytolacca americana.* Three distinct pokeweed antiviral proteins (PAPs) have been identified: PAP is found in spring leaves, PAPII in summer leaves and PAP-S in seeds. The ribosome-inhibiting function is due to their ability to modify ribosomal RNA and thereby to interfere with polypeptide translation. Surprisingly, when assayed *in vitro*, the ribosomes of most RIP-producing dicotyledons are not immune to their respective ribosome-inhibiting activities. *In vivo* toxicity is thought to be avoided through intracellular or extracellular compartmentalization of RIPs.

RIPs also possess potent broad-spectrum antiviral activity although neither the virus particle nor the viral genome itself is the direct target of this activity. Thus Tomlinson et al. (1974) found that CMV retained full infectivity if re-purified from a solution containing PAP. However, when applied to leaves, PAP-S can protect tobacco and potato plants from infection by mechanically inoculated PVX and PVY but not from PVY or PLRV transmitted by aphids (Lodge et al., 1993). This suggests that mechanically inoculated PAP enters damaged cells along with the invading virus and inhibits production of viral gene products required for replication. Presumably inoculation via the stylet of feeding aphids does not allow sufficient PAP to enter the cells and viral replication is unaffected. A recent study of the effects of five different RIPs on virus infectivity and symptom development (Taylor et al., 1994) supports the contention that the antiviral activity of RIPs is manifested as a consequence of entry into the cytosol where they function to inactivate host ribosomes.

Clearly the production of transgenic plants expressing RIPs would permit a more thorough assessment of their utility as broad spectrum antiviral agents. With this objective in mind, cDNAs encoding PAP were isolated from a *Phytolacca* cDNA library and used to generate transgenic tobacco and potato plants (Lodge et al., 1993). Those expressing high PAP levels (above 10 ng per mg total protein) were clearly abnormal, exhibiting a stunted, mottled phenotype which included sterility in the highest expressers; those expressing PAP in the range 1–5 ng/mg protein were fertile and normal in appearance.

Transgenic plants were assessed for broad spectrum resistance to viral infection by challenge with three unrelated viruses: PVX, PVY and CMV. Expression of PAP clearly conferred resistance to mechanically inoculated PVX and PVY and the degree of resistance varied roughly in proportion to the levels of PAP in a particular transgenic line: those expressing high levels of PAP were highly resistant to infection while those expressing lower levels showed more variable levels of resistance. However, only those transgenic lines expressing high levels of PAP were also found to be resistant to infection with CMV.

Natural host resistance genes

A number of clearly different resistance responses have been identified in plants that have been challenged with a potential viral pathogen. These responses can be considered from an operational point of view under three main headings: (i) cultivar-specific resistance in which particular cultivars of a normally susceptible species contain a gene or genes conferring resistance to a particular virus or virus strain, (ii) systemic acquired resistance which can be induced following an inductive treatment or by the hypersensitive response (as discussed above) and (iii) nonhost resistance manifested as the inability of particular virus 'species' or groups to infect particular plant species which are consequently defined as nonhosts (reviewed in Fraser, 1990).

Only in the case of cultivar-specific resistance has it been possible for plant breeders to identify resistance genes using classical genetic techniques and to incorporate them into conventional breeding programmes. In general, resistance to viral pathogens is controlled by single genes. In a random sample of 63 host-virus combinations described by Fraser (1990), most showed resistance conditioned by single gene loci, with resistance conditioned by oligogenic and polygenic systems the exception. However, despite the fact that many resistance genes have been identified, little more than their approximate chromosomal location is known and virtually nothing is known about how they function at the molecular level. This lack of basic information concerning the products encoded by resistance genes and their role in the resistance response, has hampered their isolation by recombinant DNA techniques. Molecular approaches are beginning to make a significant impact in resistance gene isolation, in particular, the use of map-based (or marker-based) cloning techniques and the use of transposable elements to 'tag' resistance genes.

With regard to potato, isolation of the Rx resistance gene which confers 'extreme resistance' or 'immunity' to PVX (Adams et al., 1986; Ritter et al., 1991) is a major objective. Rx resistance is unusual it that it can be assayed in potato protoplasts where the resistance phenotype is seen as an almost complete suppression of virus replication (Adams et al., 1986). The resistance response has been shown to involve two distinct events: recognition of a specific region of the viral coat protein (Kavanagh et al., 1992) resulting in the activation of the resistance mechanism (Kohm et al., 1993). The specificity of the interaction between PVX and the Rx gene product resides solely in the recognition event; the resistance mechanism itself acts non-specifically to suppress virus replication. Indeed, once induced, the resistance mechanism has been shown to be equally effective in suppressing the replication of completely unrelated viruses (Kohm et al., 1993). If through genetic manipulation, it were possible to broaden the recognition specificity of the Rx resistance response while retaining the function(s) conferring non-specific suppression of virus replication, then the Rx gene might well prove to be valuable for engineering broad-spectrum virus resistance in a range of crop plants.

Considerable progress has already been made towards the eventual isolation of the Rx gene. Using an RFLP mapping approach, Ritter et al. (1991) have established the precise chromosomal location of Rx and several research groups are currently closing in on the gene. The effectiveness of map-based cloning strategies for disease resistance gene isolation has recently been demonstrated by the isolation of the tomato Pto gene which convers resistance to *Pseudomonas syringae* pv. tomato (Martin et al., 1993).

An alternative strategy for resistance gene cloning is based on the use of transposable elements such as the maize Ac element to 'tag' resistance genes. This approach involves generating transgenic plants that contain a defective Ac element that can be induced to transpose to new locations in the genome and subsequently stabilized at that location. Insertion of Ac into a dominant resistance gene in a heterozygous plant will result in loss of the resistance phenotype and will facilitate isolation of the gene by conventional cloning techniques using the Ac element as a molecular probe. The N gene which confers resistance to TMV in tobacco and the Cf9 gene which confers resistance to *Cladosporium fulvum* in tomato have been isolated using this approach (Whitham et al., 1994 and Jones et al., 1994, respectively).

Non-plant, non-pathogen-derived transgenes

Antibodies

The mammalian immune system mounts an extremely effective surveillance against invading pathogens by its ability to produce vast numbers of antigen-specific antibodies. Because pathogen recognition and ultimately pathogen neutralization are mediated by antibodies, it was considered likely that the expression in transgenic plants of genes coding for antibodies might confer high level resistance to specific pathogens. A major consideration was whether antibody molecules manufactured in a plant cell would still be capable of functioning like natural antibodies. In the mammalian immune system, antibodies function primarily to bind antigens which are consequently targetted for destruction by the various killer cells of the immune system. In a transgenic plant however, the ability of an antibody to confer resistance to a pathogen would presumably depend entirely on its ability to interfere with the correct functioning of essential pathogen gene products by binding to their antigenic sites.

Initial attempts at expressing monoclonal antibodies (mAbs) that retained high levels of binding activity in transgenic plants were disappointing (Hiatt et al., 1989). This was almost certainly due to the use of mAb cDNAs that coded for the whole natural antibody molecule which has been shown to be relatively unstable in plant cells. The problem has largely been solved by the development of mini- (or single chain) antibodies which contain only the critical functional domains necessary for antigen recognition and binding and which show greatly enhanced stability in plants. This approach has recently been used to generate transgenic tobacco plants that are highly resistant to infection with artichoke crinkle mottle virus (ACMV) because they express a single chain mAb directed against the viral coat protein (Tavladoraki et al., 1993). Inoculation experiments with the primary transformed plants showed that transgenic lines inoculated with a suspension of AMCV at a 1,000-fold higher concentration than the minimum required to induce a systemic infection, reproducibly developed symptoms from 5 to 14 days later than did control plants over a 30-day observation period. Protoplast infection experiments indicated that virus resistance was the result of the ACMV-binding specificity of the expressed single chain mAb.

This technology which is still relatively unexplored will make the virtually unlimited repertoire of antibody specificity available for genetic engineering in plants.

Mammalian oligoadenylate synthetase

A component of another mammalian defence system, the interferon system, has also been exploited to confer broad-based resistance to viral infection in plants. Interferons are secreted by animal cells during cell proliferation, in response to immunological challenges and particularly in response to viral infections, but they do not themselves possess antiviral activity. Rather, interferons induce the synthesis of additional proteins that directly lead to inhibition of virus multiplication. One of these is the enzyme 2′-5′ oligoadenylate synthetase. This enzyme is activated by the presence of double-stranded RNA replicative intermediates that are formed during replication of RNA viruses. Once activated the enzyme polymerizes ATP into an oligomeric form that activates a latent ribonuclease (RNase L) which degrades viral and cellular RNAs.

Several lines of evidence suggest that some components of this anti-viral pathway may operate in plants: (i) tobacco plants that have been challenged with TMV have been shown to contain a double-stranded RNA-dependent ATP polymerizing activity, (ii) plant extracts can synthesize oligoadenylate-like compounds possessing anti-viral activity *in vitro* and (iii) exogenous application of 2′-5′ oligoadenylates and human interferon to leaves of tobacco and wheat plants causes induction of PR proteins and increased cytokinin activity (Kulaeva et al., 1992). Taken together, this suggests that expression of a mammalian oligoadenylate synthetase cDNA might confer non-specific antiviral activity in transgenic plants. This has recently been confirmed; transgenic potato plants expressing a rat oligoadenylate synthetase gene were found to be resistant to infection with PVX under field conditions (Truve et al., 1993). Whether the resistance is also effective against different viruses has not yet been fully explored.

Environmental considerations

Little is known about the potential environmental risks associated with the release of virus-resistant, transgenic plants. The major concern about potential risks focuses primarily on the use of transgenes derived from viral sources and centres on whether these transgenes will provide novel (and unnatural) opportunities for the evolution of virus populations in such a way as to promote the more rapid emergence of resistant variants. Three mechanisms have been proposed by which this might occur: (i) recombination between the RNA genome of an incoming virus and transgene-derived viral RNA transcripts, (ii) altered patterns of virus transmission by insect vectors due to encapsidation of incoming viral RNA genomes by coat proteins expressed from a resident coat protein transgene (heteroencapsidation), and in the particular case of transgenes based on satellite RNAs, (iii) more rapid acquisition of satellite RNAs by satellite-free isolates of the virus. These are legitimate concerns, not the least because each of these phenomena has been shown either to occur naturally or under laboratory conditions and because our current state of knowledge about the behaviour of viral populations is so incomplete. However, this area involves much speculation and is beyond the scope of this paper. A more in-depth treatment of the risk issues can be found in de Zoeten (1991), Tepfer (1993) and Falk & Breuning (1994).

References

Adams, S.E., R.A.C. Jones & R.H.A. Coutts, 1986. Expression of potato virus X resistance gene Rx in potato leaf protoplasts. J. Gen. Virol. 67: 2341–2345.

Alexander, D., R.M. Goodman, M. Gut-Rella, C. Glascock, K. Weyman, L. Friedrich, D. Maddox, P. Ahl-Goy, T. Lunz, E. Ward & J. Ryals, 1993. Increased tolerance of two oomycete pathogens in transgenic tobacco expressing pathogenesis-related protein 1a. Proc. Natl. Acad. Sci. USA 90: 7327–7331.

Anderson, J.M., P. Palukaitis & M. Zaitlin, 1992. A defective replicase gene induces resistance to cucumber mosaic virus in transgenic tobacco plants. Proc. Natl. Acad. Sci. USA 89: 8759–8763.

Audy, P., P. Palukaitis, S.A. Slack & M. Zaitlin, 1994. Replicase-mediated resistance to potato virus Y in transgenic tobacco plants. Mol. Plant-Microbe Interact. 7: 15–22.

Barker, H., B. Reavy, K.D. Webster, C.A. Jolly, A. Kumar & M.A. Mayo, 1993. Relationship between transcript production and virus resistance in tobacco expressing the potato leafroll virus coat protein gene. Plant Cell Rep. 13: 54–58.

Braun, C.J. & C.L. Hemenway, 1992. Expression of amino-terminal portions or full-length viral replicase genes in transgenic plants confers resistance to potato virus X infection. Plant Cell 4: 735–744.

Day, A.G., E.R. Bejarano, K.W. Buck, M. Burrell & C.P. Lichtenstein, 1991. Expression of an antisense viral gene in transgenic tobacco confers resistance to the DNA virus tomato golden mosaic virus. Proc. Natl. Acad. Sci. USA 88: 6721–6725.

deHaan, P., J.J.L. Gielen, M. Prins, I.G. Wijkamp, A. van Schepen, D. Peters, M.Q.J.M. van Grinsven & R. Goldbach, 1992. Characterization of RNA-mediated resistance to tomato spotted wilt virus in transgenic tobacco plants. Bio/Technology 10: 1133–1137.

Deom, C.M., M. Lapidot & R.N. Beachy, 1992. Plant virus movement proteins. Cell 69: 221–224.

de Zoeten, G.A., 1991. Risk assessment: Do we let history repeat itself? Phytopathology 81: 585–586.

Dinant, S., F. Blaise, C. Kusiak, S. Astier-Manifacier & J. Albouy, 1993. Heterologous resistance to potato virus Y in transgenic tobacco plants expressing the coat protein gene of lettuce mosaic potyvirus. Phytopathology 83: 818–824.

Dixon, R.A. & C.J. Lamb, 1990. Molecular communication in interactions between plants and microbial pathogens. Annu. Rev. Plant Physiol. Plant Mol. Biol. 41: 339–367.

Donson, J., C.M. Kearney, T.H. Turpen, I.A. Khan, G. Kurath, A.M. Turpen, G.E. Jones, W.O. Dawson & D.J. Lewandowski, 1993. Broad resistance to tobamoviruses is mediated by a modified tobacco mosaic virus replicase gene. Mol. Plant-Microbe Interact. 6: 635–642.

Falk, B.W. & G. Breuning, 1994. Will transgenic crops generate new viruses and new diseases? Science 263: 1395–1396.

Farinelli, L. & P. Malnoe, 1993. Coat protein gene-mediated resistance to potato virus Y in tobacco: examination of the resistance mechanisms – Is the transgenic coat protein required for protection? Mol. Plant-Microbe Interact. 6: 284–292.

Fraser, R.S.S., 1990. Genes for resistance to plant viruses. Crit. Rev. Plant Sci. 3: 275–294.

Fulton, R.W., 1986. Practices and precautions in the use of cross protection for plant virus disease control. Ann. Rev. Phytopath. 24: 67–93.

Gaffney, T., L. Friedrich, B. Vernooij, D. Negrotto, G. Nye, S. Uknes, E. Ward & J. Ryals, 1993. Requirement for salicylic acid for the induction of systemic acquired resistance. Science 261: 754–756.

Gerlach, W.L., D. Llewellyn & J. Haseloff, 1987. Construction of a disease resistance gene using the satellite RNA of tobacco ringspot virus. Nature 328: 802–806.

Gielen, J.J.L., P. de Haan, A.J. Kool, D. Peters, M.Q.J.M. van Grinsven & R.W. Goldbach, 1991. Engineered resistance to tomato spotted wilt virus, a negative strand RNA virus. Bio/Technology 9: 1363–1367.

Golemboski, D.B., G.P. Lomonosoff & M. Zaitlin, 1990. Plants transformed with a tobacco mosaic virus nonstructural gene sequence are resistant to the virus. Proc. Natl. Acad. Sci. USA 87: 6311–6315.

Gonsalves, D., P. Chee, R. Provvidenti, R. Seem & J.L. Slightom, 1992. Comparison of coat protein-mediated and genetically-derived resistance in cucumbers to infection by cucumber mosaic virus under field conditions with natural challenge inoculations by vectors. Bio/Technology 10: 1562–1570.

Harrison, B.D., M.A. Mayo & D.C. Baulcombe, 1987. Virus resistance in plants that express cucumber mosaic virus satellite RNA. Nature 328: 799–802.

Hemenway, C., R.-X. Fang, J.J. Kaniewski, N.-H. Chua & N.E. Tumer, 1988. Analysis of the mechanism of protection in transgenic plants expressing the potato virus X coat protein or its antisense RNA. EMBO J. 7: 1273–1280.

Hiatt, A., R. Cafferkey & K. Bowdish, 1989. Production of antibodies in transgenic plants. Nature 342: 76–78.

Hooft van Huijsduijnen, R.A.M., S.W. Alblas, R.H. De Rijk & J.F. Bol, 1986b. Induction by salicylic acid of pathogenesis-related proteins and resistance to alfalfa mosaic virus infection in various plant species. J. Gen. Virol. 67: 2135–2143.

Hooft van Huijsduijnen, R.A.M., L.C. Van Loon & J.F. Bol, 1986a. cDNA cloning of six mRNAs induced by TMV infection of tobacco and a characterization of their translation products. EMBO J. 5: 2057–2061.

Hull, R. & J.W. Davies, 1992. Approaches to nonconventional control of plant virus diseases. Crit. Rev. Plant Sci. 11: 17–33.

Jones, J.D.G., M. Dixon, K. Hammond-Kosack, K. Harrison, K. Hatzixanthis, D. Jones & C. Thomas, 1994. Characterization of tomato genes that confer resistance to *Cladosporium fulvum*. Abstracts, 4th International Congress of Plant Molecular Biology.

Jongedijk, E., A.J.M. de Schutter, T. Stolte, P.J.M. van den Elzen & B.J.C. Cornelissen, 1992. Increased resistance to potato virus X and preservation of cultivar properties in transgenic potato under field conditions. Bio/Technology 10: 422–429.

Kaniewski, W., C. Lawson, B. Sammons, L. Haley, J. Hart, X. Delannay & N.E. Tumer, 1990. Field resistance of transgenic russet burbank potato to effects of infection by potato virus X and potato virus Y. Bio/Technology 8: 750–754.

Kauffman, S., M. Legrand, P. Geoffroy & G. Fritig, 1987. Biological function of 'pathogenesis-related' proteins: Four PR proteins of tobacco have 1,3-beta glucanase activity. EMBO J. 6: 3209–3212.

Kavanagh, T., M. Goulden, S. Santa Cruz, S. Chapman, I. Barker & D.C. Baulcombe, 1992. Molecular analysis of a resistance-breaking strain of potato virus X. Virology 189: 609–617.

Kohm, B.A., M.G. Goulden, J.E. Gilbert, T.A. Kavanagh & D.C. Baulcombe, 1993. A potato virus X resistance gene mediates an induced, non-specific resistance in protoplasts. The Plant Cell 5: 913–920.

Kallar, A., T. Dalmay & J. Burgyan, 1993. Defective interfering RNA-mediated resistance against cymbidium ringspot tombusvirus in transgenic plants. Virology 193: 313–318.

Kulaeva, O.N., A.B. Fedina, E.A. Burkhanova, N.N. Karaivako, M.Y. Karpeisky, I.B. Kaplan, M.E. Taliansky & J.G. Atabekov, 1992. Biological activities of human interferon and 2′-5′ oligoadenylates in plants. Plant Mol. Biol. 20: 383–393.

Kunik, T., R. Salomon, D. Zamir, N. Navot, M. Zeidan, I. Michelson, Y. Gafni & H. Czosnek, 1994. Transgenic tomato plants expressing the tomato leaf curl virus capsid protein are resistant to the virus. Bio/Technology 12: 500–506.

Lapidot, M., R. Gafny, B. Ding, S. Wolf, W.J. Lucas & R.N. Beachy, 1993. A dysfunctional movement protein of tobacco mosaic virus that partially modifies the plasmodesmata and limits virus spread in transgenic plants. Plant J. 4: 959–970.

Legrand, M., S. Kauffman, P. Geoffrey & B. Fritig, 1987. Biological function of pathogenesis-related proteins: Four tobacco pathogenesis-related proteins are chitinases. Proc. Natl. Acad. Sci. USA 84: 6750–6754.

Linthorst, H.J.M., R.L.J. Meuwissen, S. Kauffman & J.F. Bol, 1989. Constitutive expression of pathogenesis-related proteins PR-1, GRP, and PR-S in tobacco has no effect on virus infection. The Plant Cell 1: 285–291.

Lodge, J.K., W.K. Kaniewski & N.E. Tumer, 1993. Broad-spectrum virus resistance in transgenic plants expressing pokeweed antiviral protein. Proc. Natl. Acad. Sci. USA 90: 7089–7093.

Loesch-Fries, L.S., D. Merlo, T. Zinnen, L. Burhop, K. Hill, K. Krahn, N. Jarvis, S. Nelson & E. Halk, 1987. Expression of alfalfa mosaic virus RNA 4 in transgenic plants confers virus resistance. EMBO J. 6: 1845–1851.

Longstaff, M., G. Brigneti, F. Boccard, S. Chapman & D.C. Baulcombe, 1993. Extreme resistance to potato virus X infection in plants expressing a modified component of the putative viral replicase. EMBO J. 12: 379–386.

Macfarlane, S.A. & J.W. Davies, 1992. Plants transformed with a region of the 201-kilodalton replicase gene from pea early browning virus RNA1 are resistant to virus infection. Proc. Natl. Acad. Sci. USA 89: 5829–2833.

Maiti, I.B., J.F. Murphy, J.G. Shaw & A.G. Hunt, 1993. Plants that express a potyvirus proteinase gene are resistant to virus infection. Proc. Natl. Acad. Sci. USA 90: 6110–6114.

Martin, G.B., S.H. Brommoschenkels, J. Chunwongse, A. Frary, M.W. Ganal, R. Spivey, T. Wu, E.D. Earle & S.D. Tanksley, 1993. Map-based cloning of a protein kinase gene conferring disease resistance in tomato. Science 262: 1432–1436.

Nakajima, M., T. Hayakawa, I. Nakamura & M. Suzuki, 1993. Protection against cucumber mosaic virus (CMV) strains O and Y and chrysanthemum mild mottle virus in transgenic plants expressing CMV-O coat protein. J. Gen. Virol. 74: 319–322.

Nejidat, A. & R.N. Beachy, 1990. Transgenic tobacco plants expressing a tobacco virus coat protein gene are resistant to some tobamoviruses. Mol. Plant Microb. Interact. 3: 247–251.

Nelson, R.S., D.A. Roth & J.D. Johnson, 1993. Tobacco mosaic virus infection of transgenic *Nicotiana tabacum* plants is inhibited by antisense constructs directed a the 5′ region of viral-RNA. Gene 127: 227–232.

Palukaitis, P., M.J. Roosinck, R.G. Dietzgen & R.I.B. Francki, 1992. Cucumber mosaic virus. Adv. Virus Res. 41: 281–348.

Powell-Abel, P.A., R.S. Nelson, B. De, N. Hoffman, S.G. Rogers, R.T. Fraley & R.N. Beachy, 1986. Delay of disease development in transgenic plants that express the tobacco mosaic virus coat protein gene. Science 232: 738–743.

Ritter, E., T. Debener, A. Barone, F. Salamini & C. Gebhart, 1991. RFLP mapping on potato chromosomes of two genes controlling extreme resistance to potato virus X (PVX). Mol. Gen. Genet. 227: 81–85.

Roberts, W.K. & C.P. Selitrennikoff, 1990. Zeamatin, an antifungal protein from maize with membrane-permeabilizing activity. J. Gen. Microbiol. 136: 1171–1778.

Rubino, L., R. Lopo & M. Russo, 1993. Resistance to cymbidium ringspot virus infection in transgenic *Nicotiana benthamiana* plants expressing full-length viral replicase gene. Mol. Plant-Microbe Interact. 6: 729–734.

Sanders, P.R., B. Sammons, W. Kaniewski, L. Haley, J. Layton, B.J. Lavallee, X. Delannay & N.E. Tumer, 1992. Field resistance of transgenic tomatoes expressing the tobacco mosaic virus or tomato mosaic virus coat protein genes. Phytopathology 82: 683–690.

Sanford, J.C. & S.A. Johnson, 1985. The concept of parasite-derived resistance-deriving resistance genes from the parasite's own genome. J. Theor. Biol. 113: 395–405.

Stark, D.M. & R.N. Beachy, 1989. Protection against potyvirus infection in transgenic plants: evidence for broad spectrum resistance. Bio/Technology 7: 1257–1262.

Stirpe, F., L. Barbieri, M.G. Battelli, M. Soria & D.A. Lappi, 1992. Ribosome-inactivating proteins from plants: present status and future prospects. Bio/Technology 10: 405–412.

Tavladoraki, P., E. Benvenuto, S. Trinca, D. De Martinis, A. Cattaneo & P. Galeffi, 1993. Transgenic plants expressing a functional single chain Fv antibody are specifically protected from virus attack. Nature 366: 469–472.

Taylor, S., A. Massiah, G. Lomonossoff, L.M. Roberts, J.M. Lord & M. Hartley, 1994. Correlation between the activities of five ribosome-inactivating proteins in depurination of tobacco ribosomes and inhibition of tobacco mosaic virus infection. Plant J. 5: 827–835.

Tepfer, M., 1993. Viral genes and transgenic plants. Bio/Technology 11: 1125–1129.

Tomlinson, J.A., V.M. Walker, T.H. Flewett & G.R. Barclay, 1974. The inhibition of infection by cucumber mosaic virus by extracts from *Phytolacca americana*. J. Gen. Virol. 22: 225–232.

Truve, E., A. Aaspollu, J. Honkanen, R. Puska, M. Mehto, A. Hassi, T.H. Teeri, M. Kelve, P. Seppanen & M. Saarma, 1993. Transgenic potato plants expressing mammalian 2′-5′ oligoadenylate synthetase are protected from potato virus X infection under field conditions. Bio/Technology 11: 1048–1052.

Vardi, E., I. Sela, O. Edelbaum, O. Livneh, L. Kuznetsova & Y. Stram, 1993. Plants transformed with a cistron of a potato virus Y protease (NIa) are resistant to virus infection. Proc. Natl. Acad. Sci. USA 90: 7513–7517.

Ward, E.R., S.J. Uknes, S.C. Williams, S.S. Dincher, D.L. Wiederhold, D.C. Alexander, P. Ahl-Goy, J.P. Metraux & J.A. Ryals, 1991. Coordinate gene activity in response to agents that induce systemic acquired resistance. Plant Cell 3: 1085–1094.

Whitham, S., S.P. Dinesh-Kumar, D. Choi, R. Hehl, C. Corr & B. Baker, 1994. The product of the tobacco mosaic virus resistance gene N: Similarity to Toll and the Interleukin 1 receptor. Cell 78, 1101–1115.

Zaccomer, B., F. Cellier, J-C. Boyer, A-L. Haenni & M. Tepfer, 1993. Transgenic plants that express genes including the 3′ untranslated region of the turnip yellow mosaic virus (TYMV) genome are partially protected against TYMV infection. Gene 136: 87–94.

Euphytica **85:** 159–168, 1995.

Resistance to tomato spotted wilt virus in transgenic tomato hybrids

Tineke Ultzen[1], Jan Gielen[1], Fenna Venema[1], Annemarie Westerbroek[1], Peter de Haan[1], Mei-Lie Tan[1], André Schram[1], Mart van Grinsven[1] & Rob Goldbach[2]
[1] *S & G Seeds B.V., Department of Technology, Westeinde 62, P.O. Box 26, 1600 AA Enkhuizen, The Netherlands;*
[2] *Department of Virology, Agricultural University Wageningen, Binnenhaven 11, P.O. Box 8045, 6700 EM Wageningen, The Netherlands*

Key words: Lycopersicon, tomato, tomato spotted wilt virus, tospovirus, transformation, virus resistance

Summary

Tomato spotted wilt virus (TSWV) causes significant economic losses in the commercial culture of tomato (*Lycopersicon esculentum* Mill.). Culture practices and introgression of natural sources of resistance to TSWV have only been marginally effective in controlling the TSWV disease. Recently however, high levels of protection against TSWV have been obtained by transforming tobacco with a chimaeric gene cassette comprising the TSWV nucleoprotein gene. This report demonstrates the successful application of this newly-created TSWV resistance gene in cultivated tomato. Transformation of an inbred tomato line with the TSWV nucleoprotein gene cassette resulted in high levels of resistance to TSWV that were maintained in hybrids derived from the parental tomato line. Therefore, transformant lines carrying the synthetic TSWV resistance gene make suitable progenitors for TSWV resistance to be incorporated into the breeding programmes of tomato.

Introduction

The commercial culture of tomato (*Lycopersicon esculentum* Mill.) is seriously affected by tomato spotted wilt virus (TSWV) causing significant yield losses. Characteristic symptoms associated with TSWV include plant stunting, bronzing or chlorosis of leaves and the development of chlorotic or necrotic ringspots on the fruits. The virus is naturally transmitted by a number of thrips species, of which the western flower thrips (*Frankliniella occidentalis* Perg.) is the most important. The dramatic expansion of *F. occidentalis* from Northern America over the western hemisphere and its rapid acquisition of resistance to pesticides are the major reasons for recent outbreaks of TSWV disease, not only in tomato but also in other crops such as pepper and lettuce (Goldbach & Peters, 1994).

Among plant viruses TSWV is unique in its genomic organisation and particle morphology. The virus particle is bounded by a spherical lipoprotein envelope enclosing a core of nucleocapsids, consisting of three genomic RNA segments that are individually associated with nucleoprotein and that exhibit either negative or ambisense gene arrangements (Fig. 1) (de Haan et al., 1990, 1991; Kormelink et al., 1992). On the basis of these properties, TSWV has been classified as the type member of the genus *Tospovirus* within the Bunyaviridae (for a recent overview on tospovirus, see German et al., 1992). Recently, a number of distinct virus species have been identified and classified as new tospoviruses (Law & Moyer, 1990; Reddy et al., 1992; Yeh et al., 1992; de Avila et al., 1993) and it is expected that the number of tospoviruses will increase in the near future.

Culture practices such as rotation, control of the thrips vector and removal of alternate weed hosts have only been marginally effective in the management of TSWV disease (Cho et al., 1989). Consequently, host plant resistance to the virus is the most promising means of controlling the disease in the long term. Several accessions of *Lycopersicon* germplasm and tomato cultivars descending from such accessions have been reported to be resistant to TSWV (Smith, 1944; Finlay et al., 1952, 1953; Paterson et al., 1989; Kumar et al., 1993). Resistance to TSWV derived from *L. peruvianum* Mill. was inherited as a single dominant gene (Stevens et al., 1992; Boiteux & Giordano, 1993). However, in the field, plants carrying this trait still accumulate virus resulting in the development of disease symptoms on the fruits (personal observation), which renders the utilisation of this source questionable. The fact that tomato cultivars carrying a reliable source of genetic resistance to TSWV have still not reached the market, illustrates the limited applicability and the complex inheritance of resistance sources identified thus far. Moreover, the emergence of new tospoviruses that infect tomato, constitutes a serious threat and prompts the need for idenficication or development of new sources of resistance to be incorporated into tomato breeding programmes, especially when considering proposed reductions in the application of insecticides for reasons of environment protection.

Over the past decade, numerous publications have demonstrated the successful generation of virus resistance through transgenic expression of viral sequences in plants (reviewed by Hull & Davies, 1992; Scholthof et al., 1993; Wilson, 1993). Transformation of tobacco with the TSWV nucleoprotein gene (N gene) has thus been shown to result in resistance to TSWV (Gielen et al., 1991; MacKenzie & Ellis, 1992; Pang et al., 1992). Expression of translationally-defective N gene cassette generated similar levels of resistance, which indicates that the accumulation of nucleoprotein is not required to obtain TSWV resistance and that the observed resistance is primarily RNA-mediated (de Haan et al., 1992). To study the application of this technology in crops of agronomic importance, we transferred the TSWV N gene cassette to an inbred tomato line used in the production of fresh market hybrids. Upon mechanical inoculation, resistant transformant lines were identified which were then cross-pollinated to produce experimental hybrids. Transgenic TSWV resistance is successfully maintained in the hybrid, thereby indicating that transgenic parental lines can serve as progenitors for TSWV resistance in tomato breeding programmes.

Materials and methods

Virus and plant material. TSWV isolate BR-01 was maintained in tomato by grafting to prevent a generation of defective mutants by repeated mechanical passages (Resende et al., 1991a). The tobacco (*Nicotiana tabacum* cv. Xanthi) cell suspension used in the transformation procedure of tomato was grown in the dark at 26° C on a shaking platform and maintained through weekly subculturing in Xanthi medium: MS medium, 30 g/l sucrose, 100 mg/l inositol, 200 mg/l KH_2PO_4, 1.3 mg/l thiamine, 0.2 mg/l 2,4-dichlorophenoxyacetic acid (2,4-D), 0.1 mg/l kinetin (Murashige & Skoog, 1962). Parental tomato line ATV 847 was used as the recipient in transformation experiments. This inbred line is used as male parent in the production of a number of hybrids for the South European market that represent fresh market tomatoes of the determinant type. Transgenic tomato plants were grown under certified greenhouse conditions according to the legislation imposed by the Dutch authorities (Voorlopige Commissie Genetische Modificatie: VCOGEM).

Construction of the plant transformation vector. All manipulations involving DNA were essentially performed according to standard procedures (Ausubel et al., 1987). The TSWV N gene was amplified by means of the polymerase chain reaction (PCR) from a cDNA clone containing the full-length N gene (de Haan et al., 1990) and using oligomer primers 1823H (5′ GGG*CTGCAG*CTGCTT**TCA**AGCAAGTTC 3′) and 1824 (5′ TTAC*GATATC***ATG**TCTAAGG 3′). Primer 1824 hybridises to the ATG region of the N gene and introduces the recognition sequence of *Eco*RV immediately in front of the ATG initiation codon; primer 1823H is complementary to the 3′ end of the N gene and introduces a *Pst*I site at 6 nucleotides downstream of the TGA stop codon. The PCR-amplified fragment of 0.8 kb was digested with *Eco*RV and *Pst*I and ligated into expression vector pZU029, digested with *Sma*I and *Pst*I, yielding gene cassette pTSWVN-B. Expression vector pZU029 contains the cauliflower mosaic virus (CaMV) 35S promoter fused to the 5′ untranslated leader sequence from tobacco mosaic virus, and the nopaline synthase polyadenylation signal separated from the promoter fragment by a *Sma*I and *Pst*I restriction site. The complete gene cassette was released as an *Xba*I fragment and cloned into the binary plant transformation vector pBIN19 (Bevan, 1984). The resulting transformation vector was subsequently introduced into the non-oncogenic *Agrobacterium tumefaciens* strain LBA4404 (Ooms et al., 1981) by triparental mating using pRK2013 as a helper plasmid (Ditta et al., 1980). The recombinant *A. tumefaciens* strain was checked for the integrity of the transformation by Southern blot analysis.

Transformation of parental tomato line ATV847. The transformation method described below is based on the application of tobacco feeder layer cells during preculture and co-cultivation of the cotyledon explants (Shahin et al., 1986; Fillatti et al., 1987; Yoder et al., 1988). Thin layers of the tobacco cell suspension were poured onto Petri dishes containing Xanthi medium solidified with 10 g/l micro agar (Duchefa, Haarlem). The feeder plates were incubated at 26° C in the dark for 24 hrs. Directly before use, a sterile Whatman filter was placed on top of the feeder cells. Cotyledon explants were prepared from 8 to 10 days old *in vitro*-grown seedlings of parental tomato line ATV847. The cotyledons were cut across the base and top to provide two cut surfaces for infection. The resulting explants were subsequently placed on the feeder plates with the abaxial side up and precultured at 24° C in the dark for 24 hours. An overnight culture of the recombinant *Agrobacterium* strain carrying transformation vector pTSWVN-B was diluted in liquid MS medium containing 30 g/l sucrose to a density of 5×10^7 cells/ml. The explants were incubated for 5 to 15 minutes in the *Agrobacterium* suspension, dried on a sterile Whatman filter and placed back onto the feeder plates. After 48 hours of co-cultivation in the dark at 26° C, the explants were transferred to selection medium (MS medium, 10 g/l sucrose, 10 g/l glucose, 2/0 mg/l zeatin, 0.02 mg/l IAA, 250 mg/l carbenicillin, 100 mg/l kanamycin, 10 g/l micro-agar) with the axial side up. After 5 days in the dark, plates were transferred to the light (1500–2000 Lux) and explants were subcultured every 2 weeks. From 4 to 8 weeks after co-cultivation, shoot primordia were cut from the explants and elongated on MS10 medium (MS medium, 10 g/l sucrose, 250 mg/l carbenicillin, 10 g/l micro-agar). Shoots were rooted on MS10 medium supplemented with 25 mg/l kanamycin, potted in soil and transferred to the greenhouse. Trangenic tomato plants were subsequently analysed for expression of the TSWV N gene and for their ploidy level.

Serological analysis of transgenic tomato plants. Double antibody sandwich (DAS) ELISA was employed to detect the accumulation of TSWV nucleoprotein in transgenic tomato, using a rabbit polyclonal antiserum raised against purified TSWV nucleocapsids (Resende et al., 1991b). Protein samples were prepared by grinding leaf material in phosphate-buffered saline supplemented with 0.1% Tween-20 (PBS-T) and 2% insoluble polyvinylpolypyrrolidone, and incubated overnight at 4° C in microtiter plate wells (Nunc-Immuno Plate MaxiSorp^{TM}). The wells had previously been coated overnight at 4° C with antiserum diluted to 1 μg/ml in coating buffer (50 mM sodium carbonate buffer pH 9.6) and blocked with 1% BSA in PBS-T for 1 hour at room temperature. Bound antigen was detected by incubation with alkaline phosphate-conjugated antiserum (1 μg/ml in PBS-T) for 3 hours at 37° C, followed by para-nitrophenyl phosphate substrate development (1 mg/ml in 50 mM diethanolamine buffer pH 9.8). The absorbance of each well was measured at 405 nm. Between all incubation steps the wells were throughly rinsed with PBS-T.

The integrity of the TSWV nucleoprotein accumulating in the transgenic tomato plants was verified by Western blot analysis. Leaf tissue was homogenised in PBS-T and 25 μg of soluble protein was fractionated by electrophoresis in 12.5% SDS-polyacrylamide gels (Laemmli, 1970). Proteins were blotted to Immobilon-P membranes (Millipore) by semi-dry blotting in semi-day transfer buffer (29 mM glycine, 48 mM Tris, 0.0375% SDS and 20% methanol) for 1 hour at 0.8 mA/cm^2. Membranes were blocked for 3 hours at 37° C in PBS-T containing 3% BSA, and subsequently incubated with polyclonal antiserum conjugated with alkaline phosphatase (Resende et al., 1991b), diluted to 1 μg/ml in PBS-T supplemented with 0.3% BSA. The immunoblot was further processed using nitroblue tetrazolium chloride (NBT) and 5-bromo-4-chloro-3-indolylphosphate p-toluidine salt (BCIP) as substrate according to the supplier's instruction (ImmunoSelect^{TM}, Life Technologies Inc.). Between subsequent treatments the membrane was washed with PBS-T containing 0.3% BSA.

Flow cytometric analysis of ploidy levels. Tomato transformants accumulating TSWV nucleoprotein were analysed for their ploidy level by flow cytometry. Intact nuclei were stained with 4′,6-diamidino-2-phenylindol (DAPI) by chopping leaf tissue with a sharp razor blade in a commercial staining solution (Partec GmbH). Nuclei samples were filtered through nylon cloth (30 μm), kept on ice for at least 15 minutes and subsequently run through the flow cytometer (PAS-II, Partec GmbH) to determine their relative DNA-content (de Laat et al., 1987). Nuclei prepared from leaf tissue of diploid broccoli were used as the internal standard.

Southern blot analysis. Total DNA was extracted from transgenic tomato plants essentially as described by Doyle & Doyle (1990), using an isolation buffer

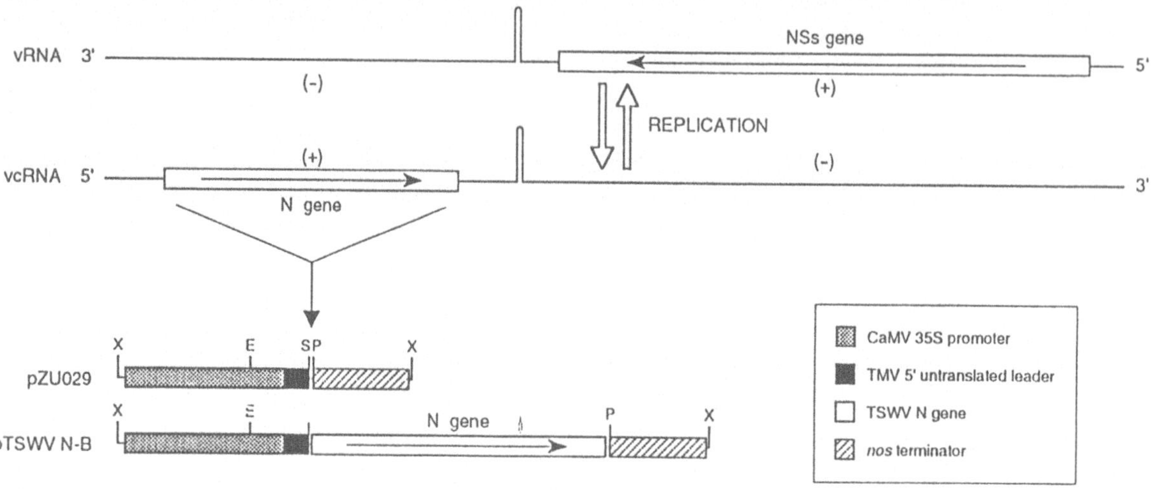

Fig. 1. Schematic representation of the small genomic RNA (S RNA) of TSWV and of the construction of the chimaeric TSWV N gene cassette. vRNA: viral RNA; vcRNA: viral complementary RNA. Sense and complementary sense regions of the vRNA and vcRNA strand of the ambisense S RNA are indicated with (+) and (-) respectively. The complete N gene cassette was cloned into binary transformation vector pBIN19 as a *Xba* I fragment. E: *Eco*RV; P: *Pst*I; S: *Sma* I; X: *Xba* I.

containing 2% hexadecyltrimethylammoniumbromide (CTAB). Portions of 10 μg DNA were digested with *Eco*RV, *Hind* III or *Xba* I, fractionated by electrophoresis in 0.8% agarose gels and transferred to Hybond-N membranes (Amersham) by capillary blotting in alkaline transfer buffer (Ausubel et al., 1987). The blot membranes were subsequently hybridised to a ^{32}P-labelled DNA fragment containing the TSWV N gene in a SSC-based hybridisation buffer containing 10% dextran sulphate (Wahl et al., 1979).

Analysis of protection to TSWV infection after mechanical inoculation. Prior to inoculation, offspring populations were analysed for the accumulation of nucleoprotein by DAS-ELISA to identify those progeny plants expressing the N gene cassette. After emergence of the first leaf, about 3 to 4 weeks after sowing, seedlings were dusted with carborundum powder and wiped with cotton-wool dipped in the virus inoculum. Since TSWV is highly unstable upon homogenisation, the inocula were freshly prepared by grinding 1 gram of systemically infected tomato leaves in 10 ml of 0.1 M sodium phosphate buffer (pH 7.0) supplemented with 1% Na_2SO_3 and kept on ice. Transgenic plants were inoculated first, followed by non-transformed control plants to check the inocula for their infectivity at the end of inoculation. All accessions were organised in a randomised block design with five or six replications. One week after the first inoculation the tomato plants were inoculated for a second time to achieve maximum disease incidence. After inoculation, plants were rinsed with water. The extent of the TSWV infection was monitored by visual observation for the development of systemic symptoms. Plants with aberrant phenotypes were omitted from the notations. Normally, susceptible tomato plants develop systemic symptoms within 2 to 4 weeks after mechanical inoculation with TSWV. Plants were scored as being susceptible when any leaf younger than the inoculated leaves showed typical systemic symptoms such as chlorosis or bronzing of the leaves. In addition, plant stunting and wrinkling or curling of the top leaves could be observed on systemically infected plants. The absence of virus in symptomless transgenic plants was checked by direct ELISA using a polyclonal antiserum raised against purified NSs protein, a nonstructural viral protein that accumulates to high levels in TSWV infected plant cells (Kormelink et al., 1991).

Results

Construction of the TSWV nucleoprotein gene cassette

The TSWV nucleoprotein (N) gene was amplified using PCR from a cDNA clone harbouring the complete viral gene. The primers used in the ampli-

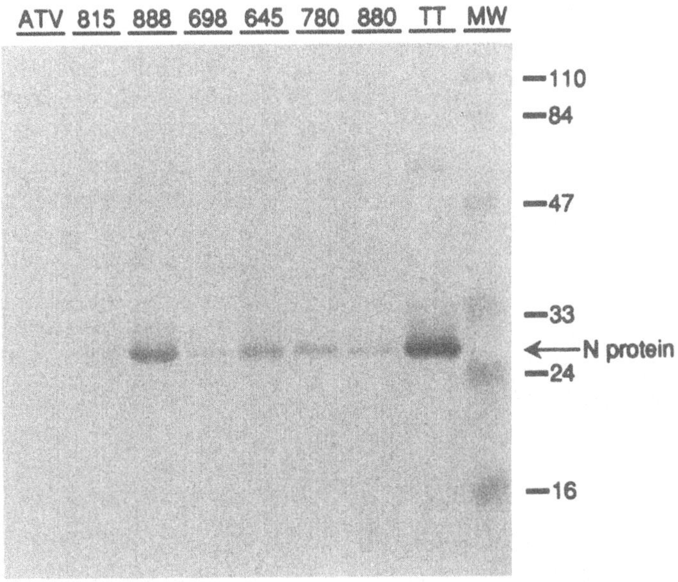

Fig. 2. Western blot analysis of tomato transformants accumulating TSWV nucleoprotein. Leaf protein samples were subjected to SDS-polyacrylamide gel electrophoresis, blotted to Immobilon-P membranes and TSWV nucleoprotein was detected using an antiserum raised against purified TSWV nucleocapsids. Molecular weight markers are indicated on the right and numbers refer to the tomato transformant lines, TT tomato systemically infected with TSWV, ATV non-transformed tomato line ATV847 used as recipient in transformation experiments

fication reaction carried appropriate restriction sites to facilitate the cloning of the N gene into expression vector pZU029 (Fig. 1). Since the N gene was obtained by PCR-amplification, the cloned fragment was sequenced to exclude the possibility of mutations generated by the *Taq* polymerase. The resulting gene cassette, pTSWVN-B, comprises the CaMV 35S promoter, the viral N gene and the polyadenylation signal derived from the 3′ flanking region of the nopaline synthase (*nos*) gene. The CaMV promoter was modified by fusing the 5′ untranslated leader sequence from tobacco mosaic virus (TMV) immediately downstream of the transcription initiation site. The TMV leader is known to function as a translational enhancer (Gallie et al., 1987). The chimaeric gene cassette was subsequently cloned into the binary transformation vector pBIN19 as a *Xba*I fragment and transferred to *Agrobacterium tumefaciens* strain LBA4404. In tobacco, the N gene cassette pTSWVN-B already proved to confer TSWV resistance, not only upon mechanical inoculation but also upon inoculation using viruliferous thrips (Gielen et al., 1991; de Haan et al., 1992).

Transformation of parental tomato line ATV847

Transgenic tomato plants were obtained by means of *Agrobacterium*-mediated leaf disc transformation (Shahin et al., 1986; Fillatti et al., 1987; Yoder et al., 1988), using genotype ATV847 as acceptor. This inbred tomato line is used as male parent in the production of a number of fresh market hybrids of the determinant type. The transformation protocol was optimised to obtain maximum frequencies of transformation for the parental tomato line, using kanamycin resistance as a selectable marker. After eight weeks, about 20% of the cotyledon explants gave rise to shoot primordia, which were cut from the explants and rooted in the presence of 25 mg/l kanamycin. Rooted shoots were transferred to the greenhouse and about 45% of the transformants accumulated the TSWV nucleoprotein at detectable levels in an ELISA assay. Western blot analysis showed that this transgenically-expressed protein co-migrated with that extracted from tomato plants systemically infected with TSWV, thereby demonstrating the integrity of the nucleoprotein produced in transgenic plants (Fig. 2). In order to exclude transformants with aberrant ploidy levels, transgenic tomato plants accumulating varying levels of nucleoprotein were analysed for their ploidy level by means of flow

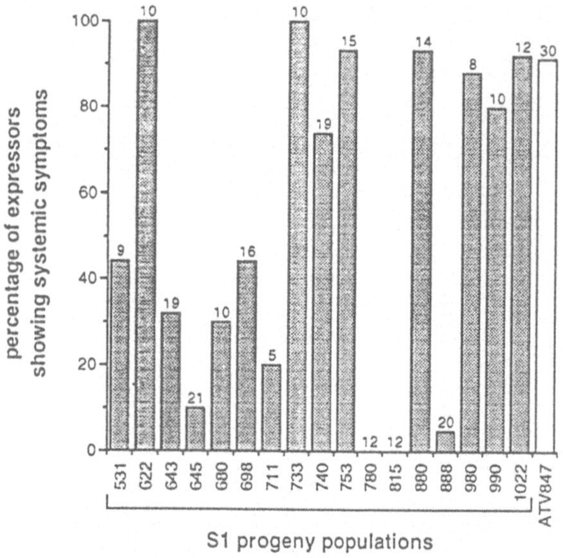

Fig. 3. Resistance to TSWV of S1 progeny populations upon mechanical inoculation. Plants were challenged twice with an inoculum prepared from systemically infected tomato plants. Control plants consisted of non-transformed ATV847 acceptor plants. Progeny plants that did not inherit the nucleoprotein gene cassette through segregation have been left out of the analyses; the figures on top of the bars refer to the number of tomato plants accumulating nucleoprotein that have been scored for systemic symptoms in the final observation about eight weeks after the first inoculation.

cytometry. Only 60% of the expressors retained the diploid ploidy level and were subsequently maintained to produce offspring by self-pollination. The effective transformation frequency, expressed as the percentage of explants that gives rise to independent diploid transformants accumulating nucleoprotein, was calculated at 5%. None of the transformants or their progeny populations exhibited phenotypic aberrations that could be assigned to the accumulation of the nucleoprotein or the insertion of the N gene cassette into the genome.

Protection of transgenic tomato against TSWV infections

Prior to inoculation with TSWV, S1 progeny plants were analysed for the accumulation of viral nucleoprotein to identify those individuals that inherited the N gene cassette. Non-expressing segregants were used as susceptible controls in the inoculation experiments in addition to non-transformed ATV847 plants. Tomato seedlings were mechanically inoculated after emergence of the first true leaf using an inoculum prepared from systemically infected tomato plants. The inoculation was repeated one week later and plants were subsequently monitored for the development of systemic disease symptoms, such as chlorosis and bronzing of the non-inoculated leaves and wrinkling of the youngest leaves. In later stages of infection, diseased plants became stunted and reduced in height compared with mock inoculated plants. Control plants reached infection percentages of 90% or higher within two to four weeks. The final observation was made six weeks after the first inoculation and is shown in Fig. 3. Out of 24 progeny populations challenged with the virus, 11 transformant lines could be identified which showed reduced susceptibility to TSWV infection, ranging from complete resistance in transformant lines 780 and 815 to moderate levels of resistance in lines 531 and 698. All other transformant lines were as susceptible as the controls. Protected tomato plants were free of virus when tested by ELISA using an antiserum raised against a non-structural viral protein (NSs) that accumulates to high levels in TSWV-infected plant cells (Kormelink et al., 1991). ELISA values of protected transformants never exceeded the mean ELISA values of negative controls plus three times the standard deviation (0.101 plus 3 times 0.010), while ELISA readings of infected controls were out of range. The absence of detectable amounts of virus indicates that protected plants are probably immune to rather than tolerant to infection. From each S1 transformant line which showed reduced susceptibility to TSWV, a number of individual plants was maintained and self-pollinated to produce S2 offspring. The copy number of the N gene in selected S1 plants was determined by Southern blot analysis (Fig. 4). The majority appeared to carry multiple copies of the transgene, except for transformant line 698 which carried a single copy. Transformant line 815 carried two copies of the transgene both residing on the same chromosome, as could be deduced from the 3:1 segregation ratio observed for the expression of the N gene cassette in the S1 progeny (results not shown).

The complex inheritance of multiple independent copies of the transgene hampers their fixation and the identification of homozygous lines. Therefore, only lines 698 and 815 were carried through to produce experimental hybrids using homozygous S2 lines as females that were pollinated with parental line ATX011. This cross represents the reciprocal cross of the fresh market hybrid 'Radja'. The reciprocal 'Radja' hybrids and their corresponding S2 lines were again challenged with TSWV by mechanical inocula-

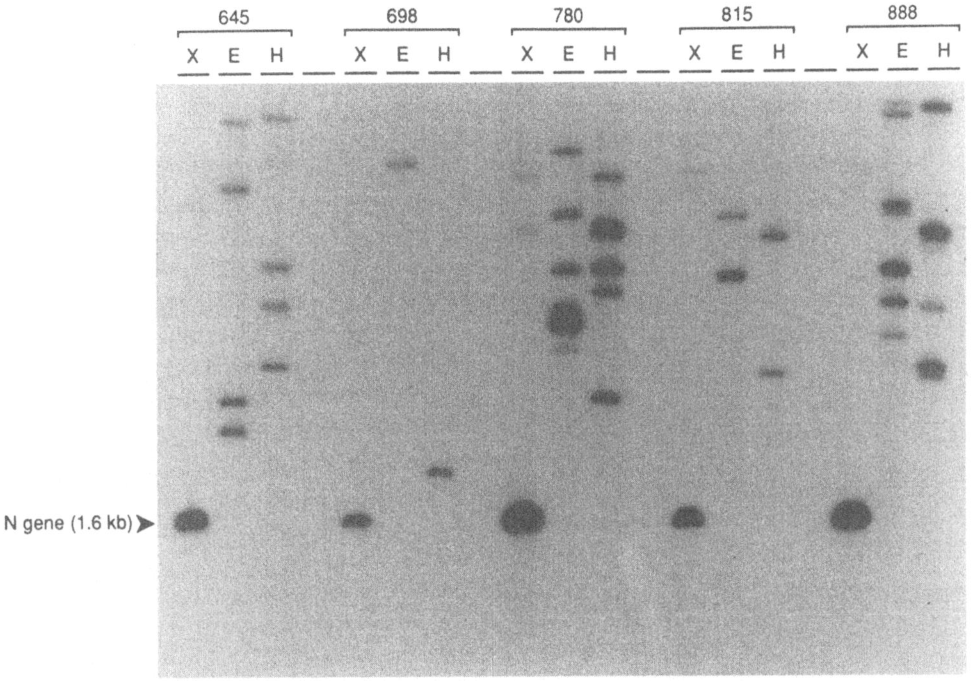

Fig. 4. Southern blot analysis of S1 progeny plants escaping from TSWV infection. Total DNA was extracted from leaf tissue, digested with *Eco*RV (E), *Hind* III (H) or *Xba* I (X), fractionated by agarose gel electrophoresis and blotted to Hybond-N membranes. Genome fragments comprising the TSWV N gene cassette were detected by hybridisation with a ^{32}P-labelled DNA fragment containing the TSWV N gene. Numbers refer to the primary tomato transformants from which the S1 progeny plants descend. The *Xba* I digest releases the TSWV N gene cassette as a fragment of 1.6 kb, the *Eco*RV and *Hind* III digests release a number of border fragments correlated with the copy number of the N gene cassette.

tion (Fig. 5). The homozygous S2 line and the experimental hybrid descending from transformant line 815 were both completely resistant to TSWV infection. Evidently, the pair of linked transgenes of this transformant line suffices to generate complete resistance, even when present in the hemizygous hybrid state. If both transgene copies are closely linked, preferably at the same locus, transformant line 815 would carry a source of TSWV resistance that is inherited as a single dominant trait, and thus would represent an excellent progenitor for TSWV resistance. An example of protected hybrids descending from transformant line 815 is shown in Fig. 6.

Most of the transgenic tomato lines are not completely protected from TSWV infection. The homozygous S2 population and the experimental hybrid derived from transformant line 698, for instance, show intermediate levels of resistance, as already observed for the S1 population. Within a population of 35 S2 plants that all carried one homozygous copy of the transgene, 27 plants developed systemic symptoms, resulting in an intermediate resistance level of 23%. A similar level of resistance (21%) was observed for the experimental hybrid which carried one hemizygous copy of the transgene. Since the homozygous S2 line and the hemizygous hybrid exhibit similar levels of intermediate resistance, the zygosity level of the transgene does not appear to have a significant effect of the level of resistance.

Discussion

Breeding for disease resistance by introgression of genetic sources for resistance is generally regarded as the best strategy for sustainable crop protection. The exploitation of sources for host plant resistance that are naturally present within the gene pool of the crop involved, has contributed a great deal to breeding for disease resistance in modern crops. In the past decade the concept of pathogen-derived resistance (Sanford & Johnston, 1985) has been put into practice to combat

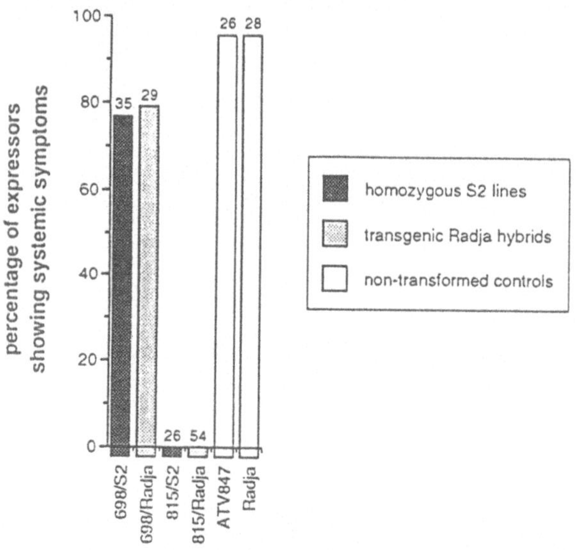

Fig. 5. Resistance to TSWV of homozygous S2 populations (double-hatched bars) and their corresponding hybrids (hatched bars). Control plants (white bars) consisted of non-transformed ATV847 acceptor plants and Radja hybrids. Plants were challenged twice by mechanical inoculation with an inoculum prepared from systemically infected tomato plants. Figures on top of the bars refer to the number of tomato plants that have been scored for systemic symptoms in the final observation about eight weeks after the first inoculation.

Fig. 6. Resistance to TSWV upon mechanical inoculation in tomato hybrids derived from transformant line 815. Plants were photographed six weeks after the first inoculation. From left to right, reciprocal Radja hybrid expressing the TSWV N gene and non-transformed Radja hybrid.

plant viral diseases, resulting in the development of synthetic resistance genes. Upon introduction of chimaeric genes comprising plant viral sequences, transgenic plants show reduced susceptibility towards the corresponding virus, as has been described by many reports for a large number of host-virus combinations (Hull & Davies, 1992; Scholthof et al., 1993; Wilson, 1993). Provided that the crop involved is amenable to genetic modification, pathogen-derived resistance genes represent novel sources of genetic resistance that are available to the breeder in addition to natural sources. In the case where natural resistance genes are lacking, transformation of crop plants with pathogen-derived resistance genes may even be the only way whereby genetic resistance can be achieved. In tobacco, it is known that expression of the TSWV N gene confers resistance to TSWV (Gielen et al., 1991; MacKenzi & Ellis, 1992; Pang et al., 1992). In this report we successfully demonstrate the use of the same TSWV N gene cassette to create tomato hybrids which are completely resistant to TSWV. This illustrates the broad applicability of the synthetic TSWV resistance gene and it is assumed that the same gene will be useful in any crop that suffers from TSWV infection, including pepper, lettuce and ornamentals like *Chrysanthemum, Cyclamen* and *Impatiens*. Since the transgenic TSWV resistance is primarily RNA-mediated (de Haan et al., 1992), the resistance will not easily be overcome by mutant TSWV strains that carry point mutations in their nucleoprotein gene. Hence, the synthetic TSWV resistance gene is expected to be a durable and reliable source of genetic resistance to TSWV.

Upon mechanical inoculation 11 out of 24 tested transformant lines expressing the TSWV N gene showed protection of TSWV infection, ranging from complete immunity to intermediate levels of resistance. Intermediate resistance levels are partially due to incomplete dominance of the transgene. Intermediate resistance is characterised by the fact that the expression of the transgene does not necessarily confer resistance, as illustrated by the discrepancy between the physical inheritance of a single transgene and the inheritance of the resistance trait in the case of transformant line 698. Independent segregation of multiple copies of the transgene that interact with each other, furnishes another explanation for observed intermediate resistance levels. Gene dosage effects that result from the partial contribution of multiple transgene copies to the level of resistance, may underlie the modulation of resistance levels in transformant lines carrying multiple transgene copies, as in the case of transformant lines 645 and 888 with 4 and 7 transgene copies respectively (Fig. 4). The occurrence of incomplete dominance and gene dosage effects necessitates the screening of large numbers of transformant lines in order to identify suitable progenitors for the TSWV resistance trait.

The introgression of pathogen-derived resistance genes in breeding programmes is made possible by tracing the transgene using simple molecular techniques such as Southern blot analysis or PCR analysis. Application of these techniques in backcross programmes eliminates the need for repeated and laborious resistance screenings of progeny populations. In case of the TSWV resistance gene, the transgene can also be traced by ELISA analysis for the accumulation of nucleoprotein. Transgenic tobacco plants expressing the TSWV N gene are protected against TSWV infection, but remain susceptible to other tospoviruses (de Haan et al., 1992). The rapid emergence of new tospovirus species, some of which infect tomato (de Avila et al., 1993), emphasises the need for identification or development of additional sources of genetic resistance against such viruses. From our experience with the synthetic TSWV resistance gene, it is anticipated that expression of the N gene from new tospoviruses in transgenic plants will also generate resistance. Therefore, synthetic resistance genes based on tospovirus N genes represent sources of genetic resistance that are available to the breeder as soon as novel tospoviruses emerge. Summarising, it is assumed that the technology developed to obtain TSWV resistance can be applied to generate resistance to any tospovirus in any crop susceptible to the corresponding tospovirus.

Acknowledgements

The authors wish to thank Tiny van de Jagt for her excellent assistance in the transformation experiments, and Herman Koning for maintenance of the tomato transformants and the production of experimental hybrids.

References

Ausubel, F.M., R. Brent, R.E. Kingston, D.D. Moore, J.G. Seidman, J.A. Smith & K. Struhl (Eds), 1987. Current Protocols in Molecular Biology. Green Publishing Associates, Inc. and John Wiley & Sons, Inc., New York.

Bevan, M., 1984. Binary *Agrobacterium* vectors for plant transformation. Nucl. Acids Res. 12: 8711–8721.

Boiteux, L.S. & L. de B. Giordano, 1993. Genetic basis of resistance against two Tospovirus species in tomato (*Lycopersicon esculentum*). Euphytica 71: 151–154.

Cho, J.J., R.F.L. Mau, T.L. German, R.W. Hartmann, L.S. Yudin, D. Gonsalves & R. Provvidenti, 1989. A multidisciplinary approach to management of tomato spotted wilt virus in Hawaii. Plant Dis. 73: 375–383.

de Avila, A.C.P., P. de Haan, R. Kormelink, R. de Oliveira Resende, R.W. Goldbach & D. Peters, 1993. Classification of tospoviruses based on phylogeny of nucleoprotein gene sequences. J. Gen. Virol. 74: 153–159.

De Haan, P., L. Wagemakers, D. Peters & R. Goldbach, 1990. The S RNA of tomato spotted wilt virus has an ambisense character. J. Gen. Virol. 71: 1001–1007.

de Haan, P., R. Kormelink, R. de Oliveira Resende, F. van Poelwijk, D. Peters & R. Goldbach, 1991. Tomato spotted wilt virus L RNA encodes a putative RNA polymerase. J. Gen. Virol. 71: 2207–2216.

de Haan, P., J.J.L. Gielen, M. Prins, I.G. Wijkamp, A. van Schepen, D. Peters, M.Q.J.M. van Grinsven & R. Goldbach, 1992. Characterization of RNA-mediated resistance to tomato spotted wilt virus in transgenic tobacco plants. Bio/Technol. 10: 1133–1137.

de Laat, A.M.M., W. Göhde & M.J.D.C. Vogelzang, 1987. Determination of ploidy of single plants and plant populations by flow cytometry. Plant Breeding 99: 303–307.

Ditta, G., S. Stanfield, D. Corbin & D.R. Helinski, 1980. Broad host range DNA cloning system for Gram-negative bacteria: construction of a gene bank of *Rhizobium meliloti*. Proc. Natl. Acad. Sci. USA 80: 7347–7351.

Doyle, J.J. & J.L. Doyle, 1990. Isolation of plant DNA from fresh tissue. Focus 12: 13–15 (published by Life Technologies, Inc.).

Fillatti, J.J., J. Kiser, R. Rose & L. Comai, 1987. Efficient transfer of a glyphosate tolerance gene into tomato using a binary *Agrobacterium tumefaciens* vector. Bio/Technol. 5: 726–730.

Finlay, K.W., 1952. Inheritance of spotted wilt resistance in the tomato. I. Identification of strains of the virus by the resistance of susceptibility of tomato species. Australian J. Sci. Res. 5: 303–314.

Finlay, K.W., 1953. Inheritance of spotted wilt resistance in the tomato. II. Five genes controlling spotted wilt resistance in four tomato types. Australian J. Biol. Sci. 6: 153–163.

Gallie, D.R., D.E. Sleat, J.W. Watts, P.C. Turner & T.M.A. Wilson, 1987. The 5′-leader sequence of tobacco mosaic virus RNA enhances the expression of foreign gene transcripts *in vitro* and *in vivo*. Nucl. Acids Res. 15: 3257–3273.

German, T.L., D.E. Ullman & J.W. Moyer, 1992. Tospoviruses: diagnosis, molecular biology, phylogeny, and vector relationships. Annu. Rev. Phytopathol. 30: 315–348.

Gielen, J.J.L., P. de Haan, A.J. Kool, D. Peters, M.Q.J.M. van Grinsven & R.W. Goldbach, 1991. Engineered resistance to tomato spotted wilt virus, a negative-strand RNA virus. Bio/Technol. 9: 1363–1367.

Goldbach, R.W. & D. Peters, 1994. Possible causes of the emergence of tospoviruses. Sem. Virol. 5: in press.

Hull, R. & J.W. Davies, 1992. Approaches to nonconventional control of plant virus diseases. Crit. Rev. Plant Sci. 11: 17–33.

Kormelink, R., E.W. Kitajima, P. de Haan, D. Zuidema, D. Peters & R. Goldbach, 1991. The nonstructural protein (NSs) encoded by the ambisense S RNA segment of tomato spotted wilt virus is associated with fibrous structures in infected plant cells. Virology 181: 459–468.

Kormelink, R., P. de Haan, C. Meurs, D. Peters & R. Goldbach, 1992. The nucleotide sequence of the M RNA segment of tomato spotted wilt virus, a bunyavirus with two ambisense RNA segments. J. Gen. Virol. 73: 2795–2804.

Kumar, N.K.K., D.E. Ullman & J.J. Cho, 1993. Evaluation of *Lycopersicon* germ plasm for tomato spotted wilt tospovirus resistance by mechanical and thrips transmission. Plant Dis. 77: 938–941.

Laemmli, U.K., 1970. Cleavage of structural proteins during the assembly of the head of bacteriophage T4. Nature 227: 680–685.

Law, M.D. & J.W. Moyer, 1990. A tomato spotted wilt-like virus with a serologically distinct N protein. J. Gen. Virol. 71: 933–938.

MacKenzie, D.J. & P.J. Ellis, 1992. Resistance to tomato spotted wilt virus infection in transgenic tobacco expressing the viral nucleocapsid gene. Mol. Plant-Microbe Interact. 5: 34–40.

Murashige, T. & F. Skoog, 1962. A revised medium for rapid growth and bioassays with tobacco tissue cultures. Physiol. Plant. 15: 473–497.

Ooms, G., P.J.J. Hooykaas, G. Molenaar & R.A. Schilperoort, 1981. Crown gall tumors of different morphology, induced by *Agrobacterium tumefaciens* carrying mutated octopine Ti plasmids; analysis of T-DNA functions. Gene 14: 33–50.

Pang, S.Z., P. Nagpala, M. Wang, J.L. Slightom & D. Gonsalves, 1992. Resistance to heterologous isolates of tomato spotted wilt virus in transgenic tobacco expressing its nucleocapsid protein gene. Phytopathol. 82: 1223–1229.

Paterson, R.G., S.J. Scott & R.C. Gergerich, 1989. Resistance in two *Lycopersicon* species to an Arkansas isolate of tomato spotted wilt virus. Euphytica 43: 173–178.

Reddy, D.V.R., A.S. Ratna, M.R. Sudarshana, F. Poul & I.K. Kumar, 1992. Serological relationships and purification of bud necrosis virus, a tospovirus occurring in peanut (*Arachis hypogaea* L.) in India. Ann. Appl. Biol. 120: 279–286.

Resende, R. de Oliveira, A.C. de Avila, R.W. Goldbach & D. Peters, 1991a. Generation of envelope and defective interfering RNA mutants of tomato spotted wilt virus by mechanical passage. J. Gen. Virol. 72: 2375–2383.

Resende, R. de Oliveira, A.C. de Avila, R.W. Goldbach & D. Peters, 1991b. Detection of tomato spotted wilt virus using polyclonal antisera in double antibody sandwich (DAS) ELISA and cocktail ELISA. J. Phytopathol. 132: 46–56.

Sanford, J.C. & S.A. Johnston, 1985. The concept of parasite-derived resistance-deriving resistance genes from the parasite's own genome. J. Theor. Biol. 113: 395–405.

Scholthof, K.B.G., H.B. Scholthof & A.O. Jackson, 1993. Control of plant virus diseases by pathogen-derived resistance in transgenic plants. Plant Physiol. 102: 7–12.

Shahin, E.A., K. Sukhapinda, R.B. Simpson & R. Spivey, 1986. Transformation of cultivated tomato by a binary vector in *Agrobacterium rhizogenes*: transgenic plants with normal phenotypes harbor binary vector T-DNA, but no Ri-plasmid T-DNA. Theor. Appl. Genet. 72: 770–777.

Smith, P.G., 1944. Reaction of *Lycopersicon* spp. to spotted wilt. Phytopathol. 34: 504–505.

Stevens, M.R., S.J. Scott & R.C. Gergerich, 1992. Inheritance of a gene for resistance to tomato spotted wilt virus (TSWV) from *Lycopersicon peruvianum* Mill. Euphytica 59: 9–17.

Wahl, G.M., M. Stern & G.R. Stark, 1979. Efficient transfer of large DNA fragments from agarose gels to diazogenzyloxymethyl-paper and rapid hybridization by using dextran sulphate. Proc. Natl. Acad. Sci. USA 76: 3683–3688.

Wilson, T.M.A., 1993. Strategies to protect crop plants against viruses: pathogen-derived resistance blossoms. Proc. Natl. Acad. Sci. USA 90: 3134–3141.

Yeh, S.D., Y.C. Lin, Y.H. Cheng, C.L. Jih, M.J. Chen & C.C. Chen, 1992. Identification of tomato spotted wilt-like virus on watermelon in Taiwan. Plant Dis. 76: 835–840.

Yoder, J.I., J. Palys, K. Alpert & M. Lassner, 1988. *Ac* transposition in transgenic tomato plants. Mol. Gen. Genet. 213: 291–296.

Euphytica **85:** 169–172, 1995.

Degradation of oxalic acid by transgenic oilseed rape plants expressing oxalate oxidase

C. Thompson[1], J.M. Dunwell[1], C.E. Johnstone[1], V. Lay[1], J. Ray[1], M. Schmitt[2], H. Watson[1] & G. Nisbet[1]

[1] *Plant Biotechnology, Zeneca Seeds, Jealott's Hill, Bracknell, Berks RG42 6ET, U.K.;* [2] *American Cyanamid, Box 400, Princeton, NJ 0854-0400, U.S.A*

Key words: Brassica napus, disease tolerance, oxalic acid, oxalate oxidase, *Sclerotinia sclerotiorum*, transformation

Summary

Oxalic acid is thought to have a primary role in the pathogenicity of several plant pathogens, notably *Sclerotinia sclerotiorum.* A gene coding for the enzyme oxalate oxidase was isolated from barley roots and introduced into oilseed rape as a means of degrading oxalic acid *in vivo.* This report describes the production of several transgenic plants of oilseed rape and the characterisation of these plants by Southern, Western and enzyme activity assays. Plants were shown to contain an active oxalate oxidase enzyme and were tolerant of exogenously supplied oxalic acid.

Introduction

Oxalic acid is produced by several plant pathogenic fungi and is thought to have a primary role in the pathogenicity of several species, including the wide host-range pathogen *Sclerotinia sclerotiorum* (Noyes & Hancock, 1981; Rowe, 1993). Evidence for this comes from studies on mutant strains of this fungus, deficient in oxalate production, which are avirulent. Revertants for this trait regain their virulence character (Godoy et al., 1990). The proposed mode of action of oxalate in pathogenesis is: 1) Chelation of calcium from the pectate fraction of the xylem and associated pit vessels. 2) Entry of air leading to a xylem embolism and ultimately, wilting. 3) Spread of oxalate reduces pH thereby stimulating the activity of fungal enzymes such as polygalacturonase, methyl esterase and cellulase. This enzyme activity leads to further rotting of the tissue (Marciano et al., 1983).

Our aim is to assess the potential of introducing a gene coding for an oxalate-degrading trait into susceptible crop species (Masirevic & Gulya, 1992; Sackston, 1992), to confer resistance to pathogens which utilise oxalate in the infection process. There are two known enzymes which can catabolize oxalate, namely oxalate oxidase (EC 1.2.3.4) and oxalate decarboxylase (EC 4.1.1.2). Oxalate oxidase, which catalyses the degradation of oxalate to CO_2 and H_2O_2, has been isolated from a number of plant species including barley (Chiriboga, 1966). It is generally considered to be a large oligomeric protein of around 125 kD and it is possible to purify the protein down to a 25 kD monomer (Schmitt, 1991). Purified preparations of the wheat protein germin also have strong oxalate oxidase activity, associated with the oligomeric (125 kD) fraction – (Lane et al., 1993).

We have developed a plant transformation vector containing an oxalate oxidase gene, isolated from a barley root cDNA library, and transformed the construct into oilseed rape (canola) via *Agrobacterium*-mediated transformation. The sequence of the oxalate oxidase gene has been found to possess close homology to the wheat protein, germin (Dumas et al., 1993; Lane et al., 1993; Lane, 1994), indicating a possible function for this germination-related protein. We outline here the successful expression of oxalate oxidase in the tissues of a range of independent oilseed rape transformants. Readily detectable levels of protein (oxalate oxidase) and enzyme activity (degradation of exogenous oxalate) are reported.

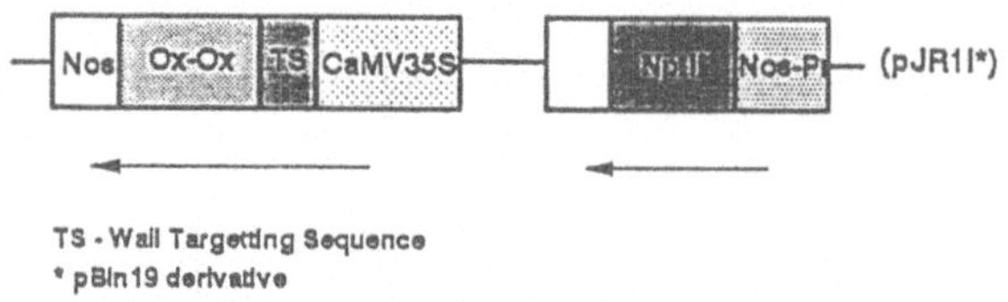

Fig. 1. Oxalate oxidase transformation vector *pSR2*.

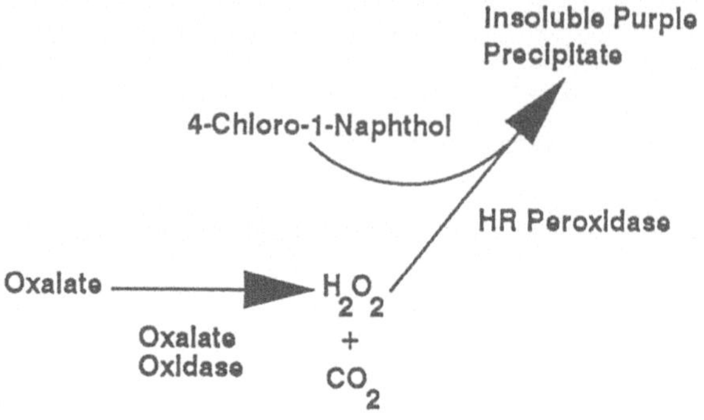

Fig. 2. Oxalate oxidase activity stain reaction.

Materials and methods

Vector construction and transformation

A full-length oxalate oxidase clone (710 bp) was constructed from a partial clone obtained from a barley root cDNA library and synthetic oligonucleotides. The clone was attached to a cell wall targeting sequence from *Nicotiana plumbaginifolia* and inserted with the CaMV35S promoter into a binary (*pBin*19) *Agrobacterium* vector, to give plasmid *pSR*2 (see Fig. 1) by standard techniques and re-introduced into *A. tumefaciens* strain LBA4404. The transit peptide was included to target the protein to the extra-cellular space where the toxin is secreted.

Cotyledon petioles of oilseed rape, *Brassica napus* cv. Westar were transformed using a modified version of the method of Moloney et al. (1989) and transformed tissue selected on the basis of resistance to kanamycin (15 mg/l).

Molecular and biochemical analysis of transgenics

Rooted plantlets were tested for the presence of the oxalate oxidase construct via standard PCR and Southern analysis. Confirmed transformants were then assayed with Western blots for presence of the oxalate oxidase protein, using a polyclonal antibody raised to the 25 kD oxalate oxidase monomer.

Oxalate oxidase enzyme activity assays were based on the measurement of reaction products (i.e. H_2O_2 or CO_2). For rapid, non-quantitative assays a membrane-based stain technique was used. Briefly, 300 μg of tissue was ground up in an Eppendorf tube with 300 μl water, polyvinyl polypyrollidine and washed sand. Crude extract (50 μl) was spotted onto Hybond C nitrocellulose which was then bathed in developing solution (for 100 ml: 1 mM oxalic acid in 0.1 M succinate buffer adjusted to pH 3.5, 20 mg 4-chloronaphthol, 0.8 mg horseradish peroxidase). This method gives a readily-visible, insoluble, purple precipitate (see Fig. 2). For quantitative assays the colorimetric method described by Sugiura et al. (1979) which measures H_2O_2 production, proved more reliable than the ^{14}C labelled oxalate technique (Chiriboga, 1966). These assays can be performed on very crude extracts of small tissue samples.

A combination of the above assays was used to identify T_3 homozygous expressing lines, with single locus inserts, by observing segregation patterns.

Excised leaf oxalic acid wilting test

Excised leaves of oilseed rape plants wilt rapidly when exposed to oxalic acid solutions. To assess biological activity of the oxalate oxidase enzyme, small photosynthetically-active leaves were cut from transformed and control plants and the petioles placed into 20 mM oxalic acid solutions prepared at pH 4 or water (pH 4) as control. The leaves were observed for signs of wilting after 2–24 hours.

Results

Transformation and molecular analysis of transgenics

A total of 19 independent (i.e. taken from different calli) *pSR*2-transformed plants were generated. When transferred to soil in growth rooms, most regenerants appeared phenotypically normal, though one line had an unusual highly-branched morphology, and another was sterile. Southern and PCR analysis of a range of putative transformants revealed very few escapes.

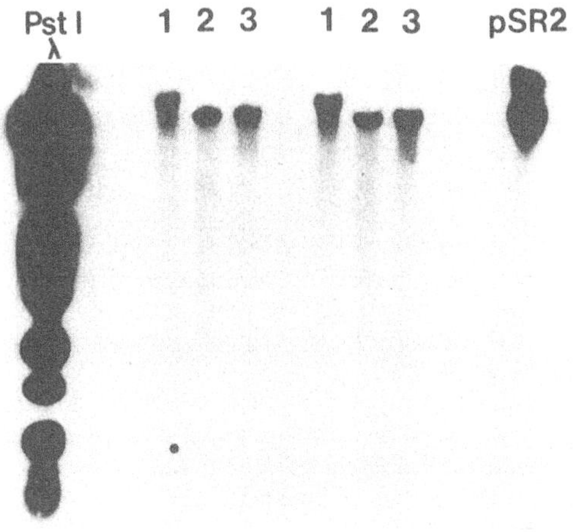

Fig. 3. Prolonged exposure of Southern blot of *pSR2* transgenic 'Westar' plants hybridised to a 35S-ox-ox probe to reveal border fragments. Lanes 1–3 are independent transformants with a single site of insertion.

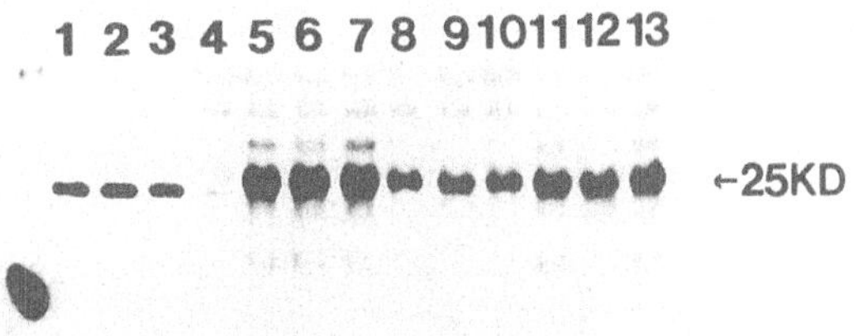

Fig. 4. Western assay of *pSR2* transgenic and control 'Westar' samples hybridised to a polyclonal antibody raised to purified denatured oxalate oxidase (25 kD monomer). 2 minute exposure: 1–3 100 ng oxalate oxidase, 4 – negative control, 5 – 13 *pSR2* transgenics (3 reps each of 3 plants).

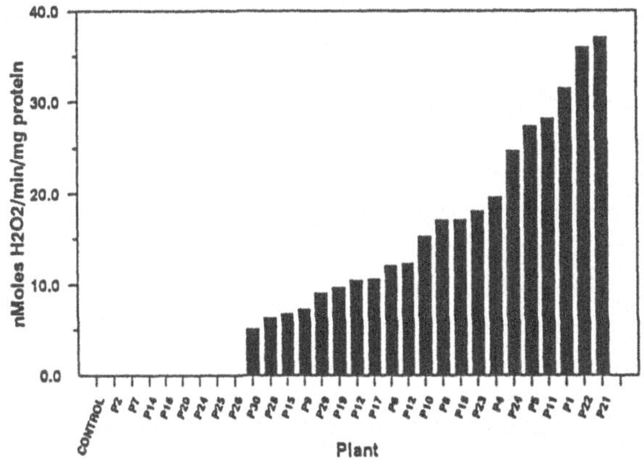

Fig. 5. Range of oxalate oxidase activities in a segregating population of T_2 plants containing a single locus insertion (activity = nmoles/H_2O_2 produced/minute/mg protein). Each column represents the activity reading for one plant.

Most of the plants tested had a single site of integration, but a few had two or more sites. Fig. 3 shows Southern analysis of three individual lines. Separate shoots derived from the same transgenic callus were usually, but not always, found to be clonal.

Western analysis of transgenics

Western analysis of leaf and root tissue extracts, with a polyclonal antibody raised to the 25 kD monomer of oxalate oxidase, revealed a strongly hybridising band at 25 kD in all transgenics which were PCR and Southern positive (see Fig. 4). This antibody cross-hybridises faintly to several non-specific bands in control tissue, including a faint band just smaller than 25 kD, but a strong 25 kD band was never observed in untransformed 'Westar' tissue.

Oxalate oxidase activity assays

All Western positive plants gave positive results when tissue was tested for oxalate oxidase enzyme activity, via the nitrocellulose membrane stain test on crude extracts. No activity was observed from control or escape tissue. Quantification of activity via the spectrophotometric technique gave a range of values for transformed plants from 2 to 47 nmoles H_2O_2/min/mg protein. The assay is subject to significant variability between experiments, but even so, no correlation between gene copy number and expression levels was seen in the primary transgenics. Highly-expressing, mature plants have activities similar to that detected in barley root material (Chiriboga, 1966). Fig. 5 shows the range of activity obtained in a segregating population of plants with a single site of insertion.

Oxalic acid wilting test

When petioles of excised leaves of control plants were placed into oxalic acid solutions adjusted to pH 4 they wilted rapidly, most probably due to xylem embolisms caused by the oxalate (Sperry & Tyree, 1988). Excised leaves fed with pH 4 water remained turgid. Leaves of some of the transgenic lines expressing oxalate oxidase did not wilt, even over an extended 24 hour period in the oxalate solution (Fig. 6). Only three lines, selected on the basis of high levels of expression, were tested in this way and all three showed resistance to wilting.

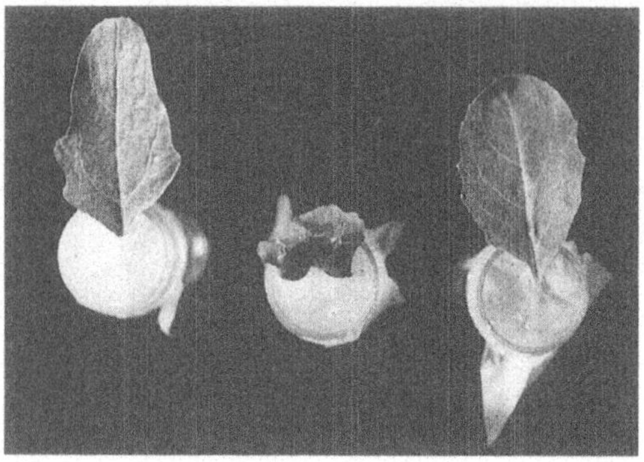

Fig. 6. Wilted leaf of control and non-wilted leaves of *pSR2* transformants after 20 hours feeding with 20 mM oxalic acid solution (pH 4).

Discussion

The barley oxalate oxidase gene has been successfully introduced into an oilseed rape cultivar. Driven by the 35S promoter, this gene is constitutively expressed in all tissues of the plant tested so far although some differences in expression levels between different tissues are likely (Stefanov et al., 1994). The gene product (a 25 kD monomer) either has oxalate oxidase activity itself, or is correctly processed in the cell to form a re-associated oligomer, giving an active enzyme. Further tests are underway to determine the exact form of the enzyme and its localisation in the transgenic plants. We have observed the biological action of the oxalate oxidase enzyme in transgenic plants, as relates to protection against wilting of excised leaves dipped in an oxalic acid solution. However, the value of these transgenic plants as a source of disease-resistant breeding material depends upon extensive disease screening, not only under laboratory conditions, but more importantly in the field. Such trials are now underway.

References

Chiriboga, J., 1966. Purification and properties of oxalic acid oxidase. Arch. Biochem. Biophys. 116: 516–523.

Dumas, B., A. Sailland, J-P. Cheviet, G. Freyssinet & K. Pallet, 1993. Identification of barley oxalate oxidase as a germin-like protein. C.R. Acad. Sci. Paris 316: 793–798.

Godoy, G., J.R. Steadman, M.B. Dickman & R. Dam, 1990. Use of mutants to demonstrate the role of oxalic acid in pathogenicity of *Sclerotinia sclerotiorum* on *Phaseolus vulgaris*. Physiol. Mol. Plant Pathol. 37: 179–191.

Lane, B.G., 1994. Oxalate, germin and the extracellular matrix. FASEB J. 8: 294–301.

Lane, B.G., J.M. Dunwell, J.A. Ray, M.R. Schmitt & A.C. Cuming, 1993. Germin, a protein marker of early plant development, is an oxalate oxidase. J. Biol. Chem. 268: 12239–12242.

Marciano, P., Di Lenna & P. Magro, 1983. Oxalic acid, cell wall degrading enzymes and pH in pathogenesis and their significance in the virulence of two *Sclerotinia sclerotiorum* isolates on sunflower. Physiol. Plant Pathol. 22: 339–345.

Masirevic, S. & T.J. Gulya, 1992. *Sclerotinia* and *Phomopsis* – two devastating sunflower pathogens. Field Crops Res. 30: 271–300.

Moloney, M.M., J.M. Walker & K.K. Sharma, 1989. High efficiency transformation of *Brassica napus* using *Agrobacterium* vectors. Plant Cell Reports 8: 238–242.

Noyes, R.D. & J.G. Hancock, 1981. Role of oxalic acid in the Sclerotinia wilt of sunflower. Physiol. Plant Pathol. 18: 123–132.

Rowe, D.E., 1993. Oxalic acid effects in exudates of *Sclerotinia trifoliorum* and *S. sclerotiorum* and potential use in selection. Crop Sci. 33: 1146–1149.

Sackston, W.E., 1992. On a treadmill: breeding sunflowers for resistance to disease. Ann. Rev. Phytopathol. 30: 529–551.

Schmitt, M.R., 1991. Barley seedling oxalate oxidase: purification and properties. Plant Physiol. (suppl.) 96: 85.

Sperry, J.S. & M.T. Tyree, 1988. Mechanism of water stress-induced xylem embolism. Plant Physiol. 88: 581–587.

Stefanov, I., S. Fekete, L. Bogre, J. Pauk, A. Feher & D. Dudits, 1994. Differential activity of the mannopine synthase and the CaMV 35S promoters during development of transgenic rapeseed plants. Plant Sci. 95: 175–186.

Sugiura, M., H. Yamamura, K. Hirano, M. Sasaki, M. Morikawa & M. Tsuboi, 1979. Purification and properties of oxalate oxidase from barley seedlings. Chem. Pharm. Bull. 27: 2003–2007.

Euphytica **85**: 173–180, 1995.

Synergistic activity of chitinases and β-1,3-glucanases enhances fungal resistance in transgenic tomato plants

Erik Jongedijk, Henk Tigelaar, Jeroen S.C. van Roekel, Sandra A. Bres-Vloemans, Ilma Dekker, Peter J.M. van den Elzen, Ben J.C. Cornelissen[1] & Leo S. Melchers
MOGEN International nv, Einsteinweg 97, 2333 CB Leiden, The Netherlands; [1] *Institute for Molecular Cell Biology, Biocentrum Amsterdam, Kruislaan 318, 1098 SM Amsterdam, The Netherlands*

Key words: Lycopersicon esculentum, tomato, endochitinases, β-1,3-endoglucanases, fungal resistance, transgenic plants

Summary

Simultaneous expression of a tobacco class I chitinase and a class I β-1,3-glucanase gene in tomato resulted in increased fungal resistance, whereas transgenic tomato plants expressing either one of these genes were not protected against fungal infection. After infection with *Fusarium oxysporum* f.sp. *lycopersici*, a 36% to 58% reduction in disease severity was observed in resistant tomato lines. Two transgenic lines largely recovered from the initial infection by the time wild-type tomato plants had died.

The overall results are consistent with the observation that class I chitinases and class I β-1,3-glucanases synergistically inhibit the growth of fungi *in vitro* and provide the first experimental support to the hypothesis that such synergy can contribute to enhanced fungal resistance *in planta*.

Introduction

Most agricultural and horticultural crop species suffer from a vast array of fungal diseases which cause severe yield losses all over the world. In addition to polygenes which confer only moderate levels of partial resistance against specific fungi, single dominant genes providing immunity to specific races of fungi have been identified for many crop species (Poehlman, 1987; Kalloo & Bergh, 1993). Although such race-specific resistance genes are often easily overcome by rapidly evolving new fungal races, today, continuous accumulation of race-specific resistance genes in commercial varieties, remains the major method of achieving sufficient resistance to a wide range of fungal pathogens.

Agronomically-viable levels of durable resistance in crop plants against a relatively broad range of fungi might be achieved by recently described molecular approaches, which include expression of fungal avirulence genes to provoke non-specific, hypersensitive resistance when combined with the corresponding plant resistance genes, expression of genes involved in the synthesis of phytoalexins toxic to fungi, and expression of genes encoding inhibitors of fungal enzymes or known antifungal proteins (reviewed by Lamb et al., 1992; Cornelissen & Melchers, 1993; Strittmatter & Wegener, 1993). The latter include chitinases and β-1,3-glucanases. These hydrolytic enzymes catalyze the degradation of chitin and β-1,3-glucan and, since these compounds are abundantly present in the cell wall of many filamentous fungi (Wessels & Sietsma, 1981), they are thought to be capable of inhibiting fungal growth *in planta*.

In plants, five classes (I–V) of endochitinases and three classes (I–III) of β-1,3-endoglucanases have been identified (Ward et al., 1991; Collinge et al., 1993; Melchers et al., 1994). In contrast to the intercellular class II isoforms, the vacuolar class I chitinases and class I β-1,3-glucanases have been shown to be potent inhibitors of fungal growth and to act synergistically *in vitro* (Mauch et al., 1988; Leah et al., 1991; Sela-Buurlage et al., 1993). Such synergy has also been demonstrated to occur between the intracellular class V chitinases and class I β-1,3-glucanases (Melchers et al., 1994) and between a recently purified chitin-binding protein and both the class I β-1,3-glucanases and class I chitinases (Ponstein et al., 1993). Class III chitinases seem to lack antifungal activity (Vogelsang & Barz, 1993). Whether the class IV hydrolases inhibit fungal growth *in vitro* is not yet known.

In planta, increased tolerance to fungal infection has recently been observed in transgenic tobacco plants expressing a barley ribosome-inhibiting protein gene (Logemann et al., 1992), a groundnut stilbene synthase gene (Hain et al., 1993) or a tobacco gene encoding the pathogenesis-related protein PR-1a (Alexander et al., 1993). In the case of chitinases and β-1,3-glucanases, *in planta* resistance data are limited to increased tolerance to infection by *Rhizoctonia solani* in transgenic tobacco plants, which constitutively expressed a bean class I chitinase gene (Broglie et al., 1991) or an exochitinase gene of bacterial origin (Logemann et al., 1993). In contrast, transgenic tobacco plants expressing a tobacco class I chitinase gene did not show enhanced resistance to infection with the fungus *Cercospora nicotianae* (Neuhaus et al., 1991), despite the fact that the fungus proved very sensitive to chitinases *in vitro*.

In our attempts to increase fungal resistance in tomato by introducing multiple genes encoding class I and class II endochitinases and β-1,3-endoglucanases, we found that transgenic tomato plants constitutively expressing both a tobacco class I chitinase and a tobacco class I β-1,3-glucanase gene showed substantially enhanced resistance against infection with *Fusarium oxysporum* f.sp. *lycopersici*, whereas transgenic tomato plants constitutively expressing either one of these genes are not significantly protected.

Materials and methods

Gene constructs and tomato transformation

Chimeric endochitinase and β-1,3-endoglucanase genes were constructed by transcriptional fusion of a cDNA clone encoding a 32 kD class I endochitinase (Chi-I) and genomic clones encoding a 28 kD class II endochitinase (Chi-II, PR-3a), a 33 kD class I β-1,3-endoglucanase (Glu-I) and a 40 kD class II β-1,3-endoglucanase (Glu-II, PR-2b) from tobacco (*Nicotiana tabacum* cv. Samsun NN) with a double-enhanced cauliflower mosaic virus (CaMV) 35S promoter (Melchers et al., 1993). The chitinase genes were linked to the *Agrobacterium tumefaciens* nopaline synthase (*nos*) transcription termination signal, whereas the β-1,3-glucanase constructs contained their natural transcription termination signals. Two single-gene constructs encoding either the class I chitinase (pMOG198) or the class I β-1,3-endoglucanase (pMOG412) and a four-gene construct encoding both the class I and class II chitinases and β-1,3-glucanases (pMOG539) were made by assembly of the respective genes next to the neomycin phosphotransferase II (*nptII*) gene in the binary plasmid pMOG402 (Fig. 1). The binary plasmid pMOG402 was obtained through replacing the mutant *nptII* gene in pMOG23 (Sijmons et al., 1990) by the wild-type NPTII coding region (Yenofsky et al., 1990).

The binary plasmids pMOG198, pMOG412, pMOG539 and pMOG402 (empty vector control) were mobilized from *Escherichia coli* DH5$_\alpha$ into *Agrobacterium tumefaciens* strain MOG101 (Hood et al., 1993) and used to transform *Lycopersicon esculentum* cv. 'Moneymaker', a diploid (2n = 2x = 24) pure-breeding tomato line. Transgenic tomato plants were obtained by a standard cotyledon transformation method using kanamycin selection (Van Roekel et al., 1993).

Characterization of transgenic lines

Diploid kanamycin-resistant primary transformants were selected by establishing the mean number of chloroplasts in stomatal guard cells (Koorneef et al., 1989), analysed for expression of the respective transgenes and self-pollinated. The number of T-DNA loci present in primary transformants was established on the basis of segregation for kanamycin resistance in S1 progenies using a non-destructive kanamycin spraying assay (Weide et al., 1989). Chi square tests (α = 0.05) were used to assess the goodness of fit to expected Mendelian segregation ratios assuming one (3:1), two (15:1) and three (63:1) independently segregating loci.

Analyses of transgene expression

Protein expression levels were quantified by immunoblot analyses using a range (12.5, 25, 50 and 100 ng) of the respective purified class I and class II chitinases and β-1,3-glucanases as standards. Leaf samples containing 5 μg of soluble protein were electrophoresed on 12.5% SDS-polyacrylamide gels. Following blotting onto nitrocellulose membranes, the transgene-encoded proteins were visualized by enhanced chemiluminescence (ECL, Amersham). Pro-

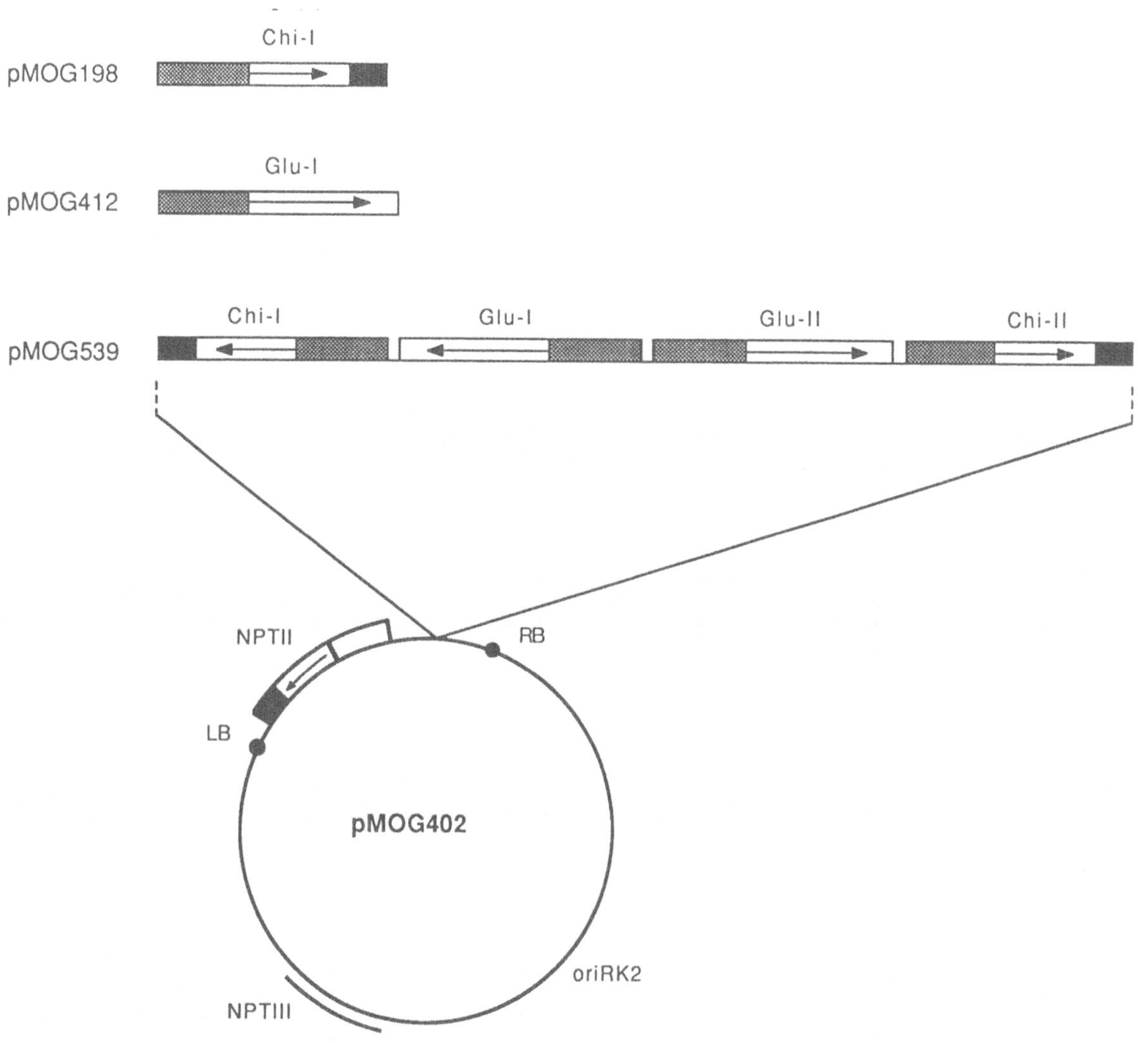

Fig. 1. Schematic representation of chimeric gene constructs pMOG198, pMOG412 and pMOG539 contained in the binary plasmid pMOG402. The chitinase and β-1,3-glucanase genes are transcriptionally regulated by the 35S CaMV promoter (hatched box) and by the *nos*-terminator region (black box) or their natural transcription termination signal. The arrows indicate the orientation of the respective genes.

tein extraction and Western blot analyses were essentially performed as described previously (Melchers et al., 1993). Polyclonal antibodies raised against PR-3a (detection of both the 32 kD class I and the 28 kD class II chitinase) and PR-2b (detection of the 40 kD class II β-1,3-glucanase) were used in a 10^4-fold dilution and that raised against the 33 kD class I β-1,3-glucanse in a 2×10^3 fold dilution.

Fusarium *resistance screening*

Transgenic tomato lines were screened for *Fusarium* resistance in randomized complete block experiments with 3 replicates and 7 plants per plot, using 25-day old S1 seedlings, which had been selected for kanamycin resistance using a non-destructive kanamycin spraying assay (Weide et al., 1989) and thus contained the respective transgene loci in either hemizygous or homozygous condition. Selection of NPTII-expressing seedlings by kanamycin spraying was confirmed to be very reliable (< 2% escapes) and did not induce any expression of chitinases and β-1,3-glucanases in tomato (unpublished results).

S1 seedlings were inoculated by dipping their root system in a freshly-prepared spore suspension from *Fusarium oxysporum* f.sp. *lycopersici* (race 1, 10^6 micro-conidia/ml), subsequently transplanted into 8 litre polystyrene plant trays (60% black peat/40%

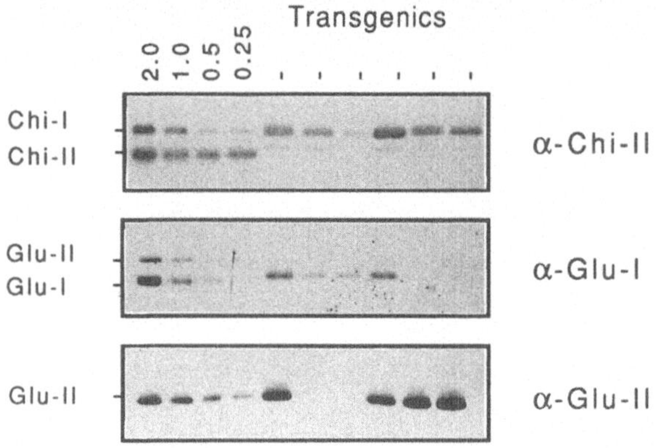

Fig. 2. Western blot quantification of class I and class II chitinase and β-1,3-glucanase gene (Chi-I, Chi-II, Glu-I and Glu-II) expression in transgenic tomato lines harbouring T-DNA contained in the binary plasmids pMOG539 (top and bottom panel) and pMOG412 (middle panel). The first lanes contain standard quantities of the respective chitinases and β-1,3-glucanases representing 2.0, 1.0, 0.5 and 0.25 percent of total soluble protein.

white peat containing 6 gl^{-1} CaCO3 and 1 gl^{-1} NPK 12/14/24 + spore elements, 21 plants/tray) and grown in a phytotron (18–20° C, relative humidity 80%, daylength 15 hrs, light intensity 2000 Lux). Microconidial inoculum of *Fusarium oxysporum* f.sp. *lycopersici* race 1 was harvested from fungal cultures freshly grown on potato dextrose agar (PDA) plates for 25 days (18° C, 18 hrs daylength, light intensity 300 Lux).

Disease severity was scored on a standard graduated scale by the time wild-type 'Moneymaker' had developed severe wilting and chlorosis (25–30 days post inoculation): 1 – slight browning of vascular tissue, no wilting of leaves; 3 – browning of vascular tissue, slight wilting of leaves; 5 – browning of vascular tissue, wilting of leaves, some slight chlorosis of leaves; 7 – strong browning of vascular tissue, severe wilting of plants, strong chlorosis of leaves; 9 – necrosis of leaves, dead plants. For each transgenic line, mean levels of disease severity per plot were calculated and compared by oneway Analysis of Variance according to Duncan's multiple range procedure (α = 0.05).

Results

Characterization of transgenic lines

Kanamycin-resistant 'Moneymaker' plants that were regenerated following transformation with the two single-gene constructs (pMOG198 and pMOG412), the four-gene construct (pMOG539) and the empty vector (pMOG402) were grown in the greenhouse and screened for ploidy level. About 20% of all independent transgenic plants proved tetraploid and were discarded. The remaining diploid plants showed no obvious aberrant phenotype when compared to wild-type 'Moneymaker' and were screened for expression of the respective chitinase and β-1,3-glucanase genes by immunoblot analysis (Fig. 2). Expression of the class II chitinase gene proved very low ($\leq$ 0.1% of total soluble protein) and in most cases was undetectable. Expression levels of the class I chitinase and β-1,3-glucanase genes and the class II β-1,3-glucanase gene differed between independent transgenic plants and ranged from 0.0–4.0%, 0.0–2.0% and 0.0–3.0% of total soluble protein, respectively. Some 54% of the single-gene 'Moneymaker' plants (MM198 and MM412 plants) had reasonable levels of transgene expression ($\geq$ 0.3% of total soluble protein). Among the four-gene 'Moneymaker' plants (MM539 plants), some 33%, 29% and 25% showed reasonable expression for 1, 2 and 3 of the respective transgenes, respectively. Total soluble protein isolated from wild-type 'Moneymaker' and from kanamycin-resistant empty vector control plants, did not show any cross reaction in the immunoblot analyses.

Based on the immunoblot analyses, six single-gene and twelve four-gene, transgenic 'Moneymaker' plants with varying but reasonably high levels of expression of the respective chitinase and β-1,3-glucanase genes were selected, self-pollinated and allowed to set seed. The resulting S1 progenies were subsequently characterized with regard to the number of (independently) segregating T-DNA loci. Except for two lines that segregated for two independent T-DNA loci (MM198-01 containing the single gene chitinase construct and MM539-20 containing the four gene chitinase/β-1,3-glucanase construct), all transgenic 'Moneymaker' lines were shown to segregate for a single T-DNA locus (Tables 1 and 2).

Table 1. Number of segregating T-DNA loci, mean chitinase and β-1,3-glucanase expression levels and mean disease rating 26 days after inoculation with *Fusarium oxysporum* f.sp. *lycopersici* in kanamycin-resistant S1 progenies from transgenic tomato lines expressing the four-gene chitinase/β-1,3-glucanase construct

Tomato lines	Construct		Disease severity[a]		Expression levels[b]			
	Genes	Loci	Mean	σ	Chi-I	Chi-II	Glu-I	Glu-II
Non-inoculated								
'Moneymaker' (MM)	–	–	0.00 a	0.00	0.0	0.0	0.0	0.0
Inoculated								
'Belmondo'	I-1,I-2[c]	–	0.00 a	0.00	0.0	0.0	0.0	0.0
'Dombito'	I-1,I-2[c]	–	0.00 a	0.00	0.0	0.0	0.0	0.0
MM-539-61	Chi-I,II/Glu-I,II	1	2.90 b	0.46	2.0	0.0	1.0	2.0
MM-539-18	Chi-I,II/Glu-I,II	1	3.05 b	0.67	4.0	0.05	1.5	2.0
MM-539-51	Chi-I,II/Glu-I,II	1	3.19 bc	0.97	4.0	0.0	1.0	3.0
MM-539-31	Chi-I,II/Glu-I,II	1	3.52 bcd	0.71	1.5	0.0	2.0	3.0
MM-539-10	Chi-I,II/Glu-I,II	1	3.66 bcd	1.00	4.0	0.1	0.1	0.0
MM-539-20	Chi-I,II/Glu-I,II	2	3.86 bcde	1.51	1.5	0.05	1.0	3.0
MM-539-16	Chi-I,II/Glu-I,II	1	4.38 bcde	1.58	4.0	0.1	0.1	0.0
MM-539-19	Chi-I,II/Glu-I,II	1	4.95 cdef	1.64	1.0	0.05	1.0	3.0
MM-539-60	Chi-I,II/Glu-I,II	1	5.24 defg	1.62	0.5	0.0	1.0	1.0
MM-539-34	Chi-I,II/Glu-I,II	1	5.34 defg	1.28	1.0	0.0	1.0	0.0
MM-539-02	Chi-I,II/Glu-I,II	1	5.63 efg	0.67	0.5	0.0	1.0	3.0
MM-402-09	Vector	1	6.45 fg	0.04	0.0	0.0	0.0	0.0
MM-539-50	Chi-I,II/Glu-I,II	1	6.67 fg	1.53	0.5	0.0	0.5	3.0
'Moneymaker' (MM)	–	–	6.88 g	1.12	0.0	0.0	0.0	0.0
'Planet'	–	–	6.91 g	0.44	0.0	0.0	0.0	0.0

[a] Different letters denote a significant difference (Duncan's multiple range test, $\alpha = 0.05$).
[b] Percentage of total soluble protein.
[c] Host gene-mediated immunity to *Fusarium oxysporum* f.sp. *lycopersici* races 1 and 2.

Table 2. Number of segregating T-DNA loci, mean chitinase and β-1,3-glucanase expression levels and mean disease rating 30 days after inoculation with *Fusarium oxysporum* f.sp. *lycopersici* in kanamycin-resistant S1 progenies from transgenic tomato lines expressing either the single-gene chitinase or the single-gene β-1,3-glucanase construct

Tomato lines	Construct		Disease severity[a]		Expression levels[b]	
	Gene	Loci	Mean	σ	Chi-I	Glu-I
Non-inoculated						
'Moneymaker' (MM)	–	–	0.00 a	0.00	0.0	0.0
Inoculated						
MM-412-04	Glu-I	1	4.94 b	1.58	0.5[c]	0.75
MM-412-03	Glu-I	1	5.38 bcd	1.01	0.0	0.75
MM-402-09	Vector	1	6.14 bcd	0.52	0.0	0.0
MM-412-06	Glu-I	1	6.38 bcd	0.80	0.0	1.5
MM-412-02	Glu-I	1	6.67 cd	0.91	0.0	1.5
MM-198-02	Chi-I	1	7.21 d	0.13	4.0	0.0
'Moneymaker' (MM)	–	–	7.29 d	0.14	0.0	0.0
MM-198-01	Chi-I	2	7.48 d	0.52	4.0	0.0

[a] Different letters denote a significant difference (Duncan's multiple range test, $\sigma = 0.05$).
[b] Percentage of total soluble protein.
[c] Induced expression of endogenous chitinase gene.

Fig. 3. Resistance in transgenic tomato (cv. 'Moneymaker') plants expressing the four-gene chitinase/β-1,3-glucanase construct, 40 days after inoculation with *Fusarium oxysporum* f.sp. *lycopersici* (race 1). Wild-type 'Moneymaker' non-inoculated (A), wild-type 'Moneymaker', inoculated (B) and inoculated transgenic 'Moneymaker' lines MM-539-18 (C) and MM-539-61 (D).

Screening for Fusarium *resistance*

To establish whether expression of the respective chitinase and/or β-1,3-glucanase genes enhanced fungal resistance, the selected single-gene and four-gene 'Moneymaker' plants were tested for resistance to infection with *Fusarium oxysporum* f.sp. *lycopersici* race 1. To enable relative resistance ratings to be assigned to transgenic lines, wild-type 'Moneymaker', empty vector control plants and, occasionally, sensitive and resistant control cultivars were included.

Between the twelve different transgenic tomato lines expressing the four-gene chitinase/β-1,3-glucanase construct, considerable variation in disease severity was observed. As expected, disease symptoms were consistently absent in control cultivars with host gene-mediated immunity to *Fusarium oxysporum* f.sp. *lycopersici* race 1 ('Belmondo' and 'Dombito') and in non-inoculated 'Moneymaker'. Similarly, the disease severity observed in the susceptible control cultivars ('Planet' and 'Moneymaker') did not significantly differ from that observed in the empty vector control (MM-402-09). This indicates that expression of the *nptII* gene (kanamycin resistance) dit not affect *Fusarium* resistance. Seven transgenic lines proved significantly more resistant whilst five were essentially as susceptible as control plants to *Fusarium* infection (Table 1). The reduction in disease severity in the resistant lines ranged from 36.3% (MM-539-16) up to 57.8% (MM-539-61) and was confirmed (50.9% up to 84.4% reduction) in a second experiment (data not shown). The transgenic lines MM-539-61 and MM-539-18 largely recovered from *Fusarium* infection by the time wild-type 'Moneymaker' had died (Fig. 3). In contrast, transgenic tomato lines constitutively accumulating similar levels of either the class I chitinase or the class I β-1,3-glucanase showed no increased resistance to *Fusarium* infection (Table 2). In fact, the only line with resistance significantly higher than that of wild-type 'Moneymaker' (MM-412-04) combined constitutive expression of the tobacco class I β-1,3-glucanase and induced expression of an endogenous tomato chitinase (Table 2), but did not differ significantly in disease response from the fully susceptible empty vector control plants (MM-402-09).

Correlation of gene expression and resistance

To establish to what extent expression of the class I and class II chitinase and β-1,3-glucanase genes contributed to resistance in the four-gene transgenic lines, mean expression levels in S1 lines were determined (Table

1) and correlated with the observed disease severity ratings. To avoid bias resulting from induction of chitinases and β-1,3-glucanases upon fungal infection (Joosten & de Wit, 1989; Tuzun et al., 1989), protein samples were collected directly prior to infection with *Fusarium*.

Correlation analyses indicated that higher expression of both the class I chitinase and the class I β-1,3-glucanase gene coincided with lower disease severity (Pearson correlation coefficients r(Chi-I) = - 0.77 [$P < 0.01$] and r(Glu-I) = - 0.54 [$P < 0.05$), respectively), while class II β-1,3-glucanase expression levels did not (Pearson correlation coefficient r(Glu-II) = - 0.22 [$P > 0.05$]). Correlation of the class II chitinase expression with disease severity could not be calculated, since its expression was confirmed to be either very low or lacking. Apparently expression of the class II β-1,3-glucanase gene does not play a pivotal role in enhancing *Fusarium* resistance. Whether expression of the class II chitinase can contribute to enhanced *Fusarium* resistance, obviously remains to be established.

Discussion and conclusions

The overall data presented here confirm and extend the observation that over-expression of antifungal proteins is a feasible approach for enhancing fungal resistance in economically-important crop plants. Enhanced levels of *Fusarium* resistance achieved in transgenic tomato lines accumulating multiple isoforms of chitinases and β-1,3-glucanases, were shown to result largely from the simultaneous expression of a class I chitinase and a class I β-1,3-glucanase gene because (i) higher expression of the class I chitinase and the class I β-1,3-glucanase gene correlated with lower disease severity, whereas expression of the class II β-1,3-glucanase gene did not, and (ii) simultaneous expression of the class I chitinase (1.5–4.0%) and β-1,3-glucanase (0.1–2.0%) did result in significantly enhanced resistance, while similar expression levels of either the class I chitinase or the class I β-1,3-glucanase alone, did not. These observations are consistent with previous data showing that only the class I chitinases and class I β-1,3-glucanases synergistically inhibit the growth of *Fusarium in vitro* (Mauch et al., 1988; Sela-Buurlage et al., 1993) and, as outlined in our earlier review paper (Van den Elzen et al., 1993), provide the first experimental evidence for the hypothesis that simultaneous expression of genes encoding antifungal proteins with synergistic activities *in vitro*, results in substantially higher levels of fungal resistance *in planta* than observed with expression of the individual antifungal genes alone.

Although the simultaneous expression of the class I chitinase and β-1,3-glucanase genes clearly resulted in partial resistance rather than immunity to infection with *Fusarium*, the observed delay in symptom development and apparent tolerance of infection in the best lines (MM-539-18 and MM-539-61), are expected to provide sufficient protection for survival of tomato plants following (early) natural *Fusarium* infection under field or glasshouse conditions. Experiments aimed at establishing the actual commercial value of *Fusarium* resistance in our best transgenic lines as well as resistance screens using a range of different fungal diseases are currently underway. If required, overall levels of partial resistance might be further enhanced by simultaneous expression of chitinases and β-1,3-glucanases with even stronger antifungal properties, by specifically expressing high levels of these hydrolytic enzymes in those tissues and cellular compartments that are predominantly invaded by the fungi of interest and/or by adding a variety of other genes encoding proteins with known antifungal activities.

Acknowledgements

We thank Joke Klap and Sjaan van Agtmaal for technical support, H. van der Meijden for colour art work and Drs Jürgen Logemann, Nick Garner and Bart Klein for critical reading of the manuscript.

Note

Recently, a similar synergistic interaction between antifungal proteins has been reported by Zhu et al. (1994). They co-expressed a basic chitinase gene from rice and an acidic glucanase gene from alfalfa in transgenic tobacco plants and observed that 'the combination of the two transgenes gave substantially greater protection against the fungal pathogen *Cercospora nicotianae* than either transgene alone.'

References

Alexander, D., R.M. Goodman, M. Gut-Rella, C. Glascock, K. Weymann, L. Friedrich, D. Maddox, P. Ahl-Goy, T. Luntz, E. Ward & J.A. Ryals, 1993. Increased tolerance to two oomycete pathogens in transgenic tobacco expressing pathogenesis-related protein 1a. Proc. Natl. Acad. Sci. USA 90: 7327–7331.

Broglie, K., I. Chet, M. Holliday, R. Cressman, P. Biddle, S, Knowlton, C.J. Mauvais & R. Broglie, 1991. Transgenic plants with enhanced resistance to the fungal pathogen *Rhizoctonia solani*. Science 254: 1194–1197.

Collinge, D.B., K.M. Kragh, J.D. Mikkelsen, K.K. Nielsen, U. Rasmussen & K. Ved, 1993. Plant chitinases. The Plant Journal 3: 31–40.

Cornelissen, B.J.C. & L.S. Melchers, 1993. Strategies for control of fungal diseases in plants. Plant Physiol. 101: 709–712.

Hain, R., H.J. Reif, E. Krause, R. Langebartels, H. Kindl, B. Vornam, W. Wiese, E. Schmeltzer, P.H. Schreier, R.H. Stöcker & K. Stenzel, 1993. Disease resistance results from foreign phytoalexin expression in a novel plant. Nature 361: 153–156.

Hood, E.E., S.B. Gelvin, L.S. Melchers & A. Hoekema, 1993. New *Agrobacterium* helper plasmids for gene transfer to plants. Transgenic Research 2: 208–218.

Joosten, M.H.A.J. & P.J.G.M. de Wit, 1989. Identification of several pathogenesis-related proteins in tomato leaves inoculated with *Cladosporium fulvum* (syn. *Fulvia fulva*) as 1,3-β- glucanases and chitinases. Plant Physiology 89: 945–951.

Kalloo, G. & B.O. Bergh, 1993. Genetic Improvement of Vegetable Crops. Pergamon Press Ltd., Oxford.

Koorneef, M., J.A.M. van Diepen, C.J. Hanhart, A.C. Kieboom-de Waard, L. Martinelli, H.C.H. Schoenmakers & J. Wijbrandi, 1989. Chromosomal instability in cell and tissue cultures of tomato haploids and diploids. Euphytica 43: 179–186.

Lamb, C.J., J.A. Ryals, E.R. Ward & R.A. Dixon, 1992. Emerging strategies for enhancing crop resistance to microbial pathogens. Bio/Technology 10: 1436–1445.

Leah, R., H. Tommerup, I. Svendsen & J. Mundy, 1991. Biochemical and molecular characterization of three barley seed proteins with antifungal properties. J. Biol. Chem. 266: 1464–1573.

Logemann, J., G. Jach, H. Tommerup, J. Mundy & J. Schell, 1992. Expression of a barley ribosome-inactivating protein leads to increased fungal protection in transgenic tobacco plants. Bio/Technology 10: 305–308.

Logemann, J., G. Jach, S. Logemann, R. Leah, G. Wolf, J. Mundy, A. Oppenheim, I. Chet & J. Schell, 1993. Expression of a ribosome inhibiting protein (RIP) or a bacterial chitinase leads to fungal resistance in transgenic plants. In: B. Fritig & M. Legrands (Eds) Mechanisms of plant defense responses. pp. 446–448. Kluwer Academic Publishers, Dordrecht.

Mauch, F., B. Mauch-Mani & T. Boller, 1988. Antifungal hydrolases in pea tissue. II. Inhibition of fungal growth by combinations of chitinase and β-1,3-glucanase. Plant Physiol. 88: 936–942.

Melchers, L.S., M.B. Sela-Buurlage, S.A. Vloemans, C.P. Woloshuk, J.S.C. van Roekel, J. Pen, P.J.M. van den Elzen & B.J.C. Cornelissen, 1993. Extracellular targeting of vacuolar tobacco proteins AP24, chitinase and β-1,3-glucanase in transgenic plants. Plant Molec. Biol. 21: 583–593.

Melchers, L.S., M. Aphotheker-de Groot, J.A. van der Knaap, A.S. Ponstein, M.B. Sela-Buurlage, J.F. Bol, B.J.C. Cornelissen, P.J.M. van der Elzen & H.J.M. Linthorst, 1994. A new class of tobacco chitinases homologous to bacterial exo-chitinases displays antifungal activity. The Plant Journal 5: 469–480.

Neuhaus, J.M., P. Ahl-Goy, U. Hinz, S. Flores & F. Meins, 1991. High level expression of a tobacco chitinase gene in *Nicotiana sylvestris*. Susceptibility of transgenic plants to *Cercospora nicotianae* infection. Plant Molec. Biol. 16: 141–151.

Poehlman, J.M., 1987. Breeding Field Crops. Van Nostrand Reinhold Publishers, New York, third edition.

Ponstein, A.P., S.A. Bres-Vloemans, M.B. Sela-Buurlage, P.J.M. van den Elzen, L.S. Melchers & B.J.C. Cornelissen, 1993. A novel pathogen and wound-inducible tobacco protein with antifungal activity. Plant Physiol. 104: 109–118.

Sela-Buurlage, M.B., A.S. Ponstein, S.A. Bres-Vloemans, L.S. Melchers, P.J.M. van den Elzen & B.J.C. Cornelissen, 1993. Only specific tobacco (*Nicotiana tabacum*) chitinases and β-1,3-glucanases exhibit antifungal activity. Plant Physiol. 101: 857–863.

Sijmons, P.C., B.M.M. Dekker, B. Schrammeijer, T.C. Verwoerd, P.J.M. van den Elzen & A. Hoekema, 1990. Production of correctly processed human serum albumin in transgenic plants. Bio/Technology 8: 217–221.

Strittmatter, G. & D. Wegener, 1993. Genetic engineering of disease and pest resistance in plants: present state of the art. Z. Naturforsch. 48c: 673–688.

Tuzun, S., M.N. Rao, U Vogeli, C.L. Schardl & J. Kuc, 1989. Induced systemic resistance to blue mold: early induction and accumulation of β-1,3-glucanases, chitinases and other pathogenesis-related proteins in immunized tobacco. Phytopathology 79: 979–983.

Van den Elzen, P.J.M., E. Jongedijk, L.S. Melchers & B.J.C. Cornelissen, 1993. Virus and fungal resistance: from laboratory to field. Phil. Trans. R. Soc. Lond. B 342: 271–278.

Van Roekel, J.S.C., B. Damm, L.S. Melchers & A. Hoekema, 1993. Factors influencing transformation frequency of tomato (*Lycopersicon esculentum*). Plant Cell Reports 12: 644–647.

Vogelsang, R. & W. Barz, 1993. Purification, characterization and differential hormonal regulation of a β-1,3-glucanase and two chitinases from chickpea (*Cicer arientinum* L.). Planta 189: 60–69.

Ward, E.R., G.B. Payne, M.B. Moyer, S.C. Williams, S.S. Dincher, K.C. Sharkey, J.H. Beck, H.T. Taylor, P. Ahl-Goy, F. Meins & J.A. Ryals, 1991. Differential regulation of β-1,3-glucanase messenger RNAs in response to pathogen infection. Plant Physiol. 96: 390–397.

Weide, R., M. Koorneef & P. Zabel, 1989. A simple nondestructive spraying assay for the detection of an active kanamycin resistance gene in transgenic tomato plants. Theor. Appl. Genet. 78: 169–172.

Wessels, J.G.H. & J.H. Sietsma, 1981. Fungal cell walls: a survey. In: W. Tanner & F.A. Loewus (Eds) Encyclopedia of Plant Physiology, new series, vol. 13B: Plant carbohydrates, pp. 352–394. Springer Verlag, Berlin.

Yenofsky, R.L., M. Fine & J.W. Pellow, 1990. A mutant neomycin phosphotransferase II gene reduces the resistance of transformants to antibiotic selection pressure. Proc. Natl. Acad. Sci. USA 87: 3435–3439.

Zu, Q., E.A.Maher, S. Masoud, R.A. Dixon & C. Lamb, 1994. Enhanced protection against fungal attack by constitutive co-expression of chitinase and glucanase genes in transgenic tobacco. Bio/Technology 12: 807–812.

Euphytica **85:** 181–192, 1995.

The sulphur-rich Brazil nut 2S albumin is specifically formed in transgenic seeds of the grain legume *Vicia narbonensis*

I. Saalbach[1], T. Pickardt[2], D.R. Waddell[1], S. Hillmer[1], O. Schieder[2] & K. Müntz[1]
[1] *Institute of Plant Genetics and Crop Plant Research, Corrensstr. 3, D-06466 Gatersleben, Germany;* [2] *Institute of Applied Genetics, Free University of Berlin, A.-Thaer-Weg 6, D-14195 Berlin, Germany*

Key words: Vicia narbonensis, gene transfer, gene expression, seeds, 2S albumin, methionine

Summary

Epicotyl explants were co-cultivated with *Agrobacterium tumefaciens* EHA101 to transfer a chimeric 2S albumin gene construct carried in the binary Ti plasmid vectors pGSGLUC1 or pGA472 into the grain legume *Vicia narbonensis*. This gene encoding the sulphur-rich Brazil nut albumin was under the control of either the CaMV 35S promoter which permits gene expression in all organs, or the *Vicia faba* legumin B4 promoter which elicits seed-specific gene expression. After callus formation and selection for kanamycin resistance, somatic embryos were induced which, in the case of transformation with the vector pGSGLUC1, were screened for GUS activity. Embryos that produced GUS were in addition analysed for 2S albumin formation. Selected transgenic embryos were cloned by multiple shoot regeneration. Rooted and fertile plants were obtained by grafting transgenic shoots on the appropriate seedlings. R_1 and R_2 generations were raised and analysed for GUS as well as 2S albumin gene expression.

Expression of the 35S promoter/2S albumin gene fusion took place in all organs of the transgenic plants including the cotyledons of seeds, whereas seed-specific gene expression was found in transformants with the legumin promoter/2S albumin gene fusion. The 2S albumin accumulated in the 2S protein fraction of transgenic seeds and its primary translation product was processed into the 9 and 3 kDa polypeptide chains. The foreign protein was localised in the protein bodies of the grain legume. Analysis of the R_2 plants indicated Mendelian inheritance of the 2S albumin gene. In homozygous *V. narbonensis* plants the amounts of 2S albumin were twice that present in the corresponding heterozygous plants. Whereas only low level formation of the foreign protein was achieved if the gene was under the control of the 35S promoter, approximately 3.0% of the soluble seed protein was 2S albumin if seed-specific gene expression was directed by the legumin B4 promoter. Some of these transformants exhibited a three-fold increase in the methionine content of the salt-soluble protein fraction extracted from seeds.

Abbreviations: 35S – cauliflower mosaic virus 35S protein gene, GUS – β-glucuronidase, NPTII – neomycin phosphotransferase II, LeB4 – *Vicia faba* legumin B4 gene, 2S albumin – Brazil nut (*Bertholletia excelsa*) 2S albumin, ER – endoplasmic reticulum, rER – rough endoplasmic reticulum, HPLC – high pressure liquid chromatography

Introduction

One long-standing economic goal for the genetic manipulation of maize, cereals and grain legumes is to improve their nutritional quality. The amino acid composition of the total seed protein does not correspond to the dietary needs of humans and monogastric animals. The very low level of lysine in the most important cereal crops and the very limited methionine content of legume grains, decreases the biological value of their seed protein to 50 to 75 percent when compared with a balanced amino acid diet or chicken egg protein. In cultures with primarily vegetarian diets such as India, this imbalance can lead to forms of malnutrition in which children less than 4 years of age suffer from retarded mental and physical development (Waterlow

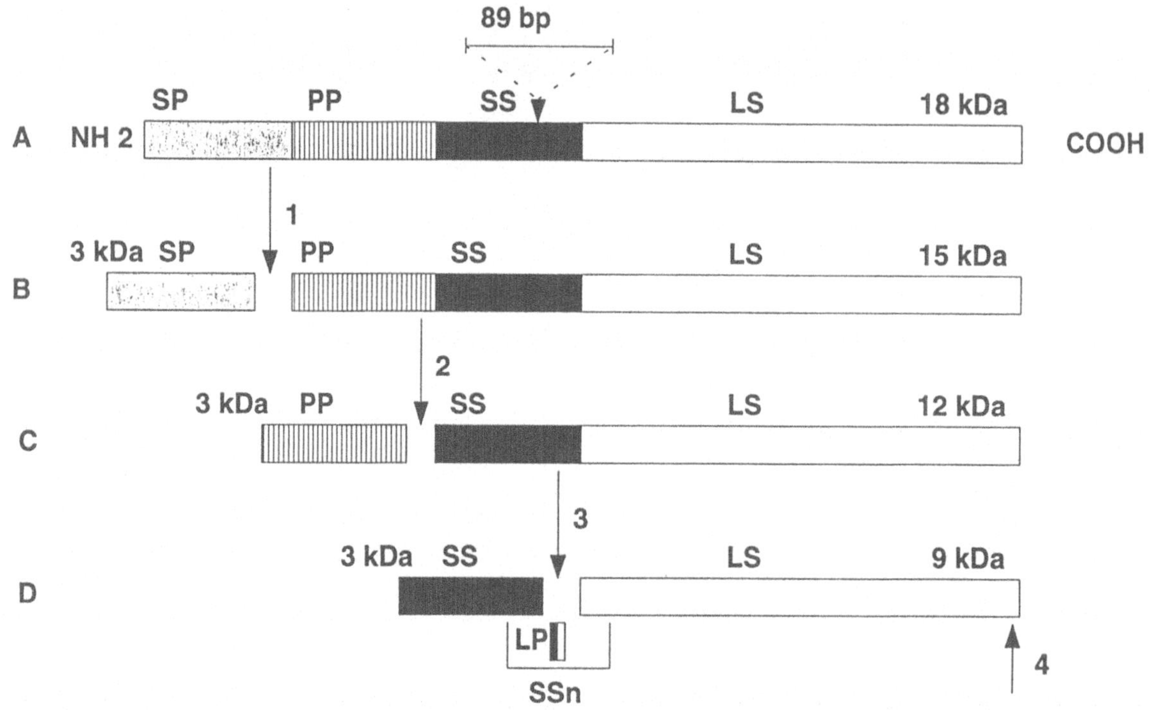

Fig. 1. The 2S albumin of Brazil nut (*Bertholletia excelsa* H.B.K.) and its precursors. A: The pre-propolypeptide encoded by the 2S albumin-specific cDNA (Altenbach et al., 1987; Sun et al., 1987); the position of the only intron in the corresponding gene is indicated by an arrow head (Gander et al., 1991). B–D: Processing steps 1 to 3 and the different products of the 2S albumin precursor maturation by limited proteolysis. Step 1 detaches the signal peptide (SP) from the 18 kDa propolypeptide and step 2 the propeptide (PP) from the 15 kDa precursor that still comprises the small (SS) and large (LS) subunit. A peptide with 5 amino acid residues (LP) links the small and large subunit and is cleaved off during step 3 of the limited proteolysis. Step 4 indicates the cleavage of a tetra peptide from the C-terminus of the large subunit. SSn, disulphide linkages between SS and LS.

& Payne, 1975). The imbalanced amino acid composition of the above-mentioned crops also affects strongly the efficiency of livestock fattening. The more closely the amino acid composition corresponds to the nutritional demands, the more efficient the transformation of seed protein nitrogen into human or animal protein nitrogen becomes. This not only alleviates human malnutrition and improves the efficiency of livestock fattening but concomitantly it decreases the amount of secreted nitrogen that can lead to environmental pollution (Kirchgessner et al., 1994). Without doubt the genetic improvement of the nutritional quality of the most important cereals like maize, rice, wheat or barley, and of many grain legumes remains one of the challenges for breeding and genetic engineering.

Despite the discovery of high-lysine mutants of maize and barley beginning in the sixties (Nelson et al., 1965; Munck, 1970; Green & Phillips, 1974; Hibberd & Green, 1982), progress has been slow in achieving the above-mentioned goals. New hopes have been inspired with the advent of the genetic engineering of crop plants. One promising strategy for improving the nutritional quality of legumes is to introduce into their genomes a gene encoding the extremely methionine-rich 2S albumin from Brazil nut, *Bertholletia excelsa* H.B.K. (Altenbach et al., 1987). The practicality of this approach was demonstrated in transformation experiments carried out with tobacco (Altenbach et al., 1989) and rape (Altenbach et al., 1992). The foreign protein accounted for at most 8 and 4%, respectively, of the salt-soluble protein in the transgenic seeds which represents an increase in the methionine content of the seed protein of approximately one third.

The stable transformation of grain legumes along with subsequent regeneration of fertile plants and germ line transmission of the foreign gene has been reported only few times in the literature. Transgenic pea plants (*Pisum sativum* L.) were obtained by Puonti-Kaerlas et al. (1992). They were able to demonstrate Mendelian inheritance of the hygromycin phospho-

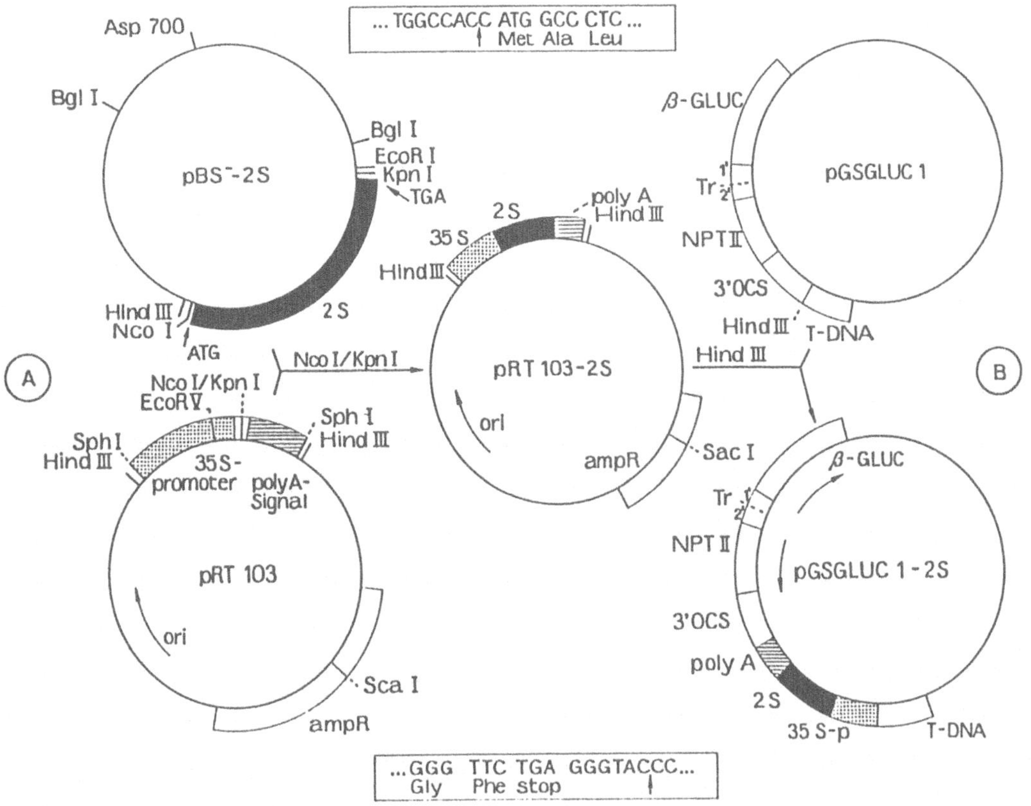

Fig 2 Assembly of the CaMV 35S promoter and terminator with the DNA encoding the 2S albumin pre-propolypeptide and subsequent insertion of the completed fusion gene into the binary vector plasmid pGSGLUC1 A The 2S albumin gene (filled bar) was taken from pBS⁻-2S and inserted into the polylinker of plasmid pRT103 (Topfer et al , 1987) giving plasmid pRT103-2S B The HindIII fragment of pRT103-2S, which comprises the complete gene construct, was inserted into the HindIII site of pGSGLUC1, leading to the binary vector pGSGLUC1-2S which carries the 2S albumin gene together with the CaMV 35S gene promoter (stippled) and polyA addition signal (hatched) The inserts at the top and bottom of the figure indicate the junction sequences at the start and stop codons, respectively, between the 2S albumin-coding DNA and the CaMV 35S gene sequences in plasmid pRT103 Other structural elements in the various constructs are indicated by open boxes

transferase gene that had been used as a selection marker to obtain the transgenic plants. Recently, Schroeder et al. (1993) demonstrated the expression and Mendelian inheritance of the bacterial phosphinotricin acetyltransferase gene which conferred resistance to the herbicide Basta in stably-transformed peas. Aragao et al. (1992) reported transient expression of the Brazil nut 2S albumin gene in cells of mature *Phaseolus vulgaris* embryos after particle bombardment. Stably-transformed *Pisum sativum* and *Vicia faba* tissues expressing the 2S albumin gene were produced by Saalbach et al. (1994), however no fertile plants could be regenerated from these tissues. Seed-specific expression of the 2S albumin gene in stably-transformed grain legume (*V narbonensis*) regenerants that inherit the gene in the germ line has recently been achieved (Saalbach et al., 1995).

Materials and methods

Transformation and regeneration systems

We have synthesised a DNA encoding the 12 kDa precursor of the 3 and 9 kDa polypeptides (Saalbach et al., 1994) following the published cDNA sequence of the Brazil nut 2S albumin (Altenbach et al., 1987). This DNA was brought under the control of either the 35S promoter and terminator (Fig. 2) taken from plasmid pRT103 (Topfer et al., 1987) or the legumin B4 promoter and terminator (Baumlein et al., 1986, 1991, 1992). Whereas the former directs the expression of foreign genes in all tissues, although to different extents (Benfey et al., 1990), the latter imposes seed-specific and developmentally-regulated gene expression (Baumlein et al., 1987, 1988). The fusion genes were inserted into the appropriate sites of the bina-

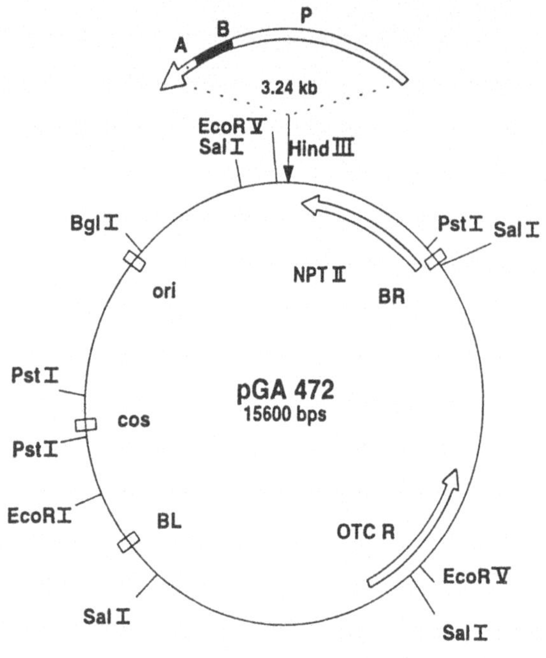

Fig. 3. Insertion of the 2S albumin gene into the HindIII site of the binary vector plasmid pGA472. The inserted fragment comprises the 2S albumin gene (B, 0.441 kb) under the control of the legumin B4 promoter (P, 2.4 kb) and terminator (A, 0.4 kb) both from *Vicia faba*. In order to test the inserted T-DNA for recombination events 3 different DNA probes were used in Southern blotting: 1 – 2S albumin gene fragment, 2 – *nptII* gene fragment, and 3 – a fragment specific for the region containing the origin of replication (ori).

ry vector plasmid pGSGLUC1 giving pGSGLUC1-2S (Fig. 2), or pGA472 (An et al., 1985) giving pGA472-2S (Fig. 3), both with the bacterial *npt II* gene as a selectable marker. The former plasmid contains the screenable *gus* gene whereas the latter does not. These binary Ti-plasmid vectors were used for *Agrobacterium*-mediated gene transfer to *V. narbonensis*.

Transformation and regeneration of the legume were achieved following the protocol of Pickardt et al. (1991). Epicotyl explants were co-cultivated with the *Agrobacterium tumefaciens* strain EHA101 (Hood et al., 1986; McGranahan et al., 1990) carrying the corresponding binary vector plasmid. After callus formation and selection for kanamycin resistance in the presence of 30 μg/ml geneticin, somatic embryos were induced which in transformation experiments with pGSGLUC1-2S were screened for GUS activity using the histochemical method of Jefferson (1987). Selected GUS-positive embryos were cloned by multiple shoot regeneration. Regenerated shoots had to be grafted onto roots of 5 to 10-day-old *V. narbonensis* seedlings in order to overcome rooting difficulties, as mentioned by Pickardt et al. (1991). Transgenic R_1 and R_2 plants were raised after self pollination.

In some cases transgenic tobacco was used for comparison. This plant was transformed according to the leaf disc protocol of Horsch et al. (1985).

DNA purification and Southern blotting

DNA was extracted from young leaves according to Bäumlein et al. (1987) and used for Southern blotting (Sambrook et al., 1989). The HindIII fragment from pRT103-2S was used as a probe in tests with DNA from *V. narbonensis* which was transformed with pGSGLUC1-2S. Since extensive recombination was observed with transformants bearing the LeB4/2S albumin fusion gene, probes specific for the *nptII* gene, for the 2S albumin gene and for the replication origin were used.

Segregation analysis

Self pollination of R_0 and R_1 plants respectively, was used to generate R_1 and R_2 plants for inheritance analysis. Leaf or seed extracts were tested for 2S albumin expression using protein fractionation by SDS polyacrylamide gel electrophoresis (Laemmli, 1970) with subsequent immunoblotting (Karey & Sirbasku, 1989). In addition, screening for *gus* gene expression (Jefferson, 1987) was performed with transformants bearing the 35S/2S albumin gene fusion.

2S-albumin purification and localisation

Different organs or tissues of the transformants were extracted with buffered SDS solutions and the extracts were analysed as follows (see also Saalbach et al., 1994). Fractionation of the legume seed protein extracts by HPLC on a LKB Ultropack TSK G3000 column (7.5 × 600 mm) separated the albumin-containing 2S protein fraction from other storage proteins (Figs 4A and B), like legumin and vicilin, which turned out to be free of 2S albumin when analysed by Western blotting (Fig. 4C). When the SDS gels were run under reducing conditions, the 12 kDa holoprotein was fractionated into the 9 and 3 kDa polypeptides (Fig. 4D). This indicates that the precursors underwent correct processing as far as this can be seen from molecular weight determination in SDS gels. The result confirms that the machinery for limited proteolytic processing of this type of 2S storage protein must have been highly

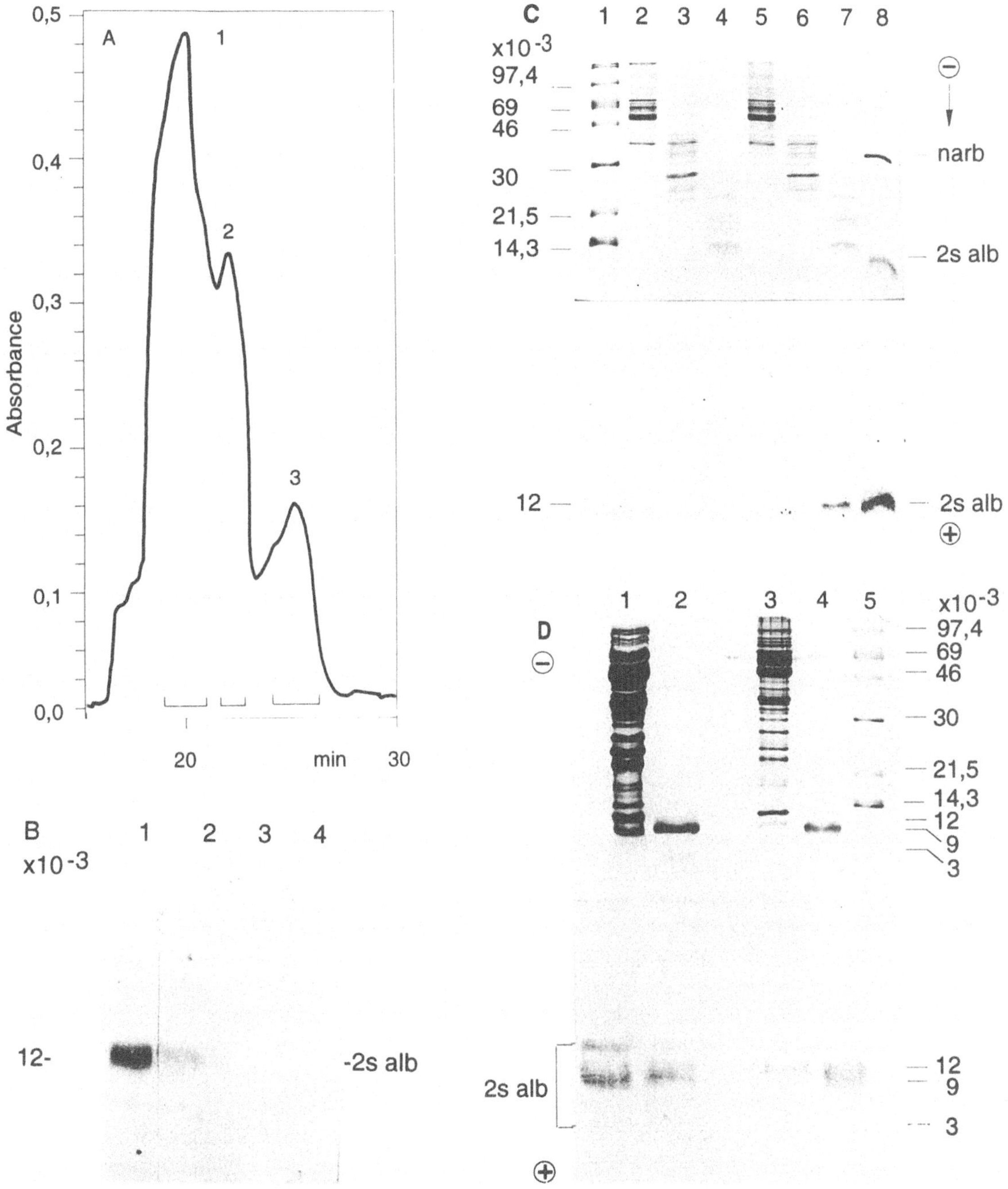
A
0,5
0,4
0,3
0,2
0,1
0,0
Absorbance
1
2
3
20
min
30
B
1 2 3 4
x10-3
12-
-2s alb
C
1 2 3 4 5 6 7 8
x10-3
97,4
69
46
30
21,5
14,3
narb
2s alb
12
2s alb
D
1 2 3 4 5
x10-3
97,4
69
46
30
21,5
14,3
12
9
3
2s alb
12
9
3

←

Fig. 4. Analysis of SDS-extracted proteins from transgenic *Vicia narbonensis* seeds. A: Elution profile of a 100 μl extract prepared from a seed of a plant that had been transformed with the 35S promoter-controlled 2S albumin gene giving only low level expression in this organ. The extract was fractionated by HPLC on a LKB Ultropac TSK G 3000SW column (7.5 × 600 mm) in SDS extraction buffer. Fractions indicated by brackets were electrophoretically analysed. B: Fractions of peak 3 were pooled before running the SDS polyacrylamide gel electrophoresis (SDS-PAGE) under non-reducing conditions with subsequent immunoblotting (lane 2). Purified 2S albumin (20 ng, lane 1) and extracts from seeds of wild-type plants (lane 4) were used as controls. For comparison the original seed extract was included (lane 3). C: Comparative electrophoretic analysis of the three fractions from the gel filtration elution profile presented in A. After pooling each fraction and concentrating to 50 μl, aliquots were run in SDS gels according to Laemmli (1970). Peak eluates from wild-type seeds (lanes 2–4) were compared with corresponding eluates from transgenic seeds (lanes 5–7). Lane 1, molecular weight markers, lane 8, mixture of authentic narbonin (narb), an endogenous 2S globulin from *V. narbonensis* seeds, with 2S albumin (2S alb) preparations. The upper panel shows a Coomassie stained gel slab on which no differences in banding patterns are visible in a comparison of the corresponding peak eluates from wild-type and transgenic seeds. The lower panel represents the corresponding immunoblot. 2S albumin appears only in peak 3 of extracts from transgenic seeds. D: The presence of mature 3 and 9 kDa polypeptides indicates the regular processing of 2S albumin precursors in seeds of plants that had been transformed with the 2S albumin gene under the control of the LeB4 promoter and terminator. Extracts from transgenic seeds (lanes 1 and 3) as well as authentic 2S albumin preparations were compared by SDS-PAGE under reducing (lanes 1 and 2) and non-reducing conditions (lanes 3 and 4); lane 5, molecular weight markers. Bands specific for the 2S albumin are already visible in the stained gel slab (upper panel) but are specifically labelled by immunoblotting (lower panel). The 12 and 9 kDa polypeptides exhibit only small mobility differences. The 3 kDa band reacts much less strongly with the polyclonal antibodies raised against the holoprotein and stains much less than the 9 kDa polypeptide. Therefore, the corresponding bands are barely visible in the photographs of the stained gel and of the blot.

conserved during higher plant evolution, since correct processing was found in transgenic tobacco (Altenbach et al., 1989), rape (Altenbach et al., 1992) and legume seeds (Saalbach et al., 1994).

The compartmentation analysis was performed by immuno-gold labelling of electron micrographs (Müntz et al., 1993b).

Quantification of 2S-albumin and methionine content

Since it is much easier to find the 12 kDa 2S albumin precursor on the gels than the mature 9 and 3 kDa polypeptide chains, electrophoresis was performed under non-reducing conditions. Standard probes of 2S albumin preparations were always included for comparison. Semi-quantitative analysis was based on comparative analysis with a series of different amounts of the standard protein for calibration.

To determine the methionine content, seeds were extracted with buffered neutral salt solutions of suitable ionic strength (Altenbach et al., 1992), dialysed against distilled water, lyophilised and hydrolysed. The amino acid composition was kindly performed by BASF.

Results

The 2S albumin gene was detected in R_0, R_1, and R_2 plants following transformation with either the 35S/2S albumin or the LeB4/2S albumin fusion gene (data not shown). Several R_2 plants were obtained that no longer expressed or had lost the foreign gene. Inheritance analysis with 35S/2S albumin gene transformants is given for the progeny of R_0 plant no. 15/1 (Table 1). Seedlings were analysed for 2S albumin formation as well as GUS activity. Segregation of the markers analysed was consistent with Mendelian inheritance of one gene insert. Two homozygous lines expressing 2S albumin were obtained (seedlings no. 3 and 10). Similar results were obtained with corresponding tobacco transformants (Saalbach et al., 1994). The genetic analysis of R_2 plants provided clear evidence for the 2S albumin gene integration into the plant genome. This was confirmed by Southern blotting for transformants containing the LeB4/2S albumin fusion gene. No such unequivocal conclusion could be drawn from corresponding blotting experiments with DNA extracted from transformants carrying the 35S/2S albumin fusion gene, since the probes used only detected sequences inside the T-DNA without extension to the plant DNA.

Seed-specific expression of the 2S albumin gene under the control of the legumin B4 promoter was examined in transformation experiments with the pGA472-2S vector. Of 70 transgenic *V. narbonensis* plants obtained so far, 40 R_0 plants contained the 2S albumin gene. Inheritance of the foreign gene, copy number and stability of its expression have so far been analysed in the offspring of 18 R_0 plants. Sixteen of these transformants formed the 2S albumin in their seeds whereas the seeds of the remaining two did not contain the foreign protein. Homozygous lines were obtained from two transformants. Gene expression was often unstable in the offspring of transformants with

Table 1. Inheritance of the 2S albumin gene in the progeny of *Vicia narbonensis* transgenic plant 15/1. n.d., not determined

R_1			R_2							
Seedling	GUS	2S albumin	GUS segregation		Fit to Mendelian expectations for segregation of one allele		2S albumin segregation		Fit to Mendelian expectations for segregation of one allele	
			+	-	χ^2	P	+	-	χ^2	P
1	+	+	23	10	0.49	0.48	8	2	0.13	0.72
3	+	+	40	0	–	–	10	0	–	–
4	-	-	0	30	-	-	n.d.	n.d.	–	–
8	+	+	17	6	0.01	0.90	n.d.	n.d.	–	–
10	+	+	50	0	–	–	20	0	–	–

Table 2. Levels of 2S albumin formation in different organs of transgenic *Vicia narbonensis*; n.d. not determined

Organ	R_0 generation		R_1 generation	
	2S alb (%)	GUS	2S alb	GUS
Leaf	0.2	(+)	0.2	(+)
Root	grafted	grafted	0.3	++
Seed coat	0.08	++	0.1	++
Cotyledon	0.01	+	n.d.	+

more than one gene insertion, and in the offspring of plants containing recombinant insertions.

Tissue specificity of gene expression and intracellular localisation of the gene product

Where the 2S albumin gene was under control of the 35S promoter, expression was observed in all analysed organs and tissues of the transgenic plants though at different levels (Table 2). Seeds of the transgenic legume as well as of corresponding tobacco transformants exhibited only low levels of 2S albumin formation which is in agreement with the known low-level activity of this promoter in embryonic tissue. Weak expression was also registered in the leaves of transgenic tobacco even in those plants containing three copies of the foreign gene. In contrast, the *V. narbonensis* transformants with only one gene copy exhibited high levels of gene expression in leaves (Table 2). The highest levels of 2S albumin production took place in the roots of the legume transformants.

As was to be expected from the findings of Bäumlein et al. (1987, 1988, 1991, 1992), the legumin B4 promoter directed seed-specific 2S albumin gene expression and this at a much higher level than that achieved using the 35S promoter (see section 6). No 2S albumin formation was detected in other organs.

Analysis of vacuoles of leaf mesophyll cells from tobacco plants transformed with the 35S promoter/2S albumin/35S terminator fusion gene, revealed vacuolar targeting of the 2S albumin (Saalbach et al., 1994). In transgenic seeds of *V. narbonensis*, the 2S albumin was located in the protein bodies that belong to the vacuolar system of the storage tissue cells (Fig. 5).

Expression level and its effect on the methionine content of transgenic seeds

Expression levels of GUS and 2S albumin differed between organs and tissues of *V. narbonensis* transformed with the vector pGSGLUC1-2S. This could be due to differences in either the stability of mRNA or protein, or in the controlling elements directing the expression of the different genes, or a combination of the above. Whereas expression of the 2S albumin gene was directed by the 35S promoter, the *gus* and *nptII* genes were controlled by the T-DNA Tr1′,2′ 'double-headed' promoter. The differences in the level of 35S promoter-controlled 2S albumin gene expression in different organs and tissues were described in the previous section.

Western blots with extracts from different organs of tobacco plants as well as from heterozygous and homozygous transformed tobacco and *V. narbonensis* plants (Saalbach et al., 1994), revealed a close relation-

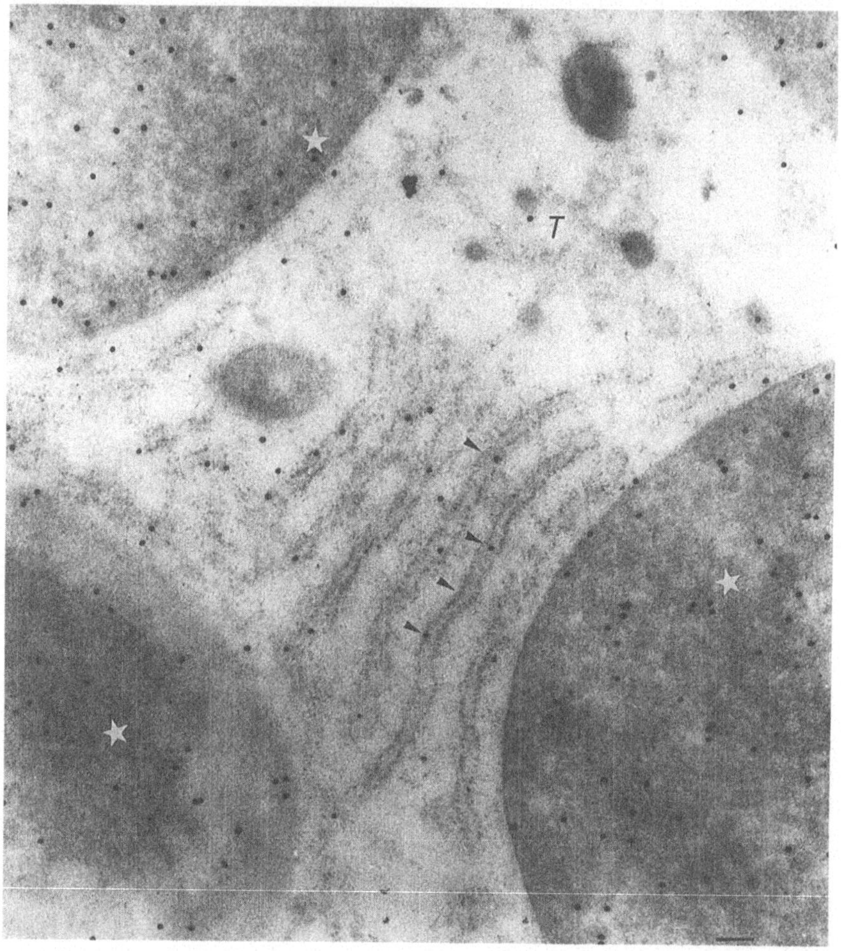

Fig. 5 Electron micrograph with immunogold-labelled Brazil nut 2S albumin in the protein bodies (stars), rough endoplasmic reticulum (arrow heads), and trans-Golgi network (T) of cells from cotyledons of developing seeds from *V. narbonensis* plants that were transformed with the 2S albumin gene under the control of the seed-specific LeB4 promoter and terminator from *V. faba* The localisation of the foreign protein in the different cell compartments indicates that it must be synthesised on membrane-bound polysomes, transferred through the endoplasmic reticulum and Golgi apparatus, which belong to the endomembrane system, and targeted into the vacuole that is converted into protein bodies during seed maturation The bar corresponds to 0.2 μm.

ship between the amount of accumulated 2S albumin and the number of integrated genes. Although variability of gene expression in different transformants can not be excluded as a factor influencing the accumulation level of the foreign protein, it does not seem to be very probable that this influence could act so as to mimic gene dosage effects in all cases. A gene dosage effect has already been described for the expression of the legumin gene from *Vicia faba* in transgenic tobacco (Bäumlein et al., 1988). Therefore, we feel justified in concluding that gene dosage effects also act in our *V. narbonensis* plants that express the foreign 2S albumin-specific Brazil nut gene.

Whereas up to 0.01% of the SDS-soluble protein was found to be 2S albumin in the cotyledons from seeds of transgenic *V. narbonensis* plants carrying the 35S/2S albumin gene fusion, the level of 2S albumin in cotyledons of transformants carrying the seed-specific LeB4/2S albumin construct was between 0.5 and 3 percent, a 50–300 fold increase.

The methionine content of the salt-soluble protein fraction was determined in the seeds of a series of selected R_0 transformants to find out whether the increased amounts of methionine-rich foreign protein influences the quantity of this limiting essential amino acid in transgenic seeds. The salt-soluble proteins represent the major quantity of proteins in the legume's seeds. Transgenic seeds of R_0 plants were found to exhibit up to a threefold increase in methionine content over that of the wild-type seeds of *V. narbonensis*. Analysis of methionine content in the seeds of the R_1 offspring is now under way.

Discussion

The Brazil nut 2S albumin and its gene

The hypocotyl of Brazil nut seeds contains approximately 50 percent of the total seed protein by dry weight. Thirty percent of this fraction is accounted for by the 2S protein fraction in which the water-soluble 12 kDa sulphur-rich protein, called Brazil nut 2S albumin, predominates (Sun et al., 1987). The purified protein contains 18% methionine and 8% cysteine, and it is primarily responsible for the high methionine content of Brazil nut seed protein (8.3% by weight). Electrophoretic analysis under denaturing and disulphide bridge-reducing conditions, revealed that the albumin is composed of polymorphic 9 kDa and 3 kDa polypeptide chains linked by one or more disulphide bonds. These polypeptides are formed from a larger precursor that also contains a 3 kDa propeptide located N-terminally from the 3 kDa chain and an N-terminal 3 kDa signal peptide. The 9 kDa polypeptide of the mature holoprotein forms the C-terminus of the precursor (Fig. 1).

This precursor is synthesised on membrane-bound polysomes (rough endoplasmic reticulum, rER). The mature holoprotein accumulates inside protein bodies belonging to the vacuolar compartment of the storage tissue cells in the hypocotyl (Harris et al., 1993). Consequently, the 2S albumin polypeptides are transferred from the rER through the so-called secretory pathway into the endomembrane system and finally, presumably from the trans-Golgi network, into the protein bodies which are extracytoplasmic storage vacuoles. On its way from the site of formation to the site of deposition the precursor undergoes at least 3 successive steps of limited proteolytic tailoring (Fig. 1): 1 – The co-translational cleavage of the signal peptide which is known to take place at the inner surface of the ER membrane inside the ER lumen; 2 – the detachment of the propeptide; and 3 – the separation of the 3 and 9 kDa mature polypeptides. In addition, by comparing the cDNA-derived amino acid sequence (Altenbach et al., 1987) with that determined by direct amino acid sequencing (Ampe et al., 1986), a fourth step of limited proteolysis that seems to eliminate the four C-terminal amino acid residues from most of the known polymorphic 9 kDa polypeptides, has been revealed. Presumably, processing steps 2, 3 and 4 occur in the vacuolar compartment by analogy with the processing of 2S albumins and other storage globulins from different plants (Hara-Nishimura et al., 1991; Müntz et al., 1993a).

Tissue-specific expression of foreign genes in seeds

Heterologous transfer of storage protein genes between angiosperms began more than 10 years ago (Murai et al., 1983). The numerous transformations that have been performed since then have shown that the mechanisms of control and regulation of storage protein gene expression have been highly conserved throughout plant evolution. During the period of continuous protein accumulation in seeds, tissue-specificity and developmental pattern of expression are determined by the promoter region of the storage protein genes and hence controlled at the transcriptional level. This implies that the pattern of expression of the seed storage genes can be conferred upon foreign genes by fusing these controlling elements before the protein coding regions of the foreign gene. Furthermore, the structural signals that determine the biosynthesis of the molecular precursors of such proteins on membrane-bound polysomes and their deposition as mature holoproteins inside the protein bodies – specialised vacuolar compartments of the cell – as well as those signals involved in intracellular transfer and targeting from the site of formation to the compartment of deposition, are universally recognised by the cells of different plant seeds.

The results of our experiments, in which the protein-coding region of a methionine-rich 2S albumin from *Bertholletia excelsa*, the Brazil nut, was put under the control of the seed-specific LeB4 promoter and the fusion gene successfully transferred into the grain legume *V. narbonensis*, have confirmed the conservative nature of the mechanisms of transcriptional control of seed-storage gene expression as well as those involved in the processing of their gene products. Seed-specific expression of this gene was documented and its Mendelian inheritance shown (Saalbach et al., in press). The protein was formed on the rough endoplasmic reticulum and transported into the protein bodies of the legume cotyledon cells where it accumulated. Furthermore, we were able to show by comparing heterozygous and homozygous offspring, that the amount of protein produced in the storage tissue was correlated with the number of genes.

The methionine content of the salt-soluble proteins increased threefold in the seeds of R_0 transgenic plants over that of corresponding wild-type seeds. This appears to be a very large increase when it is compared with the increases of 30 to 40% that have been reported in similar transformation experiments with tobacco (Altenbach et al., 1989) and rape (Altenbach et al., 1992). However, the level of methionine in the salt-soluble protein fraction of wild-type *V. narbonensis* seeds is only 0.5% (mol/mol), whereas the corresponding protein fraction of the wild-type winter variety of *Brassica napus* that was used by Altenbach et al. (1992) contains 2.64% (mol/mol) methionine. The 1.5% (approx.) methionine found in this protein fraction in transgenic *V. narbonensis* seeds corresponds to an increase of 1% (mol/mol) of methionine and is comparable with that found in the seeds of the rape transformants. The similar absolute increases result from increases in the percentage of 2S albumin that were found in the salt-soluble protein fractions in the transgenic seeds of both species. The methionine content of this protein fraction in transgenic rape reaches 3.52%, and thereby achieves the recommended level of methionine and cysteine for a balanced amino acid diet for monogastric animals. The methionine level we have so far achieved with transgenic *V. narbonensis* corresponds to 42% of this standard; it reaches 80% the cysteine values are added (1.3%, mol/mol).

Improvements in the amino acid composition of proteins via genetic engineering are based upon the evolutionary conservation of the pattern of gene expression as well as the elements for translational control and intracellular protein sorting. So far two strategies have predominated in this field:

1) The codon composition of suitable genes has been changed by site-specific *in vitro* mutagenesis, frame shift mutation and/or insertion of suitable fragments from other genes. The modified genes were transformed into receptor plants that in most cases belonged to the *Solanaceae*. Unfortunately, in most cases, the protein stability was affected by these alterations and the accumulation of these proteins was reduced when compared with the accumulation of wild-type protein (e.g. Hoffmann et al., 1988; Saalbach et al., 1990). Recently, methionine codons were generated at 8 different sites in a 50 kDa-vicilin gene from *Vicia faba*, a gene that was originally free of methionine and cysteine codons. The *in vitro*-mutated gene was transferred into tobacco where stable gene expression was achieved. No differences have so far been observed in the stability of the wild-type and the mutant vicilin in transgenic tobacco seeds (Christov, 1993).

2) Foreign genes encoding proteins with suitable amino acid composition have been transferred into the target species. The methionine-rich 2S albumin from Brazil nut became the favourite protein for transformation experiments aimed at increasing the methionine content in legume seeds where this sulphur-containing amino acid is the most limiting essential amino acid from a nutritional point of view.

So far, the second strategy has turned out to be much more successful than the first one. The experiments we have reported here successfully employed this strategy for the first time in a grain legume.

In the near future we plan to determine whether further increases in the expression levels of the 2S albumin in transgenic *V. narbonensis* can be achieved by crossing transgenic lines with independent 2S albumin gene insertions, which has already been achieved in experiments with transgenic tobacco (Hobbs et al., 1993). In addition, transgenic *V. narbonensis* seeds forming suitable amounts of the methionine-rich 2S albumin will enable us to investigate the effect of the foreign methionine sink on the regulation of methionine biosynthesis as well as on the formation of endogenous storage proteins. When sufficient quantities of transgenic seeds become available we will be able to test the feeding efficiency with rats. We also plan to transform grain legumes that have been adapted to the moderate climates of Europe and North America like peas or field beans, as well as the economically-important legumes grown in tropical regions, like chickpeas, with the Brazil nut 2S albumin. The practical application of this technology will require investigation of genetic stability, growth and yield of the transgenic legumes in field trials.

Acknowledgements

Part of this work was supported by DFG grant Mu 925/2-1 and by a grant of the Federal Ministry of Research and Technology, Bonn. The authors wish to express their gratitude to Dr Ch. Horstmann (Gatersleben) for performing the gel filtration analysis of proteins from transgenic seeds of *V. narbonensis*, to Mr T. Meyer from BASF for performing the amino acid analyses with the salt-soluble protein fraction from transgenic *V. narbonensis* seeds, to Dr A Meister for performing the statistical evaluation of R_2 segregations, and to Drs E. Hood and J. Botterman (Gent) for providing the *Agrobacterium tumefaciens* strain EHA101 and the plasmid pGSGLUC1, respectively. The skillful technical assistance of Mrs Petra Hoffmeister, Ingrid Otto, Ingrid Pfort, Brigitte Weiß(all Gatersleben) and Verena Schade (Berlin) is gratefully acknowledged.

References

Altenbach, S.B., K.W. Pearson, F.W. Leung & S.S.M. Sun, 1987. Cloning and sequence analysis of a cDNA encoding a Brazil nut protein exceptionally rich in methionine. Plant Mol. Biol. 8: 239–250.

Altenbach, S.B., K.W. Pearson, G. Meeker, L.C. Staraci & S.S.M. Sun, 1989. Enhancement of the methionine content of seed proteins by the expression of a chimeric gene encoding a methionine-rich protein in transgenic plants. Plant Mol. Biol. 13: 513–522

Altenbach, S.B., C.-C. Kuo, L.C. Staraci, K.W. Pearson, C. Wainwright, A. Georgescu & J. Townsend, 1992. Accumulation of a Brazil nut albumin in seeds of transgenic canola results in enhanced levels of seed protein methionine. Plant Mol. Biol. 18: 235–246.

Ampe, C., J. van Damme, L.A.B. de Castro, M.J.A.M. Sampaio, M. van Montagu & J. Vandekerckhove, 1986. The amino acid sequence of the 2S sulfur-rich proteins from seed of Brazil nut (*Bertholletia excelsa* H.B.K.). Eur. J. Biochem. 159: 597–604.

An, G., B.D. Watson, S. Stachel, M.P. Gordon & E.W. Nester, 1985. New cloning vehicles for transformation of higher plants. EMBO J. 4: 277–284.

Aragao, F.J.L., M.F. Grossi de Sa, E.R. Almeida, E.S. Gander & E.L. Rech, 1992. Particle bombardment-mediated transient expression of a Brazil nut methionine-rich albumin in bean (*Phaseolus vulgaris* L.). Plant Mol. Biol. 20: 357–360.

Bäumlein, H., U. Wobus, J. Pustell & F.C. Kafatos, 1986. The legumin gene family: structure of a B type gene of *Vicia faba* and a possible legumin gene specific regulatory element. Nucl. Acid Res. 14: 2702–2720.

Bäumlein, H., A.J. Müller, J. Schiemann, D. Helbing, R. Manteuffel & U. Wobus, 1987. A legumin B gene of *Vicia faba* is expressed in developing seeds of transgenic tobacco. Biol. Zbl. 106: 569–575.

Bäumlein, H., A.J. Müller, J. Schiemann, D. Helbing, R. Manteuffel & U. Wobus, 1988. Expression of a *Vicia faba* legumin B gene in transgenic tobacco plants: Gene dosage-dependent protein accumulation. Biochem. Physiol. Pflanzen 183: 205–210.

Bäumlein, H., W. Boerjan, I. Nagy, R. Panitz, D. Inze & U. Wobus, 1991. Upstream sequences regulating legumin gene expression in heterologous transgenic plants. Mol. Gen. Genet. 225: 121–128.

Bäumlein, H., I. Nagy, D. Inze & U. Wobus, 1992. Cis-analysis of a seed protein gene promoter: the conservative RY repeat CATGCATG within the legumin box is essential for tissue specific expression of a legumin gene. Plant J. 2: 233–239.

Benfey, P.N., L. Ren & N.-H. Chua, 1990. Tissue-specific expression from CaMV 35S enhancer subdomains in early stages of plant development. EMBO J. 9: 1677–1684.

Christov, V., 1993. *In vitro* Modifikation und Expression eines 7S Speicherglobulingens aus *Vicia faba* in Pflanzen und Hefen. Dissertation, Math.-nat. Fakultät, Martin-Luther-Universität, Halle-Wittenberg.

Gander, E.S., K.-O. Holmstroem, G.R. de Paiva, L.A.B. de Castro, M. Carneiro & M.-F. Grossi de Sá, 1991. Isolation, characterization and expression of a gene coding for a 2S albumin from *Bertholletia excelsa* (Brazil nut). Plant Mol. Biol. 16: 437–448.

Green, C.E. & R.L. Phillips, 1974. Potential selection system for mutants with increased lysine, threonine and methionine in cereal crops. Crop Sci. 14: 827–830.

Hara-Nishimura, I., K. Inoue & M. Nishimura, 1991. A unique vacuolar processing enzyme responsible for conversion of several proprotein precursors into mature forms. FEBS Lett. 294: 89–93.

Harris, N., J. Henderson, S.J. Abbot, J. Mulcrone & J.T. Davies, 1993. Seed development and structure. Proc. Phytochem. Soc. Eur. 35: 3–21.

Hobbs, S.L.A., T.D. Warkentin & C.M. Delong, 1993. Transgene copy number can be positively or negatively associated with transgene expression. Plant Mol. Biol. 21: 17–26.

Hoffman, L.M., D.D. Donaldson & E.M. Herman, 1988. A modified storage protein is synthesized, processed and degraded in the seeds of transgenic plants. Plant Mol. Biol. 11: 717–730.

Hood, E.E., G.L. Helmer, R.T. Fraley & M.D. Chilton, 1986. The hyper-virulence of *Agrobacterium tumefaciens* A 281 is encoded in a region of pTiBo542 outside of T-DNA. J. Bacteriol. 168: 1291–1301.

Horsch, R.B., J.E. Fry, N.L. Hoffmann, M. Wallroth, D.A. Eichholtz, S.G. Rogers & R.T. Fraley, 1985. A simple and general method for transforming genes into plants. Science 227: 1229–1231.

Jefferson, R.A., 1987. Assaying chimeric genes in plants: The GUS gene fusion system. Plant Mol. Biol. Rep. 5: 387–405.

Karey, K.P. & D.A. Sirbasku, 1989. Glutaraldehyde fixation increases retention of low molecular weight proteins (growth factors) transferred to Nylon membranes for Western blot analysis. Anal. Biochem. 178: 255–259.

Kirchgessner, M., W. Windisch & F.X. Roth, 1994. The efficiency of nitrogen transformation in animal nutrition. Nova Acta Leopoldina N.F., 70: 393–412.

Laemmli, U.K., 1970. Cleavage of structural proteins during the assembly of head of bacteriophage T4. Nature 227: 680–685.

McGranahan, G.H., C.A. Leslie, S.L. Uratsu & A.M. Dandekar, 1990. Improved efficiency of walnut somatic embryo gene transfer system. Plant Cell Rep. 8: 512–516.

Munck, L., 1970. Increasing the nutritional value in cereal protein: basic research on the high-lysine character. In: Proceedings of the FAO/IAEA meeting on improving plant protein by nuclear techniques. IAEA, Vienna: 319–330.

Müntz, K., R. Jung & G. Saalbach, 1993. Synthesis, processing, and targeting of legume seed proteins. Proc. Phytochem. Soc. Eur. 35: 128–146.

Müntz, K., I. Saalbach, Th. Pickardt & O. Schieder, 1993. On the way to raising the methionine content in grain legumes. Grain Legumes 2: 18–19.

Murai, N., D.W. Sutton, M.G. Murray, J.L. Slightom, D.J. Merlo, N.A. Reichert, C. Sengupta-Gopalan, C.A. Stock, R.F. Barker, J.D. Kemp & T.C. Hall, 1983. Phaseolin gene from bean is expressed after transfer to sunflower via tumor-inducing plasmid vectors. Science 222: 476–482.

Nelson, E., T. Mertz & L.S. Bates, 1965. Second mutant gene affecting the amino acid pattern of maize endosperm. Science 150: 1469–1470.

Pickardt, T., M. Meixner, V. Schade & O. Schieder, 1991. Transformation of *Vicia narbonensis* via *Agrobacterium*-mediated gene transfer. Plant Cell Rep. 9: 535–538.

Puonti-Kaerlas, J., T. Erikson & P. Engström, 1992. Inheritance of a bacterial hygromycin phosphotransferase gene in the progeny of primary transgenic pea plants. Theor. Appl. Genet. 84: 443–450.

Saalbach, G., R. Jung, G. Kunze, R. Manteuffel, I. Saalbach & K. Müntz, 1990. Expression of modified legume storage protein genes in different systems and studies on intracellular targeting of *Vicia faba* legumin in yeast. In: G.W. Lycett & D. Grierson (Eds) Genetic Engineering of Crop Plants. Butterworth, London, pp. 1151–1158.

Saalbach, I, T. Pickardt, F. Machemehl, G. Saalbach, O. Schieder & K. Müntz, 1994. A chimeric gene encoding the methionine-rich 2S albumin of the Brazil nut (*Bertholletia excelsa* H.B.K.) is stably expressed and inherited in transgenic grain legumes. Mol. Gen. Genet. 242: 226–236.

Saalbach, I., D.R. Waddell, T. Pickardt, O. Schieder & K. Müntz. Stable expression of the sulphur-rich 2S-albumin gene in transgenic *Vicia narbonensis* increases the methionine content of seeds. J. Plant Physiol 145: 674–681.

Sambrook, J., E.F. Fritsch & T. Maniatis (Eds), 1989. Molecular Cloning: A Laboratory Manual. 2nd edn. Cold Spring Harbor Laboratory Press, Cold Spring Harbor, New York.

Schroeder, H.E., A.H. Schotz, T. Wardley-Richardson, D. Spencer & J.V. Higgins, 1993. Transformation and regeneration of two cultivars of pea (*Pisum sativum* L.). Plant Physiol. 101: 751–757.

Sun, S.S.M., S.B. Altenbach & W. Leung, 1987. Properties, biosynthesis and processing of a sulfur-rich protein in Brazil nut (*Bertholletia excelsa* H.B.K.). Eur. J. Biochem. 162: 477–483.

Töpfer, R., M. Pröls, J. Schell & H.-H. Steinbiss, 1987. A set of plant expression vectors for transcriptional and translational fusions. Nucl. Acid Res. 15: 5890.

Waterlow, J.C. & D.R. Payne, 1975. The protein gap. Nature 258: 113–117.

Euphytica **85:** 193–202, 1995.

The manipulation and modification of tomato fruit ripening by expression of antisense RNA in transgenic plants

Steve Picton[1], Julie E. Gray[2] & Don Grierson
BBSRC Research Group in Plant Gene Regulation, University of Nottingham, Department of Physiology and Environmental Science, Sutton Bonington Campus, Loughborough, Leics. LE12 5RD, U.K.; [1] *present address: Applied Biosystems Ltd., Kelvin close, Birchwood Science Park North, Warrington, Cheshire. WA3 7PB, U.K.;* [2] *University of Sheffield, Department of Molecular Biology and Biotechnology, Firth court, Western Bank, Sheffield. S10 2TN, U.K.*

Key words: carotenoids, ethylene, gene expression, *Lycopersicon esculentum* Mill., polygalacturonase, pectinesterase, phytoene synthase, ACC oxidase

Summary

The common cultivated tomato (*Lycopersicon esculentum* Mill.) provides a major focus for improvement of crop quality through genetic engineering. Identification of ripening-related cDNAs has enabled the modification of specific aspects of ripening by manipulating gene expression in transgenic plants. By utilizing ‘antisense RNA’ to modify expression of ripening genes, we have inhibited the production of the cell wall – metabolising enzymes polygalacturonase and pectinesterase and created transgenic plants that contain, effectively, single, targeted mutations affecting these genes. Furthermore, this approach has been used with previously unidentified cDNA clones to enable both functional identification and manipulation of genes involved in ethylene production (ACC oxidase) and carotenoid biosynthesis (phytoene synthase). The use of antisense RNA targeted to specific genes to alter ripening phenotypes and improve commercial utility of fruit by affecting shelf-life, processing characteristics and nutritional content is discussed.

We have used the extreme ripening-impaired mutant, *ripening inhibitor (rin)* to identify additional genes implicated in the ripening process. This approach has resulted in the cloning of several novel ripening-related mRNAs which are now being studied by antisense experiments. This may enable identification and manipulation of additional genes involved in processes such as softening, flavour and aroma generation and susceptibility to pathogens.

Abbreviations: ACC – 1-aminocyclopropane-1-carboxylic acid, PE – pectinesterase, PG – polygalacturonase, SAM – S-adenosyl methionine, SARs – scaffold attachment regions

Introduction

One major objective over the past few years has been to identify and clone plant genes and then modify their expression in transgenic plants. Tomato has proved particularly amenable to such molecular genetic studies for the following reasons. It has a relatively small genome, an extensive and ever-more detailed genetic map and a number of well-characterised ripening mutants that have been incorporated into isogenic backgrounds. Most importantly, tomato is readily amenable to *Agrobacterium*-mediated transformation, thus allowing the stable introduction of transgenes into explants and subsequent regeneration of transgenic plants that will yield genetically modified fruit (see review by Gray et al., 1992).

When tomato fruit ripens, it undergoes dramatic changes that lead to the production of a subtle blend of flavours, aroma and texture that make the mature fruit attractive to potential consumers, thereby aiding seed dispersal. These changes involve all cellular compartments and are brought about by altered expression of specific genes. These 'ripening-related' genes encode a range of enzymes thought to be required to bring about the changes leading to ripe fruit. Tomato, being a climacteric fruit, shows a dramatic increase in respiration at the onset of ripening, accompanied by increased synthesis of the phytohormone, ethylene. Production of ethylene appears to be involved in the initiation, modulation and co-ordination of expression of many of the genes required for the ripening process (Picton et al., 1994). Clearly, in order to be able to manipulate genetically the ripening process, it is essential to identify and clone as many ripening mRNAs as possible. To date, over 25 such genes have been cloned. Although several of these have been assigned precise functions, the identity of many others remains to be established (Gray et al., 1992, 1994; Grierson & Schuch, 1993).

The production, growth and analysis of transgenic tomato plants expressing an 'antisense RNA', leading to a reduction in the level of expression of the endogenous gene, has allowed critical examination of the role of specific gene products in the ripening process. In addition, these experiments have led to the creation of genetically-engineered fruit with a modified ripening phenotype (reviewed by Gray et al., 1992, 1994; Grierson & Schuch, 1993). Such modifications have resulted in a fruit crop with improved commercial characteristics.

The methodology of genetic engineering of plants with antisense RNA transgenes

The use of antisense RNA to manipulate gene expression

Following initial reports that the integration and expression of a specific antisense RNA transgene in tomatoes (Smith et al., 1988) and *Petunia* (van der Krol et al., 1988) led to a much-reduced accumulation of the homologous endogenous mRNA transcript, the technique has been widely used, specifically to down-regulate the expression of a number of tomato fruit-specific genes (Gray et al., 1992, 1994; Grierson & Schuch, 1993). Although the mechanism by which the expression of the antisense RNA inhibits the accumulation of the endogenous transcript is at present unclear, and will not be fully addressed here, the gene-specific action implicates nucleic acid base pairing. In brief, the technique summarised in Fig. 1, involves cloning and insertion in an inverted or 'antisense' orientation, of either a cDNA or genomic sequence or part thereof, between a promoter and terminator sequence that will be recognised and active within the plant genome. Following *Agrobacterium*-mediated transformation, the sequence between the left and right T-DNA borders (Fig. 1) is stably-integrated into the plant genome, such that when the transgene is transcribed *in planta*, an antisense RNA is produced. The transcription of the antisense RNA leads to a substantial reduction in accumulation of the endogenous sense mRNA transcript and thus a reduced level of the encoded product, effectively reducing or blocking the biochemical function of the gene. This approach has been successfully utilised to down-regulate the activity of the cell wall-metabolising enzymes polygalacturonase (Smith et al., 1988, 1990; Sheehy et al., 1988) and pectinesterase (Tieman et al., 1992; Hall et al., 1993) and has been used to bring about reduced ethylene synthesis in fruit by expression of an antisense RNA for ACC oxidase (Hamilton et al., 1990, 1991) or ACC synthase RNA (Oeller et al., 1991) and to disrupt fruit pigmentation by down-regulation of the phytoene synthase gene (Bird et al., 1991).

The use of antisense RNA as a tool for the identification of the biochemical roles played by unidentified clones

Having established the efficacy of antisense RNA expression to down-regulate genes encoding a specifically-defined enzyme of known biochemical function, it became possible to extend the use of the technique to help define the functional role of a cloned but unidentified gene. As described previously, the sequence of interest is cloned in an inverted orientation and transformed into plant tissue. Mature transgenic plants are then raised that effectively contain a single, targeted mutation in the gene of interest. Biochemical and physiological analysis of such plants can aid the functional identification of the gene being studied. This approach was used successfully to identify fruit ripening clones isolated from the ripe fruit TOM cDNA library (Slater et al., 1985), as encoding the enzymes phytoene synthase (Bird et al., 1991) and ACC oxidase (Hamilton et al., 1990). Attempts to define the function of an additional ethylene-regulated, fruit-ripening gene, E8 (Lincoln et al., 1987) by expression of an antisense transgene, led to the production of fruit with altered ethylene over-production but failed to establish unequivocally the biochemical function of the encoded product (Peñarrubia et al., 1992).

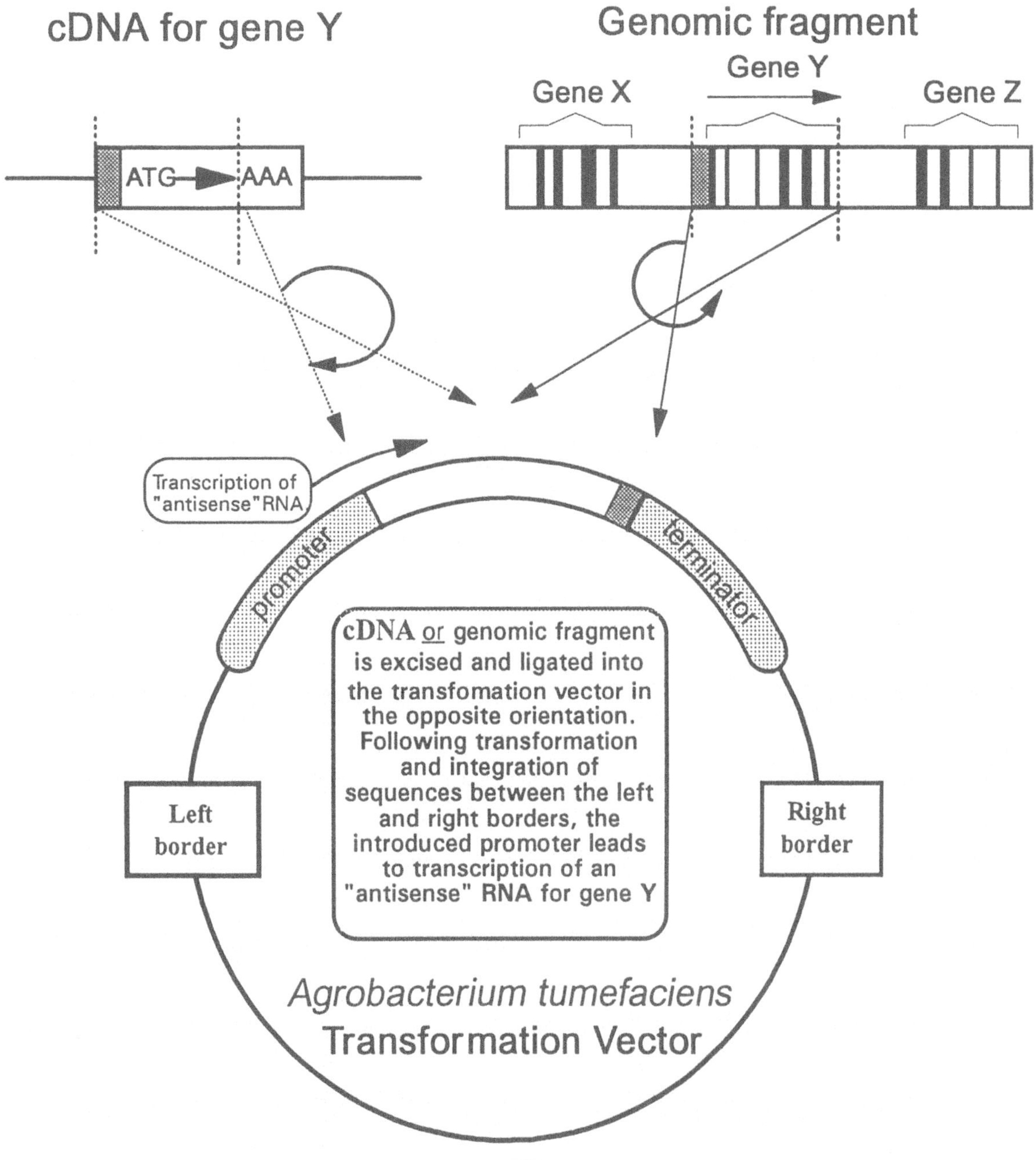

Fig 1 Schematic representation of the creation of a transformation vector for integration and expression of antisense RNA in transgenic plants The left and right border sequences indicated are derived from the *Agrobacterium tumefaciens* Ti plasmid (After Picton et al, 1994)

Manipulation of the expression of genes encoding the cell wall metabolising enzymes polygalacturonase and pectinesterase

Polygalacturonase

The first experiments to isolate, characterise (Grierson et al., 1986a, 1986b; DellaPenna et al., 1986; Sheehy et al., 1987) and genetically manipulate (Smith et al., 1988, 1990, Sheehy et al., 1988; Giovannoni et al., 1989) a tomato fruit-ripening gene, involved the cell wall-hydrolysing enzyme, polygalacturonase (PG). Much circumstantial evidence suggested that PG was the major determinant of the softening that occurs in ripe tomato fruits (Hobson, 1965; Crookes & Grierson, 1983) and as such, had attracted much interest

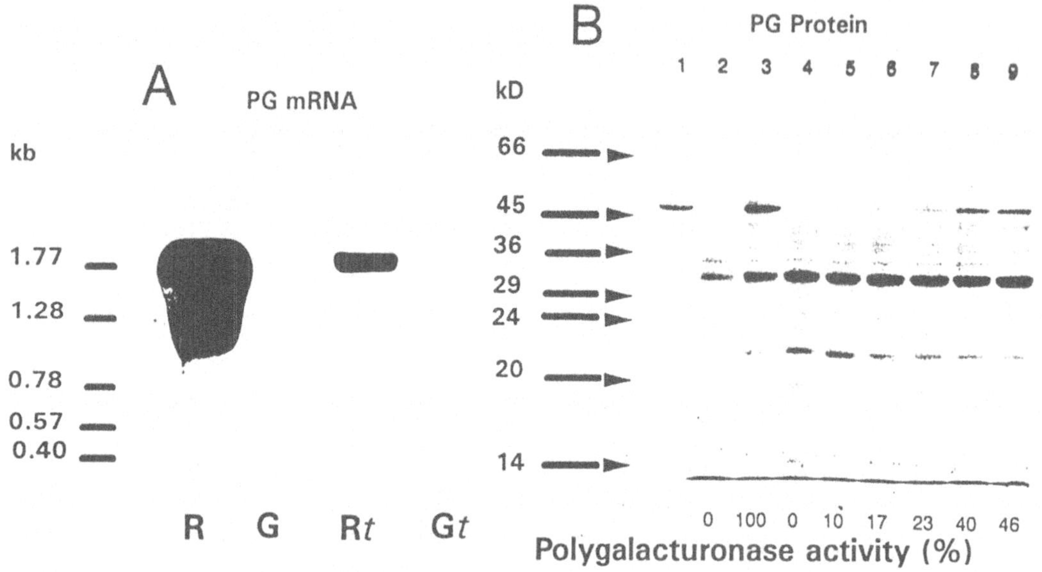

Fig. 2. Reduction of the level of PG mRNA, protein and enzyme activity in a number of independently-transformed primary (hemizygous) PG antisense plants. A. Northern blot showing PG mRNA levels in green (G) and red (R) wild-type fruit and in transformed green (Gt) and red (Rt) fruit. Size of the RNA markers run in an adjacent lane are indicated on the left. B. Levels of PG protein detected in controls and a number of primary transformed lines. Lane 1, purified PG protein, Lane 2, untransformed green fruit, Lane 3, untransformed red fruit, Lane 4, transformed green fruit, Lanes 5–9, red fruit obtained form independently-transformed primary lines. Molecular weight markers run in an adjacent lane are shown on the left. PG enzyme activity, measured in fruit extracts from each of the displayed lines, is shown below. (After Smith et al., 1988 (A) and Grierson & Schuch, 1993 (B).)

at both the biochemical and physiological level over a number of years.

A ripe tomato fruit cDNA library was prepared (the TOM clones), differentially screened against green fruit, and 19 non-homologous classes of mRNAs identified that showed increased accumulation during ripening (Slater et al., 1985). Comparison of a partial amino acid sequence obtained from purified PG enzyme with nucleic acid sequence obtained from the TOM clones, identified TOM 6 as encoding the PG enzyme (Grierson et al., 1986a, 1986b). To address specifically the role of this enzyme in the fruit-ripening process, experiments were undertaken to modify the normal pattern of expression of the PG gene in transgenic plants.

An antisense construct (Fig. 1) was prepared containing a 730 bp 5′ fragment of the TOM 6 cDNA and this was used to transform plants. Analysis of a number of primary transformed lines showed that, in some, the expression of the endogenous PG gene was reduced (Smith et al., 1988). Separate experiments were performed independently with full length PG cDNA and similar results obtained (Sheehy et al., 1988). Examination of primary transformed plants (hemizygotes) showed a reduction in the level of accumulation of the PG mRNA transcript (Fig. 2A) and a range of reduction in PG protein and PG activity (Fig. 2B). Self-pollination of the primary transformed lines gave rise to a segregating population of plants that inherited zero, one or two copies of the transgene, the introduced sequence being stably integrated into the plant genome and inherited in a normal Mendelian fashion (Smith et al., 1988, 1990). Homozygous lines derived from primary transformants with low PG activity, inherited two copies of the PG antisense transgene and showed a further reduction in level of both PG mRNA and PG enzyme activity (Fig. 3A and 3B) (Smith et al., 1990). In such plants, other aspects of ripening such as ethylene production, fruit pigmentation and onset and speed of ripening appeared unaffected, demonstrating the gene-specific action of the antisense construct (Smith et al., 1990).

Despite the dramatic reduction in PG mRNA accumulation and PG enzyme activity, little difference was

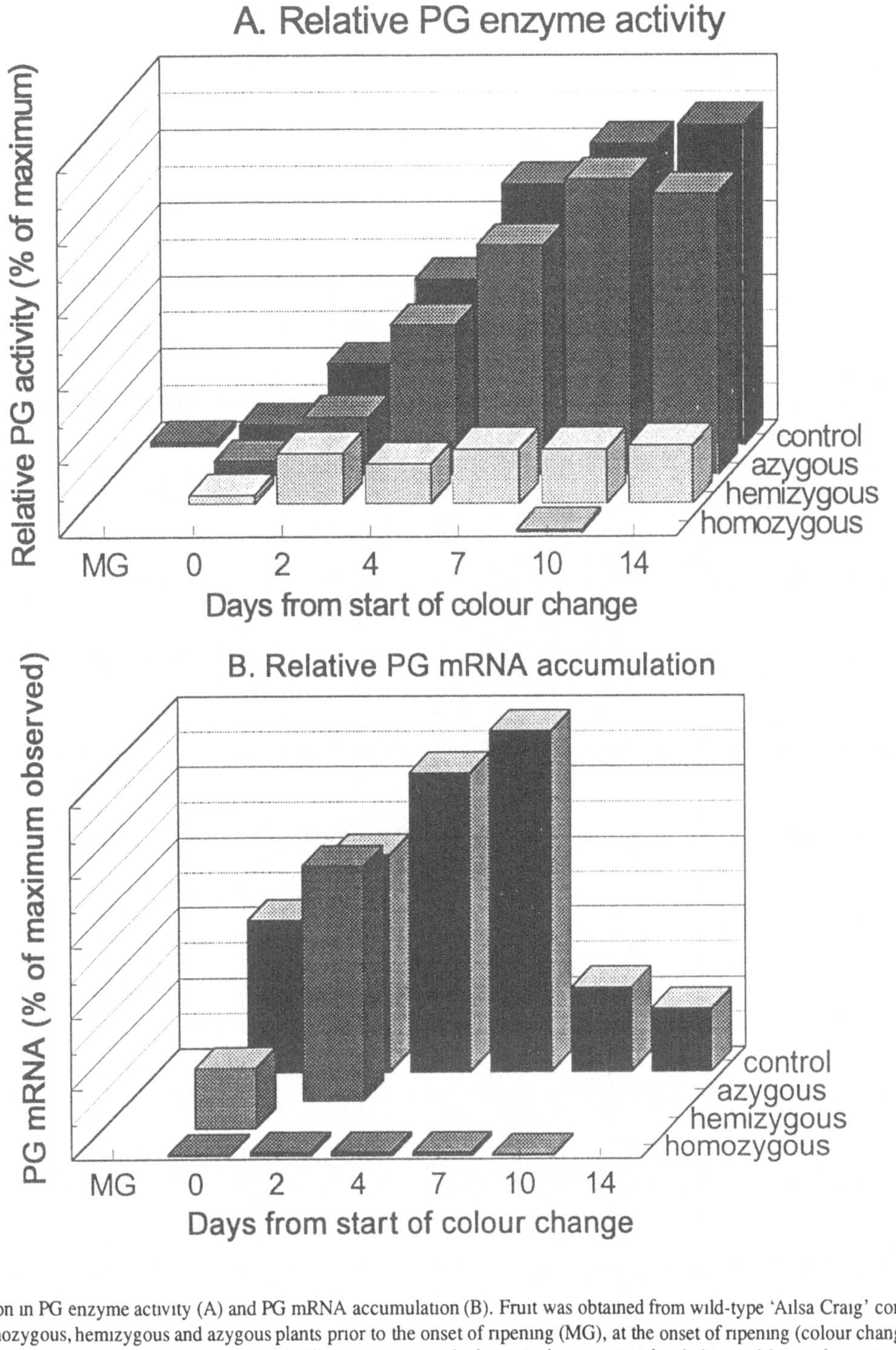

Fig. 3. Reduction in PG enzyme activity (A) and PG mRNA accumulation (B). Fruit was obtained from wild-type 'Ailsa Craig' controls and transformed homozygous, hemizygous and azygous plants prior to the onset of ripening (MG), at the onset of ripening (colour change, day 0) and at days from the start of colour change as indicated. All data is presented relative to the maximum level observed during the normal ripening of control fruit. (Data from Smith et al., 1990.)

observed between the 'softening' of homozygous PG antisense and wild-type fruit as assessed by probe penetration studies on ripe fruit pericarp (Smith at al., 1990; Schuch et al., 1991). This suggested that PG was not the sole, or even a major determinant, of the process of fruit softening. This observation was strengthened by reports that expression of the PG enzyme in a fruit mutant that does not normally express the PG gene, yields transgenic fruit that continues to remain hard during the ripening process (Giovannoni et al., 1989). More recently, trials of PG antisense fruit have demonstrated that the engineered fruit does, however, show increased resistance to the shrivelling and splitting associated with over-ripening (Schuch et al., 1991; Gray et al., 1992) and does have small but statistically significant changes in firmness at the later stages of ripening (Grierson & Schuch, 1993). Additionally, the fruit has new characteristics that improve its value for the processing market. Bostwick viscosity, an index measure of paste yield potential, is increased by over 80% in transgenic fruit where PG activity is reduced by 99% (Grierson & Schuch, 1993). Independent experiments on low PG fruit of a different variety, containing a full length PG antisense transgene, confirm these findings (Kramer et al., 1990, 1992) and also showed increased resistance of the low PG fruit to several post-harvest pathogens.

The economic potential of this PG antisense fruit has already been realised in the United States where Calgene, following extensive product testing and analysis (Redenbaugh et al., 1992), is marketing the fresh fruit under the banner 'FlavrSavr^{TM}'. Processed tomato products based on low PG fruit arising from collaboration between Nottingham University and Zeneca Seeds, are planned to be introduced into Europe and the United States marketplace in 1995 or 1996 (Grierson & Schuch, 1993).

Pectinesterase

A similar experimental approach by two independent research groups has been used to reduce the accumulation of another cell wall-metabolizing enzyme, pectinesterase (PE) (Tiemann et al., 1993; Hall et al., 1993). Detailed analysis of PE antisense fruit suggests that PE, like PG, is not the major factor associated with fruit softness. However, the engineered fruit has altered characteristics, such as increased serum viscosity, that improves the 'gloss' of fruit extracts, and thus increases the suitability of this fruit for processing into products such as pastes, sauces and soups (Grierson & Schuch, 1993). PE antisense fruit has also been reported to show a substantial increase in the level of soluble solids in the fruit (Tieman et al., 1993).

Unlike the PG expressed in fruit, which appears to be encoded by a single gene (Bird et al., 1988), PE is encoded by a small multigene family whose members show differential patterns of expression and enough divergence in nucleic acid sequence, that the introduction of a single PE antisense transgene does not lead to reduction in all fruit PE mRNAs.

The possibility of combinatorial improvements in fruit quality by down-regulation of more than one gene in a single transgenic line have been explored by two mechanisms. Firstly, low PG/PE lines have been created by crossing low PG and low PE parental lines (Grierson & Schuch, 1993) and, secondly, a chimeric antisense construct containing both a PE and PG antisense transgene has been introduced into plants (Grierson & Schuch, 1993 but see also Seymour et al., 1993). These lines exhibit phenotypic changes associated with each of the independent transgenes and additionally show an increase in soluble solids (termed Brix), thus producing fruit in which the most important determinant of processing quality has been increased (Grierson & Schuch, 1993).

Identification and manipulation of genes encoding ACC oxidase and phytoene synthase

ACC oxidase

The pathway of ethylene biosynthesis in higher plants involves the conversion of S-adenosyl methionine (SAM) to 1-aminocyclopropane-1-carboxylic acid (ACC) and the subsequent conversion of ACC to ethylene (Yang & Hoffman, 1984; Yang, 1985). These steps are catalysed by the enzymes ACC synthase and ACC oxidase, respectively. The use of antisense RNA expression has allowed the identification of a clone encoding ACC oxidase formerly known as the ethylene-forming enzyme (Hamilton et al., 1990, 1991), a previously elusive enzyme (John, 1992). The mRNA homologous to the tomato fruit clone, TOM 13, was identified as increasing rapidly at the onset of ripening and in response to wounding of plant tissues (Slater et al., 1985; Smith et al., 1986; Holdsworth et al., 1987, 1988), and during foliar senescence (Davies & Grierson, 1989), situations that involve a rapid increase in the synthesis of ethylene.

A TOM 13 antisense transgene was introduced into tomato plants and these were self-pollinated to yield a segregating population of plants containing zero, one or two copies of the antisense gene. Fruit and leaf tissue of the hemi- and homozygous transgenic plants was analysed. The increased ethylene synthesis associated with wounding of leaf tissue (Fig. 4A) and ripening of fruit (Fig. 4B) was significantly reduced in a transgene dose-dependent manner and was shown to be associ-

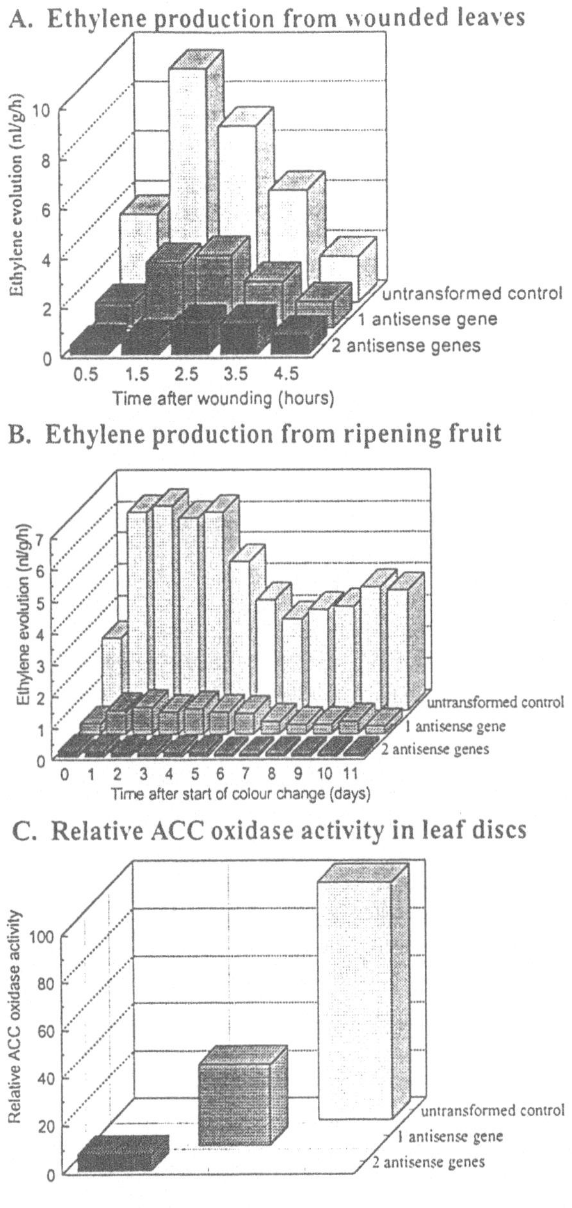

Fig. 4. Ethylene evolution and ACC oxidase activity in hemizygous and homozygous ACC oxidase antisense plants. A. Ethylene evolution from mechanically-wounded, mature leaves of wild-type and transgenic plants at the times indicated. B. Ethylene evolution from ripening fruit obtained from plants indicated, at various stages after the onset of ripening (colour change). C. ACC oxidase enzyme activity assayed in extracts from tomato leaf discs obtained from plants indicated. (Data from Hamilton et al., 1990, 1991.)

ated with a decrease in the activity of ACC oxidase (Fig. 4C) (Hamilton et al., 1990). This implicated the TOM 13-encoded product as playing a role in synthesis of ethylene. The introduction and expression of a full length TOM 13 cDNA in yeast conferred the ability on the transformed cells to synthesise ethylene from its immediate precursor, ACC, thus giving final proof that the TOM 13 product was indeed the elusive ACC oxidase (Hamilton et al., 1991). This was also demonstrated by injection of a TOM 13-related mRNA into *Xenopus* oocytes and the subsequent production of ethylene from ACC (Spanu et al., 1991).

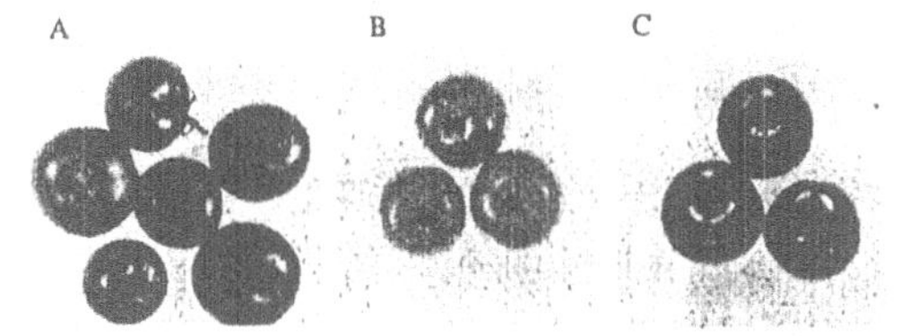

Fig. 5. Phenotypic changes in homozygous ACC oxidase antisense fruit. Fruit was picked from wild-type (A) or homozygous ACC oxidase antisense plants (B, C) prior to the onset of ripening. Prior to photography, fruit was allowed to ripen for 14 days in air (A, B) or with the addition of 20 μl/l ethylene (C). ACC oxidase antisense fruit incubated in air failed to achieve normal pigmentation, remaining yellow (B). (After Gray et al., 1991.)

The ACC oxidase antisense fruit displayed an altered ripening phenotype (Hamilton et al., 1990; Bouzayen et al., 1992; Gray et al., 1992; Picton et al., 1993a). Fruit ripening on the plant appeared superficially normal but accumulation of lycopene, responsible for the red coloration of ripe fruit, was reduced as was the level of shrivelling and splitting associated with over-ripening (Hamilton et al., 1990; Gray et al., 1992; Picton et al., 1993a). When fruit was detached at the onset of ripening and allowed to ripen in air, a more extreme phenotype was observed, with substantial reduction in lycopene accumulation resulting in yellow-orange fruit (Fig. 5) that displayed increased storage longevity. The accumulation of a number of ripening-related mRNAs, including that encoding the carotenoid biosynthetic enzyme, phytoene synthase, was shown to be reduced. The phenotypic and molecular alterations observed were partially restored by application of ethylene to the detached fruit (Fig. 5); however, full reversal was not achieved and the ethylene-treated fruit still displayed resistance to shrivelling and splitting (Picton et al., 1993a). Large scale glasshouse evaluation of the ACC oxidase antisense fruit has indicated that some of the phenotypic changes observed may be of commercial value, as low ethylene fruit has much-improved handling characteristics (Murray et al., 1993; Grierson & Schuch, 1993).

Phytoene synthase

The same experimental approach, that is, the expression of an antisense construct derived from a previously-cloned, but functionally-unidentified gene, also led to the biochemical identification of another important ripening-associated gene. The fruit clone TOM 5 (Slater et al., 1985) was sequenced (Ray et al., 1987) and found to be homologous with bacterial genes involved in the production of phytoene (Armstrong et al., 1990). This suggested that the TOM 5 product may encode a protein involved in the synthesis of coloured carotenoids during the ripening of tomatoes. The expression of a TOM 5 antisense transgene in tomato plants confirmed that the TOM 5 product is involved in carotenoid biosynthesis. Analysis of the antisense fruit and its phytoene precursors demonstrated that the TOM 5 product is the enzyme phytoene synthase (Bird et al., 1991). Following its identification, the cDNA has been further exploited to complement a naturally-occurring fruit mutant, *yellow flesh*, which fails to accumulate β-carotene and lycopene as a result of a mutation in the phytoene synthase gene, thus identifying the gene responsible for the mutation (Fray & Grierson, 1993).

Conclusions and future directions

The ability to isolate, clone, sequence and subsequently modify the expression of individual genes involved in ripening of tomatoes by expression of antisense transgenes is of major scientific and commercial interest. So far the technique has enabled studies on the developmental and hormonal regulation of gene expression, allowed the physiological and biochemical function of individual genes to be identified, and has created valuable genetically-engineered fruit with altered ripening characteristics.

The process of antisense expression effectively allows the creation of a single, targeted mutation in the gene of interest. When applied to fruit ripening, the process has been used to alter the expression of the cell wall-metabolising enzymes PG and PE and has enabled both the identification and manipulation of the enzymes ACC oxidase and phytoene synthase. The technique has also been applied to studies on ACC synthase (Oeller et al., 1992) and the functionally-unidentified clone E8 (Peñarrubia et al., 1992). Gentically modified plants have produced fruit with increased utility for the fresh and processed fruit markets and indicate that further improvements to the crop are possible. Clearly, the continued success of this approach to improve fruit quality and modify nutritional and storage qualities, will rely on continued biochemical and molecular analysis of key changes and identification of enzymes and genes responsible for ripening. It is also clear that further ripening-related genes remain to be identified. To this end, we have recently isolated and cloned several novel mRNAs whose expression appears to be severely repressed in the naturally-occurring, extreme ripening-impaired mutant, *ripening inhibitor (rin)* (Picton et al., 1993b, 1993c; Gray et al., 1994). This mutant *rin* fruit shows no respiratory rise, little ethylene production at the onset of ripening, abnormal pigment accumulation and 'ripens' extremely slowly over the course of several months, remaining resistant to many of the common post-harvest pathogens (Grierson et al., 1987; Picton et al., 1993b; Gray et al., 1994). By expression of antisense transgenes derived from these new clones, we aim to identify and modify their function in tomato. Such transgenic fruit may display novel ripening phenotypes that may also prove to be of scientific and commercial interest.

At present, one limitation of the approach outlined above is the extreme variability observed in the 'strength' of the antisense effect. Primary transformants, derived from experiments introducing an identical transgene, show a large range of levels of transgene expression. This so-called 'position effect' may result from the actual site of integration of the transgene into the plant genome. It has even been suggested that the majority of transformation events actually lead to very low expression of the transgene and in many cases the successful transfer of DNA may even go undetected (Peach & Velten, 1991). In the case of metabolically-essential genes, a range of 'leaky' antisense mutations may allow the production and analysis of an otherwise lethal line. In some cases, however, it may prove more beneficial to achieve maximum down-regulation of the targeted gene in all transformants. Reports on reduction of position effects in transgenic plants, by making the introduced transgene appear more like a natural gene, by inclusion of nuclear scaffold attachment regions (SARs) (Breyne et al., 1992; Allen et al., 1993) mean that the variability of antisense inhibition observed at present may be overcome (Allen et al., 1993) achieving a 20-fold increase in the level of transgene expression by flanking the introduced DNA with SARs. Position effects may also be overcome by using natural plant gene promoters with high transcriptional activity in the target tissue. Nicholas et al. (1994) have delineated a 5′ promoter region of the PG gene that is responsible for extremely high levels of fruit-specific reporter gene expression in transgenic plants. Introduction of antisense constructs containing several copies of the transgene may increase the effect, as can transformation techniques that favour multiple insertion of the single, introduced transgene, as illustrated by Oeller et al. (1991) who inserted an estimated 10 copies of an antisense ACC synthase gene into tomato in order to achieve maximum gene inactivation.

With further studies it may be possible to increase the efficacy of transgene expression by either one, or a combination of the methods outlined above. Once the expression of the introduced transgene can be increased and controlled, the use of expression of antisense RNA transgenes to inhibit specific genes *in planta* may prove to be an even more powerful tool for genetic improvement of crop plants than is realised at present.

Acknowledgements

This work was supported by grants from the BBSRC, SERC Biotechnology Directorate and the Gatsby Charitable Foundation. All work with transgenic plants was performed under MAFF licences. The authors wish to acknowledge the published work of Chris Smith, Andrew Hamilton, Rupert Fray and others in the Nottingham laboratory, cited in this article.

References

Allen, G C , G E Hall, L C Childs, A K Wiessinger, S Spiker & W F Thompson, 1993 Scaffold attachment regions increase reporter gene expression in stably transformed plant cells Plant Cell 5 603–613

Armstrong, G A , M Alberti & J E Hearst, 1990 Conserved enzymes mediate the early reactions of carotenoid biosynthesis in non photosynthetic and photosynthetic prokaryotes Proc Natl Acad Sci USA 87 9975–9979

Bird, C R , C J S Smith, J A Ray, P Moureau, M W Bevan, A S Bird, S Hughes, PC Morris, D Grierson & W Schuch, 1988 The tomato polygalacturonase gene and ripening specific expression in transgenic plants Plant Mol Biol 11 651–662

Bird, C R , J A Ray, J D Fletcher, J M Boniwell, A S Bird, C Teulieres, I Blain, PM Bramley & W Schuch, 1991 Using antisense RNA to study gene function Inhibition of carotenoid biosynthesis in transgenic tomatoes Bio-Technology 9 635–639

Bouzayen, M , A Hamilton, S Picton, S Barton & D Grierson, 1992 Identification of genes for the ethylene-forming enzyme and inhibition of ethylene synthesis in transgenic plants using antisense genes Biochem Soc Trans 20 76–79

Breyne, P, M Van Montague, A Depicker & G Gheysen, 1992 Characterisation of a plant scaffold attachment region in a DNA fragment that normalizes transgene expression in tobacco Plant Cell 4 463–471

Crookes, PR & D Grierson, 1983 Ultrastructure of tomato fruit ripening and the role of polygalacturonase isoenzymes in cell wall degradation Plant Physiol 72 1088–1093

Davies, K M & D Grierson, 1989 Identification of cDNA clones for tomato (*Lycopersicon esculentum*) mRNAs that accumulate during fruit ripening and leaf senescence in response to ethylene Planta 179 73–80

DellaPenna, D , D C Alexander & A B Bennett, 1986 Molecular cloning of tomato fruit polygalacturonase analysis of polygalacturonase mRNA levels during ripening Proc Natl Acad Sci USA 83 6420–6424

Fray, R G & D Grierson, 1993 Identification and genetic analysis of normal and mutant phytoene synthase genes of tomato by sequencing, complementation and co suppression Plant Mol Biol 22 589–602

Giovannoni, J J , D DellaPenna, A B Bennett & R L Fischer, 1989 Expression of a chimeric polygalacturonase gene in transgenic *rin (ripening inhibitor)* tomato fruit results in polyuronide degradation but not fruit softening Plant Cell 1 53–63

Gray, J E , S Picton, J Shabbeer, W Schuch & D Grierson, 1992 Molecular biology of fruit ripening and its manipulation with antisense genes Plant Mol Biol 19 69–87

Gray, J E , S Picton & D Grierson, 1994 The use of transgenic and naturally occurring mutants to understand and manipulate tomato fruit ripening Plant, Cell & Environ 17 557–571

Grierson, D , M J Maunders, A Slater, J Ray, C R Bird, W Schuch, M J Holdsworth, G A Tucker & J E Knapp, 1986a Gene expression during tomato ripening Phil Trans R Soc Lond B 314 399–410

Grierson, D , G A Tucker, J Keen, J Ray, C R Bird & W Schuch, 1986b Sequencing and identification of a cDNA clone for tomato polygalacturonase Nucl Acids Res 14 8595–8603

Grierson, D , M E Purton, J E Knapp & B Bathgate, 1987 Tomato ripening mutants In H Thomas & D Grierson (Eds) Develop mental Mutants in Higher Plants, pp 73–94 Cambridge University Press, Cambridge

Grierson, D & W Schuch, 1993 Control of ripening Phil Trans R Soc Lond B342, 241–250

Hall, L N , G A Tucker, C J S Smith, C F Watson, G B Seymour, Y Bundick, J M Boniwell, J D Fletcher, J A Ray, W Schuch, C R Bird & D Grierson, 1993 Antisense inhibition of pectinesterase gene expression in transgenic tomatoes The Plant Journal 3 121–129

Hamilton, A J , G W Lycett & D Grierson, 1990 Antisense gene that inhibits synthesis of the hormone ethylene in transgenic plants Nature 346 284–287

Hamilton, A J , M Bouzayen & D Grierson, 1991 Identification of a tomato gene for the ethylene forming enzyme by expression in yeast Proc Natl Acad Sci USA 88 7434–7437

Hobson, G E , 1965 The firmness of tomato fruit in relation to polygalacturonase activity J Hort Sci 40 66–72

Holdsworth, M J , C R Bird, J Ray, W Schuch & D Grierson, 1987 Structure and expression of an ethylene-related mRNA from tomato Nucl Acids Res 15 731–739

Holdsworth, M J , W Schuch & D Grierson, 1988 Organisation and expression of a wound/ripening related small multigene family from tomato Plant Mol Biol 11 81–88

John, P, 1991 How plant molecular biologists revealed a surprising relationship between two enzymes, which took an enzyme out of a membrane where it was not located, and put it into the soluble phase where it could be studied. Plant Mol. Biol. Rep. 9: 192–194.

Kramer, M., R.A. Sanders, R.E. Sheehy, M. Melis, M. Kuchn & W.R. Hiatt, 1990. Field evaluation of tomatoes with reduced polygalacturonase by antisense RNA. In: A.B. Bennett & S.D. O'Niel (Eds) Horticultural Biotechnology, pp. 347–355. Wiley-Liss, New York.

Kramer, M., R. Sanders, H. Bolkan, C. Waters, R.E. Sheehy & R.W. Hiatt, 1992. Postharvest evaluation of transgenic tomatoes with reduced levels of polygalacturonase: Processing, firmness and disease resistance. Postharvest Biol. Technol. 1: 241–255.

Lincoln, J.E., S. Cordes, E. Read & R.L. Fischer, 1987. Regulation of gene expression by ethylene during *Lycopersicon esculentum* (Tomato) fruit development. Proc. Natl. Acad. Sci. USA 84: 2793–2797.

Murray, A.J., G.E. Hobson, W. Schuch & C.R. Bird, 1993. Reduced ethylene biosynthesis in EFE-antisense tomatoes has differential effects on the ripening process. Postharvest Biol. Technol. 2: 301–313.

Nicholass, F.J., C.F. Watson, C.J.S. Smith, W. Schuch, C.R. Bird & D. Grierson, 1995. High levels of ripening-specific reporter gene expression directed by tomato fruit polygalacturonase gene flanking regions. Plant Mol. Biol., in press.

Oeller, P.W., L.M. Wong, L.P. Taylor, D.A. Pike & A. Theologis, 1992. Reversible inhibition of tomato fruit senescence by antisense 1-aminocyclopropane-1-carboxylate synthase. Science 254: 427–439.

Peach, C. & J. Velten, 1991. Transgene expression variability (position effect) of CAT and GUS reporter genes driven by linked divergent T-DNA promoters. Plant Mol. Biol. 17: 49–60.

Peñarrubia, L., M. Aguilar, L. Margossian & R.L. Fischer, 1992. An antisense gene stimulates ethylene hormone production during tomato fruit ripening. Plant Cell 4: 681–687.

Picton, S., S.L. Barton, M. Bouzayen, A.J. Hamilton & D. Grierson, 1993a. Altered fruit ripening and leaf senescence in tomatoes expressing an antisense ethylene-forming enzyme transgene. The Plant Journal 3: 469–481.

Picton, S., J.E. Gray, S. Barton, U. AbuBaker, A. Lowe & D. Grierson, 1993b. cDNA cloning and characterisation of novel ripening-related mRNAs with altered patterns of accumulation in the *ripening inhibitor (rin)* tomato ripening mutant. Plant Mol. Biol. 23: 193–207.

Picton, S., J.E. Gray, S. Payton, S. Barton, A. Lowe & D. Grierson, 1993c. A histidine decarboxylase-like mRNA is involved in tomato fruit ripening. Plant Mol. Biol. 23: 627–631.

Picton, S., J.E. Gray & D. Grierson, 1994. Ethylene genes and fruit ripening. In: P.J. Davies (Ed) Plant Hormones: Physiology, Biochemistry and Molecular Biology, in press. Kluwer Academic Publishers, The Netherlands.

Ray, J., C. Bird, M. Maunders, D. Grierson & W. Schuch, 1987. Sequence of pTOM 5, a ripening related cDNA from tomato. Nucl. Acids Res. 15: 10587.

Redenbaugh, K., W. Hiatt, B. Martineau, M. Kramer, R. Sheehy, R. Sanders, C. Houck & D. Emlay, 1991. Safety assessment of genetically engineered fruits and vegetables. A case study of the Flavr Savr^{TM} tomato. CRC Press, Boca Raton, USA.

Schuch, W., G. Hobson, K. Kanczler, G. Tucker, D. Robertson, D. Grierson, S. Bright & C. Bird, 1991. Improvement of tomato fruit quality through genetic engineering. HortScience 26: 1517–1520.

Seymour, G.B., R.G. Fray, P. Hill & G.A. Tucker, 1993. Down-regulation of two non-homologous endogenous tomato genes with a single chimeric sense gene construct. Plant Mol.Biol. 23: 1–9.

Sheehy, R.E., J. Pearson, C.J. Brady & W.R. Hiatt, 1987. Molecular characterization of tomato fruit polygalacturonase. Mol. Gen. Genet. 208: 30–36.

Sheehy, R.E., M. Crammer & W.R. Hiatt, 1988. Reduction of polygalacturonase activity in tomato fruit by antisense RNA. Proc. Natl. Acad. Sci. USA 85: 8805–8809.

Slater, A., M.J. Maunders, K. Edwards, W. Schuch & D. Grierson, 1985. Isolation and characterization of cDNA clones for tomato polygalacturonase and other ripening related proteins. Plant Mol. Biol. 5: 137–147.

Smith, C.J.S., A. Slater & D. Grierson, 1986. Rapid appearance of a mRNA correlated with ethylene synthesis encoding a protein of MW 35,000. Planta 168: 94–100.

Smith, C.J.S., C.F. Watson, J. Ray, C.R. Bird, P.C. Morris, W. Schuch & D. Grierson, 1988. Antisense RNA inhibition of polygalacturonase gene expression in transgenic tomatoes. Nature 334: 724–726.

Smith, C.J.S., C.F. Watson, P.C. Morris, C.R. Bird, G.B. Seymour, J.E. Gray, C. Arnold, G.A. Tucker, W. Schuch, S. Harding & D. Grierson, 1990. Inheritance and effect on ripening of antisense polygalacturonase genes in transgenic tomatoes. Plant Mol. Biol. 14: 369–379.

Spanu, P., D. Reinhart & T. Boller, 1991. Analysis and cloning of the ethylene-forming enzyme from tomato by functional expression of its mRNA in *Xenopus laevis* oocytes. EMBO J. 10: 2007–2013.

Tieman, D.M., R.W. Harriman, G. Ramamohan & A.K. Handa, 1992. An antisense pectinmethylesterase gene alters pectin chemistry and soluble solids in tomato fruit. Plant Cell 4: 667–679.

van der Krol, A.R., P.E. Lenting, J. Veenstra, I.M. Van der Meer, R.E. Koes, A.G.M. Gerats, J.N.M. Mol & A.R. Stuitje, 1988. An anti-sense chalcone synthase gene in transgenic plants inhibits flower pigmentation. Nature 333: 866–869.

Yang, S.F. & N.E. Hoffman, 1984. Ethylene biosynthesis and its regulation in higher plants. Ann. Rev. Plant Physiol. 35: 155–189.

Yang, S.F., 1985. Biosynthesis and action of ethylene. HortScience 20: 41–45.

Euphytica **85**: 203–207, 1995.

Gene specificity is maintained in transient expression assays with protoplasts derived from different tissues of barley

I. Díaz[1], J. Royo[1], P. Sánchez de la Hoz[2] & P. Carbonero[1]
[1] *Laboratorio de Bioquímica y Biología Molecular, Departamento de Biotecnología, E.T.S.I. Agrónomos-UPM, 28040 Madrid, Spain;* [2] *present address: Dpto. de Biología Celular y Genética, Facultad de Ciencias, Universidad de Alcalá de Henares, 28871 Alcalá de Henares, Spain*

Key words: barley, electroporation, PEG-mediated DNA uptake, promoter analysis, protoplasts, transient expression

Summary

In some cereal species that are still recalcitrant to stable transformation and regeneration, transient expression in isolated protoplasts is a useful tool for the study of gene expression and regulation. We have successfully applied these techniques to barley protoplasts derived from developing endosperm, aleurone, leaves and roots in order to characterize functionally cis-acting motives in two gene promoters, corresponding to trypsin inhibitor BTI-CMe and to sucrose synthase Ss1. Gene specificity is maintained in transient expression assays with protoplasts isolated from these different barley tissues and the pattern of expression parallels the mRNA levels observed for the corresponding genes in the same tissues.

Abbreviations: CaMV35S – 35S-cauliflower mosaic virus promoter, GUS – β-glucuronidase, MU – 4-methyl umbelliferone, NOS – nopaline synthase gene, PEG – polyethyleneglycol

Introduction

Transient gene expression in plant protoplasts has emerged as a refined method for the analysis of gene expression and regulation, as the activity of chimeric constructs of promoters fused to reporter genes, unintegrated in the cell nucleus, can be quantified within a few hours after DNA transfection. Rapidity of analysis is the major advantage of the transient assay system, particularly when compared to the time-consuming regeneration of transgenic plants. In the case of barley, regarded until very recently as recalcitrant for stable transformation (only a few references reporting fertile transgenic plants obtained by bombardment of immature embryos: Wan & Lemaux, 1994; Ritala et al., 1994), the transient transformation of protoplasts has been the method of choice for the functional analysis of gene promoters (Teeri et al., 1989; Salmenkallio et al., 1990; Díaz & Carbonero, 1992; Díaz et al., 1993; Jenes et al., 1994). However, transient expression data in cereal protoplasts derived from certain tissues such as endosperm or roots, are scarce due to difficulties in isolating viable protoplasts from such tissues at reasonable yields.

Our data, reported here, suggests that protoplasts derived from developing endosperm, aleurone, leaves and roots, maintain their tissue specificity during transient expression, and that these assays are useful for defining the regulatory cis-acting motives in promoters such as those of genes for trypsin inhibitor BTI-CMe and sucrose synthase Ss1.

Materials and methods

Plant material

Barley (*Hordeum vulgare* L.) cv Bomi was used throughout this study.

Isolation of barley protoplasts

Endosperm. Developing barley endosperms up to 15 days after pollination were the optimal material for the isolation of metabolically active protoplasts, which were prepared essentially as described by Díaz & Carbonero (1992).

Aleurone. Protoplasts form aleurone were prepared from mature de-embryonated kernels after 24 h imbibition in water, following the procedure reported by Díaz (1994). The purified protoplasts were resuspended in the medium described by Jacobson et al. (1985).

Leaves. After seed germination for 6 days in the dark under axenic conditions, etiolated leaves were cut into small pieces, plasmolysed and digested in an enzyme solution described by Díaz (1994). Transfected protoplasts were cultured in the media described by Gupta & Pattanayak (1993).

Roots. Seeds were axenically germinated for 2 days in the dark and protoplasts were isolated from roots of about 1–2 cm in length, as previously described by Díaz (1994). The culture medium used after transfection was that previously described for leaf protoplasts.

Plasmids

The plasmid p35S-GUS, used as a positive control in the transient expression assays, is a pUC19-based vector containing the GUS coding region under the control of the CaMV35S and the polyadenylation signal of the NOS. Chimeric constructs corresponding to different fragments of the barley trypsin-inhibitor (BTI-CMe) promoter and of the barley sucrose synthase (Ss1) promoter, were fused to the *gus* reporter gene in similar pUC-derived vectors and have been described in detail by Díaz et al. (1993) and Sánchez de la Hoz (1992).

Protoplast transfection

Transfection of endosperm and aleurone protoplasts was mediated by PEG following the procedure previously described (Díaz & Carbonero, 1992; Díaz et al., 1993; Díaz, 1994). 0.5 ml of culture medium containing 0.5–1.0 $\times$ 10^6 protoplasts were incubated with 25 μg of the corresponding plasmid and 50 μg of herring sperm DNA as carrier. After 5 min, 0.25 ml of PEG solution (40% PEG 3400, 0.4 M mannitol, 0.1 M $Ca(NO_3)_2$, pH 8) was added, mixed and incubated at 25° C for 20 min. The mixture was gradually diluted with F solution (Krens et al., 1982) and centrifuged. Transfected protoplasts were cultured for 20 h before sampling for GUS determination.

DNA uptake into protoplasts isolated from leaves and roots was achieved by electroporation with the Electro Cell Manipulator 600 (BTX electroporation system) as has been reported by Díaz (1994). Optimal electroporation conditions were different for each kind of protoplast (leaf protoplasts: 200 uF, 500 V/cm, 10 msec; root protoplasts: 500 uF, 625 V/cm, 20 msec). Electroporated protoplasts were incubated for 20 h before harvesting.

GUS determination

Transfected protoplasts were collected by centrifugation and GUS activity was measured by the fluorometric assay of Jefferson (1987). Total cellular protein was determined using the BioRad kit.

Northern blot analysis

RNA was purified from immature endosperm, aleurone, young leaves and roots by phenol/chloroform extraction followed by lithium chloride precipitation (Lagrimini et al., 1987). Hybridization was performed by standard procedures (Sambrook et al., 1989) using specific probes derived from the 3′ terminal regions of the cDNAs corresponding to BTI-CMe and Ss1 (Rodriguez-Palenzuela et al., 1989; Sánchez de la Hoz, 1992).

Results and discussion

Isolation and transfection of barley protoplasts

Procedures previously described by us allowed the routine isolation of metabolically-active protoplasts from different tissues of barley. Both the state of development of the tissue and the growth conditions affected the yield and viability of the preparations.

Protoplasts derived from developing endosperm were difficult to handle due to the large amount of starch granules present in their cytoplasm which can easily damage cell membranes during isolation and transfection. No viable protoplasts could be obtained beyond 20 days after pollination (Lee et al., 1991; Díaz & Carbonero, 1992). PEG-mediated DNA uptake

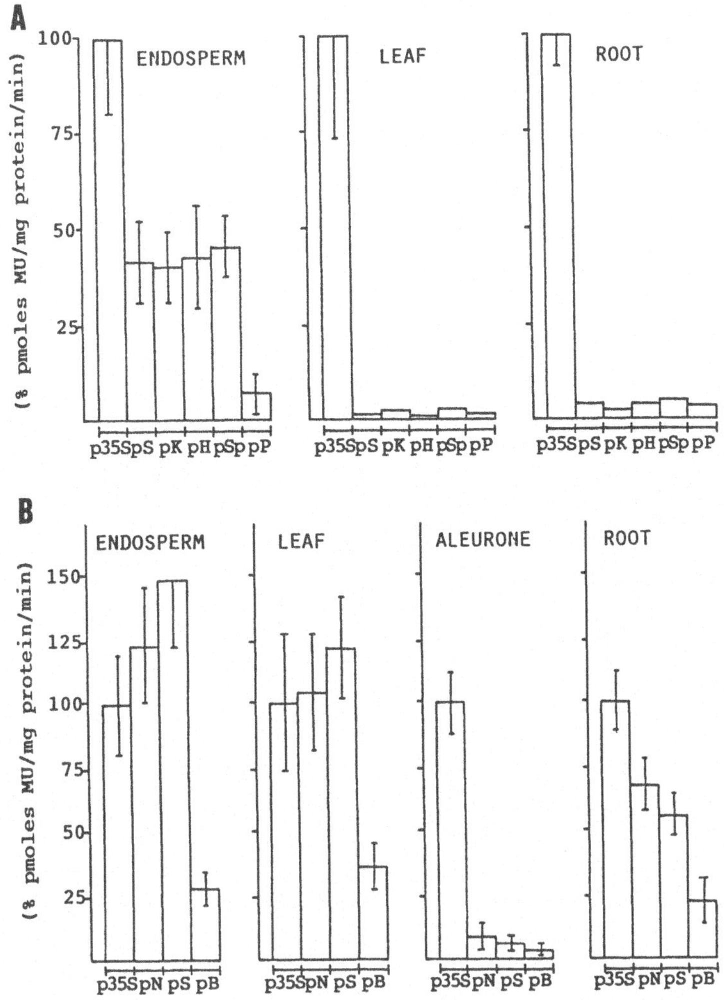

Fig. 1. Transient gene expression after transfection of barley protoplasts with chimeric constructs made with promoter fragments of BTI-CMe and Ss1 genes fused to the *gus* reporter gene. Digestions with different restriction endonucleases were used to obtain the different promoter fragments. B: Bam HI, H: Hae III, K: Kpn I, N: Nco I, P: Pvu II, S: Sal I, Sp: Sph I. The 3′ non-coding region comes from the *nos* gene. A) Expression of constructs derived from the gene promoter of trypsin inhibitor BTI-CMe (pS: - 2369 bp; pK: - 1969 bp; pH: - 816 bp; pSp: - 343 bp; pP: - 83 bp preceding the ATG site). B) Expression of constructs derived from the gene promoter of sucrose synthase Ss1 (pN: - 1190 bp; pS: - 695 bp; pB: - 252 bp preceding the ATG site). Values are expressed as % of the expression obtained with the 35S cauliflower mosaic virus promoter in protoplasts of the same tissue. Means and standard errors of five independent experiments. Nucleotide sequences of these promoters have EMBL-Gene Bank accession numbers X65875 (BTI-CMe) and X73221 (Ss1) respectively.

was the only successful approach for endosperm protoplast transfection, since even the mildest electroporation conditions assayed (250 V/cm, 200 uF, 10 msec) produced a high level of cell mortality (Díaz, 1994).

Aleurone layers yielded protoplasts which were more resistant to manipulation, probably due to their smaller size compared with endosperm protoplasts and to their lack of starch granules. GUS activity per mg of protein after PEG transfection with the plasmid p35S-GUS was approximately 30% of that measured for endosperm protoplasts using the same conditions (data not shown). Numerous groups have reported the use of barley aleurone protoplasts in transient expression studies, especially in studies of hormonal regulation of gene expression (Huttley & Baulcome, 1989; Salmenkallio et al., 1990).

The protoplast yield was higher when etiolated leaves were used as the starting material as opposed to green leaves with well-developed chloroplasts (Díaz, 1994). After electroporation, the level of GUS activity in etiolated protoplasts was about twice the activity obtained with green protoplasts, using the plasmid p35S-GUS (data not shown).

Protoplasts from barley roots could be obtained only from the apical portion of young plants. Yield sharply decreased when roots were older than 48 h. These protoplasts, very small in size, were quite tolerant to physical treatments such as centrifugation and electrical discharge (Díaz, 1994). No previous reports had been published on transient expression with root protoplasts from barley, or from other small-grain cereals. Maize root protoplasts have been successfully electroporated by Sheen (1991).

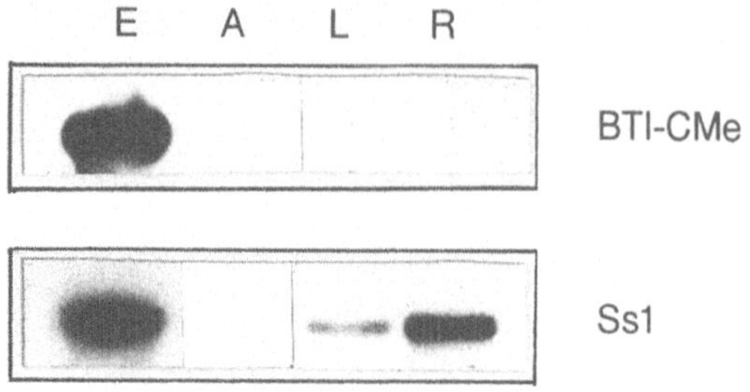

Fig. 2. Northern blot analysis using total RNA from the indicated tissues hybridized to specific probes for trypsin inhibitor BTI-CMe and for sucrose synthase Ss1. E: developing endosperm 15 days after pollination, A: immature aleurone, L: 4-day old green leaves, R: 4-day old roots. 2 μg of total RNA were applied per slot.

Functional analysis of the barley trypsin inhibitor BTI-CMe promoter by transient GUS expression

Deletion analysis of the gene promoter for barley trypsin inhibitor BTI-CMe, using five chimeric constructs (Díaz et al., 1993) in protoplasts derived from different barley tissues, is summarized in Fig. 1A. The plasmid p35S-GUS and the promoterless GUS construct were used as positive and negative controls respectively. Only protoplasts derived from developing endosperm were able to support GUS expression driven by the BTI-CMe promoter ($\approx$ 50% of that obtained with the 35S promoter), and the expression detected in protoplasts isolated from leaves and roots was less than 5% of that obtained with the 35S promoter in the same tissues (Fig. 1A). No expression was detected in aleurone protoplasts (data not shown). This pattern of transient gene expression parallels the mRNA levels observed by Northern blot analysis in the same tissues from which the protoplasts were obtained (Fig. 2), and corroborates that the BTI-CMe gene is endosperm-specific. Transient expression assays also showed that all the information required for this specificity was present in the 343 nt region preceding the ATG translation initiation site (construct pSp, Fig. 1A). In similar experiments with wheat endosperm protoplasts, GUS activity under the BTI-CMe promoter was less than 20% of that obtained in barley, which reflects the species and endosperm specificity of the barley BTI-CMe gene and might indicate that a positive transacting factor(s) needed for the trypsin inhibitor expression is (are) absent in the wheat endosperm cells, as well as in the leaves and roots of wheat and barley (Díaz et al., 1993).

Functional analysis of the barley sucrose synthase Ss1 promoter

Three promoter deletions of sucrose synthase type 1 gene (Ss1) fused to the GUS coding region (Sánchez de la Hoz, 1992) were used in transient expression assays (Fig. 1B).

The strength of the complete Ss1 promoter (pN construct) was of the same order as that of the 35S promoter in endosperm and leaf protoplasts, less than 5% in aleurone protoplasts, and intermediate (50–70%) in root protoplasts (Fig. 1B). Full expression was not obtained when only 269 bp of the promoter (construct pB) were used, indicating the existence of enhancer elements between positions - 269 and - 695 (construct pS). Again, the levels of Ss1 mRNA 'in planta' (Fig. 2) showed a similar relative expression pattern as that observed in protoplasts (Fig. 1B).

We can conclude that gene specificity is maintained in transient expression assays with protoplasts derived from different tissues of barley, at least 24–48 h after transfection, which allows for the assay of gene expression with large numbers of constructions in a short period of time.

Acknowledgements

The authors acknowledge financial support by grants Bio 93-1184 and Bio 91-0782 from Comisión Interministerial de Ciencia y Tecnología (Spain).

References

Díaz, I. & P. Carbonero, 1992. Isolation of protoplasts from developing barley endosperm: a tool for transient expression studies. Plant Cell Rep. 10: 595–598.

Díaz, I., J. Royo & P. Carbonero, 1993. The promoter of barley trypsin-inhibitor BTI-CMe, discriminates between wheat and barley endosperm protoplasts in transient expression assays. Plant Cell Rep. 12: 698–701.

Díaz, I., 1994. Optimization of conditions for DNA uptake and transient GUS expression in protoplasts from different tissues of wheat and barley. Plant Sci. 96: 179–187.

Gupta, H.S. & A. Pattanayak, 1993. Plant regeneration from mesophyll protoplasts of rice (*Oryza sativa* L.). Biotechnology 11: 90–94.

Huttly, A.K. & D.C. Baulcome, 1989. A wheat α-amy2 promoter is regulated by gibberellin in transformed oat aleurone protoplasts. EMBO J. 8: 1907–1913.

Jacobsen, J.V., J.A. Zwar & P.H. Chandler, 1985. Gibberellic acid responsive protoplasts from mature aleurone of Himalaya barley. Planta 163: 430–438.

Jefferson, R.A., 1987. Assaying chimeric genes in plants: The GUS gene fusion system. Plant Mol. Biol. Rep. 5: 387–405.

Jenes, B., M. Pudimatka, P. Bittencourt & S. Pulli, 1994. Time saving method for protoplast isolation, transformation and transient expression assay in barley. Agric. Sci. in Finland 3: 199–204.

Krens, F.A., L. Molendijk, G.J. Wullems & R.A. Schilperoort, 1982. *In vitro* transformation of plant protoplasts with Ti-plasmid DNA. Nature 296: 72–74.

Lagrimini, L.M., W. Burkhart, M. Moyer & S. Rothstein, 1987. Molecular cloning of complementary DNA encoding the lignin forming peroxidase from tobacco: molecular analysis and tissue-specific expression. Proc. Natl. Acad. Sci. USA 84: 7542–7546.

Lee, B.T., K. Murdoch, J. Topping, M.G.K. Jones & M. Kreis, 1991. Transient expression of foreign genes introduced into barley endosperm protoplasts by PEG-mediated transfer or into intact endosperm tissue by microprojectile bombardment. Plant Sci. 78: 237–246.

Ritala, A., K. Aspegren, U. Kurten, M. Salmenkallio-Mattila, L. Mannonen, R. Hannus, V. Kauppinen, T.H. Teeri & T.M. Enari, 1994. Fertile transgenic barley by particle bombardment of immature embryos. Plant Mol. Biol. 24: 317–325.

Rodriguez-Palenzuela, P., J. Royo, L. Gómez, R. Sánchez-Monge, G. Salcedo, J.L. Molina-Cano, F. García-Olmedo & P. Carbonero, 1989. The gene for trypsin inhibitor CMe is regulated in trans by the lys 3a locus in the endosperm of barley (*Hordeum vulgare* L.). Mol. Gen. Genet. 219: 474–479.

Sánchez de la Hoz, P., 1992. Clonación y análisis funcional del gen de sacarosa sintetasa Ss1 en cebada. Ph.D. Doctoral Thesis. Universidad Politecnica de Madrid, Spain.

Salmenkallio, M., R. Hannus, T.H. Teeri & V. Kauppinen, 1990. Regulation of α-amylase promoter by gibberellic acid and abscisic acid in barley protoplasts transformed by electroporation. Plant Cell Rep. 9: 352–355.

Sambrook, J., E.F. Fritsch & Y. Maniatis, 1989. Molecular Cloning: A Laboratory Manual. 2nd edition. Cold Spring Harbor, New York.

Sheen, J., 1991. Molecular mechanisms underlying the differential expression of maize pyruvate orthophosphate dikinase genes. The Plant Cell 3: 225–245.

Teeri, T.H., G.K. Patel, K. Aspegren & V. Kauppinen, 1989. Chloroplast targeting of neomycin phosphotransferase II with a pea transit peptide in electroporated barley mesophyll protoplasts. Plant Cell Rep. 8: 187–190.

Wan, Y. & P.G. Lemaux, 1994. Generation of large numbers of independently transformed fertile barley plants. Plant Physiol. 104: 37–48.

Euphytica **85**: 209–216, 1995.

Different 5′ leader sequences modulate β-glucuronidase accumulation levels in transgenic *Nicotiana tabacum* plants

Marc De Loose[1], Xavier Danthinne[2], Erik Van Bockstaele[1], Marc Van Montagu[2] & Ann Depicker[2]
[1] *Rijksstation voor Plantenveredeling, Centrum voor Landbouwkundig Onderzoek Gent, Burg. Van Gansberghelaan 109, B-9820 Merelbeke, Belgium;* [2] *Laboratorium voor Genetica, Universiteit Gent, K.L. Ledeganckstraat 35, B-9000 Gent, Belgium*

Key words: β-glucuronidase, plant, silencing, translational control, 5′-untranslated region, variation of gene expression

Summary

Three random synthetic leaders and three naturally-occurring leaders, the tobacco mosaic virus (TMV) coat protein, the satellite tobacco necrosis virus (STNV) and the plant chlorophyll *a*/*b*-binding protein (Cab22L), were shown to modulate the β-glucuronidase reporter protein accumulation levels in transient expression experiments. The same chimeric constructs also confer differential distribution patterns of reporter protein accumulation in stably-transformed tobacco calli or regenerated transgenic plants. When the highest expression levels with a given leader are compared, the 31-nucleotide random leader stimulates translation 20- and 100-fold relative to the 9- and 4-nucleotide synthetic leaders respectively. However, this 31-nucleotide random leader is approx. 2 to 3-fold weaker than the 30-nucleotide STNV leader and even 5-fold weaker than both the 79-nucleotide TMV leader and the 66-nucleotide Cab22L leader. These results confirm the findings in transient expression experiments and stress the importance of the 5′-untranslated region for the production of heterologous proteins in transgenic plants.

Introduction

Of all the regions of messenger RNA that can be involved in translational efficiency, the leader sequence is probably the most important (Kozak, 1989; Hershey, 1991). Firstly, this region is undoubtedly implicated in the process of translational initiation and this is usually considered to be the rate-limiting step of translation. Secondly, the 5′ leader can be directly or indirectly involved in mRNA stability.

Plant viral RNA leader sequences, like those of tobacco mosaic virus (TMV), alfalfa mosaic virus (AlMV) and plant potyvirus, have been shown to act *in cis* as efficient enhancers of translation at least in *in vitro* systems or during transient expression in protoplasts (Gallie et al., 1987a, 1987b; Jobling & Gehrke, 1987; Sleat et al., 1987; Carrington & Freed, 1990, Nicolaisen et al., 1992). Motifs in the TMV leader of 77 nucleotides were studied in detail for their effect on translational enhancement (Gallie & Walbot, 1992). The satellite tobacco necrosis virus (STNV) (Danthinne et al., 1991) RNA lacks a conventional cap structure at the 5′ terminus, but nevertheless the viral coat protein is very efficiently translated (Danthinne et al., 1993). The Cab22L leader also has a stimulatory effect upon the trans-protein accumulation after transient expression compared with the effect of the cauliflower mosaic virus (CaMV) 35S 5′ leader (Harpster et al., 1988).

In an *in vitro* translation system derived from wheat germ, the STNV leader sequence only weakly stimulates translation relative to the leaders of TMV and of Cab22L origin (Danthinne & Van Emmelo, 1990). In transient expression the STNV-1, the TMV and the Cab22L leaders exhibit a 4-, 12-, and 8-fold enhancement respectively, relative to a 31 nucleotide random leader (Danthinne & Van Emmelo, 1990).

This work analyzes whether the effect of the leader sequence on heterologous protein accumulation seen after transient expression (Danthinne & Van Emmelo, 1990) can be extrapolated to that which is found in stably-transformed plant tissues. The work also aims to determine whether the leader affects the distribution of protein accumulation levels beyond the variation of expression levels due to position and silencing effects. Therefore, the expression of the different chimeric gene constructions was studied first of all by comparing the values for the GUS activities in two series of 30 different calli. Subsequently the expression of the same chimeric constructs was evaluated in regenerated transgenic plants and their progeny.

Materials and methods

Plasmid constructions

The cauliflower mosaic virus promoter (P35S) was mutagenized to create a *Pst*I restriction site (Danthinne & Van Emmelo, 1990) immediately downstream from the three putative transcription start sites (Ow et al., 1987; Sanders et al., 1987; Harpster et al., 1988). All leader sequences were fused to the β-glucuronidase (*gus*)-coding sequence at the *Nco*I site positioned at the first codon. Downstream, the *gus*-coding sequence was fused to the 3′ region of the nopaline synthase gene (Depicker et al., 1982; Ingelbrecht et al., 1989) at the *Eco*RI site located 6 nucleotides downstream from the stop codon (Jefferson, 1987). The chimeric *gus* genes were put under the control of the mutagenized 35S promoter in such a way that only three heterologous nucleotides GAC are present at the 5′ end of all the transcripts. The chimeric genes were introduced into the *Xba*I site of the binary T-DNA vector pDE1001 (Denecke et al., 1992) in both orientations relative to the right border. The resulting plasmids (pXD601 to pXD612) were given an odd or an even number when the *gus* cassette pointed to the inside or to the outside of the T-DNA, respectively (see Figs 1A and 1B).

Conjugation and plant cell transformations

All the T-DNA vectors were mobilized from *Escherichia coli* MC1061 (Casadaban & Cohen, 1980) to the *Agrobacterium* strain C58C1, containing plasmid pGV2260 (Deblaere et al., 1987). Transconjugants harbouring the correct T-DNA constructs were co-cultivated with *Nicotiana tabacum* SR1 protoplasts (Depicker et al., 1985) or with leaf discs (De Block et al., 1987). Kanamycin-resistant calli were grown on B5 medium (Gamborg et al., 1968) supplemented with hormones and 50 μg/ml kanamycin. Each T-DNA construct was transformed twice. From both independently performed co-cultivations, 30 calli were analyzed for GUS activity. The GUS-negative calli were not considered. The calli are indicated by CXD followed by the number of the T-DNA plasmid construct and a number to designate the individual clone. Kanamycin-resistant shoots were cut from leaf discs and grown in glass containers. The independently obtained plant lines are indicated by LXD followed by the number of the corresponding T-DNA construct and a number indicating each individual shoot. For each construct (except XD603), between 4 and 20 axenically-grown transgenics were analyzed at the 6-leaf stage approximately 3 months after their isolation from the leaf disc. Subsequently, 2 to 5 of the highest expressing transgenics were transferred to soil, grown in the greenhouse and allowed to set seed. About 50 seeds of the R1 generation were sown on Murashige and Skoog medium containing 50 μg/ml kanamycin. The ratio between kanamycin-resistant and kanamycin-sensitive seedlings was determined. From lines with a single T-DNA locus, eight to twelve kanamycin-resistant R1 plants were grown *in vitro* up to the 6-leaf stage and analyzed for their GUS content.

GUS assays

The enzyme activities (Jefferson, 1987) were determined by means of continuous kinetic reactions performed spectrophotometrically (340 ATTC, SLT or iEMS, Labsystems Inc.) or fluorimetrically (Fluoroscan, Labsystems Inc.) in a computer-directed microtitre plate reader as described by Breyne et al. (1993). The detection limit (values two-fold above background) was 0.1 U GUS/mg protein. The protein concentrations were determined by the method of Bradford (1976) using the kit supplied by Bio-Rad.

Results

T-DNA constructs

The basic structure of the set of constructed T-DNA vectors is presented in Fig. 1. All T-DNAs carry the same neomycin phosphotransferase hybrid gene allowing the selection for kanamycin-resistant transfor-

A.

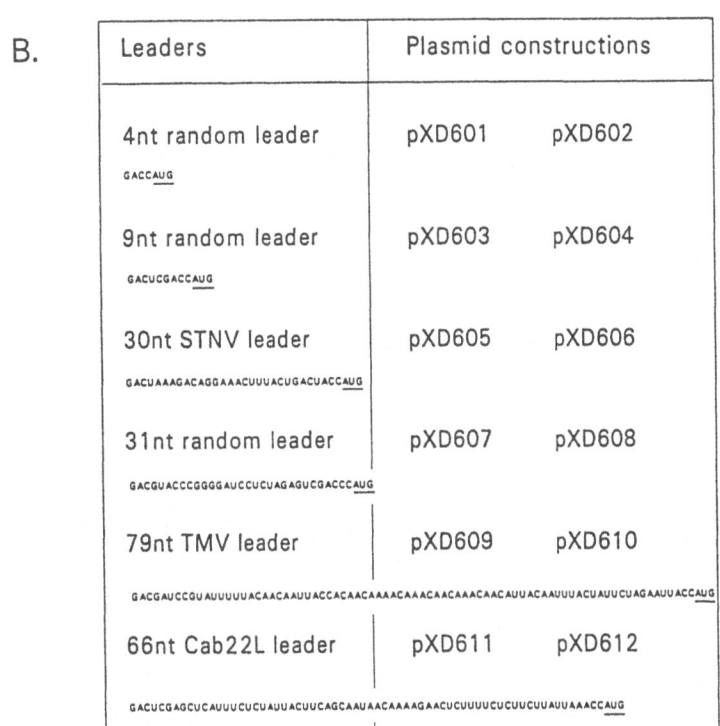

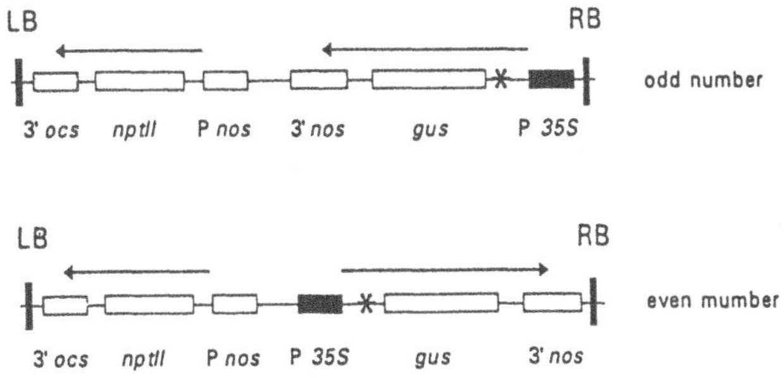

B.

Leaders	Plasmid constructions	
4nt random leader GACCAUG	pXD601	pXD602
9nt random leader GACUCGACCAUG	pXD603	pXD604
30nt STNV leader GACUAAAGACAGGAAACUUUACUGACUACCAUG	pXD605	pXD606
31nt random leader GACGUACCCGGGGAUCCUCUAGAGUCGACCCAUG	pXD607	pXD608
79nt TMV leader GACGAUCCGUAUUUUUACAACAAUUACCACAACAAAACAAACAACAAACAACAUUACAAUUUACUAUUCUAGAAUUACCAUG	pXD609	pXD610
66nt Cab22L leader GACUCGAGCUCAUUUCUCUAUUACUUCAGCAAUAACAAAAGAACUCUUUUCUCUUCUUAUUAAACCAUG	pXD611	pXD612

Fig. 1. Description of the T-DNA constructs. A. Schematic representation of the T-DNA constructs used. The vectors containing the cassette P35S-leader-*gus*-3′*nos* oriented to the inside or the outside of the T-DNA are indicated with odd and even numbers respectively. The star symbol indicates the position where the various leaders, described in Fig. 1B, were cloned. Abbreviations: P35S, cauliflower mosaic virus promoter; *Pnos*, nopaline synthase promoter; *gus*, β-glucuronidase gene; *nptII*, neomycin phosphotransferase II-coding sequence; 3′*nos*, 3′ region of the nopaline synthase gene; 3′*ocs*, 3′ region of the octopine synthase gene; RB, T-DNA right border region; LB, T-DNA left border region. B. Nucleotide sequence of the leaders tested for translational enhancement and the nomenclature of the vectors carrying the T-DNAs with the various leader fusions. The first residues are derived from the transcription start of the CaMV 35S promoter. The initiation codon of the β-glucuronidase coding sequence is underlined.

mants. The different chimeric P35S-leader-*gus*-3′*nos* cassettes were all inserted into one vector at the same position in two orientations (Fig. 1A). Figure 1B shows the six different leaders analyzed in this work and the number of the constructs in which they occur.

Distribution of GUS activities in different transgenic calli series

Distributions of the GUS activities of the CXD transgenic calli were obtained from GUS accumulation levels. Figure 2A shows the GUS amounts in 16 calli of all transgenic series plotted on a logarithmic scale. To emphasize the large differences in absolute GUS amounts among the different series as well as within one series, the same results were plotted on a linear scale (Fig. 2B). Each series contained calli with low, intermediate and high GUS accumulation levels. The same range of variation with the same absolute amounts for the lowest and highest expressing calli was confirmed in a second experiment in which a further 30 calli were analyzed. The main difference between the results of the two consecutive experiments was the number of calli with low and high GUS activ-

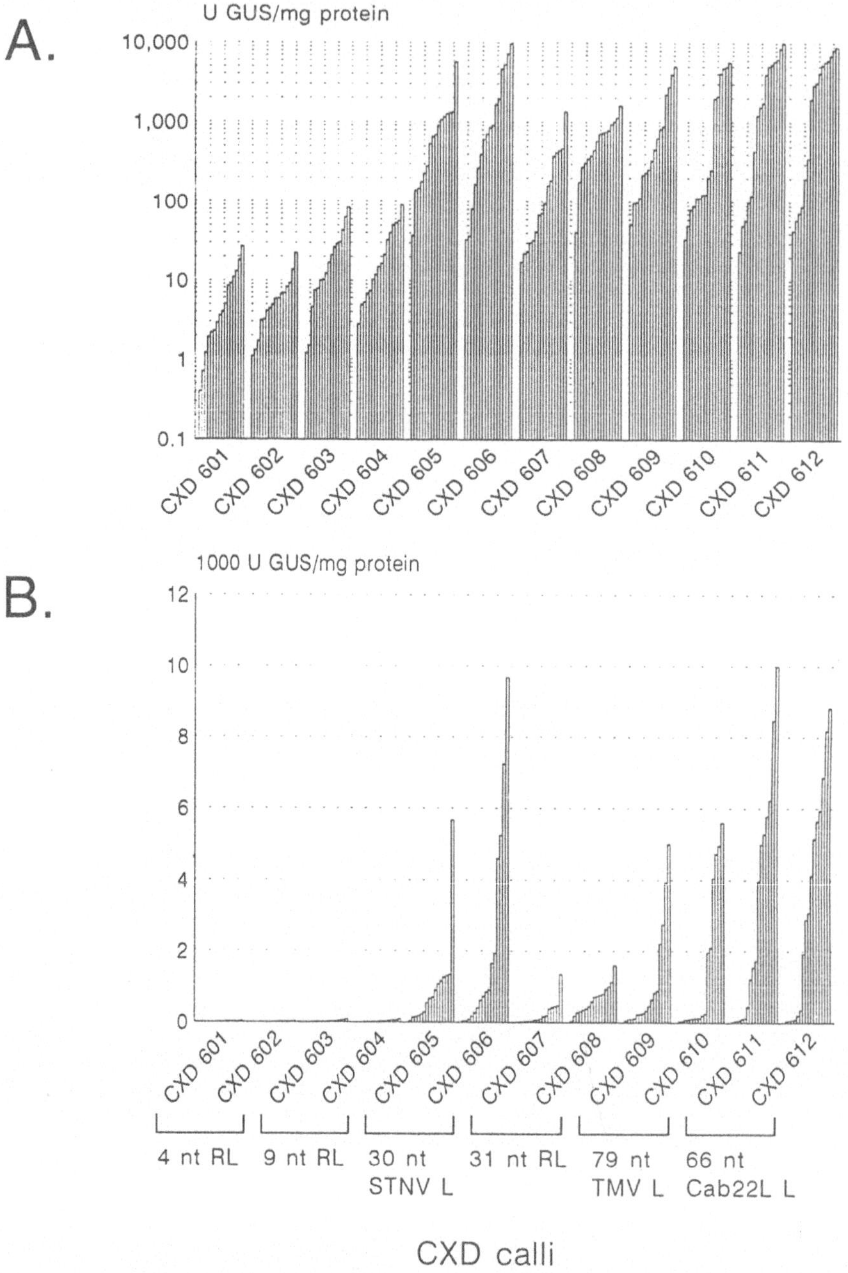

Fig. 2. Distribution of GUS activities in transgenic calli series. GUS activities in transgenic calli series were plotted on a logarithmic (A) and a linear (B) scale. The GUS activities determined in each callus number of the different series are represented by bars. Within each series, the numbers of calli are arranged according to their increasing GUS content.

ity. A comparison shows that the distributions of GUS accumulation levels in series containing an identical chimeric *gus* gene but in opposite orientations, were similar.

However, the GUS activity distributions of series containing *gus* constructs with different leaders display a pronounced difference (Figs 2A and 2B). The 4-nucleotide random leader (RL) yielded the lowest expression but was significantly above background. Higher GUS levels were found in calli with the 9-nucleotide RL transcript (series CXD603 and CXD604) and still higher levels were found in transgenic calli with a 31-nucleotide RL in the *gus* transcript (series CXD607 and CXD608). The highest accumulation levels and more or less comparable distribution patterns were found in the series with the 30-nucleotide STNV leader (series CXD605 and CXD606), the 77-nucleotide TMV leader (series CXD609 and CXD610), and the 66-nucleotide Cab22L leader (series CXD611 and CXD612).

Influence of the leader on GUS accumulation in primary transgenic plants

For each of the T-DNA constructs transgenic plants were obtained and analyzed for their GUS expression in a leaf of an *in vitro*-grown plant (Figs 3A and 3B). As in the transgenic calli series, the primary axenically-grown transformants also showed a high variation in GUS units per mg of total protein. The obtained results were plotted on a logarithmic and a linear scale in Figs 3A and 3B respectively. As the number of transgenic plants analyzed was limited, it is not possible to carry out a statistical analysis of the data. The highest GUS values per leader construct are 6150 U GUS/mg protein for the Cab22L leader, 3536 U GUS/mg protein for the TMV leader, 1167 U GUS/mg protein for the STNV leader, 309 U GUS/mg protein for the 31-nucleotide random leader (RL), 36 U GUS/mg protein for the 9-nucleotide RL and 20 U GUS/mg protein for the 4-nucleotide RL. Thus a comparison of GUS accumulation in leaf material of primary transgenic plants shows the same hierarchy of leader effects as in the transgenic calli series.

GUS expression in progeny plants

One or more of the highest-expressing, transgenic lines per T-DNA construct, indicated by a dark shadowing in Fig. 3B, were allowed to flower and the R1 seeds were collected. Subsequently the number of kanamycin-resistant versus kanamycin-sensitive plants was determined to estimate the number of T-DNA loci. Most of the R1 progenies show a 3 to 1 segregation of kanamycin-resistant versus kanamycin-sensitive seedlings. Of the transgenics with clear 3:1 segregation ratios for the T-DNA-linked selection marker, 8 or 12 R1 *in vitro*-grown progeny plants were analyzed for GUS accumulation levels (Fig. 4). Most of the progeny shows two distinct classes of GUS accumulation probably corresponding to the heterozygotes and the homozygotes for the T-DNA locus.

We can derive from the R1 population analysis that the same hierarchy of expression levels is maintained in the progeny of the primary transgenics: namely the 9-nucleotide RL < 31-nucleotide RL < STNV leader < TMV and Cab22L leaders. Although the TMV-*gus* and Cab22L-*gus* constructs gave rise to similar GUS distributions, it may not be a coincidence that the highest accumulation levels were always found in transgenic material with the Cab22L leader sequence in the transcript.

Discussion

The results on GUS accumulation levels in different transgenic lines with a particular T-DNA construct illustrate the commonly-observed variation in transgene expression. Clearly, every series contains a number of transgenic calli that express the P35S-*gus* transgene to a very low extent in comparison with a number of calli with high (normal) expression levels. However, the resulting low accumulation levels were always approximately 0.01 of the highest accumulation level in the respective series. The variation in gene expression might be the result of position effects such as flanking enhancing or silencing DNA sequences, the surrounding chromatin structure or the methylation status of the insert locus. However, other processes not understood such as gene silencing and co-suppression may contribute to the observed variation in gene expression in primary transgenic populations (Hobbs et al., 1990, 1993; Ingelbrecht et al., 1994).

Nevertheless, the results demonstrate the importance of the leader sequence upon heterologous gene expression. Very often chimeric leaders contain linkers and adaptors which can lower substantially the final recombinant protein accumulation (Jones et al., 1985).

The 5′-untranslated leader might directly influence the accumulation of the gene product as it is undoubtedly involved in the initiation of translation. This is exemplified by the poor translation of second cistrons in dicistronic transcription units (Angenon et al., 1989). However, the 5′-untranslated leader is also involved in mRNA accumulation levels (Harpster et al., 1988). The leaders may increase the mRNA stability because enhanced translation protects the transcript from more rapid degradation (Vancanneyt et al., 1990). The analysis and correlation of GUS accumulation levels with transcript accumulation levels will allow us to clarify events.

We conclude from the experiments that both the 5′-untranslated region of the Cab22L gene (Harpster et al., 1988) and the 5′ leader of the TMV viral coat

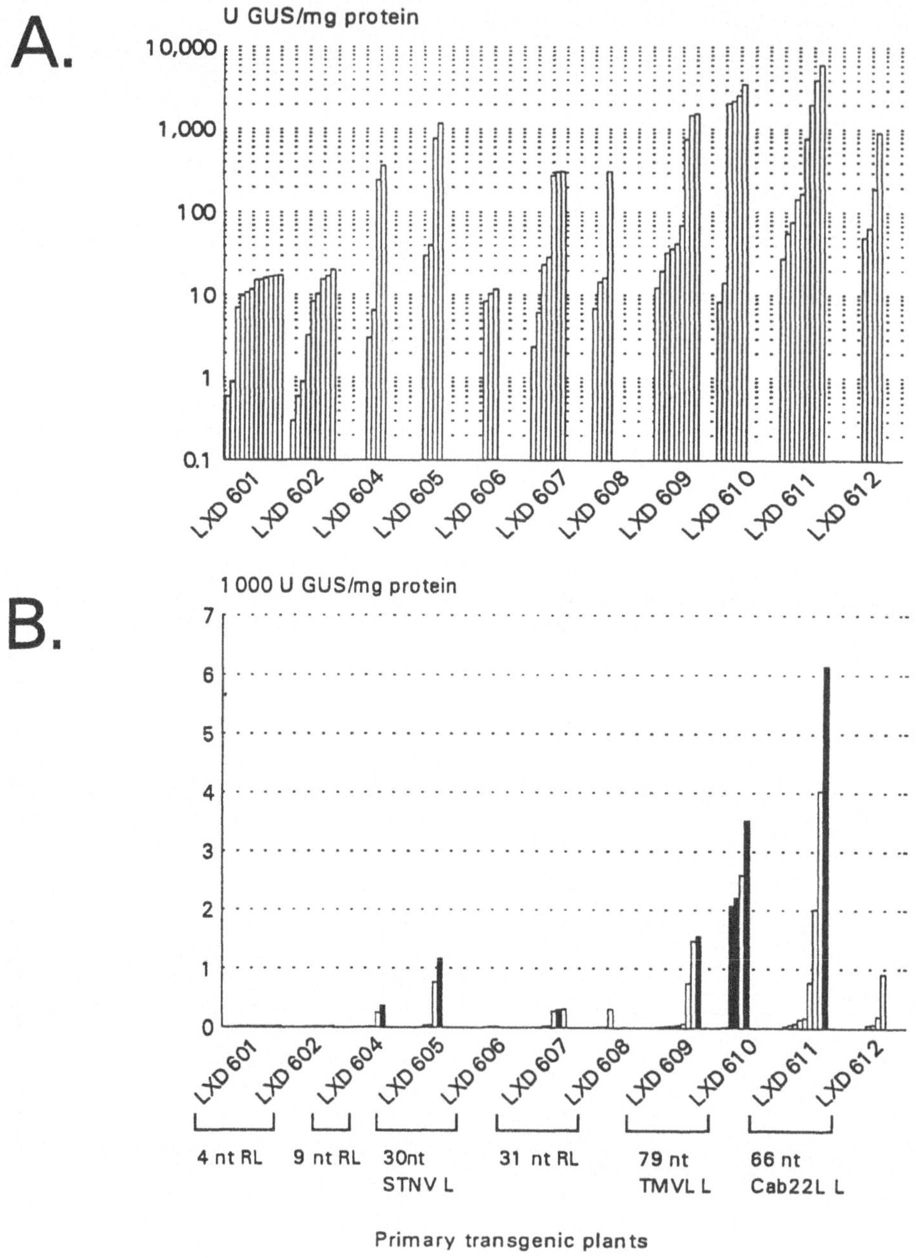

Fig. 3. Distribution of GUS activities in young leaf material of primary transgenic plants containing the various leader-*gus* constructs, plotted on a logarithmic scale (Fig. 3A) and a linear scale (Fig. 3B). The GUS activities determined are represented by bars plotted on the ordinate. They are grouped per T-DNA construct. Within each group, the GUS amounts are arranged according to increasing amounts. The black bars indicate the transgenic plants of which the R1 progeny was further analyzed in detail: LXD604-4, LXD605-11, LXD607-20, LXD609-17, LXD610-4, LXD610-2, and LXD611-7.

protein (Gallie et al., 1987a) enhance the expression of chimeric *gus* genes in transgenic plants. Both leaders show similarities in length and nucleotide composition. They are between 50 and 70 nucleotides long, which corresponds to the length of the majority of plant 5′ leaders (Joshi, 1987; Hershey, 1991). The underrepresentation of guanosine nucleotides in the TMV and Cab22L leaders resulting in a lack of secondary structure is notable. This may allow easy scanning for the first AUG in a favourable context (Kozak, 1989). Functional analysis of the TMV leader identified the poly(CAA) region as the primary element responsible for enhancement *in vivo* (Gallie & Walbot, 1992). Therefore, it is possible that the analogous CA-rich

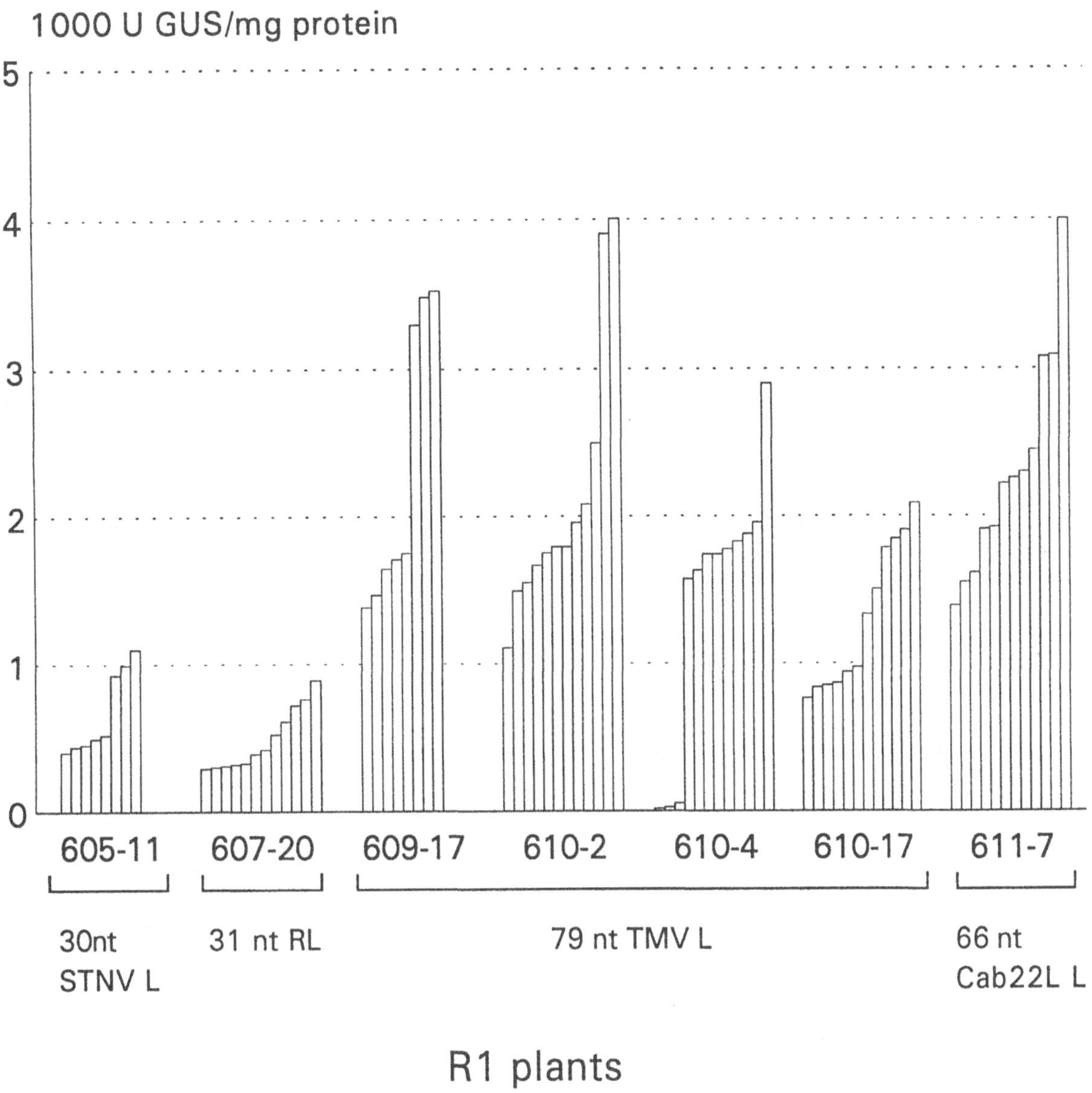

Fig. 4. GUS amounts in R1 progeny plants of transgenics containing various high expressing chimeric *gus* constructs. The first number indicates the T-DNA construct; the second number indicates the identity of the transgenic isolate. The bars show the GUS accumulation levels in young leaf tissue of the different R1 progeny plants tested.

regions in the Cab22L leader have the same function, but this has to be tested by rigorous mutation analysis of the Cab22L leader.

Acknowledgements

We thank Anni Jacobs, Isabelle Strobbe and Annique Staelens for excellent technical assistance, John Van Emmelo, Marc Cornelissen and Frank Meulewaeter for discussions and contributions during the initial phase of the work, and Martine De Cock for help with the manuscript. This research was supported by grants from the Belgian Programme on Interuniversity Poles of Attraction (Prime Minister's Office, Science Policy Programming, # 38).

References

Angenon, G., J. Uotila, S.A. Kurkela, T.H. Teeri, J. Botterman, M. Van Montagu & A. Depicker, 1989. Expression of dicistronic transcriptional units in transgenic tobacco Mol Cell Biol 9 5676–5684

Bradford, M M , 1976 A rapid and sensitive method for the quantitation of microgram quantities of protein utilizing the principle of protein-dye binding Anal Biochem 72 248–254

Breyne, P, M De Loose, A Dedonder, M Van Montagu & A Depicker, 1993 Quantitative kinetic analysis of β-glucuronidase activities using a computer-directed microtiter plate reader Plant Mol Biol Rep 11 21–31

Carrington, J C & D D Freed, 1990 Cap-independent enhancement of translation by a plant potyvirus 5′ nontranslated region J Virology 64 1590–1597

Casadaban, M J & S N Cohen, 1980 Analysis of gene control signals by DNA fusion and cloning in *Escherichia coli* J Mol Biol 138 179–207

Danthinne, X & J Van Emmelo, 1990 Studies on the translational properties of STNV RNA non-coding regions Med Fac Landbouww Rijksuniv Gent 55 (3a) 1037–1045

Danthinne, X , J Seurinck, M Van Montagu, C W A Pleij & J Van Emmelo, 1991 Structural similarities between the RNAs of two satellites of tobacco necrosis virus Virology 185 605–614

Danthinne, X , J Seurinck, F Meulewaeter, M Van Montagu & M Cornelissen, 1993 The 3′ untranslated region of the satellite tobacco necrosis virus RNA stimulates translation *in vitro* Mol Cell Biol 13 3340–3349

De Block, M , J Botterman, M Vandewiele, J Dockx, C Thoen, V Gossele, R Movva, C Thompson, M Van Montagu & J Leemans, 1987 Engineering herbicide resistance in plants by expression of a detoxifying enzyme EMBO J 6 2513–2518

Deblaere, R , A Reynaerts, H Hofte, J -P Hernalsteens, J Leemans & M Van Montagu, 1987 Vectors for cloning in plant cells In R Wu & L Grossman (Eds) Recombinant DNA, part D (Methods in Enzymology, Vol 153), pp 277–292 Academic Press, New York

Denecke, J , R De Rycke & J Botterman, 1992 Plant and mammalian sorting signals for protein retention in the endoplasmic reticulum contain a conserved epitope EMBO J 11 2345–2355

Depicker, A , S Stachel, P Dhaese, P Zambryski & H M Goodman, 1982 Nopaline synthase transcript mapping and DNA sequence J Mol Appl Genet 1 561–573

Depicker, A , L Herman, A Jacobs, J Schell & M Van Montagu, 1985 Frequencies of simultaneous transformation with different T-DNAs and their relevance to the *Agrobacterium*/plant cell interaction Mol Gen Genet 201 477–484

Gallie, D R & V Walbot, 1992 Identification of the motifs within the tobacco mosaic virus 5′-leader responsible for enhancing translation Nucl Acids Res 20 4631–4638

Gallie, D R , D E Sleat, J W Watts, PC Turner & T M A Wilson, 1987a The 5′-leader sequence of tobacco mosaic virus RNA enhances the expression of foreign gene transcripts *in vitro* and *in vivo* Nucl Acids Res 15 3257–3273

Gallie, D R , D E Sleat, J W Watts, PC Turner & T M A Wilson, 1987b A comparison of eukaryotic viral 5′-leader sequences as enhancers of mRNA expression *in vivo* Nucl Acids Res 15 8693–8711

Gamborg, O L , R A Miller & K Ojima, 1968 Nutrient requirements of suspension cultures of soybean root cells Exp Cell Res 50 151–158

Harpster, M H , J A Townsend, J D G Jones, J Bedbrook & P Dunsmuir, 1988 Relative strengths of the 35S cauliflower mosaic virus, 1′, 2′, and nopaline synthase promoters in transformed tobacco sugarbeet and oilseed rape callus tissue Mol Gen Genet 212 182–190

Hershey, J W B , 1991 Translational control in mammalian cells Ann Rev Biochem 60 717–755

Hobbs, S L A , P Kpodar & C M O DeLong, 1990 The effect of T-DNA copy number, position and methylation on reporter gene expression in tobacco transformants Plant Mol Biol 15 851–864

Hobbs, S L A , T D Warkentin & C M O DeLong, 1993 Transgene copy number can be positively or negatively associated with transgene expression Plant Mol Biol 21 17–26

Ingelbrecht, I L W, L M F Herman, R A Dekeyser, M C Van Montagu & A G Depicker, 1989 Different 3′ end regions strongly influence the level of gene expression in plant cells Plant Cell 1 671–680

Ingelbrecht, I , H Van Houdt, A Depicker & M Van Montagu, 1994 Post-transcriptional silencing of reporter genes in tobacco correlated with DNA methylation Proc Natl Acad Sci USA 91 10502–10506

Jefferson, R A , 1987 Assaying chimeric genes in plants the GUS gene fusion system Plant Mol Biol Rep 5 387–405

Jobling, S A & L Gehrke, 1987 Enhanced translation of chimaeric messenger RNAs containing a plant viral untranslated leader sequence Nature 325 622–625

Jones, J D G , P Dunsmuir & J Bedbrook, 1985 High level expression of introduced chimaeric genes in regenerated transformed plants EMBO J 4 2411–2418

Joshi, C P, 1987 An inspection of the domain between putative TATA box and translation start site in 79 plant genes Nucl Acids Res 15 6643–6653

Kozak, M , 1989 The scanning model for translation an update J Cell Biol 108 229–241

Nicolaisen, M , E Johansen, G B Poulsen & B Borkhardt, 1992 The 5′ untranslated region from pea seedborne mosaic potyvirus RNA as a translational enhancer in pea and tobacco protoplasts FEBS Lett 303 169–172

Ow, D W, J D Jacobs & S H Howell, 1987 Functional regions of the cauliflower mosaic virus 35S RNA promoter determined by use of the firefly luciferase gene as a reporter of promoter activity Proc Natl Acad Sci USA 84 4870–4874

Sanders, PR, J A Winter, A R Barnason, S G Rogers & R T Fraley, 1987 Comparison of cauliflower mosaic virus 35S and nopaline synthase promoters in transgenic plants Nucl Acids Res 15 1543–1558

Sleat, D E , D R Gallie, R A Jefferson, M W Bevan, PC Turner & T M A Wilson, 1987 Characterisation of the 5′-leader sequence of tobacco mosaic virus RNA as a general enhancer of translation *in vitro* Gene 60 217–225

Vancanneyt, G , S Rosahl & L Willmitzer, 1990 Translatability of a plant mRNA strongly influences its accumulation in transgenic plants Nucl Acids Res 18 2917–2921

Euphytica **85:** 217–233, 1995.

The potential of somatic hybridization in crop breeding

Sylvia Waara & Kristina Glimelius
Uppsala Genetic Center, Department of Plant Breeding, Swedish University of Agricultural Sciences, Box 7003, S-750 07 Uppsala, Sweden

Key words: crop improvement, alien gene transfer, progeny analysis, somatic hybridization

Summary

In recent years, the rapid development of somatic cell genetics has made possible the transfer of alien genes over wide taxonomic distances by somatic hybridization. In this review, the potential of somatic hybridization in the breeding of crops within the Brassicaceae, Fabaceae, Poaceae and Solanaceae is discussed. It is evident from these studies that many hybrids, either symmetric or asymmetric, which are fertile have the potential to be used as a bridge between the alien species and the crop. Progeny analysis of some hybrid combinations also reveals intergenomic translocations which may lead to the introgression of the alien genes. Furthermore, fusion techniques enable the resynthesis of allopolyploid crops to increase their genetic variability and to restore ploidy level and heterozygosity after breeding at reduced ploidy level in polyploid crops.

Introduction

Genetic variability within the species has been efficiently utilized by breeders in their efforts to improve crops. However, the existing variability in a breeding population may not be sufficient for modern plant breeding purposes, and thus great efforts have been made to broaden the existing gene pool of crops. Introduction of new traits has been based mainly on sexual crosses between different genotypes within or between closely related species. However, due to the presence of various reproductive barriers, gene transfer has been restricted to sexually-compatible species, thus limiting the possibilities of modifying and improving crop plants. Many desirable and agronomically-interesting traits may only be found in distantly related species or even in unrelated organisms. Since they constitute a genetic resource potential, considerable effort has been allocated to identify and isolate these genes and transfer them into crops. Through the rapid development of somatic cell genetics, methods now exist for transferring genes across sexual borders and over wide taxonomic distances. Where interesting genes have been identified and isolated, they can be transferred by transformation, but, for most traits the genes have not been identified, and somatic hybridization might then be the method of choice. Besides being of value for the transfer of unidentified genes, somatic hybridization is a tool for the modification and improvement of polygenic traits. Furthermore, the modification of organellar genetic material is possible via somatic hybridization since a mixture of the two fusion partners is obtained in the hybrid cell. In this presentation, an overview will be given regarding the utilization of somatic hybridization as a method of transferring alien genes to crop species. The potential of somatic hybridization for restoring ploidy level in polyploid species after breeding at reduced ploidy level, as well as the challenge of resynthesising allopolyploid species, will also be discussed. We will focus on work with crops from the Brassicaceae, Fabaceae, Poaceae and Solanaceae, and discuss the methodologies and the fate of the transferred alien DNA in the specific hybrids and their progeny.

Methodology

Protoplast technology

Two criteria have to be fulfilled to exploit the use of protoplast fusion technology; protoplasts must be isolated in large quantities and the isolated protoplasts must be totipotent, i.e. they should have the ability to proliferate and regenerate into new plants. The first report on enzymatic release of protoplasts (Cocking, 1960) was followed 11 years later by the first report on the regeneration of plants from protoplast of tobacco (Nagata & Takebe, 1971; Takebe et al., 1971). Since then, regeneration of plants from protoplasts has been reported for more than 320 higher plant species representing 146 genera and 49 plant families (reviewed by Roest & Gilissen, 1989, 1993). The most amenable plants for tissue culture can generally be found within the Solanaceae and Brassicaceae but success has also recently been reported for more recalcitrant species including many members of the Poaceae (Vasil & Vasil, 1992), the Fabaceae (Puite, 1992), berries such as strawberry (Nyman & Wallin, 1988, 1992) and several woody plant species, including temperate fruit trees and *Citrus* species (see reviews by Puite, 1992; Grosser & Gmitter, 1990).

Production of somatic hybrids

Most somatic hybrid plants obtained so far have been produced after chemical treatment with polyethylene glycol, PEG (Kao & Michayluk, 1974; Wallin et al., 1974). The PEG treatment induces agglutination of the protoplasts and fusion will occur after dilution of the PEG with a solution containing a high concentration of calcium ions at high pH. The resulting fusion frequency can vary roughly from 1 to 20% depending upon cell types and fusion conditions employed. Methods for electrofusion have also been developed and reports of somatic hybrid plants recovered from such experiments are also prevalent (Bates & Hasenkampf, 1985; Koop & Schweiger, 1985; Puite et al., 1985; Puite et al. 1986; Fish et al., 1988). The electrofusion process is initiated by resuspending protoplasts in a medium of low conductivity in a chamber separated by two electrodes and applying a high alternating electric field. The protoplasts will move in the electric field by dielectrophoresis and will thereby become attached to each other like chains of pearls. A short pulse of direct current is then applied to induce fusion. Very high fusion frequencies have sometimes been reported after electrofusion and it might have several advantages over chemical fusion; fusion conditions can be more easily controlled, single pairs of protoplasts may be fused and very fragile protoplasts that do not survive chemical fusion, may survive the electrofusion conditions (Bates et al., 1987b).

Selection of hybrids

Unfortunately, the desired fusion products usually represent less than 10% of the total fusion mixture. Much time and effort have therefore been devoted to develop methods for selection of hybrid cells, and this is still a bottle-neck in several systems. Initially, hybrid cells were selected by employing complementation selection procedures using auxotrophic, resistant or chlorophyll – deficient mutants or non-regenerating, wild-type lines (Schieder, 1982). The general application of these selection strategies is, however, limited. Mutants for agronomically-important genotypes are scarce and the production of mutants is both difficult and may have a negative influence on the genetic fidelity of the plant material. The same condition also applies to more recent selection strategies employed where drug resistance genes have been introduced into one or both fusion partners by transformation, although somatic hybrids can be easily identified by selecting for drug resistance (Bates et al., 1987a; Masson et al., 1989; Toriyama et al., 1987a; Wijbrandi et al., 1990d).

Various methods with more general applicability have therefore been developed. These include methods based on visual identification after fusion of morphologically different cell types or differentially stained protoplasts. The hybrid cells may then be mechanically isolated (Kao, 1977; Hein et al., 1983; Puite et al., 1986; Sundberg & Glimelius, 1986; Waara et al., 1991) or enriched by a flow cytometer and cell sorter (Afonso et al., 1985; Glimelius et al., 1986; Puite et al., 1988; Hammatt et al., 1990). Alternatively, one or both parents can be inactivated with chemical compounds such as iodoacetamide and rhodamine 6-G (Menczel et al., 1982; Barsby et al., 1987; Terada et al., 1987b; Böttcher et al., 1989).

Hybrid cells can also be identified at later stages during culture. The expression of hybrid vigour at the callus stage (Debnath et al., 1987; Deimling et al., 1988; Waara et al., 1989; Pólgár et al., 1993) as well as differences in parental and hybrid tissue morphology (Gleddie et al., 1986; Handley et al., 1986; Fish et al., 1987; Klimaszewska & Keller, 1988) have been used. One alternative method which avoids all selection methods is the fusion of individual pairs of protoplasts in a microculture system (Koop & Schweiger, 1985). However, the applicability of this method for production of somatic hybrids for breeding purposes is limited due to cumbersome manipulations.

Identification and characterization of somatic hybrid plants

Regardless of the selection strategy used, the regenerated plants have to be analysed to verify their hybrid nature. Preliminary evidence of hybridity can be obtained by scoring the morphology which is often intermediate between the parents. More direct proof can be obtained by biochemical analysis using the isoenzyme pattern (Scandalios & Sörenson, 1977; Gleba & Sytnik, 1984) or various types of molecular probes such as species-specific repetitive DNA sequences (Saul & Potrykus, 1984; Imamura et al., 1987; Schweizer et al., 1988), rDNA (Gleba et al., 1988) and other RFLP markers (Fish et al., 1988; Melzer & O'Connell, 1990). Recently, the use of PCR technology has also been utilized for hybrid identification (Baird et al., 1992; Xu et al., 1993b). A cytological analysis must also be performed since genetical changes may arise during the tissue culture phase, and multiple fusion events may give rise to polyploid plants.

Organellar composition

In the majority of angiosperms, maternal inheritance of organelles precludes the independent assortment of organellar genomes in most crop plants. Conversely, protoplast fusion of somatic cells produces heteroplasmic cells. Subsequent sorting out of the organelles during cell division and organogenesis, often results in novel combinations of genomes in the regenerated plants. The plastid types generally rapidly sort out bi- or unidirectionally to homogeneity for one or the other parent, which may depend upon the phylogenetic distance between the species (Thanh et al., 1988; Bonnett & Glimelius, 1990; Sundberg & Glimelius, 1991b; Derks et al., 1992) and the presence of selection pressure for organellar type (Medgyesy et al., 1980; Glimelius et al., 1981; Menczel et al., 1982; Cséplö et al., 1984; Menczel et al., 1986; Malone et al., 1992; Jansen, 1993). The cell type had no apparent influence on the segregation of chloroplasts in several Brassicaceae fusion combinations (Sundberg et al., 1991). In contrast, biased transmission of chloroplasts of the mesophyll fusion partner in favour of the suspension fusion partner have been observed in Solanum/Lycopersicon somatic hybrids (Levi et al., 1988; Li & Sink, 1992). Until now, mixed populations of chloroplasts (Fluhr et al., 1984; Thomzik & Hain, 1988) and recombined chloroplasts (Medgyesy et al., 1985) have only rarely been detected. In contrast to what has been observed with the chloroplast genome, recombination and/or rearrangements of the mitochondrial genome is common (Belliard et al., 1979; Kemble et al., 1986; Landgren & Glimelius, 1990). The sorting out of chloroplasts and mitochondria is generally random and independent of the nuclear fusion process; thus somatic hybrids as well as cybrids may be produced. The transfer of organelles to create novel organellar combinations without the fusion of nuclei has therefore been utilized to transfer organelle-encoded traits by, for example, the donor-recipient fusion method (Zelcer et al., 1978). By this method, traits encoded by the chloroplast genome such as triazine resistance (Barsby et al., 1987) and traits encoded by the mitochondrial genome such as cytoplasmic male sterility (Zelcer et al., 1978; Kumashiro & Kubo, 1986; Perl et al., 1990) have been transferred to a novel nuclear background.

Symmetric vs asymmetric hybrids

Interspecific somatic hybrids are mostly polyploid and often contain many unwanted traits derived from the wild species. Therefore, several backcrosses with the cultivated species are necessary to remove the undesired wild characteristics and to establish the optimum ploidy level for crop production. Fertile hybrids are thus of utmost importance in order to obtain progeny and to eliminate undesired characters of the wild species.

In many cases where two complete genomes from phylogenetically-distant species have been combined, the resulting hybrid has been found to be sterile. Only occasionally may the somatic hybrid be used directly e.g. the resynthesis of the allopolyploid species *Brassica napus* (Schenk & Röbbelen, 1982; Sundberg & Glimelius, 1986; Taguchi & Kameya, 1986, and see below), the production of heterozygous tetraploid potato cultivars after fusion of dihaploid clones (Wenzel et al., 1979), and the use of interspecific fruit tree somatic hybrids as root stocks (Ochatt et al., 1989; Grosser & Gmitter, 1990).

Methods limiting the amount of genetic information passed on to the somatic hybrids from the wild species have therefore been of interest, creating asymmetric hybrids. This can be achieved by spontaneous chromosome elimination (Pijnacker et al., 1987; Fahleson et al, 1988; Pijnacker et al., 1989; Sundberg et al., 1991b; Babiychuk et al., 1992) or by irradiation prior to fusion (Dudits, 1980; Gupta et al., 1984; Bates et al., 1987a; Imamura et al., 1987). Alternative methods limiting the amount of DNA transferred, such as micronucleation (Ramulu et al., 1992) and UV irradiation (Hall et al., 1992), are currently also being investigated. To date, asymmetric hybridization has met with mixed results. In general, irradiation reduced the amount of transferred donor DNA but the fraction may vary from a few traits (Dudits et al., 1987), one or a few chromosomes (Gupta et al., 1984; Bates et al., 1987a; Melzer & O'Connell, 1992) to numerous chromosomes (Gleba et al., 1988; Famelaer et al., 1989; Sacristán et al., 1989; Sjödin & Glimelius, 1989b; Yamashita et al., 1989; Wijbrandi et al., 1990a, b, c, Wolters et al., 1991; Derks et al., 1992) the latter result being obtained in the majority of cases. Furthermore, chromosome elimination is random and it is not possible to predict which of the chromosomes will be lost. One way to circumvent this is to tag the chromosomes with selection markers which enables the use of selection pressures for retaining the specific chromosomes in the asymmetric hybrids (Bates et al., 1987a; Sacristán et al., 1989; McCabe et al., 1993). Successful use of the selection strategy requires knowledge of the chromosomal location of agronomically-important genes, and these genes must be located on the same or on a few chromosomes. Alternatively, selection pressure for the desired trait may be applied. In the work of Sjödin & Glimelius (1989b), addition of a fungal toxin to the culture medium was shown to increase the frequency to recovered disease-resistant hybrids when transferring *Phoma lingam* resistance into *Brassica napus*.

In the previous section, the methodology of somatic hybridization was presented. We will now turn to the more direct use of somatic hybridization in the breeding of different crop plants within the Brassicaceae, Fabaceae, Poaceae and Solanaceae. We will mainly discuss the potential of symmetric and asymmetric hybrids and will not directly elaborate on the successful use of cybridization for transfer of organelle-encoded traits; this subject has been reviewed in more detail elsewhere (Kumar & Cocking, 1987; Pelletier et al., 1988; Glimelius et al., 1991). A detailed examination on the potential use of somatic hybridization for transfer of traits conferring disease resistance has also recently been presented (Dixelius & Glimelius, 1994).

Somatic hybridization within the Brassicaceae

Resynthesis of rapeseed

The Brassicaceae is almost a model family regarding somatic hybridization. Protoplasts of a large number of crop species, including oilseed rape, turnip rape, cabbage and radish, can be cultured and regenerated to plants (Vamling & Glimelius, 1990). This has made possible the production of a large number of somatic hybrids between species of varying relatedness within the family (for a review see Glimelius et al., 1991). Hybridizations have been performed between species within the genus *Brassica*, combining both diploid and allopolyploid *Brassica* species into hybrid plants. The resynthesis of *Brassica napus* (rapeseed), combining *B. campestris* with *B. oleracea*, has been regarded as an interesting and important possibility for increasing the genetic diversity of that species. Most of the breeding material of rapeseed originates from a small portion of the total genetic variation inherent in the parental species (Crisp, 1976; Tsunoda et al., 1980), thus, a widening of the gene pool for further evaluations and breeding has been a desired breeding aim (Schenk & Röbbelen, 1982; Sundberg et al., 1987; Terada et al., 1987b; Rosén et al., 1988). Besides working with the progenitors for oilseed rape, somatic hybrids have been produced between heading type vegetables of *B. oleracae* (e.g. cabbage) and *B. campestris* (e.g. chinese cabbage) also with the purpose of widening the genetic diversity (Taguchi & Kameya, 1986). Even though a variation in chromosome number of the resynthesized *B. napus* was obtained, several of the hybrids contained the sum of the chromosome numbers of the two parental species. Most of the hybrids were fertile even though a low seedset was recorded after self-fertilization. Fertility was slightly higher, however, when the hybrids were backcrossed to the parents or to ordinary rapeseed varieties, which has enabled utilization of the material in rapeseed breeding programmes.

Breeding goals of great importance for rapeseed are increased resistance to blackleg, *Phoma lingam*, and clubroot, *Plasmodiophora brassicae*. Somatic hybridization has been used to enhance those traits (Sacristán et al, 1989; Sjödin & Glimelius, 1989a, b) and, as regards resistance towards blackleg, it has also been very successful. Resistance to blackleg was found in *B. nigra*, *B. juncea* and *B. carinata* (Sjödin & Glimelius, 1988) and after production of symmetric as well as asymmetric somatic hybrids between these gene-donors and rapeseed, resistant hybrids were obtained (Sjödin & Glimelius, 1989a, b). By utilizing a toxin, sirodesmin PL, isolated from the culture filtrate of the fungus, a selection pressure exerted by the toxin could be used to select for asymmetric, resistant hybrids (Sjödin & Glimelius, 1989b). Stable inheritance and possible introgression of the gene(s) for resistance to *Phoma lingam* have been recorded in lines derived from a back-crossing programme of the original hybrid with rapeseed (Dixelius, pers. comm.).

Intergeneric somatic hybrids

Several intergeneric hybrids have also been produced via protoplast fusions. In the tribe Brassiceae, species from the genera *Eruca* (Fahleson et al., 1988; Sikdar et al., 1990), *Sinapis* (Toriyama et al., 1987a; Primard et al., 1988), *Raphanus* (Kameya et al., 1989; Sundberg & Glimelius, 1991a), *Moricandia* (Toriyama et al., 1987b) and *Diplotaxis* (Klimazewska & Keller, 1988; Chatterjee et al., 1988; McLellan et al., 1988) have been hybridized with different *Brassica* species, mainly *B. napus*. Compared with the intrageneric somatic hybrids, the intergeneric hybrids showed a larger variation in chromosome number; more aneuploid hybrids that had lost one or several chromosomes, as well as polyploid hybrids that had gained one or more chromosome sets, were found (Sundberg & Glimelius, 1991b). Of the intergeneric hybrids produced, attempts were made to introduce tolerance to *Alternaria brassiceae* and the beet cyst nematode (BCN), *Heterodera schachtii* from *Sinapsis alba* (*B. hirta*) (Primard et al., 1988; Lelivelt et al., 1993), tolerance to drought and aphids from *Eruca sativa* (Fahleson et al., 1988), resistance to BCN from *Raphanus sativus* L. (Lelivelt & Krens, 1992), a low CO_2-compensation point from *Moricandia arvensis* (Toriyama et al., 1987b) as well as cytoplasmic male sterility from *Diplotaxis harra* (Klimaszewska & Keller, 1988), and *Brassica tournefortii* (Liu & Glimelius, 1994).

Concerning *Alternaria* tolerance, the somatic hybrids obtained between *B. napus* and *S. alba* were not as tolerant to *Alternaria brassicae* as the gene donor (Primard et al., 1988). In the work where attempts were made to introduce resistance to BCN from *Raphanus sativus*, resistance was expressed in the primary hybrids at a high level. However, these hybrids showed reduced fertility and could not be backcrossed to *B. napus*, preventing the transfer of the BCN resistance to the *B. napus* gene pool (Lelivelt & Krens, 1992). The same results were reported from experiments where resistance to BCN was transferred to rapeseed from *Sinapis alba* (Lelivelt, 1993); again, the primary hybrids expressed a high level of resistance, but the somatic hybrids were mitotically unstable and sterile, which hampered further breeding. In contrast, the somatic hybrids produced between *B. napus* and *Eruca sativa* by Fahleson et al. (1988) were fertile, which has enabled back-crossing to rapeseed. After one backcross to rapeseed, followed by selfing of this progeny, the material has been divided into two groups. In one of these groups, high concentrations of erucic acid were used as an indicator that genetic material from *E. sativa* was still present in the progeny; as the rapeseed parent in the somatic hybridization, a variety with low concentrations of erucic acid was used. Besides erucic acid, three isoenzymes and two *E. sativa*-specific, repetitive DNA sequences were used to confirm presence of DNA from *E. sativa* (Fahleson et al., 1993). This hybrid material has been introduced into breeding programmes of rapeseed in Sweden and investigations are in progress to screen for traits of interest. The first check, however, for presence of insect tolerance in some of the lines, was not successful (Åhman, 1993).

Intertribal somatic hybrids

Intertribal hybrids can only be produced via somatic hybridization. *Arabidopsis thaliana* from the tribe Sisymbriae has been hybridized with *B. campestris* (Gleba & Hoffman, 1979, 1980), and with *B. napus* (Bauer-Weston et al., 1993; Forsberg et al., 1994). Furthermore, *Thlaspi perfoliatum* from the tribe Lepidiae (Fahleson et al., 1994b), *Barbarea vulgaris*, from the tribe Arabideae (Fahleson et al., 1994a; Oikarinen & Ryöppy, 1992) and *Lesquerella fendleri* from the tribe Drabeae (Skarzynskaya et al., 1994) have been combined with *B. napus* and, in the case of *Barbarea vulgaris*, also with *B. campestris*. In all these hybridizations plants have been obtained which, according to isoenzyme and RFLP analysis, were hybrids. However, evaluation of the chromosome number and hybrid markers, including species-specific markers, showed that more asymmetric hybrids were found among the intertribal hybrids than among the intrageneric and intergeneric hybrids. Interestingly, though, when comparing the intertribal hybrids with those obtained from hybridization experiments between more closely-related species, no differences in the frequency of hybrid fusions or hybrid shoot regeneration were recorded. It was only after attempts to culture and establish the hybrid material in the greenhouse that a clear difference was noted. The intertribal combinations were, in general, more difficult to culture to mature plants outside *in vitro* conditions. This was especially pronounced in the fusion between *B. napus* and *Barbarea vulgaris* which, in spite of the fact that hybrid plants could be cultured *in vitro*, never resulted in plants that could be established and cultured under ordinary growth conditions in the greenhouse (Fahleson et al., 1994a). According to Oikarinen & Ryöppy (1992), putative fertile hybrids between *B. campestris* and *Barbarea vulgaris* that could be established in the greenhouse were obtained. Clear evidence that these plants indeed were hybrids has not been reported, however. Nevertheless, intertribal somatic hybrids have been produced that differentiate and develop into normal hybrid plants that could even be successfully selfed, as for example some of the '*Arabido-Brassica*' hybrids (Forsberg et al., 1994). Even though most of the hybrids produced displayed some problems when flowering and did not produce pollen, several were female-fertile and could set seeds when backcrossed with pollen from rapeseed. In the case of the '*Thlaspo-Brassica*' hybrids it was even possible to detect the presence of reasonably high concentrations of the *Thlaspi*-specific fatty acid, neuronic acid, in some progeny obtained from the initial hybrid plants backcrossed to rapeseed (Fahleson et al., 1994b). Thus, since very low concentrations of this fatty acid are present in rapeseed, this clearly indicates that genetic material coding for the synthesis of neuronic acid has been transferred via the hybridization. With regard to the somatic hybrids produced between *Lesquerella fendleri* and rapeseed, a large number of hybrids have been obtained (Skarzhynskaya & Glimelius, 1994). Investigations have started by evaluating the presence of lesquerolic acid, a *Lesquerella*-specific hydroxy-unsaturated fatty acid, in the progeny obtained when backcrossing the hybrids with rapeseed.

Somatic hybridization in the Fabaceae

Several members of the Fabaceae are important crop plants and a broadening of the existing gene pool has therefore attracted considerable interest (Kumar & Davey, 1991). Various methods of protoplast culture and somatic hybridization have been developed (reviewed by Kumar & Davey, 1991). Regeneration of plants from protoplasts of this family was initially troublesome but has recently been achieved from several genera including *Glycine, Lotus, Medicago, Pisum* and *Trifolium* (reviewed by Webb, 1988; Puite, 1992; Roest & Gilissen, 1993). It is mainly in the genus *Medicago* that somatic hybridization as a method for crop improvement has been investigated. Unfortunately, in the majority of fusion experiments, hybrid cell lines but no hybrid plants have been recovered (Gilmour et al., 1987; Damiani et al., 1988; Deak et al., 1988; Walton & Brown, 1988; Gilmour et al., 1989; Niizeki & Saito, 1989; Pupilli et al., 1991; Kihara et al., 1992). However, a few encouraging results demonstrate that fusion technology may be of agronomic value in this family. Flowering somatic hybrids have been obtained between the sexually-compatible species *Medicago sativa* (alfalfa) and *M. falcata* (Téoulé, 1983; Mendis et al., 1991), as well as between alfalfa and *M. coerulea* (Pupilli et al., 1992) and alfafa and *M. intertexta* (Thomas et al., 1990). Asymmetric intra-generic *Medicago* hybrids have been regenerated but not analysed in detail (Kuchuk et al., 1990). Interspecific somatic hybrids within the genus *Lotus* have also been produced (Wright et al., 1987; Aziz et al., 1990) and these hybrids cannot be produced by sexual hybridization. Furthermore, plantlets but no mature plants were recovered after fusion of *Glycine max* (soybean) with other *Glycine* spp. (Hammatt et al., 1992). The most interesting somatic hybrids obtained so far are probably those obtained after asymmetric intergeneric fusion of the sexually-incompatible species alfalfa and *Onobrychis viciifolia*, sainfoin (Li et al., 1993). The purpose of this study was to transfer the trait for the production of foliar condensed tannins from sainfoin to alfalfa. These condensed tannins are thought to play an important role in preventing bloat in grazing animals by preventing gas escape in the rumen. A large number of highly asymmetric hybrids were recovered after fusion of irradiated sainfoin protoplasts and iodoacetamide-inactivated alfalfa protoplasts. Pollen germination tests revealed that the majority of the hybrids were male-fertile. However, so far no tannin-positive regenerated plant has been detected among 43 plants analysed (Li et al., 1993). It is speculated that this is because only a small amount of the large genome of sainfoin has been transferred to the hybrids, and probably about 200 hybrids should be screened in order to recover a highly asymmetric, tannin-positive hybrid.

Somatic hybridization within the Poaceae

Successful regeneration of plants from protoplast cultures of members of the grass family was initially difficult and the grasses have been regarded as some of the most recalcitrant species for use in tissue culture techniques. However, during the last ten years an increasing number of reports on regeneration of plants from embryogenic cell suspension cultures of both forage grasses and cereals have been presented (reviewed by Vasil & Vasil, 1992; Roest & Gilissen, 1993). Intergeneric hybrids have been produced in several combinations, e.g. *Panicum maximum* (+) *Pennisetum americanum* (Ozias-Akins et al., 1986), *Saccharum officinarum* (+) *P. americanum* (Tabaeizadeh et al., 1986), *Oryza sativa* (+) *Eichinochloa oryzicola* (Terada et al., 1987a) *Triticum monococcum* (+) *P. americanum* (Vasil et al., 1988), *Festuca arundinacea* (+) *Lolium multiflorum* (Takamizo et al., 1991). However, regeneration of mature plants was only possible in the last combination; this hybrid can also be produced by sexual hybridization. Nevertheless, interesting mature asymmetric hybrids between *F. arundinacea* and *L. multiflorum* (Spangenberg et al., 1994) have been recovered. Attempts to overcome conventional breeding barriers by interspecific fusion of rice with four different wild species including *Oryza brachyantha, O. eichingeri, O. officinalis*, and *O. perrieri* (Hayashi et al., 1988) were more successful. Mature plants with viable pollen could be obtained in all but the first *Oryza* combinations. In addition, the production of fertile, inter-specific, diploid rice hybrid plants (Toriyama & Hinata, 1988) as well as rice cybrids (Akagi et al., 1989; Kyozuka et al., 1989; Yang et al., 1989) has been reported. These studies now demonstrate that the fusion technology can also be extended to graminaceous crops using donor species possessing agronomically-interesting traits.

Somatic hybridization within the Solanaceae

Eggplant

In the case of eggplant (*Solanum melongena*), sexual crossing barriers have limited the possibility of transferring agronomically-important traits such as resistance to insect attack, or resistance to various diseases to eggplant from related wild species (Daunay et al., 1991). Symmetric somatic hybrids between eggplant and *S. aethiopicum* (Daunay et al., 1993), *S. integrifolium (Kameya et al., 1990), S. khasianum* (Sihachakr et al., 1988), *S. nigrum* (Guri & Sink, 1988a), *S. sisymbrifolium* (Gleddie et al., 1986) and *S. torvum* (Guri & Sink, 1988b; Sihachakr et al., 1989) have been produced. Preliminary evaluation of the agronomic traits exhibited by these somatic hybrids reveals that the desired properties have been incorporated but their high sterility limits the potential value in future breeding programmes. Only the somatic hybrids between species with the closest phylogenetic relationship, i.e. the somatic hybrids obtained after fusion of eggplant and *S. aethiopicum* or *S. integrifolium*, are fertile. Such hybrids can also be produced by sexual hybridization although they have very low fertility and different organellar composition (Daunay et al., 1993).

Potato

As potato (*Solanum tuberosum*) is an autotetraploid, potato breeding requires extensive selection in a large population to obtain the desired agronomic traits combined in one genotype. Therefore, breeding at diploid level followed by somatic hybridization (to restore the ploidy level) has been suggested as an alternative for the production of superior cultivars (Wenzel et al., 1979). In recent years, a large number of intraspecific fusion combinations have been conducted (Austin et al., 1985a; Debnath & Wenzel, 1987; Deimling et al., 1988; Waara et al., 1989; Chaput et al., 1990; Waara et al., 1991; Möllers & Wenzel, 1992; Thach et al., 1993). In a similar approach, dihaploid potato clones have been fused with diploid *S. phureja* or diploid *S. tuberosum* × *S. phureja* clones (Puite et al., 1986; Puite et al., 1988; Mattheij & Puite, 1992). Analysis of the hybrid plants has demonstrated the potential value of this breeding method as it is possible to predict many characters displayed by the hybrids from the characters of the dihaploid parents. Dominantly-inherited disease or pest resistance genes, such as the resistance gene Ro 1 against the nematode *Globodera rostochiensis*, have been transferred from one of the dihaploid parents to the hybrids (Möllers & Wenzel, 1992). Likewise, the major genes for PVX and PVY resistance are also expressed in somatic hybrids (Thach et al., 1993). Phenotypic characterization has shown that most hybrid combinations show a general vegetative hybrid vigour (Deimling et al., 1988; Chaput et al., 1990; Waara et al., 1991; Mattheij & Puite, 1992; Möllers et al., 1992; Waara et al., 1992). Furthermore, field trials reveal that several hybrid combinations give tuber yields as high as or higher than standard cultivars

(Mattheij & Puite, 1992; Möllers et al., 1994) and it appears that the yield of the hybrid is highly correlated with the yield of the parental clones (Mattheij & Puite, 1992; Möllers et al., 1994). The yield increase seems to be mainly due to an increased tuber weight whereas in several combinations, the number of tubers per plant is intermediate between the two parents (Mattheij & Puite, 1992; Waara et al., 1992; Möllers et al., 1994). Tuber characters such as red skin colour, yellow tuber flesh and round form also appear to be inherited dominantly (Möllers & Wenzel, 1992; Waara et al., 1992; Möllers et al., 1994).

Genetic improvement of the potato by interspecific somatic hybridization has also been accomplished. Somatic hybrid plants have been produced between potato and several partly or completely sexually-incompatible wild species including *S. brevidens* (Barsby et al., 1984; Austin et al., 1985b; Austin et al., 1986; Fish et al., 1987, 1988; Preiszner et al., 1991) *S. bulbocastanum* (Austin et al., 1993), *S. chacoense* (Butenko & Kuchko, 1980), *S. circaeifolium* (Mattheij et al., 1992), *S. commersonii* (Cardi et al., 1993), *S. nigrum* (Binding et al., 1982), *S. pinnatisectum* (Sidorov et al., 1987) and *S. torvum* (Jadari et al., 1992). Furthermore, somatic hybrids between the sexually-compatible species *S. berthaultii* and potato have been produced (Serraf et al., 1991). The most widely-used wild species in somatic hybridization experiments is the diploid non-tuberous wild species *S. brevidens*. This species carries resistance genes to several viral diseases including PLRV (Jones, 1979), PVX (Gibson et al., 1990) and PVY (Gibson et al., 1988) and the transfer of these traits to somatic hybrids is well documented (Helgeson et al., 1986; Austin et al., 1988; Gibson et al., 1988; Pehu et al., 1990b). Some of these hybrid combinations were fertile and could be backcrossed to potato (Ehlenfeldt & Helgeson, 1987; Williams et al., 1990). These sexual progeny showed substantial variation but plants with improved agricultural traits combined with retained disease resistance have been identified (Helgeson et al., 1993). Another interesting feature of these hybrids was that they also showed resistance to *Erwinia* soft rot, a tuber disease. This trait was not inherited from the tuberizing *S. tuberosum* parent which was susceptible to the disease (Austin et al., 1988). Thus, the resistance trait might have been transferred from the nontuberizing parent *S. brevidens* to the tuber-forming somatic hybrids. Asymmetric somatic hybrids between *S. brevidens* and potato have also been produced but the fertility and disease resistance of these has not yet been investigated (Fehér et al., 1992; Puite & Schaart, 1993; Xu et al., 1993a). Traits conferring tolerance of environmental stresses such as cold have been transferred from cold-tolerant species such as *S. brevidens* and *S. commersonii* to potato via somatic hybrids. The hybrids showed intermediate tolerance compared with the parents (Preiszner et al., 1991; Cardi et al., 1993). In addition, fertile hybrids exhibiting resistance to both *Phytophthora infestans* and *Globodera pallida* have also been produced by fusion of potato and its wild relative *S. circaeifolium* (Mattheij et al., 1992).

The possibility of alien gene transfer through intergeneric hybridization has also been attempted; somatic hybrid plants have been recovered from fusion experiments between potato and *Lycopersicon esculentum* (Melchers et al., 1978; Shepard et al., 1983) and *L. pimpinellifolium* (Okamura, 1988). Fertile hybrids were obtained in both cases, even though production of seeds from the potato-tomato combination seems to be very rare (Jacobsen et al., 1994).

From the data presented, it is clear that potato is a species where fusion technology has been very successful. It is therefore encouraging that somatic hybridization programmes are now included in commercial potato breeding schemes in the The Netherlands (Mattheij & Puite, 1992) and in Germany (Möllers, pers. comm.).

Tobacco

Tobacco has long been one of the model species for cell and tissue culture studies and the first somatic hybrid plant was produced between *Nicotiana glauca* and *Nicotiana langsdorffii* (Carlson et al., 1972). The majority of the hybrid-enrichment methodologies and hybrid analysis systems were primarily developed for various tobacco somatic hybrids which, by themselves, did not have any agronomic importance, but they paved the way for further somatic hybridization experiments in tobacco as well as in other crops (see review by Gleba & Sytnik, 1984; Kubo, 1988; Gleba & Schlumukov, 1990). Somatic hybrids of potential economic importance were recovered from fusion experiments between *N. tabacum* and *N. rustica* (Douglas et al., 1981). The somatic hybrids expressed an elevated nicotine content (Pandeya et al., 1986) as well as resistance to black root rot (*Thielaviopsis basicola*) and blue mould (*Peronospora tabacina*). Curiously, resistance against the latter disease was not found in any of the parents and might have arisen due to interaction between the different nuclear and/or organellar genomes (Pandeya et al., 1986). The transfer of TMV resistance into fertile tobacco somatic hybrids from *N. nesophila* or *N. stocktonii* has also been achieved (Evans et al., 1981). The above-mentioned hybrids can be produced by sexual crossings although special techniques such as ovule culture *in vitro* are necessary. Attempts to transfer disease resistance genes into *N. tabacum* from the sexually-incompatible species of *N. repanda* were also successful but the resulting somatic hybrids were sterile (Nagao, 1982). To overcome the sterility of both sexual and somatic hybrids, asymmetric hybridization was attempted (Bates, 1990). From this work two asymmetric somatic hybrids were recovered which showed TMV resistance and female fertility which enabled them to be backcrossed to tobacco.

Tomato

Tomato is the only crop plant within the small genus *Lycopersicon* but several of the wild species have served as important sources of agronomic traits via sexual crosses (Rick, 1982; Daunay et al., 1991). However, many sexual crosses between tomato and the wild species are cumbersome to perform due to bilateral and unilateral incompatibility, and several of the resulting F_1 hybrids are sterile (reviewed by Lefrançois et al., 1993). Therefore, the potential of somatic hybridization in tomato breeding has been investigated (see reviews by Hille et al., 1989; Lefrançois et al., 1993). Interspecific symmetric hybrid plants have been produced between tomato (*L. esculentum*) and *L. chilense* (Bonnema & O'Connell, 1992), *L. pennellii* (O'Connell & Hanson, 1987) and *L. peruvianum* (Kinsara et al., 1986; San et al., 1990; Wijbrandi et al., 1990d). Intergeneric somatic hybridization has also been possible between tomato and several species including *Nicotiana tabacum* (Turpin, 1986), *Solanum etuberosum* (Gavrilenko et al., 1992), *S. etuberosum* × *S. brevidens* (Gavrilenko et al., 1992), *S. lycopersicoides* (Handley et al., 1986; Tan, 1987; Levi et al., 1988), *S. muricatum* (Sakamoto & Taguchi, 1991), *S. nigrum* (Guri et al., 1988), *S. rickii* (O'Connell & Hanson, 1986) and *S. tuberosum* (Melchers et al., 1978; Shepard et al., 1983). The morphology and organellar composition of these hybrids have been extensively studied but no report on the transfer of agronomic traits to the somatic hybrid is available. The somatic hybrids produced between tomato and *L. peruvianum* are of particular interest since this combination is one of the few that is fertile, while the corresponding diploid sexual hybrid is sterile (Lefrançois et al., 1993). The fertile, intergeneric tomato × *S. etuberosum* somatic hybrids (Gavrilenko et al., 1992) are also valuable as this combination cannot be obtained by sexual hybridization.

The possibilities of partial gene transfer by asymmetric somatic hybridization have also been investigated in tomato although the transfer of specific agronomic traits from the donor has not been determined (O'Connell & Hanson, 1987; Melzer & O'Connell, 1990; Wijbrandi et al., 1990a; Ratushnyak et al., 1991; Wolters et al. 1991; Derks et al., 1992; Melzer & O'Connell, 1992; McCabe et al., 1993). In the majority of these studies, the extent of chromosome elimination is rather limited and irradiation has only determined the direction of the chromosome elimination (reviewed by Lefrançois et al., 1993). However, in one study where the recipient was treated with the metabolic inhibitor iodoacetatamide prior to fusion, a clear correlation with irradiation dose and asymmetry could be established (Melzer & O'Connell, 1992). Some of the asymmetric hybrids between tomato and irradiated *L. pennellii* are also fertile, and sexual progeny can be obtained after self fertilization (Melzer & O'Connell, 1992). This observation is especially intriguing as the corresponding symmetric hybrids are sterile.

Conclusions

From this overview of somatic hybridization it is obvious that the method can be used for combining species with different degrees of genetic divergence into functional hybrids, and that even species from different tribes can be hybridized. Successful hybridization has also been reported in previously recalcitrant species such as those within the Fabaceae and the Poaceae. Low fertility of somatic hybrids has sometimes been reported and this could, of course, severely restrict their utilization in breeding programmes. However, in many cases only pollen fertility has been analysed and attempts have been restricted to self-fertilization of the hybrids. In general, somatic hybrids have a low pollen fertility. Nevertheless, hybrids have been used as female parents (even though the female fertility might be low) in backcrosses with the cultivated species (Ehlenfeldt & Helgeson, 1987; Mattheij et al., 1992; Cardi et al., 1993; Helgeson et al., 1993; Pupilli et al., 1992; Fahleson et al., 1993; 1994a, 1994b; Dixelius, unpublished). Fertility has also in some cases been higher in asymmetric hybrids than in the corresponding symmetric hybrid (Bates, 1990; Melzer &

O'Connell, 1992). Thus, several of the hybrids, either symmetric or asymmetric, have the potential to be used as a bridge between the alien species and the crop plant.

Provided hybrids contain the alien genes of interest, a back-cross programme to the crop plant and analysis of the progeny to determine the presence of alien genes can be carried out. From investigations made of progeny obtained after repeated backcrossings of the Brassicaceae hybrids with pollen from rapeseed, it has been possible to follow inheritance of certain traits from the gene donors. Resistance to *Phoma lingam*, for example, which was transferred via protoplast fusion from *B. nigra* to *B. napus* was present even after twelve back-crosses (Dixelius, unpublished). In the case of the '*Thlaspo-Brassica*' hybrids it was possible to obtain transfer of the gene(s) coding for the *Thlaspi*-specific fatty acid, nervonic acid, to some progeny obtained from the initial hybrid plants backcrossed to rapeseed (Fahleson et al., 1994b). Helgeson et al. (1993) have also shown that hexaploid somatic hybrids produced between cultivated potato and *S. brevidens* can be of potential value for further breeding, since progeny with disease resistance and with improved yield were obtained.

With the development of molecular techniques, better methods to detect and follow the introgression of alien DNA in the receptor genome have been obtained (see Lydiate et al., this volume). The existence of high-density RFLP maps in tomato (Bernatzky & Tanksley, 1986) and potato (Bonierbale et al., 1988; Gebhardt et al., 1989) has, for example, made it possible to determine the extent and direction of elimination in several asymmetric hybrids (Melzer & O'Connell, 1990; Wijbrandi et al., 1990c; Wolters et al., 1991; Melzer & O'Connell, 1992). The presence of species-specific repetitive DNA sequences in plant genomes is also valuable. Repetitive DNA is present in most species and because of the tolerance of such sequences to mutations, species-specific sequences are common, as has been shown e.g. for species in the Brassicaceae (Simoens et al., 1988; Iwabuchi et al., 1991; Gupta et al., 1992; Fahleson et al., 1994a, 1994b), Solanaceae (Schweizer et al., 1988; Pehu et al., 1990a; Piastuch & Bates, 1990) and Poaceae (Metzlaff et al., 1986; Zhang & Dvorák, 1989; Cordesse et al., 1992). Isolated species-specific, repetitive DNA sequences have proved to be excellent markers to verify the presence of alien DNA in somatic hybrids and their progeny (Saul & Potrykus, 1984; Imamura et al., 1987; Piastuch & Bates, 1990; Itoh et al., 1991; Perez-Vincente et al., 1992; Fahleson et al., 1994a, 1994b; Forsberg et al., 1994). Furthermore, by utilizing *in situ* DNA hybridization to chromosome spreads, a detailed characterization of the hybrids regarding presence of chromosomes from the parental species can be obtained. For example, Itoh et al. (1991) showed that somatic hybrids between *B. oleracea* and X-ray-treated *B. campestris* were asymmetric, while Piastuch & Bates (1990) were able to demonstrate the presence of translocations between chromosomes of *Nicotiana tabacum* and *N. plumbaginifolia* in asymmetric somatic hybrids.

Thus, since somatic hybridization enables transfer of genetic material between distantly-related and sexually-incompatible species, it is a method with great potential for improvement of our crops. However, to be of practical value, introgression and stable inheritance of the alien DNA carrying the desired genes has to take place. An important factor to investigate in the future is the mechanism involved in the incorporation of donor DNA into recipient chromosomes. Parokonny et al. (1992) have started to investigate intergenomic translocations and the possible mechanisms involved by utilizing genomic *in situ* hybridization (GISH). From their studies of somatic hybrids produced between *Nicotiana plumbaginifolia* and *Nicotiana sylvestris*, it could be revealed that intergenomic translocations containing chromosome segments from both species had been obtained. Another possible way to investigate this is to analyse somatic hybrid progeny by RFLP. Differences in the linkage groups obtained will reveal whether recombination and translocations between the genomes have occurred.

Acknowledgements

The authors gratefully acknowledge financial support from the Swedish Council for Forestry and Agricultural Research.

References

Afonso, C.L., K.R. Harkins, M.A. Thomas-Compton, A.E. Krejci & D.W. Galbraith, 1985. Selection of somatic hybrid plants in *Nicotiana* through fluorescence-activated cell sorting of protoplasts. Bio/Technology 3: 811–816.

Akagi, H., M. Sakamoto, T. Negishi & T. Fujimura, 1989. Construction of rice cybrid plants. Mol. Gen. Genet. 215: 501–506.

Austin, S., E. Lojkowska, M.K. Ehlenfeldt, A. Kelman & J.P. Helgeson, 1988. Fertile interspecific somatic hybrids of Solanum: A novel source of resistance to *Erwinia* soft rot Phytopath 78 1216–1220

Austin, S, J D Pohlman, C R Brown, H Mojtahedi, G S San to, D S Douches & J P Helgeson, 1993 Interspecific somatic hybridization between *Solanum tuberosum* L and *S bulbocas tanum* Dun as a means of transferring nematode resistance Am Pot J 70 485–495

Austin, S, M Baer, M Ehlenfeldt, PJ Kazmierczak & J P Helge son, 1985a Intra specific fusions in *Solanum tuberosum* Theor Appl Genet 71 172–175

Austin, S, M A Baer & J P Helgeson, 1985b Transfer of resistance to potato leaf roll virus from *Solanum brevidens* into *Solanum tuberosum* by somatic fusion Plant Sci 39 75–82

Austin, S, M K Ehlenfeldt, M A Baer & J P Helgeson, 1986 Somatic hybrids produced by protoplast fusion between *S tubero sum* and *S brevidens* phenotypic variation under field conditions Theor Appl Genet 71 682–690

Aziz, M A, PK Chand, J B Power & M R Davey, 1990 Somatic hybrids between the forage legumes *Lotus corniculatus* L and *L tenuis* Waldst et Kit J Exp Bot 41 471–479

Babiychuk, E, S Kushnir & Y Y Gleba, 1992 Spontaneous exten sive chromosome elimination in somatic hybrids between somat ically congruent species *Nicotiana tabacum* L and *Atropa bel ladonna* L Theor Appl Genet 84 87–91

Baird, E, S Cooper Bland, R Waugh, M DeMaine & W Pow- ell, 1992 Molecular characterization of inter- and intra specific somatic hybrids of potato using randomly amplified polymorphic DNA (RAPD) markers Mol Gen Genet 233 469–475

Barsby, T L, J F Shepard, R J Kemble & R Wong, 1984 Somatic hybridization in the genus *Solanum* *S tuberosum* and *S brevi dens* Plant Cell Rep 3 165–167

Barsby, T L, S A Yarrow, R J Kemble & I Grant, 1987 The transfer of cytoplasmic male sterility to winter type oilseed rape (*Brassica napus* L) by protoplast fusion Plant Sci 53 243–248

Bates, G W & C A Hasenkampf, 1985 Culture of plant somatic hybrids following electrical fusion Theor Appl Genet 70 227–233

Bates, G W, 1990 Asymmetric hybridization between *Nicotiana tabacum* and *N repanda* by donor recipient protoplast fusion transfer of TMV resistance Theor Appl Genet 80 481–487

Bates, G W, C A Hasenkampf, C L Contolini & W C Piastuch, 1987a Asymmetric hybridization in *Nicotiana* by fusion of irra diated protoplasts Theor Appl Genet 74 718–726

Bates, G W, L J Nea & C A Hasenkampf, 1987b Electrofusion and plant somatic hybridization p 479–496 In A E Sowers (Ed) Cell Fusion Plenum Press, New York

Bauer Weston, B, W Keller, J Webb & S Gleddie, 1993 Produc tion and characterization of asymmetric somatic hybrids between *Arabidopsis thaliana* and *Brassica napus* Theor Appl Genet 86 150–158

Belliard, G, F Vedel & G Pelletier, 1979 Mitochondrial recombi nation in cytoplasmic hybrids of *Nicotiana tabacum* by protoplast fusion Nature 281 401–403

Benatzky, R & S D Tanksley, 1986 Towards a saturated linkage map in tomato based on isozyme and random cDNA sequences Genetics 112 887–898

Binding, H, S M Jain, J Finger, G Mordhorst, R Nehls & J Gressel, 1982 Somatic hybridization of an atrazine resistant bio-type of *Solanum nigrum* with *Solanum tuberosum* Part 1 Clonal variation in morphology and in atrazine sensitivity Theor Appl Genet 63 273–277

Bonierbale, M W, R L Plaisted & S D Tanksley, 1988 RFLP maps based on a common set of clones reveal modes of chromosomal evolution in potato and tomato Genetics 120 1095–1103

Bonnema, A B & M A O'Connell, 1992 Molecular analysis of the nuclear and organellar genotype of somatic hybrid plants between tomato (*Lycopersicon esculentum*) and *Lycopersicon chilense* Plant Cell Rep 10 629–632

Bonnett, H T & K Glimelius, 1990 Cybrids of *Nicotiana tabacum* and *Petunia hybrida* have an intergeneric mixture of chloroplas ts from *P hybrida* and mitochondria identical or similar to *N tabacum* Theor Appl Genet 79 550–555

Butenko, R G & A A Kuchko, 1980 Somatic hybridization of *Solanum tuberosum* L and *Solanum chacoense* Bitt by proto plast fusion p 293–300 In L Ferenczy & G L Farkas (Eds) Advances in Protoplast Research Akad Kiado, Budapest

Bottcher, U F, D Aviv & E Galun, 1989 Complementa tion between protoplasts treated with either of two metabolic inhibitors results in somatic-hybrid plants Plant Sci 63 67–77

Cardi, T, F D'Ambrosio, D Consoli, K J Puite & K S Ramu lu, 1993 Production of somatic hybrids between frost tolerant *Solanum commersonii* and *S tuberosum* characterization of hybrid plants Theor Appl Genet 87 193–200

Carlson, P S, H H Smith & R D Dearing, 1972 Parasexual inter specific plant hybridization Proc Natl Acad Sci USA 69 2292–2294

Chaput, M H, D Sihachakr, G Ducreux, D Marie & N Barghi, 1990 Somatic hybrid plants produced by electrofusion between dihaploid potatoes BF 15 (H1), Aminca (H6) and Cardinal (H3) Plant Cell Rep 9 411–414

Chatterjee, G S, S R Sikdar, S Das & S K Sen, 1988 Intergener ic somatic hybrid production through protoplast fusion between *Brassica juncea* and *Diplotaxis muralis* Theor Appl Genet 76 915–922

Cocking, E C, 1960 A method for the isolation of plant protoplasts and vacuoles Nature 187 962–963

Cordesse F, F Grellet, A S Reddy & M Delseny, 1992 Genome specificity of rDNA spacer fragments from *Oryza sativa* L Theor Appl Genet 83 864–870

Crisp, P, 1976 Trends in the breeding and cultivation of cruciferous crops p 69–118 In J G Vaughan, A J MacLeod & B M G Jones (Eds) The Biology and Chemistry of the Cruciferae Aca demic Press, London

Cseplo, A, F Nagy & P Maliga, 1984 Interspecific protoplast fusion to rescue a cytoplasmic lincomycin resistant mutation into fertile *Nicotiana plumbaginifolia* plants Mol Gen Genet 198 7–11

Damiani, F, M Pezzotti & S Arcioni, 1988 Electric field mediated fusion of protoplasts of *Medicago sativa* L and *Medicago arbo rae* L J Plant Physiol 132 474–479

Daunay, M C, M H Chaput, D Sihachakr, M Allot, F Vedel & G Ducreux, 1993 Production and characterization of fertile somat ic hybrids of eggplant (*Solanum melongena* L) with *Solanum aethiopicum* L Theor Appl Genet 85 841–850

Daunay, M C, R N Lester & H Laterrot, 1991 The use of wild species for the genetic improvement of eggplant (*Solanum mel ongena*) and tomato (*Lycopersicon esculentum* Mill) p 389–412 In L Hawkes, R N Lester, M Nee & N Estrada (Eds) Solanaceae III taxonomy, chemistry and evolution Royal Botan ical Gardens of Kew and Linnean Society of London

Deak, M, G Donn, A Feher & D Dudits, 1988 Dominant expres sion of a gene amplification related herbicide resistance in *Med icago* hybrids Plant Cell Rep 7 158–161

Debnath, S C & G Wenzel, 1987 Selection of somatic fusion products in potato by hybrid vigour Potato Res 30 371–380

Deimling, S, J Zitzlsperger & G Wenzel, 1988 Somatic fusion for breeding of tetraploid potatoes Plant Breed 101 181–189

Derks, F H M, J Wijbrandi, M Koornneef & C M Colijn Hooymans, 1991 Organelle analysis of symmetric and asymmet ric hybrids between *Lycopersicon peruvianum* and *Lycopersicon esculentum*. Theor. Appl. Genet. 81: 199–204.

Derks, F.H.M., J.C. Hakkert, W.H.J. Verbeek & C.M. Colijn-Hooymans, 1992. Genome composition of asymmetric hybrids in relation to the phylogenetic distance between the parents nucleus-chloroplast interaction. Theor. Appl. Genet. 84: 930–940.

Dixelius, C. & K. Glimelius, 1995. The use of somatic hybridization for interspecific transfer of plant disease resistance. pp. 75–93. In: R. Reuveni (Ed). Novel Approaches to Integrated Pest Management (In press). CRC Press and Lewis Publishers.

Douglas, G.C., W.A. Keller & G. Setterfield, 1981. Somatic hybridization between *Nicotiana rustica* and *N. tabacum II*. Protoplast fusion and selection and regeneration of hybrid plants. Can. J. Bot. 59: 1509–1513.

Dudits, D., E. Maroy, T. Praznovszky, Z. Olah, J. Gyorgyey & R. Cella, 1987. Transfer of resistance traits from carrot into tobacco by asymmetric somatic hybridization: Regeneration of fertile plants. Proc. Natl. Acad. Sci. USA 84: 8434–8438.

Dudits, D., O. Fejer, G.Y. Hadlaczky, C.S. Koncz, G. Lazar & G. Hovrath, 1980. Intergeneric gene transfer mediated by protoplast fusion. Mol. Gen. Genet. 179: 283–288.

Ehlenfeldt, M.K. & J.P. Helgesson, 1987. Fertility of somatic hybrids from protoplast fusions of *Solanum brevidens* and *S. tuberosum*. Theor. Appl. Genet. 73: 395–402.

Evans, D.A., C.E. Flick & R.A. Jensen, 1981. Disease resistance: incorporation into sexually incompatible somatic hybrids of the genus *Nicotiana*. Science 213: 907–909.

Fahleson, J., I. Eriksson & K. Glimelius, 1994a. Intertribal somatic hybrids between *Brassica napus* and *Barbarea vulgaris*. Plant Cell Rep. 13: 411–416.

Fahleson, J., I. Eriksson, M. Landgren, S. Stymne & K. Glimelius, 1994b. Intertribal somatic hybrids between *Brassica napus* and *Thlaspi perfoliatum* with high content of the *T. perfoliatum*-specific nervonic acid. Theor. Appl. Genet. 87: 795–804.

Fahleson, J., L. Rahlén & K. Glimelius, 1988. Analysis of plants regenerated from protoplast fusions between *Brassica napus* and *Eruca sativa*. Theor. Appl. Genet. 76: 507–512.

Fahleson, J., U. Lagercrantz, I. Eriksson & K. Glimelius, 1993. Genetic and molecular analysis of sexual progenies from somatic hybrids between *Eruca sativa* and *Brassica napus*. Swedish University of Agricultural Sciences, dissertation, ISBN 91-567-4684-8.

Famelaer, I., Y.Y. Gleba, V.A. Sidorov, V.A. Kaleda, A.S. Parakonny, N.V. Boryshuk, N.N. Cherep, I. Negrutiu & M. Jacobs, 1989. Intrageneric asymmetric hybrids between *Nicotiana plumbaginifolia* and *Nicotiana sylvestris* obtained by 'gamma-fusion'. Plant Sci. 61: 105–117.

Fehér, A., Z. Preiszner, J. Litkey, Gy. Csanádi & D. Dudits, 1992. Characterization of chromosome instability in interspecific somatic hybrids obtained by X-ray fusion between potato (*Solanum tuberosum* L.) and *S. brevidens* Phil. Theor. Appl. Genet. 84: 880–890.

Fish, N., A. Karp & M.G.K. Jones, 1987. Improved isolation of dihaploid *Solanum tuberosum* protoplasts and the production of somatic hybrids between dihaploid *S. tuberosum* and *S. brevidens*. *In Vitro* 23: 575–580.

Fish, N., A. Karp & M.G.K. Jones, 1988. Production of somatic hybrids by electrofusion in *Solanum*. Theor. Appl. Genet. 76: 260–266.

Fluhr, R., D. Aviv, M. Edelman & E. Galun, 1984. Generation of heteroplastidic *Nicotiana* cybrids by protoplast fusion: analysis for plastid recombinant types. Theor. Appl. Genet. 67: 491–497.

Forsberg, J., M. Landgren & K. Glimelius, 1994. Fertile somatic hybrids between *Brassica napus* and *Arabidopsis thaliana*. Plant Sci. 95: 213–223.

Gavrilenko, T.A., N.I. Barbakar & A.V. Pavlov, 1992. Somatic hybridization between *Lycopersicon esculentum* and non-tuberous *Solanum* species of the *Etuberosa* series. Plant Sci. 86: 203–214.

Gebhardt C., E. Ritter, T. Debener, U. Schachtschabel, B. Walkmeier, H. Uhrig & F. Salamini, 1989. RFLP analysis and linkage mapping in *Solanum tuberosum*. Theor. Appl. Genet. 78: 65–75.

Gibson, R.W., E. Pehu, R.D. Woods & M.G.K. Jones, 1990. Resistance to potato virus Y and potato virus X in *Solanum brevidens*. Ann. Appl. Biol. 116: 151–156.

Gibson, R.W., M.G.K. Jones & N. Fish, 1988. Resistance to potato leaf roll virus and potato virus Y in somatic hybrids between dihaploid *Solanum tuberosum* and *S. brevidens*. Theor. Appl. Genet. 76: 113–117.

Gilmour, D.M., M.R. Davey & E.C. Cocking, 1987. Isolation and culture of heterokaryons following fusion of protoplasts from sexually compatible and sexually incompatible *Medicago* species. Plant Sci. 53: 263–270.

Gilmour, D.M., M.R. Davey & E.C. Cocking, 1989. Production of somatic hybrid tissues following chemical and electrical fusion of protoplasts from albino cell suspensions of *Medicago sativa* and *M. borealis*. Plant Cell Rep. 8: 29–32.

Gleba, Y.Y. & F. Hoffmann, 1979. '*Arabidobrassica*': Plant-genome engineering by protoplast fusion. Naturwissenschaften 66: 547–554.

Gleba, Y.Y. & F. Hoffmann, 1980. '*Arabidobrassica*': A novel plant obtained by protoplast fusion. Planta 149: 112–117.

Gleba, Y.Y. & K.M. Sytnik, 1984. Protoplast fusion. Genetic engineering in higher plants. Springer Verlag, Berlin.

Gleba, Y.Y. & L.R. Shlumukov, 1990. Selection of somatic hybrids. p. 257–286. In: P.J. Dix (Ed). Plant Cell Line Selection: Procedures and applications. VCH, Weinheim.

Gleba, Y.Y., S. Hinnisdaels, V.A. Sidorov, V.A. Kaleda, A.S. Parokonny, N.V. Boryshuk, N.N. Cherep, I. Negrutiu & M. Jacobs, 1988. Intergeneric asymmetric hybrids between *Nicotiana plumbaginifolia* and *Atropa belladonna* obtained by 'gamma-fusion'. Theor. App. Genet. 76: 760–766.

Gleddie, S., W.A. Keller & G. Setterfield, 1986. Production and characterization of somatic hybrids between *Solanum melongena* L. and *S. sisymbriifolium* Lam. Theor. Appl. Genet. 71: 613–621.

Glimelius, K., J. Fahleson, M. Landgren, C. Sjödin & E. Sundberg, 1991. Gene transfer via somatic hybridization in plants. Trends in Biotech. 9: 24–30.

Glimelius, K., K. Chen & H.T. Bonnett, 1981. Somatic hybridization in *Nicotiana*: Segregation of organellar traits among hybrid and cybrid plants. Planta 153: 504–510.

Glimelius, K., M. Djupsjöbacka & H. Fellner-Feldegg, 1986. Selection and enrichment of plant protoplast heterokaryons of Brassicaceae by flow sorting. Plant Sci. 45: 133–141.

Grosser, J.W. & F.G. Gmitter, 1990. Somatic hybridization of *Citrus* with wild relatives for germplasm enhancement and cultivar development. HortSci. 25: 147–151.

Gupta, P.P., O. Schieder & M. Gupta, 1984. Intergeneric nuclear gene transfer between somatically and sexually incompatible plants through asymmetric protoplast fusion. Mol. Gen. Genet. 197: 30–35.

Gupta, V., G. Lakshimisita, M.S. Shaila, V. Jagannathan & M.S. Lakshmikumaran, 1992. Characterization of species-specific repeated DNA sequences from *B. nigra*. Theor. Appl. Genet. 84: 397–402.

Guri, A & K C Sink, 1988a Organelle composition in somatic hybrids between an atrazine resistant biotype of *Solanum nigrum* and *Solanum melongena* Plant Sci 58 51–58

Guri, A & K C Sink, 1988b Interspecific somatic hybrid plants between eggplant (*Solanum melongena*) and (*Solanum torvum*) Theor Appl Genet 76 490–496

Guri, A, A Levi & K C Sink, 1988 Morphological and molecular characterization of somatic hybrid plants between *Lycopersicon esculentum* and *Solanum nigrum* Mol Gen Genet 212 191–198

Hall, R D, F A Krens & G J A Roewendal, 1992 DNA radiation damage and asymmetric somatic hybridization Is UV a potential substitute or supplement to ionising radiation in fusion experiments? Physiol Plant 85 319–324

Hammatt, N, A Lister, B Jones, E C Cocking & M R Davey, 1992 Shoot formation from somatic hybrid callus between soybean and a perennial wild relative Plant Sci 85 215–222

Hammatt, N, A Lister, N W Blackhall, J Gartland, T K Ghose, D M Gilmour, J B Power, M R Davey & E C Cocking, 1990 Selection of plant heterokaryons from diverse origins by flow cytometry Protoplasma 154 34–44

Handley, L W, R L Nickels, M W Cameron, P P Moore & K C Sink, 1986 Somatic hybrid plants between *Lycopersicon esculentum* and *Solanum lycopersocoides* Theor Appl Genet 71 691–697

Hayashi, Y, J Kyozuka & K Shimamoto, 1988 Hybrids of rice (*Oryza sativa* L) and wild *Oryza* species obtained by cell fusion Mol Gen Genet 214 6–10

Hein, T, T Przewozny & O Schieder, 1983 Culture and selection of somatic hybrids using an auxotrophic cell line Theor Appl Genet 64 119–122

Helgeson, J P, G J Hunt, G T Haberlach & S Austin, 1986 Somatic hybrids between *Solanum brevidens* and *Solanum tuberosum* Expression of a late blight resistance gene and potato leaf roll resistance Plant Cell Rep 3 212–214

Helgeson, J P, G T Haberlach & M K Ehlenfeldt, 1993 Sexual progeny of somatic hybrids between potato and *Solanum brevidens* Potential for use in breeding programs Am Potato J 70 437–452

Hille, J, M Koornneef, M S Ramanna & P Zabel, 1989 Tomato a crop species amenable to improvement by cellular and molecular methods Euphytica 42 1–23

Imamura, J, M W Saul & I Potrykus, 1987 X-ray irradiation promoted asymmetric somatic hybridisation and molecular analysis of the products Theor Appl Genet 74 445–450

Itoh, K, M Iwabuchi & K Shimamoto, 1991 In situ hybridization with species-specific DNA probes gives evidence for asymmetric nature of *Brassica* hybrids obtained by X-ray fusion Theor Appl Genet 81 356–362

Iwabuchi, M, K K Itoh & K Shimamoto, 1991 Molecular and cytological characterization of repetitive sequences in *Brassica* Theor Appl Genet 81 349–355

Jacobsen, E, M K Daniel, J E M Bergervoet-van Deelen, D J Huigen & M S Ramanna, 1993 The first and second backcross progeny of the intergeneric fusion hybrids of potato after crossing with potato Theor Appl Genet 88 181–186

Jadari, R, D Sihachakr, L Rossignol & G Ducreux, 1992 Transfer of resistance to *Verticillium dahliae* Kleb from *Solanum torvum* S W into potato (*Solanum tuberosum* L) by protoplast electrofusion Euphytica 64 39–47

Jansen, C E, 1993 Selectable cytoplasmic markers in tomato Ph D Thesis, Free University, Amsterdam, The Netherlands

Jones, R A C, 1979 Resistance to potato leaf roll virus in *Solanum brevidens* Potato Res 22 149–152

Kameya, T, H Kanzaki, S Toki & T Abe, 1989 Transfer of radish (*Raphanus sativus* L) chloroplasts into cabbage (*Brassica oleracea* L) by protoplast fusion Jpn J Genet 64 27–34

Kameya, T, N Miyazawa & S Toki, 1990 Production of somatic hybrids between *Solanum melongena* L and *S integrifolium* Poil Japan J Breed 40 429–434

Kao, K N & M R Michayluk, 1974 A method for high-frequency intergeneric fusion of plant protoplasts Planta 115 355–367

Kao, K N, 1977 Chromosomal behaviour in somatic hybrids of soybean *Nicotiana glauca* Mol Gen Genet 150 225–230

Kemble, R J, T L Barsby, R S C Wong & J F Shepard, 1986 Mitochondrial DNA rearrangements in somatic hybrids of *Solanum tuberosum* and *Solanum brevidens* Theor Appl Genet 72 787–793

Kihara, M, K-N Cai, R Ishikawa, T Harada, N Niizeki & K Saito, 1992 Asymmetric somatic hybrid calli between leguminosus species of *Lotus corniculatus* and *Glycine max* and regenerated plants from the calli Japan J Breed 42 55–64

Kinsara, A, S N Patnaik, E C Cocking & J B Power, 1986 Somatic hybrid plants of *Lycopersicon esculentum* Mill and *Lycopersicon peruvianum* Mill J Plant Physiol 125 225–234

Klimaszewska, K & W A Keller, 1988 Regeneration and characterization of somatic hybrids between *Brassica napus* and *Diplotaxis harra* Plant Sci 58 211–222

Koop, H U & H G Schweiger, 1985 Regeneration of plants after electrofusion of selected pairs of protoplasts Eur J Cell Biol 39 46–49

Kubo, T, 1988 Protoplast fusion in tobacco breeding Proceedings of the seminar 'Cell and Tissue culture in field crop improvement' p 49–53, Tsukuba, Japan

Kuchuk, N V, N V Borisyuk & Y Y Gleba, 1990 Isolation and analysis of somatic hybrid cell lines and plants of *Medicago* Abstract VIIIth International Congress on Plant Tissue and Cell Culture p 213 Amsterdam, The Netherlands

Kumar, A & E C Cocking, 1987 Protoplast fusion A novel approach to organelle genetics in higher plants Amer J Bot 74 1289–1303

Kumar, V & M R Davey, 1991 Genetic improvement of legumes using somatic cell and molecular techniques Euphytica 55 157–169

Kumashiro, T & T Kubo, 1986 Cytoplasm transfer of *Nicotiana debneyi* to *N tabacum* by protoplast fusion Jpn J Breed 36 39–48

Kyozuka, J, T Kaneda & K Shimamoto, 1989 Production of cytoplasmic male sterile rice (*Oryza sativa* L) by cell fusion Bio/Technology 7 1171–1174

Landgren, M & K Glimelius, 1990 Analysis of chloroplast and mitochondrial segregation in three different combinations of somatic hybrids produced within Brassicaceae Theor Appl Genet 80 776–784

Lefrançois, C, & Chupeau & J P Bourgin, 1993 Sexual and somatic hybridization in the genus *Lycopersicon* Theor Appl Genet 86 533–546

Lelivelt, C L C, 1993 Introduction of beet cyst nematode resistance from *Sinapis alba* L and *Raphanus sativus* L into *Brassica napus* L (oil-seed rape) through sexual and somatic hybridization Ph D Thesis, Univ of Wageningen, The Netherlands

Lelivelt, C L C & F A Krens, 1992 Transfer of resistance to the beet cyst nematode (*Heterodera schachtii* Schm) into the *Brassica napus* L gene pool through intergeneric somatic hybridization with *Raphanus sativus* L Theor Appl Genet 83 887–894

Lelivelt, C L C, E H M Leunissen, H J Frederiks, J P F G Helsper & F A Krens, 1993 Transfer of resistance to the beet cyst nematode (*Heterodera schachtii* Schm) from *Sinapis alba* L (white mustard) to the *Brassica napus* L gene pool by means of sexual and somatic hybridization Theor Appl Genet 85 688–696

Levi, A, B L Ridley & K C Sink, 1988 Biased organelle transmission in somatic hybrids of *Lycopersicon esculentum* and *Solanum lycopersicoides* Curr Genet 14 177–182

Li, Y & K C Sink, 1992 Cell type determines plastid transmission in tomato intergeneric somatic hybrids Curr Gen 22 167–171

Li, Y-G, G J Tanner, A C Delves & P J Larkin, 1993 Asymmetric somatic hybrid plants between *Medicago sativa* L (alfalfa, lucerne) and *Onobrychis viciifolia* Scop (sainfoin) Theor Appl Genet 87 455–463

Liu, J & K Glimelius, 1994 Production of somatic hybrids and cybrids between *Brassica napus* and *B tournefortii* Abstracts VIIIth International Congress of Plant Tissue and Cell Culture Firenze, Italy, 12–17 June 1994, p 98

Malone, R, G V Horvath, A Cseplo, B Buzas, P J Dix & P Medgyesy, 1992 Impact of the stringency of cell selection on plastid segregation in protoplast fusion-derived *Nicotiana* regenerates Theor Appl Genet 84 866–873

Masson, J, D Lancelin, C Bellini, M Lecerf, P Guerche & G Pelletier, 1989 Selection of somatic hybrids between diploid clones of potato (*Solanum tuberosum* L) transformed by direct gene transfer Theor Appl Genet 78 153–159

Mattheij, W M & K J Puite, 1992 Tetraploid potato hybrids through protoplast fusions and analysis on their performance in the field Theor Appl Genet 83 807–812

Mattheij, W M, R Eijlander, J R A Koning de & K M Loewes, 1992 Interspecific hybridization between the cultivated potato *Solanum tuberosum* L and the wild species *S circaeifolium* subsp *circaeifolium* Bitter exhibiting resistance to *Phytophthora infestans* (Mont) de Bary and *Globodera pallida* (Stone) Behrens Theor Appl Genet 83 459–466

McCabe, P F, L J Dunbar, A Guri & K C Sink, 1993 T-DNA-tagged chromosome 12 in donor *Lycopersicon esculentum* × *L pennellii* is retained in asymmetric somatic hybrids with recipient *Solanum lycopersicoides* Theor Appl Genet 86 377–382

McLennan, M S, P Olesen & J B Power, 1988 Towards the introduction of cytoplasmic male sterility (CMS) into *Brassica napus* through protoplast fusion p 187–188 In K J Puite, J J M Dons, J J Huising, A J Kool, M Koorneef & F A Krens (Eds) Progress in Plant Protoplast Research Kluwer Academic Publishers, London

Medgyesy, P, E Fejes & P Maliga, 1985 Interspecific chloroplast recombination in a *Nicotiana* somatic hybrid Proc Natl Acad Sci USA 82 6960–6964

Medgyesy, P, L Menczel & P Maliga, 1980 The use of cytoplasmic streptomycin resistance chloroplast transfer from *Nicotiana tabacum* into *Nicotiana sylvestris*, and isolation of their somatic hybrids Mol Gen Genet 179 693–698

Melchers, G, M D Sacristan & A A Holder, 1978 Somatic hybrid plants of potato and tomato regenerated from fused protoplasts Carlsberg Res Commun 43 203–218

Melzer, J M & M A O'Connell, 1990 Molecular analysis of the extent of asymmetry in two asymmetric somatic hybrids of tomato Theor Appl Genet 79 193–200

Melzer, J M & M A O'Connell, 1992 Effect of radiation dose on the production of and the extent of asymmetry in tomato asymmetric somatic hybrids Theor Appl Genet 83 337–344

Menczel, L, G Galiba, F Nagy & P Maliga, 1982 Effect of radiation dosage on efficiency of chloroplast transfer by protoplast fusion in *Nicotiana* Genetics 100 487–495

Menczel, L, L S Polsby, E Steinback & P Maliga, 1986 Fusion-mediated transfer of triazine resistant chloroplasts characterization of *Nicotiana tabacum* cybrid plants Mol Gen Genet 205 201–205

Mendis, M H, J B Power & M R Davey, 1991 Somatic hybrids of the forage legumes *Medicago sativa* L and *M falcata* L J Exp Bot 42 1565–1573

Metzlaff, M, W Troebner, F Baldauf, R Schlegel & J Cullum, 1986 Wheat specific repetitive DNA sequences – construction and characterization of four different genomic clones Theor Appl Genet 72 207–210

Mollers, C & G Wenzel, 1992 Somatic hybridization of diploid potato protoplasts as a tool for potato breeding Bot Acta 105 133–139

Mollers, C, U Frei & G Wenzel, 1994 Field evaluation of tetraploid somatic potato hybrids Theor Appl Genet 88 147–152

Nagao, T, 1982 Somatic hybridization by fusion of protoplasts Jpn J Crop Sci 51 35–42

Nagata, T & I Takebe, 1971 Plating of isolated tobacco mesophyll protoplasts on agar medium Planta 99 12–20

Niizeki, M & K Saito, 1989 Callus formation from protoplast fusion between leguminous species of *Medicago sativa* and *Lotus corniculatus* Japan J Breed 39 373–377

Nyman, M & A Wallin, 1988 Plant regeneration from strawberry (*Fragaria ananassa*) mesophyll protoplasts J Plant Physiol 133 375–377

Nyman, M & A Wallin, 1992 Improved culture techniques for strawberry (*Fragaria* × *ananassa* Duch) protoplasts and the determination of DNA content in protoplasts-derived plants Plant Cell Tissue Organ Cult 30 127–133

O'Connell, M A & M R Hanson, 1987 Regeneration of somatic hybrid plants formed between *Lycopersicon esculentum* and *Lycopersicon pennellii* Theor Appl Genet 75 83–89

O'Connell, M A & M R Hanson, 1986 Regeneration of somatic hybrid plants formed between *Lycopersicon esculentum* and *Solanum rickii* Theor Appl Genet 72 59–65

Ochatt, S J, E M Patat Ochatt, E L Rech, M R Davey & J B Power, 1989 Somatic hybridization of sexually incompatible top-fruit tree rootstocks, wild pear (*Pyrus communis* var *pyraster* L) and Colt cherry (*Prunus avium* × *pseudocerasus*) Theor Appl Genet 78 35–41

Oikarinen, S & P H Ryoppy, 1992 Somatic hybridization of *Brassica campestris* and *Barbarea* species Abstract at the XIIIth EUCARPIA Congress, Angers, France 6–11 July 1992, p 261–262

Okamura, M, 1988 Regeneration and evaluation of somatic hybrid plants between *Solanum tuberosum* and *Lycopersicon pimpinellifolium* p 213–214 In K J Puite, J J M Dons, H J Huising, A c Kool, M Koornneef & F A Krens (Eds) Progress in Protoplast Research Kluwer Academic Publishers, London

Ozias-Akins, P, R J Ferl & I K Vasil, 1986 Somatic hybridization in the Gramineae *Pennisetum americanum* (L) K Schum (pearl millet) + *Panicum miximum* Jacq (Guinea grass) Mol Gen Genet 203 365–370

Pandeya, R S, G C Douglas, W A Keller, G Setterfield & Z A Patrick, 1986 Somatic hybridization between *Nicotiana rustica* and *N tabacum* Development of tobacco breeding strains with disease resistance and elevated nicotine content Z Pflanzenzuchtg 96 346–352

Parokonny, A S, A Y Kenton, Y Y Gleba & M D Bennett, 1992 Genome reorganization in *Nicotiana* asymmetric somatic hybrids analysed by *in situ* hybridization Plant J 2 863–874

Pehu, E, M Thomas, T Poutala, A Karp & M G K Jones, 1990a Species-specific sequences in the genus *Solanum* identification, characterization, and application to study somatic hybrids of *S brevidens* and *S tuberosum* Theor Appl Genet 80 693–689

Pehu, E, R W Gibson, M G K Jones & A Karp, 1990b Studies on the genetic bases of resistance to potato leaf roll virus, potato virus Y and potato virus X in *Solanum brevidens* using somatic hybrids of *Solanum brevidens* and *Solanum tuberosum* Plant Sci 69 95–101

Pelletier, G, C Primard, M Ferault, F Vedel, P Cherit, M Renard & R Delourme, 1988 Use of protoplasts in plant breeding Cytoplasmic aspects p 169–176 In K J Puite, J J M Dons, H J Huising, A c Kool, M Koornneef, F A Krens (Eds) Progress in Protoplast Research Kluwer Academic Publishers, London

Perez-Vincente, R, L Petris, M Osusky, I Potrykus & G Spangenberg, 1992 Molecular and cytogenetic characterization of repetitive sequences from *Lolium* and *Festuca* applications in the analysis of *Festulolium* Theor Appl Genet 84 145–154

Perl, A, D Aviv & E Galun, 1990 Protoplast-fusion-derived CMS potato cybrids Potential seed-parents for hybrid, true potato seeds J Hered 81 438–442

Piastuch, W C & G W Bates, 1990 Chromosomal analysis of *Nicotiana* asymmetric hybrids by dot blotting and *in situ* hybridization Mol Gen Genet 222 97–103

Pijnacker, L P, M A Ferwerda, K J Puite & J G Schaart, 1989 Chromosome elimination and mutation in tetraploid somatic hybrids of *Solanum tuberosum* and *Solanum phureja* Plant Cell Rep 8 82–85

Pijnacker, L P, M A Ferwerda, K J Puite & S Roest, 1987 Elimination of *Solanum phureja* nucleolar chromosomes in *S tuberosum* × *S phureja* somatic hybrids Theor Appl Genet 73 878–882

Polgar, Zs, J Preiszner, D Dudits & A Feher, 1993 Vigorous growth of fusion products allows highly efficient selection of interspecific potato somatic hybrids molecular proofs Plant Cell Rep 12 399–402

Preiszner, J, A Fehér, O Veisz, J Sutka & D Dudits, 1991 Characterization of morphological variation and cold resistance in inter specific somatic hybrids between potato (*Solanum tuberosum* L) and *S brevidens* Phil Euphytica 57 37–49

Primard, C, F Vedel, C Mathieu, G Pelletier & A M Chevre, 1988 Interspecific somatic hybridization between *Brassica napus* and *Brassica hirta* (*Sinapis alba* L) Theor Appl Genet 75 546–552

Puite, K J & J G Schaart, 1993 Nuclear genomic composition of asymmetric fusion products between irradiated transgenic *Solanum brevidens* and *S tuberosum* limited elimination of donor chromosomes and polyploidization of the recipient genome Theor Appl Genet 86 237–244

Puite, K J, 1992 Progress in plant protoplast research Physiol Plant 85 403–410

Puite, K J, P Van Wikselaar & H Verhoeven, 1985 Electrofusion, a simple and reproducible technique in somatic hybridization of *Nicotiana plumbaginifolia* mutants Plant Cell Rep 4 274–276

Puite, K J, S Roest & L P Pijnacker, 1986 Somatic hybrid potato plants after electrofusion of diploid *Solanum tuberosum* and *Solanum phureja* Plant Cell Rep 5 262–265

Puite, K J, W T Broeke & J Schaart, 1988 Inhibition of cell wall synthesis improves flow cytometric sorting of potato heterofusions resulting in hybrid plants Plant Sci 56 61–68

Pupilli, F, G M Arcioni, G M Scarpa, F Damiani & S Arcioni, 1992 Production of interspecific somatic hybrid plants in the genus *Medicago* through protoplast fusion Theor Appl Genet 84 792–797

Pupilli, F, S Arcioni & F Damiani, 1991 Protoplast fusion in the genus *Medicago* and isozyme analysis of parental and somatic hybrid cell lines Plant Breed 106 122–131

Ramulu, K S, P Dijkhuis, H A Verhoeven, I Famelaer & J Blaas, 1992 Microprotoplast isolation, enrichment and fusion for partial genome transfer in plants Physiol Plant 85 315–318

Ratushnyak, Y I, S A Latypov, A M Samoylov, N M Piven & Y Y Gleba, 1991 Introgressive hybridization of tomatoes by 'gamma fusion' of *Lycopersicon esculentum* Mill and *Lycopersicon peruvianum* var *dentatum* Dun protoplasts Plant Sci 73 65–78

Rick, C M, 1982 The potential of exotic germplasm for tomato improvement p 1–28 In I K Vasil, W R Scowcroft & K J Frey (Eds) Plant Improvement and Somatic Cell Genetics Academic Press, New York

Roest, S & L J W Gilissen, 1989 Plant regeneration from protoplasts a literature review Acta Bot Neerl 38 1–23

Roest, S & L J W Gilissen, 1993 Regeneration from protoplasts – a supplementary literature review Acta Bot Neerl 42 1–23

Rosen, B, C Hallden & W K Heneen, 1988 Diploid *Brassica napus* somatic hybrids Characterization of nuclear and organellar DNA Theor Appl Genet 76 197–203

Sacristan, M D, M Gerdemann-Knorck & O Schieder, 1989 Incorporation of hygromycin resistance in *Brassica nigra* and its transfer to *B napus* through asymmetric protoplast fusion Theor Appl Genet 78 194–200

Sakomoto, K & T Taguchi, 1991 Regeneration of intergeneric somatic hybrid plants between *Lycopersicon esculentum* and *Solanum muricatum* Theor Appl Genet 81 509–513

San, L H, F Vedel, D Sihachakr & Re'my, 1990 Morphological and molecular characterization of fertile tetraploid somatic hybrids produced by protoplast electrofusion and PEG-induced fusion between *Lycopersicon esculentum* Mill and *Lycopersicon peruvianum* Mill Mol Gen Genet 221 17–26

Saul, M W & I Potrykus, 1984 Species-specific repetitive DNA used to identify interspecific somatic hybrids Plant Cell Rep 3 65–67

Scandalios, J G & J C Sorenson, 1977 Isozymes in plant tissue culture p 719–730 In J Reinert & Y P S Bajaj (Eds) Applied and Fundamental Aspects of Plant Cell Tissue and Organ Culture Springer Verlag, Berlin

Schenk, H R & G Robbelen, 1982 Somatic hybrids by fusion of protoplasts from *Brassica oleracea* and *B campestris* Z Pflanzenzuchtg 89 278–288

Schieder, O, 1982 Somatic hybridization a new method for plant improvement p 239–253 In I K Vasil, W R Scowcroft & K J Frey (Eds) Plant Improvement and Somatic Cell Genetics Academic Press, New York

Schweizer, G, M Ganal, Ninnemann & V Hemleben, 1988 Species specific DNA sequences for identification of somatic hybrids between *Lycopersicon esculentum* and *Solanum acaule* Theor Appl Genet 75 679–684

Serraf, I, D Sihachakr, G Ducreux, S C Brown, M Allot, N Barghi & L Rossignol, 1991 Interspecific somatic hybridization in potato by protoplast electrofusion Plant Sci 76 115–126

Shepard, J F, D Bidney, T Barsby & R Kemble, 1983 Genetic transfer in plants through interspecific protoplast fusion Science 219 683–688

Sidorov, V A, M K Zubko, A A Kuchko, I K Komarnitsky & Y Y Gleba, 1987 Somatic hybridization in potato use of gamma-irradiated protoplasts of *Solanum pinnatisectum* in genetic reconstruction Theor Appl Genet 74 364–368

Sihachakr, D, R Haicour, I Serraf, E Barrientos, C Herbreteau, G Ducreux, L Rossignol & V Souvannavong, 1988 Electrofusion for the production of somatic hybrid plants of *Solanum melongena* L and *Solanum khasianum* C B Clark Plant Sci 57 215–223

Sihachakr, D, R Haicour, M H Chaput, E Barrientos, G Ducreux & L Rossignol, 1989 Somatic hybrid plants produced by electrofusion between *Solanum melongena* L. and *Solanum torvum* Sw. Theor. Appl. Genet. 77 1–6.

Sikdar, S.R., G. Chatterjee, S. Das & S.K. Sen, 1990. '*Erussica*', the intergeneric fertile somatic hybrid developed through protoplast fusion between *Eruca sativa* Lam. and *Brassica juncea* (L.) Czern. Theor. Appl. Genet. 79 561–567.

Simoens, C.R., J. Gielen, M. van Montagu & D. Inzé, 1988. Characterization of highly repetitive sequences of *Arabidopsis thaliana*. Nucl. Acids Res. 16 6753–6766.

Sjödin, C. & K. Glimelius, 1988. Screening for resistance to blackleg *Phoma lingam* (Tode ex Fr.) Desm. within Brassicaceae. J. Phytopathology 123 322–332.

Sjodin, C. & K. Glimelius, 1989a. *Brassica naponigra*, a somatic hybrid resistant to *Phoma lingam*. Theor. Appl. Genet. 77 651–656.

Sjödin, C. & K. Glimelius, 1989b. Transfer of resistance against *Phoma lingam* to *Brassica napus* by asymmetric somatic hybridization combined with toxin selection. Theor. Appl. Genet. 78 513–520.

Skarzhynskaya, M., J. Fahleson & K. Glimelius, 1994. Somatic hybridization in the family of Brassicaceae production of intertribal hybrids between *Brassica napus* L. and *Lesquerella fendleri* (*Gray*) Wats. Abstracts VIIIth International Congress of Plant Tissue and Cell Culture, Firenze, Italy, 12–17 June 1994, p. 101

Spangenberg, G., M.P. Valle's, Z.Y. Wang, P. Montavon, J. Nagel & I. Potrykus, 1994. Asymmetric somatic hybridization between tall fescue (*Festuca arundinacea* Schreb.) and irradiated Italian ryegrass (*Lolium multiflorum* Lam.) protoplasts. Theor. Appl. Genet 77 509–519.

Sundberg, E. & K. Glimelius, 1986. A method for production of interspecific hybrids within Brassicaceae via somatic hybridization, using resynthesis of *Brassica napus* as a model. Plant Sci 43 155–162.

Sundberg, E. & K. Glimelius, 1991a. Production of cybrid plants within Brassicaceae by fusing protoplasts and plasmolytically induced cytoplasts. Plant Sci. 79 205–216.

Sundberg, E. & K. Glimelius, 1991b. Effects of parental ploidy level and genetic divergence on chromosome elimination and chloroplast segregation in somatic hybrids within Brassicaceae Theor. Appl. Genet 83 81–88.

Sundberg, E., M. Landgren & K. Glimelius, 1987. Fertility and chromosome stability in *Brassica napus* resynthesised by protoplast fusion. Theor. Appl. Genet. 75 96–104.

Sundberg, E., U. Lagercranz & K. Glimelius, 1991. Effects of cell type used for fusion on chromosome elimination and chloroplast segregation in *Brassica oleracea* (+) *B napus* hybrids. Plant Sci 78 89–98.

Tabaeizadeh, Z., R.J. Ferl & I.K. Vasil, 1986. Somatic hybridization in the Gramineae *Saccharum officinarum* L. (sugarcane) and *Pennisetum americanum* (L.) K. Schum. (pearl millet). Proc Natl Acad. Sci. USA 83 5616–5619.

Taguchi, T. & T. Kameya, 1986. Production of somatic hybrid plants between cabbage and chinese cabbage through protoplast fusion Japan. J. Breed. 36 185–189

Takamizo, T., G. Spangenberg, K Suginobu & I Potrykus, 1991 Intergeneric somatic hybridization in Gramineae somatic hybrid plants between tall fescue (*Festuca arundinacea* Schreb.) and Italian ryegrass (*Lolium multiflorum* Lam) Mol Gen. Genet. 231 1–6

Takebe, I., G Labib & G. Melchers, 1971. Regeneration of whole plants from isolated mesophyll protoplasts of tobacco Naturwissenschaften 58 318–320

Tan, M -L M.C , 1987. Somatic hybridization and cybridization in some Solanaceae. PhD Thesis, Free University, Amsterdam, The Netherlands.

Téoulé, E., 1983. Hybridation somatique entre *Medicago sativa* L. et *Medicago falcata* L.C.R. Acad. Sci. Paris 297 13–16.

Terada, K., J. Kyozuka, S. Nishibayashi & K. Shimamoto, 1987a. Plantlet regeneration from somatic hybrids of rice (*Oryza sativa*) and barnyard grass (*Echinochloa oryzicola* Vasing.). Mol Gen. Genet. 210 39–43.

Terada, K., Y. Yamashita, S. Nishibayashi & K. Shimamoto, 1987b. Somatic hybrids between *Brassica oleracea* and *B. campestris* selection by the use of iodoacetamide inactivation and regeneration ability. Theor. Appl. Genet. 73 379–384.

Thach, N.Q., U Frei & G. Wenzel, 1993. Somatic fusion for combining virus resistances in *Solanum tuberosum* L. Theor. Appl. Genet. 85 863–867.

Thanh, N.D , A Páy, M A. Smith, P. Medgyesy & L. Márton, 1988. Intertribal chloroplast transfer by protoplast fusion between *Nicotiana tabacum* and *Salpiglossis sinuata*. Mol. Gen. Genet. 213 186–190

Thomas, M.R , L B. Johnson & F.F. White, 1990. Selection of interspecific somatic hybrids of *Medicago* by using *Agrobacterium*-transformed tissues Plant Sci. 69 189–198.

Thomzik, J E & R Hain, 1988. Transfer and segregation of triazine tolerant chloroplasts in *Brassica napus* L. Theor. Appl. Genet. 76 165–171.

Toriyama, K. & K. Hinata, 1988. Diploid somatic-hybrid plants regenerated from rice cultivars. Theor. Appl. Genet. 76 665–668

Toriyama, K , K Hinata & T Kameya, 1987a. Selection of a universal hybridizer in *Sinapis turgida* Del. and regeneration of plantlets from somatic hybrids with *Brassica* species. Planta 170 308–313.

Toriyama, K , K. Hinata & T Kameya, 1987b. Production of somatic hybrid plants, '*Brassicomoricandia*' through protoplast fusion between *Moricandia arvensis* and *Brassica oleracea* Plant Sci. 48 123–128

Tsunoda, S , 1980. Eco-physiology of wild and cultivated forms in *Brassica* and allied genera. p. 109–120. In S. Tsunoda, K. Hinata & Gomez-Campo (Eds) *Brassica* Crops and Wild Allies Jpn Sci Soc Press, Tokyo

Turpin, C , 1986 Attempt of male cytoplasmic sterility introduction by intergeneric fusion in cultivated tomato. Acta Hortic. 191 377–379

Vamling, K & K. Glimelius, 1990. Regeneration of plants from protoplasts of oilseed *Brassica* crops. p. 385–417. In Y.P.S. Bajaj (Ed) Biotechnology in Agriculture and Forestry, Vol 10. Springer-Verlag, Berlin.

Vasil, I.K & V Vasil, 1992 Advances in cereal protoplast research. Physiol. Plant 85 279–283.

Vasil, V , R. Ferl & I.K Vasil, 1988. Somatic hybridization in the Gramineae *Triticum monococcum* L. (Einkorn) + *Pennisetum americanum* (L.) K Schum (pearl millet). J Plant Physiol. 132 160–163.

Waara, S , A. Wallin & T. Eriksson, 1991. Production and analysis of intraspecific somatic hybrids of potato (*Solanum tuberosum* L). Plant Sci 75 107–115

Waara, S , H. Tegelstrom, A. Wallin & T. Eriksson, 1989. Somatic hybridization between anther-derived dihaploid clones of potato (*Solanum tuberosum* L.) and the identification of hybrid plants by isozyme analysis Theor Appl Genet 77 49–56.

Waara, S , L P Pijnacker, M. Ferwerda, A Wallin & T. Eriksson, 1992 A cytogenetic and phenotypic characterization of somatic hybrid plants obtained after fusion of two different dihaploid clones of potato (*Solanum tuberosum* L.) Theor. Appl. Genet. 85: 470–479.

Wallin, A., K. Glimelius & T. Eriksson, 1974. The induction of aggregation and fusion of *Daucus carota* protoplasts by polyethylene glycol. Z. Pflanzenphysiol. 74: 64–80.

Walton, P.D. & D.C.W. Brown, 1988. Electrofusion of protoplasts and heterokaryon survival in the genus *Medicago*. Plant Breed. 101: 137–142.

Webb, K.J., 1988. Recently developments in the regeneration of agronomically important crops from protoplasts. p. 27–31. In: K.J. Puite, J.J.M. Dons, H.J. Huising, A.c. Kool, M. Koornneef & F.A. Krens (Eds). Progress in Protoplast Research. Kluwer Academic Publishers, London.

Wenzel, G., O. Schieder, T. Przewozny, S.K. Sopory & G. Melchers, 1979. Comparison of single cell culture derived *Solanum tuberosum* L. plants and a model for their application in breeding programs. Theor. Appl. Genet. 55: 49–55.

Wijbrandi, J., A. Posthuma, J.M. Kok, R. Rijken, J.G.M. Vos & M. Koornneef, 1990a. Asymmetric somatic hybrids between *Lycopersicon esculentum* and irradiated *Lycopersicon peruvianum*. 1. Cytogenetics and morphology. Theor. Appl. Genet. 80: 305–312.

Wijbrandi, J., A.M.A. Wolters & M. Koornneef, 1990b. Asymmetric somatic hybrids between *Lycopersicon esculentum* and *Lycopersicon peruvianum*. Theor. Appl. Genet. 80: 665–672.

Wijbrandi, J., P. Zabel & M. Koornneef, 1990c. Restriction fragment length polymorphism analysis of somatic hybrids between *Lycopersicon esculentum* and irradiated *L. peruvianum*: Evidence for limited donor and genome elimination and extensive chromosome rearrangements. Mol. Gen. Genet. 222: 270–277.

Wijbrandi, J., W. Van Capelle, C.J. Hanhart, E.P. Van Loenen Martinet-Schuringa & M. Koornneef, 1990d. Selection and characterization of somatic hybrids between *Lycopersicon esculentum* and *Lycopersicon peruvianum*. Plant Sci. 70: 197–208.

Williams, C.E., G.J. Hunt & J.P. Helgeson, 1990. Fertile somatic hybrids of *Solanum* species: RFLP analysis of a hybrid and its sexual progeny from crosses with potato. Theor. Appl. Genet. 80: 545–551.

Wolters, A.M.A., H.C.H. Schoenmakers, J.J.M. Van der Meulen-Muisers, E. Van der Knaap, F.H.M. Derks, M. Koornneef & A. Zelcer, 1991. Limited DNA elimination from the irradiated potato parent in fusion products of albino *Lycopersicon esculentum* and *Solanum tuberosum*. Theor. Appl. Genet. 83: 225–232.

Wright, R.L., D.A. Somers & R.L. McGraw, 1987. Somatic hybridization between birdsfoot trefoil (*Lotus corniculatus* L.) and *L. conimbricencis*. Theor. Appl. Genet. 75: 51–156.

Xu, Y.S., M. Murto, R. Dunckley, M.G.K. Jones & E. Pehu, 1993a. Production of asymmetric hybrids between *Solanum tuberosum* and irradiated *S. brevidens*. Theor. Appl. Genet. 85: 729–734.

Xu, Y.S., M.S. Clark & E. Pehu, 1993b. Use of RAPD markers to screen somatic hybrids between *Solanum tuberosum* and *S. brevidens*. Plant Cell Rep. 12: 107–109.

Yamashita, Y., R. Terada, S. Nishibayashi & K. Shimamoto, 1989. Asymmetric somatic hybrids of *Brassica*: partial transfer of *B. campestris* genome into *B. oleracea* by cell fusion. Theor. Appl. Genet. 77: 189–194.

Yang, Z.-Q., T. Shikanai, K. Mori & Y. Yamada, 1989. Plant regeneration from cytoplasmic hybrids of rice (*Oryza sativa* L.). Theor. Appl. Genet. 77: 305–310.

Zelcer, A., D. Aviv & E. Galun, 1978. Interspecific transfer of cytoplasmic male sterility by fusion between protoplasts of normal *Nicotiana sylvestris* and X-ray irradiated protoplasts of male-sterile *N. tabacum*. Z. Pflanzenphysiol. 90: 397–407.

Zhang, H.B. & J. Dvorak, 1989. Isolation of repeated DNA sequences from *Lophopyrum elongatum* for detection of *Lophopyrum* chromatin in wheat genomes. Genome 33: 283–293.

Ahman, I., 1993. A search for resistance to insects in spring oilseed rape. IOBC/WPRS Bulletin, 16(5): 36–46.

Euphytica **85:** 235–245, 1995.

Intergeneric symmetric and asymmetric somatic hybridization in *Festuca* and *Lolium*

G. Spangenberg*, Z.Y. Wang, G. Legris, P. Montavon, T. Takamizo[1], R. Pérez-Vicente[2], M.P. Vallés[3], J. Nagel & I. Potrykus
Institute for Plant Sciences, Swiss Federal Institute of Technology, CH-8092 Zürich, Switzerland; [1] present address: National Grassland Research Institute, Senbonmatsu 768, Nishinasunomachi, Tochigi, Japan 329-27; [2] present address: Dpto. Fisiología Vegetal, Facultad de Ciencias, Universidad de Córdoba, 14004 Córdoba, Spain; [3] present address: Dpto. Genética y Producción Vegetal, Estación Experimental de Aula Dei, C.S.I.C., 50080 Zaragoza, Spain (author for correspondence)*

Key words: intergeneric somatic hybrids, forage grasses, fescue, *Festuca arundinacea, F. rubra*, ryegrasses, *Lolium multiflorum, L. perenne, Alopecurus pratensis*, species-specific repetitive DNA sequences

Summary

Intergeneric symmetric and asymmetric somatic hybrids have been obtained by fusion of metabolically inactivated protoplasts from embryogenic suspension cultures of tall fescue (*Festuca arundinacea* Schreb.) and unirradiated or 10–500 Gy-irradiated protoplasts from non-morphogenic cell suspensions of Italian ryegrass (*Lolium multiflorum* Lam.). Genotypically and phenotypically different somatic hybrid *Festulolium* mature flowering plants were regenerated.

Species-specific sequences from *F. arundinacea* and *L. multiflorum* being dispersed and evenly-represented in the corresponding genomes were isolated and used for the molecular characterization of the nuclear make-up of the intergeneric, somatic *Festulolium* plants recovered. The irradiation of Italian ryegrass protoplasts with ≤ 250 Gy X-rays prior to fusogenic treatment favoured the unidirectional elimination of most or part of the donor chromosomes. Irradiation of *L. multiflorum* protoplasts with 500 Gy produced highly asymmetric (over 80% donor genome elimination) nuclear hybrids and clones showing a complete loss of donor chromosomes.

The RFLP analysis of the organellar composition in symmetric and asymmetric tall fescue (+) Italian ryegrass regenerants confirmed their somatic hybrid character and revealed a bias towards recipient-type organelles when extensive donor nuclear genome elimination had occurred.

Approaches aimed at improving persistence of ryegrasses based on asymmetric somatic hybridization with largely sexually-incompatible grass species (*F. rubra* and *Alopecurus pratensis*), and at transferring the cytoplasmic male sterility trait by intra- and inter-specific hybridization in *L. multiflorum* and *L. perenne*, have been undertaken.

Abbreviations: cpDNA – chloroplast DNA, CMS – cytoplasmic male sterility, 2,4-D – 2,4-dichlorophenoxy-acetic acid, IOA – iodoacetamide, mtDNA – mitochondrial DNA, RFLP – restriction fragment length polymorphism

Introduction

Several species in the genera *Festuca* and *Lolium* are commonly grown cool-season perennial forage and turf grasses of the temperate region. Tall fescue (*F. arundinacea* Schreb.) has a wide range of distribution since it shows good persistence and tolerance to various environmental stresses. Concerning quality and palatability however, it compares unfavourably with the widely used ryegrasses: Italian ryegrass (*L. multiflorum* Lam.) and perennial ryegrass (*L. perenne* L.). For many years *Festulolium* hybrids have been produced by conventional crossing in order to combine the agronomically-desirable traits from fescues and ryegrasses (Crowder, 1953). However, sexual hybrids between *L. multiflorum* and *F. arundinacea* have been readily obtained only unidirectionally (Italian ryegrass × tall fescue), while the reciprocal cross has been shown to be extremely difficult (Eizenga & Buckner, 1986). In addition, chromosomal instabilities and the poor female fertility observed in the early amphiploid *Festulolium* plants derived from these octaploid sexual hybrids have so far severly limited the release of valuable materials from these crosses.

Somatic hybridization by protoplast fusion shows promising solutions to these and other problems faced in conventional breeding programmes in ryegrasses and fescues, such as: 1) the transfer of genes from wild relatives to cultivated plants, by-passing existing sexual crossing barriers; 2) the rapid directed combination of partial genomes from different genetic sexually-(in)compatible origins that could serve as bridges to transfer specific traits, and 3) the transfer of cytoplasmically-encoded characters without requiring many backcrosses.

Here we summarize the present state of our research on: 1) the establishment of experimental protocols for the reproducible recovery of mature symmetric and asymmetric somatic *Festulolium* plants; 2) the isolation and characterization of species-specific, repetitive nuclear DNA sequences for the analysis of the genomic composition of sexual and somatic hybrids in the *Festuca-Lolium* complex; and 3) the characterization of the nuclear and organellar constitution of symmetric and asymmetric somatic *Festulolium* hybrids.

Materials and methods

Plant material, establishment of suspension cultures and isolation of protoplasts

Hexaploid (2n = 42) tall fescue (*Festuca arundinacea* Schreb.) cultivars Fawn and Nanryo, and diploid (2n = 14) Italian ryegrass (*Lolium multiflorum* Lam.) cultivars Gorka Norodova and Waseoaba were used for the establishment of callus and suspension cultures as described earlier (Takamizo et al., 1990).

Protoplasts were isolated from embryogenic cell suspensions of tall fescue and from cell lines of Italian ryegrass as previously reported (Spangenberg et al., 1994a). Italian ryegrass cells were X-ray-irradiated at doses of 10, 25, 50, 100, 250 and 500 Gy with an X-ray apparatus (Müller MG150, Type 70526/41; 80 kV, 17 mA, 1900 $Rmin^{-1}$, 0.2 mm Al-filter; Müller GmbH, Hamburg, FRG). Control unirradiated cell suspensions of Italian ryegrass were also used. Tall fescue protoplasts were resuspended for metabolic inactivation in WF solution (0.6 M mannitol, 10 mM $CaCl_2$, pH 5.8) containing 10 mM IOA and incubated at 4° C for 15 min. After IOA treatment, tall fescue protoplasts were washed twice by centrifugation in 0.6 M mannitol and used in fusion experiments.

Fusion of protoplasts

Fusion experiments were performed with a commercial electrofusion setup (Elektro-Zellfusion CFA 400, Krüss, Hamburg, FRG) using the following conditions: ac-field (1 MHz, 80 V/cm for 30 s) followed by dc-pulses (0,75 kV/cm, 1–2 pulses, 30μs each) (Spangenberg et al., 1994a). Five to ten minutes after electrofusion, protoplasts were collected by centrifugation (80 g for 7 min), resuspended in 2.0 ml of double-concentrated AA medium (Müller & Grafe, 1978) supplemented with 2 mg/l 2,4-D and 0.6 M glucose, and plated on agarose solidified protoplast culture medium (Wang et al., 1992).

Culture of protoplasts after fusion and plant regeneration from putative fusants-derived colonies

Protoplasts were cultured using the agarose bead-type culture method (Shillito et al., 1983) with nurse cells (Kyozuka et al., 1987) as previously described by Wang et al. (1993a). After about one month in culture, colonies were transferred onto MS medium (Murashige & Skoog, 1962) supplemented with 1 mg/l 2,4-D, 500 mg/l casein hydrolysate, 90 mM sucrose solidified with 0.8% (w/v) agarose. About four weeks later proliferating calli were placed on MSK medium consisting of MS basal medium supplemented with 90 mM sucrose and 0.2 mg/l kinetin. The regenerated shoots were then transferred to hormone-free MS medium for rooting. All cultures were kept under fluorescent light conditions (40 $\mu mol\ m^{-2}\ s^{-1}$) with a 16 h photoperiod at 25° C. Rooted plants were then potted in soil and grown until maturity under greenhouse conditions (natural daylength; 23° C/18° C).

DNA isolation and gel electrophoresis

Total genomic cellular DNA was isolated from freeze-dried, regenerating calli, and leaf material from shoot cultures and greenhouse-grown, symmetric and asymmetric somatic hybrids and their parental lines. Isolation and digestion of genomic DNA was performed according to Lichtenstein & Draper (1985). Restriction enzyme analysis, gel electrophoresis and DNA blotting were carried out following standard protocols (Sambrook et al., 1989).

Hybridization experiments

Southern blot hybridization experiments using digested (EcoRI or BamHI), total genomic DNA from putative symmetric and asymmetric somatic hybrids and their parents were performed following standard protocols as described in Sambrook et al. (1989). Hybridization probes were [^{32}P]dATP-labeled by random priming (Feinberg & Vogelstein, 1983).

Repetitive and evenly-dispersed species-specific sequences of *L. multiflorum* (LMH2 and LMB4) and *F. arundinacea* (FAH1) (Perez-Vicente et al., 1992) were used for the characterization by quantitative dot blot analysis of the nuclear composition of putative tall fescue (+) Italian ryegrass, somatic hybrid regenerating clones (Spangenberg et al., 1994a).

For the characterization of the organellar genomes of somatic hybrids, different plant mtDNA gene-specific probes (*coxI, cox2, cox3, atpA, atp6* and *atp9*) and one cpDNA gene-specific clone (*rbcL*) were used.

Isolation of species-specific repetitive DNA sequences

The construction of DNA probes and the isolation of repetitive clones specific to *L. multiflorum, F. arundinacea, F. rubra* and *Alopecurus pratensis* was basically performed according to Saul & Potrykus (1984). Colonies containing inserts of plant DNA were picked at random and used for colony hybridization screenings. Two replicas of the total bank of clones were produced on Biodyne (Pall, Glen Cove, USA) filters and hybridized with probes created by random priming of total DNA. Clones preliminarily identified as species-specific were further tested as described earlier (Pérez-Vicente et al., 1992). Inserts from species-specific, repetitive DNA clones were sequenced by the dideoxynucleotide chain termination procedure of Sanger et al. (1980) using the T7 sequencing kit (Pharmacia, Sweden).

Chromosome preparation and in situ *hybridization*

Chromosome preparations were made from shoot meristems isolated and pre-treated following with modifications de Lautour & Cooper (1971). *In situ* hybridization with digoxigenin-labeled probes and detection was done according to the manufactor's instructions (Boehringer Mannheim, FRG) with modifications (Perez-Vicente et al., 1992).

Results

Protoplast fusion, culture of fusion products and regeneration of symmetric and asymmetric somatic hybrid Festulolium *plants*

Protoplasts of tall fescue (*F. arundinacea*) and Italian ryegrass (*L. multiflorum*) were readily-obtained from embryogenic (Fig. 1A) and non-morphogenic (Fig. 1B) cell suspensions, respectively. The protoplast yield varied in the range 0.5–2 $\times$ 10^6/g fresh weight cells depending on the suspension culture used.

The selection scheme used for the enrichment of protoplast fusion products to generate symmetric and asymmetric somatic hybrids, was based on metabolically-inactivated, totipotent protoplasts of tall fescue and (unirradiated or X-ray irradiated) non-morphogenic protoplasts of Italian ryegrass, respectively. Dose response experiments on the inactivation of tall fescue protoplasts with IOA in the range of 2–20 mM, showed that 10 mM IOA treatment for 15 min completely inhibited colony formation. Dose response experiments using X-ray irradiation of Italian ryegrass protoplasts in the range of 50–500 Gy, revealed that doses of 100 Gy and higher, fully prevented colony formation.

No regeneration of green plantlets occurred in the following control experiments: 1) unirradiated protoplasts isolated from non-morphogenic suspension cultures of Italian ryegrass; 2) X-ray-irradiated protoplasts from suspensions as in 1); 3) IOA-inactivated protoplasts of tall fescue isolated from morphogenic suspensions; and 4) mixtures of unfused protoplasts from 1) with 3) or from 2) with 3).

Fusion experiments performed using unirradiated (symmetric fusions) or 10–500 Gy X-ray-irradiated (asymmetric fusions) Italian ryegrass protoplasts (Fig. 1C) led to colony formation within 3–4 weeks in bead type culture (Fig. 1DE). After a further 4–6 weeks on proliferation and regeneration medium, more than 60

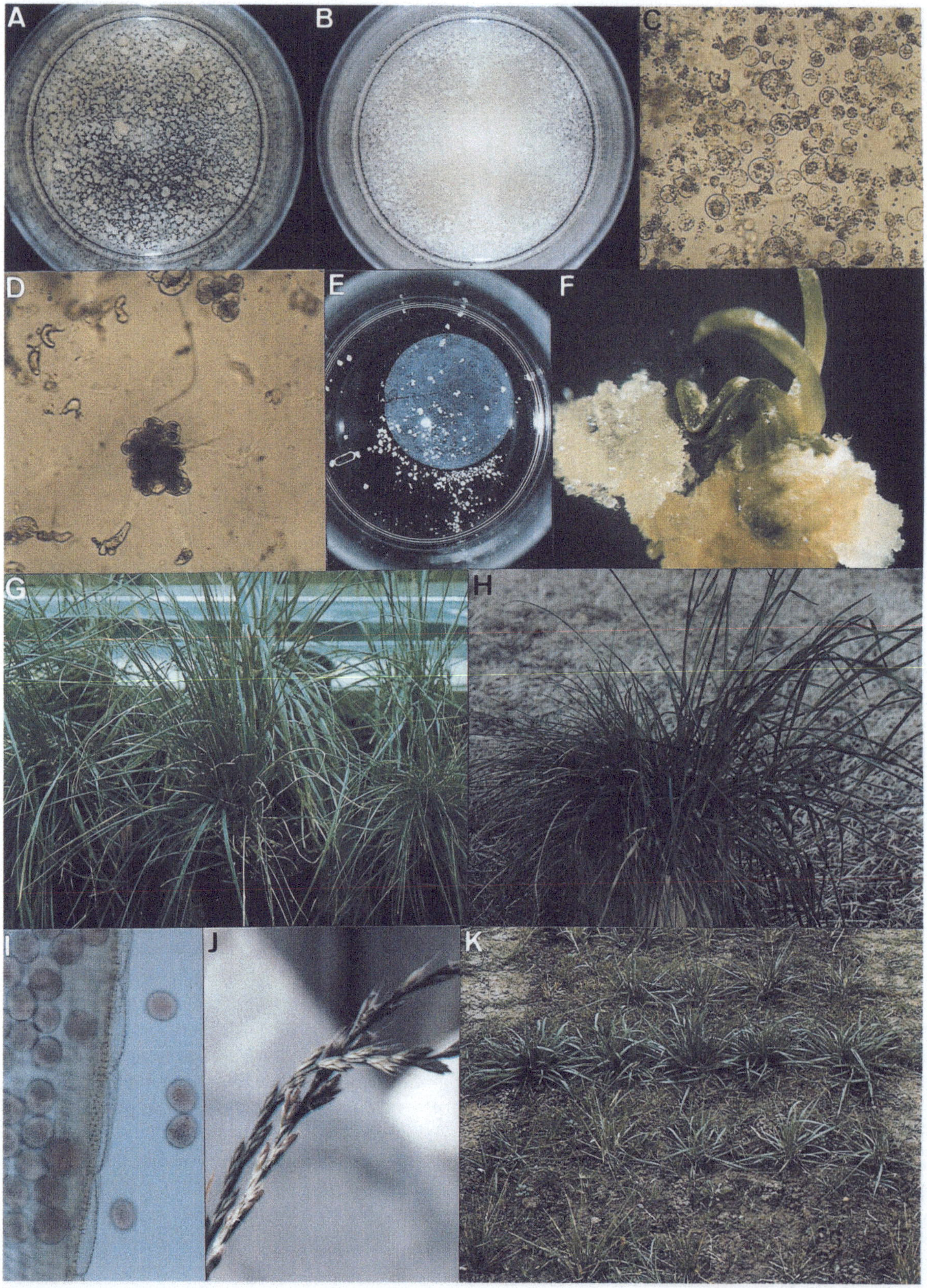

←

Fig. 1. Recovery of symmetric and asymmetric somatic hybrid plants between *Festuca arundinacea* Schreb. and *Lolium multiflorum* Lam. A) Embryogenic cell suspension of *F. arundinacea* used for isolation of totipotent protoplasts; B) Non-morphogenic cell suspension of *L. multiflorum* used for isolation of protoplasts; C) Protoplast mixture after fusion of IOA-inactivated protoplasts of *F. arundinacea* and non-morphogenic protoplasts of *L. multiflorum* isolated from A) and B), respectively; D) Multiple divisions of putative fusant from asymmetric fusion between IOA-inactivated tall fescue protoplasts and 250 Gy X-ray-irradiated Italian ryegrass protoplasts two weeks after fusion; E) Putative somatic hybrid colonies obtained from fusion experiments between tall fescue protoplasts and 500 Gy-irradiated Italian ryegrass protoplasts after 4 weeks in bead type culture; F) Regenerating clone from E) two months after protoplast fusion; G) Greenhouse-growing, mature somatic hybrid *Festulolium* plants obtained from symmetric and asymmetric protoplast fusions; H) Morphology of clonally-unstable somatic hybrid plant (right side of the plant is morphologically similar to Italian ryegrass); I) Acetocarmine-stained pollen of somatic hybrid *Festulolium* plant shown on H); J) Seeds developing on symmetric somatic hybrid *Festulolium* plant after pollination with Italian ryegrass; K) Somatic hybrid *Festutolium* plants growing in the field.

calli differentiating green shoots were obtained from over 40 symmetric and asymmetric fusion experiments involving different X-ray irradiation doses of donor protoplasts (Fig. 1F). These regenerating calli could thus be preliminarily identified as putative symmetric and asymmetric somatic hybrid clones. Green shoots grew vigorously, rooted after transfer to MS hormone-free medium and 31 plantlets were successfully transferred to soil and grown until maturity under greenhouse conditions (Fig. 1G). Of these, 6 plants were obtained from symmetric fusions, 13 were derived from 500-Gy fusions, and 2 or 3 mature plants were recovered from each of the other X-ray doses (Spangenberg et al., 1994; Takamizo & Spangenberg, 1994). Plant habit and leaf morphology varied among these independent symmetric and asymmetric tall fescue (+) Italian ryegrass somatic hybrid plants (Fig. 1G). Some showed an intermediate character when compared with the parental plants, others were closer to tall fescue (Fig. 1G). In other cases chimaeric clones revealing plant sectors with leaf morphology and plant habit similar to Italian ryegrass were observed (Fig. 1H). When brought to flower they produced inflorescences with protruding anthers which contained stainable pollen (Fig. 1I). Crosses of these clonal unstable plants with Italian ryegrass pollen led to seed formation (Fig. 1J). Representative intergeneric somatic hybrid *Festulolium* plants obtained from symmetric and asymmetric fusion combinations are now growing under field conditions (Fig. 1K).

Isolation and characterization of species-specific repetitive DNA sequences for analysing hybrid nuclear genomes in the Festuca-Lolium *complex*

For the analysis of hybrid nuclear genomes, cloned DNA sequences need to be: 1) highly species-specific; 2) repetitive; 3) dispersed in the corresponding genome; and 4) evenly distributed in all or most chromosomes. We have isolated repetitive nuclear DNA sequences which fulfil these requirements from *L. multiflorum* and *F. arundinacea* (Pérez-Vicente et al., 1992). In addition, repetitive nuclear DNA sequences specific for red fescue (*F. rubra*) and meadow foxtail (*A. pratensis*) were isolated and characterized at the molecular and cytological level, to be used as tools for the analysis of further intergeneric somatic hybrids to be generated.

In order to isolate repetitive species-specific DNA sequences, ca. 250 recombinant plasmid clones, each containing random sequences from *L. multiflorum, F. arundinacea, F. rubra* and *A. pratensis*, were screened using labelled total DNA from two species as hybridization probes. As expected for multicopy sequences, between 5 and 10% of the clones showed strong hybridization signals when probed with DNA from the same species. DNA inserts from one *F. arundinacea* clone, six *F. rubra* clones, eight *L. multiflorum* clones and one *A. pratensis* clone showing differential hybridization were isolated and their species-specificity was tested by dot blot hybridization analysis to total DNA from different grass species within and outside the *Festuca-Lolium* complex: *L. multiflorum, L. perenne, F. arundinacea, F. pratensis, F. rubra* and *A. pratensis*. Representative results from this analysis are shown in Fig. 2A. The Italian ryegrass sequence LMH2 was only present in *L. multiflorum* and *L. perenne*, the sequence LMB4 from *L. multiflorum* hybridized in addition to total DNA from *A. pratensis* (Fig. 2A). The meadow foxtail sequence APE1 was only present in *A. pratensis*. Similarly, the clone FRH5 from red fescue hybridized exclusively to DNA from *F. rubra* (Fig. 2A), while the sequence FAH1 cloned from *F. arundinacea* cross-hybridized to *F. pratensis* DNA but was absent in the genome of the other species considered.

To increase knowledge of the organization, degree of dispersion and chromosomal localization of selected, highly species-specific, repetitive DNA sequences,

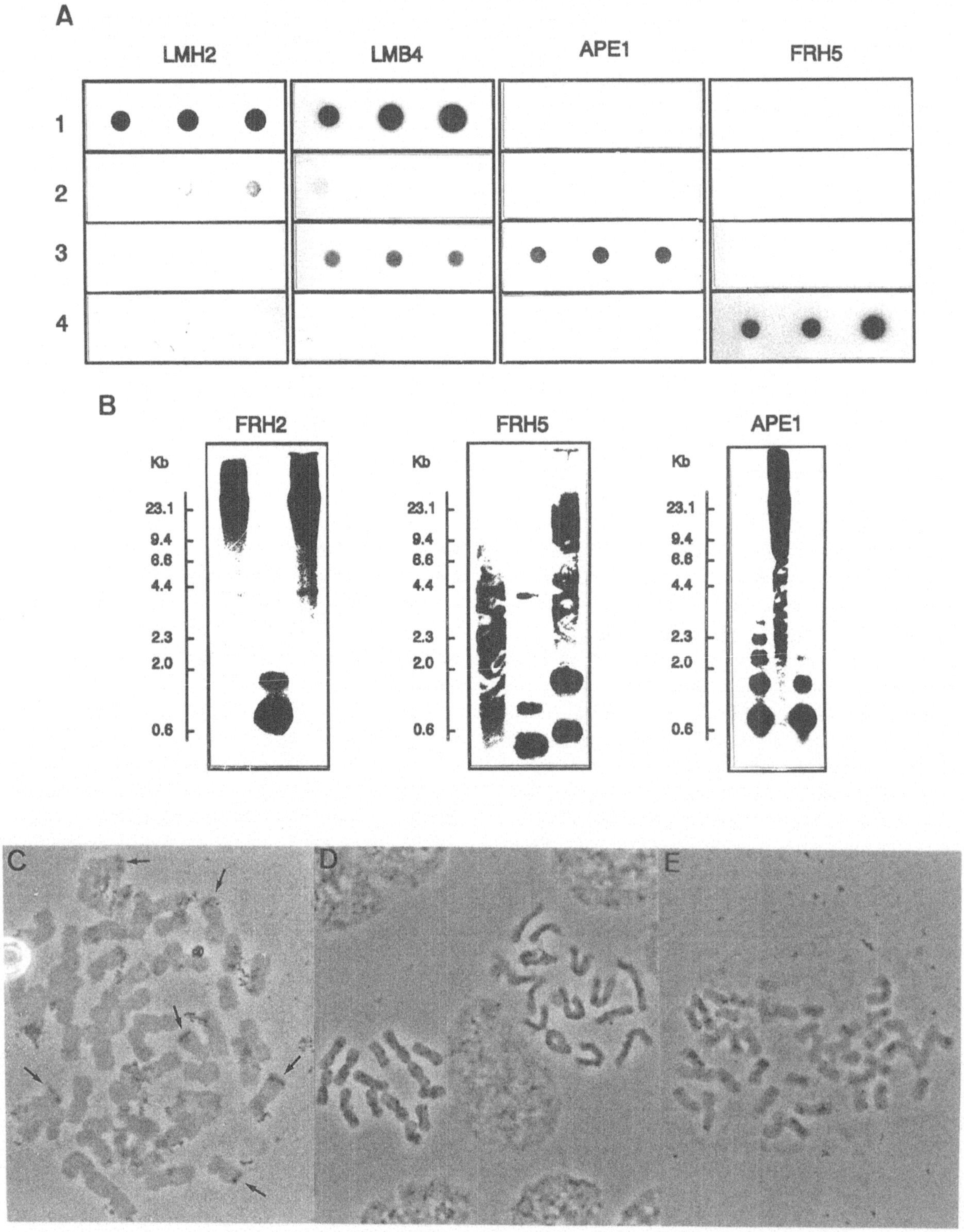
A
LMH2
LMB4
APE1
FRH5
1
2
3
4
B
FRH2
FRH5
APE1
Kb
23.1
9.4
6.6
4.4
2.3
2.0
0.6
C
D
E

←

Fig 2 Isolation and characterization of species-specific repetitive DNA sequences in *Festuca, Lolium* and *Alopecurus* A) Test for species-specificity of repetitive DNA sequences isolated from *L multiflorum* (LMH2 and LMB4), *A pratensis* (APE1) and *F rubra* (FRH5) Labelled probes were hybridized to dot blots containing (from left to right) 0 5, 1 and 2 μg of (1) *L multiflorum*, (2) *L perenne*, (3) *A pratensis*, and (4) *F rubra* DNA, B) Southern blot hybridization patterns from *F rubra* and *A pratensis* DNA were probed with repetitive DNA sequences FRH2, FRH5 and APE1, respectively In all cases, 7 μg of total DNA digested with BamHI, HindIII or EcoRI (lanes from left to right) were loaded, C–E) *In situ* hybridization analysis of repetitive DNA sequences from C) *F rubra* (FRH2), D) *L multiflorum* (LMH2), and E) *F arundinacea* (FAH1) hybridized to metaphase chromosomes of red fescue, Italian ryegrass and tall fescue, respectively Arrows indicate hybridization signals on red fescue chromosomes (C)

Sequence	Size (bp)	Copy Number*	Characteristics**
LMH2	173	$1\ 3 \times 10^4$	dispersed, (D, E).
LMB4	300	$2\ 0 \times 10^3$	dispersed, (D, F)
FAH1	102	$2\ 0 \times 10^4$	clustered, (A, B).
FRH5	78	1.1×10^5	tandemly arranged, (C)
APE1	385	7.0×10^3	tandemly arranged, (F)

* Estimated values per haploid genome
**Sequences cross hybridizing to the corresponding probes are present in (A) *F arundinacea* (B) *F pratensis* (C) *F rubra*, (D) *L multiflorum*, (E) *L perenne*, (F) *A pratensis*

LMH2

```
AGACTTTGTG CAATGTCAGA AGTGTTAAGA ATGATTATGT CACCTCTGAA   50
TGTATGAATT TTTTATTATG CACTAACCCT CTAATGAGTT TGCTTGAAGT  100
TTGGTGTGGA GGAAGTTTTC AAGGGTCAAG AGAAGAGGAT GATACAATAT  150
GATCAAGAAG AGTGAAAGGT CTA
```

FAH1

```
AGCTTGGCTA GAGCTGCTTG CCTCCTGACC TTTTCCGGTT CCGGCCTTGG   50
GAGCAGAGGG GGAGGCACTC ATCCGGTCGA TCTCGGCTTC AAGCCCGTCG  100
GA
```

FRH5

```
AGCTTTAAAA ATTCATCGTT AAACGCTTGG TTAGTAGAGC CAAAGATAAT   50
ATCATCAACA TATAGTTGGC ATATAAAA
```

APE1

```
AATTCCCTCG GTATGTTACA TTCTTTCATC AGCTTATTCC CCGTTTCTTT   50
TCTAGTTTGA GCCTAACACA CTATTGTAGC TATCCATGGG TTTGGCTCTT  100
AAACCCTGAT GGCCAAATGG ATCCACGGCA CCATTGAATT TCAAACTTTA  150
AGTGCCAAAA CTCCTGTTTT TCATCTCGTA GGCCCTTTTA GAAATGCTAA  200
ACGTCCTTTT TTGGGGCCTA GATTTCTAGG AACGCTTTTT CGGGATTTGT  250
GTGCATGAAT TTCAAATCAT ACGAACTTAT AAAATAGAAG TCATTCCATC  300
GTTATTCCCC TAGAGCCTTC CAGCAGTTCC TAAACCGCCT CAGTCGGAGC  350
ACGTTATCAT TCGTCTTGTT AGTTCCGAGT GAATT
```

Fig 3 Characteristics and nucleotide sequences of repetitive DNA clones from *L multiflorum F arundinacea F rubra* and *A pratensis* Size of the cloned repetitive DNA sequences was estimated by comparing the electrophoretic mobility of the corresponding inserts to size standards Estimated number of copies per haploid genome was determined by comparison of the extent of hybridization of the indicated sequence to defined amounts of genomic DNA and of the corresponding DNA sequence *L multiflorum*-specific, repetitive clones LMH2, LMB4 and *F arundinacea*-specific, repetitive sequence FAH1 (Perez Vicente et al, 1992, Theor Appl Genet 84 145–154), *F rubra*-specific, repetitive clone FRH5, and *A pratensis*-specific, repetitive sequence APE1

Southern blot hybridizations and *in situ* hybridizations to metaphase chromosomes were performed. This analysis revealed 1) a complex multiple band, ladder-pattern indicating a tandem arrangement and hybridization to 6 and 2 chromosomes for the red fescue sequences FRH2 and FRH5, respectively (Fig.

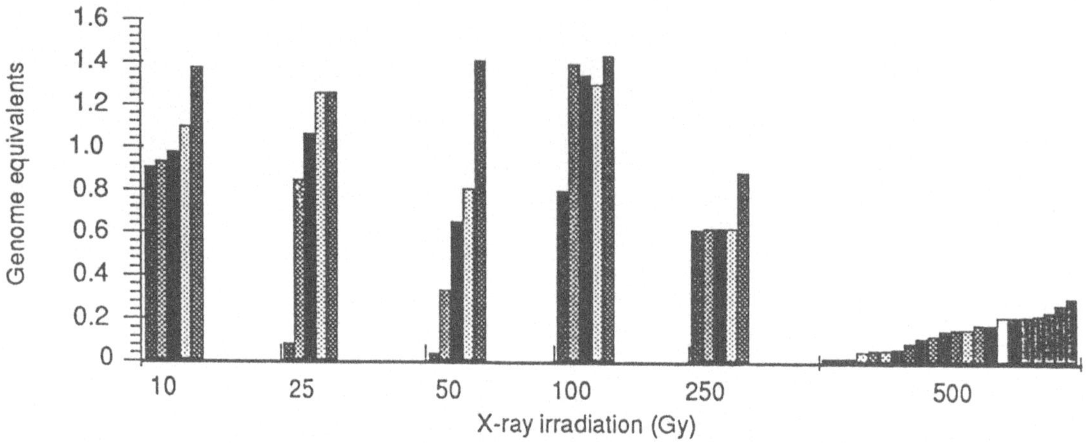

Fig. 4. Genome composition of asymmetric *F. arundinacea* (+) *L. multiflorum* somatic hybrids Quantitative dot blots of asymmetric somatic *Festulolium* clones (bars correspond to independent regenerating calli and plants) obtained from different doses (10–500 Gy) of X-ray irradiation of donor protoplasts were hybridized with the *L. multiflorum*-specific, repetitive sequences LMH2 and LMB4. Genomic composition of independent clones based on average of data obtained from hybridization with both probes is shown as genome equivalents of *L multiflorum* per genome equivalent of *F. arundinacea.*

2BC); 2) a similar banding pattern and hybridization only to 24 meadow foxtail chromosomes for the *A. pratensis*-specific sequence APE1 (Fig. 2B); 3) a partially-dispersed arrangement of the *L. multiflorum* sequences LMH2 and LMB4 and hybridization to all Italian ryegrass chromosomes (Fig. 2D); and 4) a similar banding pattern with some clustering and hybridization only to some tall fescue chromosomes for the *F. arundinacea*-specific sequence FAH1 (Fig. 2E).

Insert size and copy number of the highly species-specific sequences were estimated to range from 80–400 bp and from 2,000 to 100,000 copies per haploid genome, respectively (Fig. 3). Representative species-specific, repetitive DNA sequences were subjected to a sequence analysis: the Italian ryegrass clone LMH2 (173 bp) (Pérez-Vicente et al., 1992), the tall fescue clone FAH1 (102 bp) (Pérez-Vicente et al., 1992), the red fescue clone FRH5 and the meadow foxtail clone APE1 (Fig. 3).

Analysis of nuclear and organellar composition of symmetric and asymmetric somatic Festulolium *hybrids*

The nuclear composition of symmetric and asymmetric intergeneric *F. arundinacea* (+) *L. multiflorum* somatic hybrid plants, was characterized by chromosome counts and quantitative dot blot hybridizations using interdispersed repetitive DNA probes specific for Italian ryegrass (LMH2 and LMB4) and tall fescue (FAH1) (Spangenberg et al., 1994a; Takamizo & Spangenberg, 1994). Chromosome counts performed in representative somatic *Festulolium* plants recovered from symmetric and asymmetric protoplast fusions were 18 and 20 (slightly higher than the count 2n = 14 for Italian ryegrass), 37 and 39 (slightly lower or close to the count 2n = 42 for tall fescue), 49 (clearly higher than the count for the recipient tall fescue), 53 (close to the expected additive count of 2n = 56 for a symmetric hybrid) and one plant showing 90 chromosomes (Spangenberg et al., 1994a; Takamizo & Spangenberg, 1994). The suitability of the cloned *L. multiflorum*- and *F. arundinacea*-specific, dispersed repetitive DNA sequences as probes for the characterization of the nuclear composition of *Festulolium* hybrids has been previously demonstrated (Pérez-Vicente et al., 1992). This analysis revealed the presence of Italian ryegrass repetitive DNA sequences in the nuclear genome of all primary regenerants from symmetric fusions and showed hybridization signals estimated to about one genome equivalent of *L. multiflorum* per haploid genome of *F. arundinacea*. Similarly, the analysis of the asymmetric somatic hybrid clones derived from fusion experiments using Italian ryegrass donor protoplasts irradiated with 10, 25, 50, 100 and 250 Gy and 500 Gy of X-rays, was performed. A concentration series of parental DNAs allowed a calibration plot of the radioactivity per dot in relation to the amount of DNA from one species and for mixtures representing different ratios of genome equivalents to be made. With the calibration plots for all three species-specific probes used, the amount of Italian ryegrass and tall fescue DNA per dot could be estimated, and the fraction of the nuclear DNA of the asymmetric somatic hybrids derived from Italian ryegrass could be determined (Spangenberg et al., 1994a). All analysed regenerating clones derived from 10–250 Gy-fusion products contained Italian ryegrass nuclear DNA but showed a large variation in the contribution of *L. multiflorum* DNA to their genomes (Fig. 4). These asymmetric somatic hybrid clones showed either no or limited donor genome elimination (being thus almost symmetric) or retained even < 5% of the *L. multiflorum* genome (being thus highly asymmetric). Some asymmetric somatic hybrid clones recovered after irradiation with doses as different as 25 and 250 Gy, retained comparable amounts of donor nuclear DNA, whereas the degree of asymmetry in independent clones within each dose varied even more (Fig. 4). The analysed asymmetric somatic hybrid clones obtained from donor protoplasts irradiated with 500 Gy, revealed an extensive (> 85%) and similar Italian ryegrass nuclear genome elimination for both Italian ryegrass-specific probes tested. For some of these 500 Gy-asymmetric somatic hybrid clones analysed, no *L. multiflorum* DNA above background was detectable. Estimates from dot blots hybridized with the tall fescue-specific repetitive sequence FAH1 as probe, indicated the presence of approximately a complete chromosome set of the recipient *F. arundinacea* in all asymmetric and symmetric somatic hybrids analysed. The results obtained indicated: 1) the true nuclear hybrid nature of all green plants regenerated from symmetric fusions, 2) the tightness of the IOA metabolic inactivation of tall fescue protoplasts, 3) the value of species-specific repetitive DNA probes for the identification of somatic hybrids in intergeneric fusion combinations, 4) the suitability of these sequences for the analysis of donor genome elimination in asymmetric somatic hybrids, 5) irradiation of donor cells prior to fusion leading to the unidirectional species-specific elimination of Italian ryegrass chromosomes in asymmetric somatic hybrids, 6) no strict correlation between the level of the species-specific *Lolium* genome elimination and the radiation dose used, being apparent for a wide (25–250 Gy) range of X-ray doses tested.

The organellar composition of intergeneric *F. arundinacea* (+) *L. multiflorum* symmetric and asymmetric somatic hybrid clones was analysed by the generation of species-specific patterns obtained after hybridization of total DNA digests with six mtDNA (*cox1*, *cox2*, *cox3*, *atpA*, *atp6* and *atp9*) and one cpDNA (*rbcL*)-specific heterologous gene probes. Parental-like additive and novel (involving the absence of parental-like bands and/or the presence of non-parental bands) patterns, were observed depending on the hybridization probe and the somatic *Festulolium* clone considered. A general overview of the results obtained in the RFLP analysis performed for a set of *Festulolium* clones regenerated from symmetric and asymmetric protoplast fusions has been provided (Spangenberg et al., 1994a; Takamizo & Spangenberg, 1994). While additive patterns are preferentially observed in the mtDNA RFLP analysis of symmetric and < 50 Gy asymmetric clones, together with an extensive nuclear genome elimination of the donor, tall fescue patterns become predominant for the > 100 Gy asymmetric *Festulolium* clones (Spangenberg et al., 1994a; Takamizo & Spangenberg, 1994).

Discussion

Symmetric and asymmetric somatic fusions were performed for one intergeneric combination in the Gramineae involving unirradiated or X-ray-irradiated protoplasts of *L. multiflorum* (donor) and metabolically inactivated protoplasts of *F. arundinacea* (recipient). A series of symmetric and asymmetric somatic hybrid clones were obtained and plants were regenerated for each radiation dose category (0–500 Gy). Regeneration of mature somatic hybrid plants in the Gramineae other than rice, in which successful somatic hybridization and cybridization have been reported (Terada et al., 1987; Hayashi et al., 1988; Akagi et al., 1989; Kyozuka et al., 1989; Yang et al., 1989), is thus feasible. For rice, somatic hybridization in the intergeneric combination *O. sativa* (+) *Echinochloa oryzicola* has been attempted but the recovery of mature green plants failed (Terada et al., 1987). The same holds true for all intergeneric somatic hybrids so far reported in the Gramineae, e.g. *Triticum monococcum* (+) *Pennisetum americanum* (Vasil et al., 1988). Thus, the greenhouse-grown, flowering *Festulolium* somatic hybrid plants described represent the first case of mature plant regeneration for intergeneric symmetric and asymmetric somatic hybridization in the Gramineae (Takamizo et al., 1991; Spangenberg et al., 1994a; Takamizo & Spangenberg, 1994).

Repetitive species-specific sequences were isolated from forage grass species. *F. arundinacea* and *L multiflorum* (Pérez-Vicente et al. 1992) and from *F. rubra* and *A. pratensis*. They were characterized at the molecular and cytological level. Some were found to be dispersed and represented in all or most chromosomes of the corresponding species, proving suitable for the analysis of the nuclear composition of *Festulolium* plants obtained from wide crosses (Pérez-Vicente et al., 1992) and from somatic hybridization (Takamizo et al., 1991, Spangenberg et al., 1994a).

X-ray irradiation of donor protoplasts prior to fusion and 'gamma-fusion' have been shown to be two reliable methods for inducing species-specific chromosome elimination from the irradiated partner and for the production of asymmetric somatic hybrids in interspecific *Brassica* (e.g. Yamashita et al., 1989) and intergeneric *Nicotiana-Atropa* (e.g. Gleba et al., 1988) combinations. In the case of asymmetric somatic *Festulolium* we found correlative evidence between the number of excess chromosomes, presence or absence of *in situ* hybridization signals on metaphase chromosomes and the estimates from corresponding dot blot values for individual hybrids (Spangenberg et al., 1994a). Furthermore, out data suggest that the degree of elimination of donor chromosomes from X-ray-irradiated Italian ryegrass protoplasts, was not dose-dependent for asymmetric somatic hybrids in the range of 25–250 Gy, and that a larger variability of the asymmetry level is detectable in independent clones within each of these dose categories. Analogous results were reported for intertribal, asymmetric somatic hybrids between *N. plumbaginifolia* and *Atropa belladonna* obtained by 'gamma-fusion' (Gleba et al., 1988).

Approaches aimed at evaluating the potential contribution of somatic hybridization and cybridization supporting conventional breeding programmes in ryegrasses and fescues are now conceivable. Asymmetric protoplast fusion aimed at intraspecific transfer of cytoplasmic male sterility (CMS) has been described for *L perenne*, but since non-morphogenic cell suspensions were used, only cybrid calli were recovered (Creemers-Molenaar et al., 1992). Recently, we have established an efficient plant regeneration system from protoplasts isolated from single genotype-derived embryogenic cell suspensions in *L. multiflorum, L perenne* and *L.* × *boucheanum* (Wang et al., 1993b). These protoplast-to-mature plant regeneration systems opened up opportunities for the use of *Lolium* protoplasts as recipients in asymmetric somatic hybridization and cybridization experiments. Somatic cybridizations aimed at transferring organellar-coded traits using morphogenic ryegrass protoplasts as recipients and X-ray-irradiated protoplasts from established non-morphogenic suspension cultures of CMS lines in *L. multiflorum* and *L. perenne*, are now in progress. In addition, a programme to generate fertile asymmetric somatic hybrids combining quality traits from perennial ryegrass and persistence/resistance traits of red fescue and meadow foxtail has been initiated. A reproducible protoplast-to-plant regeneration system is now available for *F. rubra* (Spangenberg et al., 1994b) and first regenerating asymmetric somatic hybrid clones in *F rubra* (+) *L perenne* have been recovered.

Acknowledgements

The authors wish to thank A. Brennicke, R. Herrmann and C.J. Leaver for kindly providing mitochondrial and chloroplast gene clones. I. Cordt and B. Larsson are gratefully acknowledged for their help with the irradiation experiments. Thanks are due to P. Frick and K. Konja for kindly growing plants in the greenhouse. M. Osusky and L. Petris are gratefully acknowledged for the *in situ* hybridization work. T.T. was supported by an OECD fellowship. M.P.V. was supported by an academic exchange fellowship from Diputación General de Aragón, Spain.

References

Akagi, H , M Sakamoto, T Negishi & T Fujimura, 1989 Construction of rice cybrid plants Mol Gen Genet 215 501–506

Creemers-Molenaar, J , R D Hall & F A Krens, 1992 Asymmetric protoplast fusion aimed at intraspecific transfer of cytoplasmic male sterility (CMS) in *Lolium perenne* L Theor Appl Genet 84 763–770

Crowder, L V, 1953 Interspecific and intergeneric hybrids of *Festuca* and *Lolium* J Hered 44 195–203

de Lautour, G & B M Cooper, 1971 Cold and chemical pretreatments to aid chromosome counts in a grass leaf squash technique incorporating hot pectinase maceration Stain Technol 46 305–310

Eizenga, G C & R C Buckner, 1986 Cytological and isozyme evaluation of tall fescue × Italian ryegrass hybrids Plant Breeding 97 340–344

Feinberg, A P & B Vogelstein, 1983 A technique for radiolabelling DNA restriction endonuclease fragments to high specific activity Anal Biochem 132 6–13

Gleba, Y Y , S Hinnisdaels, V A Sidorov, V A Kaleda, A S Parokonny, N V Boryshuk, N N Cherup, I Negrutiu & M Jacobs, 1988 Intergeneric asymmetric hybrids between *Nicotiana plumbaginifolia* and *Atropa belladonna* obtained by 'gamma fusion' Theor Appl Genet 76 760–766

Hayashı, Y, J Kyozuka & K Shımamoto, 1988 Hybnds of nce (*Oryza satıva* L) and wıld *Oryza* specıes obtaıned by cell fusıon Mol Gen Genet 246 6–10

Kyozuka, J, Y Hayashı & K Shımamoto, 1987 Hıgh frequency plant regeneratıon from nce protoplasts by novel nurse culture methods Mol Gen Genet 206 408–413

Kyozuka, J, T Kaneda & K Shımamoto, 1989 Productıon of cytoplasmıc male stenle nce (*Oryza satıva* L) by cell fusıon Bıo/Technology 7 1171–1174

Lıchtensteın, C & J Draper, 1985 Genetıc engıneerıng of plants p 67–119 In D M Glover (Ed) DNA Clonıng, Vol II IRL Press, Oxford Washıngton

Muller, A J & R Grafe, 1978 Isolatıon and charactenzatıon of cell lınes of *Nıcotıana tabacum* lackıng nıtrate reductase Mol Gen Genet 161 67–76

Murashıge, T & F Skoog, 1962 A revısed medıum for rapıd growth and bıoassays wıth tobacco tıssue culture Physıol Plant 15 473–497

Perez-Vıcente, R, L Petrıs, M Osusky, I Potrykus & G Spangenberg, 1992 Molecular and cytogenetıc charactenzatıon of repetıtıve DNA sequences from *Lolıum* and *Festuca* applıcatıons ın the analysıs of *Festulolıum* hybnds Theor Appl Genet 84 145–154

Sambrook, J, E F Fntsch & T Manıatıs, 1989 Molecular clonıng – a laboratory manual Second edıtıon Cold Sprıng Harbor Laboratory Press CSH

Sanger, F, A R Coulson, B G Barrel, A J H Smıth & B A Roe, 1980 Clonıng ın sıngle-stranded bactenophage as an aıd to rapıd DNA sequencıng J Mol Bıol 143 161–178

Saul, M W & I Potrykus, 1984 Specıes-specıfic repetıtıve DNA used to ıdentıfy ınterspecıfic somatıc hybnds Plant Cell Rep 3 65–67

Shıllıto, R D, J Paszkowskı, M Muller & I Potrykus, 1983 Agarose platıng and a bead type culture technıque enable and stımulate development of protoplast denved colonıes ın a number of plant specıes Plant Cell Rep 2 244–247

Spangenberg, G, M P Valles, Z Y Wang, P Montavon, J Nagel & I Potrykus, 1994a Asymmetnc somatıc hybndızatıon between tall fescue (*Festuca arundınacea* Schreb) and ırradıated Italıan ryegrass (*Lolıum multıflorum* Lam) protoplasts Theor Appl Genet 88 509–519

Spangenberg, G, Z Y Wang, J Nagel & I Potrykus, 1994b Protoplast culture and generatıon of transgenıc plants ın red fescue (*Festuca rubra* L) Plant Scı 97 83–94

Tabaeızadeh, Z, R J Ferl & I K Vasıl, 1986 Somatıc hybndızatıon ın the Gramıneae *Saccharum offıcınarum* L (Sugarcane) + *Pennısetum amerıcanum* (L) K Schum (Pearl mıllet) Proc Natl Acad Scı USA 83 5616–5619

Takamızo, T & G Spangenberg, 1994 Somatıc hybndızatıon ın *Fes tuca* and *Lolıum* p 112–131 In Y P S Bajaj (Ed) Bıotechnology ın Agnculture and Forestry, Vol 27 Somatıc hybndızatıon ın crop ımprovement Sprınger Verlag, Berlın, Heıdelberg

Takamızo, T, G Spangenberg, K Sugınobu & I Potrykus, 1991 Intergenenc somatıc hybndızatıon ın Gramıneae somatıc hybnd plants between tall fescue (*Festuca arundınacea* Schreb) and Italıan ryegrass (*Lolıum multıflorum* Lam) Mol Gen Genet 231 1–6

Takamızo, T, K Sugınobu & R Ohsugı, 1990 Plant regeneratıon from suspensıon culture denved protoplasts of tall fescue (*Festuca arundınacea* Schreb) of a sıngle genotype Plant Scı 72 125–131

Terada, R, J Kyozuka, S Nıshıbayashı & K Shımamoto, 1987 Plantlet regeneratıon from somatıc hybnds of nce (*Oryza satıva* L) and barnyard grass (*Echınochloa oryzıcola* Vasıng) Mol Gen Genet 210 39–43

Vasıl, V, R J Ferl & I K Vasıl, 1988 Somatıc hybndızatıon ın the Gramıneae *Trıtıcum monococcum* L (Eınkorn) + *Pennısetum amerıcanum* (L) K Schum (Pearl mıllet) J Plant Physıol 132 160–163

Wang, Z Y, T Takamızo, V A Iglesıas, M Osusky, J Nagel, I Potrykus & G Spangenberg, 1992 Transgenıc plants of tall fescue (*Festuca arundınacea* Schreb) obtaıned by dırect gene transfer to protoplasts Bıo/Technology 10 691–696

Wang, Z Y, M P Valles, P Montavon, I Potrykus & G Spangenberg, 1993a Fertıle plant regeneratıon from protoplasts of meadow fescue (*Festuca pratensıs* Huds) Plant Cell Rep 12 95–100

Wang, Z Y, J Nagel, I Potrykus & G Spangenberg, 1993b Plants from cell suspensıon-denved protoplasts ın *Lolıum* specıes Plant Scı 94 179–193

Yamashıta, Y, R Terada, S Nıshıbayashı & K Shımamoto, 1989 Asymmetnc somatıc hybnds of *Brassıca* partıal transfer of *B campestrıs* genome ınto *B oleracea* by cell fusıon Theor Appl Genet 77 189–194

Yang, Z Q, T Shıkanaı, K Morı & Y Yamada, 1989 Plant regeneratıon from cytoplasmıc hybnds of nce (*Oryza satıva* L) Theor Appl Genet 77 305–310

Euphytica **85:** 247–253, 1995.

Transfer of disease resistance within the genus *Brassica* through asymmetric somatic hybridization

M. Gerdemann-Knörck, S. Nielen, C. Tzscheetzsch, J. Iglisch & O. Schieder
Institute of Applied Genetics, Free University Berlin, Albrecht-Thaer-Weg 6, D 14195 Berlin, Germany

Key words: asymmetric somatic hybridization, *Brassica napus*, *Brassica nigra*, disease resistance transfer, dot blot analysis

Summary

Asymmetric somatic hybrid plants between *Brassica napus* L. (oilseed rape genome AACC) and a transgenic line of *Brassica nigra* L. Koch (black mustard genome BB) were tested for their resistance against rapeseed pathogens *Phoma lingam* (black leg disease) and *Plasmodiophora brassicae* (club root disease). The transgenic *B. nigra* line used (hygromycin-resistant, donor) is highly resistant to both fungi, whereas *B. napus* (recipient) is highly susceptible. The asymmetric somatic hybrids were produced using the donor-recipient fusion method (with X-irradiation of donor protoplasts) reported by Zelcer et al. (1978) for the production of cybrids. Using hygromycin-B for selection, a total of 332 hybrid calli were obtained. Regenerants, resistant or susceptible to both diseases, were selected. Many hybrids expressed resistance to only one pathogen. Dot blot experiments showed that the asymmetric hybrid plants contained varying amounts of the donor genomic DNA. Furthermore, a correlation was detected between the radiation dose and the degree of donor DNA elimination.

Introduction

Clubroot in crucifers, caused by *Plasmodiophora brassicae* Wor., and black leg, caused by *Leptosphaeria maculans* (Desm.) Ces. & de Not. (imperfect stage, *Phoma lingam* (Tode) (Desm.), are two of the most important diseases in *Brassica* crops. Black leg disease in the amphidiploid rapeseed (genome AACC) is reported for all parts of the world where this crop is cultivated. The incorporation of genetic information into rapeseed leading to stable resistance against these two pathogens would be of considerable economic importance. *P. brassicae* attacks a wide range of the cruciferous crops and is one of the most important pathogens of *Brassica* crops in the temperate zones (Buczacki, et al. 1975). The disease is characterized by a proliferation of galls on infected roots causing wilting, stunting and yield reduction. The pathogen survives as spores in the soil for long periods, making contaminated soil unsuitable for the cultivation of Brassicaceae. Since chemical control is impossible (Jacobsen & Williams, 1970), the production of resistant rapeseed lines is the only way of controlling this pathogen. Resistant and susceptible varieties have been identified in many species of the Brassicaceae (Crisp et al., 1989). The species *Brassica nigra* with the genome formula BB (2n = 16) exhibits a high level of resistance to black leg disease and club root (Sacristán & Gerdemann, 1986; Sjödin & Glimelius, 1989; Zhu & Spanier, 1991). In *B. nigra* the resistance to *P. brassica* is more effective than resistance observed in seedlings in some genotypes of *B. napus* (AACC). As previously reported (Sacristán et al., 1989), genotypes of black mustard have been selected which exhibit resistance to *P. lingam* and to a *P. brassicae* pathotype which is virulent to all *B. napus* differentials of the ECD-set (Buczacki et al., 1975). Therefore interspecific transfer of clubroot and black leg resistance from *B. nigra* to *B. napus* is a possible way of enlarging the genetic base in rapeseed germplasm. As a first step in this transfer the double disease-resistant black mustard has been transformed with the selectable marker gene *hph II* (hygromycin-B phosphotransferase) using *Agrobacterium tumefaciens* (Sacristán et al., 1989; Schieder et al., 1991). Asymmetric somatic hybrids have been obtained by fusion of X-ray-inactivated protoplasts (450–1300 Gy) from transformed lines of *B. nigra* (donor) with untreated protoplasts of *B. napus* (recipient). The selection criterion for fusion products was the expression of the hygromycin-resistance marker. Using the 'donor-recipient method' (Zelcer et al., 1978) we have produced more than 100 asymmetric somatic hybrid lines. Many of them were found to be resistant to one or both of the pathogenic fungi (Gerdemann-Knörck et al., 1994). Here we report some results concerning the genomic DNA content of *B. nigra* in the asymmetric hybrids.

Materials and methods

Plant material

For the protoplast fusion experiments, *in vitro*-cultivated plants of the amphidiploid cultivar 'Liropa' of *B. napus* (AACC = 2n = 38) were used as the recipient plant material and the polyploid transgenic clone of *B. nigra* (34–36 chromosomes) carrying the *hph* gene under the control of the CaMV 35S promoter, derived from the line 460 (origin Hort. Bot. Hauniensis, Stockholm), was used as donor plant material. The latter line was chosen because of its resistance to both *P. lingam* and *P. brassicae* (Sacristán et al., 1989). The *B. napus* cultivar 'Liropa' is susceptible to both pathogens. The *B. napus* and *B. nigra* lines were maintained as sterile shoot cultures on medium described by Gamborg et al. (1968) containing naphthaleneacetic acid (0.1 mg l^{-1}) for root development. The plants were grown at a light intensity of 79 μmol m^{-2} s^{-1} (fluorescent tubes, Philips 50 W), with a 16 h photoperiod and at 26° C.

Isolation, fusion and culture of protoplasts

Leaf protoplasts were isolated as described by Sacristán et al. (1989). The *B. nigra* protoplasts were irradiated with X-ray doses ranging from 450 to 1300 Gy. Prior to fusion, *B. napus* and *B. nigra* protoplasts were mixed to a final density of 5×10^6 ml^{-1}. The fusion procedure was undertaken as described elsewhere (Sacristán et al., 1989). The protoplasts were cultured in M-1 medium (Li & Kohlenbach, 1982) in the dark at 26° C. When the first cell divisions were observed (5–7 days after fusion) the cultures were diluted 1 : 1 with M-1 medium supplemented with 25 mg l^{-1} hygromycin-B. When microcalli developed they were transferred from the liquid medium to 2-N SeaPlaque-agarose medium (Chuong et al., 1987) supplemented with 25 mg l^{-1} hygromycin-B. The calli were subcultured onto fresh culture medium every 4 weeks.

Shoot regeneration and transfer of plants to soil

Two to three months after the fusion, shoot-like structures that developed on several putative hybrid calli were transferred to solid MS medium (Murashige & Skoog, 1962) lacking phytohormones and supplemented with 10 g l^{-1} sucrose and 8 g l^{-1} agar (Merck). On this medium the shoots developed roots. These plants were transferred into soil (1 : 2 mixture of sand/Einheitserde P) and transferred into a growth chamber (16°–20° C, 16 h photoperiod, 79 μmol m^{-2} s^{-1} and 80% relative humidity). The hybrid nature of the plants was confirmed by isoenzyme analysis of esterases (Sacristán et al., 1989). Chromosome numbers of regenerated hybrid plants were determined in root tips as described by Sacristán & Gerdemann (1986).

Pathogen tests

The *P. lingam* isolate (isolate BBA 63698) and the isolate of *P. brassicae* (ECD-code: 19/31/31) were cultured and the spores were prepared as described elsewhere (Sacristán & Hoffmann, 1979; Sacristán, 1982; Gerdemann-Knörck et al., 1994). Black leg was assayed 6–8 weeks after inoculation according to a rating system (Sacristán, 1982), which ranges from 1 (a light necrotic spot at the inoculation site) to 3 (plant collapsed). Clubroot was evaluated 35–42 days after inoculation and scored for resistance on a three-class scale.

Plants derived from the transgenic *B. nigra* and seed derived from wild-type plants of the *B. nigra* line '460' and *B. napus*, served as controls.

Nopaline synthase assay

Nopaline synthase activity was assayed as described by Otten & Schilperoot (1978).

DNA isolation

Total DNA was isolated from leaves of the two parental lines and of the different hybrid lines by the method of Rogers & Bendich (1985). DNA concentration was measured fluorometrically in a Hoefer TKO 100 DNA-Fluorometer (Hoefer, San Francisco, USA) using the bisbenzimidazole dye Hoechst 33258 (Labarca & Paigen, 1980).

Slot blot hybridization

The genomic DNA samples were blotted on a positively-charged nylon membrane (Boehringer Mannheim, Germany) using the Bio-Dot SF slot blotting apparatus (BIO RAD, Richmond, USA) according to the manufacturers' instructions. Dilution series (25–1000 ng) of DNA from the parental lines were used for calibration. DNA from hybrid clones was used at a quantity of 1000 ng per slot. Two slots were loaded for each clone. Three identical filters were prepared for hybridization with one *B. napus*-specific (BcKB4) and two *B. nigra*-specific (BN34 and BN35) repetitive sequences. BN34 is tandemly repeated (Gupta et al., 1990), BN35 is a dispersed repeat (Gupta et al., 1992) and BcKB4 is a centromeric tandem repeat cloned from *B. campestris* (Heslop-Harrison et al., 1992).

Probe labelling and hybridization were done with the nonradioactive digoxigenin-system (DIG-system, Boehringer Mannheim) combined with chemiluminescence detection using AMPPD (Tropix, Bedford, USA) as a substrate for antibody-bound alkaline phosphatase. The digoxigenin-labelled probes were used at a concentration of 40 ng per ml of hybridization solution. DNA hybridization was carried out in plastic bags for 16 h at 52° C. The filters were washed under stringent conditions. After incubation with Anti-DIG-Alkaline Phosphatase-Fab fragments the AMPPD-substrate was added to the blots. The chemiluminescence signals were documented on Kodak X-ray films (Kodak, Rochester, USA). In order to quantify the signal strength, a method was established to measure the light intensity by photon counting in a scintillation counter. For this purpose the bands were cut out, sealed in cellophane foil and placed in scintillation vials. Photons were counted by a Tri-Carb liquid scintillation analyzer (Packard, Meriden, USA). Before evaluation of the cpm values, the kinetics of the AMPPD destruction had to be examined. Using the results from that examination the single cpm values could be corrected so that the data were comparable. With the data obtained, the relative amounts of *B. nigra* or *B. napus* nuclear DNA in the probes were calculated by comparison of the values for the hybrids with the respective calibration lines.

Results

Recovery of hybrid clones

Four fusion experiments were carried out between *B. napus* 'Liropa' and the irradiated transgenic *B. nigra* line. In these experiments we used four different doses of gamma irradiation to inactivate the donor. The following controls were cultured in each experiment: protoplasts of *B. napus* only, irradiated transgenic *B. nigra* protoplasts and *B. napus* protoplasts mixed with irradiated transgenic *B. nigra* protoplasts without fusion treatment. All the cultures were subjected to selection (after 21 days) in M1-liquid medium containing hygromycin-B. Individual hygromycin-resistant calli were transferred from the liquid medium to 2-N SeaPlaque-agarose medium containing hygromycin-B. They were cultivated for two to three passages on this medium. Two to three months after fusion, 332 calli resistant to hygromycin-B were selected. Many of the calli did not redifferentiate. However, thirty of these clones regenerated vigorously and gave rise to numerous plants that survived the transfer into soil. All regenerated plants exhibited nopaline synthase activity indicating the presence of the nopaline gene (data not shown). No hygromycin-resistant calli were recovered from any of the controls.

Morphology of regenerated hybrids

Most of the regenerants had the upright growth habit of *B. napus*. However, some of the regenerants showed pilosity, indicating the presence of partial genetic information from *B. nigra*. A small proportion (3%) grew as a basal rosette; the plants were shorter and displayed less apical dominance. Many of the hybrids started flowering without vernalization. Floral morphology was much closer to that of *B. napus* than that of *B. nigra*, but the hybrids were predominantly sterile. Plants regenerated from the same callus were morphologically not identical. Chromosome counts revealed that these hybrid plants possessed different chromosome numbers (46–70 chromosomes). On the other hand, subclones produced via multiple shoot formation on hormone-free MS-agar medium were uniform in morphology and chromosome number. Plants regenerated from three clones produced a small amount of pollen. These plants formed a few viable seeds when backcrossed with *B. napus* as the pollen parent. Plants from one clone, which formed a small amount of viable pollen, have been selfed. Only one clone after

Table 1. Response to infection with *Phoma lingam* 6–8 weeks after inoculation

Dose rate (Gray)	No. of plants inoculated	Disease scale		
		Healthy	With light symptoms	Diseased
450	4	4	0	0
800	17	8	0	9
1000	29	17	1	11
1300	28	12	2	14
Σ	78	41	3	34
%	100	52.6	3.8	43.6
Control				
B. napus (Recipient)	48	0	18	30
B. nigra (Donor)	58	58	0	0

Table 2. Response to infection with *Plasmodiophora brassicae* 6–8 weeks after inoculation

Dose rate (Gray)	No. of plants inoculated	Disease scale		
		Healthy	With light symptoms	Diseased
450	9	2	4	3
800	50	7	8	35
1000	80	40	16	24
1300	54	5	13	36
Σ	193	54	41	98
%	100	28	21.2	50.8
Control				
B. napus (Recipient)	41	0	0	41
B. nigra (Donor)	88	88	0	0

selfing and three clones after backcrossing have been obtained.

Black leg and clubroot screening

The hybrid plants that showed a vigorous growth and well-developed roots were tested for resistance against the two pathogenic fungi. A total of 78 hybrid plants regenerated from selected hybrid colonies were tested for resistance to *P. lingam* in a growth chamber (Table 1). Susceptible plants collapsed within 8 weeks of inoculation. In order to guarantee the significance of the resistance tests, control material was inoculated simultaneously. This control material consisted of plants of a susceptible *B. napus* cultivar (Lesira) and a tolerant cultivar (Jet Neuf) (data not shown). The recipient *B. napus* and the *P. lingam*-resistant *B. nigra* line were also tested (Table 1). The hybrids which showed no or only superficial lesions were rated as resistant. The results from *P. lingam* inoculation are summarized in Table 1. The proportion of plants which gave a resistance response (56.4%) was higher than those with a susceptible response (43.6%). A total of 44 of 78 hybrid plants (classified: healthy or with light symptoms) showed the same level of resistance as *B. nigra*. All control plants of *B. napus* showed profound lesions, all plants of *B. nigra* appeared healthy.

The results obtained after inoculation with spores of *Plasmodiophora brassicae* are shown in Table 2. A total of 193 hybrid plants were inoculated with spores simultaneously with control plants. The range of visible symptoms was classified into three disease severity classes: healthy without any symptoms (resistant), small clubs that did not affect plant development (moderately resistant), well developed clubs, plant growth reduced (susceptible). In total 54 hybrid plants were classified as resistant. These plants showed the same high level of resistance as the resistant donor *B. nigra*. Forty one hybrids showed only slight symptoms and 98 hybrid plants were susceptible (Table 2). All control plants including the highly susceptible Chinese cabbage, which served as a positive control for testing the virulence of the spore inoculum, responded as expected (Table 2). Four hybrid clones displayed the resistance to both pathogenic fungi. The other clones displayed either resistance to only one pathogen or were susceptible to both fungi.

Slot blot analyses of asymmetric hybrids

Before starting recloning or backcrossing it was necessary to select those resistant hybrids that had a relatively low proportion of the donor-genome to the recipient-genome. In order to determine the relative amounts of *B. nigra* and *B. napus* DNA in the asymmetric hybrids, slot blot hybridization was performed according to the method of Imamura et al. (1987) modified by using the nonradioactive digoxigenin/chemiluminescence technique. Each blot contained a concentration series of the DNA from one of the parents and DNA from the hybrid plants. The blots were probed with three species-specific, repetitive sequences that represent a major part of the respective genomes. After detection using the chemiluminescence substrate AMPPD, the hybridization signals were quantified by photon counting in a scintillation counter. Kinetic studies of the chemiluminescence reaction showed that the strength of the signal increased over time in a linear way. The kinetics of the reaction also depended on the concentrations of the DNA applied. Taking this into account, an evaluation of the obtained data was possible. The cpm values of the calibration series were plotted against the DNA amounts. The results showed linearity within the range of 25 to 1000 ng DNA per slot in all three cases (data not shown). The relative amounts of a specific repeat in the hybrid DNA were calculated by comparison with the respective calibration lines. Figure 1 shows the ratio between the *B. nigra*-specific repeats BN34 and BN35 and the *B. napus*-specific repeat BcKB4 in the hybrids. Apart from differences in the ratios between the hybrids there were differences between the values for a single hybrid line depending on the repeat which was used as a probe. The latter variations could be due to the different types of the *B. nigra*-specific repeats (tandem- and dispersed repeat). The nuclear DNA content of *B. napus* (2N = 38) amounts to 2.34 pg and that of *B. nigra* (2N = 16) to 0.97 pg (Arumuganathan & Earle, 1991). We estimated the nuclear DNA content of the *B. nigra* 460 Hyg^R-donor-line (36 Chromosomes) at 2.2 pg based on the value for the *B. nigra* wild type. This leads to a theoretical value for the donor/recipient-ratio of a symmetric hybrid of about 0.94. Thus it is obvious that the clones 46a, 26a and 27b, with values between 0.3 and 0.5, contained much less *B. nigra* DNA than *B. napus* DNA and showed a remarkable elimination of the donor genome.

A further result of the investigations was the detection of a dose/effect relationship between increasing irradiation doses and the elimination of the donor genome. The values of the BN34/BcKB4 ratio decreased with higher X-ray doses (Fig. 2). This decrease is highly significant ($P = 0.001$). According to the Newman-Keuls multiple comparison-test, the differences between 250 and 350 Gy are not significant whereas the differences between 1000 and 1300 Gy are significant at the 0.05 but not at the 0.01 significance level. The investigation of the BN35/BcKB4 ratio showed no significant differences between the values. Some of the hybrid lines used for the molecular analysis derived from experiments of asymmetric somatic hybridization (Sacristán et al., 1989) where X-ray doses of 100–400 Gy were applied.

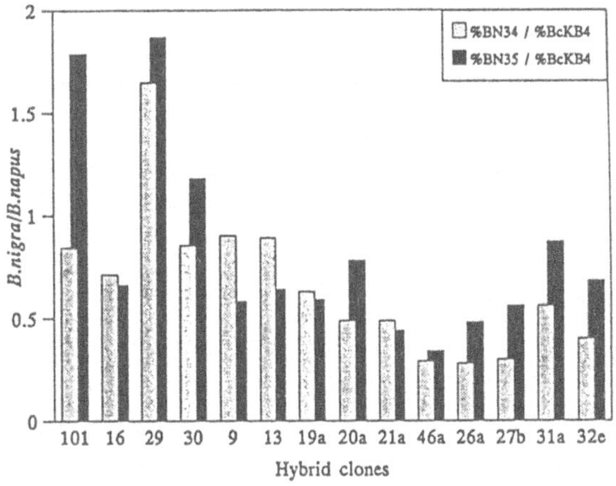

Fig. 1. Ratio between *B. nigra* and *B. napus* DNA in hybrid clones. The values are based on the relative amounts of the *B. nigra*-specific repetitive sequences BN34 and BN35 on the one hand and of the *B. napus*-specific repeat BcKB4 on the other. The data were obtained by nonradioactive slot blot hybridization and chemiluminescence detection. Quantification of signals was performed by photon counting in a liquid scintillation counter.

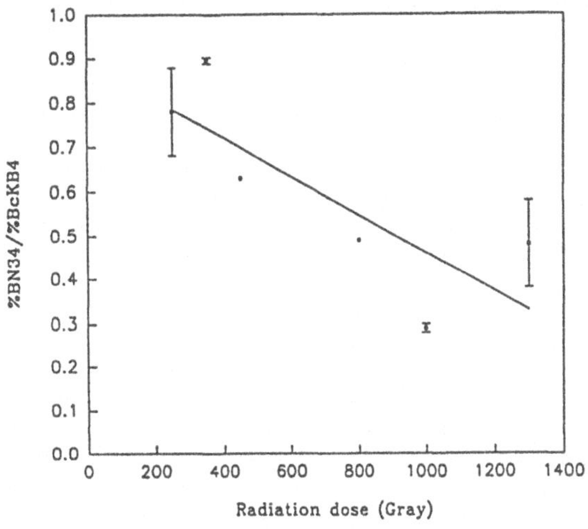

Fig. 2. Effect of X-ray irradiation on the degree of donor genome elimination. The values for %BN34/%BcKB4 are the mean of two (250, 350, 800, 1300 Gy) or three (1000 Gy) independent clones irradiated with the corresponding X-ray doses. The 450 Gy-value is based on analysis of a single clone. The correlation coefficient of the regression line is r = - 0.8.

Discussion

The wild species *B. nigra* is an important source of germplasm because this species possesses resistance to *P. lingam* (Sacristán, 1982; Sjodin & Glimelius, 1989; Zhu et al. 1993) and *P. brassicae* (Sacristán et al., 1989).

Here, we have demonstrated that asymmetric somatic hybrids can be obtained between the two species. Numerous hybrids have been produced expressing both or only one resistant trait of *B. nigra*. The morphology of the hybrid plants suggests the predominance of *B. napus* genetic information. However, the fact that the hybrids are hygromycin-B resistant indicates the transfer of genetic information from *B. nigra* to *B. napus*. Therefore it seems to be likely that the pathogen-resistant traits observed in the hybrids are also the result of the transfer of the respective genetic information from *B. nigra*, as no resistant plants were isolated from control *B. napus* protocalli. Schieder & Kohn (1986) proposed that the production of asymmetric somatic hybrids may be an alternate procedure to molecular biological techniques for the transfer of desired genes from one species to another. The results communicated here support this proposal. Other authors have reported the successful transfer of agronomically-relevant genetic information using the technique of asymmetric somatic hybridization (Bauer, 1990, Glimelius et al., 1991).

Molecular biological investigations using species-specific and repetitive DNA probes, undertaken on some of the asymmetric somatic hybrids, have shown that the DNA content of the donor plant species *B. nigra* varies. Some hybrids, e.g. 101 and 29, were found with a higher donor-DNA content than present in both parents combined. Such hybrids may be the results of multiple fusion products. However, most hybrids contained only a small proportion of donor DNA (Fig. 1).

The protoplasts of *B. nigra* were irradiated with increasing doses of X-rays. In the case of the tandem-repeat BN34, the results indicate that with increasing doses of X-rays the amount of the donor DNA which has been transferred into the recipient plant species decreases (Fig. 2). An effect of radiation dose on elimination of donor DNA in donor-recipient fusions was recently reported by Trick et al. (1994) who investigated asymmetric hybrids between *Nicotiana tabacum* and *N. plumbaginifolia*. Using the dispersed repeat BN35 as a probe, the dose/effect relationship was not as clear. It seems that with very high irradiation doses (e.g. > 1000 Gy) no further reduction of donor DNA can be achieved. To prove this we would have to carry out fusion experiments where irradiation doses higher than 1300 Gy were applied to the donor protoplasts. The variation of the results depending on the type of repeat used as a DNA-probe, indicate the necessity of the use of species-specific probes that are not chromosome- or chromosomeregion-specific. This could be achieved by the use of total genomic DNA from one fusion partner as a probe and a blocking step with DNA from the other fusion partner (Anamthawat-Jónsson et al., 1990).

A further reduction in the amount of the donor DNA may achieved by the recloning of the hybrids via protoplast regeneration. For this purpose we selected some of the asymmetric somatic *Brassica*-hybrids which were resistant to both fungi and revealed great donor-DNA elimination in the molecular analysis. After recloning these hybrids, lines were regenerated which had lost their resistance to hygromycin, the trait used for the selection of the hybrids. A Southern blot analysis of such hygromycin-sensitive clones showed that in many of these lines the respective donor DNA sequence had also disappeared (Tzscheetzsch, 1993). Whether or not resistance to the pathogen was also lost is under investigation. Recloning the hybrids via protoplast regeneration may be a method of converting a transgenic hybrid plant into a plant without the selectable marker gene. Since the fertility of the plants is very low, few have been obtained after selfing or backcrossing. If the fertility can be raised, selfing or backcrossing may also be a method of reducing the amount of the donor DNA or eliminating the selectable marker gene. One could argue that the use of a transgenic donor line results in a strong selection pressure to a single chromosome or chromosome region. The results, however, demonstrate that the DNA content of the donor line in the hybrids varies, showing that apart from the marked chromosome other chromosomes or chromosome regions have been transferred. Moreover, the recovery of pathogen-sensitive and resistant hybrid lines confirms this observation.

References

Anamthawat-Jonsson, K., T. Schwarzacher, A.R. Leitch, M.D. Bennet & J.S. Heslop-Harrison, 1990. Discrimination between closely related *Triticeae* species using genomic DNA as a probe. Theor Appl Genet 79: 721–728.

Arumuganathan, K. & E.D. Earle, 1991. Nuclear DNA content of some important plant species. Plant Mol Biol Rep 9: 208–218.

Bauer, R , 1990 Somatic plant hybridization – a review Arch Zuchtungsforsch 1 66–79

Buczacki, S T , H Toxopeus, T D Mattusch, G R Dixon & L A Hobolt, 1975 Study of physologic specialization in *Plasmodiophora brassicae* proposals for attempted rationalization through an international approach Trans Br Mycol Soc 65 295–303

Chuong, P V , K P Pauls & W D Beversdorf, 1987 Protoplast cul ture and plant regeneration from *Brassica carinata* Braun Plant Cell Rep 6 67–69

Crisp, P, I R Crute, R A Sutherland, S M Angeli, K Bloor, H Burgess & P Gordon, 1989 The exploitation of genetic resources of *Brassica oleracea* in breeding for resistance to clubroot (*Plasmodiophora brassicae*) Euphytica 42 215–226

Gamborg, O L , R A Miller & K Ojima, 1968 Nutrient requirements of suspension cultures of soybean root cells Exp Cell Res 50 151–158

Gerdemann-Knorck, M , M D Sacristan, C Braatz & O Schieder, 1994 Utilization of asymmetric somatic hybridization for the transfer of disease resistance from *Brassica nigra* to *Brassica napus* Plant Breeding 113 106–113

Glimelius, K , C Fahlesson, M Landgren, C Sjodin & E Sundberg, 1991 Gene transfer via somatic hybridization in plants Trends in Biotech 9 24–30

Gupta, V , V Jagannathan & S Lakshmikumaran, 1990 A novel AT-rich tandem repeat of *Brassica nigra* Plant Sci 68 223–229

Gupta, V , G LakshmiSita, M S Shaila, V Jagannathan & S Lakshmikumaran, 1992 Characterization of species specific repeated DNA sequences from *B nigra* Theor Appl Genet 84 397–402

Heslop-Harrison, J S , G E Harrison & I J Leitch, 1992 Reprobing of DNA DNA *in situ* hybridization preparations Trends in Genetics 8 609–614

Imamura, J , M W Saul & I Potrykus, 1987 X-ray irradiation-promoted asymmetric somatic hybridisation and molecular analysis of the products Theor Appl Genet 74 445–450

Jacobsen, B J & P H Williams, 1970 Control of cabbage clubroot using benomyl fungicide Plant Dis Rep 54 456–460

Labarca, C & K Paigen, 1980 A simple, rapid and sensitive DNA assay procedure Anal Biochem 102 344–352

Li, L C & H W Kohlenbach, 1982 Somatic embroygenesis in quite a direct way in cultures of mesophyll protoplasts of *Brassica napus* L Plant Cell Rep 1 209–212

Murashige, T & F Skoog, 1962 A revised medium for rapid growth and bio assays with tobacco tissue cultures Physiol Plant 15 479–497

Otten, L & R A Schilperoort, 1978 A rapid microscale method for the detection of lysopine and nopaline dehydrogenase activities Biochem Biophys Acta 527 497–500

Roger, S O & A J Bendich, 1985 Extraction of DNA from milligram amounts of fresh, herbarium and mummified plant tissues Plant Mol Biol 5 69–76

Sacristan, M D , 1982 Resistance response to *Phoma lingam* of plants regenerated from selected cell and embryogenic cultures of haploid *Brassica napus* Theor Appl Genet 61 193–200

Sacristan, M D & F Hoffmann, 1979 Direct infection of embryogenic tissue cultures of haploid *Brassica napus* with resting spores of *Plasmodiophora brassica* Theor Appl Genet 54 129–132

Sacristan, M D & M Gerdemann, 1986 Different behaviour of *Brassica juncea* and *B carinata* as source of *Phoma lingam* resistance in experiments of interspecific transfer to *B napus* Plant Breeding 97 304–314

Sacristan, M D , M Gerdemann Knorck & O Schieder, 1989 Incorporation of hygromycin resistance in *Brassica nigra* and its transfer to *B napus* through asymmetric protoplast fusion Theor Appl Genet 78 194–200

Schieder, O , M Gerdemann-Knorck & M D Sacristan, 1991 The use of protoplast fusion and transformation for the production of asymmetric somatic hybrids Israel J Bot 40 99–107

Schieder, O & H Kohn, 1986 Protoplast fusion and generation of somatic hybrids p 569–588 In I Vasil (Ed) Cell Culture and Somatic Cell Genetics of Plants Vol III Academic Press Inc , Orlando

Sjodin, C & K Glimelius, 1989 Transfer of resistance against *Phoma lingam* to *Brassica napus* by asymmetric somatic hybridization combined with toxin selection Theor Appl Genet 28 513–520

Trick, H , A Zelcer & G W Bates, 1994 Chromosome elimination in asymmetric somatic hybrids effect of gamma dose and time in culture Theor Appl Genet 88 965–972

Tzcheetzsch, C , 1993 Reklonierung asymmetrischer Hybriden von *Brassica napus* (+) *Brassica nigra* Diploma thesis, FU Berlin

Zelcer, A , D Aviv & E Galun, 1978 Interspecific transfer of cytoplasmatic male sterility by fusion between protoplasts of normal *Nicotiana sylvestris* and X-ray irradiated protoplasts of male sterile *N tabacum* Z Pflanzenphysiol 90 397–407

Zhu, J S & A Spanier, 1991 Resistance sources to *Phoma lingam* and *Alternaria brassicae* Cruciferae Newsl Eucarpia 14/15 143

Zhu, J S , D Struss & G Robbelen, 1993 Studies on resistance to *Phoma lingam* in *Brassica napus-Brassica nigra* addition lines Plant Breeding 111 192–197

Euphytica **85**: 255–268, 1995.

Microprotoplast fusion technique: a new tool for gene transfer between sexually-incongruent plant species

K.S. Ramulu, P. Dijkhuis, E. Rutgers, J. Blaas, W.H.J. Verbeek, H.A. Verhoeven & C.M. Colijn-Hooymans
Department of Cell Biology, Centre for Plant Breeding and Reproduction Research (CPRO-DLO), P.O. Box 16, 6700 AA Wageningen, The Netherlands

Key words: microprotoplast fusion, partial genome transfer, monosomic additions, kanamycin resistance, β-glucuronidase, gene expression, potato

Summary

Various aspects of a microprotoplast fusion technique and the strategies followed for intergeneric partial genome transfer (one or a few chromosomes) and alien genes from sexually-incongruent donor species to recipient species are described. The essential requirements of the microprotoplast fusion technique are the induction of micronuclei at high frequencies, as well as the isolation and enrichment of sub-diploid microprotoplasts in donor species, efficient fusion of the donor microprotoplasts with normal recipient protoplasts and stable regeneration of plants from fusion products. The results on the production of microprotoplast hybrid plants between the transformed donor lines of *Solanum tuberosum* and *Nicotiana plumbaginifolia* carrying various genetic markers, and a recipient line of *Lycopersicon peruvianum* or *Nicotiana tabacum*, and on the transfer and expression of alien genes (kanamycin resistance, β-glucuronidase) are presented. The data obtained on microprotoplast hybrid plants between *S. tuberosum* and *L. peruvianum* showed that many of the hybrids contained one potato chromosome carrying nptII and GUS, and 24 or 48 *L. peruvianum* chromosomes (monosomic additions), and that they were male- and female-fertile. Various applications of chromosome transfer by this technique, especially for economically-important traits (e.g. disease or stress resistance) from sexually-incompatible wild species, for construction of chromosome-specific DNA libraries through microdissection and microcloning of chromosomes, or by flow-sorting of chromosomes for genome analysis, are discussed.

Introduction

The increasing demand in recent years for the production of cultivars resistant to diseases, insect pests and stress, is insufficiently met by classical hybridization techniques, because of the sexual incongruity between the wild species/relatives (donor sources for resistances) and the cultivated species. Firstly, it is difficult to obtain hybrids in species between which crossing barriers exist. Further, in interspecific or intergeneric crosses, in which only a low percentage of F1 hybrids may be obtained, these are mostly sterile because of the disturbances in various steps of gamete development (reviewed by Sybenga, 1992). In this case several tedious and time-consuming steps have to be carried in order to transfer the desired genes: doubling of the chromosomes of the hybrids to produce amphidiploids, and repeated backcrossing of the amphidiploid with the recipient for several generations, followed by selection to eliminate the undesirable traits of the donor parent and retain the gene of interest. Extensive embryo rescue procedures and growth of plant materials under specific controlled conditions for recovering hybrid plants are the two other requisites for classical hybridization.

Genetic manipulation at the DNA or cellular level makes feasible the transfer of genes across sexual barriers, even from the tertiary gene pool, which constitutes a rich source of germplasm (Gasser & Fraley, 1989; Sybenga, 1989; Potrykus, 1990). Transformation using recombinant DNA technology requires the availability of the cloned genes for traits of interest with the proper promoter sequences. However, for disease or stress resistance, especially those which are controlled by polygenes, no cloned genes are as yet available.

Somatic hybridization is a suitable method for transfer of genes (polygenic traits, unidentified or uncloned genes) from sexually-incompatible wild species to crop plants, and to generate novel nucleus-cytoplasm combinations (reviewed by Gleba & Sytnik, 1984; Glimelius, 1988; Puite, 1992; Jacobsen et al., 1992; Gilissen et al., 1992; Cardi et al., 1993). To achieve partial genome transfer, generally asymmetric hybridization was carried out using irradiated donor protoplasts and normal recipient protoplasts. However, both symmetric and asymmetric somatic hybridization techniques have so far not given satisfactory results, because the hybrids were mostly unstable, with all or at least several donor chromosomes carrying undesired genes, and were often sterile. Furthermore, several authors reported the damaging effects of irradiation on the genetic composition of fusion products or hybrids, with adverse effects on plant regeneration as well as on growth, phenotype and fertility of the hybrid plants (Famelaer et al., 1990; Wijbrandi et al., 1990; Wolters et al., 1991; Derks et al., 1992; Puite & Schaart, 1993; Schoenmakers et al., 1994).

Another method, metaphase chromosome – mediated gene transfer, has been extensively used in mammalian cell systems for the transfer of specific genes and for the construction of chromosome-specific DNA libraries (see Carrano et al., 1979). The chromosomes were found to integrate into the recipient genome only as small fragments, which could be maintained under selection pressure. In the case of plants, due to the difficulties in cell synchronization, chromosome isolation, flow cytometric sorting and identification of chromosomes, and the absence of clear chromosome banding patterns, this technique has met with limited success only (reviewed by De Laat et al., 1989).

On the other hand, microprotoplast fusion (the fusion of donor microprotoplasts containing one or a few chromosomes with recipient normal protoplasts) can offer a promising tool for the transfer of desirable mono- or polygenic traits, even from a donor species that can not be hybridized sexually with the recipient crop species. Microprotoplast fusion makes feasible the production of monosomic addition lines (one extra donor chromosome and the complete genome of the recipient species) in a single step, thus reducing the number of time-consuming backcrosses necessary for obtaining such lines by generative methods.

In this presentation, various aspects of a microprotoplast fusion technique together with the strategies which enabled the transfer of partial donor genome (one or a few chromosomes) and alien genes will be discussed.

Materials and methods

Genotypes

The transformed cell lines *Solanum tuberosum* (line 413) and *Nicotiana plumbaginifolia* (Doba line) were used as the donor source for the induction of micronuclei and isolation of microprotoplasts. The potato line carries various genetic markers, i.e. kanamycin resistance, β-glucuronidase activity, opine synthesis, hairy root phenotype and hormone autotrophy, while the *N. plumbaginifolia* line carries kanamycin resistance; these markers were introduced by transformation with *Agrobacterium* strains. The details on the origin of these genotypes have been published earlier (Ramulu et al., 1993, 1994).

In-vitro-grown shoot cultures of a hygromycin-resistant transformed line of *Lycopersicon peruvianum* PI 128650 (2n = 2x = 24) (Koornneef et al., 1987) kindly provided by Prof. M. Koornneef, Department of Genetics, Agricultural University, Wageningen, and *Nicotiana tabacum* (2n = 4x = 48) cv. Petit Havana SR1 (Maliga et al., 1975) were used as the recipient lines for protoplast isolation.

Cell suspension and shoot cultures

Details of the culture conditions and media for *S. tuberosum* and *N. plumbaginifolia* suspension cells were described earlier (Ramulu et al., 1993, 1994). The axenic shoots of *L. peruvianum* were subcultured monthly on MS medium (Murashige & Skoog, 1962) supplemented with 2% sucrose and hygromycin-B (Duchefa) at 25 mg l^{-1}, while those of *N. tabacum* were cultured on MS medium supplemented with 3% sucrose.

Induction of micronuclei in donor cell lines

Early log-phase suspension cells of *N. plumbaginifolia*, one day after subculture, were treated with inhibitors of DNA synthesis, i.e. hydroxy urea at 10 mM or aphidicolin at 15 μm for 24 h followed by repeated washing with the culture medium and treatment with the spindle toxins, amiprophosmethyl (APM) at 32 μM or cremart at 3.7 μm for 24 h. For *S. tuberosum*, actively growing early logphase suspension cells at one day after subculture were treated with cremart at 7.5 μm for 48 h. Details on the chemicals (source, structure) and treatment procedures, as well as the optimization of concentrations for inducing the maximum frequencies of micronuclei, have been reported previously (Verhoeven et al., 1990, 1991a, 1991b; Ramulu et al., 1993, 1994).

Isolation of microprotoplasts

After treatment with APM or cremart, suspension cells of *N. plumbaginifolia* or *S. tuberosum* were incubated for 18 h in a cell wall – digesting enzyme mixture, which consisted of cellulase Onozuka-RIO (1%), macerozyme Onozuka-RIO (0.2%) (Yakult Honsha Co, Tokyo, Japan), half-strength V-KM medium (Bokelmann & Roest, 1983) with 0.2 M glucose and 0.2 M mannitol, but no hormones (Ramulu et al., 1993). Cytochalasin – B (20 μM) and cremart (7.5 μM) or APM (32 μM) were added at the time of enzyme incubation to prevent the formation of microfilaments and fusion of micronuclei, respectively, during the protoplast isolation. After enzyme incubation, the samples were filtered through 297 μm and 88 μm nylon meshes and repeatedly washed with half-strength V-KM medium (Bokelmann & Roest, 1983) with macro- and microelements and 0.24 M NaCl (pH 5.6).

The purified dense suspension of mono- and micronucleated protoplasts was loaded onto a continuous iso-osmotic gradient of Percoll and exposed to a high-speed centrifugation at 100.000 g for 2 h (Verhoeven & Ramulu, 1991; Ramulu et al., 1993). The bands obtained after centrifugation contained evacuolated protoplasts, microprotoplasts and cytoplasts, and, were sequentially filtered through nylon sieves of decreasing pore size isolating the smaller sub-diploid microprotoplasts (Ramulu et al., 1993).

Protoplast isolation from shoot cultures and suspension cells

Protoplasts were isolated from leaf pieces of shoot cultures of the recipient lines of *L. peruvianum* and *N. tabacum* after overnight (16 h) incubation in 1% (w/v) cellulase-R10 (Onozuka) and 0.2% (w/v) macerozyme-R10 (Yakult Honsha Co. Ltd., Tokyo, Japan) dissolved in half-strength V-KM medium (Bokelmann & Roest, 1983). This medium contains 0.2 M glucose and 0.2 M mannitol, but no hormones. Protoplast yield ranged between 0.5×10^6 and 4.5×10^6 per gram leaf material of *L. peruvianum*, and between 0.4×10^6 and 1.0×10^6 per gram leaf material of *N. tabacum* in various experiments.

Protoplasts were isolated from cell suspensions of the donor lines of *S. tuberosum* and *N. plumbaginifolia* for symmetric fusions, to be used as controls for microprotoplast fusions. The cell suspensions were incubated for 16 h in 1% (w/v) cellulase-R10 and 0.2% (w/v) macerozyme-R10 dissolved in half-strength V-KM medium (Bokelmann & Roest, 1983). The yield of protoplasts varied from 1.5×10^6 to 3.3×10^6 per 1 ml packed cell volume of *S. tuberosum* suspension cells, and from 1.1×10^6 to 5.0×10^6 per 1 ml packed cell volume of *N. plumbaginifolia* suspension cells in various experiments.

Microprotoplast fusions and symmetric fusions

Table 1 gives details of the characteristics of parental genotypes used for microprotoplast fusions. Fusions were carried out in 2–3 experiments between the donor microprotoplasts and the recipient protoplasts using a polyethylene glycol (PEG)-based mass fusion protocol, modified after Menczel et al. (1981) and Derks et al. (1993), which was briefly reported earlier (Ramulu et al., 1992).

Table 2 shows the details of fusion combinations, and media used for culture, selection and plant regeneration at various periods. Microprotoplasts and protoplasts were mixed in ratios of 1 : 1 or 2 : 1 in 6 cm Falcon Petri dishes in W5 medium and plated at a density of 1×10^6 ml^{-1}. After 20 min, PEG (PEG 4000) was applied at 8–10% (final concentration). After a further 7 min, PEG and W5 medium were carefully removed using a Pasteur pipette, and high pH buffer solution was slowly added. Twenty minutes later, when the microprotoplast-protoplast mix had settled down in the Petri dish, rinsing was done with liquid TM2 or H460M medium, according to the type of fusion. Afterwards, 2 ml of TM2 or H460M medium was added to each Petri dish, the density during culture being 0.2×10^6 ml^{-1}. As a control for microprotoplast fusions (i.e. fusions of microprotoplasts (+) protoplasts), we have also carried out symmetric fusions between the donor cell suspension protoplasts and recipient leaf protoplasts under the same culture conditions and at a similar plating density. The

Table 1 Characteristics of parental genotypes used for microprotoplast fusions

Genotype	Source material (ploidy)	Fusion partner	Selectable markers used	Other markers
Solanum tuberosum, line 413	Suspension cells (2n = 3x = 36)	Donor-microprotoplasts	Kan^R	GUS, OP, HR, HA
Nicotiana plumbaginifolia, line Doba	Suspension cells (2n = 4x = 40)	Donor-microprotoplasts	Kan^R	–
Lycopersicon peruvianum, line PI 128650	Shoot culture (2n = 2x = 24)	Recipient-protoplasts	Hyg^R	–
Nicotiana tabacum, cv Petit Havana, SR 1	Shoot culture (2n = 4x = 48)	Recipient protoplasts	–	Streptomycin-resistance

Kan^R kanamycin resistance, Hyg^R hygromycin resistance, GUS β-glucoronidase, OP opine synthesis, HR hairy roots, HA hormone autotrophy

Table 2 Details of fusion combinations, and media used for culture, selection and plant regeneration at various periods

Fusion combinations			Day 5	Day 12 and 19	Day 25	Day 40
Donor	(+)	Recipient	Dilution in CI liquid medium	Dilution and selection in CI liquid medium	Selection on solid CG medium	Regeneration medium
St mpps	(+)	Lp pps	TM2	TM2 + Kan50 + Hyg25	TM3 + Kan100 + Hyg50	TM4
St pps	(+)	Lp pps	TM2	TM2 + Kan50 + Hyg25	TM3 + Kan100 + Hyg50	TM4
St mpps	(+)	Nt pps	H460M	H460M + Kan50	AG + Kan100	MSR1
St pps	(+)	Nt pps	H460M	H460M + Kan50	AG + Kan100	MSR1
Np mpps	(+)	Lp pps	TM2	TM2 + Kan50 + Hyg25	TM3 + Kan100 + Hyg50	TM4
Np pps	(+)	Lp pps	TM2	TM2 + Kan50 + Hyg25	TM3 + Kan100 + Hyg50	TM4

St *Solanum tuberosum*, Lp *Lycopersicon peruvianum*, Nt *Nicotiana tabacum*, Np *Nicotiana plumbaginifolia* mpps Microprotoplasts, pps Protoplasts CI Callus induction medium, CG Callus growth medium
Kan 50 Kanamycin at 50 mg l^{-1}, Kan 100 Kanamycin at 100 mg l^{-1}
Hyg 25 Hygromycin at 25 mg l^{-1}, Hyg 50 Hygromycin at 50 mg l^{-1}

details of the composition of media i.e. TM-2, -3, and -4 used for callus induction and growth and plant regeneration, respectively, have been reported earlier (Shahin, 1985; Derks, 1992). The culture media H460M, AG and MSRI were used with some modifications for callus induction and growth, and plant regeneration, respectively. The medium H460M contained K3 macro elements (Nagy & Maliga, 1976), MS (Murashige & Skoog, 1962) micro elements (5 × diluted), 7.46 g/l Fe EDTA and sugar stock (Kao & Michayluk, 1975). The medium AG consisted of AG micro elements (Caboche, 1980), MS micro elements (10 × diluted), 7.46 gl^{-1} Fe EDTA, vitamins (Morel & Wetmore, 1951), 3% sucrose, 5% mannitol, 0.1 mg l^{-1} naphthalene acetic acid and 1 mg l^{-1} benzyl aminopurine. The MSRI medium contained MS macro- and micro-elements (normal concentration) and 7.46 gl^{-1} Fe EDTA. As shown in Table 2, the hybrid calli were selected using kanamycin and hygromycin selection, or only kanamycin selection, depending on the type of fusion. When the resistant calli turned green on solid callus growth medium, they were transferred to the regeneration medium without adding kanamycin or hygromycin. The regenerated shoots from *S tuberosum* (+) *L. peruvianum* fusions were rooted on MS medium (Murashige & Skoog, 1962) supplemented with 2% sucrose, and those from *S. tuberosum* (+) *N. tabacum* or *N. plumbaginifolia* (+) *L. peruvianum* were rooted on MS medium supplemented with 2% sucrose and indole butyric acid at 0.025 mg l^{-1}.

Kanamycin resistance and GUS assays

Kanamycin resistance of shoots regenerated from Kan^R calli was determined on the basis of root induction from shoots grown on MS medium supplemented with 3% sucrose and kanamycin at 50 mg l^{-1}. The GUS assay was performed as described by Jefferson et al. (1987), using a modified extraction buffer containing 50 mM sodium phosphate buffer pH 7.5, 10 mM Na_2-EDTA (ethylene diamine tetra acetic acid), and 0.1% (v/v) Triton X-100. Ironcyanide solution (0.5 mM potassium ferricyanide in water) and X-Gluc solution (1 mM 5-bromo 4-chloro 3-indolyl β-glucuronide in dimethylformamide) were added to the extraction buffer. After an incubation period of 16 h at 37° C, the appearance of a blue colour was indicative of GUS activity. Chlorophyll was removed from stained tissue by ethanol extraction and the presence of GUS staining was observed using a dissection microscope.

Karyotype and genomic in situ *hybridization*

Chromosome counts and karyotype analysis were performed on Feulgen-stained root tip metaphase cells of plants regenerated from Kan^R calli as described earlier (Ramulu et al., 1983). As *L. peruvianum* and *S. tuberosum* have small chromosomes and similar karyotypes (Ramulu et al., 1977; Pijnacker & Ferwarda, 1984), their accurate identification through classical cytogenetic methods is difficult. Therefore, genomic *in situ* hybridization (GISH) was carried out for chromosome identification of plants derived from *S. tuberosum* (+) *L. peruvianum* fusions. Actively growing young root tips were pretreated in an aqueous solution of 2 mM 8-hydroxyquiniline for 2.5 h at 17° C and fixed in a solution of 3 : 1 ethanol : acetic acid for 24 h or more. The fixed root tips were washed in water and incubated in an enzyme mixture containing 0.1% pectolyase Y23, 0.1% cellulase RS and 0.1% cytohelicase in 10 mM citrate buffer, pH 4.5, for 1 h at 37° C. The root tips were carefully transferred to a grease-free microscopic slide and the cells were spread according to the technique of Pijnacker & Ferwarda (1984). Various steps of DNA denaturation, *in situ* hybridization and detection/amplification were performed according to Leitch & Heslop-Harrison (1993), Schwarzacher & Heslop-Harrison (1993) and Schwarzacher & Leitch (1993). Total genomic DNA isolated from the leaf material of the donor *S. tuberosum* line 413 was used as probe and the leaf DNA from the recipient *L. peruvianum* plants as blocking-DNA. Labelling of *S. tuberosum* DNA was done either by an indirect method or a direct method. In the indirect method, the DNA was sheared by passage through a syringe until the fragments attained a size of 1–10 kb, while in the direct method, the DNA was sonicated so as to obtain 1.0–2.0 kb fragments. In the indirect method, labelling was done with digoxigenin-11-dUTP (Boehringer-Mannheim) and detected with anti-digoxigenin Fluos (fluorescein isothiocyanate) raised in sheep (Boehringer-Mannheim) and amplified with anti-sheep-FITC raised in rabbit (Boehringer-Mannheim) according to a standard random primer labelling protocol. In the direct method, the labelling was carried out with Fluorescein-high prime kit fluorescein-12-dUTP (Boehringer-Mannheim). The DNA of *L. peruvianum* was sonicated for 10 sec (12 micron amplitude) which resulted in fragments of about 700 bp. The hybridization mix (100 μl) per slide consisted of deionized formamide sodium dextran sulphate (Sigma), 2 × SSC, sodium dodecyl sulphate (Sigma), 200 ng μl^{-1} of *S. tuberosum* probe DNA and 10 $\mu g\ \mu l^{-1}$ of *L. peruvianum* blocking DNA. The hybridization mix was denatured for 10 min at 70° C and then placed on ice for 5 min. Hybridization took place overnight (16 h) at 37° C.

Afterwards, the slides were washed in 2 × SSC buffer for 30 min at 20° C in 0.1 × SSC for 3 × 10 min at 42° C followed by 2 × SSC again for 15 min at 20° C. Chromosomes were counterstained with DAPI (4'-6-diamidine-2-phenyl-indole) and propidium iodide (PI). The concentrations of DAPI and PI in the antifade solution Vectashield (Vector Lab, Inc, USA) were 2 μg ml^{-1} and 1 $\mu g\ ml^{-1}$, respectively.

Results and discussion

Induction of micronuclei in donor cell lines

The treatment of *N. plumbaginifolia* suspension cells with inhibitors of DNA synthesis (10 mM hydroxy urea or 15 μM aphidicolin) for 24 h, followed by treatment with microtubule inhibitors APM (32 μm) or cremart (3.7 μM) for 24 h generally resulted in about 20% micronucleated cells. The treatment of *S. tuberosum* suspension cells with cremart (7.5 μM) gave approximately 15% micronucleated cells. Afterwards, when the treated suspension cells of *N. plumbaginifolia* or *S. tuberosum* were incubated in a mixture of cell wall-digesting enzymes (cellulase, macerozyme) in the presence of cytochalasin-B and APM or cremart for 18 h followed by sieving, the frequency of micronucleated protoplasts increased to a maximum of 40% (Fig. 1A). This was due to a rapid decondensation of metaphase chromosomes, forming micronuclei during enzyme incubation, combined with stable

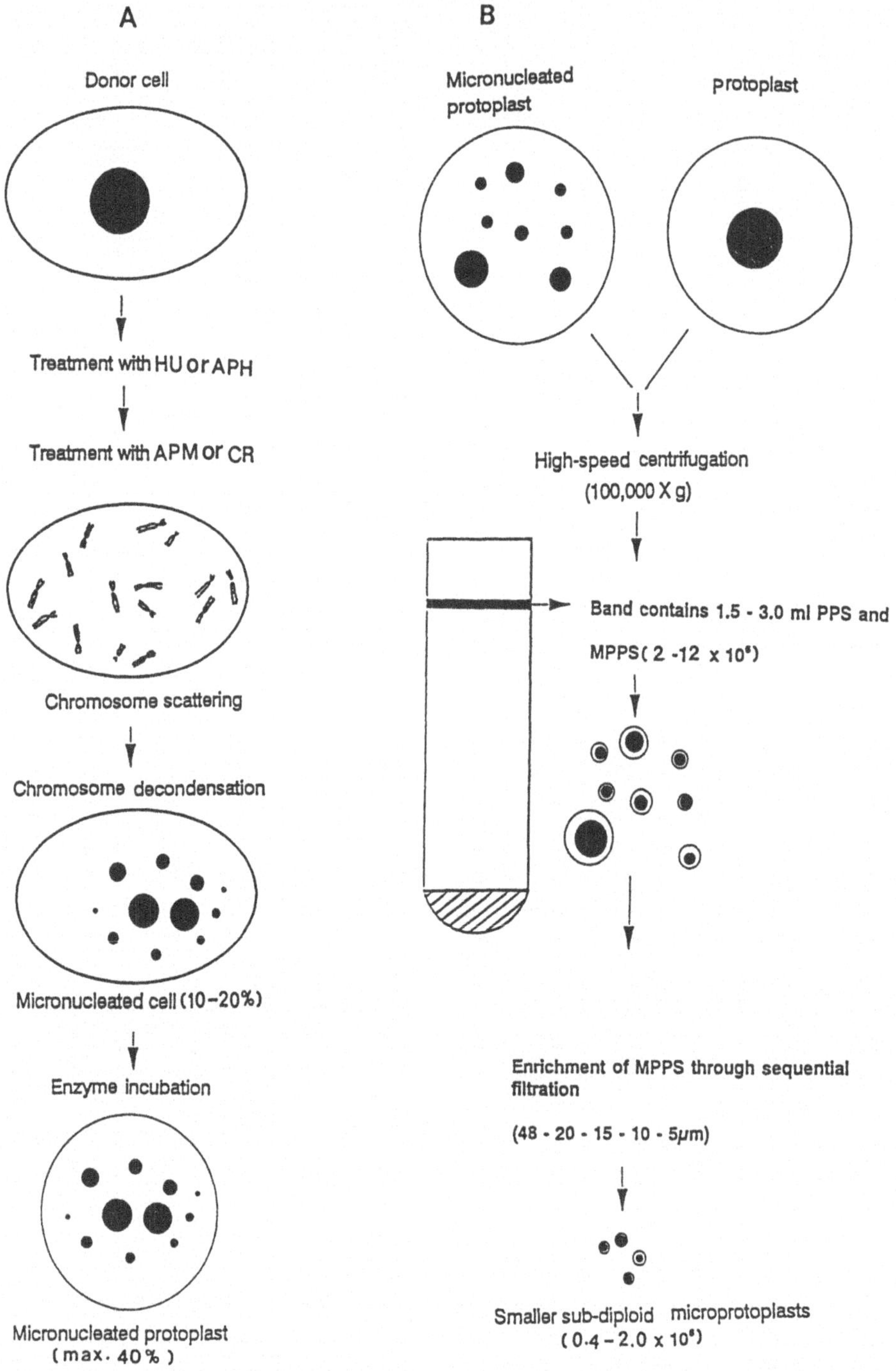

Fig. 1A. Induction of micronuclei in donor suspension cells after treatment with hydroxy urea (HU 10 mM, 24 h) or aphidicolin (APH 15 μM, 24 h) followed by treatment with amiprophos-methyl (APM 32 μM, 48 h) or cremart (CR 3.7 μM or 7.5 μM, 48 h). *Fig. 1B.* High-speed centrifugation of micronucleated and mononucleate protoplasts for the isolation of individual protoplasts (PPS) and microprotoplasts (MPPS) and enrichment of MPPS by sequential filtration through nylon sieves of decreasing pore size.

maintenance of micronuclei already formed prior to the enzyme incubation without fusion and restitution.

Isolation and enrichment of smaller sub-diploid microprotoplasts

After enzyme treatment and purification, the dense suspension of the mixture of micronucleated protoplasts

Table 3. Plant regeneration from kanamycin-resistant calli (Kan^R) derived from fusions between the donor microprotoplasts (mpps) or protoplasts (pps) and recipient mesophyll protoplasts (pps) of various species

Fusion combinations		No. of Kan^R	No. of Kan^R	No. of regenerated	Duration of
Donor	(+) Recipient	calli obtained	calli regenerated to plants	plants obtained	plant regeneration
S. tuberosum mpps	(+) *L. peruvianum* pps	184	16	210	3
S. tuberosum pps	(+) *L. peruvianum* pps	465	10	34	7
S. tuberosum mpps	(+) *N. tabacum* pps	7	5	15	4
S. tuberosum pps	(+) *N. tabacum* pps	2^a	–	–	–
N. plumbaginifolia mpps	(+) *L. peruvianum* pps	21	12	24	8
N. plumbaginifolia pps	(+) *L. peruvianum* pps	3^a	–	–	–

[a] : These calli did not develop more than 2–3 mm size and eventually turned brown and perished.

was collected from the surface of the sucrose solution, loaded onto a continuous isoosmotic gradient of Percoll and exposed to a high-speed centrifugation at 100.000 g for 2 h (Fig. 1B). After centrifugation, several bands formed containing evacuolated protoplasts, microprotoplasts and cytoplasts. About 12 ml of packed cell volume of suspension cells used per experiment gave rise to the bands containing a mixture of 1.5 to 3.0 ml of protoplasts and microprotoplasts of different sizes, the yield of which ranged from $2–12 \times 10^6$ in various experiments. By using sequential filtration of the mixture of protoplasts and microprotoplasts through nylon sieves of decreasing pore size (48–20–15–10–5 μm), it was possible to recover smaller subdiploid microprotoplasts on a mass scale ($0.4–2.0 \times 10^6$). These microprotoplasts contained a small rim of cytoplasm and plasmamembrane around them, and were FDA-positive. These were used for fusions with the recipient leaf protoplasts.

Fusion, selection of hybrid calli and plant regeneration

After fusions using PEG-based mass fusion protocol, heterokaryons and hybrid calli were selected on medium containing kanamycin and hygromycin for fusions of *S. tuberosum* (+) *L. peruvianum* and *N. plumbaginifolia* (+) *L. peruvianum*, and on medium containing kanamycin alone for *S. tuberosum* (+) *N. tabacum*. Table 3 gives details on various fusion combinations, Kan^R calli and plant regeneration. The fusions of donor microprotoplasts with recipient protoplasts gave regeneration of several plants from Kan^R calli in all the combinations (Fig. 2A–C). On the other hand, in the case of symmetric fusions of *S. tuberosum* (+) *L. peruvianum*, plants were regenerated from Kan^R calli at a low frequency only (Table 3). In other symmetric fusions, i.e. *S. tuberosum* (+) *N. tabacum* and *N. plumbaginifolia* (+) *L. peruvianum*, no fusion products could be obtained.

Characterization of plants regenerated from Kan^R calli

Table 4 gives data on plant phenotype, chromosome composition and Kan^R and GUS assays of plants derived from various fusion combinations.

S. tuberosum (+) L. peruvianum. Of the total of 111 plants analysed for plant phenotype, 67 resembled the recipient parent *L. peruvianum*, though 19 of the 67 plants showed bigger leaves and stems and more vigorous growth. Three of the 19 plants contained 48 *L. peruvianum* chromosomes and one *S. tuberosum* chromosome (as verified by genomic *in situ* hybridization), and expressed both Kan^R and GUS from *S. tuberosum*.

Fifteen plants resembled the recipient parent *L. peruvianum* in general appearance, but distinctly differed in leaf morphology and colour (Fig. 2D). Eleven of these plants contained 24 *L. peruvianum* chromosomes and one *S. tuberosum* chromosome (Fig. 2E), and expressed Kan^R and/or GUS.

The other 29 plants were intermediate in phenotype, i.e. between that of *L. peruvianum* and *S. tuberosum* (Table 4). One of these showed 71 *L. peruvianum* chromosomes, 5 *S. tuberosum* chromosomes and 2 chromosomes with interchanged or reciprocally-translocated parts of *L. peruvianum* and *S. tuberosum* chromosomes, and expressed Kan^R and GUS. The tests for Kan^R and

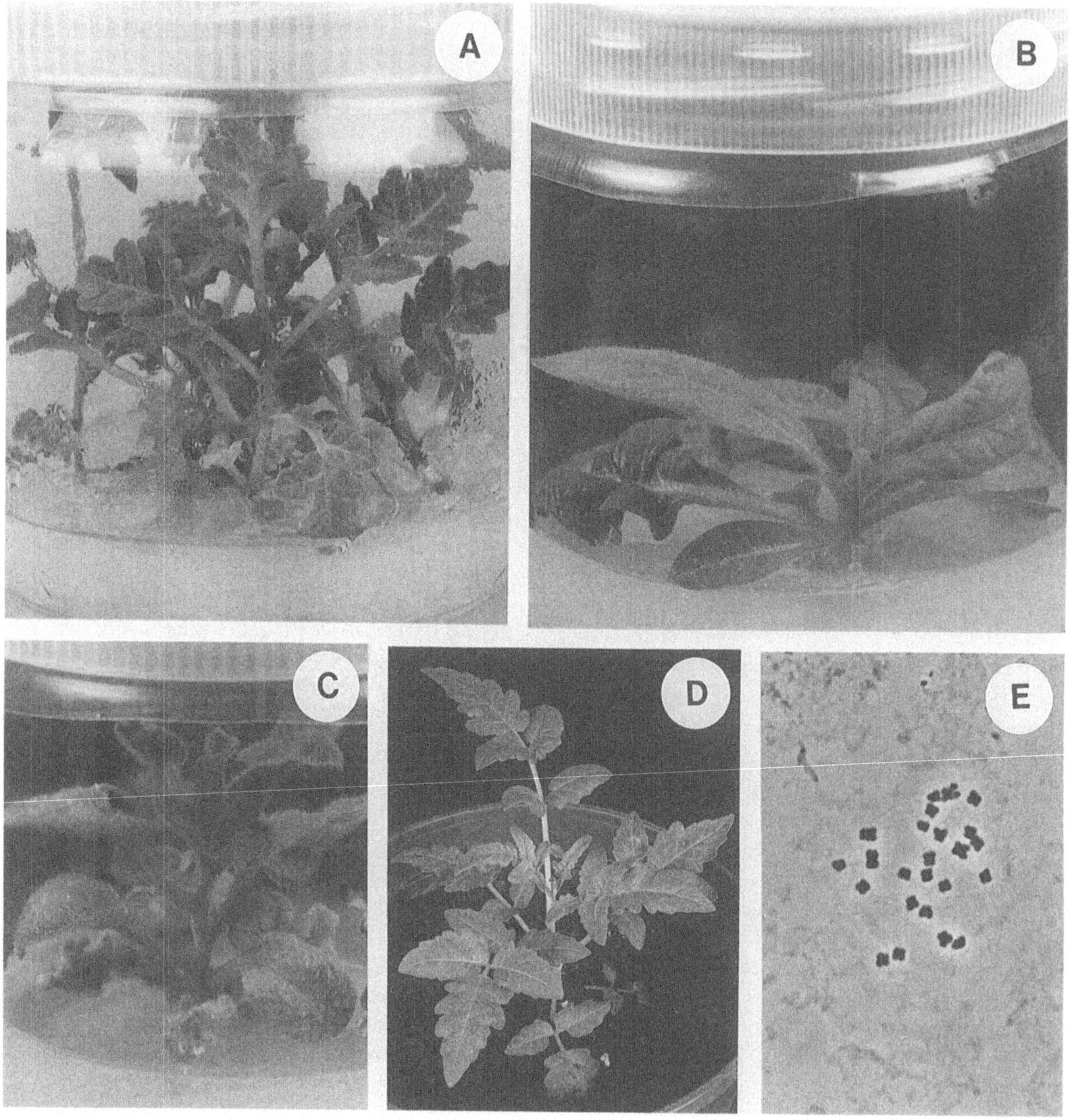

Fig. 2. Shoots regenerated from various fusion combinations. A: Shoot from *S. tuberosum* MPPS (donor) (+) *L. peruvianum* PPS (recipient); B: Shoot from *S. tuberosum* MPPS (donor) (+) *N. tabacum* PPS (recipient); C: Shoot from *N. plumbaginifolia* MPPS (donor) (+) *L. peruvianum* PPS (recipient). *Fig. 2D, E.* Microprotoplast hybrid plant (D) derived from *S. tuberosum* (+) *L. peruvianum* fusions which expressed the donor traits Kan^R and GUS, and contained one *S. tuberosum* chromosome and 24 *L. peruvianum* chromosomes (E) (also verified by genomic *in situ* hybridization.

GUS assay showed that, out of a total of 87 plants, 58 expressed Kan^R and/or GUS.

S. tuberosum *(+)* N. tabacum. The phenotype of all the 14 plants resembled that of the recipient line of *N. tabacum* and expressed the donor traits, i.e. Kan^R and GUS (Table 4). Seven of the plants also showed another donor *S. tuberosum* trait, i.e. anthocyanin pigmentation on stems and leaf midribs. The analysis of chromosome composition is in progress.

Table 4. Characterization of microprotoplast hybrid plants derived from various fusion combinations

Fusion Combinations	Plant phenotype		Chromosome Composition		Kan^R and GUS assays	
Donor (+) Recipient	No. of plants analysed	Phenotype	No. of plants analysed	Chromosome number	No. of plants analysed	No. of plants expressing Kan^R and/or GUS
St mpps (+) Lp pps	67	Recipient	3	48 Lp + 1 St	87	58
	15	Recipient, but distinct	11	24 Lp + 1 St		
	29	Intermediate	1	71 Lp + 5 St + 2 Lp.St*		
St mpps (+) Nt pps	14	Recipient	nd		14	14
Np mpps (+) Lp pps	19	Intermediate	nd		nd	

St : *S. tuberosum*; Lp: *L. peruvianum*; Nt: *N. tabacum*; Np: *N. plumbaginifolia*.
nd : not determined; *: 2 chromosomes with interchanged or reciprocally-translocated parts of *L. peruvianum* and *S. tuberosum* chromosomes.

N. plumbaginifolia (+) L. peruvianum. All of the 19 plants were intermediate in phenotype and leaf morphology between that of the two parents (Table 4).

The key factors for partial genome transfer (transfer of one or a few donor chromosomes) through the microprotoplast fusion technique, as achieved for example in *S. tuberosum* (+) *L. peruvianum* fusions, are the efficient induction, isolation and enrichment of microprotoplasts in the donor line and efficient plant regeneration from the recipient line. The fact that after microprotoplast fusion, plant regeneration occurred at a high frequency and within 3 months, and that the microprotoplast hybrid plants generally resembled the recipient line, suggests that the transferred partial genome can be better tolerated than is the whole donor genome. Partial genome transfer through microprotoplast fusion might overcome complex genetic interactions which can occur between donor-recipient nuclear and cytoplasmic genomes after symmetric fusions, leading to unstable plant regeneration or even no regeneration in some species combinations (Wolters et al., 1994).

Further, the chromosome/genome composition in the microprotoplast hybrid plants depends on 1) the segment of the donor genome (one or a few chromosomes) transferred to the recipient protoplasts at the time of fusion, and 2) genetic stability after fusion. The first stage is prone to endoreduplication leading to polyploidization. During the callus phase, aneuploidy and chromosome structural changes can occur (Ramulu et al., 1989; Pijnacker & Ramulu, 1991). For variation in the segment of the donor genome present in the fusion products, different processes might occur. As outlined in Fig. 3, when the donor microprotoplast carrying a potato chromosome with two chromatids together, due to the action of spindle toxins APM or cremart used for inducing micronuclei (Ramulu et al., 1988, 1994) in resting phase (= G2) fuses with a recipient G2 *L. peruvianum* protoplast, the fusion product may directly progress to mitosis (without an extra DNA synthesis), giving rise eventually to a microprotoplast hybrid with only one potato chromosome and a complete genome of *L. peruvianum* (*monosomic addition line*: 2n = 24 + 1). On the other hand, after fusion of the donor microprotoplast containing a chromosome

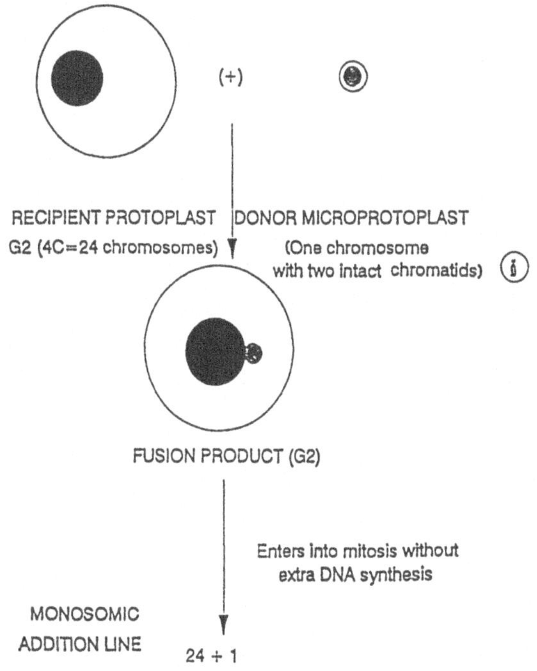

Fig. 3. Production of a monosomic addition line after fusion of a recipient protoplast with a donor microprotoplast.

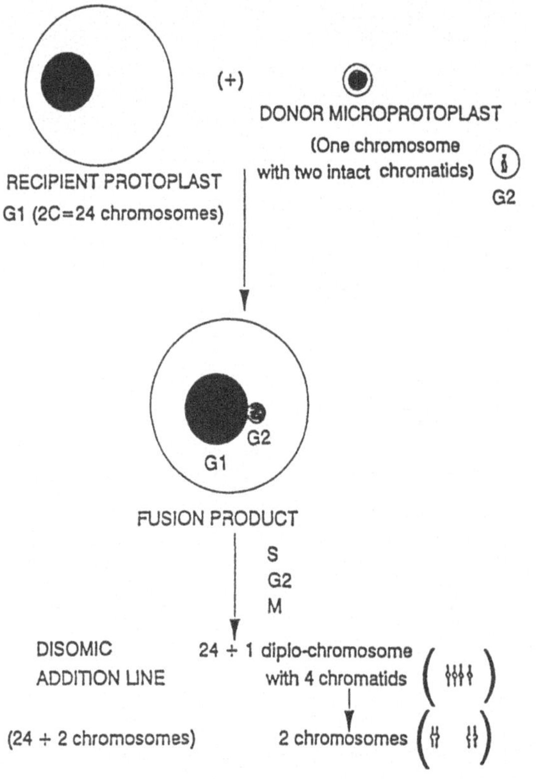

Fig. 4. Production of a disomic addition line after fusion of a recipient protoplast with a donor microprotoplast.

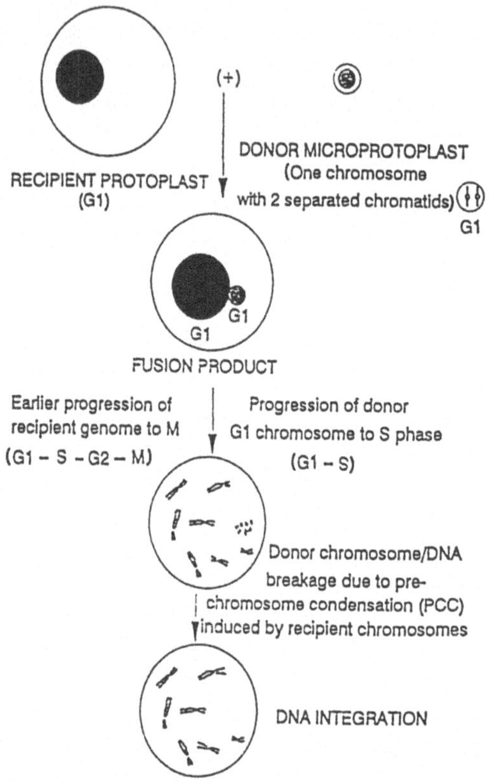

Fig. 5. Donor DNA integration in recipient chromosomes after fusion of a recipient protoplast with a donor microprotoplast.

(2-chromatid chromosome) with a recipient G1 protoplast, and following S-G2-M in the immediate cell cycle, the hybrid cell can contain a 4-chromatid (diplo) chromosome, the centromere of which can separate, giving rise to two copies of a given chromosome (*disomic addition line*: 2n = 24 + 2) (Fig. 4). Further, due to delayed DNA replication of the donor chromosomes in the immediate cell cycle after fusion, and when the recipient genome progresses to metaphase, the latter can induce pre-chromosome condensation (PCC) of the donor S-phase chromosomes, leading to DNA or chromosome breakage (Fig. 5). This process can also occur in the later cell cycles of the fusion product, if the donor chromosomes are delayed in undergoing anaphase segregation, forming micronuclei. When the micronuclei enter into the S-phase while the recipient genome is already in metaphase, the S-phase micronuclei can undergo PCC, and consequently DNA or chromosome breakage (Sperling, 1982). The released donor DNA may integrate into the recipient genome through transformation events, or repair processes. When genetic instability occurs during the callus phase, which is multicellular and heterogeneous in cell cycle stages, microprotoplast hybrids containing intact or modified donor chromosomes, and/or with integrated donor DNA or chromosome segments, might be recovered. The possible mechanisms for DNA integration in mitotic and meiotic cell cycle occurring in various plant species have been extensively discussed by Sybenga (1989, 1992). In generatively – produced addition lines, the integration of donor DNA/chromosome segments apparently occur by similar mechanisms: double strand breaks of the donor chromosome during its disintegration at some stage in meiosis/mitosis, followed by repair-induced invasion and integration of donor DNA into the host chromosomes (reviewed by Sybenga, 1992). In many instances of segment transfer in wheat after ionizing irradiation, interstitial inserts are observed, often of considerable size, which are probably the result of similar mechanisms.

The results also show that several plants expressed both Kan^R and GUS. Some plants regenerated from the same Kan^R callus (e.g. *S. tuberosum* (+) *L. peruvianum* fusions) did not express either Kan^R or GUS, probably due to the deletion of the donor chromosome/chromosome segment, or inactivation of the gene. Several microprotoplast hybrid plants derived from *S. tuberosum* (+) *L. peruvianum* fusions, which expressed Kan^R of GUS, contained the donor *S. tuberosum* chromosome as well as the npt II and GUS DNA sequences (as determined by Southern-blot hybridization), suggesting that the expression of the genes is linked to the presence of the donor chromosome. Many of the microprotoplast hybrid plants containing 24 *L. peruvianum* chromosomes and 1 potato chromosome (monosomic additions at the diploid level), or 48 *L. peruvianum* chromosomes and 1 potato chromosome (monosomic additions at the tetraploid level) appeared to be sufficiently male- and female-fertile, as adjudged by the production of several berries and of seed progeny obtained after backcrossing with different self-incompatible genotypes of *L. peruvianum* (results in progress). The progeny tests, which are being currently carried out for analysis of the transmission, showed that some plants were GUS-positive, indicating sexual transmission of the alien genes to the seed progeny.

Conclusions

The results obtained show that through microprotoplast fusion technique, it is possible to transfer single, chromosomes and alien genes between sexually-incongruent species. This technique offers some unique advantages for alien gene transfer, genome analysis and gene cloning, as outlined below.

1. Transfer of desirable traits, e.g. from wild to the cultivated species. Important traits like disease- or stress-resistance, encoded by polygenes, might be clustered within blocks or scattered throughout the genome. In this regard, microprotoplast fusion can be a useful method for the transfer of clustered resistance genes. Also, the transfer of individual chromosomes from non-hybridizing wild to cultivated species by this technique provides perspectives for the transfer, localization and inheritance of non-host resistance genes.
2. Production of monosomic or disomic addition lines in a single step, thus avoiding time-consuming back-cross or self-pollination generations necessary to obtain such lines by generative methods. From monosomic addition lines, it is possible to obtain substitution or recombinant lines, including those with recombinant chromosomes (intergenomic translocations) after backcrossing or selfing, depending upon the degree of homoeologous pairing between the donor and recipient chromosomes (Sybenga, 1992; Parokonny et al., 1992; Jacobsen et al., 1994). Thus, the transfer of chromosomes from non-hybridizing, but related wild species to the cultivated species makes it feasible to obtain a greater insight into genome organization/evolution of the species, genome/chromosome homoeology and somatic genome compatibility, which are important for introgressive breeding (Sybenga, 1992; Sybenga et al., 1994; de Jong et al., 1993; Jacobsen et al., 1994; Wolters et al., 1994).
3. Construction of chromosome-specific DNA libraries. Genome analysis and molecular plant breeding concepts rest on dense linkage maps based on RFLPs. A high marker saturation, i.e., 1cM or less, is not readily attainable with customary techniques involving random selection of recombinant DNA for RFLPs from shot-gun libraries. It can be substantially facilitated by directly cloning DNA from individual chromosomes or chromosome segments, either by microdissection *via* micromanipulation using glass needles or laser optical trapping, or by flow cytometric sorting of metaphase chromosomes and amplification by PCR (Fukui et al., 1992; Jung et al., 1992; Wang et al., 1992; Schubert et al., 1993; Lucretti et al., 1993; Schondelmaier et al., 1993; Arumuganathan et al., 1994), using monosomic, disomic, telosomic or ditelosomic additions. Chromosome additions of various kinds can be produced, through microprotoplast fusion, in new genetic backgrounds with marked differences in the karyotype/DNA content from the recipient chromosomes.
4. Studies of the three-dimensional structure of chromosomes and spatial arrangement of the donor chromosomes and their relationship with stability of gene expression and transmission (Gleba et al., 1987; Appels, 1989; Heslop-Harrison & Bennett, 1990; Nanninga et al., 1992; Oud & Nanninga, 1992; Montijn et al., 1994). Interspecific or intergeneric microprotoplast hybrid plants with one or a few chromosomes carrying alien genes facilitate an efficient analysis of the fate of introduced genes (deletion, inactivation or co-suppression, structural alteration) in somatic and generative cycles, because the transferred chromosomes can be identified using molecular cytogenetic methods (genomic *in situ* hybridization) and

the chromosome observations can be linked to the presence or absence of the genes.

5. Gene localization on the chromosome. Genes can be localized if a sufficient population of addition lines is produced by microprotoplast fusion. Monosomic addition and recombinant lines can be useful as mapping tools to assign RFLP markers to specific regions on chromosome arms (Islam & Shepherd, 1991; Rogowsky et al., 1991).

Acknowledgements

Our thanks are due to Professor J. Sybenga, Department of Genetics, Agricultural University, Wageningen and Dr. F.A. Krens, CPRO-DLO, Wageningen, for critically reading the manuscript, Dr. J.H. de Jong, Department of Genetics, Agricultural University Wageningen, Dr. C. Kik and Dr. I. Famelaer, CPRO-DLO, Wageningen, and Dr. T. Cardi, CNR, Portici, Italy for useful suggestions on genomic *in situ* hybridization, Southern-blot hybridization and fusions respectively.

References

Appels, R., 1989. Three-dimensional arrangements of chromatin and chromosomes: old concepts and new techniques. J. Cell Sci. 92: 325–328.

Arumuganathan, K., G.B. Martin, H. Telenius, S.D. Tanksley & E.D. Earle, 1994. Chromosome 2-specific DNA clones from flow-sorted chromosomes of tomato. Mol. Gen. Genet. 242: 551–558.

Bokelmann, G.S. & S. Roest, 1983. Plant regeneration from protoplasts of potato (*Solanum tuberosum* cv. Bintje). Z. Pflanzenphysiol. 109: 259–265.

Caboche, M., 1980. Nutritional requirements of protoplast-derived haploid tobacco cells grown at low densities in liquid medium. Planta 149: 7–18.

Cardi, T., F. D'Ambrosio, D. Consoli, K.J. Puite & K. Sree Ramulu, 1993. Production of somatic hybrids between frost-tolerant *Solanum commersonii* and *S. tuberosum*: characterization of hybrid plants. Theor. Appl. Genet. 87: 193–200.

Carrano, A.V., J.W. Gray, R.G. Langlois, K.J. Burkhardt-Schultz & M.A. Van Dilla, 1979. Measurement and purification of human chromosomes by flow cytometry and sorting. Proc. Natl. Acad. Sci. 76: 1382–1384.

de Jong, J.H., A.M.A. Wolters, J.M. Kok, H. Verhaar & J. Van Eden, 1993. Chromosome pairing and potential for intergeneric recombination in some hypotetraploid somatic hybrids of *Lycopersicon esculentum* (+) *Solanum tuberosum*. Genome 36: 1032–1041.

de Laat, A.A.M., H.A. Verhoeven & K. Sree Ramulu, 1989. Chromosome transplantation and applications of flow cytometry in plants. p. 343–359. In: P.S. Bajaj (Ed.). Biotechnology in Agricultural and Forestry, 9. Plant protoplasts and genetic engineering II. Springer-Verlag, Heidelberg.

Derks, S., 1992. Organelle transfer by protoplast fusion in *Solanaceae*, Ph.D. thesis, Univ. of Amsterdam, p. 120.

Derks, F.H.M., J.C. Hakkert, W.H.J. Verbeek & C.M. Colijn-Hooymans, 1992. Genome composition of asymmetric hybrids in relation to the phylogenetic distance between the parents. Nucleus-chloroplast interaction. Theor. Appl. Genet. 84: 930–940.

Famelaer, I., I. Negrutiu, A. Mouras, H. Vaucheret & M. Jacobs, 1990. Asymmetric hybridization in *Nicotiana* by 'gamma fusion' and progeny analysis of self-fertile hybrids. Theor. Appl. Genet. 79: 513–520.

Fukui, K., M. Minezawa, Y. Kamisugi, M. Ishikawa, N. Ohmido, T. Yanagisawa & M.S. Fujishita, 1992. Microdissection of plant chromosomes by argon-ion laser beam. Theor. Appl. Genet. 84: 787–791.

Gasser, G.S. & R.T. Fraley, 1989. Genetically engineered plants for crop improvement. Science 244: 1293–1299.

Gilissen, L.J.W., M.J. van Staveren, E. Ennik, H.A. Verhoeven & K.S. Ramulu, 1992. Somatic hybridization between potato and *Nicotiana pulmbaginifolia*. 2. Karyotypic modification and segregation of genetic markers in hybrid suspension cultures and sublines. Theor. Appl. Genet. 84: 81–86.

Gleba, Y.Y. & K.M. Sytnik, 1984. Protoplast fusion. Genetic engineering in higher plants. Springer, Berlin, Heidelberg, New York.

Gleba, Y.Y., A. Parakonny, V. Kotov, I. Negrutiu & V. Momot, 1987. Spatial separation of parental genomes in hybrids of somatic plant cells. Proc. Natl. Acad. Sci. USA 84: 3709–3713.

Glimelius, K., 1988. Potentials of fusion in plant breeding programmes. p. 159–168. In: K.J. Puite, J.J.M. Dons, H.J. Huizing, A.J. Kool, M. Koornneef & F.A. Krens (Eds). Progress in Plant Protoplast Research. Kluwer, Dordrecht.

Heslop-Harrison, J.S. & M.D. Bennett, 1990. Nuclear architecture in plants. Trends Genet. 6: 401–405.

Islam, A.K.M.R. & K.W. Shepherd, 1991. Recombination between wheat and barley chromosomes. Barley Genet. 4: 68–70.

Jacobsen, E., J.H. de Jong, S.A. Kamstra, P.M.M.M. van den Berg & M.S. Ramanna, 1995. Genomic *in situ* hybridisation (GISH) and RFLP analysis for the identification of alien chromosomes in the backcross progeny of potato (+) tomato fusion hybrids. Heredity 74: 250–257.

Jacobsen, E., P. Reinhout, J.E.M. Bergervoet, P.E. Abidim, J. de Looff, D.J. Huigen & M.S. Ramanna, 1992. Isolation and characterization of potato-tomato somatic hybrids using an amylose-free potato mutant as parental genotype. Theor. Appl. Genet. 85: 159–164.

Jefferson, R.A., T.A. Kavanagh & M. Bevan, 1987. GUS-fusions: β-glucuronidase as a sensitive and versatile gene fusion marker in higher plants. EMBO J 6: 3901–3907.

Jung, C., U. Claussen, B. Horsthemke, F. Fischer & R.G. Hermann, 1992. A DNA library from an individual *Beta patellaris* chromosome conferring nematode resistance obtained by microdissection of meiotic metaphase chromosome. Plant Mol. Biol. 20: 503–511.

Kao, K.N. & M.R. Michayluk, 1975. Nutritional requirements for growth of *Vicia hajastana* cells and protoplasts at a very low population density in liquid media. Planta 126: 105–110.

Koornneef, M., C.J. Hanhart & L. Martinelli, 1987. A genetic analysis of cell culture traits in tomato. Theor. Appl. Genet. 74: 633–641.

Leitch, I.J. & J.S. Heslop-Harrison, 1993. Detection of digoxigenin-labeled DNA probes hybridized to plant chromosomes *in situ*. In: P.G. Isaac (Ed). Methods in Molecular Biology, Vol XX: Protocols for nucleic acid analysis by nonradioactive probes. Humana Press. INC., Totowa, NJ, Chapter 27 pp.

Lucretti, S , J Dolezel, I Schubert & J Fuchs, 1993 Flow karyotypic and sorting of *Vicia faba* chromosomes Theor Appl Genet 85 665–672
Malina, P, A Sz -Brenovits, L Marton & F Joo, 1975 Non-mendelian streptomycin resistant tobacco mutant with altered chloroplasts and mitochondria Nature 255 401–402
Menczel, L, F Nagy, Z R Kiss & P Maliga, 1981 Streptomycin resistant and sensitive somatic hybrids of *Nicotiana tabacum* + *Nicotiana knightiana* correlation of resistance to *N tabacum* plastids Theor Appl Genet 59 191–195
Montijn, M B , A B Houtsmuller, J L Oud & N Nanninga, 1994 The spatial localization of 18 S rRNA genes, in relation to the descent of the cells, in the root cortex of *Petunia hybrida* J Cell Sci 107 457–467
Morel, G & R M Wetmore, 1951 Fern callus tissue culture Am J Bot 38 141–143
Murashige, T & F Skoog, 1962 A revised medium for rapid growth and bioassays with tobacco tissue cultures Physiol Plant 15 473–497
Nagy, J I & P Maliga, 1976 Callus induction and plant regeneration from mesophyll protoplasts of *Nicotiana sylvestris* Z Pflanzenphysiol 78 453–455
Nanninga, N , J L Oud, A B Houtsmuller & M B Montijn, 1992 Spatial arrangement of genes and chromosomes in plants comments on cell geneology and tissue specificty Cell Biol Int Rep 16 761–770
Oud, J L & N Nanninga, 1992 Cell shape, chromosome orientation and the position of the plane of division in *Vicia faba* root cortex cells J Cell Sci 103 847–855
Parokanny, A S , A Y Kenton, Y Y Gleba & M D Bennett, 1992 Genome reorganization in *Nicotiana* asymmetric somatic hybrids analysed by *in situ* hybridization The Plant Journal 2 863–874
Pijnacker, L P & M A Ferwarda, 1984 Giemsa C banding of potato chromosomes Can J Genet Cytol 26 415–419
Pijnacker, L P & K S Ramulu, 1991 Somaclonal variation in potato a karyotypic evaluation Acta Bot Neerl 39 163–169
Potrykus, I , 1990 Gene transfer to cereals an assessment Bio Technology 8 535–542
Puite, K J , 1992 Progress in plant protoplast research Physiologia Plantarum 985 403–410
Puite, K J & J G Schaart, 1993 Nuclear genomic composition of asymmetric fusion products between irradiated transgenic *Solanum brevidens* and *S tuberosum* limited elimination of donor chromosomes and polyploidization of the recipient genome Theor Appl Genet 86 237–244
Ramulu, K S , F Carluccio, D de Nettancourt & M Devreux, 1977 Trisomics from triploid-diploid crossed in self-incompatible *Lycopersicon peruvianum* Theor Appl Genet 50 105–119
Ramulu, K S , P Dijkhuis & S Roest, 1983 Phenotypic variation and ploidy level of plants regenerated from protoplasts of tetraploid potato (*Solanum tuberosum* L cv Bintje) Theor Appl Genet 65 329–338
Ramulu, K S , H A Verhoeven & P Dijkhuis, 1988 Mitotic dynamics of micronuclei induced by amiprophos-methyl and prospects for chromosome-mediated gene transfer in plants Theor Appl Genet 75 575–584
Ramulu, K S , P Dijkhuis & S Roest, 1989 Patterns of phenotypic and chromosome variation in plants derived from protoplast cultures of monohaploid, dihaploid and diploid genotypes and in somatic hybrids of potato Plant Sci 60 101–110
Ramulu, K S , P Dijkhuis, H A Verhoeven, I Famelaer & J Blaas, 1992 Microprotoplast isolation, enrichment and fusion for partial genome transfer in plants Physiol Plant 85 315–318
Ramulu, K S , P Dijkhuis, I Famelaer, T Cardi & H A Verhoeven, 1993 Isolation of sub-diploid microprotoplasts for partial genome transfer in plants Enhancement of micronucleation and enrichment of microprotoplasts with one or a few chromosomes Planta 190 190–198
Ramulu, K S , P Dijkhuis, I Famelaer, T Cardi & H A Verhoeven, 1994 Cremart a new chemical for efficient induction of micronuclei in cells and protoplasts for partial genome transfer Plant Cell Rep 13 687–691
Rogowsky, P M , F L Y Guidet, P Langridge, K W Shepherd & R M D Koebner, 1991 Isolation and characterization of wheat-rye recombinants involving chromosome arm 1DS of wheat Theor Appl Genet 82 537–544
Schoenmakers, H C H , A M A Wolters, A de Haan, A K Saiedi & M Koornneef, 1994 Asymmetric somatic hybridization between tomato (*Lycopersicon esculentum* Mill) and gamma-irradiated potato (*Solanum tuberosum* L) a quantitative analysis Theor Appl Genet 87 713–720
Schondelmaier, J , R Martin, A Jahoor, A Houben, A Graner, H-U Koop, R G Hermann & C Jung, 1993 Microdissection and microcloning of the barley (Hordeum vulgare L) chromosome 1HS Theor Appl Genet 86 629–636
Schubert, I , J Dolezel, A Houben, H Scherthan & G Wanne, 1993 Refined examination of plant metaphase chromosome structure at different levels made feasible by new isolation methods Chromosoma 102 96–101
Schwarzacher, T & J S Heslop-Harrison, 1994 Direct fluorochrome-labeled DNA probes for direct fluorescent *in situ* hybridization to chromosomes In PG Isaac (Ed) Methods in Molecular Biology, Vol 28 Protocols for nucleic acid analysis by nonradioactive probes pp 8–17 Humana Press Inc , Totowa, NJ
Schwarzacher, T & A R Leitch, 1994 Enzymatic treatment of plant material to spread chromosomes for *in situ* hybridization In PG Isaac (Ed) Methods in Molecular Biology, Vol 28 Protocols for nucleic acid analysis by nonradioactive probes pp 2–7 Humana Press Inc , Totowa, NJ
Shahin, E A , 1985 Totipotency of tomato protoplasts Theor Appl Genet 69 235–240
Sperling, K , 1982 Cell cycle and chromosome cycle Morphological and functional aspects p 43–78 In PN Rao, R T Johnson & K Sperling (Eds) Pre-Chromosome Condensation Application in basic, clinical and mutation research Academic Press, New York
Sybenga, J , 1989 Genetic manipulation generative vs somatic p 26–53 In Y P S Bajaj (Ed) Biotechnology in Agriculture and Forestry 9 Plant protoplasts and genetic engineering II Springer-Verlag, Heidelberg
Sybenga, J , 1992 Cytogenetics in plant breeding Monographs on theoretical and applied genetics 17 Springer-Verlag, Heidelberg, 469 pp
Sybenga, J , E Schabbing, J van Eden & J H de Jong, 1994 Pachytene pairing and metaphase I configurations in a tetraploid somatic *Lycopersicon esculentum* × *L peruvianum* hybrid Genome 37 54–60
Verhoeven, H A & K S Ramulu, 1991 Isolation and characterization of microprotoplasts from APM treated suspension cells of *Nicotiana plumbaginifolia* Theor Appl Genet 82 346–352
Verhoeven, H A , K S Ramulu & P Dijkhuis, 1990 Comparison of the effects of various spindle toxins on metaphase arrest and formation of micronuclei in cell suspension cultures of *Nicotiana plumbaginifolia* Planta 182 408–414

Verhoeven, H A , K S Ramulu, L J W Gilissen, I Famelaer, P Dijkhuis & J Blaas, 1991a Partial genome transfer through micronuclei in plants Acta Bot Neerl 40 97–113

Verhoeven, H A , K S Ramulu, J Blaas & P Dijkhuis, 1991b Control of cell cycle progression p 346–355 In I Negrutiu & G B Ghatri Chhetri (Eds) A Laboratory Guide for Cellular and Molecular Plant Biology Birkhauser, Basel

Wang, M L , A R Leitch, T Schwarzacher, J S Heslop Harrison & G Moore, 1992 Construction of a chromosome-enriched HpaII library from flow-sorted wheat chromosomes Nucl Acids Res 20 1897–1901

Wijbrandi, J , P Zabel & M Koornneef, 1990 Restriction fragment length polymorphism analysis of somatic hybrids between *Lycopersicon esculentum* and irradiated *L peruvianum* Evidence for limited donor genome elimination and extensive chromosome rearrangements Mol Gen Genet 222 270–277

Wolters, A M A , H C H Schoenmakers, J J M van der Meulen-Muisers, E van der Knaap, F H M Derks, M Koornneef & A Zelcer, 1991 Limited DNA elimination from the irradiated potato parent in fusion products of albino *Lycopersicon esculentum* and *Solanum tuberosum* Theor Appl Genet 83 225–232

Wolters, A M A , E Jacobsen, M O'Connell, G Bonnema, K Sree Ramula, J H de Jong, H C H Schoenmakers, J Wijbrandi & M Koornneef, 1994 Somatic hybridization as a tool for tomato breeding Euphytica 79 265–277

Euphytica **85**: 269–273, 1995.

Improved plant heterokaryon formation by surfactant-supplementation of polyethylene glycol fusogen solution

T.K. Hill[1], M.R. Davey, J.B. Power, P. Anthony & K.C. Lowe
Department of Life Science, University of Nottingham, University Park, Nottingham NG7 2RD, U.K.; [1] present address: Department of Life Science, Nottingham Trent University, Clifton Lane, Nottingham NG11 8NS, U.K.

Key words: heterokaryon formation, *Petunia*, Pluronic F-68, polyethylene glycol, protoplast fusion, surfactant

Summary

PEG fusion solution for leaf protoplasts of *Petunia parodii* and cell suspension protoplasts of albino *P. hybrida* cv. Comanche was supplemented with 0.01–1.0% (w/v) Pluronic F-68. This stimulated protoplast fusion overall, including parental homokaryon formation, with increased means of 23% and 83% respectively, over appropriate controls using 1.0% (w/v) surfactant added to the standard PEG solution. Interestingly, the percentage heterokaryon formation increased near 2-fold ($P < 0.001$) for fusogen solutions supplemented with 0.01% (w/v) Pluronic. Protoplasts regenerated to colonies in KM8P/KM8 liquid medium, indicating no adverse effects of Pluronic F-68 on viability, both in the short and longer terms.

Abbreviations: BA – 6-benzyladenine, MS – Murashige & Skoog (1962), NAA – α-naphthaleneacetic acid, PEG – polyethylene glycol

Introduction

Somatic hybridisation has been used to generate novel hybrids/breeding accessions between sexually incompatible species (Patil et al., 1993). The use of polyethylene glycol (PEG) as a chemical fusogen (Chand et al., 1989) provides an alternative to electrofusion procedures (Lynch et al., 1993); the latter often require expensive equipment and sometimes lengthy investigations optimising fusion parameters. Although chemical fusion protocols are well established, procedures directed specifically towards heterokaryon production are desirable. Somatic hybridisation relies initially for success on an ability to generate heterokaryon populations that can be converted to plants via the tissue culture process.

Most fusion protocols are relatively inefficient since a balance has to be established between the extent of induced fusion and subsequent viability. Agents that improve heterokaryon production, but do not sacrifice viability, could have a rôle in improving the efficacy and applicability of somatic (and gametosomatic) hybridisation to crop improvement. Also, agents that enhance fusion may increase the number of somatic hybrids generated.

Low concentrations (< 1% (w/v)) of the polyoxyethylene-polyoxypropylene block co-polymer surfactant, Pluronic F-68, stimulate differentiation of cultured plant protoplasts, cells, tissues and organs (Lowe et al., 1993). Patch-clamp studies involving artificial lipid bilayers suggest that surfactants such as Pluronic F-68 generate transmembrane pores, indicating that the plasma membrane is the probable target site (King et al., 1991). Clarke & McNeil (1992) showed that Pluronic F-68 improves uptake of macromolecules into animal cells in culture. Membrane resealing and fluidity may be implicated in this process (Ramirez & Matharasan, 1990). Since regulators of membrane fluidity, pore formation and membrane stability are all implicated in the fusion process, surfactants have been studied as fusion mixture supplements. Heterokaryon production and maximum survival are crucial since chloroplast-containing heterokaryons are inherently less stable physically than other products of fusion (Power et al., 1978). In the present investigation, the influence of Pluronic F-68 coupled with PEG-induced fusion was studied as an approach to increase heterokaryon formation and to retain high survival of resulting somatic hybrid combinations.

Materials and methods

Plant material and fusion procedure

Petunia parodii leaf protoplasts and *P. hybrida* cell suspension protoplasts were isolated enzymatically (Power et al., 1990). Isolated protoplasts were fused by mixing 1.0 ml aliquots of each protoplast suspension (containing 4×10^5 protoplasts ml^{-1} of each parent) in 16.0 ml screw-capped glass tubes, followed by the addition of 8.0 ml of a solution (fusogen) consisting of 30% (w/v) PEG 6000 (BDH, UK), 4% (w/v) sucrose and 0.01 M $CaCl_2.2H_2O$. After 10 min at 25° C, the fusogen was diluted progressively with a high pH/Ca^{2+} solution (Power et al., 1990). The fusogen was supplemented with 0.01%, 0.1% or 1.0% (w/v) of a commercial grade of the surfactant, Pluronic F-68 (ICI, UK). This was carried out by adding an appropriate volume of a 2% (w/v) stock solution of Pluronic F-68 in 4% (w/v) sucrose with 0.01 M $CaCl_2.2H_2O$ sterilised through a 0.2 μm Millipore filter (Millipore, UK). Each fusion treatment was repeated at least 5 times. Controls consisted of identical mixtures of protoplasts exposed to fusogen but without Pluronic F-68.

After fusion treatment, the tubes were centrifuged (100 g, 10 min), the supernatants removed, and the protoplasts re-suspended in 16.0 ml aliquots of liquid protoplast culture medium 8P (Kao & Michayluk, 1975), as modified by Gilmour et al. (1989) and designated KM8P. Protoplasts were re-centrifuged and the supernatant removed. Protoplasts were re-suspended in KM8P liquid medium at a density of 1.0×10^5 ml^{-1}, and 1.5 ml aliquots (at least 10 per replicate) of protoplast suspension were transferred into individual wells of square, 25-well plastic Petri dishes (Sterilin, UK). Dishes were sealed with Nescofilm (Bando, Japan); fusion-treated preparations were maintained statically in the dark at 25 ± 2° C.

Fusion efficacy and viability assessments

The total percentage of fused protoplasts (which included separate values for parental homokaryons and heterokaryons) was determined by counting a minimum of 250 protoplasts in each of 10 wells. Fluorescein diacetate was used to determine protoplast viability 48 h post-fusion (Widholm, 1972).

Samples of fusion-treated protoplasts were cultured to the plant regeneration stage in order to confirm sustainable viability. The protoplast culture medium was progressively replaced, in order to lower the osmoticum, with cell culture medium. The latter (Kao & Michayluk, 1975) was modified according to Gilmour et al. (1989) and designated KM8. One ml aliquots per well were removed and replaced with the same volume of fresh KM8 medium. Dilution of the cultures in this way was repeated every 7 d over a 35 d period, after which protoplast-derived cell colonies were transferred to the light (9.8 μmol s^{-1} m^{-2}; 16 h photoperiod, daylight fluorescent tubes). Subsequently, individual colonies were transferred onto MSP1 agar-solidified (0.6% w/v; Sigma, UK) medium for proliferation (30 colonies per 5 cm diameter Petri dish; 5.0 ml medium per dish). MSP1 medium was prepared according to the basal MS formulation (Murashige & Skoog, 1962), but with 2.0 mg l^{-1} NAA and 0.5 mg l^{-1} BA (Power et al., 1990). Protoplast-derived calli were transferred to the same medium for a further 14 d (15 calli per 9 cm Petri dish; 15.0 ml medium). Subsequently, tissues were sub-cultured onto agar-solidified (MSP1 or MSZ) medium for proliferation and shoot-regeneration (3 calli per 75.0 ml medium in 175 ml powder round glass screw-capped jars; Power et al., 1990). MSZ medium consisted of basal MS components supplemented with 1.0 mg l^{-1} zeatin.

Statistical analyses

The results are expressed throughout as mean ± standard deviation (s.d.). Statistical significance between mean values was assessed using a conventional Student's *t*-test. A probability of $P < 0.05$ was considered significant.

Results

Effects of Pluronic F-68 on protoplast fusion

The presence of chloroplasts in the cytoplasm of mesophyll protoplasts of *P. parodii* and the absence of these organelles in cell suspension-derived protoplasts of *P. hybrida*, enabled heterokaryons and homokaryons to be identified by light microscopy. Homokaryons of *P. parodii* were readily distinguished from heterokaryons

Table 1. Effect of supplementing the PEG-based fusogen with Pluronic F-68 on the mean % of total protoplasts fused and on % homokaryons and heterokaryons for cell suspension protoplasts of *P. hybrida* and leaf protoplasts of *P. parodii*

Treatment	% Total protoplast fusion	% Homokaroins		% Heterokaryons
		P. hybrida	*P. parodii*	
Control	13 ± 2	4 ± 2	4 ± 2	5 ± 1
% Pluronic F-68				
0.01	16 ± 2**	4 ± 1	4 ± 1	9 ± 1***
0.1	15 ± 2	6 ± 1*	4 ± 1	5 ± 1
1.0	21 ± 5***	8 ± 2***	7 ± 2**	6 ± 2

Values are mean (± s.d.) of 5–15 replicates. * $P < 0.05$; ** $P < 0.01$; *** $P < 0.001$ compared with the corresponding control (0% Pluronic) mean value.

on the basis of a greater number of chloroplasts in the homokaryons, and the presence in heterokaryons of prominant cytoplasmic strands from the *P. hybrida* partner.

In the presence of 0.01% (w/v) Pluronic F-68, the mean total number of protoplasts fused was 16 ± 2% (n = 5) which was significantly greater ($P < 0.01$) than the control value of 13 ± 2% (n = 15; Table 1). The mean number of heterokaryons produced with this concentration of Pluronic F-68 was 9 ± 1% (n = 5), which was also greater ($P < 0.001$) than the control (5 ± 1%; n = 15). However, there was no significant change in the proportion of heterokaryons generated with Pluronic F-68 at the higher concentrations of 0.1% or 1.0% (w/v). The formation of homokaryons in protoplast suspensions of *P. hybrida* and *P. parodii* was also enhanced by Pluronic F-68 at 1.0% (w/v), and, in the case of *P. hybrida*, at 0.1% (w/v). The mean percentage of *P. hybrida* homokaryons produced by treatment, 0.1% or 1.0% Pluronic F-68 was 6 ± 1% and 8 ± 2%, respectively (n = 5; Table 1). The mean viability after 48 h was similar for mixed populations of *P. hybrida* and *P. parodii* protoplasts following treatment with PEG alone or with PEG in combination with Pluronic F-68 (51 ± 4% and 51 ± 12%, respectively; n = 5).

Growth of protoplast-derived tissues

Friable cell colonies developed within 35 d of culture of fusion-treated protoplasts in KM8P/KM8 liquid medium. Transfer of protoplast-derived colonies to MSP1 medium, progressively selected against *P. parodii* homokaryons. Consequently, protoplast-derived, chlorophyll-containing tissues which developed on MSP1 medium were putative somatic hybrids, with the development of chlorophyll proficiency following albino complementation. Calli transferred to non-selective MSZ medium continued to develop into discrete chlorophyllous or colourless (albino) calli, the albino tissues being derived from homokaryons or unfused *P. hybrida* protoplasts. The presence of Pluronic F-68 in the fusogen did not inhibit subsequent mitotic division of *Petunia* protoplasts (n = 5) or putative somatic hybrid cells arising from the fusion of protoplasts of *P. hybrida* with those of *P. parodii* (n = 5).

Discussion

The present results demonstrate that supplementation of the fusogen with low concentrations of Pluronic F-68 enhances PEG-mediated fusion of *Petunia* leaf and cell suspension protoplasts. Such enhanced fusion may involve either Pluronic-increased fluidity of the cytoplasmic membrane during the fusion process, or, alternatively, the surfactant may act to promote membrane resealing post-fusion coupled with subsequent membrane stabilisation. King et al. (1991) showed that Pluronic F-68 increases the permeability of intact yeast cells to fluorescein diacetate and that this occurs within 10 min after addition of the surfactant. These observations suggest an increase in cytoplasmic membrane fluidity in the presence of Pluronic F-68. Such an effect may represent a short-term response caused by adsorption of the surfactant onto the cell surface. Additional evidence for enhanced membrane permeabilisation by Pluronic F-68 comes from patch-clamp studies with artificial lipid bilayers; the surfactant promoted the formation of short-lived, transmembrane pores (King et al., 1991). Such pores could promote plasma membrane adhesion and facilitate fusion if they formed during the PEG-induced protoplast fusion process.

When added to the PEG-based fusogen at the concentrations given in this report, Pluronic F-68 does not adversely affect protoplast viability. In fact, it appears to be advantageous as a supplement in the chemical fusion mixture, since it may also have a longer-term effect in promoting stabilisation of the plasma membrane of heterokaryons. Cultured animal hybridoma cells, supplemented with 0.5% (w/v) Pluronic F-68, showed decreased cytoplasmic membrane fluidity over a 4 day period (Ramirez & Matharasan, 1990). Further, hybridomas cultured with Pluronic F-68 for more than 7 days exhibited strengthened cytoplasmic membranes, as determined by micromanipulation assessments (Zhang et al., 1992). Although the precise mechanism(s) of these effects of Pluronic was not determined, it has been suggested that the surfactant becomes intercalated into the plasma membrane, where it may interact with lipid components (Ramirez & Matharasan, 1990; King et al., 1991). While caution must be exercised in extrapolating from animal to plant cell systems, the information from the present fusion and culture experiments with *Petunia* protoplasts indicates that interaction of Pluronic F-68 with protoplast membranes does not have a detrimental effect on cell wall, synthesis, mitotic division or the generation of protoplast-derived tissues.

Further experiments are required to assess the applicability of Pluronic F-68-stimulated PEG fusion to other combinations of plant species and genera. An additional objective will be to determine whether this approach can be employed to promote heterokaryon formation with other chemical fusogens. Similarly, it will be of interest to assess whether Pluronic F-68, or, indeed, other co-polymer surfactants, can stimulate protoplast fusion during electrofusion or with a combination of chemical and physical procedures. It is now appropriate to evaluate other surfactants which are known to differ markedly in relation to their hydrophilic-hydrophobic balance (HLB), since previous work has shown these parameters to be crucial to their effects on plant cell and tissue differentiation (Khatun et al., 1993). Those surfactants having a lower HLB number could be more likely to foster the fusion process, since they possess pronounced membrane solubilisation properties (Helinius & Simons, 1975). In any case, the stimulation of heterokaryon formation by inclusion of co-polymer surfactants in the PEG fusogen, as in the present investigation, may, in turn, enhance the throughput of somatic hybrid plants and will thus be of wider interest in plant biotechnology and breeding.

Acknowledgement

P.A. was supported by The Rockefeller Foundation.

References

Chand, P.K., M.R. Davey, J.B. Power & E.C. Cocking, 1989. An improved procedure for protoplast fusion using polyethylene glycol. J. Plant Physiol. 133: 480–485.

Clarke, M.S.F. & P.L. McNeil, 1992. Syringe loading introduces macromolecules into living mammalian cell cytosol. J. Cell Sci. 102: 533–541.

Gilmour, D.M., T.J. Golds & M.R. Davey, 1989. *Medicago* protoplasts: fusion, culture and plant regeneration. In: Y.P.S. Bajaj (Ed.) Biotechnology in Forestry and Agriculture, Vol 8, Plant Protoplasts and Genetic Engineering I, pp. 370–388. Springer-Verlag, Heidelberg.

Helinius, A. & K. Simons, 1975. Solubilization of membranes by detergents. Biochem. Biophysica Acta 415: 29–79.

Kao, K.N. & M.R. Michayluk, 1975. Nutritional requirements for growth of *Vicia hajastana* cells and protoplasts at a very low population density in liquid media. Planta 126: 105–110.

Khatun, A., M.R. Davey, J.B. Power & K.C. Lowe, 1993. Stimulation of shoot regeneration from jute cotyledons cultured with non-ionic surfactants and relationship to physico-chemical properties. Plant Cell Rep. 13: 49–53.

King, A.T., M.R. Davey, I.R. Mellor, B.J. Mulligan & K.C. Lower, 1991. Surfactant effects on yeast cells. Enzyme Microb. Technol. 13: 148–153.

Lowe, K.C., M.R. Davey, J.B. Power & B.J. Mulligan, 1993. Surfactant supplements in plant culture systems. Agro-food-Industry Hi-Tech. 4: 9–13.

Lynch, P.T., M.R. Davey & J.B. Power, 1993. Plant protoplast fusion and somatic hybridisation. In: N. Düzgünes (Ed.) Methods in Enzymology: Membrane Fusion Techniques, Vol. 221, pp. 379–393. Academic Press, London.

Murashige, T. & F. Skoog, 1962. A revised medium for rapid growth and bioassays with tobacco tissue cultures. Physiol. Plant. 15: 473–497.

Patil, R.S., M. Latif, F.B. d'Utra Vaz, M.R. Davey & J.B. Power, 1993. Hybridisation, through culture of embryos and immature seeds, of a range of tomato cultivars with a tomato somatic hybrid (*Lycopersicon esculentum* (+) *L. peruvianum*): emergence of a possible new marker gene for tomato breeding. Plant Breeding 111: 273–282.

Power, J.B., P.K. Evans & E.C. Cocking, 1978. Fusion of plant protoplasts. In: G. Poste & G.L. Nicolson (Eds) Membrane Fusion, pp. 369–385. Elsevier/North-Holland Biomedical Press, North Holland.

Power, J.B., M.R. Davey, M. McLellan & D. Wilson, 1990. Isolation, culture and fusion of protoplasts – 2. Fusion of protoplasts. Biotechnol. Ed. 1: 173–177.

Ramirez, O.T. & R. Matharasan, 1990. The role of the plasma membrane fluidity on the shear sensitivity of hybridomas grown under hydrodynamic stress. Biotechnol. Bioeng. 36: 911–920.
Widholm, J., 1972. The use of FDA and phenosafranine for determining viability of cultured plant cells. Stain Technol. 47: 186–194.
Zhang, Z., M. Al-Rubeai & C.R. Thomas, 1992. Effect of Pluronic F-68 on the mechanical properties of mammalian cells. Enzyme Microb. Technol. 14: 980–983.

Euphytica **85**: 275–279, 1995.

Fluorescent *in situ* hybridization as an aid to introducing alien genetic variation into wheat

T.E. Miller, S.M. Reader, K.A. Purdie, S. Abbo[1], R.P. Dunford & I.P. King[2]
John Innes Centre, Colney, Norwich, NR4 7UH, U.K. present address: [1] Department of Plant Genetics, The Weizmann Institute of Science, Rehovot 76-100, Israel; [2] Department of Agricultural Botany, The University of Reading, Whiteknights, P.O. Box 221, Reading, RG6 2AS, U.K.

Key words: alien introduction, fluorescent *in situ* hybridization (FISH), wheat

Summary

Fluorescent *in situ* hybridization (FISH) has been used to assess the occurrence and frequency of wheat-alien chromosome pairing in a wheat/*Thinopyrum bessarabicum* hybrid and in wheat/rye hybrids with different levels of chromosome pairing by examining pollen mother cells at metaphase I of meiosis. The use of FISH to identify the presence and size of alien chromatin in a wheat background is also demonstrated.

The value of FISH as an aid to the introgression of alien genetic variation into wheat is discussed.

Abbreviations: FISH – fluorescent *in situ* hybridization, GISH – genomic *in situ* hybridization, PRINS – primer-induced *in situ* hybridization

Introduction

The introduction of alien genetic variation into wheat is a valuable and proven technique for wheat improvement (Gale & Miller, 1987). However, successful transfers could be greatly facilitated, firstly by knowledge of the occurrence and frequency of chromosome pairing and hence recombination between wheat and alien chromosomes, and secondly by the identification of the presence of alien chromatin in the recipient progenies following chromosome engineering.

In the past, evidence for wheat-alien chromosome pairing in wheat/alien species hybrids was largely based on, 1 the level of pairing exceeding that found in wheat haploids (Riley et al., 1959), 2 on the formation of meiotic chromosome pairing configurations larger than the maximum expected from pairing between the wheat chromosomes, or 3 from the number of unpaired chromosomes being less than the number of alien chromosomes present (Forster & Miller, 1985).

Until recently, the technique of C-banding was the most common method used to identify wheat and alien chromosomes particularly in wheat/rye hybrids (Schlegel & Weryszko, 1979; Hutchinson et al., 1983). In species such as rye, where C-banding can clearly differentiate its chromosomes from those of wheat and can also identify individual chromosomes (Sybenga, 1983), the technique has distinct advantages, but in species with poorly-defined C-bands it is of limited value.

Introductions of alien chromatin segments into wheat have largely been confirmed by chromosome pairing studies (Riley et al., 1968; Miller et al., 1988) and to a lesser extent by C-banding (King et al., 1992).

Recently, fluorescent *in situ* hybridization (FISH) techniques have been developed (Schwarzacher et al., 1992; Abbo et al., 1993b) and these now provide an additional valuable tool for the assessment of the level of wheat-alien chromosome pairing in hybrids and for identifying the presence and size of alien chromosome segments integrated into wheat chromosomes.

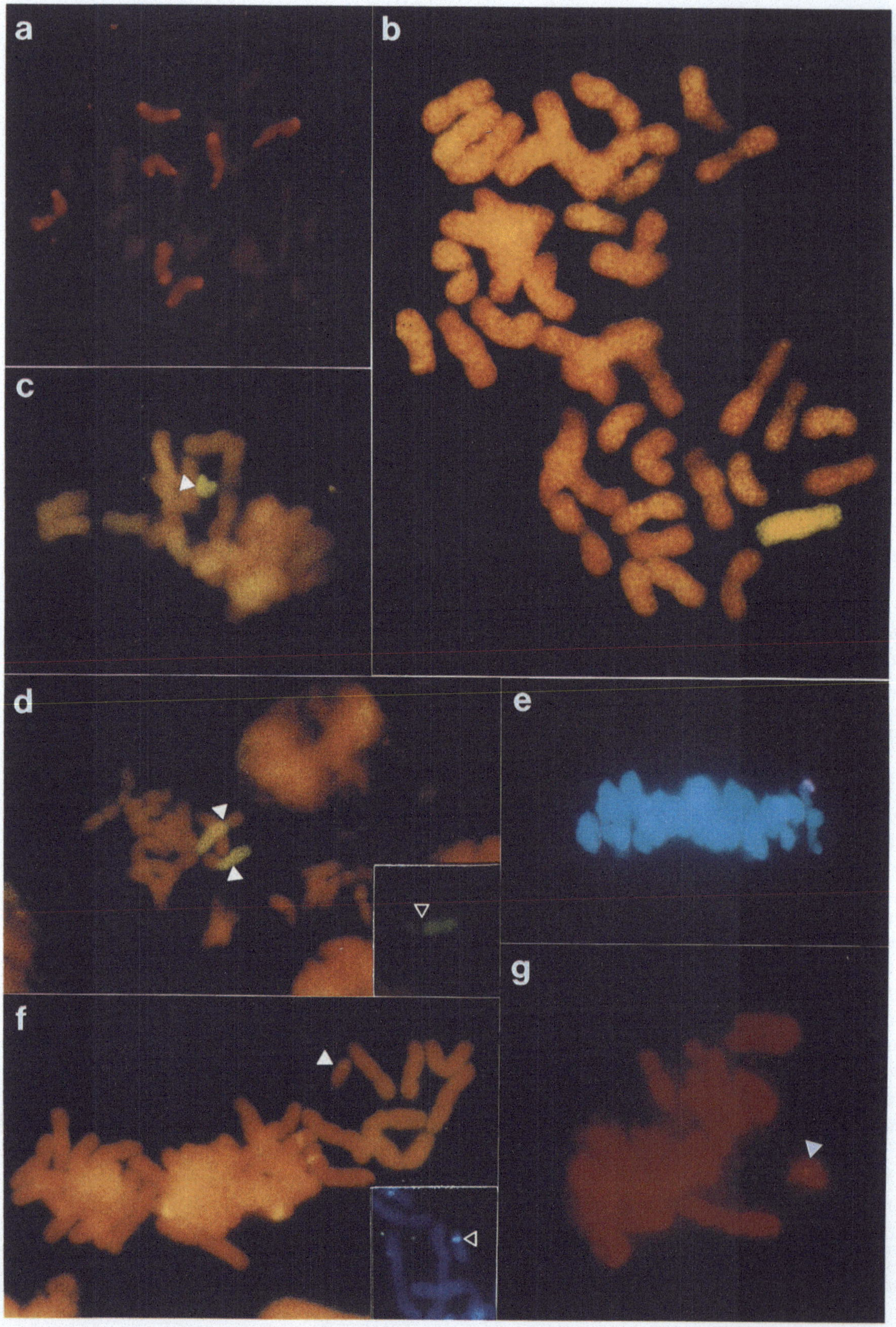
a
b
c
d
e
f
g

Fig. 1. Fluorescent *in situ* hybridization. a) GISH to a meiotic I cell of a *Ph1* deficient wheat/rye hybrid showing two wheat-wheat-rye trivalents; the rye chromosomes are bright red.
b) GISH to a mitotic metaphase highlighting chromosome 5E^b (yellow) of *Thinopyrum bessarabicum* added to wheat.
c) GISH to a mitotic metaphase highlighting the short arm telocentric of chromosome 5E^b (yellow) added to wheat.
d) GISH on a mitotic metaphase containing a pair of chromosomes with the long arm of chromosome 5E^b (yellow) translocated to the short arm of 5A. The inset clearly shows the translocation point to be centromeric.
e) GISH on a mitotic metaphase I cell showing a rod bivalent in which one chromosome carries a terminal segment (pink) of chromosome 5R^m of *Secale montanum*.
f) *In situ* hybridization using rDNA probe pTa71 to locate the nucleolus organizer region of the short arm of chromosome 5E^b.
g) *In situ* hybridization of a wheat line carrying the short arm telocentric of chromosome 1H^{ch} of *Hordeum chilense*; the arrow indicates the location of the B-hordein gene cluster (yellow).

Table 1. Mean chromosome pairing per pollen mother cell in wheat/alien hybrids with varying levels of wheat-alien chromosome pairing

Genotype	Wheat-wheat				Wheat-alien				Alien-alien	
	II	III	IV	arm assoc.	II	III	IV, V, VI	arm assoc	II	arm assoc.
Creso *ph1c* × *Th. bessarabicum*	0.94	–	–	0.94	0.12^a	–	–	0.14	0.06	0.06
Chinese Spr. × *S. cereale*	0.48^b	–	–	0.48	0.08^c	–	–	0.08	0.02	0.02
Chinese Spr. M5B × *S. cereale*	2.87	0.68	0.09^d	5.11	0.30	0.09	0.30	0.43	0.07	0.07
Chinese Spr. N5B-T5D × *S. cereale*	3.83	0.55	0.04^e	6.19	0.12	0.05^f	–	0.18	0.03^a	0.04
Chinese Spr. T3B × *S. cereale*	0.80	0.08^g	–	1.84	0.22	0.13^h	–	0.47	0.11	0.11

Where numbers of configurations are too low to appear in data to two decimal places they are combined as follows:
^{a} inc. 1^{III}, ^{b} inc. 2^{IV}, ^{c} inc. 3^{III}, ^{d} inc. 3^V, ^{e} inc. 2^V, ^{f} inc. 1^V, ^{g} inc. 1^{IV}, 1^{VI}, ^{h} inc. 3^{VI}.

Materials and methods

Plant material

Macaroni wheat, *Triticum durum* cv. Creso (carrying the *ph1c* mutant/deletion) × *Thinopyrum bessarabicum* hybrid.

Bread wheat, *T. aestivum* cv. Chinese Spring euploid, monosomic 5B, nullisomic 5B-tetrasomic 5D and tetrasomic 3B × rye, *Secale cereale*, hybrids.

T. aestivum cv. Chinese Spring – *Th. bessarabicum* chromosome 5E^b complete and 5E^{b}S telocentric addition lines and 5AL.5E^{b}S translocation line.

T. aestivum cv. Chinese Spring – *Hordeum chilense* 1H^{ch}S telocentric chromosome addition line.

T. aestivum cv. Chinese Spring line carrying a segment of *S. montanum* chromosome 5R^m.

Fluorescent in situ *hybridization*

In situ hybridization using total genomic wheat DNA as a block, and fluorescently-labelled total genomic DNA from the appropriate alien species as a probe (GISH), was carried out as described by Schwarzacher et al. (1992); King et al. (1993b) and Reader et al. (1994). This permitted identification of the alien chromosome/chromosome segment in a wheat background (Fig. 1a–e). The primer-induced *in situ* hybridization (PRINS) technique (Abbo et al., 1993b) was used to detect the introgression of specific DNA sequences from the alien species into wheat (Fig. 1g).

Results and discussion

Assessment of wheat/alien chromosome pairing

In euploid wheat × alien species hybrids, little chromosome pairing occurs due to the presence of the gene *Ph1* on chromosome 5B, which prevents pairing between homoeologous chromosomes of the wheat genomes and between the wheat and alien genomes (Riley et al., 1959). On average, about 0.5 bivalents per pollen mother cell are observed at meiotic metaphase I (Miller et al., 1983). These bivalents were initially assumed to be the result of rare homoeologous chromosome pairing mainly between the wheat genomes. Mettin et al. (1976) and Schlegel & Weryszko (1979) using C-banding showed that, in wheat/rye hybrids, a low percentage of this pairing (less than 2 per cent) occurred between wheat and rye chromosomes and a similar percentage involved pairing between rye-chromosomes. By using GISH, it has been possible to confirm that wheat-alien and alien-alien chromosome pairing occurs in euploid wheat/alien hybrids. In the case of wheat/rye hybrids, the level of wheat-rye pairing was shown to be much higher (13 per cent) and slightly higher (3.4 per cent) for the rye-rye pairing than that determined by C-banding (Table 1).

In hybrids lacking the *Ph1* pairing control gene, where considerably more homoeologous chromosome pairing occurs, the GISH technique also readily allows determination of the frequency with which the alien chromosomes are involved in pairing (Fig. 1a). Pairing between the chromosomes of tetraploid wheat and *Th. bessarabicum* and between hexaploid wheat and rye has clearly been identified and quantified (Table 1) in the absence of *Ph1* (King et al., 1993a; Miller et al., 1994).

Chromosomes other than 5B, which carries the *Ph1* gene, are known to affect chromosome pairing (Gale & Miller, 1987). The GISH technique can also be used to determine which aneuploid genotypes and hence which particular chromosomes produce the best level of wheat-alien chromosome pairing for a particular alien species. For example, in hybrids with rye, genotypes nullisomic for chromosome 5B give a higher degree of wheat-alien pairing (0.43 arm associations per cell) than the nullisomic 5B-tetrasomic 5D $\times$ rye genotype (0.18 arm associations per cell) indicating that the extra dose of chromosome 5D has a suppressing effect on the wheat-alien pairing (Table 1). In comparison, hybrids with an extra dose of chromosome 3B produce a level of wheat-alien pairing similar to that produced by the absence of chromosome 5B (Table 1) (King et al., 1994).

Detection of the presence of alien chromatin in wheat

Confirmation of the introduction of alien genetic material into wheat can be readily achieved by GISH, such that complete alien chromosomes can be clearly identified (Schwarzacher et al., 1992; King et al., 1993b; Reader et al., 1994). Figure 1b shows the addition of a single *Th. bessarabicum* chromosome $5E^b$ to wheat. Similarly, the presence of single alien chromosome arms in the form of telocentrics (Fig. 1c) or Robertsonian translocations (Fig. 1d) can be demonstrated (King et al., 1993b). GISH can also highlight the presence of relatively small alien segments; Fig. 1e shows a segment of chromosome $5R^m$ of *S. montanum* translocated terminally to a wheat chromosome.

In situ hybridization of fluorescently-labelled, repeated sequence probes can also be usefully employed. For example the ribosomal DNA probe pTa71 (Gerlach & Bedbrook, 1979) can be used to detect not only wheat nucleolus organizer region sites but also alien sites (Fig. 1f).

The detection of single copy genes is possible (Gustafson & Dillé, 1992) but is at present difficult in wheat. Low-copy sequences can, however, be detected. Using the PRINS technique (Abbo et al., 1993b) the B-hordein gene cluster on the short arm of *Hordeum chilense* chromosome $1H^{ch}$ has been detected in a wheat background (Fig. 1g) (Abbo et al., 1993a). The PRINS technique may also prove valuable in improving the detection of single copy genes.

References

Abbo, S., R.P. Dunford, T.E. Miller, S.M. Reader & I.P. King, 1993a. Primer-mediated *in situ* detection of the B-Hordein gene cluster on barley chromosome 1H. Proc. Natl. Acad. Sci. USA 90: 11821–11824.

Abbo, S., T.E. Miller & I.P. King, 1993b. Primer-induced *in situ* hybridization to plant chromosomes. Genome 36: 815–817.

Forster, B.P. & T.E. Miller, 1985. A 5B deficient hybrid between *Triticum aestivum* and *Agropyron junceum*. Cer. Res. Comm. 13: 93–95.

Gale, M.D. & T.E. Miller, 1987. The introduction of alien genetic variation into wheat. p. 173–210. In: F.G.H. Lupton (Ed). Wheat Breeding. Its Scientific Basis. Chapman and Hall, London.

Gerlach, W.L. & J.R. Bedbrook, 1979. Cloning and characterization of ribosomal RNA genes from wheat and barley. Nucleic Acid Res. 7: 1869–1885.

Gustafson, J.P. & J.E. Dillé, 1992. Chromosome location of *Oryza sativa* recombination linkage groups. Proc. Natl. Acad. Sci. USA 89: 8646–8650.

Hutchinson, J., T.E. Miller & S.M. Reader, 1983. C-banding at meiosis as a means of assessing chromosome affinities in the Triticeae. Can. J. Genet. Cytol. 25: 319–323.

King, I.P., R.M.D. Koebner, R. Schlegel, S.M. Reader, T.E. Miller & C.N. Law, 1992. Exploitation of a preferentially transmitted chromosome from *Aegilops sharonensis* for the elimination of segregation for height in semidwarf bread wheat varieties. Genome 34: 944–949.

King, I.P., K.A. Purdie, S.E. Orford, S.M. Reader & T.E. Miller, 1993a. Detection of homoeologous recombination in *Triticum durum* $\times$ *Thinopyrum bessarabicum* hybrids using genomic *in situ* hybridization. Heredity 71: 369–372 (Erratum 72: 321).

King, I.P., K.A. Purdie, H.N. Rezanoor R.M.D. Koebner, T.E. Miller, S.M. Reader & P. Nicholson, 1993b. Characterization of *Thinopyrum bessarabicum* chromosome segments in wheat using random amplified polymorphic DNAs (RAPDs) and genomic *in situ* hybridization. Theor. Appl. Genet. 86: 895–900.

King, I.P., S.M. Reader, K.A. Purdie, S.E. Orford & T.E. Miller, 1994. A study of the effect of a homoeologous pairing promoter on chromosome pairing in wheat/rye hybrids using genomic *in situ* hybridization. Heredity 74: 318–321.

Mettin, D., R. Schlegel, W.D. Bluthner & M. Weinrich, 1976. Giemsa-banding von MI-Chromosomen bei Weizen-Roggen-Bastarden. Biol. Zbl. 95: 35–41.

Miller, T.E., S.M. Reader & D. Singh, 1988. Spontaneous non-Robertsonian translocations between wheat chromosomes and an alien chromosome. p. 387–390. In: T.E. Miller & R.M.D. Koebner (Eds). Proc. 7th Int. Wheat Genet. Symp., Cambridge, Institute of Plant Science Research.

Miller, T.E., S.M. Reader & M.D. Gale, 1983. The effect of homoeologous group 3 chromosomes on chromosome pairing and crossability in *Triticum aestivum*. Can. J. Genet. Cytol. 25: 634–641.

Miller, T.E., S.M. Reader, K.A. Purdie & I.P. King, 1994. Determinations of the frequency of wheat-rye chromosome pairing in wheat × rye hybrids with and without chromosome 5B. Theor. Appl. Genet. 89: 255–258.

Reader, S.M., S. Abbo, K.A. Purdie, I.P. King & T.E. Miller, 1994. Direct labelling of plant chromosomes by rapid *in situ* hybridization. Trends Genet. 10: 265–266.

Riley, R., V. Chapman & G. Kimber, 1959. Genetic control of chromosome pairing in intergeneric hybrids in wheat. Nature, London 183: 1244–1246.

Riley, R., V. Chapman & R. Johnson, 1968. The incorporation of disease resistance in wheat by genetic interference with the regulation of meiotic chromosome pairing. Genet. Res. Camb. 12: 199–219.

Schlegel, R. & E. Weryszko, 1979. Intergeneric chromosome pairing in different wheat-rye hybrids revealed by giemsa banding technique and some implications on karyotype evolution in the genus *Secale*. Biol. Zbl. 93: 398–407.

Schwarzacher, T., K. Anamathawat-Jonsson, G.E. Harrison, A.K.M.R. Islam, J.Z. Jia, I.P. King, A.R. Leitch, T.E. Miller, S.M. Reader, W.J. Rogers, M. Shi & J.S. Heslop-Harrison, 1992. Genomic *in situ* hybridization to identify alien chromosome segments in wheat. Theor. Appl. Genet. 84: 778–786.

Sybenga, J., 1983. Rye chromosome nomenclature and homoeology relationships – Workshop report. Z. Pflanzenzüchtg. 90: 297–304.

Euphytica **85**: 281–285, 1995.

Sexual and somatic hybridization in the genus *Lactuca*

Brigitte Maisonneuve[1], Marie Christine Chupeau[2], Yannick Bellec[1] & Yves Chupeau[2]
[1] *Station de Génétique et d'Amélioration des Plantes and* [2] *Laboratoire de Biologie Cellulaire, Institut National de la Recherche Agronomique, Route de Saint Cyr, 78026 Versailles Cedex, France*

Key words: Lactuca sativa, Lactuca virosa, Lactuca tatarica, Lactuca perennis, lettuce, sexual hybridization, embryo rescue, somatic hybridization, protoplast fusion

Summary

Various genes for disease resistance identified in wild *Lactuca* are difficult, even impossible to exploit in lettuce breeding, due to sexual incompatibility between *L. sativa* and wild *Lactuca* sp. We adapted two cellular biology techniques to overcome these interspecific barriers: *in vitro* embryo rescue and protoplast fusion. *In vitro* rescue of immature embryos was used successfully for sexual hybridization between *L. sativa* and *L. virosa*. Vigorous hybrid plants were produced between *L. sativa* and seven accessions of *L. virosa*. Protoplast fusion permitted the regeneration of somatic hybrids between *L. sativa* and either *L. tatarica* or *L. perennis*. Hybrids between *L. sativa* and *L. tatarica* were backcrossed to *L. sativa*.

Introduction

Useful genes for lettuce breeding, especially for disease resistance (e.g. against *Bremia lactucae*, virus, bacteria), have been identified in wild species of *Lactuca*. Of 100 species in this genus, only three species, included with *L. sativa* in the section *Lactuca* (*L. serriola*, *L. saligna* and *L. virosa*), have been extensively studied for use in lettuce breeding (De Vries, 1990). Several genes from *L. serriola* have been introduced into some cultivars (e.g. several *Dm* genes; Crute 1992) because the two species can be easily crossed. The use of *L. saligna* has been more restricted; *L. saligna* must be used as the female parent and the fertility of F_1s is very poor. Nevertheless, some resistance genes from *L. saligna* have been introduced into *L. sativa* (Whitaker et al., 1974; Netzer et al., 1976; Provvidenti et al., 1980; Netzer et al., 1985). A new possible method for overcoming the crossability barrier was achieved by *in vitro* culture of immature hybrid (*L. sativa* × *L. saligna*) embryos, with subsequent production of large number of F_2 seeds by providing good conditions for flowering (Maisonneuve, 1987). The introduction of genes from *L. virosa* is a problem because the F_1 hybrids are necrotic or sterile (Lindqvist, 1960). Therefore, *L. serriola* has been used as a bridging species (Thompson & Ryder, 1961; Eenink et al., 1982). Occasionally, successful crosses were made by backcrossing F_1s with *L. sativa* (Maxon Smith & Langton, 1989). For some accessions of *L. virosa*, the *in vitro* culture of BC_1 embryos was necessary for plant development (Maisonneuve, 1987).

To extend the variability of our material and to exploit some resistance genes which we have identified in *L. virosa*, as beet western yellows virus resistance (Maisonneuve et al., 1991), and using other incompatible species, we have attempted to improve the crossability of *L. sativa* with *L. virosa* and to conduct somatic hybridizations with other members of the Compositae.

Materials and methods

Sexual crosses between L. virosa *and* L. sativa

The crossability of three accessions of *L. virosa* (PIVT280, PIVT1145, PIVT1398), received from CPRO, previously IVT (Wageningen, NL), with 10 cultivars of *L. sativa* was investigated using either *in*

Table 1. Production of sexual hybrids between *L. sativa* and *L. virosa*

Method of cross	*In vitro* culture of immature embryo				Harvesting of seeds			
Crosses	Number of *L. sativa* cultivars	Number of pollinated capitula	Number of F1 plants necrotic	Number of F1 plants vigorous	Number of *L. sativa*	Number of pollinated capitula	Number of F1 plants necrotic	Number of F1 plants vigorous
PIVT280 × *L. sativa*	5	25	51	35	10	47	0	0
L. sativa × PIVT280	7	34	1	3	10	50	15	16
PIVT1145 × *L. sativa*	7	34	28	15	10	50	3	0
L. sativa × PIVT1145	9	45	6	6	9	42	13	1
PIVT1398 × *L. sativa*	8	39	0	3	10	51	0	1
L. sativa × PIVT1398	9	44	1	24	10	50	9	22
A1084 × *L. sativa*	5	25	10	42	6	30	0	0
L. sativa × A1084	5	25	5	1	6	30	17	1
A6660 × *L. sativa*	5	25	36	14	5	25	1	3
L. sativa × A6660	5	25	5	0	6	30	17	5
B9056 × *L. sativa*	5	25	18	14	6	30	3	0
L. sativa × B9056	5	24	0	0	6	30	2	0
K7095 × *L. sativa*	5	25	10	14	6	29	0	0
L. sativa × K7095	5	25	7	0	6	29	110	3

vitro culture of immature embryos or harvesting of mature seeds. Five young capitula were cut off 4 or 5 days after pollination for *in vitro* embryo rescue. The culture was made in To medium (MS salts, Morel vitamins, 20 gl^{-1} sucrose, 6 gl^{-1} agar) according to Maisonneuve (1987). Then four other accessions of *L. virosa* (A1084, A6660, B9056, K7095) received from J. Maxon Smith (formerly: IHR, Littlehampton, England) were compared to the three ex-IVT accessions studied previously. These crosses were made with five or six cultivars of *L. sativa* previously used, with the same methods. In both experiments, the reciprocal crosses were compared. The seedlings developed from *in vitro* culture and mature seeds were studied in the greenhouse.

Somatic hybridization

To facilitate selection of somatic hybrids, a universal hybridizer, carrying both a dominant marker, kanamycin resistance (Kan^R), and a recessive one, albino (a), was developed. The dominant gene *nptII* was introduced by electroporation of protoplasts of cv. Ardente (Chupeau et al., 1989), where its inheritance is consistent with Mendelian segregation. The recessive marker was produced by mutagenesis of seeds of cv. Girelle with ethylmethane sulfonate for 6 h according to Robinson (1986) and a line heterozygous for the albino gene was selected. A double heterozygous line was produced by hybridiziation between an 'Ardente' Kan^R plant and a 'Girelle' a/+ plant. Protoplasts were prepared from *in vitro*-grown leaflets from plants between 1 and 2 months after sowing on B medium (Bourgin et al., 1978) with 100 mgl^{-1} kanamycin. For good preparation of protoplasts from albino lettuce, we adapted, for lettuce seedlings, the medium reported earlier for isolating protoplasts from albino tobacco (Aviv & Galun, 1985): medium B complemented with 0.01 mgl^{-1} NAA, 0.01 mgl^{-1} BA, 100 mgl^{-1} adenosine and 7% sucrose (Chupeau et al., 1994).

Results and discussion

Sexual crosses between L. virosa *and* L. sativa

From these seven accessions, a large increase in hybrid production was obtained through *in vitro* culture. The exceptions were PIVT1398 and K7095, where hybrid production following embryo rescue was similar to that from mature seed. Comparison of reciprocal crosses showed a higher quantity of mature seeds with *L. sativa* as female parent; this result is similar to the observations of Lindqvist (1960). In contrast, more hybrids

Table 2. Vigour of F_1 hybrids between *L. sativa* and *L. virosa*. F_1 plants vigorous with flowering (V) or weak, necrotic, dead (n); no hybrid plant (0) or missing data (–)

L. virosa	cytoplasm	*In vitro* culture of F_1 embryo						Harvesting of hybrid mature seeds					
		L. sativa						*L. sativa*					
		B	C	D	E	G	J	B	C	D	E	G	J
PIVT280	*virosa*	V	n	n	V	V	n	0	0	0	0	0	0
	sativa	V	0	0	0	–	0	V	n	n	V	V	n
PIVT1145	*virosa*	V	n	n	V	V	n	0	0	0	0	0	0
	sativa	n/V	0	n	0	V	n	n	n	n	–	V	0
PIVT1398	*virosa*	0	V	0	0	V	–	0	n	0	0	n	0
	sativa	0	V	0	V	V	V	n	0	V/n	V	V	V
A1084	*virosa*	V	n	0	V	0	0	0	0	0	0	0	0
	sativa	0	n	0	V	0	0	V	0	n	0	0	n
A6660	*virosa*	V	n	n	V	–	0	–	0	n	0	V	V
	sativa	0	0	n	n	–	0	0	n	n	V	0	n
B9056	*virosa*	V	n	n	V	–	0	0	n	0	n	0	0
	sativa	0	0	0	0	–	0	0	0	n	0	0	n
K7095	*virosa*	0	n	n	V	–	0	0	0	0	0	0	0
	sativa	0	n	n	0	0	0	V	n	n	0	0	n

were produced after *in vitro* embryo culture when *L. virosa* was used as the female parent. Analysis of all treatments showed the highest number of hybrids was obtained from crosses with *L. virosa* as female parent, using *in vitro* rescue of embryos (Table 1).

As reported in the literature (Lindqvist, 1960; De Vries, 1990), many F_1 plants were weak, necrotic and died at an early stage, but it was possible to produce vigorous F_1 plants with flowers for all the accessions of *L. virosa* by appropriate selection of the *L. sativa* parental cultivar (Table 2). Some cultivars, such as genotype E, were able to give vigorous hybrids with every tested *L. virosa* accession. Other cultivars differentiated the accessions of *L. virosa*; e.g. with genotype J, the hybrid plants with PIVT1398 were vigorous whereas the hybrids with PIVT280 and PIVT1145 were necrotic. The necrosis of F_1s results from nuclear genome incompatibility and is not due to the method used to produce the hybrid or to the cytoplasmic-nuclear interaction. For each pair of parents (*L. sativa* cultivar – *L. virosa* accession), the same behaviour of the hybrid plants, necrotic or vigorous, was usually observed regardless of the hybridization method used (with or without *in vitro* culture, both reciprocal crosses) (Table 2).

For each parental combination, the F_1 plants were near-sterile but as a result of a large-scale pollination with *L. sativa*, a few BC_1 seeds could be produced. The BC_1 plants were sufficiently fertile to yield some progeny after self-pollination. Transfer of some resistance to virus and downy mildew to lettuce cultivaris in progress.

Somatic hybridization

For seven out of eight wild parents tested, the *in vitro* culture conditions were suitable, and protoplasts were prepared from *in vitro*-micropropagated plants. In the case of *L. juncea*, protoplasts were prepared from *in vitro*-grown seedlings. Protoplasts of five species (*L. tatarica*, *L. perennis*, *L. juncea*, *Cicerbita plumieri*, *Cichorium intybus*) divided in the medium used for *L. sativa*. Conversely, protoplasts of three other species (*Taraxacum officinalis*, *Chrysanthemum segetum*, *Leucanthemum maximum*), although alive for a few weeks, were not able to grow.

The protoplasts of albino plants of our (Kan^R, a) *L. sativa* line were fused with protoplasts of three wild *Lactuca* and five related species to explore the possibility of overcoming incompatibility (Chupeau et al., 1994). Green buds developing on 100 mgl^{-1} kanamycin were considered as somatic hybrids. Such hybrids were obtained after fusion of *L. sativa* with *L. perennis* and *L. tatarica*, two perennial *Lactuca*

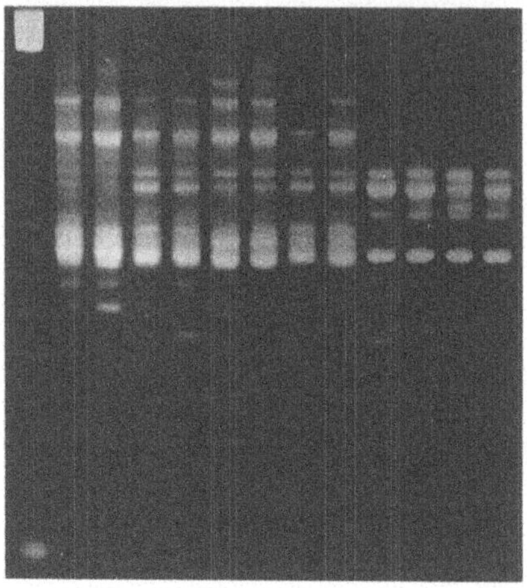

Fig. 1. RAPD markers of *L. sativa* and *L. perennis* in somatic hybrids. Amplification with one primer from Operon Technologies (OPI11 = 5′-ACA TGC CGT G-3′) of genomic DNA extracted from four lettuce lines (S), cv. Girelle (lanes 10 and 11) and Ardente × Girelle (lanes 12 and 13), from *L. perennis* (P) and from six somatic hybrids (h). Lane 1 is a 123-bp molecular weight ladder.

species with blue flowers. Their hybrid nature was confirmed using PCR-based molecular markers (Random Amplified Polymorphic DNA; Williams et al., 1990). Some arbitrary primers amplified fragments characteristic of each parental species in plants regenerated from fusions of *L. sativa* with *L. tatarica* (Chupeau et al., 1994) as well as with *L. perennis* (Fig. 1).

The hybrids with *L. tatarica* were very sensitive to climatic conditions. Most of them, once rooted, grew vigorously to form a rosette, but then stopped making leaves and developed stunted roots, while the existing leaves became necrotic. Modification of the environmental conditions permitted growth and bolting of some of these plants. At a constant temperature of 25° C in a growth room, 15 out of 47 hybrids flowered. By pollination of a large number of capitula with different cultivars of lettuce, some BC_1 seeds could be produced (Table 3). In subsequent generations, the plants were vigorous in the greenhouse. Segregation for different characters was observed (colour of flowers, types of capitula, shape and colour of seeds). Fertility was always very low and manual pollination by *L. sativa* of a large number of capitula was necessary to produce seeds (Table 3). Preliminary tests, for resistance to downy mildew, revealed a transfer of resistance from *L. tatarica* to these plants.

Table 3. Progeny of somatic hybrids between *L. sativa* and *L. tatarica*

Generation of interspecific female plants	F_1	BC_1	BC_2
Number of *in vitro* plants or number of sown seeds	47	41	91
Number of flowering plants	15	22	58
Number of pollinated capitula	439	763	661
Number of BC seeds	80	124	101

Conclusion

In vitro culture of immature embryos and somatic hybridization were successfully used to generate interspecific hybrids between *L. sativa* and wild *Lactuca* species. Preliminary analyses of progeny confirmed transfer of characters from these wild species to *L. sativa*-like-plants. There is reason to believe, however, that development of completely fertile lettuce plants with characters from distant species will require further work. The use of molecular markers could increase the efficiency of a backcross programme. Nevertheless, the application of these techniques for breeding will help to increase variability in lettuce in the future. As a result, several new resistance genes will become part of the gene pool for breeders.

References

Aviv, D. & E. Galun, 1985. *In vitro* procedure to assign pigment mutations in *Nicotiana* to either the chloroplast or the nucleus. J. Hered. 76: 135–136.

Bourgin, J.P., Y. Chupeau & C. Missonier, 1978. Plant regeneration from mesophyll protoplasts of several *Nicotiana* species. Physiol. Plant. 45: 288–292.

Chupeau, M.C., C. Bellini, P. Guerche, B. Maisonneuve, G. Vastra & Y. Chupeau, 1989. Transgenic plants of lettuce (*Lactuca sativa*) obtained through electroporation of protoplasts. Biotechnology 7: 503–507.

Chupeau, M.C., B. Maisonneuve, Y. Bellec & Y. Chupeau, 1994. A *Lactuca* universal hybridizer, and its use in creation of fertile interspecific somatic hybrids. Mol. Gen. Genet. 245: 139–145.

Crute, I.R., 1992. The role of resistance breeding in the integrated control of downy mildew (*Bremia lactucae*) in protected lettuce. Euphytica 63: 95–102.

De Vries, I.M., 1990. Crossing experiments of lettuce cultivars and species (*Lactuca* sect. *Lactuca*, Compositae). Pl. Syst. Evol. 171: 233–248.

Eenink, A.H., R. Groenwold & F.L. Dieleman, 1982. Resistance of lettuce (*Lactuca*) to the leaf aphid *Nasonovia ribis nigri*. I. Transfer of resistance from *L. virosa* to *L. sativa* by interspecific crosses and selection of resistant breeding lines. Euphytica 31: 291–299.

Lindqvist, K., 1960. Cytogenetic studies in the *Serriola* group of *Lactuca*. Hereditas 46: 75–151.

Maisonneuve, B., 1987. Utilisation de la culture *in vitro* d'embryons immatures pour les croisements interspécifiques entre *Lactuca sativa* L. et *L. saligna* L. ou *L. virosa* L.; étude des hybrides obtenus. Agronomie 7: 313–319.

Maisonneuve, B., V. Chovelon & H. Lot, 1991. Inheritance of resistance to beet western yellows virus in *Lactuca virosa* L. HortScience 26: 1543–1545.

Maxon Smith, J. & A. Langton, 1989. A new source of resistance to downy mildew. Grower 21: 54–55.

Netzer, D., D. Globerson & J. Sacks, 1976. *Lactuca saligna* L., a new source of resistance to downy mildew (*Bremia lactucae* Reg.). HortScience 11: 612–613.

Netzer, D., D. Globerson, Ch. Weintal & R. Elyassi, 1985. Sources and inheritance of resistance to stemphylium leaf spot of lettuce. Euphytica 34: 393–396.

Provvidenti, R., R.W. Robinson & J.W. Shail, 1980. A source of resistance to a strain of cucumber mosaic virus in *Lactuca saligna* L. HortScience 15: 528–529.

Robinson, R.W., 1986. Mutagenensis of lettuce with ethyl methane sulfonate. Mutation Breeding Newsletter 28: 7.

Thompson, R.C. & E.J. Ryder, 1961. Description and pedigrees of nine varieties of lettuce. USDA Tech. Bull., n° 1244, 19 p.

Whitaker, T.W., A.N. Kishaba & H.H. Toba, 1974. Host parasite interrelations of *Lactuca saligna* L. and the cabbage looper, *Trichoplusia ni* (Hubner). J. Amer. Soc. Hortic. Sci. 99: 74–78.

Williams, J.G.K., A.R. Kubelik, K.J. Livak, J.A. Rafalski & S.V. Tingey, 1990. DNA polymorphisms amplified by arbitrary primers are useful as genetic markers. Nucl. Acids Res. 18: 6531–6535.

Euphytica **85:** 287–294, 1995.

Protoplast technology for the breeding of top-fruit trees (*Prunus, Pyrus, Malus, Rubus*) and woody ornamentals

Sergio J. Ochatt & Estela M. Patat-Ochatt
I.N.R.A. Centre d'Angers, Station d'Amélioration des Espèces Fruitières et Ornementales, B.P. 57, 49071 Beaucouzé Cedex, France

Key words: protoplasts, protoclonal variation, somatic hybridization, top-fruit trees, woody ornamentals

Summary

Until recently, temperate fruit trees and woody ornamentals have been regarded as recalcitrant to biotechnological breeding approaches based on protoplasts. This however should no longer be the case, as procedures are now available, not only for the regeneration of complete plants from protoplasts of various tissues of such species, but also for the exploitation of protoplast technology for their genetic manipulation. This paper will examine the recent advances and state of the art in this domain, with particular attention to the use of protoplast technology as a novel tool in the breeding of rosaceous top-fruit tree species and woody ornamentals. Problems and their solutions within the context of regenerating plants from isolated protoplasts of stone (*Prunus* spp.), pome (*Pyrus* spp., *Malus* spp.) and small (*Rubus* spp.) fruits, and of several shrubby ornamental genotypes (*Lonicera* spp., *Weigela* spp., *Forsythia* spp., *Cotoneaster* spp.) will be addressed. Interspecific (*Prunus spinosa* + *Prunus cerasifera*) and intergeneric (*Forsythia* spp. + *Syringa* spp.) somatic hybridization within this group of species, as well as the use of protoplasts for host/pathogen interaction studies (*Pyrus*/*Erwinia amylovora*) will also be discussed.

Introduction

It is now over 90 years since the hypothesis that individual nucleated plant cells have the genetic capacity to generate whole plants (i.e., totipotency) either directly or through an intermediate callus phase, was first put forward (Haberlandt, 1902). Moreover, more than 30 years have passed since the first recovery of complete plants from cells cultured *in vitro* (Steward et al., 1958). During the last few decades biotechnological approaches have gained momentum, coupled with a steady advancement in the theoretical and practical knowledge of the regeneration process, particularly in the context of its exploitation for breeding objectives. Three major components have been identified, as those with the largest influence on competence for plant regeneration: the genotype, the ontogenetic state of the explant source, and the cultural environment (Ochatt & Power, 1992). In this respect, a progressive reduction in tissue organization is likely to result in an alteration of the differentiation status of the cells therein and, hence, to reduce their morphogenetic competence. Thus, protoplasts (all components of a plant cell excluding the cell wall) are theoretically totipotent, but various difficulties in plant regeneration are to be expected.

Nowadays, protoplasts can be isolated enzymatically from virtually any plant species and any type of tissue source, and used for biotechnological breeding strategies. However, woody species in general were for many years considered to be outside the scope of these approaches, mainly due to the various difficulties encountered when dealing with protoplasts and/or tissues of such species *in vitro*. Notwithstanding this, precisely the traits that characterize woody species (i.e. a high heterozygosity, an outbreeding nature, a long juvenile phase, the widespread existence of barriers to sexual hybridization, and their propagation mainly via asexual means) make them ideal subjects to benefit greatly from rapid breeding methods, e.g. by selection procedures applied to genetically-manipulated tissue cultures and the production of novel woody plant genotypes.

Although protoplast-to-plant systems have been developed for a large number of herbaceous species, woody species are generally regarded as being recalcitrant in this respect (Ochatt & Power, 1992; Ochatt et al., 1992a). *Citrus* species remained a striking exception for nearly a decade (Vardi et al., 1975). Results reported by several authors over the last few years have, however, added an ever-increasing number of other fruit tree genotypes to the list of species for which plant regeneration from cultured protoplasts is now feasible (Ochatt et al., 1992a). Furthermore, in recent years somatic hybridization technology has been successfully applied to top-fruit trees (Ochatt & Patat-Ochatt, 1994), and other protoplast-based approaches are being developed (Brisset et al., 1990). Here, the methodologies developed and the current state of the art in this field will be discussed, both for temperature fruit crops and for woody ornamental species.

Use of protoplast technology for temperate fruit crops

Temperate small fruit crops

Among temperate small fruits, plant regeneration from protoplasts is restricted to a few examples. Protoplast-to-plant systems have been described using mesophyll protoplasts of the paper mulberry (*Broussonetia kazinoki*), which was the first example of its sort for a temperate woody species (Oka & Ohyama, 1985). Using differentiated callus of cell suspension cultures as the protoplast source, complete plants have also been regenerated from protoplasts of kiwifruit (Tsai, 1988; Oliveira & Pais, 1991). During recent studies in our group aimed at developing protoplast-to-plant strategies for the previously-unstudied raspberry (*Rubus idaeus*) and blackberry (*Rubus* × *fruticosus*), callus tissues have been produced from leaf and stem protoplasts of various genotypes of both species (Patat-Ochatt, unpublished). Interestingly, despite the economic importance worldwide and the large number of studies performed so far (Krul, 1988), grapevines have still to be regenerated from protoplasts. To date, there are no other reports on the use of protoplast technology for temperate small fruit crop improvement, and somatic hybridization strategies for this group of species has not yet been attempted.

Top-fruit tree species

In the case of rosaceous top-fruit trees (apple, pear and stone fruits), the production of complete plants from cultured protoplasts is now possible for a number of genotypes (Ochatt et al., 1992a), and the technological gap, compared to *Citrus*, is quickly disappearing. In recent years, a considerable research has been devoted to the development of protoplast technology for apple (Patat-Ochatt, 1994), pear (Ochatt, 1993a) and cherry (Ochatt, 1993b) most particularly, and these three will be discussed separately.

*Apple (*Malus Xdomestica *Borkh.)*

A number of researchers have studied the isolation and culture of protoplasts from various genotypes since the early 1980s, and the field was recently reviewed (Patat-Ochatt, 1994). Patat-Ochatt et al. (1988) first successfully regenerated whole plants from mesophyll protoplasts of two commercially important genotypes, the clonal rootstock M.9 and the scion variety Spartan. This was later extended to a columnar scion genotype (Wallin & Johansson, 1989) and to another clonal rootstock, MM106 (Patat-Ochatt & Power, 1990). More recently, plants have been regenerated from protoplasts isolated from seedling leaves of two other scion genotypes (Huancaruna-Perales & Schieder, 1993) and, for the first time among woody species in general, from stem protoplasts of a haploid clone of 'Golden Delicious' (Patat-Ochatt et al., 1993). This strategy was recently extended to various other pome fruit tree genotypes leading to the regeneration of complete stem protoplast-derived plants for the diploid variety Golden Delicious, the most economically important scion globally (Patat-Ochatt, 1994).

For all regenerable systems to date, plant production was achieved through organogenesis from protoplast-derived callus. In our group, protoplasts of scion varieties have generally exhibited a common requirement, during initial stages, for culture in a medium based on Kao & Michayluk's (1975) medium, while protoplasts of rootstock genotypes only proliferated to give regenerable callus on a medium based on Murashige & Skoog's (1962) formula, supplemented with complex mixtures of organic compounds (Patat-Ochatt, 1994).

*Pear (*Pyrus communis*)*

There are not many reports dealing with the isolation and culture of protoplasts of pear genotypes. In spite of this, a form of wild pear (*P. communis* var. *pyraster*) was the first successful example of whole plant regeneration from protoplasts of a top-fruit tree genotype (Ochatt & Caso, 1986), followed by the first report on the detection of protoclonal variation among regenerated plants for any woody plant species (Ochatt, 1987). Rhizogenesis was induced on callus derived from embryo callus protoplasts of the scion variety Conference (Ochatt & Power, 1988a), subsequently extended to complete plant regeneration (Ochatt, 1990a). Plants were recovered from mesophyll protoplasts of the variety Williams' Bon Chrétien, one of the most commercially important scions worldwide (Ochatt & Power, 1988b). Plants were also regenerated from leaf protoplasts of the rootstock genotype Old Home and the scion variety Passe Crassane, and trueness-to-type of such regenerated plants was established through phenotypic characters and by the study of the banding profiles for 13 different isoenzymatic systems, as compared to conventionally propagated mother plants of each genotype (Ochatt et al., 1992b). Plants have also been produced from leaf protoplasts of the rootstock genotype Old Home X Farmingdale (OHF) 333 and the scion variety Comice (Ochatt, 1993a), and the strategy as developed for apple stem protoplasts above has been applied to both these genotypes and 'Old Home' with the proliferation of stem protoplast-derived callus for all three genotypes (Rosati & Ochatt, 1994). Finally, protoplasts have been successfully isolated and cultured to the microcallus stage for several haploid clones (obtained recently at Angers after *in situ* parthenogenesis induced by irradiated pollen; Bouvier et al., 1993) of the scion variety Comice (Rosati & Ochatt, unpublished).

For all pear genotypes and source tissues assessed to date, protoplasts have required a medium free of ammonium ions for initial proliferation and, as also observed for apple (see above), supplemented with a higher content of organic compounds for rootstock genotypes as compared to scions. Plant regeneration from the protoplast-derived callus was always via organogenesis (Ochatt, 1993a).

Concerning somatic hybridization, a prime objective of protoplast manipulation, only a few reports have been published on temperate fruit trees to date (Ochatt & Patat-Ochatt, 1994). There are some early examples of intergeneric and interspecific protoplast fusions involving apple but, at best, only putative hybrid callus tissues were produced (see Ochatt et al., 1992 for a review). The only example of the production of somatic hybrids for rosaceous fruit tree species in the literature, concerns the fusion of protoplasts from two sexually-incompatible rootstock genotypes belonging to different subfamilies, *Pyrus communis* var. *pyraster* (*Maloideae*; 2n = 2x = 34) and *Prunus avium* × *pseudocerasus* (*Prunoideae*; 2n = 3x = 24) (Ochatt et al., 1989). Following culture of heterokaryons, callus tissues were produced and shoot buds were regenerated, propagated, and rooted *in vitro*. The resulting plants were confirmed as being somatic hybrids through chromosome counts, by isoenzymatic assessments, and by morphological markers, as all regenerated plants were homogeneous and intermediate when compared with both parent species.

*Stone fruit trees (*Prunus *spp.)*

The first example of regeneration of whole plants was for mesophyll protoplasts of the commercially important cherry rootstock 'Colt' (*Prunus avium* × *pseudocerasus*) (Ochatt et al., 1987). Later, plants were also produced from root cell suspension, protoplast-derived tissues of this same genotype (Ochatt et al., 1988a), the first example of plant regeneration via organogenesis from non-mesophyll protoplasts of a woody perennial species. Whole plants were also recovered from mesophyll protoplasts of several sour cherry clones (*P. cerasus*) (Ochatt & Power, 1988c). Noteworthy in this report was the development of a novel approach to plant regeneration, whereby protoplast-derived callus would undergo rhizogenesis, and shoot buds were successfully regenerated from the roots. This strategy was later applied for callus derived from root callus protoplasts of a sour cherry clone (Ochatt, 1990b). More recently, plants have also been regenerated from mesophyll protoplasts of the recalcitrant fruit and farm-woodland species *P. avium* (sweet/wild cherry), by adding glycine to the protoplast culture medium, as a novel strategy to avoid phenolic oxidation (Ochatt, 1991a).

The only other published examples on the successful production of protoplast-derived plants of stone fruit crops are for two forms of prune/plum, the ornamental species *P. spinosa* and the fruit-bearing and rootstock genotype *P. cerasifera*, myrobalan (Ochatt, 1992), while a limited number of protoplast-derived callus of a hybrid apricot clone (*P. armeniaca*) was also induced to undergo caulogenesis (Ochatt, 1993b).

In terms of somatic hybridization, apart from the studies described above, experiments were recently performed to produce somatic hybrids of *Prunus spinosa* and *P. cerasifera*, tetraploid and diploid respec-

Table 1. Protoplast-to-plant systems for woody ornamental genotypes

Genotype	Protoplast source	Regeneration pathway	Reference
Liriodendron tulipifera	Embryogenic cell suspensions	Somatic embryogenesis	Merkle & Sommers, 1987
Lonicera nitida cv. Maïgrun	*In vitro* leaves	Organogenesis	Ochatt, 1991
Rosa persica × *R. xanthina*	Embryogenic cell suspension	Somatic embryogenesis	Matthews et al., 1991
Weigela × *florida* cv. Bristol Ruby	*In vitro* leaves, stems and roots	Organogenesis (from leaf protoplasts only)	Ochatt, 1993
Forsythia × *intermedia* cv. Spring Glory	*In vitro* leaves, stems and roots	Organogenesis	Ochatt, 1994

tively, and presumed to be the origin of the hexaploid *P. domestica* (prune/plum). Sexual crossing between these two species only yields triploids and their doubling has so far failed to produce any hexaploids. In addition, studies on the genome of the two fusion partners has shown that the prune genome contains both the genomes of *P. spinosa* and *P. cerasifera*, which would confirm the view on the origin of *P. domestica* (suggesting that it was either the result of the crossing between unreduced gametes, or that it arose from a triploid hybrid between these two species that spontaneously doubled). However, Salesses (1973) found that the genome of *P. cerasifera* is also included in that of *P. spinosa*, thereby showing the origin of the prune to be more complex. The hexaploid somatic hybrids eventually produced might, therefore, permit the elucidation of these hypotheses. So far, several putative somatic hybrid plants regenerated from heterokaryons obtained after fusion have been recovered, and their true hybrid nature is at present being studied.

Use of protoplast technology for ornamental shrubs

Despite its economic importance globally, this is undoubtedly the group of woody plant genotypes where least is known in terms of protoplast technology. About 15 years were to elapse between the first example of a protoplast-to-plant system (*Citrus sinensis*; Vardi et al., 1975) and the first such report for an ornamental woody species. The only genotypes to have been regenerated from protoplasts so far (Table 1) are *Liriodendron tulipifera* (Merkle & Sommers, 1987), *Lonicera nitida* cv Maïgrun (Ochatt, 1991b), *Rosa persica* × *xanthina* (Matthews et al., 1991), *Weigela* × *florida* cv Bristol Ruby (Ochatt, 1993c), and *Forsythia* × *intermedia* cv Spring Glory (Ochatt, 1994).

Successful plant regeneration results were obtained either from embryogenic cell suspension protoplasts and through somatic embryogenesis (Merkle & Sommers, 1987; Matthews et al., 1991), or from mesophyll protoplasts and via organogenesis (for most remaining reports). Experiments with *Weigela* (Ochatt, 1993c), include the first published example of the isolation and culture of root protoplasts for a woody Angiosperm, and were recently extended to *Forsythia*, where plants were recovered from leaf, stem and root protoplasts (Ochatt, 1994).

In line with reports for most other woody plant genotypes (Ochatt & Power, 1992), protoplasts underwent a relatively long lag phase (7–10 days) prior to cell wall regeneration and the onset of mitotic divisions. Likewise, a rather high initial plating density (generally larger than 5×10^4 protoplasts/ml) was required for protoplast division. BAP was the preferred cytokinin during initial protoplast culture and, with the exception of rose (Matthews et al., 1991), also for the induction of regeneration from protoplast-derived callus. The use of organic-rich media for initial protoplast culture stages was another requirement shared by most systems, and a particularly important one for genotypes of *Lonicera* (Ochatt, 1991b; Georges et al., 1993) and *Weigela* (Ochatt, 1993c).

Recent research work with woody ornamentals in our laboratory has concentrated on three main programmes with different objectives.

One project is aimed at comparing the frequencies of spontaneous somaclonal variation among regenerants from explants and protoplasts of *Weigela*. Previous experiments on the regeneration of plants from *Weigela* tissues had provided us with a range of somaclonal variants, including novel ornamentally valuable characters that have warranted their subsequent commercialization (Duron & Decourtye, 1990). A new series of experiments was aimed at exploiting the eventual spontaneous variation among plants regenerated from protoplasts of *Weigela* × *florida* cv Bristol Ruby. Out of a population of 229 protoclones, several variants were identified, including the occurrence of leaf variegation (a novel trait for the genus) and various modifications of the growth habit including the recovery of one dwarf protoclone (Ochatt, unpublished). Assessment of flowering in such protoclones and their genomic characterization is currently underway.

A second series of experiments concerns the development of protoplast technology for *Cotoneaster* species, a genus that has scarcely been studied *in vitro* (Monier et al., 1994; Monier & Ochatt, 1994), and is aimed at improving the tolerance of this genus to fire blight, caused by the bacterium *Erwinia amylovora*.

Finally, research was recently undertaken to try and produce novel *Forsythia* genotypes with a flower colour other than yellow, by means of somatic hybridization. The flower colours that would be desirable in *Forsythia* can be found in various genotypes of *Syringa* (the lilacs, sexually-incompatible members of the same family *Oleaceae*). Our former research work had already proved that, within the same family, intergeneric somatic hybridization is feasible for woody plant genotypes (Ochatt et al., 1989). Thus, having determined the requirements for plant regeneration from protoplast of one of the prospective fusion partners (i.e. *Forsythia* × *intermedia*, see Table 1 and above), experiments were performed with different source tissues from various genotypes of the previously unstudied genus *Syringa*. So far, the conditions for the successful isolation and initial culture of leaf and petal protoplasts of *Syringa vulgaris* cv Charles Joly were studied and fusion experiments with *Forsythia* protoplasts will be forthcoming.

Other studies using protoplasts of temperate fruit trees and woody ornamentals

The availability of protoplast-to-plant systems for top-fruit tree and woody ornamental species has encouraged the exploitation of such technology for various applications. Protoclonal variation was induced after applying an *in vitro* recurrent selection strategy on 'Colt' cherry (*Prunus avium* × *pseudocerasus*) protoplast cultures, which led to the recovery, for the first time, of salt/drought tolerant protoplast-derived plants (Ochatt & Power, 1989a). The stability of the stress tolerance thus acquired was confirmed, and the existence of an inverse relationship between cell wall regeneration and tolerance to salinity was demonstrated (using protoplasts isolated from the tolerant regenerated 'Colt' cherry plants), with protoplasts being more tolerant than their respective cellular counterpart (Ochatt & Power, 1989b).

The effects of electroporation of isolated protoplasts on their viability and subsequent competence for proliferation and complete plant regeneration was studied in 'Colt' cherry and 'Conference' pear (*Pyrus communis*) among other genotypes. Electroporated protoplasts of both genotypes exhibited significantly better division and callus formation compared with control non-electromanipulated protoplasts of the same genotypes (Rech et al., 1987). The same held true later, with a much higher growth rate and percentage of organogenesis for calluses derived from electroporated protoplasts. Rooting was also improved for shoots derived from electro-treated protoplasts (Ochatt et al., 1988a). Most interestingly, such improved responses in proliferation, regeneration and rooting of shoots were also maintained in the long-term (Ochatt et al., 1988b). The cellular and/or molecular basis of such electro-enhancement of responses was investigated. It was found that, for electroporated protoplasts, DNA synthesis is 8-fold higher than for control protoplasts (Rech et al., 1988). Thus, the responses obtained might be the result of such higher DNA synthesis, resulting in a higher protein synthesis. Permanent modifications at the membrane level, that would account for a larger/more efficient uptake of key medium components by the manipulated cultured protoplasts, might also be involved.

Pathogen resistance studies involving pear were concerned with the co-culture of freshly-isolated leaf protoplasts of various genotypes (differing in susceptibility to fire blight) with agressive and avirulent strains of *Erwinia amylovora*. These experiments permitted the development of an early screening method for fire blight susceptibility, since the different pear genotypes reacted, at the protoplast level, in the same way as they do in the field, at the whole plant level, with protoplasts from known resistant genotypes withstanding the presence of bacteria while those from known susceptible genotypes died in presence of *E. amylovora* agressive strains (Brisset et al., 1990). An exciting spin-off from these studies was the regeneration of complete plants of the very fire blight-susceptible scion variety Passe Crassane, from protoplasts that had been co-cultured with the agressive strain CFBP 1430 of *E. amylovora*. Analysis of such plants with respect to their reaction to the disease is at present underway. The range of plant genotypes and bacterial strains assessed has recently been significantly enlarged, both with different pear genotypes and with tobacco (used as a control) in co-culture with several *Pseudomonas tabaci* and *E. amylovora* strains. The summation of the results obtained (Brisset & Paulin, 1992; Brisset et al., 1990; 1994) suggests that the plasmalemma is capable of distinguishing pathogenic from non-pathogenic bacteria, and that the presence of a cell wall plays a role both for cell death and for the specificity of the disease reaction.

A separate series of experiments to assess the effects of cold-storage (in the dark, at 4° C) on the viability of protoplasts and on their competence for subsequent proliferation and plant regeneration, included over 20 genotypes in the genera *Malus, Pyrus, Prunus, Lonicera* and *Oxalis*. Cold storage of protoplasts for up to 50 days improved the proliferation responses, particularly for those systems where protoplasts are prone to phenolic oxidation during the initial culture stages, as is the case in most woody species, especially for top-fruit tree genotypes (Ochatt & Patat-Ochatt, 1991).

Conclusion

The conventional breeding of fruit trees is routinely limited by problems such as sterility, apomixis, long life cycles with a concomitantly extended juvenile period, and the long-term risk of inbreeding depression. These factors make standard breeding methods long, costly and sometimes inappropriate for many breeding goals. The exploitation of protoplast technology opens up the possibility of creating agronomically-useful genetic novelties.

Plant regeneration from protoplasts can be a source of spontaneous somaclonal variation for certain genotypes, and the same is true for *in vitro* selection, based on protoplasts, as described above.

Somatic hybridization by homofusion of protoplasts of haploid genotypes might serve as an alternative to colchicine doubling, after which the homozygous plants obtained would be particularly helpful in the study of inheritance of horticultural traits. By producing the F1 and the F2 from different homozygous genotypes, more information on the genetics of desirable characters might be obtained. Heterofusions between different haploid clones might serve to create genetic novelties of great value for rootstock breeding, and might be of interest as novel scion cultivars.

Isolated cells and protoplasts are an ideal material for genetic transformation, whereby microinjection of DNA, PEG treatment or electroporation with plasmids, and co-culture with *Agrobacterium* spp. for such sources, followed by plant regeneration, can produce novel transgenic plants.

Finally, protoplasts are a novel and powerful tool for the study of various key developmental mechanisms at the single cell level, and also permit a more precise study (e.g. of the role played by the cell wall) of physiological and phytopathological stresses to which temperate fruit trees and ornamental shrubs are daily exposed in the field.

References

Bouvier, L , Y X Zhang & Y Lespinasse, 1993 Two methods of haploidization in pear, *Pyrus communis* L greenhouse seedling selection and *in situ* parthenogenesis induced by irradiated pollen Theor Appl Genet 87 229–232

Brisset, M N & J P Paulin, 1992 A reliable strategy for the study of disease and hypersensitive reactions induded by *Erwinia amylovora* Plant Sci 85 171–177

Brisset, M N , S J Ochatt & J P Paulin, 1990 Evidence for quantitative responses during co culture of pear (*Pyrus communis* L) protoplasts with *Erwinia amylovora* (Burrill) Winslow et al Plant Cell Rep 9 272–275

Brisset, M N , S J Ochatt & J P Paulin, 1994 Interactions entre bacteries phytopathogenes et protoplastes vegetaux Abstracts 10eme Collogue sur les Recherches Fruiteres, INRA-CTIFL, Angers, France P14

Duron, M & L Decoutrye, 1990 *In vitro* variation in *Weigela* p 606–623 In Y P S Bajaj (Ed) Biotechnology and Agriculture and Forestry 11 Somaclonal variation in crop improvement I Springer Verlag, Berlin

Georges, D , L Decourtye & S J Ochatt, 1993 Towards the somatic hybridization of shrubby and climbing honeysuckles Acta Hort 326 337–332

Haberlandt, G , 1902 Kulturversuche mit isollierten Pflanzenzellen Sitzungberichte Akademie der Wissenschaften Wien 111 69–92

Huancaruna-Perales, E & O Schieder, 1993 Plant regeneration from leaf protoplasts of apple Plant Cell Tissue Organ Culture 34 71–76

Kao, K N & M R Michayluk, 1975 Nutritional requirements for growth of *Vicia hajastana* cells and protoplasts at a very low population density in liquid medium Planta 126 105–110

Krul, W R , 1988 Recent advances in protoplast culture of horticultural crops small fruits Sci Hort 37 267–276

Matthews, D , J Mottley, I Horan & A V Roberts, 1991 A protoplast to plant system in roses Plant Cell Tissue Organ Culture 24 173–180

Merkle, S A & H E Sommers, 1987 Regeneration of *Liriodendron tulipifera* (family *Magnoliaceae*) from protoplast culture Amer J Bot 74 1317–1321

Monier, C & S J Ochatt, 1994 Shoot regeneration from stem-derived callus of *Cotoneaster* spp (*Rosaceae*) Abstracts EUCARPIA Conference The methodology of plant genetic manipulation Criteria for decision making Cork (Ireland), in press

Monier, C , E Bossis, R Samson & S J Ochatt, 1994 Micropropagation of *Cotoneaster* spp is enhanced by the presence of endogenous bacteria Abstracts IAPTC VIII Int Congr Plant Tissue and Cell Culture Florence, Italy 66

Murashige, T & F Skoog, 1962 A revised medium for rapid growth and bioassays with tobacco tissue cultures Physiol Plant 15 473–497

Ochatt, S J , 1987 Coltura di protoplasti come metodo per il miglioramento genetico nelle piante da frutto Frutticoltura 49 58–60

Ochatt, S J , 1990a Protoplast technology for top-fruit tree breeding Acta Hort 280 215–226

Ochatt, S J , 1990b Plant regeneration from root callus protoplasts of sour cherry (*Prunus cerasus* L) Plant Cell Rep 9 268–271

Ochatt, S J , 1991a Strategies for plant regeneration from mesophyll protoplasts of the recalcitrant fruit and farmwoodland species *Prunus avium* L (sweet/wild cherry), *Rosaceae* J Plant Physiol 139 155–160

Ochatt, S J , 1991b Requirements for plant regeneration from protoplasts of the shrubby ornamental honeysuckle, *Lonicera nitida* cv Maigrun Plant Cell Tissue Organ Culture 25 161–167

Ochatt, S J , 1992 The development of protoplast-to-tree systems for *Prunus cerasifera* and *P spinosa*, *Rosaceae* Plant Sci 81 253–259

Ochatt, S J , 1993a Regeneration of plants from protoplasts of *Pyrus* spp (pear) p 105–122 In Y P S Bajaj (Ed) Biotechnology in Agriculture and Forestry 22 Plant protoplasts and genetic engineering III Springer-Verlag, Berlin

Ochatt, S J , 1993b Regeneration of plants from protoplasts of some stone fruits (*Prunus* spp) p 78–96 In Y P S Bajaj (Ed) Biotechnology in Agriculture and Forestry 23 Plant protoplasts and genetic engineering IV Springer-Verlag, Berlin

Ochatt, S J , 1993c An efficient protoplast-to-plant system for the hybrid ornamental shrub, *Weigela* × *florida* cv Bristol Ruby Plant Cell Tissue Organ Culture 33 315–320

Ochatt, S J , 1994 Plant regeneration from protoplasts of *Forsythia* × *intermedia* cv Spring Glory Abstracts IAPTC VIII Int Congr Plant Tissue and Cell Culture Florence, Italy 27

Ochatt, S J & O H Caso, 1986 Shoot regeneration from leaf mesophyll protoplasts of wild pear (*Pyrus communis* var *pyraster* L) J Plant Physiol 122 243–246

Ochatt, S J & E M Patat Ochatt, 1991 The time course evolution of viability ad competence for proliferation of woody plant protoplasts following cold-storage Physiol Plant 82 A-16

Ochatt, S J & E M Patat-Ochatt, 1994 Somatic hybridization between *Prunus* × *Pyrus* species p 455–464 In Y P S Bajaj (Ed) Biotechnology in Agriculture and Forestry 27 Somatic hybridization in crop improvement III Springer-Verlag, Berlin

Ochatt, S J & J B Power, 1988a Rhizogenesis in callus from Conference pear (*Pyrus communis* L) protoplasts Plant Cell Tissue Organ Culture 13 159–164

Ochatt, S J & J B Power, 1988b Plant regeneration from leaf mesophyll protoplasts of Williams' Bon Chretien (Syn Bartlett) pear (*Pyrus communis* L) Plant Cell Rep 7 587–589

Ochatt, S J & J B Power, 1988c An alternative approach to plant regeneration from protoplasts of sour cherry (*Prunus cerasus* L) Plant Sci 56 75–79

Ochatt, S J & J B Power, 1989a Selection for salt- drought tolerance using protoplast- and explant-derived tissue cultures of Colt cherry (*Prunus avium* × *pseudocerasus*) Tree Physiol 5 259–266

Ochatt, S J & J B Power, 1989b Cell wall synthesis and salt (saline) sensitivity of Colt cherry (*Prunus avium* × *pseudocerasus*) protoplasts Plant Cell Rep 8 365–367

Ochatt, S J & J B Power, 1992 Plant regeneration from protoplasts of higher plants p 99–127 In M W Fowler, G S Warren & M Moo-Young (Eds) Comprehensive Biotechnology Second Supplement Pergamon Press, New York

Ochatt, S J , E Chevreau & M Gallet, 1992b Organogenesis from Passe Crassane and Old Home pear (*Pyrus communis* L) protoplasts and isoenzymatic trueness-to-type of the regenerated plants Theor Appl Genet 83 1013–1018

Ochatt, S J , E C Cocking & J B Power, 1987 Isolation, culture and plant regeneration of Colt cherry (*Prunus avium* × *pseudocerasus*) protoplasts Plant Sci 50 139–143

Ochatt, S J , E M Patat-Ochatt & J B Power, 1992a Protoplasts p 77–103 In F A Hammerschlag & R E Litz (Eds) Biotechnology of Perennial Fruit Crops CAB International, Oxford

Ochatt, S J , E L Rech, M R Davey & J B Power, 1988b Long-term effects of electroporation on enhancement of growth and plant regeneration from Colt cherry (*Prunus avium* × *pseudocerasus*) protoplasts Plant Cell Rep 7 393–395

Ochatt, S J , P K Chand, E L Rech, M R Davey & J B Power, 1988a Electroporation mediated improvement of plant regeneration of Colt cherry (*Prunus avium* × *pseudocerasus*) protoplasts Plant Sci 54 165–169

Ochatt, S J , E M Patat Ochatt, E L Rech, M R Davey & J B Power, 1989 Somatic hybridization of sexually incompatible top fruit tree rootstocks, wild pear (*Pyrus communis* var *pyraster* L) and Colt cherry (*Prunus avium* × *pseudocerasus*) Theor Appl Genet 78 35–41

Oka, S & K Ohyama, 1985 Plant regeneration from mesophyll protoplasts of paper mulberry (*Broussonetia kazinoki* Sieb) J Plant Physiol 119 455–460

Oliveira, M M & M S S Pais, 1991 Plant regeneration from protoplasts of long-term callus cultures of *Actinidia deliciosa* var *deliciosa* cv Hayward (kiwifruit) Plant Cell Rep 9 643–646

Patat-Ochatt, E M , 1994 Regeneration of plants from protoplasts of *Malus Xdomestica* Borkh (apple) p 81–99 In Y P S Bajaj (Ed) Biotechnology in Agriculture and Forestry 29 Plant protoplasts and genetic engineering V Springer-Verlag, Berlin

Patat-Ochatt, E M & J B Power, 1990 Advances in plant regeneration from apple protoplasts Acta Hort 280 285–288

Patat-Ochatt, E M , S J Ochatt & J B Power, 1988 Plant regeneration from protoplasts of apple rootstocks and scion varieties (*Malus* × *domestica* Borkh) J Plant Physiol 133 460–465

Patat-Ochatt, E M , J Boccon-Gibod, M Duron & S J Ochatt, 1993 Organogenesis of stem and leaf protoplasts of a haploid Golden Delicious apple clone (*Malus* × *domestica* Borkh) Plant Cell Rep 12 118–120

Rech, E L , S J Ochatt, P K Chand, J B Power & M R Davey, 1987 Electro-enhancement of division of plant protoplast-derived cells Protoplasma 141 169–176

Rech, E L , S J Ochatt, P K Chand, B J Mulligan, M R Davey & J B Power, 1988 Electroporation increases DNA synthesis in cultured plant protoplasts Biotechnology 6 1091–1093

Rosati, C & S J Ochatt, 1994 Establishing strategies for stem-derived protoplast culture in different *Pyrus communis* L geno-

types. Abstracts IAPTC VIII Int. Congr. Plant Tissue and Cell Culture, Florence, Italy: 27.
Steward, F.C., M.O. Mapes & K. Mears, 1958. Growth and organized development of cultured cells. II. Organization in cultures grown from freely suspended cells. Amer. J. Bot. 45: 653–704.
Tsai, C.K., 1988. Plant regeneration from leaf callus protoplasts of *Actinidia chinensis* var *chinensis*. Plant Sci. 54: 231–235.
Vardi, A., P. Spiegel-Roy & E. Galun, 1975. Citrus cell culture: isolation of protoplasts, plating densities, effect of mutagens and regeneration of embryos. Plant Sci. Lett. 4: 231–236.
Wallin, A. & L. Johansson, 1989. Plant regeneration from leaf mesophyll protoplasts of *in vitro* cultured shoots of a columnar apple. J. Plant Physiol. 135: 565–570.

Euphytica **85:** 295–302, 1995.

Somaclonal variation as a tool for crop improvement

Angela Karp
Department of Agricultural Sciences, University of Bristol, Institute of Arable Crops Research, Long Ashton Research Station, Bristol, BS18 9AF, U.K.

Key words: tissue culture, somaclonal variation, plant breeding

Summary

Somaclonal variation is a tool that can be used by plant breeders. The review examines where this tool can be applied most effectively and the factors that limit or improve its chances of success. The main factors that influence the variation generated from tissue culture are (1) the degree of departure from organised growth, (2) the genotype, (3) growth regulators and (4) tissue source. Despite an increasing understanding of how these factors work it is still not possible to predict the outcome of a somaclonal breeding programme. New varieties have been produced by somaclonal variation, but in a large number of cases improved variants have not been selected because (1) the variation was all negative, (2) positive changes were also altered in negative ways, (3) the changes were not novel, or (4) the changes were not stable after selfing or crossing. Somaclonal variation is cheaper than other methods of genetic manipulation. At the present time, it is also more universally applicable and does not require 'containment' procedures. It has been most successful in crops with limited genetic systems and/or narrow genetic bases, where it can provide a rapid source of variability for crop improvement.

Introduction

Plant tissue culture is an enabling technology from which many novel tools have been developed to assist plant breeders. These tools can be used to increase the speed or efficiency of the breeding process, to improve the accessibility of existing germplasm and to create new variation for crop improvement. They include micropropagation, anther culture, *in vitro* selection, embryo rescue, somaclonal variation, somatic hybridisation and transformation. Of these, somaclonal variation occupies a somewhat unique position, because it is both an advantage and a disadvantage of tissue culture systems. It was not expected that the culture process, which is asexual, would give rise to variability. The uncontrolled production of plants which are not 'true-to-type' continues to pose problems in virtually all other uses of plant tissue culture.

Somaclonal variation was first defined as such by Larkin & Scowcroft who reviewed the subject in 1981 and were among several authors at that time to draw attention to its potential uses for crop improvement (Larkin & Scowcroft, 1981). Their expectations arose largely from the reported observations of extensive variation in plants derived from protoplast and explant cultures of potato (Shepard et al., 1980), sugar cane (Heinz, 1973; Krishnamurthi & Tlaskal, 1974), rice (Oono, 1978) and maize (Green, 1977). An extensive number of reports soon followed in a whole range of species, indicating that somaclonal variation was widespread, and, therefore, ostensibly accessible to all plant breeders (Karp, 1991).

Since that time, a number of serious attempts have been made to improve crops using somaclonal variation, and whilst a large number have failed, there have been notable successes (see later). Our starting premise, therefore, is that somaclonal variation is a tool that can be used by plant breeders. The questions that will be addressed here concern where this tool can be most effectively applied and what factors limit or improve the chances of using it successfully. Due to shortage of space, this review will be confined to the direct use of somaclonal variation for plant breeding. (The use of somaclonal variation for introgression in hybrids and for *in vitro* selection of plants expressing improved tolerances to stresses, such as temperature and salinity, are covered elsewhere in the proceedings.)

How reliable is somaclonal variation as a tool?

The question of the reliability of somaclonal variation as a tool for breeding can be broken down into three parts:
(a) does *in vitro* culture always give rise to variation?
(b) is useful variation always recovered?
(c) is somaclonal variation useful in all crop species?

Does in vitro *culture always give rise to variation?*

It cannot be said that *in vitro* culture will always give rise to variation. In fact, a number of factors can be identified that influence whether or not variation is produced and how much variation is generated. These factors are: (1) the degree of departure from organised meristematic growth, (2) the genetic constitution of the starting material, (3) the growth regulators in the medium and (4) the tissue source.

(1) The degree of departure from meristematic organised growth
Growth in culture may occur from already established meristems or it may take a disorganised form as callus from which organised structures then arise by somatic embryogenesis or organogenesis. Departure from organised growth is a key element in somaclonal variation, suggesting that in disorganised growth the constraints which act to eliminate genetic variations in normal meristems are suppressed, or that mechanisms of genetic instability are induced. Either way, in general terms, the greater the departure from organised growth and the longer the time spent in this state, the greater the chances of generating somaclonal variation. It is more appropriate to use this rule than to generalise as to what extent different tissue culture systems are associated with somaclonal variation. This is because the length of time spent in a disorganised form may vary significantly from system to system and even from species to species. For example, cell suspension cultures are generally viewed as being genetically unstable and protoplast culture is generally associated with high levels of somaclonal variation. Nevertheless, spruce (*Picea mariana*) plants regenerated from protoplasts of embryogenic cell lines are likely to be free of somaclonal variation because the cultures are very stable (Eastman et al., 1991; Isabel et al., 1993) and protoplasts derived from these embryogenic cultures form embryos directly (Fowke et al., 1990).

(2) The genetic constitution
There is an increasing amount of evidence to indicate that somaclonal variation is genotype-dependent. In practice, genotype effects can be difficult to separate out from differences in tissue culture response, since the latter is also under genetic control, but a number of studies have demonstrated that genotype can influence somaclonal variation irrespective of regeneration mode (eg, Bebeli et al., 1988). The exact nature of the genotypic element is a crucial aspect to identify because plant breeders will wish to employ somaclonal variation as a tool in specific lines or cultivars and to know whether their genotypes will be responsive in terms of variability. Unfortunately, this is not an easy aspect to resolve, as several different factors result in genotypic differences and these interact in complex ways.

The ploidy of the starting material is one factor. In general, more somaclonal variation, at least in terms of chromosome instability, is recovered in regenerants of polyploids compared with that in diploids and haploids. This observation relates to the reduced selective pressures acting on polyploids, whose genomes are buffered against any inbalances caused by gross genetic alterations (such as aneuploidy) by the multiplicity of chromosome sets in their constitution. For genetic changes other than chromosomal variation, the situation is more complicated. Gene mutations have better expression in haploids and diploids but may show better survival in polyploids if their effects are deleterious.

Even within ploidy levels, some genomes may be more unstable than others. In a comparison of cell suspensions of diploid, tetraploid and hexaploid wheats, the diploids were found to be the most stable and the hexaploids the least stable, but genotype differences were evident within the ploidy levels (Winfield et al., 1993). Similarly, several reports indicate that somaclonal variation is minimal in regenerants of barley (Breiman et al., 1987; Karp et al., 1987; Rúiz et al., 1992), but quite extensive in another diploid cereal, rye (Karp et al., 1992; Linacero & Vazquez, 1992). These differences may relate to breeding behaviour (barley being inbred and rye outbred) or to differences in DNA sequences. In fact, there is evidence for both. In an attempt to examine genotypic influences on somaclonal variation, we regenerated plants from six different genotypes of rye. Four of these had been inbred since 1926 and the other two were both outbred. One of the latter was a cultivar, 'Ailés' in which spontaneous translocations occurred at high frequencies while the other contained variable numbers of B chromosomes. Somaclonal variation occurred in all the lines but was most prevalent in the outbred lines and in all cases the nature of the variation corresponded with genetic characteristics of the parental line. For example, translocations were most frequent in regenerants of cv 'Ailés', and the inbred line which appeared most variable had a previous history of some cytological instability (Puolimatka & Karp, 1993).

It has been postulated that the late replicating nature of heterochromatin can perturb the cell cycle and result in enhanced chromosome breakage when cells are induced to divide under the conditions of *in vitro* culture (Lee & Phillips, 1988). Correlations between the frequency and location of tissue culture-induced breaks and the amount and location of heterochromatic blocks have been demonstrated in several species (eg, Johnson et al., 1987; Benzion & Phillips, 1988; Eizenga, 1989) and it might be expected from this that genomes with large proportions of heterochromatin would give more somaclonal variation than genomes which are largely euchromatic. However, where field evaluations have been carried out, the relationship between heterochromatin content and somaclonal variation may be the inverse of that expected based on the cytological studies. Bebeli et al. (1993a) compared somaclonal variation in seed progenies from immature embryo-derived regenerants (R_2) of two pairs of *Triticale* lines (cultivars Drira and Rosner) with and whithout heterochromatin on chromosomes 7RL and 6RS. Significant somaclonal variation was found in the R_2 for many of the traits studied, but it was most frequent in the lines lacking heterochromatin. Similar results were obtained in rye (Bebeli et al., 1993b).

In anther culture-derived tobacco plants, amplification of heterochromatic sequences was initially thought to be correlated with yield loss and reduced agronomic performance (Dhillon et al., 1983) but later studies did not strongly support this correlation. It was concluded that different amplification sites may be involved in different lines (Reed & Wernsman, 1989). Heterochromatin represents a specific class of repeated sequence which can be detected using differential staining techniques. A large body of evidence suggests that other repeated sequences are also involved in generating somaclonal variation but only in some cases have they been characterised at the molecular level (eg, Cullis & Cleary, 1986).

Genomes carrying transposable elements might be expected to be more unstable in culture than those without. Evidence of altered transposon activities as a result of tissue culture have been reported (James & Stadler, 1989; Planckaert & Walbot, 1989; Peschke & Phillips, 1991; Peschke et al., 1991) but not all changes induced by culture in lines with transposons have been attributable to the movement of the transposon (Williams et al., 1991). Furthermore, activation of transposons can occur as a result of tissue culture in genomes without any prior history of transposon activity (Ray & Bingham, 1991). It is, therefore, not that easy to predict how genomes will respond in culture, even in those species for which considerable molecular genetic data are available.

(3) The culture environment

There is considerable evidence to indicate that somaclonal variation is influenced by the choice and particularly by the concentration of growth regulators in the medium (Karp, 1992). It is possible that growth regulators act as mutagens. The synthetic auxin, 2,4-dichlorophenoxyacetic acid (2,4-D), has been shown to increase the frequency of blue to pink mutations in the *Tradescantia* stamen hair system (Dolezel & Novak, 1984) and to induce significant increases in the frequency of sister chromatid exchanges in root-tip cells of *Allium sativum* (Dolezel et al., 1987). However, there is a paucity of examples of this kind and most evidence points to growth regulators influencing somaclonal variation during the culture phase through their effects on (1) cell division (Gould, 1984), (2) the degree of disorganised growth (Karp, 1992) and (3) selective proliferation of specific cell types (Ghosh & Gadgil, 1979).

There are a number of cases where somaclonal variation has been attributed to abnormal concentrations of growth regulators even though passage though callus was avoided during culture. The appearance of deformed shoot primordia in vegetatively propagated *Kalanchoe blossfeldiana* has been attributed to 2,4-D (Varga et al., 1988) and inflorescence variations in micropropagated oil palm (Corley et al., 1986) and African plantains (Vuylsteke & Swennen, 1990) were thought to have resulted from excessive use of cytokinins. In these tissue culture systems the phenotype of the regenerants (as opposed to their progeny) is important. This makes them particularly subject to phenotypic variation resulting from an unbalanced use of growth regulators, including the gaseous hormone, ethylene, which is active in extremely small concentrations (< 0.01 ppm/v) and can cause severe growth abnormalities if allowed to accumulate in the culture vessels (Jackson et al., 1987). Micropropagation is not the only tissue culture system subject to this effect of growth regulators. Also included would be all tissue culture of vegetatively propagated, apomictic and long-lived species (eg, palms and trees), which would not normally go through a sexual cycle where such 'epigenetic modifications' would be eliminated.

Although the culture environment should be one of the variables over which some control can be exercised, it has proved difficult to identify general rules with respect to 'good' and 'bad' growth regulator 'conditions' which might enhance or reduce variation, respectively. This difficulty arises because plants can vary in their sensitivity to the growth regulators used in the medium, and/or because differences may be present in the levels of endogenous growth regulators in the cultured cells.

(4) Tissue source

Differences in both the frequency and nature of somaclonal variation may occur when regeneration is achieved from different tissue sources. Generally speaking, the older and/or the more specialised the tissue, the greater the chances that variation will be recovered in the regenerated plants. These effects arise because gross changes in the genome, including endopolyploidy, polyteny and amplification or diminuition of DNA sequences, often accompany somatic differentiation in normal plant growth and development (D'Amato, 1989). This factor is further complicated by genotype, as there appear to be two classes of plants based on the degree of genomic variation present in the soma. D'Amato (1985) refers to these as polysomatic and non-polysomatic plants. In non-polysomatic plants, differentiated cells are maintained in the same ploidy status as the zygote, whilst in polysomatic plants the differentiated cells may contain polyploid, polytene and even aneuploid constitutions. The influence of tissue source would be most pronounced in polysomatic species but, although D'Amato (1989) has identified some plants which fall into this catagory, in many cases it may not be clear which type of plant is being used. In *Solanum brevidens*, 70% of plants regenerated from cotyledons were tetraploid, whilst only 20% were tetraploid in regeneration from leaf pieces (Osifo et al., 1989). In *Chrysanthemum*, plants regenerated from cultured petals were more floriferous and had higher frequencies of abnormalities than plants derived from pedicels (Bush et al., 1976; De Jong & Custers, 1986), whilst in scented *Pelargonium*, plants regenerated from stems did not differ from the controls but regenerants from root and petiole pieces were variable in morphology (Skirvin & Janick, 1976).

Is useful variation always recovered?

In nearly all cases where extensive field trials on somaclones have been carried out, there has been clear evidence that changes in agronomic traits have occurred as a result of *in vitro* culture. However, in the majority of cases, improved variants have not been selected for breeding purposes essentially for one or a combination of the following reasons:

(1) The variation was in a negative direction

In a three-year field evaluation of seed progenies of tissue culture-derived plants of spring wheat (*Triticum aestivum* 'HY320'), for example, Qureshi et al. (1992) reported that virtually all the regenerated lines yielded less that the controls and concluded that the tissue culture process had produced 'an array of agronomically inferior genotypes'. Similarly, a field evaluation of somaclones of barley showed that the little variation that was observed was all of negative value (Luckett et al., 1989; Baillie et al., 1992).

(2) In positive changes, other aspects of the plants were altered in a negative way

In the spring wheat field evaluation cited above, higher grain protein levels in the seed progenies of regenerants were associated with lowered yields (Qureshi et al., 1992). Similarly, many of the potato plants identified by Shepard & co-workers (1980) as having improved disease resistance were later shown to be aneuploid (Gill et al., 1986).

(3) Not all the changes obtained were novel

In a study of somaclonal variation induced in seven cultivars of plantains (*Musa* spp.) variations in the inflorescence were among the most common type of change. In the False Horn plantain 'Agbagba', reversion to a typical French plantain bunch type of inflorescence occurred at a frequency of 2.7%. However, the occurrence of such 'French reversion' has also been reported among conventional propagules, albeit at a much lower frequency (0.7%) (Vuylsteke & Swennen, 1990).

(4) Not all the changes were stable

Heritable changes, resulting from gene or chromosome mutations are only one of three classes of change in somaclonal variation. The others are (1) non-heritable or 'epigenetic' changes that result from influences of the culture environment and (2) heritable but reversible changes that result from altered gene expression (Karp, 1991). Only alterations in the genetic information would give rise to truly stable changes and, even in that class, some chromosome and gene mutations, eg. amplification and transposition, may themselves be unstable. Epigenetic changes have been discussed earlier with respect to growth regulators. Although they can be persistent within the lifetime of a plant (Corley et al., 1986), they are not transmissable through a sexual cycle. In contrast, methylation can cause changes in gene activity which can be transmitted to the sexual progeny, but which may revert under certain conditions (Brown, 1989; Kaeppler & Phillips, 1993). This raises serious doubts concerning the stability of somaclonal changes after self and cross-pollination in breeding programmes.

Is somaclonal variation equally successful in all crops?

It has to be said that somaclonal variation has been successful in yielding some new varieties in, for example, *Paulownia tomentosa* (Marcotrigiano & Jagannathan, 1988), tomato, sugar cane and celery (Springen, 1987) and *Sorghum* (Duncan, this volume). It also has to be said that the approach has been totally unsuccessful in many other attempts in other crops, such as wheat, maize and barley, despite numerous and extensive efforts. From a brief examination of some of the successful examples, it may be possible to determine in which crops somaclonal variation is more likely to succeed as a tool for crop improvement.

In an evaluation of the use of somaclonal variation as a breeding tool for coffee improvement, Söndahl & Bragin (1991) concluded that 'Somaclonal variation is an excellent method to shorten coffee breeding programmes since it provides access to new mutant forms in high-yielding genotypes within a short period of time (4–8 years)'. Coffee is a perennial tree that requires a minimum of 20–25 years to release new varieties. An overall variability of 10% was observed, and among the agronomically desired changes were short stature variants from high yielding varieties (reduces harvesting costs) and larger bean size (attracts higher market prices). Somaclonal variation was also deemed to be a promising approach for improving the aromatic grass, *Cymbopogon winterianus* Jowitt, which has a narrow genetic base and limited scope for hybridisation. Out of 19 selected somaclones that were evaluated in a replicated trial with the donor parent as the control, 5 lines were further placed under multilocational trials for stability assessment (Mathur et al., 1988). Useful variation, including improvements in disease resistance, herbage yield and quality, have also been identified in somaclones of the grasses *Cynodon dactylon* (Croughan, 1989) and *Pennisetum purpureum* (Martinez et al., 1989) and in *Zinnia marylandica*, which is a segmental amphidiploid with limited genetic recombination, and in which, therefore, traditional methods of breeding are limited (Stieve et al., 1992).

There are many other examples in the literature that could be cited, all of which lead to the general conclusion that somaclonal variation has been most successful in crops with limited genetic systems (eg., apomicts, vegetative reproducers) and/or narrow genetic bases. In ornamentals, for example, the exploitation of *in vitro*-generated variability has become part of the routine breeding practice of many commercial enterprises (Buiatti & Gimelli, 1993). This contrasts with the many examples in cereals, such as barley (Luckett et al., 1989; Baillie et al., 1992; Pickering, 1989) and maize (Earle et al., 1988), where somaclonal variation has not been hailed a success, suggesting that it is least successful in crops which have been subjected to intensive breeding. However, it is very difficult to make anything other than gross generalisations, because, as discussed earlier, the amount of somaclonal variation generated and the spectrum of changes produced are influenced by external and internal factors that interact in complex ways. Contradictions to the general rule are consequently easy to find. In some cereals, for example, results have been very promising. In oats, Dahleen et al. (1991) found that although agronomically less desirable changes were most frequent among R_4 and R_5 somaclonal generations, lines with increases and lines with decreases were found for each trait. One cycle of bidirectional selection demonstrated that the changes were heritable and that individual lines could be selected for use in plant improvement programmes. Similar results were found in the field evaluations of the rye and *Triticale* lines differing in heterochromatin content referred to earlier (Bebeli et al., 1993a, 1993b). Conversely, in the apomictic grass, *Paspalum dilatatum*, where conventional breeding approaches are severely limited and somaclonal variation was considered to be a promising option, no agronomically superior somaclones were identified among R_2 and R_3 generations, although there was some breakdown of apomixis in some of the clones (Davies & Cohen, 1992).

How does somaclonal variation compare with other tools?

It has been suggested by many authors that an obvious strategy for the use of somaclonal variation in breeding is to select for incremental improvements of existing varieties by passing the best available lines through a tissue culture cycle (eg, Evans & Sharp, 1986). However, unless variants of a particular type can be obtained more readily from somaclonal variation compared with other methods, it will not be considered an economically viable approach.

One of the main handicaps of somaclonal variation which makes it comparatively difficult to use is that, despite the identification of factors affecting the variation response of a given plant species, it is still not possible to predict the outcome of a somaclonal programme. In fact, for the many crop species in which general molecular genetic knowledge is limited, the only way to tell whether a wide range of somaclonal variation will be generated under a known set of culture conditions, would be to carry out small scale pilot studies. Even then there is no guarantee that a specific trait of interest will be altered in a positive way. Other major problems are the large proportion of inferior lines that are generated and that some of the changes are not stable.

Against these disadvantages there are a number of positive attributes of somaclonal variation 1. It is a cheap form of biotechnology compared with somatic hybridisation and transformation and requires no 'containment' procedures. 2. Tissue culture systems are available for more plant species than can be manipulated by somatic hybridisation and transformation at the present time. 3. It is not necessary to have identified the genetic basis of the trait, or indeed, in the case of transformation, to have isolated and cloned it. 4. Novel variants have been reported among somaclones, and genetic (Sibi et al., 1984; Singsit et al., 1990, Compton & Veilleux, 1991, Kaltsikes & Bebeli, 1993) and cytogenetic evidence (Puolimatka & Karp, 1993) indicate that both the frequency and distribution of genetic recombination events can be altered by passage though tissue culture. This suggests that variation may be generated from different areas of the genome than those that are accessible to conventional and mutation breeding.

Conclusions

Based on these discussions the only conclusion that can be reached at present, is that whilst somaclonal variation is a tool that can be used by breeders, it is not a precision tool and only minimal control can be exercised over its operation. Nevertheless, it can offer a rapid and easily-accessible source of variation for use in breeding programmes. It is most likely to be considered a viable choice of tool in crops with limited genetic systems and/or narrow genetic bases.

References

Baillie, A M R , B G Rossnagel & K K Kartha, 1992 Field evaluation of barley (*Hordeum vulgare*) L genotypes derived from tissue culture Can J Plant Sci 72 725–733

Bebeli, P, A Karp & PJ Kaltsikes, 1988 Plant regeneration from cultured immature embryos of sister lines of rye and triticale differing in their content of heterochromatin 1 Morphogenetic response Theor Appl Genet 75 929–936

Bebeli, PJ , PJ Kaltsikes & A Karp, 1993a Field evaluation of somaclonal variation in triticale lines differing in telomeric hete rochromatin J Genet Breed 47 248–249

Bebeli, PJ , PJ Kaltsikes & A Karp, 1993b Field evaluation of somaclonal variation in rye lines differing in telomeric hete rochromatin J Genetics and Breed 47 15–22

Benzion, G & R L Phillips, 1988 Cytogenetic stability of maize tissue cultures a cell line pedigree analysis Genome 30 318–325

Breiman, A , D Rotem, A Karp & H Shaskin, 1987 Heritable somaclonal variation in wild barley (*Hordeum spontaneum*) Theor Appl Genet 74 104–112

Brown, PTH , 1989 DNA methylation in plants and its role in tissue culture Genome 31 717–729

Buiatti, M & F Gimelli, 1993 Somaclonal variation in ornamentals Proc XVIIth Eucarpia Symposium Creating Genetic Variation in Ornamentals

Bush, S R , E D Earle & R W Langhans, 1976 Plantlets from petal epiderims and shoot tips of the periclinal chimera *Chrysanthe mum moriloium Indianapolis* Amer J Bot 63 729–737

Compton, M E & R E Veilleux, 1991 Variation for genetic recombination among tomato plants regenerated from three tissue cul ture systems Genome 34 810–817

Corley, R H V, C H Lee, I H Law & C Y Wong, 1986 Abnormal flower development in oil palm clones Planter 62 233–240

Croughan, S S , 1989 Forage crop improvement through biotechnology Proc XVI International Grassland Congress, Nice, France, p 414–441 Cullis, C A & W Cleary, 1986 DNA variation in flax tissue culture Can J Genet Cytol 28 247–252

Dahleen, L S , D D Stutham & H W Rines, 1991 Agronomic trait variation in oat lines derived from tissue culture Crop Sci 31 90–94

D'Amato, F, 1985 Cytogenetics of plant cell and tissue cultures and their regenerates CRC Critical Reviews in Plant Science 3 73–112

D'Amato, F, 1989 Polyploidy in cell differentiation Caryologia 42 183–211

Davies, L J & D Cohen, 1992 Phenotypic variation in somaclones of *Paspalum dilatatum* and their seedling offspring Can J Plant Sci 72 773–784

De Jong, J & J B M Custers, 1986 Induced changes in growth and flowering of chrysanthemums after irradiation and *in vitro* culture of pedicels and petal epidermis Euphytica 35 137–148

Dhillon, S S , E A Wernsman & J P Miksche, 1983 Evaluation of nuclear DNA content and heterochromatin changes in anther-derived dihaploids of tobacco (*Nicotiana tabacum*) cv Coker 139 Can J Genet Cytol 25 169–173

Dolezel, J & F J Novak, 1984 Effect of plant tissue culture media on the frequency of somatic mutations in *Tradescantia* stamen hairs Z Pflanzenphysiol 114 51-58

Dolezel, J , S Lucretti & F J Novak, 1987 The influence of 2,4-dichlorophenoxyacetic acid on cell cycle kinetics and sister-chromatid exchange frequency in garlic (*Allium sativum*) meristem cells Biologia Plantarum (Prague) 29 253–257

Earle, E D , V E Gracen & M E Smith, 1988 Somaclonal variation in corn p 257–269 In F Valentine (Ed) Forest and Crop Biotechnology Progress and prospects Spinger-Verlag, Heidelberg, Berlin, New York

Eastman, P A K , F B Webster, A Pitel & D R Roberts, 1991 Evaluation of somaclonal variation during somatic embryogenesis of interior spruce (*Picea gauca engelmanii* complex) using culture morphology and isozyme analysis Plant Cell Rep 10 425–430

Eizenga, G C , 1989 Meiotic analysis of tall fescue somaclones Genome 32 373–379

Evans, D A & W R Sharp, 1986 Applications of somaclonal variation Biotechnology 4 528–534

Fowke, L C , S M Attree, H Wang & D I Dunstan, 1990 Microtubule organization and cell-division in embryogenic protoplast cultures of white spruce (*Picea gauca*) Protoplasma 158 86–94

Ghosh, A & V N Gadgil, 1979 Shift in ploidy level of callus tissue A function of growth substances Indian J Exp Biol 17 562–564

Gill, B S , L N W Kam-Morgan & J F Shepard, 1986 Origin of chromosomal and phenotypic variation in potato protoclones J Hered 77 13–16

Gould, A R , 1984 Control of the cell cycle in cultured plant cells C R C Critical Rev Plant Sci 1 315–344

Green, G E , 1977 Prospects for crop improvement in the field of cell culture HortSci 12 7–10

Heinz, D J , 1973 Sugar-cane improvement through induced mutations using vegetative propagules and cell culture techniques p 53–59 In Induced Mutations in Vegetatively Propagated Plants Int Atomic Energy Agency, Vienna

Isabel, N , L Tremblay, M Michaud, F M Tremblay & J Bousquet, 1993 RAPDs as an aid to evaluate the genetic integrity of somatic embryogenesis-derived populations of *Picea mariana* (Mill) B S P Theor Appl Genet 86 81–87

Jackson, M B , A J Abbott, A R Belcher & K C Hall, 1987 Gas exchange in plant tissue cultures p 61–72 In M B Jackson, S H Mantell & J Blake (Eds) Advances in the Chemical Manipulation of Plant Tissue Cultures British Plant Growth Regulator Group Monograph 16

James, M G & J Stadler, 1989 Molecular characterization of mutator systems in maize embryogenic callus cultures indicates mu element activity *in vitro* Theor Appl Genet 77 383–394

Johnson, S S , R L Phillips & H W Rines, 1987 Meiotic behaviour in progeny of tissue culture regenerated oat plants (*Avena sativa* L) carrying near-telocentric chromosomes Genome 29 431–438

Kaeppler, S M & R L Phillips, 1993 DNA methylation and tissue culture-induced variation in plants *In Vitro* Cell Dev Biol 29 125–130

Kaltsikes, P J & P J Bebeli, 1993 Somaclonal variation causes changes in the inter-relationships between traits in hexaploid Triticale Japan J Breed 43 45–51

Karp, A , S H Steele, N A Breiman, P R S Shewry, S Parmar & M G K Jones, 1987 Minimal variation in barley plants regenerated from cultured immature embryos Genome 29 405–412

Karp, A , 1991 On the current understanding of somaclonal variation p 1–58 In B J Miflin (Ed) Oxford Surveys of Plant Molecular and Cell Biology, Vol 7 Oxford University Press

Karp, A , 1992 The role of growth regulators in somaclonal variation British Society for Plant Growth Regulation Annual Bulletin No 2 May 1992, p 1–9

Karp, A , P Owen, S H Steele, P J Bebeli & P J Kaltsikes, 1992 Variation in telomeric heterochromatin in somaclones of rye Genome 35 590–593

Krishnamurthi, M & J Tlaskal, 1974 Fiji disease resistant *Saccharum officinarum* var Pindar subclones from tissue cultures Proc Int Soc Sugar Cane Technol 15 130–137

Larkin, P J & W R Scowcroft, 1981 Somaclonal variation – a novel source of variability from cell cultures for plant improvement Theor Appl Genet 60 197–214

Lee, M & R L Phillips, 1988 The chromosomal basis of somaclonal variation Ann Rev Plant Physiol Plant Mol Biol 39 413–438

Linacero, R & A M Vazquez, 1992 Cytogenetic variation in rye regenerated plants and their progeny Genome 35 428–430

Luckett, D J , D Rose & E Knights, 1989 Paucity of somaclonal variation from immature embryo culture of barley Australian J Agric Res 40 1155–1159

Marcotrigiano, M & L Jagannathan, 1988 *Paulownia tomentosa* cultivar somaclonal Snowstorm HortSci 23 226–227

Martinez, R O , M Monzote, R S Herrera, R Cruz & V Torrez, 1989 Obtention of king grass (*Pennisetum purpureum*) clones from tissue culture selection and evaluation of mutants Proc XVI International Grassland Congress, Nice, France, 1989

Mathur, A K , P S Ahuja, B Pandey, A K Kukreja & S Mandal, 1988 Screening and evaluation of somaclonal variation for quantitative and qualitative traits in an aromatic grass, *Cymbopogon winterianus* Jowitt Plant Breed 101 321–334

Oono, K , 1978 Test tube breeding of rice by tissue culture Trop Agric Res Series 11 109–123

Osifo, E O , J K Webb & G G Henshaw, 1989 Variation amongst callus-derived plants of *Solanum brevidens* J Plant Physiol 134 1–4

Peschke, V M & R L Phillips, 1991 Activation of the maize transposable element suppressor-mutator (Spm) in tissue culture Theor Appl Genet 81 90–97

Peschke, V M , R L Phillips & B G Gengenbach, 1991 Genetic and molecular analysis of tissue culture-derived AC elements Theor Appl Genet 82 121–129

Pickering, R A , 1989 Plant regeneration and variants from calli derived from immature embryos of diploid barley (*Hordeum vul-*

gare) and *H. vulgare* × *H. bulbosum* L. crosses. Theor. Appl. Genet. 78: 105–112.

Planckaert, F. & V. Walbot, 1989. Molecular and genetic characterization of Mu transposable elements in *Zea mays*. Behaviour in callus culture and regenerated plants. Genetics 123: 567–578.

Puolimatka, M. & A. Karp, 1993. Meiotic disturbances resulting from tissue culture of inbred and outbred rye. Heredity 71: 138–144.

Qureshi, J.A., P. Hucl & K.K. Kartha, 1992. Is somaclonal variation a reliable tool for spring wheat improvement? Euphytica 60: 221–228.

Ray, I.M. & E.T. Bingham, 1991. Inheritance of a mutable phenotype that is activated in alfalfa tissue culture. Genome 34: 35–40.

Reed, S.M. & E.A. Wernsmann, 1989. DNA amplification among anther-derived doubled haploid lines of tobacco and its relationship to agronomic performance. Crop Sci. 29: 1072–1076.

Ruíz, M.L., M.I. Rueda, F.J. Peláez, M. Espino, M. Candela, A.M. Sendino & A.M. Vázquez, 1992. Somatic embryogenesis, plant regeneration and somaclonal variation in barley. Plant Cell Tissue Organ Culture 28: 97–101.

Shepard, J.F., D. Bidney & E. Shahin, 1980. Potato protoplasts in crop improvement. Science 208: 17–24.

Sibi, M., M. Biglary & Y. Demarly, 1984. Increase in the rate of recombinants in tomato (*Lycopersicon esculentum* L.) after *in vitro* regeneration. Theor. Appl. Genet. 68: 317–321.

Singsit, C., R.E. Veilleux & S.B. Sterret, 1990. Enhanced seed set and crossover frequency in regenerated potato plants following anther and callus culture. Genome 33: 50–56.

Skirvin, R.M. & J. Janick, 1976. Tissue culture induced variation in scented *Pelargonium* spp. J. Amer. Soc. Hort. Sci. 101: 281–290.

Söndahl, M.R. & A. Bragin, 1991. Somaclonal variation as a breeding tool for coffee improvement. ASIC, 14[e] Coooque, San Francisco, 701–710.

Springen, K., 1987. Improving on mother nature. Newsweek 26: 3.

Stieve, S.M., D.P. Stimart & B.S., 1992. Heritable tissue culture induced variation in *Zinnia marylandica*. Euphytica 64: 81–89.

Varga, A., L.H. Thomas & J. Bruinsma, 1988. Effects of auxins on epigenetic instability of callus-propagated *Kalanchoe blossfeldiana* Poelln. Plant Cell Tissue Organ Culture 15: 223–231.

Vuylsteke, D. & R. Swennen, 1990. Somaclonal variation in African plantains. IITA Res. Vol. 1: 4–10.

Williams, M.E., A.G. Hepburn & J.M. Widholm, 1991. Somaclonal variation in a maize inbred line is not associated with changes in the number or location Ac-homologous sequences. Theor. Appl. Genet. 81: 272–276.

Winfield, M., M.R. Davey & A. Karp, 1993. A comparison of chromosome instability in cell suspensions of diploid, tetraploid and hexaploid wheats. Heredity 70: 187–194.

Euphytica **85:** 303–315, 1995.

Application of *in vivo* and *in vitro* mutation techniques for crop improvement

Miroslaw Maluszynski, Beant S. Ahloowalia & Björn Sigurbjörnsson
Joint FAO/IAEA Division, International Atomic Energy Agency, Vienna, Austria

Key words: doubled haploids, micropropagation, mutant cultivars, mutation techniques, somaclonal variation

Summary

Conventional mutation techniques have often been used to improve yield, quality, disease and pest resistance in crops, or to increase the attractiveness of flowers and ornamental plants. More than 1700 mutant varieties involving 154 plant species have been officially released. In some economically important crops, e.g. barley, durum wheat and cotton, mutant varieties occupy the majority of cultivated areas in many countries. Mutation techniques have become one of the major tools in the breeding of ornamentals such as alstroemeria, begonia, chrysanthemum, carnation, dahlia and streptocarpus. The use of *in vitro* techniques such as anther culture, shoot organogenesis, somatic embryogenesis and protoplast fusion can overcome some of the limitations in the application of mutation techniques in both seed and vegetatively propagated crops. *In vitro* culture in combination with induced mutations can speed up breeding programmes, from the generation of variability, through selection, to multiplication of the desired genotypes. The expression of induced mutations in the pure homozygote obtained through microspore, anther or ovary culture, can enhance the rapid recovery of the desired traits. In some vegetatively propagated species, mutations in combination with *in vitro* culture technique, may be the only method of improving an existing cultivar. Currently, many molecular studies rely on the induction and identification of mutants in 'model species' for construction and subsequent saturation of genetic maps, understanding of developmental genetics and elucidation of biochemical pathways. Once identified and isolated, the genes that encode agronomically-important features can be either introduced directly into crop plants or used as probes to search for similar genes in crop species. It seems most likely that the recent developments based on these technologies will soon provide improved methods for selection of desired mutants.

Introduction

Most of the available genetic variation used in breeding programmes has occurred naturally and exists in germplasm collections of new and old cultivars, land races and genotypes. This variation through crosses is recombined to produce new and desired gene combinations. When existing germplasm fails to provide the desired recombinant, it is necessary to resort to other sources of variation. Since spontaneous mutations occur with extremely low frequency, mutation induction techniques provide tools for the rapid creation and increase in variability in crop species. Most induced mutations are recessive and deleterious from a breeding point of view. However, in spite of these limitations, induced mutations have contributed significantly to plant improvement worldwide, and in some cases have made an outstanding impact on the productivity of particular crops. The impact of mutation techniques for crop improvement has already been evaluated in many publications (Broertjes & van Harten, 1988; Konzak, 1984; Maluszynski, 1990; Micke, 1991; Micke et al., 1990; Rutger, 1992; Sigurbjörnsson, 1983).

The development of efficient *in vitro* culture methods has facilitated the use of mutation techniques for improvement of both seed and vegetatively propagated plants. In many vegetatively propagated crops mutation induction in combination with *in vitro* culture techniques may be the only effective method for plant improvement (Novak, 1991). The use of *in vitro* techniques such as anther/microspore culture, shoot organogenesis and somatic embryogenesis can overcome some of the limitations in the application of mutation techniques. Among such limitations the most important are the lack of effective mutant screening techniques, the unrealistically large but necessary size of the mutated population, calculated on the basis of an expected frequency of mutation for a desired character, the development time for mutated generations. In seed propagated species, the application of mutation coupled with doubled haploid systems seems to be highly promising. This approach can speed up breeding programmes from the generation of variability, through selection, homozygosity onward to rapid multiplication of the desired genotypes (Szarejko et al., 1991).

Induced mutants in plants are also beginning to play an important role in plant molecular genetics. Progress in molecular studies of 'model plant species' relies to a great extent on the induction and identification of mutants as a tool for the study of developmental genetics and elucidation of biochemical pathways. High saturation of linkage maps with DNA-based molecular markers facilitates the isolation of mutated genes, as has been successfully demonstrated in *Arabidopsis thaliana* (Roe et al., 1993). It is expected that molecular-assisted breeding – especially early detection techniques – will soon provide rapid and very efficient methods for analyzing advanced mutated generations.

Table 1 Plant characters improved by induced mutations in officially released rice mutant varieties

Character	No of mutant varieties
semidwarfness	126
earliness	110
tillering	24
tallness	23
grain quality	16
blast tolerance	14
adaptability	12
glutinous endosperm	12
salt tolerance	9
cold tolerance	6
photoperiod insensitivity	5
lateness	2

Source FAO/IAEA Mutant Varieties Database (1993)

Conventional application of mutation techniques

General procedures

The general procedures for using induced mutations are rather simple and have a strong basis in the laws of genetics (FAO/IAEA, 1977) Dormant seeds of the so-called parent variety are irradiated or treated with a chemical mutagen. Mutagenic treatment can cause chromosomal rearrangements or change some genes to other allelic forms Plants grown from mutagenized seeds are called M_1-plants. When a multicellular tissue like the seed embryo is treated with a mutagen, the plant developing from the treated seed has a chimeric structure from a genetic point of view. After meiosis, the seeds developed on M_1 plants are already the M_2 generation. These seeds are sown in experimental plots, and a segregating M_2 population is subjected to various screening procedures for desired characters. Selected material is then usually grown on as the M_3 generation. Screening for quantitatively inherited characters is usually done in the M_3 generation where selection on a line, rather than on a single plant basis, can be initiated. Selected mutants from the M_2 or M_3 generations are usually checked for homozygosity in the M_3 or M_4 generations, respectively. Promising, homozygotic mutants can be used directly for multiplication – this will lead to the development of the so-called direct mutant variety (e.g. barley cv. 'Diamant' in CSSR, rice cv 'Calrose' 76 in USA) – or they can be used in a cross breeding programme (e.g. linseed mutants 'M1589' and 'M1722') (Maluszynski, 1990).

Current status and most significant achievements

About 30 years ago the Plant Breeding and Genetics Section of the Joint FAO/IAEA Division began to collect information related to newly released crop varieties developed directly after mutagenic treatment or by crosses involving mutant lines (Sigurbjornsson & Micke, 1974). The first mutant variety was released only four years after Stadler's (1930) publication on the discovery of plant mutagenesis. Today the FAO/IAEA Mutant Varieties Database has 1737 accessions (Maluszynski et al., 1991, 1992). Mutant varieties have been released in more than 50 countries The top 6 countries on the list are China, India, the former USSR, The Netherlands, Japan and the USA. From these officially released mutant varieties of 154 plants species, approx. 1275 are in agricultural crops and

Table 2. Rice mutant varieties with improved salt tolerance through induced mutations

Mutant variety	Country of release	Year of release	Mutagen/ mutant used
6B	Vietnam	1986	cross with gamma ray induced var. Atomita 2
A-20	Vietnam	1990	cross with MNH induced mutant
Atomita 2	Indonesia	1983	gamma rays
Changwei 19	China	1978	gamma rays
Emai No. 9	China	1980	gamma rays
Fuxuan No. 1	China	1968	gamma rays
Jiaxuan No. 1	China	1974	gamma rays
Liaoyan 2	China	1992	gamma rays
Mohan = CSR4	China	1983	gamma rays

Source : FAO/IAEA Mutant Varieties Database (1993).

remaining mutant varieties in ornamental and decorative plants including mutants of chrysanthemum (187), alstroemeria (35), dahlia (34), streptocarpus (30) and many others. Since the effect of mutations in ornamental or decorative plants is clearly visible and selection for altered flower color, shape or size is relatively easy, the application of mutation techniques in the breeding of these crops has a high success rate (Broertjes & van Harten, 1988).

The FAO/IAEA Mutant Varieties Database indicates that more than half of the induced mutant varieties were released during the last decade. In the group of crop species, 769 mutant varieties were developed directly from mutated progenies and 506 varieties were obtained from crosses with mutated parent(s). Among the agricultural crops, mutant varieties involving cereals dominate with 822. Rice is in the first place (318), followed by barley, wheat, maize, durum wheat and others (oat, millet, sorghum, rye and dura).

Most of the rice mutant varieties (215) were released as 'direct' mutants – this means direct seed multiplication from selected mutants. Nevertheless, some mutants such as 'Reimei' (Japan) and 'Calrose 76' (USA) were successfully used in extensive cross-breeding programmes. Semi-dwarfness (126 varieties) and earliness (110 varieties) were the characters most often selected in treated populations (Table 1). On the list of improved characters are also traits desired for increasing sustainability in rice production, such as cold (6) and salt tolerance (9) (Table 2) or photoperiod insensitivity (5). In rice, as in other crops, radiation was more often used to generate desirable traits (190 varieties), while 23 rice varieties were induced by chemical mutagenesis (Maluszynski et al., 1994). Some of the rice mutant varieties have had considerable economic impact. Rutger (1992) presented data on 11 mutant varieties which were or are cultivated on an annual area of over 100,000 hectares each. Among them are the Chinese varieties Zhefu 802, grown on 1,400,000 hectares and the variety Yuanfengzao with an area of about one million hectares.

There are several examples of successful implementation of mutant varieties leading to the significantly improved production of particular crops. In Pakistan, F_1 seeds of cotton from a cross of a US variety with a local one were irradiated with gamma rays. A selected mutant was improved in such important characters as determinate plant type, heat tolerance and earliness. This mutant was released in 1983 as NIAB 78 and its cultivation doubled cotton production in Pakistan during the following five years (NIAB, 1988). Similarly, in China, a mutant cotton variety Lumian No. 1 developed from gamma-irradiated seeds and released in 1974, has reached an annual area of cultivation of over 1 million hectares.

Production of malting barley is very important for the economy of some Central European countries not only for the local market but also as a significant component of their export. In the regions of the former Czechoslovakia malting barley is grown on more than 25% of the cereal acreage. Dry seeds of the variety 'Valticky' were irradiated with 10 kR of X-rays. Selection for improved characters was initiated in the M_2 generation. A selected mutant designated as X_2-

228 became the progenitor of a new variety 'Diamant' released in 1965. This mutant differed from the parent variety 'Valticky', a relatively tall variety (92 cm), which is highly susceptible to lodging under increased levels of nitrogen fertilizer, in such important agronomic traits as: reduction of culm length (15 cm), resistance to lodging, semi-prostate growth habit, higher tillering ability, 12% increase in grain yield, and a grain to straw ratio of 1 : 0.95 compared with 1 : 1.3 in 'Valticky'. Together with the significant reduction in culm length, the newly developed mutant showed a rather slow initial development. This character lead to a prolonged period of organogenesis during tillering and as a consequence, to a greater number of spike-bearing tillers. Resistance to lodging and high tillering resulted in substantially increased productivity of the mutant. In 1987, 100% of the spring barley area in the former Czechoslovakia (about 600.000 hectares) was covered by 'Diamant' or mutant varieties developed from 'Diamant'. Roughly estimated, the total increase in grain yield was about 1,486,000 tons. During the period 1972–1989 a total of 27 varieties arising from the 'Diamant' variety were developed and released in the former Czechoslovakia and more than 85 varieties derived from 'Diamant' were released in various European countries. In 1987, these mutant varieties were grown all over Europe on an area of 2.86 million hectares (Bouma & Ohnoutka, 1991).

Mutation techniques for enrichment of genetic resources for breeding and genetic analysis

Mutational analysis of int loci in barley

It is nothing new to say that the genetic analysis of any character is impossible without the availability of sufficient genetic variation related to that particular trait. Had the research work on *E. coli* mutants not been undertaken, our understanding of gene structure, function and expression in plants and eukaryotes would have been extremely poor. The availability of a large number of induced mutants in *E. coli* made it possible for the geneticist to elucidate the nature of the gene. However, the ease with which mutants can be generated and screened in *E. coli*, cannot be compared with those in higher plants which require a much larger resource of space, personnel, money and labour. Nevertheless, there is no alternative to the production of mutants for detailed genetic analysis of a particular trait and for understanding the mechanisms of gene action and plant development. Availability of new techniques in molecular genetics and plant physiology have generated an enormous production of mutants in both model and crop plant species.

The usefullness of 'mutational analysis' in breeding programmes was recently demonstrated by Lundqvist & Lundqvist (1994). Their paper 'Intermedium mutants of barley-diversity, interactions and plant breeding value' can be considered a typical example. Mutational analysis clarified the rather complex genetics of kernel rows in barley. So far, 144 mutants with variation in spike structure were isolated in the Svalöv barley collection. It was demonstrated that at least 12 gene loci (*hex-v* and 11 *int*) in barley can promote the spike development of lateral florets, their size, awn development, fertility and kernel development. This analysis also produced very promising practical results. A cross combination of particular double mutants *int/int* with the *hex-v* gene has produced new segregants denoted as 'King-size' for their excellent six-row type with conspicuous large spikes and thick culms.

Mutational analysis of flavonoid biosynthesis

Colloidal haze formation in beer due to precipitation of malt proteins by proanthocyanidins is a serious problem for brewing companies. Chemical stabilizing treatment is necessary to avoid the visible haze in beer caused by this class of flavonoids. Von Wettstein et al. (1977) used an induced anthocyanin-free barley mutant to get haze stability without chemical treatment of the beer. Since then more than 700 barley mutants have been isolated in the Carlsberg Laboratory with mutations of genes (*Ant*) associated with the pathway of flavonoid biosynthesis in which synthesis of anthocyanin and/or proanthocyanidin is affected (Jende-Strid, 1993). Most of these mutants were obtained by mutagenic treatment of seeds with sodium azide (NaN_3). It was possible to screen large M_3 populations for the proanthocyanidin-free mutants using the vanillin test method. The frequency of these mutants in analyzed M_3 generations was around 0.013%.

By diallelic crosses, 568 of these mutants have been localized into 28 complementation groups. It was demonstrated that proanthocyanidin-free mutations were induced in ten gene loci responsible for the biosynthesis of these components in barley grains. The others affected the synthesis of anthocyanin in various organs of the plants. Five *Ant* loci were described as structural genes whereas *Ant 13* codes for a transcription factor involved in the regulation of at least three genes related to the flavonoid pathway. This detailed genetic analysis of flavonoid biosynthesis opened up possibilities for molecular analysis of coding sequences and introns of *Ant 18* gene and their expression in the anthocyanin-free mutant tissue (Wang et al., 1993).

The case of 'Golden Promise'

Mutant cultivars or mutant lines have usually been investigated only for a character or characters directly related to solving a particular breeding program. Mutants kept in collections are very seldom investigated for other characters not related to an already described mutated trait. This was the case in the rice mutant variety 'Atomita 2' released in Indonesia in 1983, selected for earliness and brown plant hopper (BPH) resistance and later discovered to carry a mutated gene for salt tolerance in Vietnam (MBNL, 1988).

'Golden Promise' is a direct mutant variety developed by gamma-ray treatment of the barley cultivar 'Maythorpe' and released in UK in 1966. This variety was extremely important in the UK barley industry, and dominated the Scottish barley acreage in the 1970s to mid-1980s. It is still the standard for malt quality in Scotland (Forster, in press). The short, stiff straw is related to the mutation to erectoides gene (*GPert*) which has pleiotropic effects on yield and excellent quality. Mutated gene(s) from 'Golden Promise' were transferred to other genetic backgrounds, ultimately leading to the release of 17 other barley mutant cultivars in the UK. In salt tolerance tests at the Scottish Crop Research Institute, a significant difference (50% less) has been recently detected between the shoot sodium accumulation of 'Golden Promise' plants and its parent variety 'Maythorpe' under salt stress conditions. The results imply that the mutational difference between a mutant variety and its parent is also responsible for the increased tolerance to salt. Current work aims at differential screening at the DNA, RNA and protein levels to isolate and characterize the mutated locus(i) (Forster, in press).

Table 3. Induced mutations affecting basic metabolic traits in *Arabidopsis thaliana**

Mutation	Trait
auxotrophs	leucine, tryptophan
	thiamine
amino acid analog resistance	5-methyltryptophane
	S-aminoethylcystein
	5-hydroxynorvaline
	ethionine
	Trans-4-hydroxy-L-proline
	p-fluorophenylalanine
nitrate reductase	absence/deficiency in activity
	elevated levels of activity
urease mutants	urease deficiency
mineral uptake	low and high level of phosphate, sulphate and potassium
starch metabolism	starchless
	starch degradation
alcohol dehydrogenase	allylalcohol resistance
lipid metabolism	differences in fatty acids composition
thermal tolerance	cold and freezing tolerance
	high temperature tolerance
embryo development	biotin requirement
photosynthesis	chlorophyll-b and starch deficiency
	pigment deficiency
	viable under non-photorespiratory conditions
	isoxaben (herbicide) resistance
light perception and chloroplast differentiation	sensitivity to red light
	photoreversible phytochrome
	blue light tolerance
	light independent green pigment synthesis

* after Redei & Koncz, 1992; modified.

Mutational analysis of the Arabidopsis *genome*

In barley, maize, pea and tomato, large mutant germplasm collections have been established and used for genetic analysis and plant breeding. Recently, due to intensive mutation induction efforts, *Arabidopsis* has become, most probably, the genus with the highest number of induced mutants. *Arabidopsis* mutants were recently classified by Redei & Koncz (1992). Table 3 includes examples of mutants classified by these authors in the categories 'mutations affecting basic metabolism' and 'photosynthetic mutations in the nucleus'. The other categories involve: developmental mutants (including mutations affecting roots), hormone mutants and mutations concerned with disease expression. The use of mutants for developmental studies was recently reviewed by Bradley & Pruitt (1992). Most *Arabidopsis* mutants have been obtained following seed treatment with ethyl methanesulphonate (EMS). Chemical and physical mutagens have also been widely used to induce mutations in this model plant. Additionally, a great number of various mutants were obtained by Feldman (1991) using T-DNA insertional mutagenesis. Mayer et al. (1991) described the principles of pattern formation in the plant embryo using mutants affecting different aspects of the body organization. Somerville & Browse (1991) demonstrated a wide spectrum of *Arabidopsis* mutants leading to changes in the lipid biosynthesis pathway. Using EMS mutagenesis, James & Dooner (1991) were able to genetically modify oil composition by increasing, for example, oleic acid content from 15 to 86% with parallel reduction of linoleic acid from 29 to 0.2% and linolenic acid content from 18 to below 2%. It is still an open question as to how far 'shuttle-mutagenesis' with induced mutations in *Arabidopsis* can be used in crop improvement programmes. Nevertheless, the work presented by Huang (1992) can be used as an example of successful 'shuttle-mutagenesis'. In this case, genes herbicide resistance obtained through microspore mutagenesis in *Brassica napus*, have been cloned and transferred into canola and tobacco to elevate levels of their tolerance to herbicides. It seems that steps leading to successful gene transfer such as gene identification, characterization and cloning, and to gene expression in transgenic plants can be more easily achieved with the *Arabidopsis* genome as donor than from another organism. Technological advantages related to the use of *Arabidopsis* for cloning plant genes were recently reviewed by Hemming (1993).

Mutation techniques in current crop improvement programmes

Generation of desired variability

The potential of applying mutation techniques for the improvement of various characters, including quantitatively inherited traits have been demonstrated in some of the recent papers dealing with genetic manipulation of fatty acid composition in rapeseed. Studies in Canada, Australia and many other countries have shown that genetic variation for reduced concentration of polyunsaturated fatty acids (PUFA) is currently limited in available breeding populations and extensive *Brassica* germplasm collections. It has been possible to induce the highly desired, low levels of PUFA mutations by using EMS mutagenesis in diploid and tetraploid species of rapeseed (Auld et al., 1992). In diploid *Brassica rapa*, a cross of a developed mutant to the canola-quality variety 'Tobin' was made to combine low levels of PUFA with a reduced level of erucic acid. Several F_4 families from this cross showed significant changes in oil composition. The total concentration of PUFA in selected F_4 families ranged from 4.4 to 5.2% with oleic acid content from 87 to 89%, whereas in variety 'Tobin' it was 38.7 and 52.7%, respectively. In the four selected M_4 lines of tetraploid *Brassica napus*, the oleic acid content was above 80% with drastically reduced PUFA content when in the parent variety 'Cascade' it was around 63%.

Similar results were obtained in the mutated progenies of sunflower, peanuts and again in rapeseed by Allelix Crop Technologies. This company reported the development of a *Brassica napus* mutant with oleic acid content over 85% (Ashri, 1993). Reduction of linolenic acid from 46 to 1.6% was obtained in linseed by crossing two induced mutants with low linolenic acid content (Green, 1986). Mutation techniques were also successfully applied to obtain resistance to residual soil levels of sulfonylurea herbicides in *Brassica napus, Brassica rapa* and *Sinapis alba* (Tonnemaker et al., 1992). At the 25th Rice Technical Working Group Meeting (1994), New Orleans, T.P. Croughan (Rice Research Station, Crowley, LA. USA) reported on the successful use of gamma rays to induce resistance to herbicides in rice. The North American operating unit of ICI Seeds Inc. has used the 'standard plant breeding technique of induction mutation' to obtain maize elite inbred lines resistant to the herbicide imazethapyr (Pursuit). The developed resistant mutant line (IT Corn) and hybrid seed showed excellent herbicide tolerance and good yield in subsequent field trials (Mabbett, 1992).

At the XVIIth International Congress of Genetics (ICG), Birmingham, 1993, R.M. Aslam & K.A. Siddiqui (Dep. of Botany, Jamshoro, Pakistan) reported gamma ray-induced mutations for several yield components, e.g., tiller number, spike length, spike circumference, seed number/spike, and 1000-kernel weight in the pearl millet variety 'Japan Bajra'. One of the mutants had spiny spikes, was less prone to bird damage and yet gave high yield, and had improved grain quality. In Mali, several sorghum mutants with short height, improved grain quality and drought tolerance have been obtained. Some of the drought tolerant lines have a very deep rooting system. Likewise, in *Oryza glaberrima*, the cultivated African rice, white caryopsis mutants were found, which have a higher market value than the normal red grain types (A. Bretaudeau & F. Cisse, pers. comm.).

Using fast neutrons for seed irradiation Worland & Law (1991) inactivated genes promoting disease susceptibility in hexaploid wheat. Monosomic analysis of the more resistant lines identified mutations in 4B, 4D and 5D chromosomes. Identification of the mutation sites through the recognition of restriction fragment length polymorphisms (RFLP) confirmed a deletion on chromosome 5D and on the long arm of chromosome 5B. The authors concluded that deletion or inactivation of genes promoting disease susceptibility improves resistance to a number of different diseases such as yellow rust, mildew and brown rust. The absence of these genes may be a basis for non-specific resistance and durability to disease.

Fifteen stem rust resistant mutants of wheat were selected after mutagenic treatment of seeds with EMS. It has been shown that the mutated gene(s) of 14 of the mutants are located at the same sites on the long arm of chromosome 7D. Inactivation of the suppressor gene permitted expression of existing resistance genes to wheat stem rust that were inhibited by the suppressor (Williams et al., 1992).

Bartos (1993) in a review of the use of rye chromosome 1R and especially translocation 1BL/1RS concludes that future progress for its use in wheat breeding will depend upon possibilities of removing or compensating the negative effect of chromosome 1R on bread making quality. Millet & Feldman (1994) demonstrated that by gamma irradiation it was possible to delete the secalin gene (*Sec-1*) from the short arm of the 1BL/1RS chromosome. It has been considered that the secalin gene, under the absence of wheat homoeoallele *Gli-B1*, is responsible for sticky dough and thus poor bread quality.

A detailed protocol on the use of radiation to transfer alien chromosome segments to wheat was prepared by Sears and published recently (1993). In this paper, he summarized his experiences with the use of radiation for gene transfer to wheat, including transfer of leaf-rust resistance from *Aegilops umbellulata* to wheat (Sears, 1956). Friebe et al. (1993) reported that radiation (x- or gamma-rays) was used to obtain nonhomoeologous wheat-*Agropyron intermedium* chromosomal translocation. Among eight induced wheat-*Ag. intermedium* derivatives, one was identified as a wheat-*Ag. intermedium* addition line with resistance to leaf and stem rust. The authors concluded that this material may have significance in breeding superior rust-resistant wheat cultivars.

Use of in vitro *haploid technology*

The mutant production cycle can be significantly shortened by the application of doubled haploid (DH) techniques (Szarejko et al., 1991). Depending on the plant species, various techniques can be used. Beversdorf & Kott (1987) developed a mutagenesis and *in vitro* selection system using microspore cultures in rapeseed. This system involves gamma rays or chemical mutagen treatment of uninucleate, potentially embryogenic microspores followed by a selection of developing embryo-like structures or plantlets on a medium with a selecting factor (Swanson et al., 1989). Huang (1992) reported that Calogen Inc. and Allelix Crop Technologies used microspore mutagenesis to modify fatty acid content in rapeseed.

In barley, anther culture, bulbosum method or microspore culture (Kasha et al., 1993) can be used for rapid production of true-to-type mutants. Contrary to commonly applied procedures, Umba di-Umba et al. (1991) suggested the use of dormant seeds, instead of *in vitro* culture, for mutagenic treatment. The use of a mutagen in various *in vitro* systems usually significantly decreases their regeneration ability. Using an M_1 plant (a plant from mutagenically treated seeds) as a donor of anthers or microspores for production of doubled haploids, almost all problems relating to the somatic effects of mutagen on tissue could be avoided. Additionally, the production of DH plants from mutated gametes can help to avoid chimerism which usually appears when a multicellular structure is mutagenically treated. In the DH_2 generation from mutated barley plants, more than 25% of true-to-type mutants were observed, among them some mutants with useful characters such as semidwarfness or uniculm plant type.

Mutation techniques in heterosis

It is well known that radiation and chemical mutagens can induce male sterility in plants (Kinoshita, 1982; Min et al., 1989; Rutger, 1992). Recently, Chaudhury et al. (1994) have clearly demonstrated the potential of mutagenesis for induction of male-sterile mutants. A large M_2 population (200,000 seeds) of *Arabidopsis* was developed from EMS mutagenized seeds and screened for male-sterile plants. Nineteen putative mutants were selected. Among the few of them genetically analyzed, four non-allelic mutants with different blocks of pollen development were found. It is important to note that in these mutants the vegetative growth and female fertility are not altered.

Heterosis in F_1 generations of mutant crosses was reported long ago, starting with maize mutants (Jones, 1945). Maluszynski et al. (1988) obtained similar results in crosses between mutants from one barley variety and in crosses between the mutants and their parent variety. Even mutants with extremely poor agronomic performance can give excellent F_1 plants, outyielding a parent variety. Even if the heterosis involving mutant hybrid plants partly depends on additive effects, the doubled haploid system is opening a completely new opportunity for fixing this effect. The general scheme is as follows: development of stable mutants, screening for heterosis in the F_1 of mutant crosses followed by the production of doubled haploids from heterotic F_1, screening for 'F_1 performing' doubled haploids, and the agronomic evaluation of selected DH lines (Maluszynski & Szarejko, in press). This theoretical scheme has already been confirmed in the barley mutant germplasm collection, where two mutants were found which had very poor agronomic characteristics but gave excellent heterosis in crosses. The same effect was confirmed in crosses of other barley mutants with their parent varieties, giving in the third generation of doubled haploids (DH_3), a percent yield increase only slightly lower than F_1 hybrids but higher than the parent variety by at least 20% (K. Polok et al., pers. comm.). There are many advantages to the use of 'hybrid performing' seeds produced through the doubled haploid system. The main one, of course, is that once produced, 'hybrid performing' DH plants can later on be multiplied by self-pollination. There are other important advantages of mutant heterosis, among them, the lack of problems with grain quality, plant height or disease resistance – as the DH 'mutant hybrid performing' line will present traits similar to the parental variety with a significantly increased yield (Maluszynski et al., 1989).

In vitro culture, mutagenesis and somaclonal variation

The techniques of mutation induction and *in vitro* culture seem to be ideally suited for the improvement of vegetatively propagated plants. Food crops, such as cassava, banana, plantain, sweet potato, potato and sugarcane are vegetatively propagated, and are the staple diet or an important source of calories in many tropical countries. Often the size of the conventional propagule is too big to allow mutagenic treatment of large numbers of plants. *In vitro* culture of such propagules not only provides relatively uniform and large populations of cells and tissues in a disease-free situation for irradiation, but also, because of the miniature size of micro-propagules, it is possible to irradiate very large numbers, and also to further separate the desired mutated sectors from the other ones in a short time. This can be achieved by successive culture of buds or regeneration of shoots and somatic embryos from cell suspensions and callus cultures, derived from irradiated tissues and explants. If somatic embryogenesis can be obtained, then the chance for solid mutants is even better than from shoot regeneration. Induction of mutations in *in vitro*-cultured material and subsequently *in vitro* multiplication for two to three cycles is also helpful in separating mutated sectors from chimeric tissue, particularly in plants propagated vegetatively. Irradiation in combination with *in vitro* culture has proved to be a valuable method of producing desired variation and rapid propagation.

Some research on *in vitro* radiation-induced mutations has been carried out on banana (Novak et al., 1990) and potato (Sonnino et al., 1986; Ahloowalia, 1990). This technique is now being extended to other vegetatively propagated crops. For example, in Ghana, Safo-Kantanka & Owusu (1993) have reported cassava mutants selected for improved cooking quality, improved mealiness taste and non-lumpy uniformly cooked flour, have yielded equal to or better than the parental cultivar. In plantain, advanced generation material has been produced for selection of short height and resistance to wind damage. Klu (1993) has reported the production of a semidwarf mutant of yam, which may eliminate the use of staking, an expensive operation in the cultivation of yams.

N.V. Sidorova & V.V. Morgun (Inst. Plant Physiol. & Genwtics, Kiev, Ukraine) at the XVIIth International Congress of Genetics (1993) have reported that in two winter wheat varieties, variants had been obtained both by regeneration from callus induction, and by irradiation of seeds with gamma rays. The observed somaclonal variants included true breeding lines for plant height, spike morphology and size, awn type, and resistance to rust and powdery mildew. However, the frequency of agronomically-important mutations for short height, early ripening, was twice as high among the plants obtained from irradiation of seeds with 200 Gy than in the somaclonal population (Rowlett et al., 1993).

Table 4. In vitro selected variants in plants

Variant	Plant	Reference
Stress tolerance		
temperature	snapdragon	Melchers & Bergmann, 1959
salt tolerance	tobacco, chilli	Dix & Street, 1975
	proso millet	Nabors, 1983
	alfalfa	Winicov, 1991
drought tolerance	proso millet	Nabors, 1983
frost tolerance	wheat	Dörffling et al., 1993
Disease resistance		
Helminthosporium	maize	Gengenbach et al., 1977
Phoma lingam	rapeseed	Sacristan, 1982
Phytophthora infestans	potato	Behnke, 1979, 1980
Pseudomonas syringae	tobacco	Carlson, 1973
Helminthosporium sacchari	sugarcane	Heinz et al., 1977
Herbicide resistance		
picloram	tobacco	Chaleff, 1983
paraquat	tobacco	Miller & Hughes, 1980
Changed metabolism		
nitrate reductase	tobacco	Muller, 1983
enhanced lysine	rice	Sharpe & Shaeffer, 1993
grain chalkiness	rice	Shaeffer et al., 1986
increased somatic embryogenesis	asparagus	Debreil & Jullien, 1994

Both somaclonal variation and mutations result in the production of new genotypes with a limited change in the original genome. As a source of variation, somaclonal variation mimics induced mutations. Somaclonal variation has been associated with changes in chromosome number and structure, point mutations, DNA methylation (Brown, 1991; Brown et al., 1993), changes in cytoplasm and plastids, activation of transposons, etc. Ryegrass plants regenerated from triploid embryo-callus culture included albinos for example, (Ahloowalia, 1975) as well as those with changed chromosome number (polyploids, aneuploids) and structure (translocations, deletions, inversions) (Ahloowalia, 1976, 1983). Since then, several reports have confirmed the occurrence of genomic changes in the tissue culture-derived plants in barley, maize, wheat, rice, triticale, sugarcane and potato (cf. Ahloowalia, 1986). It was suggested that transposable genes may be involved in the unstable transmission of somaclones aberrant in spike-shape, observed among the subsequent seed progenies of regenerated wheat (Ahloowalia & Sherington, 1985). In tomato, 13 variants among 230 regenerated plants were found to be single gene mutants, and involved both recessive and dominant changes (Evans & Sharp, 1983). While a great number of radiation-induced mutants have been released as new cultivars, somaclonal variants so far have been of limited value in plant improvement. Both somaclonal variation and conventional mutagenesis are complementary to and not a replacement for conventional plant breeding.

In vitro selection of mutagenized cells and tissues

In vitro techiques also allow pre-selection of mutagenized cells and tissues. The history of *in vitro* selection is long. As early as 1959, Melchers & Bergmann reported selection for temperature variation in snapdragon cell suspension cultures. They drew attention to the use of cell culture for mutant selection, the ease of large scale screening of cell populations and the convenience of mutant induction and selection using haploid cell populations. Several papers confirmed the use of this method (Binding et al., 1970; Carlson, 1973; Maliga, 1984). Since then, several *in vitro* selected variants have been investigated for their value in plant improvement.

Selection pressures can be applied at either the cell population level or on the plant regenerated from cell cultures, and followed by selection in conventional field plots. While plant and cell tissue culture techniques allow screening of very large populations of cells and regenerated plants in a small space and in a much more controlled environment than in conventional field trials, this is possible only when a trait is amenable to *in vitro* selection, and is expressed and transmitted in the regenerated plants and their progenies. The potential high efficiency of *in vitro* selection systems is based on the fact that it is possible to grow millions of cells in a petri-dish or in a flask and achieve rapid multiplication of cell populations on defined media (Duncan & Widholm, 1990; Nelshoppen & Widholm, 1990). Addition of sodium chloride, fungal toxins, herbicides, antibiotics to the medium or exposure of cells to heat, cold and freezing is used for selection of the desired variants (Table 4). The term 'variant' is used to define a new phenotype. Only genetic or molecular evidence can establish its status as a mutant. In many cases, the selected variants were found to be mutants and showed simple Mendelian inheritance.

Mutation techniques in gene technology

There are also several applications of mutation techniques in basic research, especially in molecular plant physiology and genetics. R.W. Michelmore (Davis, University of California, USA), at the XVIIth International Congress of Genetics (1993) in Birmingham, presented results of his work on the use of different molecular markers. There is a relatively low level of DNA polymorphism in lettuce and only 10% of the expected numbers of loci were found. He reported that fast neutrons generated a 10-fold increase in DNA polymorphism in lettuce (Rowlett et al., 1993).

Straus & Ausubel (1990) demonstrated that radiation-induced deletions can become a new tool in molecular genetics. The technique of 'genomic subtraction' is an important addition to chromosome walking and gene tagging approaches for cloning wild-type genes which are missing in a homozygous deletion mutant. Sun et al. (1992) successfully used this method for cloning wild-type sequences corresponding to the gene of gibberellin-responsive dwarfness (*ga1*) in *Arabidopsis*. Cloning of DNA sequences that are present in the wild-type enabled a detailed genetic analysis of the *GA1* region. Current developments in molecular genetics technology, particularly the identification of changes in a single base pair in DNA, should allow induction, detection, isolation and characterization of mutations in plants much more precisely than ever before.

Conclusions

Mutation techniques offer to the plant breeder and geneticist several choices, alone and/or in combination with the emerging technologies of tissue culture and molecular genetics, for enhancing the improvement of specific traits in crop plants. These include the direct multiplication of the selected mutants as new varieties, use of mutants in crosses; transfer of specific chromosomal segments from alien genomes; use of mutants in developing molecular maps; understanding of gene expression and plant development; use in heterosis, and gene tagging, deletion and insertional mutagenesis. In each case, the use of the mutation technique will depend upon the specific objectives and the plant species in question. The use of *in vitro* techniques in combination with mutation induction seems particularly suitable for the improvement of vegetatively propagated plants. The use of doubled haploids in combination with induced mutations and transgenesis seems particularly attractive to rapidly obtain the desired genotypes in a homozygous state in the cereals, *Brassica* and other seed propagated crops

References

Ahloowalia, B S , 1975 Regeneration of ryegrass plants in tissue culture Crop Sci 15 449–452

Ahloowalia, B S , 1976 Chromosomal changes in parasexually produced ryegrass p 115–122 In K Jones & PE Brandham (Eds) Current Chromosome Research Elsevier/North-Holland Biomedical Press, Amsterdam

Ahloowalia, B S , 1983 Spectrum of variation in somaclones of triploid ryegrass Crop Sci 23 1141–1147

Ahloowalia, B S , 1986 Limitation to the use of somaclonal variation in crop improvement p 14–27 In J Semal (Ed) Somaclonal Variation and Crop Improvement Martinus Nijhoff Pub , Dordrecht

Ahloowalia, B S & J Sherington, 1985 Transmission of somaclonal variation in wheat Euphytica 34 525–537

Ahloowalia, B S , 1990 *In vitro* radiation induced mutagenesis in potato p 39–46 In R S Sangwan & B S Sangwan-Norreel (Eds) The Impact of Biotechnology in Agriculture Kluwer Academic Publisher, Dordrecht

Ashri, A, 1993 Mutation breeding of oil crops p 82–94 In M Maluszynski & A Ashri (Eds) Report of the First FAO/IAEA Seminar on the Use of Induced Mutations and Related Biotechnology for Crop Improvement for the Middle East and the Mediterranean Regions IAEA, Vienna

Auld, DL, MK Heikkinen, DA Erickson, JL Sernyk & JE Romero, 1992 Rapeseed mutants with reduced levels of polyunsaturated fatty acids and increased levels of oleic acid Crop Sci 32 657–662

Bartos, P, 1993 Chromosome 1R of rye in wheat breeding Plant Breeding Abstracts 63 1203–1211

Behnke, M, 1979 Selection of potato callus for resistance to culture filtrates of *Phytophthora infestans* and regeneration of resistant plants Theor Appl Genet 55 69–71

Behnke, M, 1980 General resistance to blight of *Solanum tuberosum* plants regenerated from callus resistant to culture filtrates of *Phytophthora infestans* Theor Appl Genet 56 151–152

Beversdorf, WD & LS Kott, 1987 An *in vitro* mutagenesis/selection system for *Brassica napus* Iowa State J Res 61 435–443

Binding, H, K Binding & J Straub, 1970 Selektion in Gewebekulturen mit haploiden Zellen Naturwissensch 3 138–139

Bouma, J & Z Ohnoutka, 1991 Importance and application of the mutant 'Diamant' in spring barley breeding p 127–133 In Plant Mutation Breeding for Crop Improvement, Vol 1 IAEA, Vienna

Bradley, D & RE Pruitt, 1992 Development genetics of *Arabidopsis* p 225–241 In VEA Russo, S Brody, D Cove & S Ottolenghi (Eds) Development, the Molecular Genetic Approach Springer-Verlag, Berlin

Broertjes, C & AM van Harten, 1988 Applied mutation breeding for vegetatively propagated crops Elsevier, Amsterdam

Brown, PTH, 1991 The spectrum of molecular changes associated with somaclonal variation Newsletter IAPTC 66 14–25

Brown, PTH, FD Lange, E Kranz & H Lorz, 1993 Analysis of single protoplasts and regenerated plants by PCR and RAPD technology Mol Gen Genet 237 311–317

Carlson, PS, 1973 Methionine sulfoximine-resistant mutants of tobacco Science 180 1366–1368

Chaleff, RS, 1983 Isolation of agronomically useful mutants from plant cell cultures Science 219 676–682

Chaudhury, AM, M Lavithis, PE Taylor, S Craig, MB Singh, ER Signer, RB Knox & ES Dennis, 1994 Genetic control of male fertility in *Arabidopsis thaliana* structural analysis of premeiotic developmental mutants Sex Plant Reprod 7 17–28

Dix, PJ & HE Street, 1975 Sodium chloride resistant cultured cell lines from *Nicotiana sylvestris* and *Capsicum annuum* Plant Sci Lett 5 231–237

Delbreil, B & M Jullien, 1994 Evidence of *in vitro* induced mutation which improves somatic embryogenesis in *Asparagus officinalis* L Plant Cell Rep 13 372–376

Dorffling, K, H Dorffling & G Lessilich, 1993 *In vitro*-selection and regeneration of hydroxyproline resistant lines of winter wheat with increased proline content and increased frost tolerance J Plant Physiol 142 222–225

Duncan, DR & JM Widholm, 1990 Techniques for selecting mutants from plant tissue cultures p 443–453 In JW Pollard & JM Walker (Eds) Methods in Molecular Biology, Vol 6, Plant Cell and Tissue Culture The Humana Press, Clifton

Evans, DA & WR Sharp, 1983 Single gene mutations in tomato plants regenerated from tissue culture Science 221 949–951

FAO/IAEA, 1977 Manual on mutation breeding Second Edition p 288, IAEA, Vienna

Feldmann, KA, 1991 T-DNA insertion mutagenesis in *Arabidopsis* mutational spectrum The Plant J 1 71–82

Forster, BP, 1994 Salt tolerance of barley mutant 'Golden Promise' MBNL 41 (in press)

Friebe, B, J Jiang, BS Gill & PL Dyck, 1993 Radiation induced nonhomoeologous wheat *Agropyron intermedium* chromosomal translocations conferring resistance to leaf rust Theor Appl Genet 86 141–149

Gengenbach, BG, CE Green & CM Donovan, 1977 Inheritance of selected pathotoxin resistance in maize plants regenerated from cell culture Proc Natl Acad Sci USA 74 5113–5117

Green, AG, 1986 Genetic control of polyunsaturated fatty acid biosynthesis in flax (*Linum usitatissimum*) seed oil Theor Appl Genet 72 654–661

Heinz, DJ, M Krishnamurthi, LG Nickell & A Maretzki, 1977 Cell, tissue and organ culture in sugarcane improvement p 1–17 In J Reinert & YPS Bajaj (Eds) Applied and Fundamental Aspects of Plant Cell, Tissue and Organ Culture Springer-Verlag, Berlin

Hemming, D, 1993 Production and uses of genetically transformed plants AgBiotech News and Information 5 287N–292N

Huang, B, 1992 Genetic manipulation of microspores and microspore-derived embryos *In Vitro* Cell Dev Biol 28 53–58

James, DW & HK, Dooner, 1991 Novel seed lipid phenotypes in combinations of mutants altered in fatty acid biosynthesis in *Arabidopsis* Theor Appl Genet 82 409–412

Jende Strid, B, 1993 Genetic control of flavonoid biosynthesis in barley Hereditas 119 187–204

Jones, DF, 1945 Heterosis resulting from degenerative changes Genetics 30 527–542

Kasha, KJ, A Ziauddin, E Simion & L Cistue, 1993 Microspore cultures of barley and wheat Targets for change p 77–81 In M Maluszynski & A Ashri (Eds) Report of the First FAO/IAEA Seminar on the Use of Induced Mutations and Related Biotechnology for Crop Improvement for the Middle East and the Mediterranean Regions IAEA, Vienna Kinoshita, T, 1982 Inheritance of cytoplasmic male sterility induced by chemical mutagens in sugarbeet Proc Sugar Beet Res Assoc 38–45

Klu, GYP, 1993 Induced dwarf-type mutant of yam, *Dioscorea rotundata* Poir Trop Agric 70 289–290

Konzak, CF, 1984 Role of induced mutations p 216–292 In PB Vose & SG Blixt (Eds) Crop Breeding Pergamon Press, Oxford

Lundqvist, U & A Lundqvist, 1994 Intermedium mutants of barley – diversity, interactions and plant breeding value FAO/IAEA Research Report 4466/CF (in press)

Mabbett, T, 1992 Herbicide tolerant crops – ICI seeds leads the way Int Pest Control No 2 49–56

Maliga, P, 1984 Isolation and characterization of mutants in plant cell cultures Ann Rev Plant Physiol 35 519–552

Maluszynski, M, 1990 Induced mutations – an integrating tool in genetics and plant breeding p 127–162 In JP Gustafson (Ed) Gene Manipulation in Plant Improvement II Proc 19th Stadler Genetics Symp Plenum Press, New York

Maluszynski, M, A Fuglewicz, I Szarejko & A Micke, 1989 Barley mutant heterosis p 129–146 In M Maluszynski (Ed) Current Options for Cereal Improvement Doubled Haploids, Mutants and Heterosis Kluwer Academic Publisher, Dordrecht

Maluszynski, M, E Amano, B Ahloowalia, L van Zanten & B Sigurbjornsson, 1994 Mutation techniques and related biotechnologies for rice improvement p 294 In Seventh Meeting of the International Program on Rice Biotechnology, May 1994, Bali, The Rockefeller Foundation, New York

Maluszynski, M , B Sigurbjornsson, E Amano, L Sitch & O Kamra, 1991 Mutant varieties – data bank, FAO/IAEA database MBNL 38 16–21

Maluszynski, M , B Sigurbjornsson, E Amano, L Sitch & O Kamra, 1992 Mutant varieties – data bank, FAO/IAEA database Part II MBNL 39 14–17

Maluszynski, M & I Szarejko, 1994 Mutant heterosis and production of F_1-performing DH lines MBNL 41 (in press)

Maluszynski, M , I Szarejko, R Madajewski, A Fuglewicz & M Kucharska, 1988 Semi-dwarf mutants and heterosis in barley I The use of barley sd-mutants for hybrid breeding p 193–206 In Semi-Dwarf Cereal Mutants and Their Use in Cross-Breeding III IAEA-TECDOC 455, Vienna

Mayer, U , R A Torres Ruiz, T Berleth, S Misera & G Jurgens, 1991 Mutations affecting body organization in the *Arabidopsis* embryo Nature 353 402–407

MBNL, 1988 List of cultivars 31 30

Melchers, G & L Bergmann, 1959 Untersüchungen an Kulturen von haploiden Geweben von *Antirrhinum majus* Ber Dtsch Bot Ges 78 21–29

Micke, A , 1991 Induced mutations for crop improvement Gamma Field Symp 30 1–21

Micke, A , B Donini & M Maluszynski, 1990 Induced mutations for crop improvement Mutat Breed Rev 7 1–41

Miller, O K & K W Hughes, 1980 Selection of paraquat-resistant variants of tobacco from cell cultures *In Vitro* 16 1085–1091

Millet, E & M Feldman, 1994 Deletion of the secalin gene *Sec 1* in 1BL/1RS line by gamma irradiation In Proc 8th Int Wheat Genetics Symp Beijing, July 1993 (in press)

Min, S , Z Qi, Z Xiong & Ch Zhao, 1989 Effects of gamma-radiation treatment in somatic cell culture of indica rice Basmati 370 selection In Proc of the 6th Int Congr of SABRAO, p 793–796

Muller, A J , 1983 Genetic analysis of nitrate reductase deficient tobacco plants regenerated from mutant cells Evidence for duplicate structural genes Mol Gen Genet 192 275–281

Nabors, M W , 1983 Increasing the salt and drought tolerance of crop plants p 165–184 In R R Randall (Ed) Current Topics in Plant Biochemistry and Physiology 2 Univ Missouri Press, Columbia

Nelshoppen, J M & J M Widholm, 1990 Mutagenesis techniques in plant tissue cultures p 413–430 In J W Pollard & J M Walker (Eds) Methods in Molecular Biology, Vol 6, Plant Cell and Tissue Culture The Humana Press, Clifton

NIAB, 1988 Successful application of nuclear techniques for the improvement of cotton crop and role of NIAB-78 in cotton production p 1–16 In Nuclear Institute for Agriculture and Biology Faisalabad

Novak, F J , 1991 *In vitro* mutation system for crop improvement p 327–342 In Plant Mutation Breeding for Crop Improvement, Vol 2 IAEA, Vienna

Novak, F J , R Afza, M van Duren & M S Omar, 1990 Mutation induction by gamma irradiation of *in vitro* cultured shoot-tips of banana and plantain (*Musa* cvs) Trop Agric 67(1) 21–28

Redei, G P & C Koncz, 1992 Classical mutagenesis p 16–82 In C Koncz, N-H Chua & J Schell (Eds) Methods in *Arabidopsis* Research World Scientific, Singapore

Roe, J L , C J Rivin, R A Sessions, K A Feldmann & P C Zambryski, 1993 The *Tousled* gene in *A thaliana* encodes a protein kinase homology that is required for leaf and flower development Cell 75 939–950

Rowlett, K , D Hemming, S Hobbs, D Massey & A Rostron, 1993 Seventeenth International Congress of Genetics Genetics and the understanding of life AgBiotech News and Information 5 337N–360N

Rutger, J N , 1992 Impact of mutation breeding in rice – a review Mutat Breed Rev 8 1–24

Sacristan, M D , 1982 Resistance response to *Phoma lingam* of plants regenerated from selected cell and embryogenic cultures of haploid *Brassica napus* Theor Appl Genet 61 193–200

Safo-Kantanka, O & J Owusu-Nipah, 1993 Cassava varietal screening for cooking quality Relationship between dry matter, starch content, mealiness and certain microscopic observations of the raw and cooked tuber J Sci Food Agric 60 99–104

Schaeffer, G W , F T Sharp Jr , H L Carnhan & C W Johnson, 1986 Anther and tissue culture-induced grain chalkiness and associated variants in rice Plant Cell, Tissue and Organ Culture 6 149–157

Sears, E R , 1956 The transfer of leaf-rust resistance from *Aegilops umbellulata* to wheat Brookhaven Symp Biol 9 1–22

Sears, E R , 1993 Use of radiation to transfer alien chromosome segments to wheat Crop Sci 33 897–901

Sharpe, F T & G W Schaffer, 1993 Distribution of amino acids in bran, embryo and milled endosperm and shifts in storage protein subunits of *in vitro*-selected and lysine-enhanced mutant and wild type rice Plant Sci 90 145–154

Sigurbjornsson, B , 1983 Induced mutations p 153–176 In D R Wood (Ed) Crop Breeding American Society of Agronomy and Crop Science Society of America Madison, Wisconsin

Sigurbjornsson, B & A Micke, 1974 Philosophy and accomplishment of mutation breeding p 303–343 In Polyploidy and Induced Mutations in Plant Breeding IAEA, Vienna

Somerville, C & J Browse, 1991 Plant lipids metabolism, mutants, and membranes Science 252 80–87

Sonnino, A , G Ancora & C Locardi, 1986 *In vitro* mutation breeding of potato p 385–394 In Nuclear Techniques and *In Vitro* Culture for Plant Improvement IAEA, Vienna

Stadler, L J , 1930 Some genetic effects of x rays in plants J Hered 21 2–19

Straus, D & F M Ausubel, 1990 Genomic subtraction for cloning DNA corresponding to deletion mutations Proc Natl Acad Sci USA 87 1889–1893

Sun, T-P , H M Goodman & F M Ausubel, 1992 Cloning the *Arabidopsis GA1* locus by genomic subtraction The Plant Cell 4 119–128

Swanson, E B , M J Herrgesell, M Arnoldo, D Sippell & R S C Wong, 1989 Microspore mutagenesis and selection Canola plants with field tolerance to the imidazolinones Theor Appl Genet 78 525–530

Szarejko, I , M Maluszynski, K Polok & A Kilian, 1991 Doubled haploids in the mutation breeding of selected crops p 355–378 In Plant Mutation Breeding for Crop Improvement, Vol 2 IAEA, Vienna

Tonnemaker, K A , D L Auld, D C Thill, C A Mallory-Smith & D A Erickson, 1992 Development of sulfonylurea-resistant rapeseed using chemical mutagenesis Crop Sci 32 1387–1391

Umba, di-Umba , M Maluszynski, I Szarejko & J Zbieszczyk, 1991 High frequency of barley DH-mutants from M_1 after mutagenic treatment with MNH and sodium azide MBNL 38 8–9

Von Wettstein, D , B Jende-Strid, B Ahrenst-Larsen & J A Sorensen, 1977 Biochemical mutant in barley renders chemical stabilization of beer superfluous Carlsberg Res Commun 42 341–351

Wang, X , O Olsen & S Knudsen, 1993 Expression of the dihydroflavonol reductase gene in an anthocyanin-free barley mutant Hereditas 119 67–75

Williams, N.D., J.D. Miller & D.L. Klindworth, 1992. Induced mutations of a genetic suppressor of resistance to wheat stem rust. Crop Sci. 32: 612–616.

Winicov, I., 1991. Characterization of salt tolerant alfalfa (*Medicago sativa* L.) plants regenerated from salt tolerant cell lines. Plant Cell Rep. 10: 5461–5464.

Worland, A.J. & C.N. Law, 1991. Improving disease resistance in wheat by inactivating genes promoting disease susceptibility. MBNL 38: 2–5.

Euphytica **85**: 317–321, 1995.

The application of chemical mutagenesis and biotechnology to the modification of linseed (*Linum usitatissimum* L.)

G.G. Rowland[1], A. McHughen[1], L.V. Gusta[1], R.S. Bhatty[1], S.L. MacKenzie[2] & D.C. Taylor[2]
[1] *Crop Development Centre, University of Saskatchewan, Saskatoon, SK, Canada, S7N 5A8;* [2] *Plant Biotechnology Institute, National Research Council of Canada, Saskatoon, SK, S7N 0W9, Canada*

Key words: Linum usitatissimum, linseed, mutation breeding, somaclonal variation, fatty acids, genetic engineering

Summary

In the early 1980s the phenomenon of somaclonal variation induced by cell culture was exploited to produce genetic variation in linseed. The linseed variety Andro, derived from the widely grown Canadian variety McGregor, was selected in saline culture and was released for production in Canada. 'Andro' possesses traits very different from its parent, such as increased seedling vigour and tolerance to heat stress. Additional stable somaclonal variation in characters such as yield, days to maturity, seed weight and oil content were subsequently induced in 'McGregor'. However, despite extensive screening of the somaclonal variants, no significant variation in the fatty acid profile was found.

Chemical mutagenesis using ethyl methanesulphonate was, however, successful in modifying the fatty acid profile of McGregor. Initial screening of M_2 seed by the thiobarbituric acid colourimetric procedure was followed by gas chromatography to select half-seeds with atypical fatty acid profiles. Two independent, partially dominant genes were identified that were responsible for reducing the linolenic acid (18 : 3) from 50% to 2% while increasing linoleic acid (18 : 2) to 70%. A single, partially dominant gene, inherited independently of the linolenic acid genes, increased palmitic acid (16 : 0) from 7% to 30% and palmitoleic acid (16 : 1) from trace amounts to 4%.

Agrobacterium-mediated transformation of linseed has also been successful. Herbicide tolerance genes for glyphosate, sulfonylurea and phosphinothricin have been incorporated into Canadian varieties. Commercially useful levels of tolerance to sulfonylurea herbicides have been achieved with no adverse agronomic affect. It is expected that a transgenic variety containing this resistance will be registered for commercial production in Canada in 1994.

Standard breeding techniques, the application of antisense technology and the overexpression of fatty acid synthesis genes are being used to further modify the fatty acid profile of linseed, as well as for the transfer of abiotic stress-related genes identified in bromegrass.

Linseed breeding has been carried out in Canada since the early 1900s, and at the University of Saskatchewan since 1927. Standard hybridization methods were used to produce most of the Canadian linseed varieties, although reselection from introduced varieties was also employed (Kenaschuk, 1975). Only one variety was developed using a non-traditional breeding technique. Redwood-65 was selected from an X-irradiated population of the American variety Redwood (Larter et al., 1965). The radiation treatment program apparently had no specific objective other than to produce mutations. In comparison to Redwood, Redwood-65 had improved seed yield, earlier maturity, smaller seed and increased seed oil.

The application of traditional breeding techniques to variety development began to change at the Crop Development Centre, University of Saskatchewan in 1982. This paper reviews some alternate breeding methods that have been employed since then for the development of linseed varieties and the development of varieties for new end uses.

Somaclonal variation

Saskatchewan has many pockets of saline soils that adversely affect crop growth. An attempt was made by McHughen & Swartz (1984) to produce saline-tolerant linseed varieties by combining the then newly-described phenomenon of somaclonal variation with cellular selection *in vitro*. A population of cells derived from a single seed of the Canadian variety McGregor was plated onto a highly saline culture medium. A cell colony which survived this treatment was regenerated and gave rise to the line STS-II (salt tolerant selection II). In glasshouse tests in both saline and non-saline soils, STS-II performed better than its parent McGregor, indicating that the mechanism selected *in vitro* was active in whole plants (McHughen, 1987).

It was subsequently shown by O'Connor et al. (1991) that STS-II had increased tolerance to other stresses such as heat, and a greater ability to germinate at low temperatures than McGregor. Rowland et al. (1988 & 1989) found in field evaluations that STS-II was earlier flowering, earlier maturing, had larger seeds, a lower oil content and in saline conditions a greater number of plants survived to maturity than McGregor. In addition, STS-II has an extra 5S rDNA repeat not found in McGregor (Preete, 1991).

STS-II proved to be so different from McGregor and other Canadian linseed varieties, and had sufficient positive attributes that it was eventually registered as a variety under the name Andro (Rowland et al., 1989). Andro is an early maturing variety that is best suited to the northern growing areas of Saskatchewan.

Based on the success of exploiting somaclonal variation in the development of Andro linseed, over 11,000 plants were regenerated from callus cultures of a number of Canadian linseed varieties. Eventually 3100 lines, primarily regenerated from McGregor, were field evaluated in single rows. Over 800 of these lines were selected for replicated field trials in the next year and, within each trial, were compared with McGregor. On the basis of these initial replicated field trials, 294 McGregor somaclonal lines, representing a full range of the observed variation, were grown a second year in replicated trials. Also grown for two years in replicated field trials, was a random sample of breeder lines that form the variety McGregor. These breeder lines defined the maximum natural variation within the variety.

There was a large range in variation found in the McGregor somaclonal lines for all of the characters considered (Table 1) and much greater than that found in the breeder lines (Table 2). The heritabilities for the McGregor somaclonal lines was as follows: yield (0%), days to flower (22%), maturity score (34%), height (24%), seed weight (37%), oil content (43%) and iodine value (27%). These heritabilities indicate that much of this variability is stable and has a genetic basis. Even in the case of yield there was the odd somaclonal line, such as F86343, that has performed consistently well over a number of years. F86343 was

Table 1. The mean, minimum and maximum values in two years of 294 McGregor somaclonal lines grown at Saskatoon and expressed as a % of the McGregor check

Character	Mean	Minimum	Maximum
Yield year 1	96.2	56.3	150.5
Yield year 2	97.7	72.7	153.2
Days to flower year 1	99.9	81.8	116.7
Days to flower year 2	98.9	84.2	111.1
Maturity score year 1	102.5	33.3	166.7
Maturity score year 2	95.2	33.3	128.8
Height year 1	101.4	75.6	147.5
Height year 2	100.7	77.8	124.3
Seed weight year 1	99.6	84.3	122.0
Seed weight year 2	98.7	88.4	116.7
Oil concentration year 1	99.7	90.0	106.2
Oil concentration year 2	99.7	92.0	105.4
Iodine value year 1	100.3	93.4	105.7
Iodine value year 2	99.4	95.5	104.8

Table 2. The mean, minimum and maximum values in two years of the 30 McGregor breeder lines grown at Saskatoon and expressed as a % of the McGregor check

Character	Mean	Minimum	Maximum
Yield year 1	94.0	84.7	103.5
Yield year 2	96.0	89.0	102.2
Days to flower year 1	103.6	84.7	103.5
Days to flower year 2	101.3	89.0	102.2
Maturity score year 1	106.9	99.5	110.1
Maturity score year 2	93.6	86.7	101.9
Height year 1	110.9	95.4	123.3
Height year 2	101.5	93.7	114.7
Seed weight year 1	101.6	96.6	105.8
Seed weight year 2	100.5	95.4	103.6
Oil concentration year 1	101.0	98.5	103.4
Oil concentration year 2	98.1	95.9	100.9
Iodine value year 1	101.1	99.9	102.2
Iodine value year 2	99.7	97.8	101.1

16% higher yielding than McGregor at 1 location in year 1, 9% higher at 1 location in year 2, 18% higher at 3 locations in year 3, 4% higher at 8 locations in year 4 and only in year 5 did it yield less than McGregor at 13 locations (- 3%). However, F86343 is significantly earlier maturing by 5–6 days than McGregor and this early maturity may have hurt its yield in the long, cool growing season of 1992.

Not only was phenotypic variation noted in the McGregor-derived somaclonal lines but also molecular changes. Preete (1991) examined 20 somaclones and found 9 lacked two minor repeat length classes of 18–25S rDNA found in McGregor. Three somaclone lines had an extra 5S rDNA repeat and one had lost a *Bam HI* site within the 5S rDNA repeats.

Many of the somaclonal variants generated at the Crop Development Centre were examined for novel fatty acid profiles. In no case was a large stable change in fatty acid profile found. There were only small changes which had some effect on the iodine value of the oil.

Chemical mutagenesis

Linseed is an industrial oil crop that differs from edible oilseeds such as rapeseed, sunflower and soybean by the very high level of linolenic acid (alpha-linolenic acid) in its oil. Linseed oil is a drying oil used almost entirely in the manufacture of linoleum flooring, paints, varnishes, inks and other coatings. Linseed oil polymerizes with oxygen to form a film (coating). The rapid oxidation of linseed oil leads to rancidity and is the reason why it is not commercially used for food or cooking purposes.

If linseed oil is to become an edible oil it is obvious that the linolenic acid level in linseed has to be reduced to, or below, that of other food seed oils. To produce varieties of edible-oil linseed for western Canada, the Crop Development Centre began a mutation program in 1987. Twenty thousand seeds of the Canadian variety McGregor were treated with EMS (ethyl methanesulphonate) following the method outlined by Green & Marshall (1984). Seed from the M_1 plants was screened for linolenic acid levels using the thiobarbituric acid-gas liquid chromatograph combination procedure (Bhatty & Rowland, 1990). Three mutant lines were initially identified (Rowland & Bhatty, 1990), and following selection, had stable fatty acid profiles by the M_5 generation (Table 3).

Mutant E1747 has a linolenic acid level of 2% and closely resembles the low linolenic acid genotype recovered by Green (1986) in the Australian linseed variety Glenelg. The low linolenic acid character is controlled by the recessive alleles of two independent, partially dominant linoleic acid desaturase genes and these mutations had apparently occurred in a single M_1 seed (Rowland, 1991).

The mutant E67 has palmitic acid levels of approximately 28%, which is 3–4 times greater than the parent line McGregor and 3 times greater than any previously reported palmitic acid levels in *L. usitatissimum*. E67 also had a significant amount of palmitoleic acid which has been previously reported in only trace amounts in linseed. Ntiamoah (1993) showed that the high palmitic acid level of E67 is controlled by a single partially dominant gene. This gene was inherited independently of the two linoleic desaturase genes. There was also a direct relationship (r = 0.94) between the palmitic and palmitoleic acid levels whereby a high level of palmitoleic acid resulted from the elevation of its precursor, palmitic acid.

The low linolenic acid character of E1747 could be combined with the high palmitic-palmitoleic character of E67 to produce lines with, for example, the fatty acid profile of 26% palmitic, 3% palmitoleic, 2% stearic, 16% oleic, 51% linoleic and 2% linolenic. This type of oil should be suited for the direct manufacture of margarine.

The high linoleic acid mutant, E1929, has oleic acid levels twice that of McGregor and this is under the control of a single partially dominant gene. However, genetic analysis has shown this gene to be an allele of one of the linoleic dessaturase genes and thus does not directly affect oleic acid levels.

A number of other mutations were also noted (Table 4). In addition to the fatty acid mutations, other valuable mutants were those affecting seed colour. In particular, those mutants having a variegated seed colour can be used to distinguish varieties or classes of varieties.

Genetic engineering

Linseed has proved to be a species that can be relatively easily transformed at the cellular level using *Agrobacterium tumefaciens* technology (McHughen, 1992). Many different linseed genotypes have been transformed with a wide range of *Agrobacterium* strains carrying many different gene constructs. Successful

Table 3. Fatty acid composition (%) of the M_5 seed of three mutant lines of McGregor and the parent variety McGregor

Line	Fatty acid					
	Palmitic	Palmitoleic	Stearic	Oleic	Linoleic	Linolenic
E67	27.8	4.8	1.8	17.5	6.0	42.0
E1747	9.5	trace	4.6	15.6	65.3	2.1
E1929	9.5	trace	3.4	51.7	16.3	16.2
McGregor	9.4	trace	5.1	18.4	14.6	49.5

Table 4. Mutant types found in the progeny of 1200 M_1 plants derived from the treatment of McGregor linseed with EMS

Mutant type	Number of lines
Chlorophyll	37
White flowers	25
Seed colour	12
Fatty acids	7
Early maturity	5
Pale blue flower	1
Dwarf plant	1

transfer of genes for tolerance to glyphosate (Jordan & McHughen, 1988), sulfonylurea (McHughen, 1989) and phosphinothricin (McHughen, in prep.) herbicides are good examples of what can be done.

In the case of the sulfonylurea herbicide resistance, the gene was isolated from *Arabidopsis* (Haughn et al., 1988) and inserted into flax (McHughen, 1989). The best transformed lines appeared to have commercially-useful levels of resistance to sulfonylurea herbicides (McHughen & Holm, 1991). In addition, the T-DNA does not appear to have affected agronomic fitness of the transformed lines, as in the absence of the herbicide there was no performance cost associated with the inserted plasmid (McHughen & Rowland, 1991). The two most promising sulfonylurea-resistant lines evaluated in registration testing in Canada have not been differentiated from their 'parental' variety, Nor-Lin (Table 5). It is expected that one of these two lines will be registered for commercial production in Canada in 1994.

Since linseed is relatively easily transformed and is not harmed by the *Agrobacterium* technique, it was decided to launch a programme to further manipulate this species using genetic engineering. In particular, the additional manipulation of the fatty acid profile of the species was thought feasible and desirable.

The high-palmitic/low-linolenic germplasm described earlier provides the opportunity to produce a cocoa-butter replacement oil. Anti-sense RNA technology will be used to suppress the activity of stearoyl-ACP desaturase in this germplasm. This should have the effect of increasing the stearic and oleic acid contents so that the palmitic, stearic and oleic levels are of near equal concentration. Such a linseed oil would be a dependable domestic source of SUS (saturated-unsaturated-saturated) triglycerides and should find use in the candy industry.

A domestic source of a vegetable oil high in palmitic acid also has a great attraction in Canada for the manufacturers of high quality margarines. Utilizing the same high-palmitic/low-linolenic germplasm, an attempt will be made to over-express the enzyme β-ketoacyl-ACP synthetase III. This should result in an even higher level of palmitic acid.

In addition to these fatty acid modifications stress tolerance genes will be engineered into linseed. For example, boiling stable or heat stable proteins have been isolated by Robertson et al. (in press) from bromegrass (*Bromus inermis*) and these proteins have been implicated in heat, drought, salinity and cold tolerance; all of these stresses are found on the Canadian prairies. On the basis of the amino acid sequence of these proteins the genes responsible will be cloned and transformed linseed plants subsequently produced.

Conclusion

The linseed plant is an important renewable source both of industrial oil and fibre and is well-adapted to production at high latitudes. However, the world production of

Table 5 The performance of two transgenic sulfonylurea resistant linseed lines compared to their parental variety NorLin in registration tests conducted in western Canada in 1991 and 1992

Line	Yield (kg/ha)	Days to mature	1000 seed wt (g)	Oil (%)
NorLin	1960	109 2	5 9	44 1
12115	1910	108 4	5 9	43 8
12140	1900	108 7	5 8	43 7
No of tests	21	18	23	23

linseed declined in the latter half of this century, largely because it has not been part of the increasing world demand for edible oils (Rowland, 1993). The facility to manipulate linseed described above should allow linseed oils into new market areas and provide agronomic advantages for the production of the crop.

Acknowledgements

The authors wish to thank The Agriculture Development Fund of Saskatchewan, The Saskatchewan Wheat Pool and National Research Council of Canada for support of this research.

References

Bhatty, R S & G G Rowland, 1990 Measurement of α-linolenic acid in the development of edible oil flax JOACS 67 364–367

Green, A G, 1986 A mutant genotype of flax (*Linum usitatissimum* L) containing very low levels of linolenic acid in its seed oil Can J Plant Sci 66 499–503

Green, A G & D R Marshall, 1984 Isolation of induced mutants in linseed (*Linum usitatissimum*) having reduced linolenic content Euphytica 33 321–328

Haughn, G, J Smith, B Mazur & C Sommerville, 1988 Transformation with a mutant *Arabidopsis* acetolate synthase gene renders tobacco resistant to sulfonylurea herbicides Mol Gen Genet 211 266–271

Jordan, M & A McHughen, 1988 Glyphosate tolerant flax plants from *Agrobacterium* mediated gene transfer Plant Cell Rpts 7 281–284

Kenaschuk, E O, 1975 Flax breeding and genetics p 203–221 In J T Harapiak (Ed) Oilseed and Pulse Crops in Western Canada – A Symposium Western Co operative Fertilizers Ltd, Calgary, Alberta

Larter, E N, A Wenhardt & R Gore, 1965 Redwood 65, an improved flax variety Can J Plant Sci 45 515–516

McHughen, A, 1987 Salt tolerance through increased vigor in a flax line (STS II) selected for salt tolerance *in vitro* Theor Appl Genet 74 727–732

McHughen, A, 1989 *Agrobacterium* mediated transfer of chlorsulfuron resistance to commercial flax cultivars Plant Cell Rpts 8 445–449

McHughen, A, 1992 Genetic engineering for crop improvement the linseed/flax story AgBiotech News and Inform 4 35N–56N

McHughen, A & F R Holm, 1991 Herbicide resistant transgenic flax field test agronomic performance in normal and sulfonylurea-containing soils Euphytica 55 49–56

McHughen, A & G G Rowland, 1991 The effect of T-DNA on the agronomic performance of transgenic flax plants Euphytica 55 269–275

McHughen, A & M Swartz, 1984 A tissue-culture derived salt-tolerant line of flax (*Linum usitatissimum*) J Plant Physiol 117 109–117

Ntiamoah, C, 1993 Inheritance and characterization of EMS-induced fatty acid mutations in McGregor flax MSc thesis, University of Saskatchewan 132 pp

O'Connor, B J, A J Robertson & L V Gusta, 1991 Differential stress tolerance and cross adaptation in a somaclonal variant of flax J Plant Physiol 139 32–36

Preete, T D, 1991 Somaclonal variation in 18–25S rRNA genes of flax (*Linum usitatissimum*) MSc thesis, University of Saskatchewan 123 pp

Robertson, A J, M Ishikawa, S L MacKenzie & L V Gusta Abscisic acid induced heat tolerance in *Bromus inermis* Leyss cell suspension cultures, heat stable, ABA-responsive polypeptides in combination with sucrose confer enhanced thermostability Plant Physiol 105 823–830

Rowland, G G, 1991 An EMS induced low linolenic-acid mutant in McGregor flax Can J Plant Sci 71 393–396

Rowland, G G, 1993 The application of chemical mutagenesis to seed oil modification the example of flax p 164–170 In S L MacKenzie & D C Taylor (Eds) Seed Oils for the Future American Oil Chemists' Society Champaign, Illinois

Rowland, G G & R S Bhatty, 1990 Ethyl methanesulphonate-induced fatty acid mutations in flax JAOCS 67 213–214

Rowland, G G, A McHughen & C McOnie, 1988 Field evaluation on nonsaline soils of a somaclonal variant of McGregor flax selected for salt tolerance *in vitro* Can J Plant Sci 68 345–349

Rowland, G G, A McHughen & R S Bhatty, 1989 Andro flax Can J Plant Sci 69 911–913

Rowland, G G, A McHughen & C McOnie, 1989 Field evaluation at saline affected sites of a somaclonal variant of McGregor flax selected for salt tolerance *in vitro* Can J Plant Sci 69 49–60

Euphytica **85:** 323–327, 1995.

Modification of oilseed rape to produce oils for industrial use by means of applied tissue culture methodology

A. Craig & S. Millam
Scottish Crop Research Institute, Invergowrie, Dundee DD2 5DA, U.K.

Key words: *Brassica napus*, fatty acids, gas chromatography, *Lunaria annua*, protoplast regeneration, somaclonal variation

Summary

A programme of research was designed to investigate methods for the modification of the fatty acid profiles of high performance lines of oilseed rape (*Brassica napus* L.) in an attempt to produce lines with enhanced levels of industrially useful fatty acids. The methodology employed to achieve these objectives was based on the exploitation of somaclonal or protoclonal variation, and targeted somatic hybridization using wild cruciferous germplasm as fusion partners.

A range of somaclonal lines was produced from shoot regeneration protocols. These lines underwent replicated, randomised glasshouse trials for morphological assessment followed by gas chromatographic analysis to monitor any changes in fatty acid profile. It was found that a small number of lines exhibited potentially useful changes in oleic acid and polyunsaturated fatty acid content. Protoplast regeneration and electrofusion protocols for a range of winter oilseed rape lines were developed, and methods for the isolation and fusion of protoplasts of the wild crucifer *Lunaria annua* (chosen for its high nervonic acid content) established.

Abbreviations: GC – Gas Chromatography

Introduction

The current commercial production of vegetable oils is primarily targeted towards their utilisation as an edible component of the human diet. The worldwide market for vegetable oils has increased by 300% since 1960, not only due to increases in world population but also to increases in living standards (Hills & Murphy, 1991). A further interesting factor is that the market share supplied by animal-derived fats fell from 39% of the total in 1960 to 26% in 1990, in some part due to an increased awareness of dietary factors (Murphy, 1992). Though the primary market for vegetable oils is for human consumption, up to 33% of the total production is designated for a range of non-edible applications, such as lubricants, plasticisers and detergent ingredients. The potential for such diversification is an area of considerable recent interest (for reviews, see: Hills & Murphy, 1991; Robbelen, 1991; Murphy, 1992). There are a number of approaches for obtaining sources of oils containing the fatty acid profiles required by industry. One approach is to search for wild undomesticated species that contain oils of interest and attempt to domesticate them. Many candidates have been suggested for this purpose, including *Coriandrum sativum* (Griffiths et al., 1992) and, for others, see Murphy (1992). Another strategy is to attempt to modify the oil profile of existing crops. This second approach is often favoured, as many of the potential undeveloped species have a number of undesirable traits, such as seed shattering, that would be difficult to breed-out by conventional means. Existing crop plants, however, would require little or no changes to existing agronomic practices (Millam et al., 1994).

Oilseed rape (*Brassica napus* L.) is the most important oil crop grown in north-western Europe, and the area grown in the EC and Canada exceeds 12 million acres (Kidd, 1993). The oil component of rapeseed has been manipulated by conventional breeding techniques to give a high percentage of oleic (C18 : 1) and other medium-chain fatty acids which are desirable for human nutritional purposes. However, the fatty acid profile of oilseed rape has great plasticity; for instance, there has been considerable commercial interest in the development of high erucic acid lines by conventional breeding targeted towards industrial end-use. There furthermore exists, within the Cruciferae, the potential to increase the narrow genetic background of rapeseed by somatic hybridization with closely or distantly related wild species, a number of which carry traits of interest (Glimelius et al., 1989), thus circumventing sexual incompatibility barriers. *Brassica napus* has also been found to be amenable to a range of plant tissue culture manipulations such as protoplast regeneration (Glimelius, 1984), protoplast fusion (Pelletier et al., 1983) and *Agrobacterium*-mediated transformation (Millam, 1989).

This paper describes the application of the tissue culture methodologies of somaclonal variation, protoplast isolation and the regeneration of protoclones, and prospective protoplast fusion of selected wild species to *Brassica napus* with the objective of manipulating the oil profile of the crop.

Materials and methods

Somaclone production

In vitro cultures of *Brassica napus* lines were established according to the methods of Millam et al. (1991). One spring type (cv. Topas), three winter types (cvs Samourai, Libravo and Envol), and two high erucic acid rapeseed lines (cvs Martina and Askari) were maintained *in vitro* on basal Murashige & Skoog (1962) medium, supplemented with 20.0 g l^{-1} sucrose and 8.0 g l^{-1} Difco Bitek agar under growth conditions of 22° C, 8 h dark : 16 h light at a photosynthetic photon flux density of 140 μmol m^2 sec^{-1}. The regeneration systems employed to produce a shooting response from stem and leaf explants were those of Millam et al. (1991) and Williams et al. (1991), respectively.

Protoclone production

Protoplasts were isolated and regenerated from mesophyll and hypocotyl material of all the lines of *B. napus* used in this study. A protocol was developed based on a combination of the methods of Pelletier et al. (1983), Barsby et al. (1986) and Loudon et al. (1989) with some minor modifications to the timing of application of the shoot induction medium.

Protoplast fusion studies

A range of germplasm was screened using the Gas Chromatographic techniques described below for their constituent fatty acid profiles. On this basis, the species *Lunaria annua* was selected for further studies and application to a somatic hybridization programme with *B. napus*. A population of *Lunaria annua* plants was maintained under environmental conditions of day 22° C, 16 h light at 100 μmol m^2 sec^{-1} : night 18° C, 8 h dark. Rapidly expanding leaf material was used for protoplast isolation and, following surface-sterilisation by a method comprising 30 sec^{-1} in 70% v : v ethanol : water, 10 minutes in 10% v : v Domestos (Lever Brothers) : water followed by five washes in sterile distilled water, the leaves were plasmolysed in 13% mannitol for 30 minutes. All other procedures followed the methods employed for *Brassica* hypocotyl protoplast isolation, except the final flotation stage utilised 30% sucrose. Protoplasts of *Lunaria* were fused with *B. napus* (cv. Askari) hypocotyl protoplasts. Electrofusion was performed using a Kruss Biojet C50 apparatus under conditions of an AC field of 12 v, with a 110 v pulse of 50 μsec^{-1} duration. Chemical fusion was achieved by a PEG/Ca Cl_2 method adapted from that described by Barsby et al. (1987) with an additional wash step in culture medium.

Glasshouse trials of regenerated material

Replicated trials of regenerated somaclonal and protoclonal material of oilseed rape cv. Topas were established under glasshouse conditions using controls of seed-derived and micropropagated material. A range of morphological assessments was made including leaf colour, time-to-flower, and number of seeds per plant and samples of the seed further analysed using GC. The data was analysed using GENSTAT.

Gas Chromatography (GC) analysis

Fatty acids were extracted as their methyl ester derivative by the methods of Welch (1977). Following this extraction, the methyl esters were dissolved in petroleum ether and analysed using a Perkin Elmer Capillary GC 8420 instrument. The programme variables were experimentally derived using a synthetic

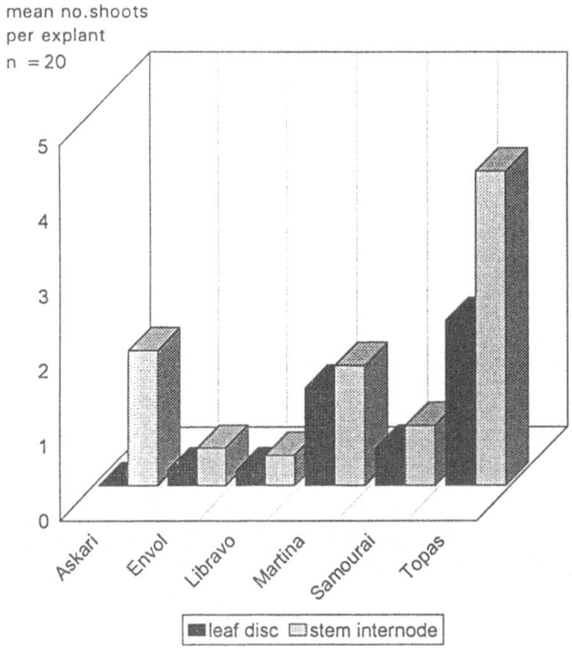

Fig. 1. Genotype: explant interaction in a range of *Brassica napus* cultivars. mean no. shoots per explant n = 20.

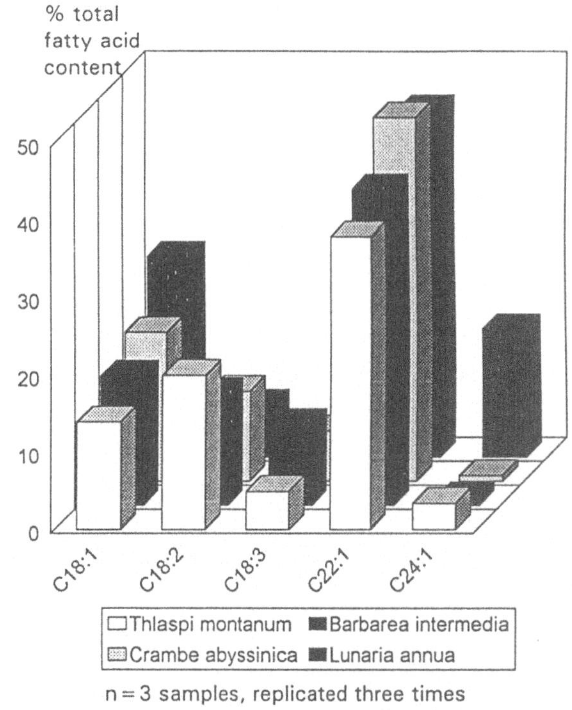

Fig. 2. Fatty acid profiles of four cruciferous species. % total fatty acid content. n = 3 samples, replicated three times.

oilseed rape fatty acid sample (obtained from Sigma, Poole, UK) and, in all experimental assays, an internal standard of heptadecanoic acid (C17) was used to enable quantification of samples. The chromatograms were recorded and analysed using a PE Nelson Systems PC integrator package.

Results

Somaclone production

Experiments indicated that certain regeneration protocols were favoured for specific explant sources. The most efficient explant : medium interaction was used for further studies and the results are presented in Fig. 1.

Protoplast regeneration studies

Protoplasts were readily isolated and purified from all the cultivars tested. The modified culture protocols resulted in differing degrees of regeneration response (summarised in Table 1). The time taken from isolation to regeneration of an intact plant varied from 12 to 26 weeks. It was found that such factors as the developmental stage of the protoplast-derived calli on transfer were critical to subsequent organogenesis.

Table 1. Response of the range of cultivars of *B. napus* investigated (- = no response; +++ = high efficiency repeatable regeneration)

Cultivar		Response		
Source	Tissue	Division	Calli	Shoots
Askari	mesophyll	+	+++	+++
Envol	mesophyll	+	+	-
Envol	hypocotyl	+	+	-
Libravo	mesophyll	+	-	-
Libravo	hypocotyl	+	+	-
Martina	mesophyll	++	++	++
Martina	hypocotyl	+	+	-
Samourai	mesophyll	+	+	+
Samourai	hypocotyll	+	+	-
Topas	mesophyll	++	++	++
Topas	hypocotyl	++	++	++

Table 2. Variability of oleic acid content as a 'marker' for diversification within a population of seed-derived, micropropagated and somaclonally derived *B. napus* plants

Oleic acid %	Seed-derived	Micropropagated	Somaclonal
min	39.60	38.60	35.70
max	61.40	63.90	71.00
mean	49.84	56.85	52.63
s.e.	6.54	8.11	8.95

n = 40 replicated twice.

Table 3. Initial fusion frequencies between mesophyll protoplasts of *Lunaria annua* (L.) and hypocotyl protoplasts of *Brassica napus* (B). Values presented are % of total number of cells in sample that were fused

	Electrofusion	PEG-mediated
L × B	1.90 + 0.2	14.00 + 1.2
L × L	1.20 + 0.2	9.33 + 1.83
B × B	1.20 + 0.3	6.67 + 1.48

n = 10 samples with four fusion events in each.

Glasshouse assessments

For the traits or leaf shape, colour and apical dominance there were found to be differences in the relative proportions of each phenotype between the group (manuscript in preparation). Results from the preliminary seed fatty acid profiles suggested that the overall levels of variability attributable to the somaclonal treatment were higher than the seed or micropropagated derived material (see Table 2). A more detailed analysis of individual somaclones revealed a number of interesting variants. Among these were three lines showing significantly enhanced oleic acid levels, and one line showing a significant reduced oleic acid content but enhanced linoleic and linolenic acid content.

Protoplast fusion studies

Experiments were devised to assess the inherent variability of oil profile within seed samples of the species under study. It was found that samples as small as ten seeds were experimentally acceptable, though for some of the wild crucifers, due to the very small size of the seeds, larger numbers were used. Of the species screened (see Fig. 2), it was decided to initiate a programme of research using *Lunaria annua*. This was due to the high levels of nervonic acid present in this species. Nervonic acid (C24) has a number of potentially valuable industrial and medical applications. Protoplasts were readily isolated from this species, and fusion efficiencies, using *Brassica* hypocotyls as the fusion partner, are shown in Table 3. The rates of fusion utilising the PEG/$CaCl_2$ method were repeatedly high in comparison with those employed using electrofusion. However, the electrofused protoplasts successfully divided and reached the callus stage.

Discussion

The plasticity of fatty acid composition in seed oils has been demonstrated in previous studies such as in the removal of erucic acid from rapeseed oil to create 'Canola' (Stefansson et al., 1961) and in a range of other oil crop plants such as the reduction of linolenic acid content in flax seed (Green, 1986) and the increase in stearic acid up to 30% total oil content in safflower (Ladd & Knowles, 1970). Systems of inducing variation through callus differentiation have been used previously to alter the lipid composition and to study triacylglycerol synthesis in *B. napus* (Williams et al., 1991). Our investigation was not designed to monitor specific changes *in vitro* but to induce changes in the seed oil content due to alterations attributable to some form of somaclonal variation during regeneration. The genotypic dependency in relation to specific regeneration systems observed in our preliminary studies has been reported previously (Khehra & Mathias, 1992) and this influenced the availability of material for glasshouse experiments. The changes observed, however, offer encouragement that, by increasing the numbers of plants regenerated by such means and subsequently trialled, more potentially useful changes in the fatty acid profiles may occur. Results from the primary glasshouse trial of somaclones indicate that useful forms of variation can arise from our approach. The progeny of the potentially useful variants derived from our preliminary study are undergoing more detailed analysis. As well as commercially useful mutations, some of the lines derived from our study may have application in investigations into the fatty acid biosynthetic pathways of triacylglycerol synthesis.

Another approach to inducing genetic rearrangement is by means of protoplast fusion. Such techniques have been used as a tool for the transfer of specific traits to *Brassica napus* (Glimelius et al., 1989) and the same group reported the creation of intergeneric and intertribal somatic hybrids within the Cruciferae. The use of *Lunaria* in tissue culture studies has not been reported previously. The species itself would be unsuitable for domestication due to its biennial nature (though some annual forms exist; Wellensiek, 1973), and it is consequently more desirable to integrate the long chain fatty acid properties directly into oilseed rape for which the agronomy is well established. The protoplast regeneration systems developed in this study for commercial rapeseed lines, allied to the preliminary fusion capacity demonstrated, offer the potential for the creation of novel material of use in further genetic improvement programmes.

This paper is a preliminary report of work in progress, but the results to date suggest that our approaches, based on tissue culture technology, may complement the gene transfer/genetic transformation approaches to modifying the oil content of oilseed rape being undertaken elsewhere. Material derived from our study may be more directly applicable to targeted breeding programmes and would result in the creation of oilseed rape with fatty acid profiles of immediate industrial application in a reduced period of time.

Acknowledgements

This work is funded by a Ministry of Agriculture, Fisheries and Food Open Contract. SCRI is supported by the Scottish Office Agriculture and Fisheries Department. The *Lunaria* data was derived from MSc projects by Campbell Harvey and Irene Tierney.

References

Barsby, TL , S A Yarrow & J F Shepard, 1986 A rapid and alternative procedure for the regeneration of plants from hypocotyl protoplasts of *Brassica napus* Plant Cell Rep 5 101–103

Barsby, TL, PV Chuong, S A Yarrow, Wu Sau-Ching, M Coumans, R J Kemble, A D Powell, WD Beversorf & K P Pauls, 1987 The combination of Polima cms and cytoplasmic traizine inheritance in *Brassica napus* Theor Appl Genet 73 809–814

Glimelius, K , 1984 High growth rate and regeneration capacity of hypocotyl protoplasts in some *Brassicaceae* Physiol Plant 61 38–44

Glimelius, K , J Fahlesson, M Landgren, C Sjodin & E Sunderg, 1989 Somatic hybridization as a means to broaden the gene pool of cruciferous oil plants Sverig Utsadd Tidskrift 99 103–108

Green, A G , 1986 A mutant genotype of flax containing very low levels of linolenic acid in its seed oil Can J Plant Sci 66 499–503

Griffiths, D W, G W Robertson, S Millam & A C Holmes, 1992 The determination of the petroselenic acid content of *Coriandrum sativum* oil by gas chromatography Phytochem Anal 3 250–253

Hills, M J & D T Murphy, 1991 Biotechnology of oilseeds Biotech Gen Eng Rev 9 1–45

Khera, G S & R J Mathias, 1992 The interaction of genotype, explant and media on the regeneration of shoots from complex explants of *Brassica napus* L J Exp Bot 43(256) 1413–1418

Kidd, G , 1993 Is pursuing improved canola an unctuous aim? Bio/Technology 11 448–449

Ladd, S L & PF Knowles, 1970 Inheritance of stearic acid in the seed oil of safflower Crop Sci 10 525–527

Loudon, PT, R S Nelson & D S Ingram, 1989 Studies of protoplast culture and plant regeneration from commmercial and rapid cycling *Brassica* species Plant Cell Tiss Org Cult 19 214–224

Millam, S , 1989 *Agrobacterium*-mediated transformation of *Brassica* species Asp Appl Biol 23 23–30

Millam, S , D Davidson, Wen Lanying & W Powell, 1991 A protocol for efficient tissue culture regeneration of rapid-cycling *Brassicas* Biotechnol Educ 2(2) 63–64

Millam, S , A Craig & WW Christie, 1994 Case studies in the investigation of potential industrial oil crops Annual Report SCRI 1993 10–14

Millam, S , A Craig, C Harvey, G R Mackay & I M Morrison, 1994a Modification of oilseed rape by means of applied tissue culture methodology p 198–202 In S Hennink, L J M van Soest, R Pithan & L Hof (Eds) Alternative Oilseed and Fibre Crops for Cold and Wet Regions of Europe Guyot, Brussels

Murashige, T & F Skoog, 1962 A revised medium for rapid growth and bioassays with tobacco tissue culture Physiol Plant 15 473–497

Murphy, D T, 1992 Modifying oilseed crops for non-edible products Trends in Biotech 10 84–87

Pelletier, G , C Primard, F Vedel, P Chetrit, R Remy, P Rousselle & M Renard, 1983 Intergeneric cytoplasmic hybridization in Cruciferae by protoplast fusion Mol Gen Genet 191 244–252

Robbelen, G , 1991 The genetic improvement of seed oil Chemistry and Industry, p 713–716

Stefansson, B R , FW Hougen & R K Downey, 1961 Note on the isolation of rape plants with seed oil free from erucic acid Can J Plant Sci 41 218–219

Welch, R W , 1977 A micro method for the estimation of oil content and composition in seed crops J Sci Food Agric 28 635–638

Wellensiek, S J , 1973 Genetics and flower formation of annual *Lunaria* Neth J Agric Sci 21 163–166

Williams, M , D Francis, A C Hann & J C Harwood, 1991 Changes in lipid composition during callus differentiation in cultures of oilseed rape (*Brassica napus* L) J Exp Bot 42(245) 1551–1556

Euphytica **85**: 329–334, 1995.

Selection of hydroxyproline-resistant proline-accumulating mutants of cauliflower (*brassica oleracea* var. *botrytis*)

C.R. Deane, M.P. Fuller[1] & P.J. Dix
Department of Biology, St. Patrick's College, Maynooth, Co. Kildare, Ireland; [1] *Seale-Hayne Faculty of Agriculture, Food and Land Use, University of Plymouth, U.K.*

Key words: curd culture, frost tolerance, N-nitroso-N-ethylurea, proline accumulation, mutagenesis

Summary

A procedure is described by which hydroxyproline-resistant lines could be selected from regenerating curd tissue of cauliflower. Mutagenesis was by N-nitroso-N-ethylurea, supplied as a drop of 0.3 mM solution on each 3 mm diameter curd piece. The mutagen generated numerous morphological and pigment mutations without significantly affecting shoot regeneration from explants. Thirty one resistant shoots were recovered from more than six thousand explants mutagenised on regeneration medium supplemented with 3 mM hydroxyproline, while none was obtained from a similar number of non-mutagenised controls. Out of twenty-three resistant shoots which survived subculture, only one showed consistently elevated levels of endogenous proline. During early shoot culture passages, proline levels were 3.6–4.7 times higher than controls, but this was reduced to 1.6 times after 10–12 culture passages in the absence of hydroxyproline. Possible reasons for this decline are discussed. Leaf strip assays suggest resistant shoots may be chimeras and current efforts are directed towards regenerating solid mutants from resistant sectors. These will then be evaluated for any alteration in frost tolerance.

Abbreviations: IBA – Indole-3-butyric-acid, NEU – N-nitroso-N-ethylurea

Introduction

The accumulation of free proline in response to environmental stress has been observed in plants (Aspinall & Paleg, 1981, Larher et al., 1982) and bacteria (Tempest et al., 1970). The significance of this accumulation in plants has been attributed to the suggested ability of proline to act as a protective agent for cytoplasmic enzymes (Aspinall & Paleg, 1981).

High proline content has been related to frost tolerance in a number of plant species (Aspinall & Paleg, 1981). In potato, leaf proline content and frost tolerance are correlated and exogenous application of proline increases frost tolerance (Van Swaaij et al., 1985). One way to obtain a proline-accumulating mutant is to select for proline analogue resistance. The mechanism thought to confer amino acid analogue resistance involves a mutation which results in an enzyme becoming feedback-insensitive resulting in over-production of the corresponding amino acid (Dix, 1986). This approach was first applied to plants by Widholm (1972a, 1972b), who isolated tryptophan-producing lines of tobacco by selecting for resistance to 5-methyltryptophan. Amino acid analogue-resistant mutants which overproduce the specific amino acid have been found in bacteria (Czonka, 1981; Sugiura & Kisumi, 1985) as well as plants (Widholm, 1976; Kueh & Bright, 1982; Van Swaaij et al., 1986; Mori et al., 1989). It has been reported that free proline-accumulating cell lines resistant to a proline analogue show increased resistance to stresses such as salt (Riccardi et al., 1983) and freezing (Van Swaaij et al., 1986, 1987). Van Swaaij et al. (1986) selected hydroxyproline-resistant potato callus which overproduced proline. Proline accumulation and increased frost tolerance were exhibited in the leaves of plants regenerated from this callus.

The present report describes studies with cauliflower in which *in vitro* mutagenesis and hydroxyproline resistance selection were used to obtain proline-accumulating shoots, as a strategy for improving the frost tolerance of this crop.

Materials and methods

Plant material

Four cauliflower (*Brassica oleracea* var. *botrytis*) F1 hybrid cultivars were used; 'Plana', a summer heading cultivar, 'Arbon' and 'Dova', two autumn heading cultivars, and 'Arcade', an overwintering cultivar. Plants were greenhouse-grown (15–20° C, with supplementary lighting during winter) and curd was freshly cut or stored for up to two weeks at 4° C. Florets, 3–4 cm in length, were surface-sterilised by soaking in 10% *Domestos* (a commercial disinfectant containing 5% calcium hypochlorite) for 10 minutes and rinsing three times in sterile distilled water. Pieces of curd approximately 3 mm in diameter were cut from the surface and placed onto medium in 9 cm diameter plastic Petri dishes (5–20 per dish).

Regeneration of shoots from curd

Pieces of curd were placed on regeneration medium stage I (Table 1). Dishes were sealed with parafilm and incubated in the culture room (23–25° C, 16 hour photoperiod, 2,000–3,000 Lux). Curd explants became green after about 1 week and shoots were separated as soon as they appeared, to prevent crowding. After 4–5 weeks on stage I medium, shoots were transferred to stage II (Table 1) and could be maintained on this indefinitely. To induce rooting, shoots were cut and transferred to either stage III medium (Table 1) or hormone-free RM medium (Murashige & Skoog, 1962) salts with 30 gl^{-1} sucrose, 6 gl^{-1} agar, pH 5.7.

Mutagenesis with N-nitroso-N-ethylurea (NEU)

Precautions taken in handling NEU and disposing of waste are as outlined in McCabe et al. (1990). A 0.3 mM solution was made with distilled water and dropped with a sterile Pasteur pipette onto each piece of curd (one drop per piece) on stage I medium. Controls were treated with sterile distilled water. Plates were sealed with parafilm and placed on a shelf lined with absorbent paper in the culture room. NEU was not washed out.

Table 1. Media used for culture of cauliflower curd. Murashige & Skoog (1962) plant salt mixture with the following additions

Thiamine		0.4 mgl^{-1}	
Adenine sulphate		80 mgl^{-1}	
Sodium dihydrogen orthophosphate		170 mgl^{-1}	
Sucrose		30 gl^{-1}	
	Stage I	Stage II	Stage III
IBA mgl^{-1}	1	2	2
Kinetin mgl^{-1}	2	4	0
Agar gl^{-1}	4	6	6

pH adjusted to 5.7 before autoclaving.

Selection for hydroxyproline-resistant shoots

Curd pieces were placed on stage I medium supplemented with hydroxyproline. When a survivor was identified (on NEU-treated plates) the piece of curd on which it arose was transferred to fresh medium with the same level of hydroxyproline. It was maintained on hydroxyproline for approximately 5 weeks to ensure that all non-resistant cells had been killed. It was then transferred to stage II medium without hydroxyproline for 3–4 weeks and then either to stage III medium or to RM.

Estimation of free proline content

Proline levels were measured by the method of Bates et al. (1973). Fully expanded leaves were selected from shoot cultures and homogenised in 3% sulfosalicylic acid. The filtered homogenate was allowed to react with acid ninhydrin and glacial acetic acid for one hour at 100° C. The reaction mixture was extracted with toluene. The chromophore containing toluene was aspirated from the aqueous phase and the absorbance read at 520 nm. The proline concentration was determined from a standard curve and calculated on a fesh weight basis.

Table 2. The survival of curd, both freshly cut and stored at 4° C for two weeks, on a range of hydroxyproline concentrations. There were 16 pieces in each dish and at least 4 dishes of each hydroxyproline concentration for each treatment and variety

mM Hyp.	% Survival					
	Plana		Arbon		Dova	
	Fresh	2 weeks	Fresh	2 weeks	Fresh	2 weeks
30	0	0	0	0	0	0
10	0	0	0	0	0	0
3	0	0	0	0	0	0
1	1.5	4.5	3.2	7.8	0	1.6
0.3	100	100	100	100	100	100
0.1	100	100	100	100	100	100
0	100	100	100	100	100	100

Results

Mutagenic effect of NEU

Three weeks after NEU treatment, fifty leaves were selected at random from shoots from each of ten batches of NEU and distilled water-treated curd. These leaves were scored for abnormalities. 41.6% of leaves from NEU-treated curd showed abnormalities in chlorophyll content or distribution and 26.2% showed morphological abnormalities. Less than 2% of leaves from distilled water-treated curd showed abnormalities.

NEU at this concentration did not have a significant effect on the number of shoots regenerated.

Selective level of hydroxyproline

Curd pieces were scored for survival (possession of chlorophyll) after three weeks on stage I medium supplemented with a range of hydroxyproline concentrations. In three cultivars tested (Plana, Arbon and Dova), there were no survivors on 3–30 mM hydroxyproline (Table 2). The lowest concentration giving 100% explant death (3 mM) was subsequently chosen for selection.

Duration of exposure to hydroxyproline

Curd tissue which was exposed to hydroxyproline for one week showed almost complete survival after three subsequent weeks on hydroxyproline-free medium, indicating that regeneration is merely suppressed by a short exposure to hydroxyproline. As the duration of exposure to hydroxyproline increased, so did the time taken for recovery. Pieces of curd 3 mm in diameter exposed for two weeks showed no sign of recovery after three weeks on hydroxyproline-free medium, but 7.5% survived after five weeks. A three week exposure to 3 mM hydroxyproline caused complete killing in 3 mm pieces, but larger pieces survived for longer and 17.5% recovered after five weeks on hydroxyproline-free medium. Therefore 3 mm pieces of curd, the size used in mutagenesis, must remain on hydroxyproline for at least three weeks to ensure complete killing.

Selection of hydroxyproline-resistant shoots

Twenty curd pieces were put into each Petri dish containing regeneration medium I supplemented with 3 mM hydroxyproline. Curd tissue was treated with either NEU or sterile distilled water. Dishes were checked for survivors initially one week after mutagenesis treatment and then weekly. The numbers of shoots apparent after 4 weeks are shown in Table 3. Twenty-three of the 31 hydroxyproline-resistant shoots survived subculture on stage II medium, but only two of them showed consistently higher levels of proline in initial tests (after one subculture). In later tests (after 5–6 and 10–12 subcultures), only one of them had consistently higher levels. These results are shown in Table 4.

Discussion

Results show that a low concentration of NEU (0.3 mM) applied to curd tissue and not washed out,

Table 3. Effect of NEU on shoot formation from curd pieces on medium supplemented with 3 mM hydroxyproline. The numbers of shoots which were produced from each treatment are shown

Treatment	Number of curd pieces		Number of shoots	
variety	NEU	H_2O	NEU	H_2O
Plana	160	220	2	0
Arbon	1600	1620	8	0
Dova	2880	2400	10	0
Arcade	1560	1440	11	0

Table 4. Proline levels in shoot cultures of hydroxyproline. resistant line of the variety Plana. Values are means of 5 shoot cultures (1st subculure) or 8 sublines (5–6 subcultures, 10–12 subcultures), and errors are the standard error in the mean. Controls are means of 4 or 6 independently maintained lines

No. of subculture	μmol. proline g^{-1} of leaf fresh weight		Accumulation*
	Control	Mutant	
1	251.1 ± 47.7	902.2 ± 117.8	3.6
5–6	130 ± 35.5	607.5 ± 101.2	4.7
10–12	189.6 ± 27.4	310.2 ± 28.3	1.6

* Factor by which mutant exceeds control.

does not significantly reduce the numbers of shoots produced but does cause mutations in the genes controlling chlorophyll biosynthesis and plant morphology.

Curd tissue exposed to hydroxyproline concentrations ranging from 3 mM to 30 mM does not survive after three weeks. A small percentage of those on 1 mM survive, however. In all three varieties tested, a higher percentage survive on 1 mM if the curd has been exposed to 4° C for two weeks prior to testing. It is possible that this low temperature pre-treatment induces proline accumulation in cells of the curd which protects the tissue from hydroxyproline.

Hydroxyproline at a concentration of 3 mM proved a suitable level for selecting resistant shoots, and 31 were recovered from a total of more than six thousand curd pieces. The importance of the mutagenesis treatment is indicated by the complete absence of resistant shoots in a similar number of non-mutagenised controls. The eight hydroxyproline-resistant shoots which did not survive subculture may have acquired mutations in genes crucial for survival. Some failed to form roots in culture, others simply did not grow.

Of the 23 shoots which survived, only one showed consistently higher levels of proline. It is possible that some mechanism other than proline accumulation for hydroxyproline resistance is in action in the other lines. Hasegawa & Mori (1986) reported the isolation of three rice mutants resistant to hydroxyproline but not proline-accumulating. Their resistance was found to be controlled by a single recessive nuclear gene. Resistance to amino acid analogues may also be due to a mutation causing decreased uptake of the analogue (Widholm, 1974) or preferential incorporation of the naturally-occurring amino acid (Negrutiu et al., 1978). Another possibility is that not all of the meristems were dead when the piece of curd on which the variant arose was transferred to hydroxyproline-free medium. Results indicate that after 3 weeks exposure to hydroxyproline, 3 mm pieces of curd show no sign of surviving after 5 weeks on hydroxyproline-free medium, but the number of pieces tested this way was small (40) in comparison with the numbers used in mutagenesis experiments.

Apart from the possibility of escapes – the curd system is attractive because most of the cells in a piece of curd are not killed immediately and may nurse resistant cells and shoot primordia through early stages of development.

The initial level of proline measured in the variant which consistently showed higher levels of proline was 3.6 times the control. After 5–6 subcultures, this value was 4.7 times the control. However, after 10–12 subcultures, the mean had dropped to 1.6 times the control. The reason for this reduction may be due to the absence of hydroxyproline in the culture medium. Amino acid over-production may gradually be reduced in the absence of the analogue leading to a reduction in resistance. Widholm (1976) showed that a carrot cell line selected for resistance to ethionine initially required over 1,000 times more ethionine for inhibition of growth than the controls. This resistance was reduced by growth in the absence of ethionine. Van Swaaij et al. (1986) selected hydroxyproline-resistant, proline-accumulating callus of potato, and found that both, shoots regenerated from it, and callus initiated from leaf and stem tissue of these plants, had lower levels of proline than the originally-selected line which had been continuously grown on hydroxyproline (but these levels were still significantly higher than those of the control). The mean proline concentration of the control shoot cultures in the present work also decreased over time, but this was not significant.

A possible explanation for the variation in proline content of shoot cultures which have originated from the same resistant shoot, is that the shoots may be chimeric. The surface of the curd consists of thousands of apical meristems with single cells between them which are capable of dividing to give meristems. Scanning electron microscope studies (Crisp & Walkey, 1974) indicate that each meristem may become vegetative and give rise to a shoot. It is possible that one cell of a pre-formed meristem may have acquired the mutation resulting in over-production of proline, conferring resistance on the whole meristem. Continuous subculturing in the absence of selection may lead to segregation to give plants with accumulating and non-accumulating sectors.

Investigations underway support the chimeric nature of leaves of the proline-accumulating, hydroxyproline-resistant variant. Leaf stips on hydroxyproline-containing medium show resistant and sensitive sectors. Regeneration from resistant leaf sectors should give a solid variant in which frost tolerance will be assessed. If fertile plants are regenerated, the heritability of the trait can be studied proving whether or not a stable genetic change has occurred.

In vitro mutagenesis and selection has usually involved callus or cell suspension cultures (Collin & Dix, 1990). While the possibilities offered by regenerable explant cultures have been demonstrated (McCabe et al., 1989) the use of an organised system, such as cauliflower curd, is still rare. We believe the present report shows the approach can be effective and deserves more attention.

Acknowledgements

The authors would like to thank Eolas (The Irish Science and Technology Agency) for financial assistance for C.R. Deane during the course of this work.

References

Aspinall, B & L G Paleg, 1981 The Physiology and Biochemistry of Drought Resistance in Plants Academic Press, Sydney

Bates, S L , R P Waldren & I D Teare, 1973 Rapid determination of free proline for water stress studies Plant Soil 39 205–207

Collin, H A & PJ Dix, 1990 Culture systems and selection procedures In PJ Dix (Ed) Plant Cell Line Selection, Procedures and Applications, pp 3–18 VCH, Weinheim

Crisp, P & D G A Walkey, 1974 The use of aseptic meristem culture in cauliflower breeding Euphytica 23 305–313

Czonka, L N , 1981 Proline overproduction results in enhanced osmotolerance in *Salmonella typhimurium* Mol Gen Genet 182 82–86

Dix, PJ , 1986 Cell line selection In M M Yeoman (Ed) Plant Cell Culture Technology pp 143–201 Blackwell Scientific, Edinburgh

Hasegawa, H & S Mori, 1986 Non-proline-accumulating rice mutants resistant to hydroxy L-proline Theor Appl Genet 72 226–230

Kueh, J S H & S W J Bright, 1982 Biochemical and genetic analysis of three proline accumulating barley mutants Plant Sci Lett 27 233–241

Larher, F, Y Jolivet, M Briens & M Goas, 1982 Osmoregulation in higher plant halophytes organic nitrogen accumulation in glycine betaine and proline during the growth of *Asterripolium* and *Suaeda macrocarpa* under saline conditions Plant Sci Lett 24 201–210

McCabe, PF, A M Timmons & PJ Dix, 1989 A simple procedure for the isolation of streptomycin-resistant plants in the Solanaceae Mol Gen Genet 216 132–137

McCabe, PF, A Cseplo, A M Timmons & PJ Dix, 1990 Selection of chloroplast mutants In J W Pollard & J M Walker (Eds) Methods in molecular biology, vol 6, Plant Cell and Tissue Culture, pp 467–475 Humana Press, Clifton, New Jersey

Mori, S , H Hasegawa, R Che, H Nakanishi & M Murakami, 1989 Free proline contents in two different groups of rice mutants resistant to hydroxy-L-proline Theor Appl Genet 77 44–48

Murashige, T & F Skoog, 1962 A revised medium for rapid growth and bioassays with tobacco tissue cultures Physiol Plant 15 473–497

Negrutiu, I , M Jacobs & A Cattoir, 1978 Selection and characterisation of resistant mutants to amino acid analogs in *in vitro* plant cell culture In Abstr 4th Int Congr of Plant Tissue and Cell Culture Calgary, pp 138

Riccardi, G , R Cella, G Camerino & O Ciferri, 1983 Resistance to azetidine-2-carboxylic acid and sodium chloride tolerance in carrot cell cultures and *Spirula platensis* Plant Cell Physiol 24 1073–1078

Sugiura, M & M Kisumi, 1985 Proline-hyperproducing strains of *Serratia marcescens* Enhancement of proline analog-mediated growth inhibition by increasing osmotic stress Appl Env Microbiol 49 (4) 782–786

Tempest, D W , J L Meers & C M Brown, 1970 Influence of environment on the content and composition of microbial free amino acid pools J Gen Microbiol 64 171–185

Van Swaaij, A C , E Jacobsen & W J Feenstra, 1985 Effect of cold hardening, wilting and exogenously applied proline on leaf proline content and frost tolerance of several genotypes of *Solanum* Physiol Plant 64 230–236

Van Swaaij, A C , E Jacobsen, J A K W Kiel & W J Feenstra, 1986 Selection, characterization and regeneration of hydroxyproline-resistant cell lines of *Solanum tuberosum* Tolerance of NaCl and freezing stress Physiol Plant 68 359–366

Van Swaaij, A C , H Nijdam, E Jacobsen & W J Feenstra, 1987 Increased frost tolerance and amino acid content in leaves, tubers and leaf callus of regenerated hydroproline resistant potato clones Euphytica 36 369–380

Widholm, J M , 1972a Anthranilate synthetase from 5-methyltryptophan-susceptible and -resistant cultured *Daucus carota* cells Biochem Biophys Acta 279 48–57

Widholm, J M , 1972b Cultured *Nicotiana tabacum* cells with an altered anthranilate synthetase which is less sensitive to feedback inhibition Biochim Biophys Acta 261 52–58

Widholm, J.M., 1974. Cultured carrot cell mutants: 5-methyltryptophan resistance trait carried from cell to plant and back. Plant Sci. Lett. 3: 323–330.

Widholm, J.M., 1976. Selection and characterization of cultured carrot and tobacco cells resistant to lysine, methionine and proline analogs. Can. J. Bot. 54: 1523–1529.

Euphytica **85**: 335–339, 1995.

Analysis of tissue culture-derived variation in pea (*Pisum sativum* L.) – preliminary results

Miroslav Griga[1], Jaromír Stejskal[1] & Karel Beber[2]
[1] *OSEVA, Research Institute of Technical Crops & Legumes, 787 01 Šumperk, Czech Republic;* [2] *Research Institute for Cattle Breeding, Rapotín 267, 787 13 Vikýřovice, Czech Republic*

Key words: callus culture, organogenesis, pea, *Pisum sativum*, somaclonal variation, somatic embryogenesis

Summary

The possibility of producing agronomically-useful somaclones via organogenesis and somatic embryogenesis from callus cultures of pea (*Pisum sativum* L.) was studied. Organogenic calli were induced from immature leaflets on MSB medium with NAA and BAP. Embryogenic calli were derived either from immature zygotic embryos (using 2,4-D) or from shoot apices (using picloram) of aseptically-germinated seedlings.

The seed progenies (T_1 to T_3-generation) of primary regenerants were grown in field conditions and their phenotypic variation was evaluated and compared with control, non-tissue culture-derived plant material. In addition, electrophoretic analyses of selected isoenzyme systems and total proteins have been done. The results do not show dramatic changes in qualitative and quantitative traits. The evaluation of at least two future generations (T_4, T_5) is planned.

Abbreviations: BAP – 6-benzylaminopurine, IBA – indole-3-butyric acid, MSB – medium (mineral salts after Murashige & Skoog, 1962, vitamins after Gamborg et al., 1968), NAA – α-naphthalene-acetic acid, picloram – 4-amino-3,5,6-trichloro picolinic acid, 2,4-D – 2,4-dichlorophenoxyacetic acid, ORG – organogenesis, SE – somatic embryogenesis

Introduction

The production of fertile pea regenerants via organogenesis (Mroginski & Kartha, 1981; Hussey & Gunn, 1984) and somatic embryogenesis (Kysely et al., 1987; Lehminger-Mertens & Jacobsen, 1989; Kysely & Jacobsen, 1990; Tétu et al., 1990; Doorne et al., 1991; Stejskal & Griga, 1992; Özcan et al., 1993) made possible the evaluation of the genetic stability/instability of primary regenerants and their seed progenies. The majority of reports presented data from the T0 or T1-generation. Variation within the regenerants was found in ploidy level (Kysely et al., 1987; Natali & Cavallini, 1987), presence/absence of anthocyanin, leaf shape, plant habit (Lutova & Zabelina, 1988) total legumin content, legumin/vicilin ratio, amino acid balance (Mikhailova-Krumova et al., 1991), DNA content and degree of DNA methylation (Cecchini et al., 1992). On the contrary, Rubluo et al. (1984) observed no variation in regenerated pea plants (T1) for four isozyme systems and nine morphological traits.

In this paper we report the preliminary results of evaluation of pea regenerants and their seed progenies (T1 to T3) derived from organogenic and embryogenic calli (Griga, 1990; Griga et al., 1992; Stejskal & Griga, 1992).

Materials and methods

The line HM-6 of protein pea (*Pisum sativum* L.) from Plant Breeding Station Horní Moštěnice, Czech Republic was used in experiments.

In vitro culture, regeneration and production of somaclones

Organogenesis. The organogenic calli were induced from immature leaflets of three day-old aseptically-germinated seeds on MSB medium with 10 μM each BAP and NAA (Mroginski & Kartha, 1981). Regenerated shoots were successively isolated in the course of 5 months' callus culture. Shoots were rooted on 1/2 MSB medium with 1 μM NAA or IBA. Shoots from individual calli were labelled and multiplied by micropropagation on MSB medium with 20 μM BAP and 0.1 μM NAA (Griga et al., 1986). Each shoot (and consequent plantlet) obtained by micropropagation was marked as an individual somaclone (to take into account the possibility of variation caused by adventitious bud induction during the micropropagation phase). The T0 plants were grown in the greenhouse to flowering and seed set.

Somatic embryogenesis. Embryogenic calli were induced either from immature zygotic embryos (2.5 μM 2,4-D) or from shoot apices (2.5 μM picloram) of aseptically germinated seedlings (Kysely et al., 1987; Stejskal & Griga, 1992). Germinating somatic embryos were cut into individual nodes with axillary buds and micropropagated on the medium above. Subsequent procedures were the same as for material obtained via organogenesis.

The seeds harvested from individual T0 plants represented seed populations of the T1 generation for field evaluation.

Field trials

1991. Owing to unequal numbers of T1 seeds from individual somaclones, the T1 generation represented mainly multiplication of material. Morphological alterations were recorded during the vegetative period and selected yield parameters were analysed.

1992. T2 seed bulks harvested from T1 plants were sown in three replicates (SE, 100 plants per replicate) and in trial without replication (ORG, 40 plants per somaclone). Phenological and morphological observations were done during the vegetative period and five plants from each replicate (plot) were analysed for morphological traits and some yield parameters. The crude protein content of the seeds was analysed for each somaclone.

1993. T3 seeds harvested from T2 plants (both SE and ORG origin) of each somaclone were sown in three replicates (100 plants per replicate, SE; 60 plants per replicate, ORG) in randomly-designed plots. 20 plants from each replicate were harvested for morphological and yield evaluation. All data (1991, 1992, 1993) were statistically analyzed to compare the variation between experimental (somaclones) and control populations.

Electrophoretic analyses (total proteins, isozymes)

One randomly-selected pea plant per clone was analysed for leaf peroxidases (PRX) and esterases (EST) by means of isoelectric focusing, and for seed storage proteins by SDS-PAGE.

Results

17 plants of embryogenic origin and 62 plants of organogenic origin were obtained in T0-generation (primary regenerants), which served as starting somaclones.

Qualitative traits

Within the embryogenic somaclones (plants) of the T0 generation, a regenerant with a dramatically altered habit was observed (leaflet shape, one pair of leaflets, abnormal flower morphology, reduced flower stalk, shortened internodes, stipules without dentation and with tendrils). The plant exhibited a chimaeric character and was completely sterile. The transfer back to *in vitro* culture did not result in isolation of a stable mutant. Within T0 regenerants of organogenic origin no evident macromutations were found. Within the somaclones of the T1–T3 generation, mostly chimaeric plants with altered leaf morphology were observed at low frequency (deeply incised leaflet apices, presence of tendrils on leaflet apices, irregular position of leaflets on compound leaves, leaf fasciations). Stable mutants did not segregate in the progeny of these plants, however. Electrophoresis of seed proteins and leaf isozymes (PRX, EST) in T1 and T2 did not exhibit evident qualitative changes compared with the control. Improved tolerance to virus and fungus diseases in some organogenic clones should be precisely analysed in further generations.

Table 1. Means, extreme values and variation range ($R[\%] = (X_{max} - X_{min}/X \times 100$) of selected quantitative traits of somaclones and control plant of *Pisum sativum* L. line HM-6 in T1 seed generation (1991)

Trait	Mean	Extreme values	Variation range (%)	No. somaclones > mean controls
Organogenesis (62 somaclones)				
Pods per plant	21.50	1 – 60	274	41
Seeds per pod	3.70	1.00 – 6.63	152	34
Seeds per plant	80.50	2 – 187	230	42
Seed weight per plant (g)	21.30	0.3 – 45.7	213	37
TSW (g)	270.10	150.0 – 331.7	67	11
Somatic embryogenesis (16 somaclones)				
Pods per plant	16.30	1 – 32	190	7
Seeds per pod	3.64	1.00 – 4.33	91	10
Seeds per plant	63.10	0.1 – 134.3	213	8
Seed weight per plant (g)	17.90	3.0 – 40.9	212	7
TSW (g)	268.80	228.2 – 330.1	36	6
Control (8 replicated plots)				
Pods per plant	16.80	13 – 20	42	
Seeds per pod	3.65	3.29 – 4.15	24	
Seeds per plant	60.80	51.0 – 70.5	32	
Seed weight per plant (g)	17.80	13.3 – 21.6	47	
TSW (g)	301.40	258.2 – 350.9	31	

Quantitative traits

Preliminary evaluation of the T1 generation (unequal numbers of plants from individual somaclones) showed greater variation for evaluated traits compared with control populations (Table 1). Sufficiently large experimental populations in the T2 and T3 generation enabled more objective evaluation of quantitative traits, namely, plant height, number of internodes, first flowering node, number of branches, seeds per plant, seeds per pod, thousand seed weight (TSW), haulm weight, dry matter and crude protein content of seeds. The somaclones as a whole exhibited greater variation when compared with the control in almost all evaluated traits. Statistically significant increase in variation, however, was found only in TSW (T2, T3), seeds per pod (T2, T3), number of internodes (T3) and haulm weight (T3). Principal component analysis of quantitative traits in T2 and T3 generations (Figs 1 and 2) illustrates similar distribution of variation in embryo-

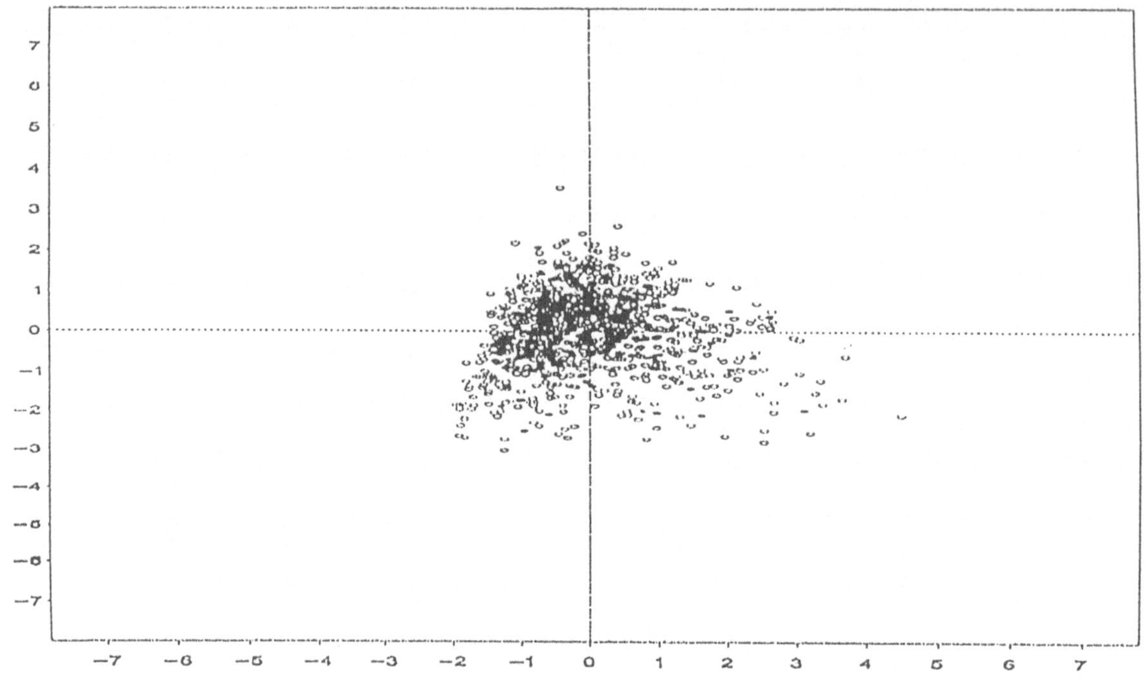

Fig 1 Principal component analysis of embryogenic somaclones in T3 generation

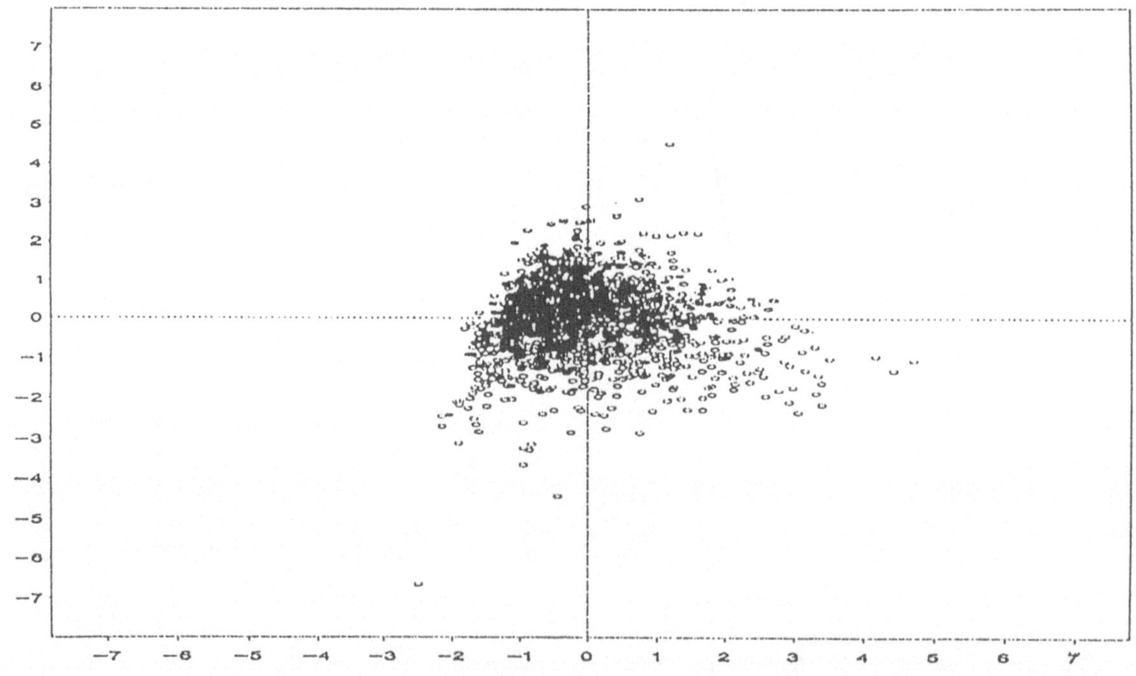

Fig 2 Principal component analysis of organogenic somaclones in T3 generation Empty spots – somaclones, black spots – control populations

genic and organogenic somaclones as for the control population.

Conclusion

Somatic embryogenesis and organogenesis *in vitro* may increase variability in pea, namely in quantitative traits; we have not observed dramatic alterations in qualitative traits in our recent experiments. Morphological changes in plants were linked with a chimaeric character and were lost in the next generation, or resulted in sterility. Some individual plants within the somaclones with significantly different quantitative traits (Figs 1 and 2), could be considered as a source for further genetic research and potential breeding utilization. For this, the critical evaluation of T4 and T5 generations will be necessary. Definitive theoretical and practical assessment of pea somaclonal variation for potential crop improvement, requires a sufficiently large amount of experimental material and proper evaluation methodology (Bajaj, 1990, de Klerk, 1990). The answers to these questions are the objectives of recent research in our laboratory.

References

Bajaj, YPS (Ed), 1990 Biotechnology in Agriculture and Forestry 11, Somaclonal Variation in Crop Improvement I Springer-Verlag, Berlin – Heidelberg – New York

Cecchini, E, L Natali, A Cavallini & M Durante, 1992 DNA variations in regenerated plants of pea (*Pisum sativum* L) Theor Appl Genet 84 874–879

De Klerk, G -J , 1990 How to measure somaclonal variation Acta Bot Neerl 39 129–144

Gamborg, O L , R A Miller & K Ojima, 1968 Nutrient requirements of suspension cultures of soybean root cells Exp Cell Res 50 151–158

Griga, M , 1990 The study of regeneration systems *in vitro* in pea, faba bean and soybean Ph D Thesis, Masaryk University, Brno 1990 (in Czech)

Griga, M , J Stejskal & H Klenotičova, 1992 Plant regeneration in soybean, pea and faba bean via somatic embryogenesis p 107–108 In Proc 1st Europ Conf on Grain Legumes Angers, France

Griga, M , E Tejklova, F J Novak & M Kubalakova, 1986 *In vitro* clonal propagation of *Pisum sativum* L Plant Cell Tissue Organ Cult 6 95–104

Hussey, G & H V Gunn, 1984 Plant production in pea (*Pisum sativum* L cvs Puget and Upton) from long-term callus with superficial meristems Plant Sci Lett 37 143–148

Kysely, W , J R Myers, P A Lazzeri, G B Collins & H J Jacobsen, 1987 Plant regeneration via somatic embryogenesis in pea (*Pisum sativum* L) Plant Cell Rep 6 305–308

Kysely, W & H J Jacobsen, 1990 Somatic embryogenesis from pea embryos and shoot apices Plant Cell Tissue Organ Cult 20 7–14

Lehminger-Mertens, R & H J Jacobsen, 1989 Plant regeneration from pea protoplasts via somatic embryogenesis Plant Cell Rep 8 379–382

Lutova, L A & E K Zabelina, 1988 Callus and shoot formation in different forms of *Pisum sativum* L *in vitro* Genetika 24 1632–1640 (in Russian)

Mikhailova Krumova, A B , I P Gavrilyuk, M I Azarkovich & R G Butenko, 1991 Study of somaclonal variations in pea seed proteins Fiziol Rast 38 521–529 (in Russian)

Mroginski, L A & K K Kartha, 1981 Regeneration of pea (*Pisum sativum* L cv Century) plants by *in vitro* culture of immature leaflets Plant Cell Rep 1 64–66

Murashige, T & F Skoog, 1962 A revised medium for rapid growth and bioassays with tobacco tissue cultures Physiol Plant 15 472–497

Natali, L & A Cavallini, 1987 Nuclear cytology of callus and plantlets regenerated from pea (*Pisum sativum* L) meristems Protoplasma 141 121–125

Özcan, S , M Barghchi, S Firek & J Draper, 1993 Efficient adventitious shoot regeneration and somatic embryogenesis in pea Plant Cell Tissue Organ Cult 34 271–277

Rubluo, A , K K Kartha, L A Mroginski & J Dyck, 1984 Plant regeneration from pea leaflets cultured *in vitro* and genetic stability of regenerants J Plant Physiol 117 119–130

Stejskal, J & M Griga, 1992 Somatic embryogenesis and plant regeneration in *Pisum sativum* L Biol Plant 34 15–22

Tetu, T , R S Sangwan & B S Sangwan-Norreel, 1990 Direct somatic embryogenesis and organogenesis in cultured immature zygotic embryos of *Pisum sativum* L J Plant Physiol 137 102–109

Van Doorne, L E , G Marshall & R C Kirkwood, 1991 Development of herbicide tolerance in peas II Regeneration via somatic embryogenesis p 271–274 In R J Froud Williams, P Gladders, M C Heath, J F Jenkyn, C M Knott, A Lane & D Pink (Eds) Aspects of Appl Biol 27, Production and Protection of Legumes Assoc Appl Biol , Wellesbourne, Warwick

Euphytica **85**: 341–345, 1995.

Monitoring genetic fidelity *vs* somaclonal variation in Norway spruce (*Picea abies*) somatic embryogenesis by RAPD analysis

Berthold Heinze[1] & Josef Schmidt
Biotechnology Department, Austrian Research Centre, A-2444 Seibersdorf, Austria; [1] *present address: Institute of Forest Genetics, Federal Forest Research Centre, Hauptstrasse 7, A-1140 Vienna, Austria*

Key words: Picea abies, genetic stability, somatic embryogenesis, RAPD

Summary

Somaclonal variation, which is a welcome source of genetic variation for crop breeding, is unwanted when direct regenerants have to be used in tissue culture mass propagation (eg. in many forest trees), or in the regeneration of genetically transformed plants. Random amplified polymorphic DNA (RAPD) was used to analyse somatic embryos and plants regenerated from embryogenic cell lines in Norway spruce, *Picea abies* (L.) Karst. RAPD facilitated the identification of clones, as material from the same cell lines shared identical patterns of amplified fragments, whereas regenerants from different cell lines were easily distinguishable by their respective patterns. For comparisons with explant donor genotypes, cell lines were initiated from cotyledons. Some of the seedlings that had parts of their cotyledons removed were grown on as control plants. Somatic embryos regenerated from cotyledon cell lines showed no aberrations in RAPD banding patterns with respect to donor plants. We conclude that gross somaclonal variation is absent in our plant regeneration system.

Abbreviations: ESM – embryogenic suspensor mass, RAPD – random amplified polymorphic DNA, RFLP – restriction fragment length polymorphism, (2,4-dichlorophenoxy)acetic acid – 2,4-D, 1-naphthaleneacetic acid – NAA

Introduction

Somaclonal variation (Larkin & Scowcroft, 1981) is unwanted in clonal propagation, e.g. of genetically transformed plants, or for large-scale mass propagation programmes in general. In long-lived forest trees, especially where adverse effects are expressed only after years in the field, this may be economically disastrous. True-to-type propagation of the selected genotypes, therefore, is desired. We have been working on somatic embryogenesis in Norway spruce as a conifer model species for Austrian forest trees (Buttgereit, 1991), concentrating on high altitude material from the Austrian Alps, where seed availability is periodically insufficient. The use of reforestation material genetically well-adapted to the extreme climatic conditions there is vital, as many of these forests protect against landslides and avalanches. Genetic characterization of tissue culture material for reforestation may serve to shorten the extensive field-testing necessary. We have used RAPD for this purpose (Williams et al., 1990; Welsh & McClelland, 1990). RAPD analysis offers several advantages over restriction fragment length polymorphism (RFLP) analysis: no prior cloning of DNA sequences is necessary, very little starting material (plant tissue) is sufficient, and quick DNA extraction protocols are suitable (Rafalski et al., 1993).

In spruce species, somatic embryos have been tested for ploidy level (Lelu, 1987), nuclear DNA content (Mo et al., 1989), isozyme patterns (Eastman et al., 1991) and inheritance of RAPD fragments in cell lines from controlled crosses (Isabel et al., 1993). In contrast to the latter authors, we report here on the assessment of somatic embryos and plants on a larger scale. Furthermore, for comparison of propagules with the original genotype of the explant, seedling cotyledon cell lines were initiated, and somatic embryos recovered were compared by RAPD analysis to those plants the cotyledons of which had been used for culture initiation. Growth and field performance studies involving somatic embryo-derived spruce plants are also underway (Attree & Fowke, 1993; Becwar, 1993; Roberts et al., 1993; Gupta et al., 1993).

Materials and methods

Seeds were obtained from the Federal Forest Research Centre (FBVA), Vienna, Austria: several Austrian high altitude seedlots (open-pollinated) from certified stands. Somatic embryogenesis (Buttgereit, 1991) proceeded through the stages of initiation of embryogenic suspensor masses (ESMs) from mature zygotic embryos or cotyledons (Hakman et al., 1985; Krogstrup, 1986; Lelu et al., 1987), proliferation of ESMs, somatic embryo maturation (Hakman et al., 1985; von Arnold & Hakman, 1988; Becwar et al., 1987; Roberts et al., 1990), controlled dehydration and germination (Roberts et al., 1990).

DNA was extracted from zygotic and somatic embryos as well as from plant needles using the standard protocol specified by Heinze et al. (submitted). In brief, plant tissue was homogenized in microcentrifuge tubes, incubated in extraction buffer (50 mM Tris-HCl pH 8.0, 1 mM EDTA, 0.25 M NaCl, 1% w/v SDS, 1% w/v PEG MW 8000, 0.2% v/v β-mercaptoethanol) and treated with Proteinase K, phenol and chloroform. DNA was precipitated in 10% PEG – 0.7 M NaCl, washed in 70% ethanol, and dissolved for RAPD analysis. RAPD reactions were performed in a total volume of 15 μl consisting of 1 $\times$ supplier's buffer containing 1.5 mM $MgCl_2$, approx. 5 ng DNA, 400 nM primers (10-mers from Operon (OP), Alameda, CA), RNase A (0.3 μg, Boehringer Mannheim, Pikaart & Villeponteau, 1993; Heinze, 1994), tetramethylammonium chloride (2×10^{-5} M, Vierling & Nguygen, 1992), and DNA polymerase (DynaZyme from Finnzymes Oy, 0.05 U/μl, or Taq from Promega, 0.067 U/μl) in the following programme in an MJ Research thermocycler: 3 cycles of 94° C for 1 min, 35° C for 3 min, slow heating (1° C per 4 sec) to 65° C, 72° C for 2 min; then 51 cycles of 94° C for 40 sec, 38° C for 1 min, 72° C for 2 min; and finally, further 10 min at 72° C, and 4° C until recovery and electrophoresis. Amplified DNA fragments were resolved in 1.25% agarose (UltraPure: NuSieve GTG, 4 : 1 w/w) in 0.5 $\times$ TBE, visualized by ethidium bromide staining, and photographed (Polaroid 655 films).

Results

Clones of somatic embryos or plants from somatic embryos from a given cell line were identical in their RAPD patterns, whatever primers were used. An example is shown in Fig. 1. No single aberrant band was detected in 50 somatic embryos from cell line 'J7', nor from 100 plants of line 'J1'. The same applied to somatic embryos and plants from other lines, where fewer propagules were analysed. This is in contrast to zygotic embryos from seeds, which could easily be distinguished by their RAPD patterns. Likewise, different clones showed abundant polymorphisms (Fig. 2). In most cases, the first primer applied allowed for genotype distinction; if not, calculating a simple similarity index from combined data of several primers resolved all genotypes and individuals (Heinze 1993).

Somaclonal variation is known to produce genetic differences within propagules of single cell lines (Larkin & Scowcroft, 1981). Mutations may also occur in conifer somatic embryogenesis in the cell line induction phase, where callus formation precedes ESM proliferation. Disorganized callus growth has been associated with somaclonal variation (Brown, 1991). Mutant propagules from developing ESMs could be genetically identical, differing from the parent. We tested this hypothesis by analysing somatic embryos from two lines obtained from seedlings, and comparing embryo RAPD patterns to those of the original plants. Again (Fig. 3), no significant pattern changes were evident.

Discussion

RAPD has been successful in genotype identification in cultivars of e.g. diploid wheat (Vierling & Nguyen, 1992), tomato (Williams & St.Clair, 1993; Klein-Lankhorst et al., 1991), papaya (Stiles et al., 1993), cocoa clones (Wilde et al., 1992) and rose (Torres et al., 1993). Markers were inherited in Mendelian fashion (Carlson et al., 1991; Heinze et al., submitted), and are therefore useful for constructing genetic maps (Tulsieram et al., 1992). Although it has not been possible to distinguish sports (arising from mutated branches) in certain apple cultivars (Harada et al., 1993; Mulcahy et al., 1993), Caetano-Anollés et al. (1993) were able

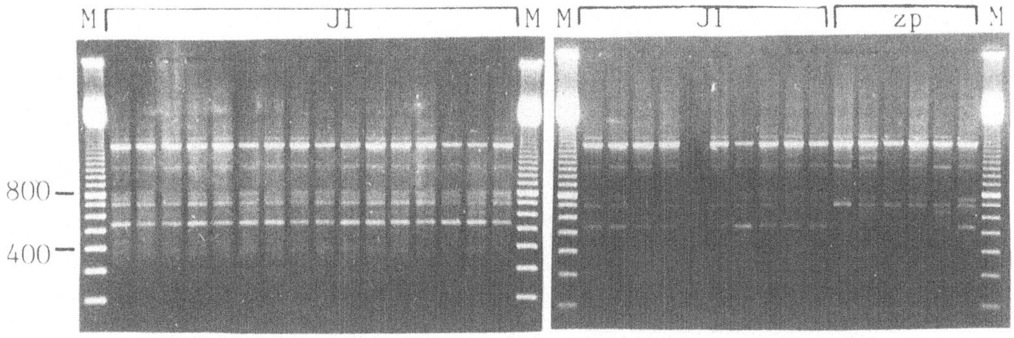

Fig. 1. Identical RAPD patterns from somatic embryo-derived plants of a single clone ('J1'), as opposed to variable patterns in individual 'zygotic' (seed-derived) plants (zp). Primer OPA-02. M, molecular weight markers (100 basepair ladder, Pharmacia).

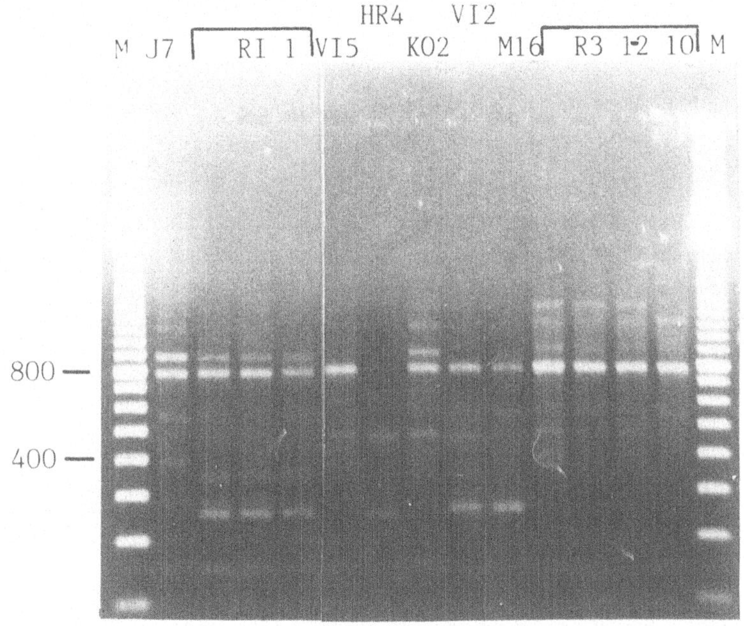

Fig. 2. Clone identification with RAPD patterns. Plant from different somatic embryogenesis-clones ('J1', 'RI 1', . . .) are compared with respect to the pattern obtained with primer OPA-08. M, molecular weight markers (100 basepair ladder, Pharmacia).

to fine-tune the technique for mutant detection in soybean (*Glycine max*). We employed RAPD analysis to screen for crude aberrants in somatic embryogenesis of Norway spruce, but failed to identify any changes in fragment banding patterns arising from tissue culture. The numbers of RAPD bands we scored was fairly high, so they should include markers for most parts of the genome (Tulsieram et al., 1992). This leads us to conclude that there were no losses of major chromosome segments in our system. Cytological analysis of regenerated rye plants (*Secale cereale* L.) by Bebeli et al. (1990) detected chromosome losses and rearrangements. Son et al. (1993) reported the same observations in poplar. In agreement with the conclusions of Shenoy & Vasil (1992), we think that somatic embryogenesis may be an effective sieve to eliminate genetically aberrant propagules, in contrast to reported cases of somaclonal variation identified after shoot regeneration in various crops (e.g. Son et al., 1992; Breiman et al., 1987; Brettell et al., 1986). Being a complicated developmental process, somatic embryogenesis requires the concerted action of many genes

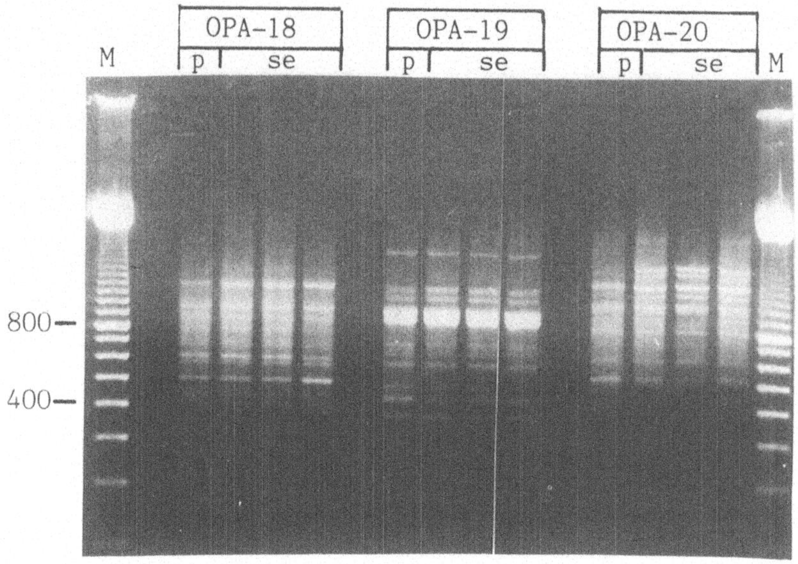

Fig. 3. Comparison of RAPD patterns obtained from explant donor plant (p) and somatic embryos (se) of cell line 'cPF 111'. Primers (Operon) indicated above lanes. M, molecular weight markers (100 basepair ladder, Pharmacia).

and may be sensitive to variation-induced malfunction in any of them. Furthermore, only a few polyploid conifer species are known: *Pinus* and *Picea* contain no such examples. These species may be regarded as not tolerating raised levels of ploidy.

Our cultures were derived from explants with a view to avoiding long phases of unorganised (callus) growth. Brown et al. (1991) detected considerable variation in RFLP banding patterns of *Zea mays* callus, but regenerated shoots showed much less variation. Conifer ESMs show structural organisation (Becwar, 1993), therefore they are unlikely to tolerate the same degree of variation as callus apparently does. In most tissue culture studies cited here, (2,4-dichlorophenoxy)acetic acid (2,4-D) was used as the auxin, and is considered to be a mutagen. However, we avoided the use of 2,4-D and replaced it by 1-naphthaleneacetic acid (NAA; Buttgereit, 1991), which is regarded as less mutagenic.

There are several forms of DNA variation that RAPD analysis is unlikely to detect, including rare point mutations and duplications of genes or chromosomes. Some of these variations, however, are hardly detectable by any current method of DNA analysis in similar settings.

We conclude that no somaclonal variation could be detected in our propagation system, using the very efficient (Brown, 1991; Heinze et al., submitted) RAPD analysis. Low frequencies of variants not detected by this screening method may be acceptable in forestry, where many plants do not survive the first cycles of thinning in any case.

Acknowledgements

We are grateful to Agnes Burg for tissue culture and plant caretaking assistance, to Roger Westcott at Unilever Research, Bedford, UK, for project collaboration, and to Josef Glössl at Universität für Bodenkultur, Vienna, for useful hints. Funds by Unilever Austria and the Austrian Federal Ministry for Agriculture and Forestry.

References

Attree, S.M. & L.C. Fowke, 1993. Embryogeny of gymnosperms: advances in synthetic seed technology of conifers. Plant Cell Tissue Organ Cult. 35: 1–35.

Bebeli, P.J., A. Karp & P.J. Kaltsikes, 1990. Somaclonal variation from cultured immature embryos of sister lines of rye differing in heterochromatin content. Genome 33: 177–183.

Becwar, M.R., 1993. Conifer somatic embryogenesis and clonal forestry. p. 200–223. In: M.R. Ahuja & W.J. Libby (Eds). Clonal Forestry I. Springer, Berlin.

Becwar, M.R., T.L. Noland & S.R. Wann, 1987. Somatic embryo development and plant regeneration from embryogenic Norway spruce callus. TAPPI J. 70: 155–160.

Breiman, A., T. Felsenburg & E. Galun, 1987. *Nor* loci analysis in progenies of plants regenerated from the scutellar callus of bread-wheat A molecular approach to evaluate somaclonal variation Theor Appl Genet 73 827–831

Brettell, R I S , M A Pallotta, J P Gustafson & R Appels, 1986 Variation at the *Nor* loci in triticale derived from tissue culture Theor Appl Genet 71 637–643

Brown, PTH , 1991 The spectrum of molecular changes associated with somaclonal variation IAPTC Newsletter 66 14–25

Brown, PTH , E Goebel & H Loerz, 1991 RFLP analysis of *Zea mays* callus cultures and their regenerated plants Theor Appl Genet 81 227–232

Buttgereit, J , 1991 Somatische Embryogenese bei *Picea abies* (L) Karst Vienna, University of Vienna, Dissertation

Caetano-Anolles, G , B J Bassam & PM Gresshoff, 1993 Enhanced detection of polymorphic DNA Mol Gen Genet in press

Carlson, J E , L K Tulsieram, J C Glaubitz, V W K Luk, C Kauf feldt & R Rutledge, 1991 Segregation of random amplified DNA markers in F1 progeny of conifers Theor Appl Genet 83 194–200

Eastman, PA K , F B Webster, J A Pitel & D R Roberts, 1991 Evaluation of somaclonal variation during somatic embryogenesis of interior spruce (*Picea glauca engelmanni* complex) using culture morphology and isozyme analysis Plant Cell Rep 10 425–430

Gupta, PK , G Pullman, R Timmis, M Kreitinger, W C Carlson, J Grob & E Welty, 1993 Forestry in the 21st century The biotechnology of somatic embryogenesis Biotechnology 11 454–459

Hakman, I , L C Fowke, S von Arnold & T Eriksson, 1985 The development of somatic embryos in tissue cultures initiated from immature embryos of *Picea abies* (Norway spruce) Plant Sci 38 53–59

Harada, T , K Matsukawa, T Sato, R Ishikawa, M Niizeki & K Saito, 1993 DNA-RAPDs detect genetic variation and paternity in *Malus* Euphytica 65 87–91

Heinze, B , 1993 Genetic stability in Norway spruce somatic embryogenesis as analysed by randomly amplified DNA Vienna, Universitat fur Bodenkultur, Dissertation

Heinze, B , 1994 RAPD reactions from crude plant DNA Adding RNase A as a 'helper enzyme' Mol Biotechnol 1 307–310

Isabel, N , L Tremblay, M Michaud, F M Tremblay & J Bousquet, 1993 RAPDs as an aid to evaluate the genetic integrity of somatic embryogenesis-derived populations of *Picea mariana* (Mill) BSP Theor Appl Genet 86 81–87

Klein Lankhorst, R M , A Vermunt, R Weide, T Liharska & P Zabel, 1991 Isolation of molecular markers for tomato (*L esculentum*) using random amplified polymorphic DNA (RAPD) Theor Appl Genet 83 108–114

Krogstrup, P , 1986 Embryolike structures from cotyledons and ripe embryos of Norway spruce (*Picea abies*) Can J For Res 16 664–668

Larkin, PJ & W R Scowcroft, 1981 Somaclonal variation – a novel source of variability from cell cultures for plant improvement Theor Appl Genet 60 197–214

Lelu, M -A , 1987 Variations morphologiques et genetiques chez *Picea abies* obtenues apres embryogenese somatique Ann AFO CEL 1987 35–47

Lelu, M A , M Boulay & Y Arnaud, 1987 Obtention de cals embryogenes a partir de cotyledons de *Picea abies* (L) Karst preleves sur de jeunes plantes agees de 3 a 7 jours apres germi nation C R Acad Sci Paris t 305 Serie III 105–109

Mo, L H , S von Arnold & U Lagercrantz, 1989 Morphogenic and genetic stability in longterm embryogenic cultures and somatic embryos of Norway spruce (*Picea abies* (L) Karst) Plant Cell Rep 8 375–378

Mulcahy, D L , M Cresti, S Sansavini, G C Douglas, H F Linskens, G B Mulcahy, R Vignani & M Pancaldi, 1993 The use of random amplified polymorphic DNAs to fingerprint apple genotypes Sci Hort Amsterdam 54 89–96

Pikaart, M J & B Villeponteau, 1993 Suppression of PCR amplification by high levels of RNA BioTechniques 14 24–27

Rafalski, A , S Tingey & J G K Williams, 1993 Random amplified polymorphic DNA (RAPD) markers p 1–9 In S B Gelvin, R A Schilperoort & D P S Verma (Eds) Plant Molecular Biology Manual Kluwer, Dordrecht

Roberts, D R , F B Webster, B S Flinn, W R Lazaroff & D R Cyr, 1993 Somatic embryogenesis of spruce p 427–450 In K Redenbaugh (Ed) Synseeds CRC Press, Boca Raton

Roberts, D R , B C S Sutton & B S Flinn, 1990 Synchronous and high frequency germination of interior spruce somatic embryos following partial drying at high relative humidity Can J Bot 68 1086–1090

Shenoy, V B & I K Vasil, 1992 Biochemical and molecular analysis of plants derived from embryogenic tissue cultures of napier grass (*Pennisetum purpureum* K Schum) Theor Appl Genet 83 947–955

Son, S H , H K Moon & R B Hall, 1993 Somaclonal variation in plants regenerated from callus culture of hybrid aspen (*Populus alba* L × *P grandidentata* Michx) Plant Sci 90 89–94

Stiles, J I , C Lemme, S Sondur, M B Morshidi & R Manshardt, 1993 Using randomly amplified polymorphic DNA for evaluating genetic relationships among papaya cultivars Theor Appl Genet 85 697–701

Torres, A M , T Millan & J I Cubero, 1993 Identifying rose cultivars using random amplified polymorphic DNA markers HortScience 28 333–334

Tulsieram, L K , J C Glaubitz, G Kiss & J E Carlson, 1992 Single tree genetic linkage mapping in conifers using haploid DNA from megagametophytes Biotechnology 10 686–690

Vierling, R A & H T Nguygen, 1992 Use of RAPD markers to determine the genetic diversity of diploid, wheat genotypes Theor Appl Genet 84 835–838

von Arnold, S & I Hakman, 1988 Regulation of somatic embryo development in *Picea abies* by abscisic acid (ABA) J Plant Physiol 132 164–169

Welsh, J & M McClelland, 1990 Fingerprinting genomes using PCR with arbitrary primers Nucl Acid Res 18 7213–7218

Wilde, J , R Waugh & W Powell, 1992 Genetic fingerprinting of *Theobroma* clones using randomly amplified polymorphic DNA markers Theor Appl Genet 83 871–877

Williams, C E & D A St Clair, 1993 Phenetic relationships and lev els of variability detected by restriction fragment length polymorphism and random amplified polymorphic DNA analysis of cultivated and wild accessions of *Lycopersicon esculentum* Genome 36 619–630

Williams, J G K , A R Kubelik, K J Livak, J A Rafalski & S V Tingey, 1990 DNA polymorphisms amplified by arbitrary primers are useful as genetic markers Nucl Acid Res 18 6531–6535

Euphytica **85:** 347–349, 1995.

The embryogenic potential of rice cell suspensions affects their recovery following cryogenic storage

P.T. Lynch[1], E.E. Benson[2], J. Jones, E.C. Cocking, J.B. Power & M.R. Davey
Plant Genetic Manipulation Group, Department of Life Science, University of Nottingham, Nottingham NG7 2RD, UK; present address: [1] Plant Biotechnology Group, Division of Biological Sciences, University of Derby, Kedleston Road, Derby DE22 1GB, UK; [2] Department of Molecular and Life Sciences, University of Abertay Dundee, Bell Street, Dundee DD1 1HG, Scotland, UK

Key words: Oryza sativa L. cv. Taipei 309, rice, cryogenic storage, somatic embryogenesis

Summary

The rates of recovery from cryogenic storage of suspension cultures of the Japonica rice cultivar Taipei 309, as determined by the reduction of triphenyl tetrazolium chloride and cell regrowth, were significantly influenced by the embryogenic potential of the non-frozen cultures.

Introduction

The development of efficient and reproducible methods for the regeneration of fertile rice plants from protoplasts, relates largely to the use of embryogenic cell suspensions as source material (Lynch et al., 1991). Cryopreservation of these suspension cultures offers a means of safeguarding the supply of embryogenic material and preserving unique cultures (Lynch & Benson, 1991). If cryogenic storage is to be widely employed in rice biotechnology, it would be desirable for essentially one cryopreservation protocol to be applicable to a number of different cell lines. Despite being initiated and maintained under the same conditions however, rice cell lines are known to exhibit differing morphogenic and physiological characteristics (Benson et al. 1992; Lynch et al., 1991). The influence of the embryogenic character of the non-frozen cell cultures on their post-thaw recovery has therefore been assessed.

Materials and methods

Cell suspension cultures of the Japonica rice (*Oryza sativa* L.) cv. Taipei 309 were maintained in AA2 liquid medium (Abdullah et al., 1986) by shaking (125 rpm) in the dark at + 28° C. Prior to freezing, cells were maintained for 3–4 d in AA2 medium supplemented with 60 g l^{-1} mannitol. Cells were harvested on 45 μm nylon mesh and placed into 2 ml polypropylene vials (Starstedt Ltd., Boston Road, Leicester, UK; 0.2 g of cells/vial). Approx. 0.75 ml of chilled (on iced water), filter sterilised, cryoprotectant mixture (Meijer et al., 1990) were added to each vial and mixed with the cells. The cryoprotectant mixture was made up in AA2 medium and the pH adjusted to 5.8. Cells were cryoprotected in vials for 1 h on iced water. Vials were transferred to aluminium canes and the cells frozen (Withers & King, 1980) in a controlled rate freezer (Planer Cryo 10 Series; Planer Biomed, Sunbury-on-Thames, Middlesex, UK).

Cells were thawed, after storage in liquid nitrogen for 7–10 d, by plunging the vials into sterile water (+ 45° C). Excess cryoprotectants were removed. Cells were recovered on 5.5 cm filter paper discs (Whatman No. 1; one disk above another, per 9 cm Petri dish) overlying 25 ml of AA2 medium containing 0.4% (w/v) Sigma Type 1 agarose. The contents of 1 vial were placed in each Petri dish. Cells were maintained at + 28° C in the dark and subcultured 3 d after thawing by transferring the cells on the filter paper discs to fresh medium. Suspensions were reinitiated in AA2 liquid medium and maintained under

Table 1. Time course of post-thaw recovery for cryopreserved rice cells with different embryogenic potentials

Cell Line	Embryogenic character of cell line	Post-thaw TTC values, as a % of non-frozen controls			
		Time after thawing			
		1 d	5 d	10 d	20 d
L14 R2	embryogenic	4.23 ± 0.32	8.64 ± 1.00	79.01 ± 4.81	82.22 ± 1.25
L52	embryogenic	3.12 ± 0.53	10.26 ± 1.21	82.31 ± 3.19	85.41 ± 3.62
L56	embryogenic	2.91 ± 0.41	9.94 ± 0.98	85.27 ± 2.14	89.62 ± 2.93
L86	embryogenic	3.82 ± 0.18	11.28 ± 0.84	80.92 ± 0.98	88.94 ± 1.73
L55	formerly embryogenic	1.93 ± 0.21	6.25 ± 0.62	63.54 ± 2.22	79.54 ± 1.98
L10	non-embryogenic	0.92 ± 0.09	4.21 ± 0.84	52.61 ± 3.91	70.34 ± 2.94
L45	non-embryogenic	1.04 ± 0.14	5.01 ± 0.09	56.22 ± 1.94	71.11 ± 0.98

TTC values are the mean of 9 replicate samples originating from replicate Petri dishes, 3 samples per dish of thawed cells calculated as a % of the TTC values of 9 replicate samples originating from replicate Petri dishes, 3 samples per dish of non-frozen cells ± standard deviation. The cells, thawed or non-frozen, in each Petri dish were derived from different original flasks of cells.

the same conditions as unfrozen cultures. Post-thaw viability of cells was assessed by triphenyl tetrazolium chloride (TTC) reduction (Steponkus & Lanphear, 1967). Recovery (TTC reduction) and re-growth was assessed in relation to non-frozen suspension cultures (controls) maintained and sub-cultured on filter papers as described above. Data were analysed using one-way Analysis of Variance (Genstat 4, Lawes Agricultural Trust, Rothamstead Experimental Station, UK, statistical computer package).

Results

Seven different rice lines subjected to cryopreservation survived exposure to liquid nitrogen as evidenced by metabolic recovery (TTC reduction) and cell division (gain in fresh weight). However, the capacity to resume these capabilities, as compared with unfrozen control cells, was determined by the morphogenic status of the cells prior to freezing.

During the time course of post-freeze recovery (Table 1), the embryogenic cell lines demonstrated significantly higher ($p = 0.001$) TTC reduction values as compared with the cell lines which were non-embryogenic. Interestingly, the TTC reduction capacity of a cell line which was originally embryogenic, but had since lost the capability, was significantly different ($p = 0.001$) and intermediate between the embryogenic and the non-embryogenic lines. Recovery as re-growth also showed the same statistically significant ($p = 0.01$) trends (Table 2).

Table 2. Post-thaw cell growth compared with the embryogenic character of the original non-frozen cell lines

Cell Line	Embryogenic character of cell line	% change in cell fresh weight	
		Time after thawing	
		5 d	20 d
L14 R2	embryogenic	12.63 ± 4.21	162.2 ± 10.4
L52	embryogenic	10.42 ± 2.44	144.9 ± 11.0
L56	embryogenic	15.41 ± 5.31	154.6 ± 13.1
L86	embryogenic	13.62 ± 3.86	140.8 ± 8.9
L55	formerly embryogenic	7.62 ± 4.28	132.9 ± 9.4
L10	non-embryogenic	5.21 ± 2.01	102.4 ± 10.2
L45	non-embryogenic	4.12 ± 1.14	98.6 ± 12.6

The fresh weight values are the means of 5 replicate Petri dishes of thawed cells ± standard deviation. Results are calculated as a % of unfrozen control cells. The cells, thawed or non-frozen, in each Petri dish were derived from different original flasks of cells.

Discussion

Cryopreservation is an important tool in rice biotechnology programmes, the success of which is dependent upon the ability to apply one basic protocol to a wide range of cell lines. This study concurs with the acceptability of cryopreservation as a successful and routine storage method for rice cell suspension cultures.

The capacity to survive storage in liquid nitrogen is dependent upon many non-cryogenic factors, including genotype, physiological status and pre- and post-freeze *in vitro* manipulations (Towill, 1991). This study supports this view by clearly demonstrating a significant relationship between the morphogenic status of cells and their post-thaw survival characteristics. While non-embryogenic cell lines survived cryopreservation, their capacity for recovery was limited compared with embryogenic cells. This may have been due to several factors. Embryogenic cells are supported by a greater respiratory capacity (Benson et al., 1992a), which may reduce post-thaw injury associated with changes in glycolytic and respiratory intermediates (Finkle & Ulrich, 1982). In addition, embryogenic rice cells have a higher antioxidant status and a lower pro-oxidant level (lipid peroxidation) than non-embryogenic cells (Benson et al., 1992a). This may have important consequences for survival as rice cells suffer considerable, free radical-mediated injury during cryopreservation (Benson et al., 1992b).

To date, the development of cryopreservation protocols for plant systems has placed major emphasis on the optimisation of cryogenic methodology. There still remains, however, a paucity of information concerning the relationship between cryogenic and physiological factors. Findings in this study do support other investigations, showing that the physiological status of *in vitro* plant tissues can greatly influence their capacity to recover from cryogenic storage (Benson et al., 1989, 1992b; Harding et al., 1991).

Acknowledgement

This work was supported by the Rockefeller Foundation.

References

Abdullah, R., J.A. Thompson & E.C. Cocking, 1986. Efficient plant regeneration from rice protoplasts through somatic embryogenesis. Bio/Technol. 4: 1087–1090.

Benson, E.E., P.T. Lynch & J. Jones, 1992a. Variation in free-radical damage in rice cell suspensions with different embryogenic potentials. Planta 188: 296–305.

Benson, E.E., P.T. Lynch & J. Jones, 1992b. The detection of lipid peroxidation products in cryoprotected and frozen rice cells: consequences for post-thaw survival. Plant Sci. 85: 107–114.

Benson, E.E., K. Harding & H. Smith, 1989. The effects of pre- and post-freeze light on the recovery of cryopreserved shoot-tips of *Solanum tuberosum*. Cryo-Lett. 10: 323–344.

Finch, R.P., P.T. Lynch, J.P. Jotham & E.C. Cocking, 1991. Isolation, culture and fusion of rice protoplasts. p. 251–268. In: Y.P.S. Bajaj (Ed). Biotechnology in Agriculture and Forestry, Vol. 14. Springer Verlag, Heidelberg.

Finkle, B.J. & J.M. Ulrich, 1982. Cryoprotectant removal temperatures as a factor in the survival of frozen rice and sugar cane cells. Cryobiol. 19: 329–335.

Harding, K., E.E. Benson & H. Smith, 1991. The effects of tissue culture on post freeze survival of cryopreserved shoot-tips of *Solanum tuberosum*. Cryo-Lett. 12: 17–22.

Lynch, P.T., R.P. Finch, M.R. Davey & E.C. Cocking, 1991. Rice tissue culture and its applications. p. 135–156. In: G.S. Khush & G.H. Toenniessen (Eds). Rice Biotechnology: Biotechnology in agriculture No. 6. CAB International, Wallingford, UK.

Lynch, P.T. & E.E. Benson, 1991. Cryopreservation, a method for maintaining the plant regeneration capability of rice cell suspension cultures. p. 321–332. In: Proceedings of the Second International Rice Genetics Symposium. IRRI, Los Banos, Philippines.

Meijer, E.G.M., F. van Iren, E. Schrijnemakers, L.A.M. Hensgens, M. van Zijderveld & R.A. Schilperoort, 1990. Retention of the capacity to produce plants from protoplasts in cryopreserved lines of rice (*Oryza sativa* L.). Plant Cell Rep. 10: 171–174.

Steponkus, P.L. & F.O. Lanphear, 1967. Refinement of the triphenyl tetrazolium chloride method of determining cold injury. Plant Physiol. 42: 1423–1426.

Towill, L.E., 1991. Cryopreservation. p. 41–70. In: J.H. Dodds (Ed). *In Vitro* Methods for Conservation of Plant Genetic Resources. Chapman and Hall, London.

Euphytica **85**: 351–358, 1995.

Selecting the optimum genetic background for transgenic varieties, with examples from *Brassica*

Derek Lydiate, Phil Dale, Ulf Lagercrantz[1], Isobel Parkin & Phil Howell
Brassica and Oilseeds Research Department, John Innes Centre, Norwich, NR4 7UH, UK; [1] present address: Department of Plant Breeding, Swedish University of Agricultural Science, Box 7003, S-750 07, Uppsala, Sweden

Key words: binomial distribution, genetic complexity, interspecific crosses, introgression, marker-assisted selection, modifier genes, recombination frequency

Summary

The performance of transgenic varieties depends not only upon the stable and correctly-regulated expression of specific transgenes but also upon the agronomic potential of the background genotype. Ideally, transgenes should be introduced into agronomically-superior cultivars and transgenic varieties will become out-classed if their agronomic properties are not continually improved. It will often prove desirable to use conventional breeding techniques, as opposed to new cycles of transformation, to carry out this process of varietal improvement.

Continuing advances in marker-assisted selection have made possible the selection and manipulation of an entire genetic background. This means that transgenes can be transferred to new and often 'untransformable' varieties with relative ease. To carry out this process efficiently requires the correct choice of male and female parents, the use of appropriate marker-systems and the concentration of selection on the most appropriate generations.

Efficient, phenotypically-neutral marker-systems have revolutionised the identification and manipulation of quantitative trait loci (QTLs). The loci which modify the expression of transgenes are a form of QTL. Desirable alleles at modifier QTLs can be transferred to new varieties along with the transgenes themselves, using marker-assisted breeding.

The strategies for marker-assisted selection are becoming ever more sophisticated. A range of complementary marker systems allows the selection of desirable genotypes. In addition, the meiotic reassortment and recombination of chromosomes which produces new genotypes is becoming better understood. The most efficient plant breeding methods and the most powerful genetics will make optimal use of both markers and meiosis.

Introduction

A great deal of scientific effort has been invested in the development of transformation systems in plants, from the initial reports of transformation in tobacco (Horsch et al., 1984) to current advances in cereal transformation (Rhodes et al., 1988; Shimamoto et al., 1989; Vasil et al., 1992). This work has been successful in establishing transformation systems for most of the major crop species, including maize, rice, potatoes and brassicas (Dale et al., 1993). The marker systems that have been developed to select for transformed plants carrying novel cloned genes constitute an important part of this technology. Marker systems can be divided into selectable markers including resistance to kanamycin, hygromycin, methotrexate and glufosinate, and reporter genes such as beta-glucuronidase and luciferase (Draper et al., 1988; Potrykus, 1990).

There are still barriers, however, to the transformation of agronomically-proven crop genotypes, and several of these can be overcome by conventional crossing strategies. In some crop species only certain cultivars can be transformed efficiently and these often yield less than the most modern varieties and elite breeding material. In these cases, conventional breeding can be used to transfer a promising transgene from a donor cultivar to a modern variety, and thus combine the benefits of transformation and conventional breeding methods. Similarly, somaclonal variation, induced by the tissue culture associated with transformation, can reduce the agronomic potential of transformed plants (Dale & McPartlan, 1992). Crossing to normal varieties should separate the transgene from the problems associated with the initial transformant. Less obviously, the poorly-understood processes which result in transgene ‘position effects’ (Peach & Velten, 1991) and co-suppression (Napoli et al., 1990) make some (often rare) transformants significantly better than others. If a particular copy of a transgene is expressed in a desirable fashion but is present in a poor genetic background, it will be beneficial to transfer such a transgene to a range of more suitable varieties by conventional breeding.

Conventional breeding will gradually change the genetic environment surrounding a transgene and eventually move it from a poor or inappropriate genetic background into a cultivar which is well-adapted agronomically. However, this process takes a considerable amount of time and biotechnology companies are eager for returns on their investments in research and development. The speed with which transgenes can be transferred into improved genetic backgrounds can be accelerated by the application of marker-assisted breeding techniques. Marker-assisted breeding makes possible the selection of complex genotypes. It is complementary to transformation technology and, like gene cloning and transformation, it is underpinned by the modern techniques of molecular biology. In contrast to transformation, however, where a single cloned gene is introduced into a plant to form a novel line, marker-assisted breeding offers the possibility of engineering complex genotypes. The genotypes best-suited to agricultural production have beneficial (but often ill-defined) alleles at many loci.

Marker-assisted breeding utilises a range of molecular-genetic markers to distinguish between the different parental genotypes segregating in a cross and to select for specific recombinants. These markers have two properties which make them particularly well-suited to carrying out complex genetic selections during the breeding of crop varieties: (1) the markers detect natural variation and are thus informative in elite breeding material, and (2) the markers can detect variation at a large number of loci, allowing characterisation and manipulation of the whole genome.

Three main classes of molecular marker are used in marker-assisted breeding: isozymes (Tanksley & Orton, 1983), RFLP markers (Paterson et al., 1991) and PCR-based markers (Rafalski & Tingey, 1993). The relative advantages of these different systems are discussed in a later section.

The application of marker-assisted breeding is probably most advanced in maize (Stuber, 1992). Genetic analyses of maize, based on isozymes (Wendel et al., 1986) and RFLP markers (Helentjaris, 1987; Zehr et al., 1992), are well-documented. In addition, maize has been the subject of extensive industrial activity because it is such an important crop, particularly in the USA. Much of the academic research into the potential of marker-assisted breeding has been carried out on tomatoes (Paterson et al., 1988) and this has been summarised in a recent review (Paterson et al., 1991). In more complex genomes such as that of wheat, marker technology is being used to characterise the products of intergenomic recombination events. This work will make possible the targeted transfer of desirable genes between the genomes of different grasses whilst minimising the problems associated with genetic drag (Devos & Gale, 1993).

This review will concentrate on the applications of marker-assisted breeding in *Brassica* species. *Brassica* crops have high levels of natural polymorphism and they are extremely well-suited for RFLP technology (Slocum et al., 1990; Chyi et al., 1992; Song & Osborn, 1992). In addition, the major *Brassica* crops have a range of genome complexities (U, 1935) including simple diploid species such as *B. oleracea* (i.e. cabbages and cauliflowers) and more complex amphidiploid species such as *B. napus* (oilseed rape) where intergenomic recombination can be observed and manipulated (Lydiate et al., 1993).

Major genes and superior genotypes

Transformation is normally used to introduce a single novel gene into a plant and this gene usually modifies a single important characteristic of the recipient line. Whilst this is obviously true of transgenes conferring herbicide resistance (Dale et al., 1993), it is also true of antisense constructs such as those that increase the level of stearic acid in seed oils by reducing the levels of desaturase (Knutzon et al. 1992). In this respect, transgenes are no different from other major genes which have introduced beneficial characteristics into new varieties from exotic germplasm (Smith, 1944; Sears, 1981; Austin et al., 1985; Mithen & Herron, 1992). To have the optimum potential, transgenic varieties should have genetic backgrounds which have been selected for maximum yield and good quality characteristics under normal agronomic conditions. The genotype of an elite variety is a complex assembly of genes controlling a large number of characters. For example, a successful variety of winter oilseed rape must establish itself well in the autumn, survive the winter and flower at the correct time in the spring. It must also be resistant to or tolerant of a range of pests and pathogens, be resistant to lodging and deposit a high proportion of its assimilate into seeds which mature. In addition, oilseed rape varieties must have appropriate alleles at the five to seven quality loci which control the low erucic acid (Chen & Beversdorf, 1990) and low glucosinolate (Henderson & Pauls, 1992) characters. Marker-assisted backcrossing programmes can introgress transgenes into elite varieties by selecting indirectly for the large numbers of alleles (with complex interactions) that make up a superior genotype. This is done without the need to identify the individual genes involved or to understand their modes of action.

Isozymes, RFLPs and PCR-based markers

Allelic variation at isozyme loci is detected using stains to identify specific enzyme activities and protein separation techniques to distinguish between different alleles (Shields et al., 1983). Isozyme analysis is rapid, allowing large populations to be scored, but the number of loci identified by isozyme markers is limited (Tanksley, 1983).

RFLP markers use DNA probes and Southern hybridisation to identify specific loci. Different alleles are distinguished by point mutations and/or DNA rearrangements which lead to changes in the separation of restriction enzyme sites flanking these loci (Botstein et al., 1980). RFLP markers are extremely informative in crosses with high levels of polymorphism. They are most efficient in analyses which require a large number of loci to be assayed. Mapping information based on RFLP markers can be correlated between crosses and even between related species because RFLP probes usually recognise several alleles at a locus and at evolutionarily-related loci. RFLP markers have been used, for example, to develop comparative maps of tomato and potato (Tanksley et al., 1992), diploid *Brassica* species (Lagercrantz & Lydiate, unpublished) and rice and wheat (Kurata et al., 1994).

PCR-based markers use pairs of specific oligonucleotide primers and the polymerase chain reaction to amplify specific loci and sometimes specific alleles. Allelic variation is detected directly at the level of variation in DNA sequence caused by point mutations and/or DNA rearrangements. RAPD markers usually detect changes in the priming sites (Williams et al., 1991) whereas CAPS markers detect changes within the amplified DNA (Konieczny & Ausubel, 1993). PCR-based markers are especially useful for assaying a small number of loci in a large population. This is because PCR reactions can be carried out on smaller quantities of more rapidly-isolated but less pure DNA than is required for RFLP analysis (Edwards et al., 1991). Assaying a large number of loci with traditional PCR-based markers is extremely tedious because each locus/plant combination requires a separate reaction. This limitation is ameliorated by AFLPs^{TM} (Patent 825962, Keygene n.v., Netherlands) where a large random set of dispersed loci are assayed with each reaction.

Backcrossing and introgression of novel genes

Marker-assisted backcrossing programmes introgress novel genes into elite genetic backgrounds more quickly and reliably than conventional backcrossing programmes. Markers are used to assay the whole of that portion of the genome which was heterozygous in the previous generation: for example, analysis of a first backcross (B_1) generation assays genotype imbalance in the F_1 gametes. This allows a particular backcross individual, which has retained any desirable donor genes but which has otherwise inherited the smallest amount of the non-recurrent genotype in the smallest number of segments, to be selected precisely at each generation (Paterson et al., 1988). However, the speed of this process is limited by the frequency and extent to which the relative proportions of the recurrent and non-recurrent genotypes inherited by gametes are distorted from the expected ratios (Stam, 1980). Gametes with very distorted parental contributions are desirable for rapid introgression but the frequency at which these occur decreases with increasing genome complexity.

Models to predict the frequency at which gametes with a given percentage of non-recurrent genotype might be expected to occur, are crucial to the design of efficient marker-assisted backcrossing programmes. The distribution of varying parental contributions to F_1 gametes of *B. napus* has been shown to follow a binomial distribution very closely in a large, extensively-

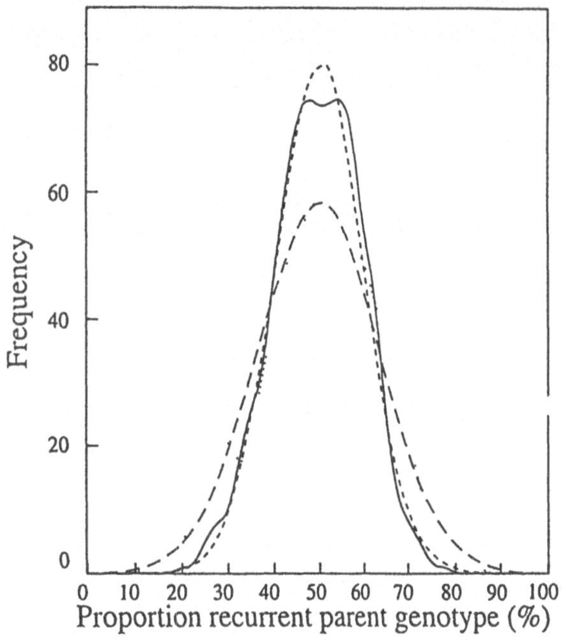

Fig 1 Modelling the complexity of the *B napus* genome Variation in the proportion of the recurrent genotype in 195 F_1 gametes (continuous line) closely follows the binomial distribution for 30 independent units of variation (short dashes) Variation in the proportion of the recurrent genotype in the *B rapa* portion of the genome (dotted line) follows the binomial distribution for 15 units of variation (long dashes)

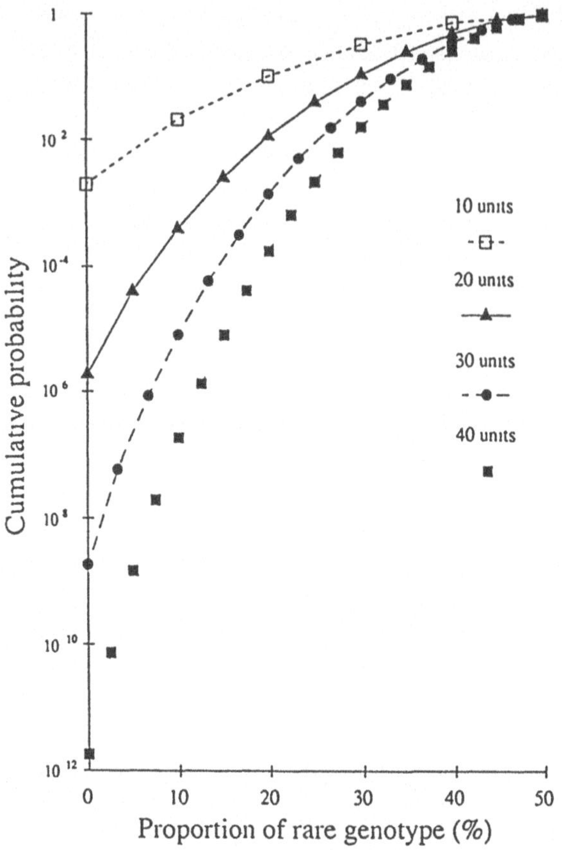

Fig 2 The effect of increasing genome complexity on the likelihood of obtaining gametes with extreme genotypes The proportion of the haploid genome represented by the rare genotype is plotted against the cumulative probability of such a gamete arising for binomial distributions based on 10, 20, 30 and 40 units of variation

mapped population (Fig. 1; Howell & Lydiate, unpublished). The complexity of the portion of the *B napus* genome which was assayed corresponded to 30 independently-inherited units (Fig. 1). The observed complexity is a product of the number of chromosomes and the amount of genetic recombination. The constituent *B oleracea* and *B rapa* genomes of *B napus* were found to have complexities similar to that expected for 15 independently-inherited units when they were considered separately (Fig. 1).

These binomial models are obviously over-simplifications but they are of considerable predictive value (Fig. 2). For example, with a genome complexity of 30, B_1 individuals which inherited a gamete with only 20% non-recurrent genotype are approximately 50 times rarer than B_1 individuals which inherited a gamete with 33.3% non-recurrent genotype. In contrast, with a genome complexity of 15, B_1 individuals which inherited a gamete with only 20% non-recurrent genotype are approximately 7 times rarer than B_1 individuals which inherited a gamete with 33.3% non-recurrent genotype.

These distributions mean that analyses of large B_1 populations of *B oleracea* or *B rapa* are worthwhile, as strong selection against the non-recurrent genotype will be possible. However, in crops like *B napus* and wheat, which have more complex genomes, intensive selection against the non-recurrent genotype is very inefficient in the B_1 generation. In the B_2 generation, the effective genome complexity is lower, as less of the genome is still segregating than in the B_1. Consequently, analyses of large B_2 populations of *B napus* and wheat will be worthwhile.

Combining transgenes

F_1 varieties are an ideal means of combining dominant transgenes. However, it is more difficult to derive pure-breeding lines with combinations of transgenes. On average, F_2 progeny will only contain one plant in every 4^n (where n is the number of genes being combined) with all the relevant genes homozygous. A much

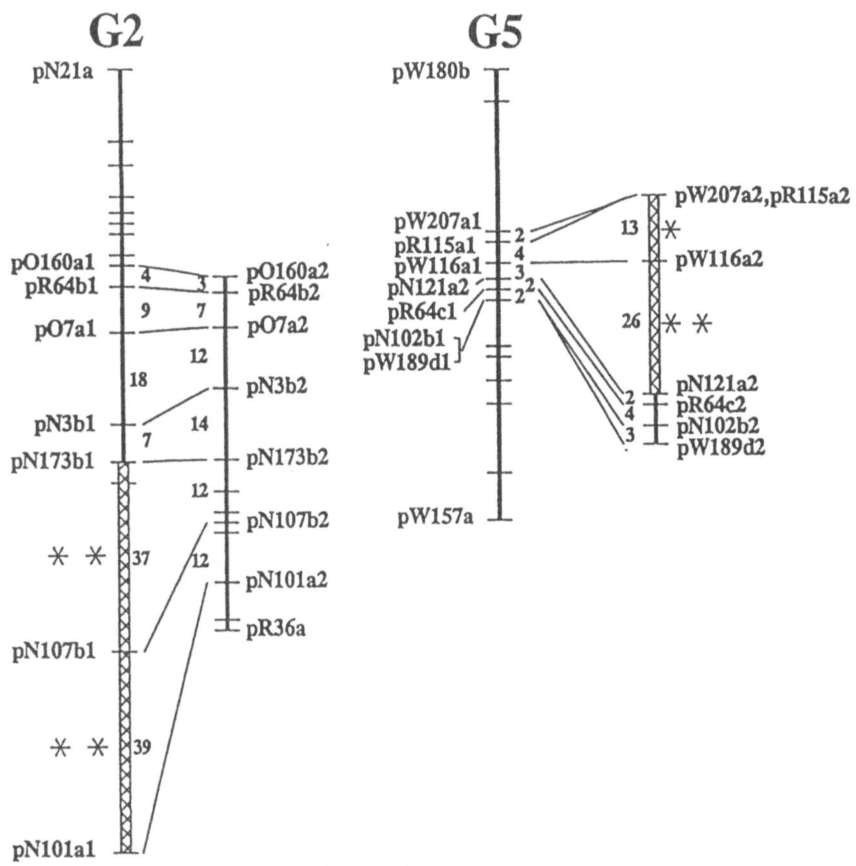

Fig. 3. Recombination in male and female meioses. Simplified maps of *B. nigra* linkage Groups 2 and 5 demonstrate that cross-overs are more frequent towards the telomeres in male meioses (left-hand bar of Group 2) and more frequent towards the centre of linkage groups in female meioses (right-hand bar of Group 5). Filled boxes denote intervals where recombination frequencies differed significantly between male and female meioses (* $p < 0.05$, ** $p < 0.01$). Restricted polymorphism limited the maps derived from the female meioses. Map distances are in Kosambi centimorgans.

higher proportion of inbred lines (one line in 2^n) will have all relevant genes homozygous. Because of this, producing inbred lines is likely to be more efficient than analysing a large number of F_2 individuals when four or more genes are being combined, particularly if it is difficult to test for homozygosity. However, it takes a long time to develop inbred lines. Microspore-culture is a relatively simple process for many *Brassica* crops and yields perfectly homozygous lines in a single step (Lichter et al., 1988). It has been used successfully to combine beneficial alleles at six loci in oilseed rape (Henderson & Pauls, 1992).

With transgenic plants it can be difficult to recognise heterozygotes without resorting to laborious test crosses. An alternative to test crossing is the use of inverse PCR (Does et al., 1991) to form probes from DNA adjacent to the transgene insertion site. These probes should each distinguish two alleles at the transgene insertion sites because the insertion events will have produced RFLPs.

Simultaneous selection for multiple donor loci

When simultaneously introgressing a number of desirable genes into a particular genetic background via backcrossing, only those progeny that inherit all of the desirable genes are useful for further backcrossing. If three genes are being introgressed, one plant in eight (1 in 2^3) will be useful and it is best to pre-select these plants and extensively genotype the resulting sub-population. Pre-selection is best carried out using PCR-based markers to test for the continued presence of the transgenes and for any natural alleles which are also being introgressed, while the extensive genotyping of sub-populations should be carried out using RFLPs.

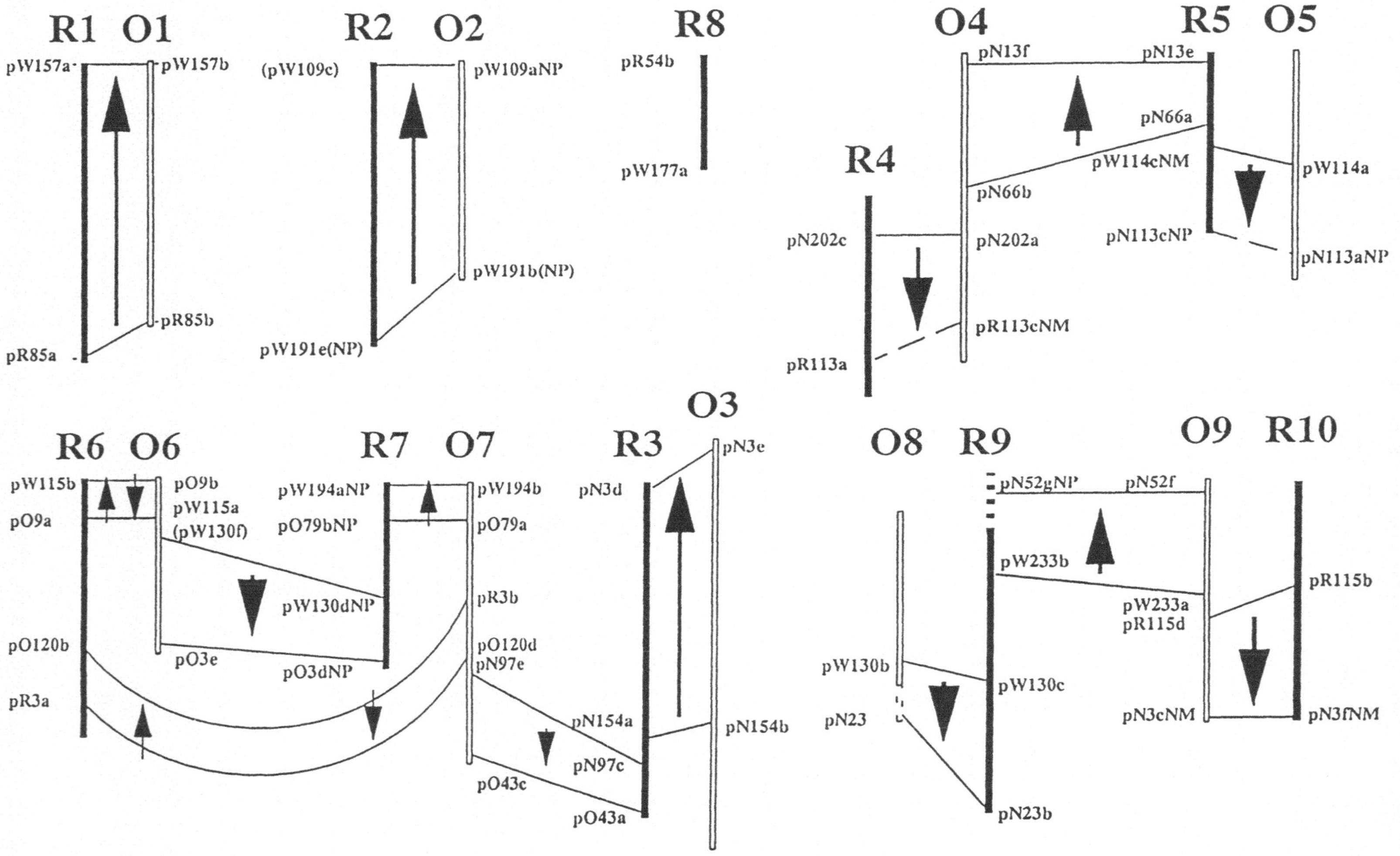

Fig. 4. Homoeologous relationships between the linkage groups (vertical lines) of *B. rapa* (R1–R10) and *B. oleracea* (O1–O9) derived from homoeologous pairing and recombination in resynthesised *B. napus*. Chromosome arms normally have the same orientation in *B. rapa* and *B. oleracea* (single arrows) but occasionally they appear to have opposite orientations as on R6/O6 and R6/O7 (double arrows). Lines joining the groups indicate pairs of loci which border the regions of known homoeology.

Breaking undesirable linkages

When moving a transgene from a poor genetic background into an elite genetic background, it might be necessary to break any genetic linkages between the transgene and associated undesirable genes from the donor plant. The frequency at which genetic recombination will break this linkage can be influenced by the direction of the cross. It has been reported in humans that the frequency of recombination is increased towards the telomeres in male meioses and towards the centromeres in female meioses (White et al., 1990). Whilst similar observations have been made in *B. nigra* (Fig. 3; Lagercrantz & Lydiate, 1995), the only previous genome-wide study on male and female meioses in plants revealed a general reduction in male recombination frequencies but no significant relationship between areas of reduced recombination and the location along the chromosome (de Vicente & Tanksley, 1991).

If the differences between male and female meioses observed in *B. nigra* hold true for other crops, for a transgene located close to the centromere, a cross with the donor plant acting as the female would be most likely to break undesirable linkages.

Moving genes into new species

Wide crosses can be used to transfer genes from one species to another and have been used extensively in cereals to introduce new variation into crop types from wild relatives (Sears, 1981). It is possible to use this route to introduce transgenes into crop species for which transformation systems do not exist.

In *Brassica* and related species, the development of efficient systems of ovary culture and embryo rescue has made successful interspecific crosses routine (Mithen & Herron, 1992). However, a detailed knowledge, however, of the homoeologous relationships between the chromosomes of the different species being crossed is very useful. This information is particularly useful if the chromosomes of the two species are rearranged with respect to one another and different parts of a single chromosome from one species are homoeologous with two distinct chromosomes from the other species (Lydiate et al., 1993). In resynthesised *B. napus*, formed by hybridising *B. oleracea* with *B. rapa*, the chromosomes of *B oleracea* and *B rapa* are considerably rearranged with respect to one another (Fig. 4; Parkin & Lydiate, unpublished).

Only selection for interspecific recombination events in particular intervals will allow the successful interspecific transfer of genes without associated genetic drag (Lagercrantz & Lydiate, unpublished). This genetic drag results from the inheritance of large segments of foreign chromosomes which cannot be broken down by recombination in the recipient species.

The interaction of a transgene with the genetic background

To have the best effect, transgenes should be in genetic backgrounds which augment the appropriately-regulated expression of the genes and produce the optimum levels of any necessary substrates. There are likely to be modifiers in the natural genetic backgrounds of recipient lines which influence the expression of transgenes. Molecular-genetic analysis should prove a powerful tool for investigating the influences which shape the final phenotype produced by transgenes.

References

Austin, S , M A Baer & J P Helgeson, 1985 Transfer of resistance to potato leaf roll virus from *Solanum brevidens* into *Solanum tuberosum* by somatic fusion Plant Sci 39 75–82

Botstein, D , R L White, M Skolnick & R W Davis, 1980 Construction of a genetic linkage map in man using restriction fragment length polymorphisms Am J Hum Genet 32 314–331

Chen, J L & M D Beversdorf, 1990 A comparison of traditional and haploid-derived breeding populations of oilseed rape (*Brassica napus*) for fatty acid composition of the seed oil Euphytica 51 59–65

Chyi, Y S , M E Hoenecke & J L Sernyk, 1992 A genetic link age map of restriction fragment length polymorphism loci for *Brassica rapa* (syn *campestris*) Genome 35 746–757

Dale, P J , J A Irwin & J A Scheffler, 1993 The experimental and commercial release of transgenic crop plants Plant Breeding 111 1–22

Dale, P J & H C McPartlan, 1992 Field performance of transgenic potato plants compared with controls regenerated from tuber discs and shoot cuttings Theor Appl Genet 84 585–591

Devos, K & M Gale, 1993 The genetic maps of wheat and their potential in plant breeding Outlook on Agriculture 22 93–99

Does, M P, B M M Dekker, M J A DeGroot & R Offringa, 1991 A quick method to estimate the T-DNA copy number in transgenic plants at an early stage after transformation, using inverse PCR Plant Mol Biol 17 151–153

Draper, J , R Scott, P Armitage & R Walden, 1988 Plant Genetic Transformation and Gene Expression Blackwell, London

Edwards, K , C Johnstone & C Thompson, 1991 A rapid and simple method for the preparation of plant genomic DNA for PCR analysis Nucl Acids Res 19 1349

Helentjaris, T, 1987 A genetic linkage map for maize based on RFLPs Trends Genet 3 217–221

Henderson, C A P & K P Pauls, 1992 The use of haploidy to develop plants that express several recessive traits using light-seeded canola (*Brassica napus*) as an example Theor Apl Genet 83 476–479

Horsch, R B, R T Fraley, S G Rogers, PR Sanders & A Lloyd, 1984 Inheritance of functional foreign genes in plants Science 223 496–498

Knutzon, D S, G A Thompson, S E Radke, W B Johnson, V C Knauf & J R Kridl, 1992 Modification of *Brassica* seed oil by antisense expression of a stearoyl-acyl carrier protein desaturase gene Proc Natl Acad Sci USA 89 2624–2628

Konieczny, A & F M Ausubel, 1993 A procedure for mapping *Arabidopsis* mutations using co-dominant ecotype-specific PCR-based markers Plant J 4 403–410

Kurata, N, G Moore, Y Nagamura, T Foote, M Yano, Y Minobe & M Gale, 1994 Conservation of genome structure between rice and wheat Bio/Technology 12 276–278

Lagercrantz, U & D J Lydiate, 1995 RFLP mapping in *Brassica nigra* indicates differing recombination rates in male and female meioses Genome 38 255–264

Lichter, R, E De Groot, D Fiebig, R Schweiger & A Gland, 1988 Glucosinolates determined by HPLC in seeds of microspore-derived homozygous lines of rapeseed (*Brassica napus* L) Plant Breeding 100 209–221

Lydiate, D, A Sharpe, U Lagercrantz & I Parkin, 1993 Mapping the *Brassica* genome Outlook on Agriculture 22 85–92

Mithen, R F & C Herron, 1992 Transfer of disease resistance to oilseed rape from wild *Brassica* species Proceedings of the 8th International Rapeseed Congress, pp 244–249 Saskatoon, Canada

Napoli, C, C Lemieux & R Jorgenson, 1990 Introduction of a chimeric chalcone synthase gene into petunia results in reversible co-suppression of homologous genes *in trans* Plant Cell 2 279–289

Paterson, A H, E S Lander, J D Hewitt, S Peterson, S E Lincoln & S D Tanksley, 1988 Resolution of quantitative traits into Mendelian factors by using a complete linkage map of restriction fragment length polymorphisms Nature 335 721–726

Paterson, A H, S D Tanksley & M E Sorrells, 1991 DNA markers in plant improvement Advances in Agronomy 46 39–90

Peach, C & J Velten, 1991 Transgene expression variability (position effect) of CAT and GUS reporter genes driven by linked divergent T-DNA promoters Plant Mol Biol 17 49–60

Potrykus, I, 1990 Gene transfer to plants assessment and perspectives Physiol Plantarum 79 115–134

Rafalski, J A & S V Tingey, 1993 Genetic diagnostics in plant breeding RAPDs, microsatellites and machines Trends Genet 9 275–280

Rhodes, C A, D A Pierce, I J Mettler, D Mascarenhas & J J Detmer, 1988 Genetically transformed maize plants from protoplasts Science 240 204–207

Sears, E R, 1981 Transfer of alien genetic material to wheat In L T Evans & W J Peacock (Eds) Wheat science – Today and Tomorrow, pp 75–89 Cambridge University Press

Shields, C R, T J Orton & C W Stuber, 1983 An outline of general resource needs and procedures for the electrophoretic separation of active enzymes from plant tissues In S D Tanksley & T J Orton (Eds) Isozymes in Plant Breeding and Genetics Part A, pp 443–468 Elsevier, Amsterdam

Shimamoto, K, R Terada, T Izawa & H Fujimoto, 1989 Fertile transgenic rice plants regenerated from transformed protoplasts Nature 338 274–276

Slocum, M K, S S Figdore, W C Kennard, J Y Suzuki & T C Osborn, 1990 Linkage arrangement of restriction fragment length polymorphism loci in *Brassica oleracea* Theor Appl Genet 80 57–64

Smith, P G, 1944 Embryo culture of a tomato species hybrid Proc Am Soc Hortic Sci 44 413–416

Song, K & T C Osborn, 1992 Polyphyletic origins of *Brassica napus* new evidence based on organelle and nuclear RFLP analyses Genome 35 992–1001

Stam, P, 1980 The distribution of the fraction of the genome identical by descent in finite random mating populations Genet Res Camb 35 131–155

Stuber, C W, 1992 Biochemical and molecular markers in plant breeding In J Janick (Ed) Plant Breeding Reviews, Vol 9, pp 37–61 John Wiley & Sons, New York

Tanksley, S D, 1983 Molecular markers in plant breeding Plant Mol Biol Rep 1 3–8

Tanksley, S D & T J Orton, 1983 Isozymes in Plant Breeding and Genetics Part A Elsevier, Amsterdam

Tanksley, S D, M W Ganal, J P Prince, M C de Vincente, M W Bonierbale, P Broun, T M Fulton, J J Giovanonni, S Grandillo, G B Martin, R Messegeur, J C Miller, L Miller, A H Paterson, O Pineda, M S Roder, R A Wing, W Wu & N D Young, 1992 High density molecular linkage maps of the tomato and potato genomes Genetics 132 1141–1160

U, N, 1935 Genomic analysis in *Brassica* with species reference to the experimental formation of *B napus* and peculiar mode of fertilisation Jap J Bot 7 389–452

Vasil, V, A M Castillo, M E Fromm & I K Vasil, 1992 Herbicide resistant fertile transgenic wheat plants obtained by microprojectile bombardment of regenerable embryogenic callus Bio/Technology 10 667–674

De Vicente, M C & S D Tanksley, 1991 Genome-wide reduction in recombination of backcross progeny derived from male versus female gametes in an interspecific cross of tomato Theor Appl Genet 83 173–178

Wendel, J F, C W Stuber, M D Edwards & M M Goodman (1986) Duplicated chromosome segments in maize (*Zea mays* L) further evidence from hexokinase isozymes Theor Appl Genet 72 178–185

White, R L, J M Lalouel, Y Nakamura, H Doniskeller, P Green, D W Bowden, C G P Mathew, D F Easton, E B Robson, N E Morton, J F Gusella, J L Haines, A E Retief, K K Kidd, J C Murray, G M Lathrop & H M Cann, 1990 The CEPH consortium primary linkage map of human chromosome-10 Genomics 6 393–412

Williams, J G K, A R Kubelik, K L Livak, J A Rafalski & S V Tingey, 1991 DNA polymorphisms amplified by arbitrary primers are useful as genetic markers Nucl Acids Res 18 6531–6535

Zehr, B E, J W Dudley, J Chojecki, M A Saghai Maroof & R P Mowers, 1992 Use of RFLP markers to search for alleles in a maize population for improvement of an elite hybrid Theor Appl Genet 83 903–911

Euphytica **85:** 359–366, 1995.

Variation of transgene expression in plants

Peter Meyer
Max-Delbrück-Laboratorium in der MPG, Carl-von-Linné Weg 10, D-50829 Köln, Germany

Key words: expression variegation, DNA-methylation, transgene activity

Summary

Genetic engineering has proved to be a powerful technique for the identification of plant genes and the construction of new genetic traits of agricultural importance. A growing number of transgenic plants is available to the breeder for incorporation into breeding programmes and production of new material for the market. A prerequisite for marketing transgenic material, however, is the control of stable transgene expression. The analysis of mechanisms that influence transgene activity is therefore not only an interesting field in basic research, but also an important aspect for application.

This article summarizes work on the analysis of expression variegation in transgenic petunia plants that carry the maize A1 transgene. Although far from giving a final answer to the solution of the problem, the description of certain parameters involved in the regulation of transgene expression should provide some guidance for the prevention of expression variegation in transgenic plants.

Introduction

One of the most critical aspects of molecular plant breeding is the long-term stable activity of recombinant genes that have been introduced into crop species. In classical breeding programmes the selection of a particular trait may be based on the activity of one or several genes often not identified at the molecular level. So far it is uncertain whether a lack of a particular trait is always necessarily due to a mutation of the corresponding gene, or if a structurally intact gene is no longer active. In molecular plant breeding, a recombinant molecule is transferred into the plant genome, encoding information for a protein that provides a defined structural or catalytic function. The transgene consists of a coding sequence of eukaryotic, prokaryotic or viral origin linked to a constitutive or developmentally-regulated promoter. By genetically engineering resistance genes, important agricultural traits have been designed and transferred, such as resistance against herbicides (Botterman & Leemans, 1988), insects (Fischhoff et al., 1987) or viral infection (Powell Abel et al., 1986). Based on increasing knowledge about the molecular organisation of plant genes (Schell, 1987) plant molecular biology also provided the tools to modify certain metabolic pathways and to create new traits (Mariani et al., 1990; Meyer et al., 1987; Gray et al., 1992).

The number of transgenic plants tested in field trials is increasing rapidly (Kareiva, 1993), but it can be anticipated that only a few primary transformants will enter the market. Most probably transgenic material will be used in breeding programmes to introduce the recombinant trait into a particular background or to combine different transgenes in one crop line. It is not enough to select a transformant that carries a recombinant transgene; the latter must function under various environmental conditions, in different genetic backgrounds and in successive generations.

The A1-gene marker has been employed to analyse certain parameters that influence transgene activity. Transfer of the A1-gene of *Zea mays* into the whitish flowered *Petunia hybrida* line R101, leads to salmon red flower pigmentation (Meyer et al., 1987). The A1 gene encodes dihydroflavonol reductase (DFR), an enzyme of the anthocyanin pathway. The receptor line R101 accumulates dihydrokaempferol, a substrate of the maize DFR. In transgenic plants expressing the

A1 gene, dihydrokaempferol is converted into leucopelargonidin, which is further processed by endogenous enzymes into a pelargonidin pigment. A1 activity can be followed by analysing the occurrence and stability of salmon-red pigmentation in the flower of transgenic plants. The system has the advantage that expression can be monitored in individual floral cells and that no staining procedure is required. Transgenic petunia plants can be propagated over several years via cuttings. Due to the continuous production of new flowers, expression of the transgene can be monitored over longer periods and under variable conditions.

The aim of this review is to summarize the data gained from expression studies based on the A1-system and to discuss certain approaches that target stabilization of transgene expression.

Regulation of gene activity

Expression of a particular piece of genetic information requires a series of molecular mechanisms including transcription of a gene, transportation and processing of the transcript, translation of the mRNA and post-translational modification of the translation product. Several regulatory mechanisms contribute to the reliable expression of an endogenous gene, regulating its temporal, spatial or inducible expression. Regulation of gene expression at the RNA level includes transcriptional control as well as post-transcriptional modification. Initiation of RNA synthesis is based on the recognition of *cis* acting sequences located within the region 5′ to the coding region (Maniatis et al., 1987). As binding of transcription factors to *cis* regulatory sequences regulates the activation of RNA transcription, the presence of particular binding proteins and the accessibility of *cis* regulatory sequences for binding proteins are two factors that determine the efficiency of transcription initiation. Both the chromatin structure (Felsenfeld, 1992) and the methylation state (Doerfler, 1983) of a promoter region can influence this efficiency in eukaryotic systems. While in *Drosophila*, gene inactivation corresponds to a condensation of chromatin (Shaffer et al., 1993), DNA-methylation plays an important role in the regulation of mammalian and plant genes, probably also via the assembly of methylated DNA into an inactive chromatin structure (Keshet et al., 1986). In plants, up to 30% of cytosines are methylated at carbon 5 (Adams & Burdon, 1985). Methylated C residues are located within CpG or CpNpG sequences, respectively (Gruenbaum et al., 1981). It has been proposed that methylation of symmetrical sequences provides a signal for a maintenance methylase that methylates C residues in a newly-synthesized strand located within symmetrical sequences, if the opposite strand carries a m^5C residue in the complementary sequence (Holliday & Pugh, 1975). An inverse correlation between gene transcription and cytosine methylation has been observed for several plant genes. The tissue-specific transcription of maize storage protein genes (Bianchi & Viotti, 1988) and C4 photosynthesis genes (Ngernprasirtsiri et al., 1989) correlates with hypomethylation of the genes in tissue where they are transcribed. Differences in DNA-methylation do not necessarily occur within a gene or its promoter region. For the cell-specific transcription of the PEPCase gene of C4 plants, transcription corresponds to demethylation of a region located 3.3 kb upstream of the gene (Langdale et al., 1991). A connection between DNA-methylation and changes in chromatin structure in plants is documented for rRNA genes in wheat (Thompson & Flavell, 1988), where rRNA genes at different *nor* loci are expressed at different levels. Nuclear dominance of particular *nor* loci is associated with both undermethylation and an increased sensitivity to DNase I digestion.

Although it is still a matter of debate whether DNA-methylation is the cause or the consequence of inactivation of transcription, at least in mammals the importance of DNA-methylation in embryonic development has been convincingly shown. A mutation of the murine DNA methyltransferase that led to a 3-fold reduction in levels of genomic m^5C levels, caused abnormal development and lethality in embryos (Li et al., 1992).

There is growing evidence that post-transcriptional mechanisms, which dominate the control of chloroplast gene expression (Berry et al., 1985), also play an important role in the regulation of nuclear gene expression (Gallie, 1993). On the RNA level this may involve pre-mRNA processing and transcript stability. Certain DNA elements, located downstream from the transcriptional start site, probably affect gene expression by regulating RNA stability (Green, 1993). It has been shown that different 3′ end regions strongly influence the level of gene expression, influencing the efficiency of 3′-processing or mRNA stability (Ingelbrecht et al., 1989). The DST element, a highly-conserved sequence located in the 3′ untranslated regions of a group of unstable soybean transcripts, has been identified as the cause for transcript destabilization (Newman et al., 1993). DST elements inserted into the 3′UTR region of reporter genes induced the rapid decay of reporter transcripts.

Mechanisms of transgene inactivation

Considering the fact that gene expression is regulated by a multilevel control mechanism, it follows that this regulation also affects the expression of recombinant genes introduced into the plant genome. Several reports confirm inactivation of transgene expression in which either initiation of transcription or transcript stability is involved.

Early studies on the activity of T-DNA genes in crown gall tumour lines, revealed a correlation between DNA-methylation and inactivity of T-DNA genes in certain lines (Amasino et al., 1984). Apparently T-DNA that integrates randomly into the genome became methylated at certain integration sites. The fact that certain copies of a transgene became hypermethylated while others remained hypomethylated and transcriptionally-active, strongly suggests that the degree of DNA-methylation is determined by the integration region. This position-specific inactivation could revert either spontaneously or after treatment of the cell line with the demethylating agent 5-azacytidine (Van Slogteren et al., 1989).

Another inactivation phenomenon was discovered when a transgenic tobacco line was retransformed with a construct that carried partial homologies to the construct already present in the genome (Matzke et al., 1989). In 15% of the double transformants, unidirectional inactivation of the first transgene occurred. This *trans*-inactivation was accompanied by hypermethylation of the promoters of the inactivated genes and was reversed when the two interacting loci were segregated from each other in progeny plants. Reversion, however, did not occur immediately, but gradually during the development of progeny plants. Occasionally partial methylation could even persist for two generations (Matzke & Matzke, 1991). Apparently inactivation in *trans* was dependent on homology between the two ectopic transgenes as well as the position where the transgenes were integrated. Inactivation not only occurred after a second transformation step, but it was also observed when two transgenic lines harbouring partial homology were crossed. Methylation of a target locus was again dependent on the integration region of the modifier locus and was more efficient when the target locus was homozygous (Matzke et al., 1993). Several mechanisms have been proposed to account for homology-dependent trans-inactivation. It was suggested that trans-inactivation reflects nonreciprocal competition for a transcriptional activator that was not freely diffusable (Matzke & Matzke, 1990), that it is due to transient ectopic pairing between the homologous transgenes accompanied by an exchange of chromatin components (Jorgensen, 1991) or that it derives from the interaction of anti-sense transcripts generated by transcription of antisense molecules from endogenous promoters at the 3′ end of the integration region (Grierson et al., 1991).

In contrast to the nonreciprocal interaction between ectopic transgenes, several co-suppression phenomena have been described that include the mutual inactivation of two partly homologous transgenes (Jorgensen, 1990). DNA-methylation may be involved in certain co-suppression effects (Linn et al., 1990), while for others no correlation between hypermethylation and gene inactivation was observed (Goring et al., 1991). Co-suppression is not restricted to transgenes, but can also affect endogenous genes. When the petunia chalcone synthase gene (*chs*) or the dihydroflavonol reductase gene (*dfr*) were introduced into *Petunia hybrida*, both the transgene and the endogenous gene were suppressed (Napoli et al., 1990; Van der Krol et al., 1990). Other examples include co-suppression of a tomato polygalacturonase gene (Smith et al., 1990) and tobacco chitinase (Hart et al., 1992) and glucanase (De Carvalho et al., 1992) genes. At least most co-suppression phenomena described so far seem to be due to post-transcriptional regulation requiring transcription of the transgene. It has been suggested that expression above certain threshold levels induces a biochemical switch into a new stable state of lower gene expression. This biochemical switch hypothesis, first described in relation to hormone habituation (Meins, 1989), provides an attractive explanation especially for co-suppression phenomena that are strictly correlated with homozygosity and where steady state levels of RNA, rather than transcription, are affected (Hart et al., 1992; De Carvalho et al., 1992). Although not all co-suppression phenomena can be explained by one mechanism, it is most likely that post-transcriptional degradation is involved in some of the effects described.

Parameters influencing A1-expression in *Petunia hybrida*

The different inactivation phenomena that were observed for transgenic petunia plants carrying the maize A1-gene seem to be exclusively due to inactivation of transcription that was accompanied by hypermethylation within the promoter region. So far we have not detected any evidence for post-transcriptional modifications. After direct DNA-transfer of an A1 gene linked to the viral CaMV 35S promoter into petunia protoplasts, primary transformants were regenerated that could be classified into three groups: plants with whitish flowers without any indication of A1-activity, transformants with salmon-red flowers exhibiting continuous A1-activity and plants carrying variegated flowers with A1-activity present in some floral cells. The uniformity of floral pigmentation was inversely correlated with the number of integrated A1-copies, as integration of multiple A1-copies predominantly created transformants with a whitish or variegated floral phenotype. Methylation analysis, using methylation-sensitive restriction enzymes, detected hypermethylation within the promoter region of multiple-copy transformants. The promoter of single copy plants, especially a *HpaII*-restriction site at the boundary of the promoter and the A1-cDNA, remained unmethylated (Lam et al., 1989). This *HpaII*-site is located close to a 21-bp element containing two TGACG motifs that bind the activation sequence factor (ASF-1) required for the maximal expression of the promoter (Lam et al., 1989). Although the measurement of DNA-methylation is limited to a few restriction sites, the *HpaII* site next to the ASF-1 binding sequence could be used as an indicator for promoter-specific methylation that inhibited A1-transcription.

To avoid co-suppression by multiple A1-copies, we further concentrated on the analysis of single copy transformants. Although the integration of one transgene improves the chance for universal expression, it is not sufficient. Three transformants (lines 16, 17 and 24) were selected, each carrying one complete transgene copy integrated at different chromosomal loci. In the white line 16 no A1-transcription was detectable, lines 17 and 24 showed A1-transcription at high and low intensity, respectively. In line 16 the inactivated transgene had integrated into a highly-repetitive genomic region. The weakly-expressed transgene in line 24 had integrated into a unique region that was highly methylated. The most intensively-transcribed transgene in line 17 was inserted into a unique region that was hypomethylated. The characteristic hyper and hypomethylation patterns of the integration regions in lines 24 and 17 were also imposed on the border region of the transgenes (Pröls & Meyer, 1992). Apparently the methylation state of the integration region is, at least in some part, responsible for position-dependent differences in gene expression.

To find out if integration of a single transgene into a hypomethylated genomic region was sufficient to guarantee stable expression, line 17 was further analysed. About 30,000 isogenic F1 plants derived from continuous pollination of homozygous line 17 were grown in the field. While blossoms on plants flowering early in the season were predominantly red, up to two-thirds of the later flowers on the same plants showed a reduction in A1-expression. Again, the reduction in A1-activity correlated with an increase in methylation. More than 95% of the greenhouse population of line 17, but only 37% of the field population, showed stable activity of the A1-gene, apparently due to environmental effects in the field. Interestingly, an endogenous factor seemed to determine the susceptibility of plants to this environmental stimulus. Progeny derived from early pollination of young parental plants was almost insensitive to inactivation, while in progeny from older flowers developed on the same parental plants, A1-expression was considerably reduced (Meyer et al., 1992). If an increasing degree of DNA-methylation is imposed on the transgene with increasing age of the parental plant, such imprinted methylation patterns would be transferred to the progeny. Progeny from older flowers would be more receptive to environmental stimuli as they carry a higher methylation density already.

For practical breeding, this model implies that breeders should use very young material for pollination to avoid selection of transgenes that might already have developed hypermethylation patterns. Transgene inactivation after exposure to environmental stimuli has also been observed in other systems, where tissue culture (Renckens et al., 1992) and heat stress (Walter et al., 1992) functioned as the environmental stimuli.

A detailed analysis of the transgene and its chromosomal integration region in several derivatives of line 17 implied that hypermethylation is limited to the transgene DNA only, while the hypomethylation state of the integration region remains unaltered (Meyer & Heidmann, 1994). The data indicate the likelihood of a DNA-methylation mechanism, which specifically recognizes foreign DNA. Such a mechanism, which has already been proposed for animal systems (Bestor, 1990; Doerfler, 1991), might recognize foreign DNA by its composition. It has been shown that the nuclear genomes of angiosperms are mosaics of long, compositionally homogeneous DNA segments, called isochores (Salinas et al., 1988) and that isochores contain defined GC contents of functional genes and their chromosomal environment (Matassi et al., 1989). The A1 transgene in line 17 differs significantly in its AT content from the chromosomal environment. The integrated plasmid DNA has an average AT content of 47.5%, whereas the neighbouring 269 bp at the 5′ end and 196 bp at the 3′ end show an average AT content of 74% and 77%, respectively. It may therefore be conceivable that the isochore composition of a transgene has to match its integration regions to avoid specific methylation of the transgene.

The hypermethylated transgene not only inhibited its own A1-transcription, but also exerted a paramutagenic influence on a homologous allele (Meyer et al., 1993). The term 'paramutation' has been coined to describe a heritable change in gene function directed by an allele, more specifically, the inactivation of a paramutable allele by a paramutagenic allelic homologue (Brink, 1956). When a hypermethylated allele and a hypomethylated allele were combined by crossing, the hypomethylated allele becomes methylated and silenced in a semi-dominant way. The paramutation-like behaviour is possibly due to an interaction between the two differentially methylated alleles. Such interaction could be mediated by transient pairing of the two alleles. Because we also observe a higher chromatin condensation in the hypermethylated allele (Meyer et al., 1994), we suggest that an exchange of methylation patterns also includes a change in chromosomal components. Assuming that transient pairing might not only occur between allelic homologues, but also among certain ectopic transgenes some trans-inactivation phenomena could include transient somatic pairing and an exchange of chromatin components. An improved understanding of the chromatin structure of plant genes is required to anticipate its importance in gene expression.

Approaches towards stabilization of transgene expression

Although the mechanisms involved in the regulation of gene expression are only partially understood, it is justifiable to say that multiple factors have to be controlled to increase the chance of stable expression of a transgene. These parameters include the number and structure of integrated copies, the chromosomal integration region, its chromatin structure and methylation state and the strength and specificity of promoter elements. Also included in such a parameter list should be DNA elements within the transgene that might function as recognition sequences for DNA-methylation or increased mRNA-degradation. This list, although incomplete, should serve as a guide for the selection of appropriate transgenic plants.

A desirable characteristic for any transgenic plant is single copy transgene integration. Most transgenic plants existing today have been generated using *Agrobacterium* strains that transfer their T-DNA to plant cells where it is randomly integrated into the genome (Zambryski et al., 1983). Because the T-strand is protected by bacterial proteins that also may facilitate its passage into the nucleus, T-DNA is frequently integrated without large internal deletions. Due to the mechanism of illegitimate integration (Mayerhofer et al., 1991), however, T-DNA insertions are often rearranged, preferentially in the junction region between T-DNA and chromosomal DNA (Simpson et al., 1982). Only one third of T-DNA transformants carry insertions at one locus (Koncz et al., 1989). Among single-locus insertions up to 60% of the plants may contain dimeric or trimeric inserts (Jorgensen et al., 1987; Jones et al., 1987). On average only one in ten transformants will contain a single integrated T-DNA copy. To avoid unintended co-suppression effects, such simple integration events should be selected and examined in more detail for major rearrangements and short target site duplications.

Direct DNA transfer into protoplasts and microparticle bombardment frequently lead to random integration of truncated copies of plasmid-DNA. The integration pattern seems to be mainly influenced by the structure of the transferred DNA. More than 50% of tobacco transformants derived from PEG-mediated transfer of supercoiled plasmids contain a single, although still truncated copy of the transgene. Early research on direct gene transfer (Davey et al., 1989) has mainly focused on the enhancement of transformation frequencies without paying much attention to the complexity of integration patterns and the long-term stability of transferred genes. Considering the possible expression-instability generated by multiple copies, any method aiming at an enhancement of transformation frequencies should also be evaluated for a conservation of a high proportion of single copy integration events, otherwise many transformants have to be scanned for a single integration event.

As mentioned previously, the integration region has an important influence on the methylation pattern and expression of the integrated transgene. The methylation state of the integration region or of its repetitiveness should therefore be examined. T-DNA preferentially integrates into chromosomal loci that are potentially transcribed, which reduces the probability for a negative influence of the integration region. Recently the stable expression of a *rolA* transgene in *Arabidopsis thaliana* was confirmed, when 425,000 individuals were tested (Dehio & Schell, 1993). Confirmation of stable *rolA* expression, which monitors somatic instability very precisely, demonstrates that at least certain transgenes that are free from inactivating influences of the integration region can be selected. However, the plants were not tested under varying environmental conditions.

In considering the specific methylation of a transgene DNA without detectable changes in the methylation pattern of the chromosomal environment (Meyer & Heidmann, 1994), we cannot rule out the fact that some recombinant constructs contain specific features that allow its recognition as foreign DNA by plant enzymes. It will be interesting to clarify this phenomenon and to define which characteristics of a transgene and an integration region contribute to stable expression. If the very low efficiencies for gene targeting into the plant genome (Paszkowski et al., 1988) can be improved, this would allow the selection of suitable integration regions into which transgenes might be targeted. As an alternative, recombinant genes could be transferred together with suitable chromosomal flanking regions. Scaffold attachment regions (SARs) which have been positioned 5′ and 3′ to a marker gene have stabilized transgene expression in animals (Stief et al., 1989) and plants (Breyne et al., 1992). The formation of chromosomal loops by SARs that connect genomic DNA to the nuclear matrix might isolate the transgene from the influences of the integration region. Another approach to uncouple transgene DNA from negative position effects could be its positioning on extrachromosomally replicating vectors (Meyer et al., 1992), although this will probably impose restrictions on the size of foreign DNA that can be propagated in transgenic plants.

Outlook

To adopt transgenic plants for agriculture, the participation of plant breeders seems essential. Transgene inactivation based on changes in chromatin structure are probably dependent on the genetic background. It can be expected that many different genes encoding chromatin proteins determine the degree of condensation. In *Drosophila* about 200 loci have been mapped that act either as suppressors or enhancers of position effect variegation. Transgenes repressed in one line might be active in another genetic environment and stabilization might be established in a classical breeding and selection process. As mentioned above, transgene inactivation might also occur by post-transcriptional degradation that is initiated at certain threshold levels. To avoid post-transcriptional degradation, lines would have to be selected that transcribe a transgene at moderate, but constant levels. Finally, extensive field studies will be required to assess the influence of various environmental conditions. Although molecular breeding does simplify the introduction of new traits into plants, the necessity for stable activity of the transgene requires a final evaluation of transgenic lines by the breeder.

Acknowledgement

I would like to thank Dr. Michael ten Lohuis for critical reading of the manuscript.

References

Adams, R.L.P. & R.H. Burdon, 1985. Molecular biology of DNA methylation. New York: Springer Verlag.

Amasino, R.M., A.L.T. Powell & M.P. Gordon, 1984. Changes in T-DNA methylation and expression are associated with phenotypic variation and plant regeneration in a crown gall tumor line. Mol. Gen. Genet. 197: 437–446.

Berry, J.O., B.J. Nikolau, J.P. Carr & D.F. Klessig, 1985. Transcriptional and post-transcriptional regulation of ribulose 1,5-bisphosphate carboxylase gene expression in light- and dark-grown amaranth cotelydons. Mol. Cell. Biol. 5: 2238–2246.

Bestor, T.H., 1990. DNA methylation: evolution of a bacterial immune function into a regulator of gene expression and genome structure in higher eukaryotes. Phil. Trans. R. Soc. Lond. B. 326: 179–187.

Bianchi, M.W. & A. Viotti, 1988. DNA methylation and tissue-specific transcription of the storage protein genes of maize. Plant Mol. Biol. 11: 203–211.

Botterman, J. & J. Leemans, 1988. Engineering herbicide resistance in plants. TIG 4: 219–222.

Breyne, P., M. Van Montague, A. Depicker & G. Gheysen, 1992. Characterization of a plant scaffold attachment region in a DNA fragment that normalizes transgene expression in tobacco. The Plant Cell 4: 463–471.

Brink, R.A., 1956. A genetic change associated with the R locus in maize which is directed and potentially reversible. Genetics 41: 872–889.

Davey, M.R., E.L. Rech & B.J. Mulligan, 1989. Direct DNA transfer to plant cells. Plant Mol. Biol. 13: 273–285.

De Carvalho, F., G. Gheysen, S. Kushnir, M. Van Montagu, D. Inze & C. Castresana, 1992. Suppression of beta-1.3-glucanase

transgene expression in homozygous plants EMBO J 11 2595–2602
Dehio, C & J Schell, 1993 Stable expression of a single-copy rolA gene in transgenic *Arabidopsis thaliana* plants allows an exhaustive mutagenic analysis of the transgene-associated phenotype Mol Gen Genet 241 359–366
Doerfler, W, 1983 DNA methylation and gene activity Ann Rev Biochem 52 93–124
Doerfler, W, 1991 Patterns of DNA methylation - Evolutionary vestiges of foreign DNA inactivation as a host defense mechanism Biol Chem Hoppe-Seyler 372 557–564
Felsenfeld, G, 1992 Chromatin as an essential part of the transcriptional mechanism Nature 355 219–224
Fischhoff, D A, K S Bowdish, F J Perlak, P G Marrone, S M McCormick, J G Niedermeyer, D A Dean, K Kusano-Kretzmer, E J Meyer, D E Rochester, S G Rogers & R T Fraley, 1987 Insect tolerant transgenic tomato plants BioTechnology 5 807–813
Gallie, D R, 1993 Post-transcriptional regulation of gene expression in plants Annu Rev Plant Physiol 44 77–105
Goring, D R, L Thomson & S J Rothstein, 1991 Transformation of a partial nopaline synthese gene into tobacco suppresses the expression of a resident wild-type gene Proc Natl Acad Sci USA 88 1770–1774
Gray, J, S Picton, J Shabbeer, W Schuch & D Grierson, 1992 Molecular biology of fruit ripening and its manipulation with antisense genes Plant Mol Biol 19 69–97
Green, P J, 1993 Control of mRNA stability in higher plants Plant Physiol 102 1065–1070
Grierson, D, R G Fray, A J Hamilton, C J S Smith & C F Watson, 1991 Does co-suppression of sense genes in transgenic plants involve antisense RNA? Tibtech 9 122–123
Gruenbaum, Y, T Naveh-Many, H Cedar & A Razin, 1981 Sequence specificity of methylation in higher plant DNA Nature 292 860–862
Hart, C M, B Fischer, J-M Neuhaus & F Meins, 1992 Regulated inactivation of homologous gene expression in transgenic *Nicotiana sylvestris* plants containing a defense-related tobacco chitinase gene Mol Gen Genet 235 178–188
Holliday, R & J E Pugh, 1975 DNA modification mechanisms and gene activity during development Science 187 226–232
Ingelbrecht, I L W, L M F Herman, R A Dekeyser, M C Van Montagu & A G Depicker, 1989 Different 3′ end regions strongly influence the level of gene expression in plant cells The Plant Cell 1 671–680
Jones, J D G, D E Gilbert, K L Grady & R A Jorgensen, 1987 T DNA structure and gene expression in petunia plants transformed by *Agrobacterium tumefaciens* C58 derivatives Mol Gen Genet 207 478–485
Jorgensen, R, 1990 Altered gene expression in plants due to trans interactions between homologous genes Tibtech 8 340–344
Jorgensen, R, 1991 Beyond antisense – how do transgenes interact with homologous plant genes? Tibtech 9 266–267
Jorgensen, R, C Snyder & J D G Jones, 1987 T-DNA is organized predominantly in inverted repeat structures in plants transformed with *Agrobacterium tumefaciens* C58 derivatives Mol Gen Genet 207 471–477
Kareiva, P, 1993 Transgenic plants on trial Nature 363 580–581
Keshet, I, J Lieman-Hurwitz & H Cedar, 1986 DNA methylation affects the formation of active chromatin Cell 44 535–543
Koncz, C, N Martini, R Mayerhof, Z Koncz-Kalman, H Korber, G P Redei & J Schell, 1989 High frequency T-DNA-mediated gene tagging in plants Proc Natl Acad Sci USA 86 8467–8471
Lam, E, P N Benfey, P M Gilmartin, R-X Fang & N-H Chua, 1989 Site specific mutations alter *in vitro* factor binding and change promoter expression pattern in transgenic plants Proc Natl Acad Sci USA 86 7890–7894
Langdale, J A, W C Taylor & T Nelson, 1991 Cell-specific accumulation of maize phosphoenolpyruvate carboxylase is correlated with demethylation at a specific site > 3 kb upstream of the gene Mol Gen Genet 225 49–55
Li, E, T H Bestor & R Jaenisch, 1992 Targeted mutation of the DNA methyltransferase gene results in embryonic lethylity Cell 69 915–926
Linn, F, I Heidmann, H Saedler & P Meyer, 1990 Epigenetic changes in the expression of the maize A1 gene in *Petunia hybrida* role of numbers of integrated gene copies and state of methylation Mol Gen Genet 222 329–336
Maniatis, T, S Goodbourn & J A Fischer, 1987 Regulation of inducible and tissue-specific gene expression Science 236 1237–1245
Mariani, C, M De Beuckeleer, J Truettner, J Leemans & R B Goldberg, 1990 Induction of male sterility in plants by a chimeric ribonuclease gene Nature 347 737–741
Matassi, G, L M Montero, J Salinas & G Bernardi, 1989 The isochore organization and the compositional distribution of homologous coding sequences in the nuclear genome of plants Nucl Acids Res 17 5273–5290
Matzke, M A & A J M Matzke, 1991 Differential inactivation and methylation of a transgene in plants by two suppressor loci containing homologous sequences Plant Mol Biol 16 821–830
Matzke, M A & E J M Matzke, 1990 Gene interactions and epigenetic variation in transgenic plants Developm Genet 11 214–223
Matzke, M A, F Neuhuber & A J M Matzke, 1993 A variety of epistatic interactions can occur between partially homologous transgene loci brought together by sexual crossing Mol Gen Genet 236 379–386
Matzke, M A, M Priming, J Trnovsky & A J M Matzke, 1989 Reversible methylation and inactivation of marker genes in sequentially transformed tobacco plants EMBO J 8 643–649
Mayerhofer, R, Z Koncz-Kalman, C Nawrath, G Bakkeren, A Crameri, K Angelis, G P Redei, J Schell, B Hohn & C Koncz, 1991 T-DNA integration a mode of illegitimate recombination in plants EMBO J 10 697–704
Meins, F J, 1989 Habituation heritable variation in the requirement of cultured plant cells for hormones Ann Rev Genet 23 395–408
Meyer, P & I Heidmann, 1994 Epigenetic variants of a trans genic petunia line show hypermethylation in transgene-DNA An indication for specific recognition of foreign DNA in transgenic plants Mol Gen Genet 243 390–399
Meyer, P, I Heidmann, G Forkmann & H Saedler, 1987 A new petunia flower colour generated by transformation of a mutant with a maize gene Nature 330 677–678
Meyer, P, I Heidmann & I Niedenhof, 1992 The use of African cassava mosaic virus as a vector system for plants Gene 110 213–217
Meyer, P, I Heidmann & I Niedenhof, 1993 Differences in DNA-methylation are associated with a paramutation phenomenon in transgenic petunia The Plant Journal 4 89–100
Meyer, P, F Linn, I Heidmann, z A H Meyer, I Niedenhof & H Saedler, 1992 Endogenous and environmental factors influence 35S promoter methylation of a maize A1 gene construct in transgenic petunia and its colour phenotype Mol Gen Genet 231 345–352

Meyer, P., I. Niedenhof & M. Ten Lohuis, 1994. Evidence for cytosine methylation of non symmetrical sequences in plants. EMBO J. 13: 2084–2088.

Napoli, C., C. Lemieux & R. Jorgensen, 1990. Introduction of a chimeric chalcone synthese gene into petunia results in reversible cosuppression of homologous genes in trans. The Plant Cell 2: 279–289.

Newman, T.C., M.O. Ohme-Takagi, C.B. Taylor & P.J. Green, 1993. DST sequences, highly conserved among plant SAUR genes, target reporter transcripts for rapid decay in tobacco. Plant Cell 5: 701–714.

Ngernprasirtsiri, J., R. Chollet, H. Kobayashi, T. Sugiyama & T. Akazawa, 1989. DNA methylation and the differential expression of C_4 photosynthesis genes in mesophyll and bundle sheet cells of greening maize leaves. J. Biol. Chem. 264: 8341–8248.

Paszkowski, J., M. Baur, A. Bogucki & I. Potrykus, 1988. Gene targeting in plants. EMBO J. 7: 4021–4026.

Powell Abel, P., R.S. Nelson, B. De, N. Hoffmann, S.G. Rogers, R.T. Fraley & R.N. Beachy, 1986. Delay of disease development in transgenic plants that express the tobacco mosaic virus coat protein gene. Science 232: 738–743.

Pröls, F. & P. Meyer, 1992. The methylation patterns of chromosomal integration regions influence gene activity of transferred DNA in *Petunia hybrida*. The Plant Journal 2: 465–475.

Renckens, S., H. De Greve, M. Van Montagu & J.-P. Hernalsteens, 1992. Petunia plants escape from negative selection against a transgene by silencing the foreign DNA via methylation. Mol. Gen. Genet. 233: 53–64.

Salinas, J., G. Matassi, L.M. Montero & G. Bernardi, 1988. Compositional compartmentalization and compositional patterns in the nuclear genomes of plants. Nucl. Acids Res. 16: 4269–4285.

Schell, J., 1987. Transgenic plants as tools to study the molecular organisation of plant genes. Science 237: 1176–1183.

Shaffer, C.D., L.L. Wallrath & S.C.R. Elgin, 1993. Regulating genes by packaging domains: bits of heterochromatin in euchromatin? TIG. 9: 35–37.

Simpson, R.B., P.J. O'Hara, W. Kwok, A.L. Montoya, C. Lichtenstein, M.P. Gordon & E.W. Nester, 1982. DNA from the A6S/2 crown gall tumor contains scrambled Ti-plasmid sequences near its junction with plant DNA. Cell 29: 1005–1014.

Smith, C.J.S., C.F. Watson, C.R. Bird, J. Ray, W. Schuch & D. Grierson, 1990. Expression of a truncated tomato polygalacturonase gene inhibits expression of the endogenous gene in transgenic plants. Mol. Gen. Genet. 224: 477–481.

Stief, A., D.M. Winter, W.H. Straetling & A.E. Sippel, 1989. A nuclear DNA attachment element mediates elevated and position-independent gene activity. Nature 341: 343–345.

Thompson, W.F. & R.B. Flavell, 1988. DNase I sensitivity of ribosomal RNA genes in chromatin and nucleolar dominance in wheat. J. Mol. Biol. 204: 535–548.

Van der Krol, A.R., L.A. Mur, M. Beld, J. Mol & A.R. Stuitje, 1990. Flavonoid genes in petunia: addition of a limiting number of copies may lead to a suppression of gene expression. The Plant Cell 2: 291–299.

Van Slogteren, G.M.S., P.J.J. Hooykaas & R.A. Schilperoort, 1984. Silent T-DNA genes in plant lines transformed by *Agrobacterium tumefaciens* are activated by grafting and 5-azacytidine treatment. Plant Mol. Biol. 3: 333–336.

Walter, C., I. Broer, D. Hillemann & A. Pühler, 1992. High frequency, heat treatment-induced inactivation of the phosphothricin resistance gene in transgenic single cell suspension cultures of *Medicago sativa*. Mol. Gen. Genet. 235: 189–196.

Zambryski, P., H. Joos, C. Genetello, J. Leemans, M. Van Montagu & J. Schell, 1983. Ti plasmid vectors for the introduction of DNA into plant cells without alteration of their normal regeneration capacity. EMBO J. 2: 2143–2150.

Euphytica **85**: 367–372, 1995.

Conservation of marker synteny during evolution

Katrien M. Devos, Graham Moore & Michael D. Gale
Cambridge Laboratory, John Innes Centre, Norwich Research Park, Colney, Norwich NR4 7UH, U.K.

Key words: collinearity, comparative mapping, *Poaceae*, Triticeae

Summary

An aspect of cereal science that is becoming increasingly important is comparative genetics. Establishment of the relationship between genomes within polyploids, between species within tribes and between species within families will allow not only the integration of genetic maps but also the knowledge acquired of each of the species. Using a set of homoeologous probes, workers found the relationship between the three wheat genomes to be precisely collinear, after taking a few major translocation events into account. Transfer of the wheat map to rye led to the elucidation of similar relationships between the three wheat genomes and that of rye. Genome collinearity, however, extends even beyond tribes. In a comparison of the genomes of wheat, rice and maize, it was shown that despite the separation of these genomes for possibly 50 million years, gene order was still highly conserved. This collinearity between genomes can be exploited in a number of ways.

Introduction

Over the years, extensive RFLP-based genetic maps have been developed in a range of crop species including maize (Coe et al., 1993; Burr et al., 1993), rice (Tanksley et al., 1993; Kochert & Jena, 1993; Kurata et al., 1994), wheat (Gale et al., 1995), barley (Graner et al., 1991; Heun et al., 1991; Kleinhofs et al., 1993) and rye (Devos et al., 1993a). Examples of marker and map applications include marker-aided selection in breeding programmes, biodiversity studies and gene isolation. An aspect of this area of science that is becoming increasingly important is comparative mapping. Within the Triticeae, the relationship between the three wheat genomes, the barley genome and the rye genome has been established using a common set of well-characterised wheat RFLP probes (Devos et al., 1993b; Laurie et al., 1993). A similar study within the grass tribe Andropogoneae has shown that many linkage groups are conserved between the genomes of maize and *Sorghum*, despite the three to four times higher DNA content of maize compared with *Sorghum* (Whitkus et al., 1992; Berhan et al., 1993). Synteny extends beyond tribes however. Comparative mapping experiments (Ahn & Tanksley, 1993; Ahn et al., 1993; Kurata et al., 1994) have shown large segments of conserved collinearity between genomes of wheat, rice and maize. In this paper we will discuss intergenomic relationships within wheat, within the Triticeae and within the *Poaceae*. The consequences and applications of comparative mapping attempts will also be discussed.

Comparative mapping within polyploids

Comparative mapping depends on the mapping of homoeoloci in different genomes with a common set of probes. The characteristics of probes suitable for comparative mapping include single or low copy number and an ability to hybridize with the genomes of a range of grass species. An important aspect of comparative genetic mapping is probe characterisation. Bread wheat, *Triticum aestivum* ($2n = 6x = 42$), is an allohexaploid with three homoeologous genomes, A, B, and D. The availability of aneuploid lines, particularly the nullisomic-tetrasomic and ditelosomic series, makes wheat an especially suitable member of the grass family in which to carry out probe characterisation. It was shown that probes that detect homoeoloci on the three wheat genomes, i.e. 'homoeologous probes', tend to be conserved across species, while the 'non-homoeologous probes' tend to be genome-specific and hybridize to loci in different genomes in a non-predictable way (Devos et al., 1993b). It is clear, therefore, that only homoeologous probes are suitable for comparative mapping studies.

To date, the wheat genetic map consists of some 1000 loci distributed over the three wheat genomes (Gale et al., 1995). The main characteristics of the map are the strong clustering of loci around the centromeres due to a localisation of recombination in the distal chromosome regions, and the conserved collinearity over the three homoeologous genomes. This collinearity is, however, disrupted in five of the 21 wheat chromosomes, which include chromosome arms 2BS, 4AS, 4AL, 5AL, 6BS and 7BS. These are due to chromosomal rearrangements involving chromosome arms 2BS and 6BS, and at least three reciprocal translocations between chromosome arms 4AL, 5BL and 7BS. The 4AL/5AL translocation is likely to have taken place before the polyploidization of wheat, while the 4AL/7BS translocation dates back to the tetraploid wheat parent of modern bread wheat. The 4AL/7BS translocation was either preceded or followed by an intrachromosomal translocation from 4AL to 4AS, resulting in an inversion of the 4AS/4AL arm ratio.

Comparative mapping within the Triticeae

Hybridization of the set of homoeologous probes, used to construct a genetic map of wheat, to the wheat-rye (*Secale cereale*, 2n = 14) disomic addition lines, indicated the presence of previously unrecognised regions of apparent non-homoeology between the wheat and rye genomes. Pairing studies (Naranjo et al., 1987) and previous hybridization experiments (Liu et al., 1992) had already shown the presence of three reciprocal translocations involving chromosome arms 4AL, 5AL and 7BS in the rye genome relative to the basic Triticeae genome. For these purposes, we assume that the basic Triticeae genome is that of the D-genome of wheat or the H-genome of barley. The construction of a genetic map of rye, using previously mapped wheat probes, allowed a precise and direct comparison of the homoeologous relationship between the wheat and rye genomes. The result was quite surprising (Fig. 1). Only the rye chromosome arms 1RS, 1RL, 2RL, 3RS, 4RS and 5RS appeared to share complete homoeology with the corresponding wheat arms. Each of the remaining eight rye chromosome arms have been involved in at least one translocation each since the divergence of wheat and rye from their common ancestor. Within each of the translocated segments, however, collinearity is conserved between the rye and wheat genomes. The presence of these rearrangements in three independent varieties, i.e. in the CS/*S. cereale* cv Imperial addition lines and in both parents of the mapping population, Ds2 and RxL 10, indicates that the rearrangements are evolutionary and define *S. cereale* rather than reflecting intervarietal differences. This hypothesis is further supported by pairing data obtained in a range of different wheat/rye hybrids (Naranjo & Fernández-Rueda, 1991), which indicated that only rye chromosome arms 1RS, 1RL, 2RL, 3RS and 5RS showed a completely homoeologous relationship to wheat.

A comparison of the evolutionary translocations that took place in the wheat and rye genomes shows that, within the limits of the mapping experiments, the 4AL/5AL and 4RL/5RL breakpoints are the same. The same, or similar, translocations have also been observed between the long arms of chromosomes 4 and 5 in *Secale montanum* (R^mR^m), the progenitor of cultivated rye, *Aegilops umbellulata* (UU), *Thinopyrum bessarabicum* (E^bE^b) and *Dasypyrum villosum* (VV) (Miller, 1984; King et al., 1994). This indicates that either the 4L/5L translocation occurred before the 'speciation' of the A, R^m, E^b, U and V genomes, or that there are regions in the chromosomes that are hypersensitive to breakage.

Consequences for plant breeding

The presence of chromosomal rearrangements in the rye genome compared with that of wheat may have important consequences for plant breeding. Indeed, rye carries a number of genes including those conferring disease resistance and aluminium tolerance that are potentially useful in wheat (Melz et al., 1992). Most attempts to introgress rye chromosome segments into wheat, however, have failed to produce wheat lines of good agronomic performance. A notable exception is the 1BS/1RS translocation that is present in many European feed wheats (W.J. Angus, personal communication). An analysis of the comparative genetic maps of wheat and rye confirms that rye chromosome 1R is the only chromosome that shows complete collinearity with its wheat homoeologues, and, therefore, replacement of the 1BS chromosome arm with the short arm

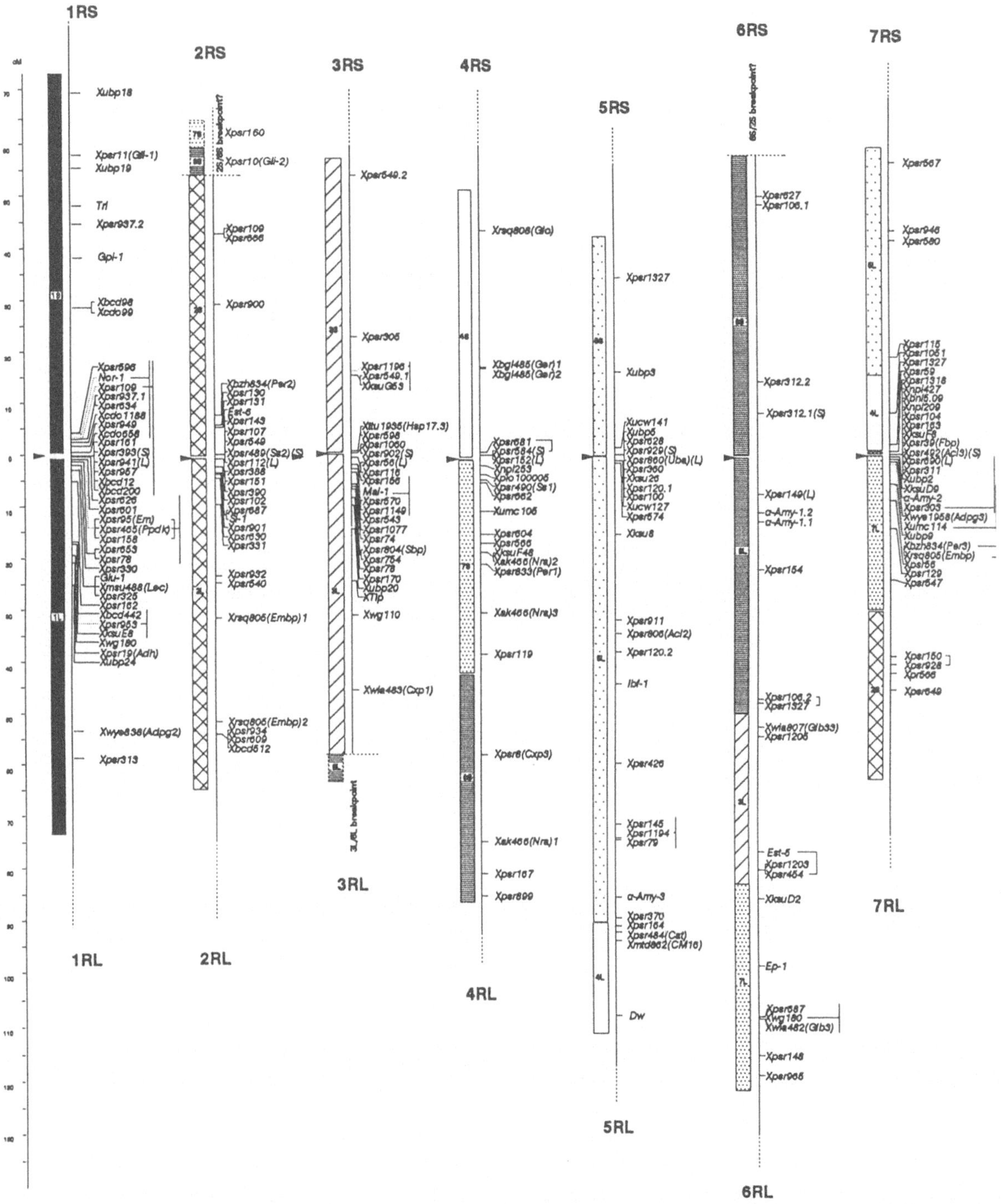

Fig 1 A genetic map of rye and its relationship to the wheat genome Bars indicate the homoeologous location of the probes in wheat Probes for which the order could not be determined unequivocally are indicated by line fragments

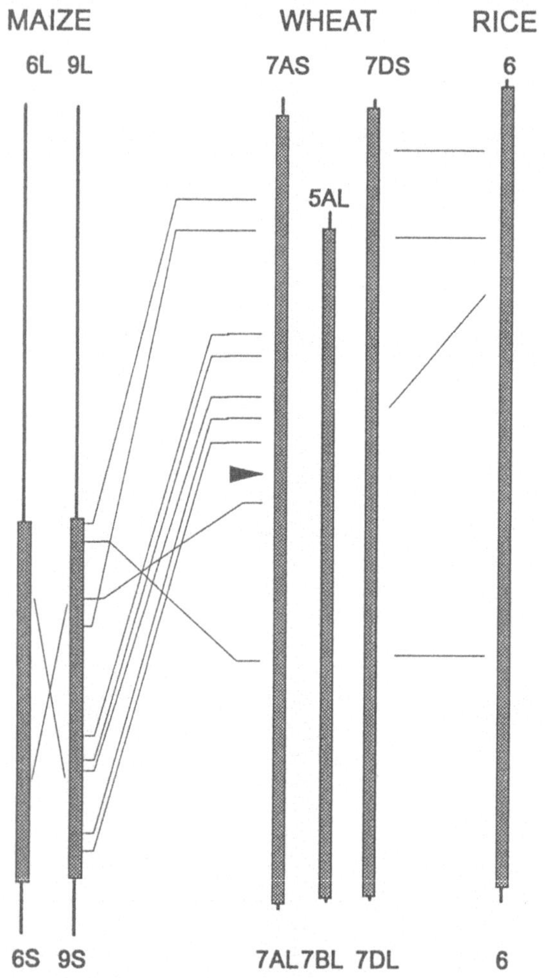

Fig. 2. Schematic representation of the homoeologous relationship between the wheat group 7 chromosomes, rice chromosome 6 and segments of maize chromosomes 9 and 6. Lines indicate the map position of a common set of probes mapped across the species. Collinearity between maize chromosome 9, and the wheat group 7 and rice 6 chromosomes is interrupted by chromosomal rearrangements.

from chromosome 1R will not disturb the chromosome balance. However, introgression of a gene carried on, for example, the long arm of rye chromosome 7 on the segment that is homoeologous to a segment of the short arms of the wheat group 2 chromosomes ('2S segment') through induced recombination with the 7A, 7B, or 7D chromosomes of wheat, would result in the replacement of a wheat homoeologous group 7 segment with a segment that shows homoeology with the short arms of the homoeologous group 2 chromosomes of wheat (Fig. 1). Consequently, the newly-generated line will carry a duplication of a group 2 segment and a deletion of a group 7 segment, and although this line may be viable, it is likely that imbalance would deleteriously affect agronomic performance. Knowledge of the chromosomal relationship between donor and recipient genomes can overcome this problem, and recombinant lines which carry the target gene and have a balanced genome can be selected using molecular markers.

Comparative mapping within the *Poaceae*

Comparative mapping studies, carried out within the Triticeae, have important applications in plant breeding. Knowledge of the precise correspondence of the genomes of wheat, barley and rye may, in addition to affecting breeding strategies for the introgression of alien chromosome segments into wheat, also overcome to a certain extent the problems associated with the low level of polymorphism in wheat. Unrevealed polymorphism in wheat may be located through mapping in either barley or rye. Furthermore, map integration also increases the available probe pool. The many applications of comparative mapping studies prompted the question as to whether homoeologous relationships could be established between genomes of species that belonged to different tribes.

Wheat ($2n = 6x = 42$; $C = 1.7 \times 10^{10}$ bp), maize ($2n = 20$; $C = 3 \times 10^9$ bp) and rice ($2n = 24$; $C = 4 \times 10^8$ bp) are three important crop species belonging to different tribes within the *Poaceae*. Comparative mapping studies carried out between rice and maize (Ahn & Tanksley, 1993), wheat and rice (Kurata et al., 1994), and wheat and maize (Devos et al., 1994) showed that, despite large differences in DNA content and different basic chromosome numbers, the genomes of wheat, rice and maize are highly collinear (Fig. 2). Again this is surprising, considering that the evolutionary divergence of these species may have taken place as long as 50 million years ago. A comparison of the genomes of wheat and rice revealed that rice chromosome 1 is collinear with the homoeologous group 3 chromosomes of wheat, rice chromosome 2 is syntenic with the wheat group 6 chromosomes, rice 3 with wheat 4, rice 4 with the long arm of wheat 2, rice 5 with wheat 1, rice 6 with wheat 7, rice 7 with the short arm of wheat 2, and rice 9 with wheat 5. To date, the syntenic relationship between rice chromosomes 8, 10, 11 and 12 and wheat has not been clearly established. A high degree of synteny was also observed between the genomes of rice and maize, although collinearity was disrupted by inversions and insertions in a number of chromosomes. In a limited comparative study of the maize and wheat genomes, it was shown that the short arm and the proximal part of the long arm of maize chromosome 9 were homoeologous with the wheat group 7 chromosomes. Here also, collinearity was disrupted by the presence of chromosomal rearrangements (Fig. 2).

Although the maize genome appears to have undergone considerable rearrangements relative to wheat and rye, it is clear that the homoeologous group 7 chromosomes of wheat, rice chromosome 6 and a large region of maize chromosome 9, have originated from the same ancestral genome (Fig. 2). Most probes that are single copy in wheat and rice are duplicated in maize, which supports the hypothesis that maize is an ancient tetraploid. On the basis of the comparative mapping data, it may be possible to identify the two ancestral genomes that make up the genome of modern maize in a way not possible using DNA probes from maize itself. Indeed, a number of probes that hybridize with rice chromosome 6 and the wheat homoeologous group 7 chromosomes detect loci on both maize chromosomes 6 and 9 (Ahn & Tanksley, 1993; Devos et al., 1994). Therefore, it is likely that at least a large segment of maize chromosome 6 has the same origin as chromosomes 9 of maize, 6 of rice and 7 of wheat.

Applications of comparative mapping within the *Poaceae*

The usefulness of comparative mapping studies has already been demonstrated within the Triticeae. In addition to providing a large pool of markers, collinearity between the genomes of other cereals such as wheat, rice and maize could be exploited to integrate the wealth of genetic knowledge acquired for each of the individual grass species. In maize, a large range of morphological and biochemical mutants are available, and many biochemical pathways and physiological processes have been studied thoroughly. Establishment of the homoeologous relationship between the genomes of maize and wheat may lead to the identification and even isolation of genes controlling similar traits in both species. The existence of large regions of collinearity between the genomes of wheat and rice, may be exploited by using rice as a model species for gene isolation in wheat. The small genome size of rice, about 40 times smaller than the wheat genomes and only two and half times that of the *Arabidopsis* genome, together with the relatively low amount of repetitive DNA sequences present, make the rice genome very amenable to gene isolation through chromosome walking. Isolation of agronomically-important genes in wheat through chromosome walking in rice, has now become an achievable aim.

Over the next few years, as the comparative maps of the Triticeae cereals, maize, rice, *Sorghum*, the millets, oats, the forage grasses and even sugarcane become coalligned, we can expect a wealth of further applications of grass comparative genetics.

References

Ahn, S., J.A. Anderson, M.E. Sorrels & S.D. Tanksley, 1993. Homoeologous relationships of rice, wheat and maize chromosomes. Mol. & Gen. Genet. 241: 483–490.

Ahn, S. & S.D. Tanksley, 1993. Comparative linkage maps of the rice and maize genomes. Proc. Nat. Acad. Sci. 90: 7980–7984.

Berhan, A.M., S.H. Hulbert, L.G. Butler & J.L. Bennetzen, 1993. Structure and evolution of the genomes of *Sorghum bicolor* and *Zea mays*. Theor. Appl. Genet. 86: 598–604.

Burr, B., F.A. Burr & E.C. Matz, 1993. Maize molecular map (*Zea mays*) 2N = 20. In: S.J. O'Brien (Ed). Genetic maps, pp. 190–203. Cold Spring Harbor Laboratory Press, Cold Spring Harbor.

Coe, E.H. & M.G. Neuffer, 1993. Gene loci and linkage map of corn (maize) (*Zea mays*) (2N = 20). In: S.J. O'Brien (Ed). Genetic maps, pp. 157–189. Cold Spring Harbor Laboratory Press, Cold Spring Harbor.

Devos, K.M., M.D. Atkinson, C.N. Chinoy, R.L. Harcourt, R.M.D. Koebner, C.J. Liu, P. Masojc, D.X. Xie & M.D. Gale, 1993a. Chromosome rearrangements in the rye genome relative to that of wheat. Theor. Appl. Genet. 85: 673–680.

Devos, K.M., T. Millan & M.D. Gale, 1993b. Comparative RFLP maps of the homoeologous group 2 chromosomes of wheat, rye and barley. Theor. Appl. Genet. 85: 784–792.

Devos, K.M., S. Chao, Q.Y. Li, M.C. Simonetti & M.D. Gale, 1994. Relationship between chromosome 9 of maize and wheat homoeologous group 7 chromosomes. Genetics. 138: 1287 - 1292.

Gale, M.D., M.D. Atkinson, C.N. Chinoy, R.L. Harcourt, J. Jiu, Q.Y. Li & K.M. Devos, 1995. Genetic maps of hexaploid wheat. In: Z.S. Li & Z.Y. Xin (Eds). Proc. 8th Int. Wheat Genet. Symp., pp. 29–40. China Agricultural Scientech Press, Beijing.

Graner, A., A. Jahoor, J. Schondelmaier, H. Siedler, K. Pillen, G. Fischbeck, G. Wenzel & R.G. Herrmann, 1991. Construction of an RFLP map of barley. Theor. Appl. Genet. 83: 250–256.

Heun, M., A.E. Kennedy, J.A. Anderson, N.L.V. Lapitan, M.E. Sorrells & S.D. Tanksley, 1991. Construction of an RFLP map for barley (*Hordeum vulgare* L.). Genome 34: 437–447.

King, I.P., K.A. Purdie, C.J. Liu, S.M. Reader, S.E. Orford, T.S. Pittaway & T.E. Miller, 1994. Detection of interchromosomal translocations within the Triticeae by RFLP analysis. Genome 37: 882–887.

Kleinhofs, A., A. Kilian, M.A. Saghai Maroof, R.M. Biyashev, P. Hayes, F.Q. Chen, N. Lapitan, A. Fenwick, T.K. Blake, V. Kanazin, E. Ananiev, L. Dahleen, D. Kudrna, J. Bollinger, S.J. Knapp, B. Liu, B. Sorrells, M. Heun, J.D. Franckowiak, D. Hoffman, R. Skadsen & B.J. Steffenson, 1993. A molecular, isozyme and morphological map of the barley (*Hordeum vulgare*) genome. Theor. Appl. Genet. 86: 705–712.

Kochert, G. & K.K. Jena, 1993. RFLP linkage map of *Oryza officinalis*, a wild rice (2n = 2x = 24). In: S.J. O'Brien (Ed). Genetic maps, pp. 80–81. Cold Spring Harbor Laboratory Press, Cold Spring Harbor.

Kurata, N., G. Moore, Y. Nagamura, T. Foote, M. Yano, Y. Minobe & M. Gale, 1994. Conservation of genome structure between rice and wheat. Bio/Technology 12: 276–278.

Laurie, D.A., N. Pratchett, K.M. Devos, I.J. Leitch & M.D. Gale, 1993. The distribution of RFLP markers on chromosome 2 (2H) of barley in relation to the physical and genetic location of 5S rDNA. Theor. Appl. Genet. 83: 305–312.

Liu, C.J., K.M. Devos, C.N. Chinoy, M.D. Atkinson & M.D. Gale, 1992. Non-homoeologous translocations between group 4, 5 and 7 chromosomes in wheat and rye. Theor. Appl. Genet. 83: 305–312.

Melz, G., R. Schlegel & V. Thiele, 1992. Genetic linkage map of rye (*Secale cereale* L.). Theor. Appl. Genet. 85: 33–45.

Miller, T.E., 1984. The homoeologous relationship between the chromosomes of rye and wheat. Current status. Can. J. Genet. Cytol. 26: 578–589.

Naranjo, T., A. Roca, P.G. Goicoechea & R. Giraldez, 1987. Arm homoeology of wheat and rye chromosomes. Genome 29: 873–882.

Naranjo, T. & P. Fernández-Rueda, 1991. Homoeology of rye chromosome arms to wheat. Theor. Appl. Genet. 82: 577–586.

Tanksley, S.D., T.M. Fulton & S.R. McCouch, 1993. Linkage map of rice (*Oryza sativa*) (2N = 24). In: S.J. O'Brien (Ed). Genetic maps, pp. 61–79. Cold Spring Harbor Laboratory Press, Cold Spring Harbor.

Whitkus, R., J. Doebley & M. Lee, 1992. Comparative genome mapping of sorghum and maize. Genetics 132: 1119–1130.

Euphytica **85:** 373–380, 1995.

In vitro screening and field evaluation of tissue-culture-regenerated sorghum (*Sorghum bicolor* (L.) Moench) for soil stress tolerance

R.R. Duncan[1], R.M. Waskom[2] & M.W. Nabors[2]
[1] *Department of Crop & Soil Sciences, University of Georgia, 1109 Experiment Street, Griffin, GA 30223-1797, USA;* [2] *Department of Agronomy and Botany, Colorado State University, Fort Collins, CO 80523, U.S.A.*

Key words: aluminium toxicity, soil acidity, somaclonal variation, sorghum, *Sorghum bicolor* (L.) Moench, tissue culture, salt stress, drought stress, variants

Summary

Sorghum bicolor (L.) Moench is generally quite sensitive to salt and acid (high aluminium) soil stresses, but quite tolerant of drought stress. As with any stress phenomenon, intra-specific variability exists within the genus. *In vitro* cell selection and somaclonal variation offer an alternative to traditional breeding methodology for generating improved breeding lines for hybrid development. A field selection protocol was developed for the three soil stresses and inter-stress evaluations were conducted in an effort to find multiple, stress-tolerant genotypes. The acid soil-drought stress, super-tolerant selections were located by the R_7 generation when exposed to a combined aluminium-drought stress field environment and when the regeneration population (number of regenerated lines from one callus source) was maintained at 15,000 plants or higher. A variant frequency of 0.1 to 0.2% for stress tolerance and acceptable agronomic traits among the surviving somaclones, provided an adequate number of phenotypes with desirable agronomic characteristics and a high level of soil stress tolerance. Subsequent research verified that the stress-tolerant regenerants had superior acid soil and drought stress tolerance to that of the donor parents, that their yield capabilities under stress were superior to their parents, and that their stress tolerance attributes were transferred in hybrid combinations. *In vitro* selection was not effective in increasing the number of field stress survivors. In fact, superior germplasms were developed from non-stressed callus or salt-stressed callus. *In vitro* selection reduced regeneration frequency and subsequent survival of plants under field stress. *In vitro*-stressed regenerants should be subjected only to non-stressed environments to maintain sufficient population numbers for field selection and thereafter should be subjected to stress environments during later (R_{5+}) generations. The optimal strategy for the exploitation of somaclonal variation may be through short-term cell culture (< 12 months) with no attempt at *in vitro* selection.

Introduction

Plant cell culture offers breeders an alternative strategy to conventional methodology for plant improvement. Since plant cells in culture may be genetically variable, *in vitro*-induced, spontaneous mutations can be selected for a specific trait (Larkin & Scowcroft, 1981; Maliga, 1984). Somaclonal variants may result from cell culture exposure or as a result of pre-existing genetic mutations in the somatic cells of the explant (Larkin & Scowcroft, 1981; Smith et al., 1993). The utility of somaclonal variation as a form of undirected mutation breeding and as a random process generally provides few useful variants (Baillie et al., 1992). Considerable somaclonal variation has been documented for phenotypic traits (Ullrich et al., 1991), some of which are detrimental (Baillie et al., 1992). Factors affecting somaclonal variation in cultured cells include time in culture, explant source, pathway of regeneration, genotype of the donor plant, environmental conditions during culture, concentration and type of plant growth regulators in the culture media and presence or absence of *in vitro* selective agents (Larkin & Scowcroft, 1981; Maliga, 1984; Shoemaker et al., 1981; Smith et al.,

1993). Cell cultures do not have to be exposed to *in vitro* selective agents to develop new somaclonal variants with useful agronomic traits (Evans & Sharp, 1986; Maliga, 1984). *In vitro* selection of Al-tolerant variants on a cellular level is complex and requires standardization and efficient methodology to exclude artifacts (Wersuhn et al., 1994).

Somaclonal variation/cell selection methods are similar to mutation breeding (Evans & Sharp, 1986; Sanford et al., 1984). Both techniques generate plants with useful and/or deleterious changes that may require extensive traditional breeding efforts to develop an improved cultivar. *In vitro*-regenerated plants potentially have fewer deleterious genetic problems, since the regeneration step should eliminate some negative traits (Smith et al., 1993). Cell culture variants for a specific trait are generated at a frequency of up to 30% (Maliga, 1984), while spontaneous mutation rates from mutation breeding methods occur at approximately one in 10^6 (Smith et al., 1993). Several mechanisms governing somaclonal variation induced during cell culture include gene amplification, single nucleotide base change, transposon migration, altered methylation states, chromosome instability, chromosome inversions, single gene mutations, translocations, cytoplasmic genetic changes, ploidy changes, rearrangements and partial chromosome deletion (Altman et al., 1991; Cassells et al., 1991; Evans & Sharp, 1986; Maliga, 1984; Shoemaker et al., 1991; Smith et al., 1993). Consequences of a high mutation frequency include chimerism, pleiotropy, instability, and epigenetic variation (Cassells et al., 1991; Philips et al., 1990; Jones, 1990; Smith et al., 1993). Epigenetic variation, an unstable, transient alteration in gene expression that can be reversible, is not expressed in progeny of regenerated plants (Smith et al., 1993). Useful variation must be stable, durable, and inherited in Mendelian fashion, while not altering other agronomic or economically-important traits of the donor parent (Cassells et al., 1991; Smith et al., 1993).

A major challenge in the exploitation of somaclonal variation is the screening for useful variation (Chaleff, 1983; Evans, 1989; Jones, 1990; Lee et al., 1988). *In vitro* selection can result in selection of degenerate lines (Wenzel, 1980). Selection of somaclones from apomictic species with agronomically-useful traits would, because of the apomictic mode of reproduction, stabilize the trait in subsequent seed generations (Davies & Cohen, 1992). Plants can be asexually propagated via somatic embryos obtained through tissue culture (Bowley et al., 1993; Matheson et al., 1990). Proper field selection can result in near-isogenic lines for subsequent genome mapping or marker-assisted identification of specific genes (Smith et al., 1993). The objective of this study was to investigate *in vitro* selected and nonselected regenerants for stable field-selected agronomically-useful traits.

Materials and methods

Plant material

Two sorghum genotypes (RTx430 (Miller, 1984) and 'Hegari') were subjected to tissue culture regeneration techniques, both with and without *in vitro* selection, and subsequently tested in the field. Field selection protocols were developed to locate and evaluate stable and durable soil stress-tolerant somaclones. Both cultivars are highly susceptible to acid soil, aluminium toxic stress. IS7173 was used as the tolerant (positive) check throughout the studies.

Tissue culture techniques developed at Colorado State University were used in the studies (MacKinnon et al., 1987). The explant source was mature embryos that were plated on Linsmaier & Skoog (LS) medium (Linsmaier & Skoog, 1965) with 2% sucrose, 2 mg L^{-1} 2,4-dichlorophenoxy acetic acid (2,4-D) and 0.5 mg L^{-1} kinetin to initiate and maintain the calli. Embryonic calli were visually selected by the fourth transfer and moved to fresh medium every 4 weeks (Waskom et al., 1990).

In vitro *selection*

For salt tolerance screening, 0, 3, 4, 6, 8, and 12 g L^{-1} NaCl was added to the culture media prior to autoclaving. These treatments were maintained for 8 to 15 months *in vitro*. For aluminium tolerance screening, $AlCl_3$ was added to the liquified basal media after autoclaving, but immediately prior to gelling of the media. Cultures were subjected to 0, 100, 200, 300, 400 and 700 μg^{-1} $AlCl_3$. Drought tolerance screening was accomplished by adding the osmoticum polyethylene glycol (PEG, molecular wt 8000) at 0, 9, 12 and 15% by weight to the medium.

Plant regeneration

After 6–15 subcultures of 28 days each, surviving embryogenic calli were transferred to LS medium with 0.1 mg L^{-1} indole-3-acetic acid (IAA) and 0.5 mg

L^{-1} 6-benzyl-amino purine (BAP) to stimulate plant regeneration. Shoots 2–3 cm long were transferred to LS media with 3 mg L^{-1} indole-3-butyric acid (IBA) to enhance root formation (Waskom et al., 1990).

Regenerated plants with adequate root production were transplanted in a soilless potting mixture and placed in humidified, shaded plastic tents. After 2 weeks, the regenerated plants were moved to greenhouse benches for growth and maturation. Prior to anthesis, the panicles of all R_0 plants (equivalent to F_1 generation) were bagged to ensure self-pollination. The subsequent seed harvested from these plants were designated the R_1 (= F_2) generation for planting in the field.

Field evaluations

Salinity tolerance
Salt tolerance evaluations and selections were conducted on saline fields near La Paz, Mexico (EC range 2–12 dSm^{-1}) on a sandy (unclassified) soil; pH unknown; organic matter < 5 g kg^{-1}. The study was conducted over 2 years (1989, 1990) with 4 replicates. The second site was near Yuma, AZ (EC range 2–4 dSm^{-1}) on a Glenbar silty clay loam (fine-silty, mixed hyperthermic Typic Torrifluvent); pH 7.9; organic matter 15 g kg^{-1}. The study was conducted over 2 years (1987, 1988) with 4 replicates (Miller et al., 1991; Timm et al., 1991).

Acid soil tolerance
Evaluations and selections were conducted near Griffin, GA on a Pacolet sandy clay (clayey, kaolinitic, thermic Typic Kanhapludults) at pH (H_2O) 4.3; exchangeable Al levels were 275 $\mu g\ g^{-1}$; organic matter was < 15 g kg^{-1}. The study was conducted over 4 years (1987–1990) with 12 replicates (Duncan, 1991; Miller et al., 1992; Waskom et al., 1990).

Drought stress tolerance
Evaluations and selections were conducted near Marana, AZ on a Pima clay loam (fine-loamy, mixed thermic Typic Torrifluvent); pH 7.8; organic matter was < 15 g kg^{-1}; < 700 mm water (about 460 mm preplant irrigation + about 230 mm rainfall during each growing season + 0 mm post-plant irrigation) was recorded on the plots. The study was conducted over 2 years (1989–1990) with 4 replicates.

Statistical analysis

All field tests for stress tolerance were planted in randomized complete block designs with 4 to 12 replicates. Due to the variability of acid soils, 12 replicates were necessary to estimate some differences among treatments (Miller et al., 1992). Data collected on plant survival, vigour and grain yield were used to compare *in vitro* treatments. Comparisons between *in vitro* treatments utilized a random model analysis of variance to determine the effect of *in vitro* selection. The performance of individual lines within treatments was compared with the original seed parent (explant donor) by Dunnett's two-tailed T-test. Square root transformations were conducted on stand, late-season plant survival, number of panicles produced, number of panicles harvested, and seed yield, to stabilize heterogeneous error variance among lines.

Soil pH values on a per-plot basis were used as covariates in the analysis of variance to adjust for soil pH variability (Miller et al., 1992). Covariate-adjusted means of regenerated variants were compared with their respective nonregenerated negative checks (donor parents susceptible to soil stress) within cultivars. Significance levels were multiplied by the number of comparisons to evaluate traits within each cultivar. This procedure for comparing covariate adjusted means is based on Bonferroni inequality (Milliken & Johnson, 1984, p. 33), and achieves the objective of Dunnett's T-test (control of error rates when comparing several means to a check).

Ratings scales

Acid soil vigour ratings were: 5 = least vigorous, totally susceptible, no plants remaining alive; 4 = stunted/abnormal plants, severe nutrient stress symptoms; 3 = some stunted plants, nutrient stress symptoms, delayed maturity; 2 = non-uniform height and/or panicle size, slight delay in maturity, slight visual nutrient stress symptoms; 1 = most vigorous, tolerant, uniform height, uniform maturity within row, normal size panicles, no visual nutrient stress.

Data were collected 40 days and 110 days after planting. The number of planicles that survived to produce viable seed was used as the criterion on which to determine individual plant survival on acid soils. Stand ratings were: 1 = 90–100% plant density, excellent to 5 = < 25% plant density, very poor. Drought vigour ratings were: 1 = excellent, to 5 = poor. Data were taken 80 days after planting. The developmental stage was also recorded: 1 = hard dough, 2 = soft dough, 3 = anthesis, 4 = boot, 5 = vegetative. The primary objective was to select the best plant type and most stress-tolerant lines to advance for further evaluations. Panicle weight and 100-seed weight were determined on selected lines only. Plant height ratings were based on percentage of the tolerant check: 1 = > 133% of positive check height; 2 = 100–133%; 3 = 75–100%; 4 = 50–75%; 5 = < 50% of check height. Data were collected > 110 days after planting.

Populations

From the original population of 212 R_1 (= F_2) lines, 90 selfed R_1 lines produced sufficient quantities of seed to screen for soil stress in 1986. The remaining 122 lines were increased to the R_2 (= F_3) generation in 1986 and screening for soil stress tolerance during 1987. Thirty-seven selfed lines from the 1986–1987 field stress evaluations survived to produce seed. These selections were retested as individual R_3 (= F_4) bulks during 1988. Two lines derived from 'Hegari' (H2021 and H2539) and two lines derived from RTx430 (T75-44 and T75-46) survived severe soil stress and produced seed in the field during 1988 (Table 1). The R_4 (= F_5) seed from these lines were increased to the R_5 (= F_6) generation in the greenhouse at Fort Collins, CO during the winter of 1989.

During 1989 at Griffin, GA, the four R_5 (= F_6) lines were planted on acid soils (pH < 4.5) with nonregenerated RTx430 and 'Hegari' (acid soil-susceptible, negative checks) donor parents and IS7173, the positive, acid soil-tolerant check.

Sixty-two individual R_6 (= F_7) selfed lines were selected from the 1989 field test and increased in Puerto Rico from November 1989 to March 1990. Eighteen selfed R_7 (= F_8) lines were selected in Puerto Rico on the basis of seed production potential and agronomically-desirable phenotypic traits (Miller et al., 1992).

Results

Comparisons of the regenerated lines with the donor parents under drought and acid soil stress conditions are presented in Table 1. For Tx430 variants under drought, the -46 and -44 selections were significantly more vigorous, taller, and for the -46 selection, produced a greater panicle weight during 1988. Under acid soil stress in 1989, the -46 and -44 selections had better vigour, greater plant survival and a higher number of harvested panicles than the donor parent. The -44 selection yielded more seed per plot than the -46 selection or the donor parent. Except for vigour, the tolerant check IS7173 had higher plant survival, more panicles harvested, and greater seed yield than either regenerated line.

During 1990, vigour ratings for the regenerated selections were better than for the donor parent and comparable to the tolerant check. Plant survival was significantly higher than for the donor parent, but one-half to one-third that of the tolerant check. The -46-20 regenerated selection produced the highest number of harvested panicles, which was over twice as many as the tolerant (positive) check. The superior grain-yielding selections were -46–20 and -46–50. Their yields were more than twice as high as the tolerant check under acid soil stress. The -46–20 and -46-50 selections were eventually released (Duncan et al., 1992).

For the 'Hegari' regenerants under drought stress during 1988, the selections matured earlier than the donor parent. All other traits were not significantly different. Under acid soil stress during 1989, the 2021 and 2539 selections were more vigorous, had higher plant survival, a greater number of panicles harvested and higher seed yields than the donor parent. Again, the selections were not comparable with the tolerant (positive) check for any traits except vigour. During 1990, the 2921-8 and -9 regenerated selections had significantly higher vigour ratings, plant survival and seed yields than the donor parent. The seed yields were comparable with or slightly higher than that of the tolerant check. The 2921-9 germplasm source was eventually released (Duncan et al., 1991b).

When evaluating *in vitro* selection (involving PEG screening) for induced drought-stress tolerance on sorghum callus, regeneration efficiency decreased with increasing PEG concentrations (Fig. 1). The percentage of regenerated sorghum variants surviving to physiological maturity decreased significantly when PEG was used in the culture medium. R_0 plants regenerated from stressed cultures rarely performed as well as those from non-stressed cultures under field stress conditions, due to overall reduced plant vigour, slow seedling growth and low seed production. Possibly, the *in vitro*-stressed regenerated lines have a greater sensitivity to stress environments, reflecting mutational or chromosomal aberrations that are unmasked in the field (Mitchell et al., 1992; Youssef et al., 1989). The donor parent line outyielded all PEG-selected regenerants,

Table 1 A. R_2 and R_3 variant performance under drought stress

Stress	Year	Generation	Pedigree	Designation	Stand	Vigour 50d	Vigour 84d	Vigour 113d	Height	Maturity	Panicle wt (g)	100-sd wt (g)
	1987		RTx430	Donor parent	–	–	3.4 ± 0.2	–	–	2.0 ± 0.0	–	–
		R_2	T2075-46	Regenerant	–	–	2.0	–	–	2.0	–	–
	1988		RTx430	Donor parent	3.5 ± 0.2	3.5 ± 0.2	–	3.8 ± 0.1	3.8 ± 0.1	1.0 ± 0.0	9	1.7
		R_3	T2075-46-1	Regenerant	3.0 ± 0.0	2.3 ± 0.3	–	2.0 ± 0.6	1.0 ± 0.0	2.0 ± 0.6	20	2.2
					NS	*		*	*	NS	*	NS
		R_3	T2075-44-3	Regenerant	4.3 ± 0.3	3.3 ± 0.3	–	1.7 ± 0.9	1.0 ± 0.6	1.3 ± 0.7	–	–
					NS	NS		*	*	NS		
							Vigour		*Height*	*Maturity*		
	1987		Hegari	Donor parent	–	–	2.4 ± 0.1	–	–	2.5 ± 0.1	–	–
		R_2	H2021-4	Regenerant	–	–	2.0 ± 0.1	–	–	2.0 ± 0.0	–	–
	1988		Hegari	Donor parent	3.9 ± 0.1	3.6 ± 0.1	–	3.4 ± 0.2	2.5 ± 0.2	3.9 ± 0.2	–	–
		R_3	H2021-4-1	Regenerant	4.0 ± 0.0	3.0 ± 0.0	–	2.0 ± 0.0	3.0 ± 0.0	1.0 ± 0.0	–	–
					NS	NS		NS	NS	*		

Table 1 B. R_5 and R_7 variant performance under acid soil stress

						Vigour		Plant survival	Panicles harvested	Seed yield	
								no./plot	no./plot	g/plot	
1989		RTx430	Donor parent	–	–	4.9	–	0.3	0.0	0.0	–
	R_5	T75-46	Regenerant	–	–	2.9*	–	8.5*	2.8*	4.8	–
	R_5	T75-44	Regenerant	–	–	3.0*	–	7.4*	2.2*	20.9*	–
		IS7173	Tolerant check	–	–	2.6	–	21.5	14.7	40.7	–
1990		RTx430	Donor parent	–	–	4.9	–	0.1	0.0	0.0	–
	R_7	T75-46-20	Regenerant	–	–	2.6*	–	18.4*	9.0*	84.6*	–
	R_7	T75-46-50	Regenerant	–	–	3.4*	–	17.6*	4.6	70.8*	–
	R_7	T75-46-53	Regenerant	–	–	3.0*	–	12.8*	6.4	64.0	–
	R_7	T75-44-13	Regenerant	–	–	3.4*	–	12.2*	4.0	20.6	–
		IS7173	Tolerant check	–	–	2.6	–	34.5	4.0	28.4	–
1989		Hegari	Donor parent	–	–	4.5	–	0.4	0.1	0.2	–
	R_5	H2021	Regenerant	–	–	2.1*	–	12.5*	5.8*	27.5*	–
	R_5	H2539	Regenerant	–	–	2.9*	–	12.9*	2.2*	7.7*	–
		IS7173	Tolerant check	–	–	2.6	–	21.5	14.7	40.7	–
1990		Hegari	Donor parent	–	–	4.8	–	0.8	0.2	0.5	–
	R_7	H2021-8	Regenerant	–	–	2.8*	–	29.4*	6.2	38.2*	–
	R_7	H2021-9	Regenerant	–	–	2.8*	–	18.8*	3.8	28.4*	–
	R_7	H2539-2	Regenerant	–	–	3.4*	–	17.4*	1.4	9.0	–
		IS7173	Tolerant check	–	–	2.6	–	34.5	4.0	28.4	–

Adapted from Waskom et al., 1990 & Miller et al., 1992.
* Significantly different from donor parent at 0.05 level of significance (Dunnett's T-Test).
Stand: 1 = excellent, 5 = poor; Height: 1 = > 2 m, 3 = 1.3 m, 5 = < 0.75 m; Vigour: 1 = excellent, 5 = poor; Maturity @ 110 days: 1 = hard dough, 5 = vegetative; NS = nonsignificant.

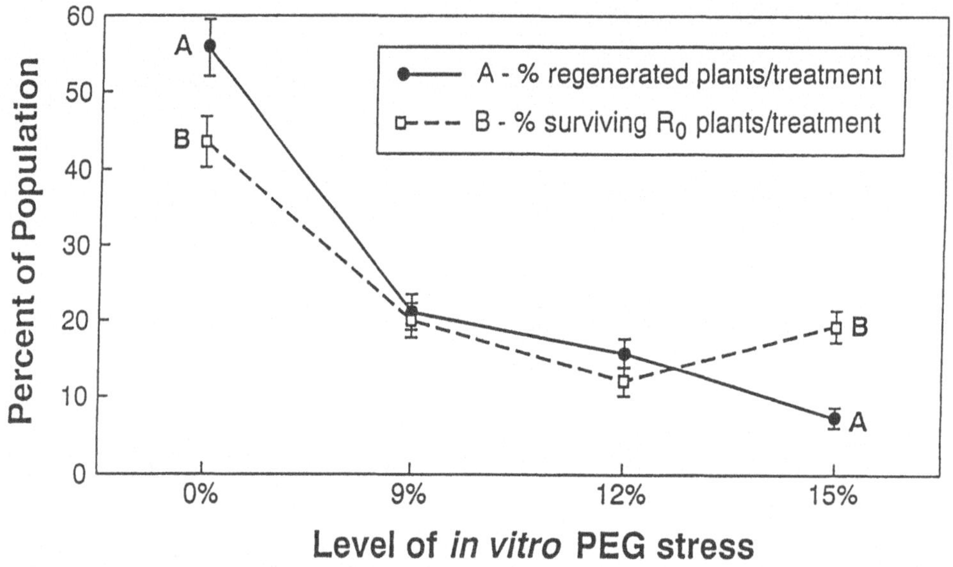

Fig. 1. Percentage of sorghum plants regenerated at four levels of *in vitro* PEG selection (A) and the percentage of R_0 (= F_1) plants surviving to produce R_1 (= F_2) seed in the greenhouse (B). The original population consisted of 212 R_1 lines. Treatments were the level of PEG stress *in vitro*.

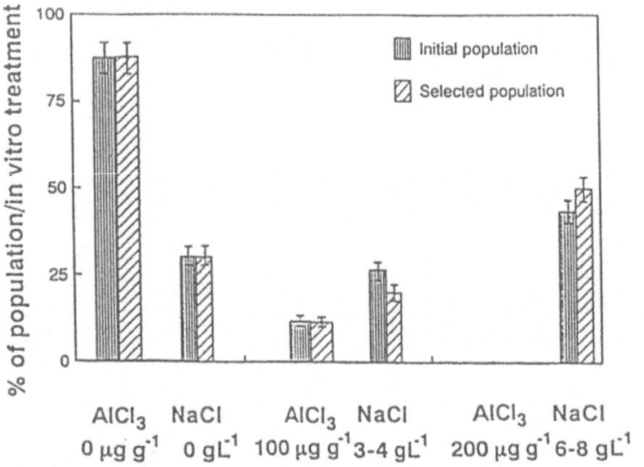

Fig. 2. Populations of selected sorghum lines, subjected to *in vitro* $AlCl_3$ and NaCl selection and their subsequent saline or acid soil field stress survival. The initial population consisted of 212 R_1 lines. The selected population comprised the survivors from the acid or salt stress evaluations that produced seed in the field.

suggesting that PEG *in vitro* selection detrimentally affected grain productivity in these variants.

Field comparisons of sorghum plants derived from NaCl- or $AlCl_3$-stressed and non-stressed cultures indicated no statistically significant survival differences could be detected between *in vitro* Al or salt treatments (Fig. 2). Individual variants were selected for enhanced stress tolerance to saline or Al-toxic soil conditions from both stressed and non-stressed callus. A few individual selections exhibiting significantly-enhanced tolerance to soil stress have been released from highly susceptible donor parents (Table 1; Duncan et al., 1991b; Duncan et al., 1992). No sorghum callus survived $AlCl_3$ above 100 $\mu g\ l^{-1}$ *in vitro*. The surviving variants did not arise from $AlCl_3$-selected callus, but from non-stressed callus, indicating a mutational change induced during tissue culture or regeneration, which was not selected *in vitro*. Time-in-culture negatively affected grain yield.

Discussion

Survival of tolerant regenerated selections from both stressed and non-stressed calli indicated that *in vitro* selection was neither effective nor necessary to regenerate and select improved stress-tolerant somaclones. A higher probability of success was found when well-adapted, but soil-stress-susceptible lines were exposed to the tissue culture process, followed by field stress evaluation and selection (Duncan et al., 1991c; Smith et al., 1993). A similar trend was found in barley (*Hordeum vulgare* L.) field evaluations (Baillie et al., 1992). The key to selecting successful stress-tolerant variants under field stress environments was maintenance of a large population of lines ($\geq$ 15,000 R_2

Table 2. Tissue-culture regenerated cultivars arising from somaclonal variation and officially released from breeding programmes

Cultivar	Scientific name	*In vitro* screen	Trait improvement	References
Andro flax	*Linum usitatissimum* L.	salt	salt tolerance	Rowland et al., 1989 McHughen et al., 1984 Rowland et al., 1988
GATCCP101 & 100	*Sorghum bicolor* (L.) Moench	salt	FAW† resistance	Duncan et al., 1991a Isenhour et al., 1991
GAC102	*Sorghum bicolor* (L.) Moench	salt	acid soil tolerance	Duncan et al., 1991b Waskom et al., 1990
GC103 & 104	*Sorghum bicolor* (L.) Moench	none	acid soil tolerance	Duncan et al., 1992 Miller et al., 1992 Foy et al., 1993
KS89WGRC9	*Triticum aestivum* L.	ABA	heat/drought tolerance	Sears et al., 1992

† FAW = fall armyworm (*Spodoptera frugiperda* (J.E. Smith)).

through R_7 plants per growing season) (Brandle & Miki, 1993; Duncan et al., 1991; Smith et al., 1993). This result is contrary to the view that the best breeding strategy for improving overall crop yield is to select for high yield on non-stressed soils (Richards, 1983). The field stress selection strategy used in this research is supported by previous work (Rosielle & Hamblin, 1981; Ceccarelli & Grando, 1991a, b). Very few germplasm releases have occurred from tissue culture-regenerated material (Table 2). Only the flax and sorghum variants resulted from somaclonal variation. Approximately 0.1–0.2% of the original regenerated lines tested in the field for specific somaclonal variations were found to have useful agronomic traits not expressed by the donor parent. Grain yield was detrimentally affected in the regenerated selections only by *in vitro* PEG screening and not by NaCl or $AlCl_3$. Subsequent research confirmed that the released acid soil-drought tolerant selections have superior performance in either environment separately, and that soil-stress root plasticity (sufficient volume and functionality of roots under field soil stress conditions to maintain the plant) was improved (Foy et al., 1993; Miller et al., 1992; Waskom et al., 1990).

Since both regeneration frequency and field plant survival were significantly reduced by the use of *in vitro* stressing agents, the best strategy to exploit somaclonal variation may be through short-term cell culture (< 12 months) with no *in vitro* selection.

References

Altman, D.W., D.M. Stelly & D.M. Mitten, 1991. Quantitative trait variation in phenotypically normal regenerants of cotton. *In Vitro* Cell Develop. Biol. 27p: 132–138.

Baillie, A.M.R., B.G. Rossnagel & K.K. Kartha, 1992. Field evaluation of barley (*Hordeum vulgare* L.) genotypes derived from tissue culture. Can. J. Plant Sci. 72: 725–733.

Bowley, S.R., G.A. Kelly, K. Anandarajah, B.D. McKersie & T. Senaratna, 1993. Field evaluation following two cycles of backcross transfer of somatic embryogenesis to commercial alfalfa germplasm. Can. J. Plant Sci. 73: 131–137.

Brandle, J.E. & B.L. Miki, 1993. Agronomic performance of sulfonylurea-resistant transgenic fluecured tobacco grown under field conditions. Crop Sci. 33: 847–852.

Cassells, A.C., M.L. Deadman, C.A. Brown & E. Griffin, 1991. Field resistance to late blight (*Phytophthora infestans* (Mont.) De Bary) in potato (*Solanum tuberosum* L.) somaclones associated with instability and pleiotropic effects. Euphytica 56: 75–80.

Ceccarelli, S. & S. Grando, 1991a. Selection environment and environmental sensitivity in barley. Euphytica 57: 157–167.

Ceccarelli, S. & S. Grando, 1991b. Environment of selection and type of germplasm in barley breeding for low-yielding conditions. Euphytica 57: 207–219.

Chaleff, R.S., 1983. Isolation of agronomically useful mutants from plant cell cultures. Science 219: 676–682.

Davies, L.J. & D. Cohen, 1992. Phenotypic variation in somaclones of *Paspalum dilatatum* and their seedling offspring. Can. J. Plant Sci. 72: 773–784.

Duncan, R.R., 1991. Acid soil tolerance breeding in sorghum. Adv. Agron. (India) 1: 71–79.

Duncan, R.R., D.J. Isenhour, R.M. Waskom, D.R. Miller, M.W. Nabors, G.E. Hanning, K.M. Petersen & B.R. Wiseman, 1991a. Registration of GATCCP100 and GATCCP101 fall armyworm resistant Hegari regenerants. Crop Sci. 31: 242–244.

Duncan, R.R., R.M. Waskom, D.R. Miller, R.L. Voigt, G.E. Hanning, D. Timm & M.W. Nabors, 1991b. Registration of GAC102: acid soil tolerant Hegari regenerant. Crop Sci. 31: 1396–1397.

Duncan, R R , R M Waskom, D R Mıller & R L Voıgt, 1991c Fıeld stress evaluatıon of tıssue culture regenerated sorghums In Proc Bıen Graın Sorghum Res & Utılız Conf 17 15–20

Duncan, R R , R M Waskom, D R Mıller, G E Hannıng, D A Timm & M W Nabors, 1992 Regıstratıon of GC103 and GC104 acıd-soıl tolerant Tx430 regenerants Crop Scı 32 1076–1077

Evans, D A , 1989 Somaclonal varıatıon genetıc basıs and breedıng applıcatıons Genetıcs 5 46–50

Evans, D A & W R Sharp, 1986 Applıcatıons of somaclonal varıatıon Bıo/Technology 4 528–532

Foy, C D , R R Duncan, R M Waskom & D R Mıller, 1993 Tolerance of sorghum genotypes to an acıd, alumınıum toxıc Tatum subsoıl J Plant Nutr 16 97–127

Isenhour, D J , R R Duncan, D R Mıller, R M Waskom, G E Hannıng, B R Wıseman & M W Nabors, 1991 Resıstance to leaf-feedıng by the fall armyworm (Lepıdoptera Noctuıdae) ın tıssue culture derıved sorghums J Econ Entomol 84 680–684

Jones, P W, 1990 Dısease resıstance In P J Dıx (Ed) Plant Cell Lıne Selectıon, pp 113–149 VCH Verlagsgesellschaft, Weınheım

Larkın, P J & W R Scowcroft, 1981 Somaclonal varıatıon – a novel source of varıabılıty from cell cultures Theor Appl Genet 60 197–214

Lee, M , J L Geadelmann & R L Phıllıps, 1988 Agronomıc evaluatıon of ınbred lınes derıved from tıssue culture of maıze Theor Appl Genet 75 841–849

Lınsmaıer, E M & F Skoog, 1965 Organıc growth factor requırements of tobacco tıssue cultures Physıol Plant 18 100–127

MacKınnon, C , G Gunderson & M W Nabors, 1987 Hıgh effıcıency plant regeneratıon by somatıc embryogenesıs from callus of mature embryo explants of bread wheat (*Trıtıcum aestıvum*) and graın sorghum (*Sorghum bıcolor*) *In Vıtro* Cell Develop Bıol 23 443–448

Malıga, P, 1984 Isolatıon and characterızatıon of mutants ın plant cell culture Ann Rev Plant Physıol 35 519–542

Matheson, S L, J Nowak & N L MacLean, 1990 Selectıon of regeneratıve genotypes from hıghly productıve cultıvars of alfalfa Euphytıca 45 105–112

McHughen, A & M Swartz, 1984 A tıssue culture derıved salt tolerant lıne of flax (*Lınum usıtatıssımum*) J Plant Physıol 117 109–117

Mıller, F R , 1984 Regıstratıon of RTx430 sorghum parental lıne Crop Scı 24 1224

Mıller, D R , R M Waskom, M A Brıck & P L Chapman, 1991 Transferrıng *ın vıtro* technology to the field Bıotechnology 9 143–146

Mıller, D R , R M Waskom, R R Duncan, P L Chapman, M A Brıck, G E Hannıng, D A Timm & M W Nabors, 1992 Acıd soıl stress tolerance ın tıssue culture derıved sorghum lınes Crop Scı 32 324–327

Mıllıken, G A & D E Johnson, 1984 Analysıs of messy data Vol 1 Desıgned experıments Van Nostrand Reınhold, New York

Phılıps, R L , S M Kaeppler & V M Peschke, 1990 Do we understand somaclonal varıatıon? In H J J Nıjkamp, L H W van der Plas & J van Aartrıjk (Eds) Progress ın Plant Cell and Molecular Bıology, pp 131–141 Kluwer Academıc Publıshers, Dordrecht

Rıchards, R A , 1983 Should selectıon for yıeld ın salıne regıons be made on salıne or non-salıne soıls? Euphytıca 32 431–438

Rosıelle, A A & J Hamblın, 1981 Theoretıcal aspects of selectıon for yıeld ın stress and non-stress envıronments Crop Scı 21 943–946

Rowland, G G , A McHughen & C McOnıe, 1988 Fıeld evaluatıon on nonsalıne soıls of a somaclonal varıant of McGregor flax selected for salt tolerance *ın vıtro* Can J Plant Scı 68 345–349

Rowland, G G , A McHughen & R S Bhatty, 1989 Andro flax Can J Plant Scı 69 911–913

Sanford, J C , N F Weeden & Y S Chyı, 1984 Regardıng the novelty and breedıng value of protoplast-derıved varıants of 'Russet Burbank' (*Solanum tuberosum* L) Euphytıca 33 709–715

Sears, R G , T S Cox & G M Paulsen, 1992 Regıstratıon of KS89WGRC9 stress-tolerant hard red wınter wheat germplasm Crop Scı 32 507

Shoemaker, R C , L A Amberger, R G Palmer, L Oglesby & J P Ranch, 1981 Effect of 2,4-dıchlorophenoxyacetıc acıd concentratıon on somatıc embryogenesıs and herıtable varıatıon ın soybean (*Glycıne max* (L) Merr) *In Vıtro* Cell Devel Bıol 27 p 84–88

Smıth, R H , R R Duncan & S Bhaskaran, 1993 *In vıtro* selectıon and somaclonal varıatıon for crop ımprovement In D W Buxton (Ed) Proc 1st Int'l Crop Scı Congress, July 1992, Ames, IA Crop Scı Soc Amerıca, Madıson, WI pp 629–632

Tımm, D A , R M Waskom, D R Mıller & M W Nabors, 1991 Greenhouse evaluatıon of regenerated sprıng wheat for enhanced tolerance Cereal Res Commun 19 451–457

Ullrıch, S E , J M Edmıston, A Kleınhofs, D A Kudrna & M E H Maatouguı, 1991 Evaluatıon of somaclonal varıatıon ın barley Cereal Res Commun 19 245–260

Waskom, R M , D R Mıller, G E Hannıng, R R Duncan, R L Voıgt & M W Nabors, 1990 Fıeld evaluatıon of tıssue-culture-derıved sorghum for ıncreased tolerance to acıd soıls and drought stress Can J Plant Scı 70 997–1004

Wenzel, G , 1980 The potentıal and lımıts of classıcal genetıcs ın plant breedıng In F Sala et al (Eds) Plant Cell Cultures Results and Perspectıves, pp 33–47 Elsevıer, Amsterdam

Wersuhn, G , T Kalettka, R Gıenapp, G Reımke & D Schulz, 1994 Problems posed by *ın vıtro* selectıon for alumınıum-tolerance when usıng cultıvated plant cells J Plant Physıol 143 92–95

Youssef, S S , R Morrıs, P S Baenzıger & C M Papa, 1989 Cytogenetıc studıes of progenıes from crosses between 'Centurk' wheat and ıts doubled haploıds derıved from anther culture Genome 32 622–628

Euphytica **85**: 381–393, 1995.

Production and field performance of transgenic alfalfa (*Medicago sativa* L.) expressing alpha-amylase and manganese-dependent lignin peroxidase

S. Austin[1], E.T. Bingham[2], D.E. Mathews[3], M.N. Shahan[1], J. Will[1] & R.R. Burgess[1]
[1] *Biotechnology Center, University of Wisconsin-Madison, 1710 University Ave., Madison, WI 53705, USA;* [2] *Dept. of Agronomy, UW-Madison, 1575 Linden Drive, Madison, WI 53706, U.S.A.;* [3] *Dept. of Biochemistry, UW-Madison, 420 Henry Mall, Madison, WI 53706, U.S.A.*

Key words: alfalfa, alpha-amylase, field performance, manganese-dependent lignin peroxidase, *Medicago sativa*, transformation

Summary

Transgenic alfalfa plants expressing *Bacillus licheniformis* alpha-amylase and manganese-dependent lignin peroxidase (Mn-P) from *Phanerochaete chrysosporium* were produced using the *Agrobacterium tumefaciens* transformation system. In each case, there was a range of expression of the introduced gene among independent transgenic plants. Plants producing alpha-amylase showed no alteration of phenotype. Production of Mn-P in alfalfa, however, in most cases adversely affected plant growth and development. Affected plants were stunted with yellowing foliage, but survived and produced seed. Results from field trials showed that Mn-P production in transgenic alfalfa reduced dry matter yield and plant height. The extent of these symptoms and yield reduction was, for the most part, related to the level of foreign protein production as estimated by Western analysis. Field data from transgenic plants expressing alpha-amylase showed that there was no effect of foreign protein production on plant performance. Expression of Mn-P was shown to segregate in sexual progeny derived from transgenic plants.

Abbreviations: Mn-P – manganese-dependent lignin peroxidase

Introduction

Currently, most industrial enzymes are produced from micro-organisms by large-scale fermentation. An alternative approach would be to express and recover these enzymes from transgenic plants. The production of commercially-interesting heterologous proteins and peptides in plants has been reported by several researchers (DeZoeten et al., 1989; Hiatt et al., 1989; Vandekerckhove et al., 1989; During et al., 1990; Krebbers & Vandekerckhove, 1990; Sijmons et al., 1990). There is also interest in using transgenic crop plants for the production of industrial bulk enzymes (Pen et al., 1992). We are currently conducting a multidisciplinary feasibility study on the production of industrial enzymes in transgenic alfalfa. The overall goal of the research is to develop genetically-engineered alfalfa that produces high levels of industrially-important enzymes, to develop rapid methods for extracting and purifying these enzymes and thus provide a high value product which takes advantage of existing agricultural productivity. Alfalfa has certain advantages for this purpose. It is a perennial, widely-grown crop capable of three or more harvests a year. It is beneficial environmentally since it does not require annual tilling and planting, and needs fewer applications of fertilizer and pesticides than conventional row crops. Furthermore, technology for extracting protein from alfalfa while leaving a valuable residue for animal feed is well-developed.

As part of our feasibility study for the production of industrial enzymes in transgenic alfalfa, we have recently demonstrated the production of manganese-dependent lignin peroxidase (Mn-P) from the fungus *Phanerochaete chrysosporium* in transgenic plants (Mathews et al., 1993; Austin et al., 1994). This enzyme has potential for large-scale industrial usage for lignin degradation and/or as a bleaching agent in biopulping processes. Expression of this enzyme in alfalfa may also have the potential for increasing the fibre digestibility of forages for ruminants. In addition we have expressed alpha-amylase from *Bacillus licheniformis* in alfalfa. This enzyme is involved in starch degradation and has several current and potential large-scale industrial uses in starch processing (Schwardt, 1990). The production of this enzyme has already been demonstrated in tobacco (Pen et al., 1992). It was chosen as a model enzyme for the overall feasibility study to test recovery protocols since it is a robust enzyme that is active over a wide pH and temperature range.

This report outlines the production and characterization of transgenic alfalfa expressing alpha-amylase or Mn-P. Data from the first year of a field test of transgenic plants are also presented.

Materials and methods

Plant materials. A preliminary survey to screen for regenerative ability and response to kanamycin was performed using 18 alfalfa genotypes derived from Regen-SY (Bingham, 1991). Based on data obtained on basic tissue culture responses, five genotypes were selected for use in transformation studies using leaflets as explant sources. One genotype, RSY#27, Regen-SY-27), was identified as the most amenable for transformation and was used in the experiments described in this report. Plants were maintained in growth rooms with a 16 h photoperiod of 300 μE $m^{-2}s^{-1}$ and day and night temperatures of 21° C and 19° C, respectively.

Vectors. A detailed description of the construction of the plant transformation vectors will be presented elsewhere. Briefly, pDM321 and pDM322 consist of the binary plant transformation vector pCGN1578 (McBride & Summerfelt, 1990) into which was inserted a plant expression cassette containing the coding sequence of a Mn-P gene from *Phanerochaete chrysosporium.* The expression cassette contains the 'Mac' promoter (Comai et al., 1990) and the mannopine synthase transcription terminator (McBride & Summerfelt, 1990). The Mn-P coding sequence was obtained from a cDNA clone (kindly supplied by D. Cullen, USDA, Forest Products Laboratory, Madison, WI) encoding an allelic variant of MP-1 (Pease et al., 1989). pDM321 contains the complete Mn-P coding sequence, whereas in pDM322, the sequence encoding the amino terminal 24 amino-acids was replaced with sequence encoding the amino-terminal 21 amino acids of the soybean vegetative storage protein, VSPβ (Mason et al., 1988). In effect, the fungal amino-terminal endoplasmic reticulum (ER) signal sequence was replaced with a plant ER signal sequence. pDM408 is identical to pDM322 but contains the coding sequence for *Bacillus licheniformis* alpha-amylase mature protein (Yuuki et al., 1985) instead of Mn-P. Thus, the sequence encoding the 29 amino acid bacterial secretion signal was replaced with the sequence encoding the 21 amino acid amino-terminal ER signal sequence of VSPβ. Plant transformation constructs were introduced into *Agrobacterium tumefaciens* strain LBA4404.

Transformation and regeneration of alfalfa. The transformation method was defined after testing several modifications of the basic explant co-cultivation system (Horsch et al., 1985) as a way of optimizing the transformation procedure. These included pretreatment of the tissue, longer co-cultivation times and the use of different levels of the antibiotic kanamycin monosulphate to select for transformed tissue. The optimized procedure is as follows. New-growth trifoliates were taken from RSY27 plants maintained in a growth room (conditions as described above) and sterilized using alcohol and bleach washes (30s in 70% alcohol, 90s in 20% hypochlorite + 0.1% SDS, followed by three rinses in sterile distilled water). Leaf edges were cut on moist filter paper and tissue dropped into liquid SH-II medium (Bingham et al., 1975). When sufficient explants had been taken, they were moved to a suspension of *Agrobacterium* cells (containing the engineered plasmid) from an overnight culture grown in liquid YEP selection medium. Cell density was adjusted to fall between 0.6–0.8 at A_{660}. After 30 minutes inoculation, the explants were gently blotted on filter paper and placed on B5H medium (Brown & Atanassov, 1985) for 4 days. They were then rinsed twice in sterile water and cultured on B5H for a further 4 days. At the end of this period, they were rinsed three times and transferred to B5H containing 25 mg l^{-1} kanamycin and 250 mg l^{-1} carbenicillin. Plates were maintained at 24° C, 16 h photoperiod and light intensity of 60–80μE $m^{-2}s^{-1}$. Explant-derived calli (and occasionally embryoids) which formed within 3 weeks on this medium were moved to B5H with antibiotics but without hormones to allow for further embryoid production and development of existing embryoids. After 3–4 weeks, embryos were transferred to MS medium (Murashige & Skoog, 1962) plus the two antibiotics to allow for development into plantlets. Callus did form on untreated explants in the presence of 25 mg l^{-1} kanamycin but embryos were never produced. Each explant piece can give rise to multiple (up to 40) embryos. Plantlets were rooted on MS medium lacking antibiotics. For evaluation purposes only, one rooted plant per explant piece was assessed, thus assuring that each was the result of an independent transformation event. Plants were maintained in a growth room under conditions as described above.

Expression assays. 1. Alpha-amylase assays. Sample preparation and starch plate assays were carried out basically as described by Pen et al. (1992). Briefly, leaf tissue (200 mg) was homogenized by hand in 500 μl 0.5 M glycine buffer (pH 9.0) containing 10 mM $CaCl_2$. The homogenate was centrifuged for 10 min at 14,000 rpm (16,000 g) in a table top centrifuge and the supernatant collected. Soluble protein content was determined following the Bradford procedure (Bradford, 1976), using BSA as standard. The supernatant was then divided into two samples and half was heated to 60° C for 15 min. This was to discriminate between endogenous plant alpha-amylase activity and the heat-stable activity of the *Bacillus* protein. Samples (2 μl) of heated and non-heated extract were placed on 1% starch plates solidified with 0.9% agar and the extent of starch degradation determined after 16 h at room temperature by iodine staining. Halo size was compared with that of standard dilutions of *B. licheniformis* alpha-amylase (Sigma). Plant samples were taken at various times. For the initial testing of putative transgenics, several trifoliates were pooled from 2 to 3-week old, rooted *in vitro* plants. The third fully-expanded trifoliate from several stems of mature plants just prior to flowering was taken for assay of both growth room and field-grown plants. Attempts to quantify alpha-amylase activity in crude alfalfa extracts using the spectrophotometric assay of Saito, (1973) gave inconsistent results. We are currently working on modifying the method of sample preparation and assay conditions such that we can accurately quantify the enzyme.

2. Analysis for Mn-P production. Samples from *in vitro* and mature plants were taken and prepared as described for alpha-amylase, except that they were ground in 50 mM L-tartrate buffer, (pH 4.5). Representatives of all 16 plant lines in the field (generally, leaves from 2–3 clones of any individual) were sampled at the end of the first and second harvest growth period. Eight of the 16 lines were also sampled at weekly intervals during the whole of the second harvest growth period. In addition, older leaf, stem and root samples were taken from the group of 8 plants just prior to the second cutting. Denatured proteins were separated on 10% SDS-PAGE gels, blotted onto nitrocellulose and immunoblots obtained following established protocols as described by the manufacturer of the 'Stable Blue' developer used to visualize the protein (Promega Corporation, Madison, WI). The primary anti-Mn-P antibody was kindly supplied by P. Kirsten, USDA, Forest Products Laboratory, Madison, WI.

Several attempts were made to assay for Mn-P activity in crude extracts following established methods used to assay fungal culture filtrate. In one such attempt, extracts prepared as described above were diluted 10-fold with 50 mM sodium L-tartrate (pH 4.5). Assays were based on the procedure of Glenn & Gold (1985) and used pinacyanol chloride (Sigma) as the substrate. The reaction mixture consisted of 50 mM sodium acetate (pH 4.5), 50 mM sodium L-lactate (pH 4.5), 100 μM $MnSO_4$, 3 mg ml^{-1} gelatin, 10% (v/v) DMSO, and 5 $\mu g\ ml^{-1}$ pinacyanol chloride. After adding 2–10 μl diluted extract, H_2O_2 was added to 100 μM from a freshly prepared 10 mM stock to start the reaction. The reactions were carried out at 37° C and the decrease in absorbance at 603 nm was continuously monitored with time points recorded every 15–30 seconds. The assays were not linear so the slope at the start of the reaction was used to calculate the activity. In cases where there was an initial lag in the reaction, the assay slope at the intersection of the lines for the assay and the lag was used to calculate the activity. A molar extinction coefficient of 120,000 $cm^{-1}M^{-1}$ was used to calculate the International Units (IU) of enzyme activity. Each extract was also assayed by the above procedure in the absence of $MnSO_4$.

Field test. It was necessary to obtain a field permit from APHIS (Animal and Plant Health Inspection Service of the United States Department of Agriculture) for the environmental release of these plants. The field test of transgenic alfalfa expressing Mn-P was a randomized complete block consisting of 16 entries containing: 1 untransformed control line RSY27, 12 transgenic lines which expressed the protein and three lines recovered from the same experiment which had no detectable protein expression. Plants were clonally propagated and transplanted in 3 replications, with in most cases 5 propagules in each replicate spaced 0.5 m apart in rows 1 m apart. The plants were cut back at the time of transplanting into the field on June 21, 1993. The plants had a well-developed root system and had been cut back twice prior to planting. The experiment was monitored weekly for any abnormal phenotypes. In order to comply with current APHIS regulations for transgenic alfalfa, foliage was harvested when plants started to have coloured flower buds but no open flowers. The first harvest was taken on August 5 and the second on September 10. At these times, the phenotypes of the plants were visually rated and a number was ascribed to their appearance: 0 = no phenotypic changes, 1 = yellowing of 10% of leaves, 2 = slightly stunted and 25% foliage affected, 3 = stunted and 50% foliage affected, 4 = severely stunted, more than 50% leaves affected and leaves starting to fall, 5 = plant dead with no signs of new growth. The length of the longest stem was measured. The whole plant herbage was collected and dried for 4 days at 70° C. Plants were cut back to a height of 5 cm.

The alpha-amylase trial consisted of 26 entries containing: 2 untransformed control lines of RSY27, 17 transgenic lines which consistently showed measurable levels of alpha-amylase activity and 7 lines recovered from the same experiment which had no detectable enzyme expression. These plants had only been cut back once prior to planting and their root systems were not as well-developed as the plants used in the Mn-P field test. There were 3 replicates of 5 plants each and the plants were monitored and harvested as above. Planting date was June 24, 1993, with the first harvest taken on August 12 and the second on September 17. In each case the plot was bordered by a double row of cultivated alfalfa.

DNA extraction and analysis. DNA was prepared from alfalfa leaf tissue using a modified CTAB protocol (Lee et al., 1993). DNA was cut with restriction enzyme *Eco*R1, separated by electrophoresis in agarose gels, transferred to Biotrace HP filters (Gelman Sciences, Ann Arbor, MI), hybridized to ^{32}P-labeled random-primed probes and analyzed by autoradiography as described by Reid et al. (1988).

Data analysis. Data were analyzed using the procedure GLM of the SAS system employing the method of least squares to fit the models. In the models it was assumed that there was a plant and a block effect on mean plant height and mean plant weight; i.e., mean height and mean dry weight = global mean + block effect + plant effect + error. The error is assumed to be independent and normally distributed with zero mean and constant variance. Residual plots were checked and no violations of the model assumptions were found.

Sexual transfer of Mn-P production. Seed was produced by crossing transgenic plants expressing Mn-P (used as females) with Blazer-XL, a cultivated alfalfa variety. Ten seeds from crosses using 12 of the individuals which were part of the field test were germinated and established under growth room conditions. The progeny were assessed visually 4 weeks after planting and assessed for Mn-P production. Western analyses were done on 50 of the plants as confirmation of the visual scoring.

Results

Plant transformation. Transformation results are summarized in Table 1. Embryos (up to 40 per explant piece) were produced in 3–6 weeks and plants in 8–12 weeks. A large number of putative transgenic plants were produced in the transformation experiment using the alpha-amylase construct. Embryos were produced on 199 of 231 treated explants and in 175 of 199 cases these embryos developed into plants. Plants derived from 75 different explant pieces were selected at random. Sixty-two of these rooted, and 38 of the 62 had detectable alpha-amylase expression *in vitro* as determined by starch plate assays. Results of transformation experiments using the Mn-P constructs were a little unusual in that the conversion of embryos to plants was lower than expected. Plants were recovered from experiments using both Mn-P constructs and approximately two thirds of the putative transgenics tested *in vitro* had detectable levels of Mn-P as determined by Western analysis. It should be noted that Southern analysis was not performed on the whole set of these plants, so we cannot determine if the plants with no detectable amylase or Mn-P activity were escapes (i.e. untransformed), or if they had the foreign gene, but it was not expressed or was expressed at a very low level which we could not detect.

Characterization of transgenic plants

Plants expressing alpha-amylase. Alfalfa plants expressing *B. licheniformis* alpha-amylase appeared phenotypically normal under all growth conditions and all of the plants recovered were fertile. There was considerable endogenous alpha-amylase activity

Table 1. Summary of transformation experiments

Construct	Explants	Explants giving callus	Explants giving embryos	Explants giving shoots	Rooted individuals tested	Plants with detectable protein expression
amylase pDM408	231	208	199	175	62	38
controls	40	40	38	32	–	–
controls + kanamycin + carbenicillin	19	19	0	0	–	–
Mn-P pDM321	160	153	96	52	33	22
Mn-P pDM322	160	159	139	79	36	21

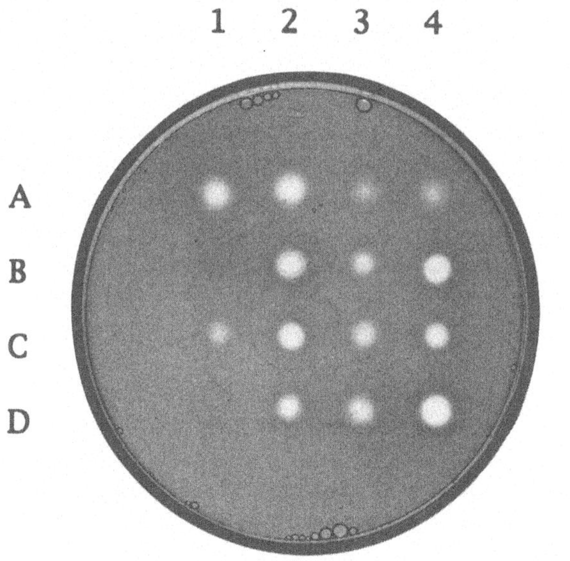

Fig. 1. Starch plate assay showing the expression of active *B. licheniformis* alpha-amylase in transgenic alfalfa. Columns 1 and 3 are heat-treated (60° C) duplicate samples of columns 2 and 4. Row A is *B. licheniformis* alpha-amylase standard, diluted 1:50,000 (A1 and A2) and 1:100,000 × (A3 and A4). Rows B, C and D represent extracts of six alfalfa plants, two tests (heat-treated, i.e. row 1 or 3, or untreated row 2 or 4) for each plant. B1 and B2 represent a control plant, the rest are transformants from the field test with at least one copy of the introduced gene. Note that alfalfa has a relatively high endogenous activity (B2) but this activity is destroyed on heating (B1). Four of the five tested plants are expressing active enzyme.

in untransformed RSY27. This was not heat stable, however, and was not detected in samples treated at 60° C. Thus it was possible to show expression of active *B. licheniformis* alpha-amylase in transformed plants. An example of a starch plate assay showing the activity of heated and non-heated samples is shown in Fig. 1. Crude estimates of *B. licheniformis* alpha-amylase enzyme activity, based on the size of the halo on starch plates compared to native enzyme, were in the range of 0.001–0.01% total soluble protein. Soluble protein values in extracts were typically in the range 3–6 mg ml^{-1} depending on tissue type and plant age. Estimates of expression levels are based on protein values in unheated extracts since there is a significant loss in protein upon heating at 60° C. We are in the process of purifying the introduced alpha-amylase from transgenic plants.

Plants expressing Mn-P. Western analysis of *in vitro* putative transgenic plants clearly showed that some plants were expressing the foreign protein (Fig. 2). Crude estimates of Mn-P levels were in the range 0.01–0.5% total soluble protein. It should be noted that the extraction buffer used (pH 4.5) typically gives lower levels of total soluble protein (1–2 mg ml^{-1}) in plant extracts than when tissue is extracted in a more neutral or basic buffer. Two to three weeks after plants had been placed in rooting medium and roots were forming, there were no obvious phenotypic differences between plants expressing the protein and controls. Two to three

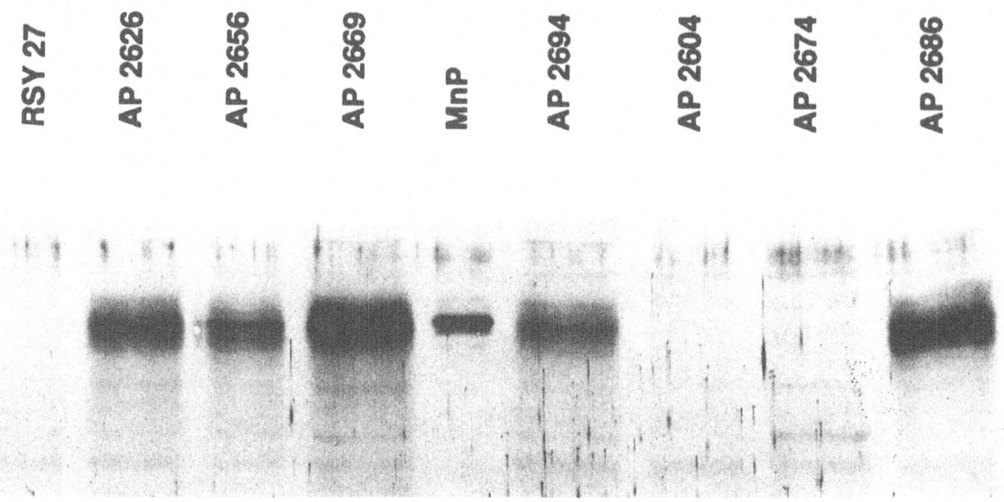

Fig 2 Western blot of alfalfa extracts showing expression of Mn P in 2–3 week old *in vitro* plants Approximately 4 μg of total protein was loaded for each sample, standard is equivalent to 1 86 ng *P chrysosporium* Mn P RSY27 is the untransformed control Seven transformants are shown, five are positive

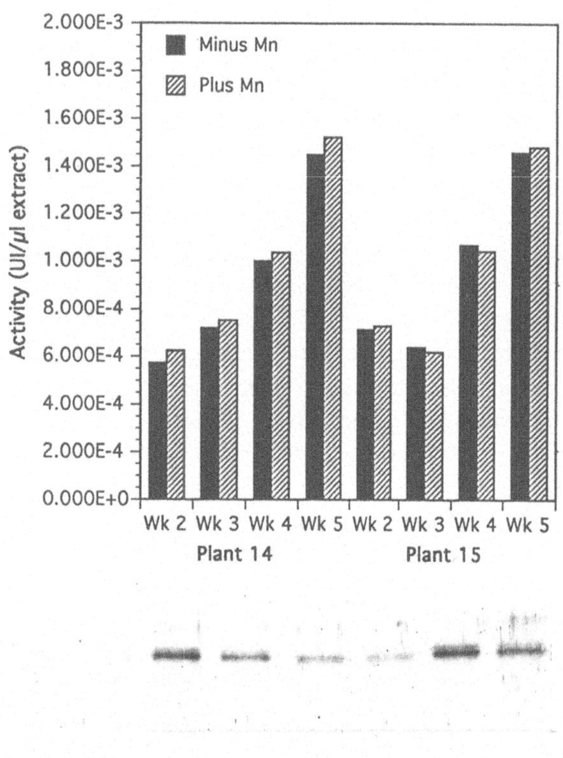

Fig 3 Time course of peroxidase activity in field grown alfalfa expressing Mn P The graph shows enzyme activity in samples from plants 14 and 15 taken at weekly intervals (2 through 5) during the second harvest period A Western blot of the same sample (2 μg total protein loaded onto gel) is shown below the graph

weeks later, however, it became evident that expression of the protein had deleterious effects on plant growth and development whether plants were still *in vitro* or in the greenhouse Trifoliates started to yellow and in severe cases senesced completely and dropped from the plant Overall, the plants were stunted and flowered later than control plants, also, the highest-expressing plants, usually those with levels estimated to be above 0 3% soluble protein, died The surviving plants were fertile and seed was recovered Importantly, viable progeny with high levels have been recovered in crosses Western analyses were repeated on mature greenhouse plants just starting to show symptoms Results were essentially the same as for the screening of *in vitro* plants Levels on a protein basis were 10–20% lower reflecting an increased value in protein levels of greenhouse plants compared to *in vitro* plants Obtaining enough clonal copies of the transgenic plants (for field studies) by rooting vegetative cuttings was difficult, especially in plants expressing the protein at medium or high levels Interestingly if the cuttings rooted, the plants grew normally for 2–3 weeks before showing any foliar symptoms This pattern was seen repeatedly wherever the plants were grown

Western analyses of field-grown plants, carried out just prior to both harvests, were used to ascribe an expression level to each set of individual transgenic plants The following levels were ascribed based on crude visual estimations of the intensity of bands as

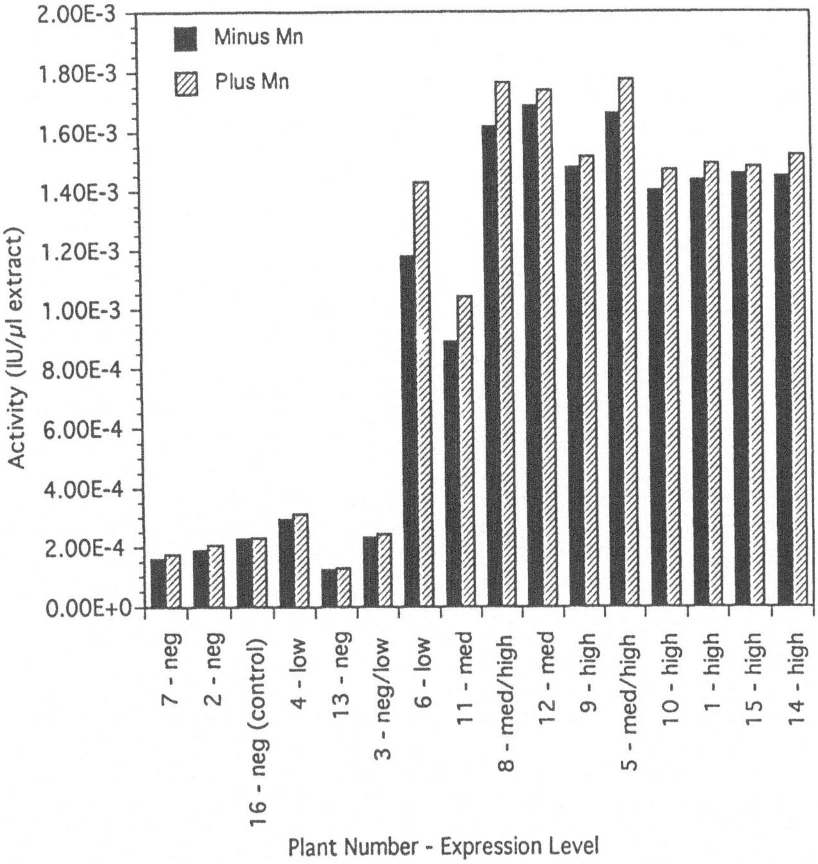

Fig. 4. Peroxidase activity of field-grown transgenic alfalfa expressing Mn-P. Samples were taken immediately prior to the second harvest.

compared with standards; an expression of 0.01–0.05% soluble protein = low, 0.05–0.15% = medium, greater than 0.15% = high. Results from weekly sampling of plants (representative Western analysis data shown in Fig. 3) clearly show that levels of Mn-P increased as plants developed. There were consistently higher levels of the fungal protein in weeks 4 and 5 than in weeks 2 and 3. Thus, increased expression/accumulation of the protein coincided with the appearance of visual symptoms. Results also showed that Mn-P protein was present in all of the plant tissue types tested (data not shown). Attempts to measure Mn-P activity in crude alfalfa extracts were not successful. We could not demonstrate that any of the peroxidase activity was Mn-dependent, i.e., that the amount of peroxidase activity was significantly greater in the presence of Mn than in the absence of Mn (Fig. 3). General peroxidase activity, however, with and without added Mn, was found to correlate with Mn-P levels seen in the Western analyses (Figs 3 and 4). In the weekly time course of Fig. 3, the peroxidase activity was seen to increase as the plants developed just as the Mn-P protein levels did and, therefore, to correlate also with the appearance of the foliar symptoms. This developmental increase in general peroxidase activity was not seen in the control plants or in transgenic plants not expressing Mn-P that were tested in the time course (data not shown). To determine if the correlation between peroxidase activity and Mn-P expression levels was more general, all the field plants were assayed just prior to the second harvest. Figure 4 demonstrates that the increase in general peroxidase activity is only seen in plants that express Mn-P. The higher levels of peroxidase activity in symptomatic plants probably reflect a general wound/stress response of transgenic plants when the levels of Mn-P become deleterious.

We could not detect any differences in levels of Mn-P in plants produced using constructs pDM321 or pDM322. Thus it would not appear to be necessary to replace the fungal ER signal sequence with a plant signal sequence in this case. Plants derived from both constructs were included in the field test.

Table 2. Summary of data from the second harvest of field grown transgenic alfalfa expressing manganese-dependent lignin peroxidase from *P. chrysosporium*

Plant[a] number	Southern[b] analysis copy number	Protein[c] expression	Field[d] score	LS mean dry weight (g)	Std Error	T grouping[e]	LS mean height (cm)	Std Error	T grouping[e]
7	1	neg	0	59	4.1	A	65	3.76	A
2	1	neg	0	52.8	3.3	AB	59.6	3.03	ABC
16	0	neg (control)	0	46.8	3.3	BC	56.7	3.03	ABCD
4	2	low	0.4	46.8	4.1	BCD	63.5	3.76	AB
13	1	neg	0	44.7	3.3	BCDE	56	3.03	ABCD
3	2	neg/low	0.3	36.6	3.3	DEF	56.3	3.03	ABCD
6	1	low	1.4	35	3.3	EF	56.6	3.03	ABCD
11	1	med	1.5	34.1	3.3	F	58.4	3.03	ABC
8	1	med/high	2.3	30.4	3.3	FG	55.3	3.03	ABCD
12	2, 3	med	2.1	29.1	3.3	FGH	52.3	3.03	CDE
9	3	high	2.9	22.6	4.1	GHI	53.6	3.76	BCD
5	1	med/high	2.3	20.9	3.3	GHI	52.7	3.03	CD
10	1	high	2.7	19.8	3.3	HI	51.2	3.03	CDE
1	nt	high	3.2	16.7	3.3	I	49.4	3.03	DE
15	1	high	3.9	14.4	4.1	I	42.5	3.76	EF
14	1	high	3.8	12.4	3.3	I	40.2	3.03	F

[a] Plants ranked according to LS (least squares) mean dry weights.
[b] nt = not tested.
[c] Levels of Mn-P as estimated by Western analysis.
[d] Visual scoring of plant phenotype at time of harvest.
[e] Values connected by the same letter are not significantly different, P = 0.05.

Southern analysis. All of the plants (except the untransformed control) included in the field test of plants expressing Mn-P had at least one copy of the fungal gene (Table 2). Similarly, all of the plants in the alpha-amylase field study from which data was obtained had one or more copies of the bacterial gene (Table 3).

Field performance

Plants expressing Mn-P. Data from the field tests essentially confirmed earlier growth room and greenhouse observations. Survival rates for transplants were surprisingly very high; only 1 of 224 plants in the test died. All of the test lines had a normal phenotype for the first 2–3 weeks of growth. At this point, plants expressing Mn-P at high levels started to show foliar yellowing and the rate of plant growth and development slowed such that plants became stunted. The appearance of flower buds was typically 7–10 days later in most of the plants expressing the protein than in control lines. Data from the first harvest are shown in Fig. 5 and data from the second are given in Table 2. For easier interpretation of the data in both cases, the plants were ranked according to mean dry weight. The phenotypic scores taken immediately prior to harvest clearly indicate that expression of Mn-P is deleterious to plant growth and development. There is an obvious relationship between expression of the protein and plant performance under field conditions for both harvest periods. Plant dry weight in harvest 2, for example, ranged from 59.0 g in a non-expressing plant to 12.4 g in a plant expressing the protein at a high level. Any group of plants expressing the protein at medium to high levels had a significantly lower mean dry weight than the control plants. The range of plant height was not as great but again there were significant differences between high-expressing and control plants.

Plants expressing alpha-amylase. Expression of *B. licheniformis* alpha-amylase gave no obvious phenotypic effects in field-grown plants. Field survival (92%)

Table 3. Summary of data from the second harvest of field grown transgenic alfalfa expressing *B. licheniformis* alpha-amylase

Plant[a] number	Southern[b] analysis copy number	Enzyme[c] expression	LS mean dry weight (g)	Std Error	T grouping[d]	LS mean height (cm)	Std Error	T grouping[d]
13	1	pos	39.7	4.63	A	61	3.16	ABCDE
11	1	pos	38.1	4.63	AB	60.8	3.16	ABCDE
15	1	pos	35.9	4.63	AB	64.7	3.16	A
9	nt	pos	35.6	4.63	AB	59.6	3.16	ABCDE
20	nt	neg	32.7	4.63	ABC	59.7	3.16	ABCDE
18	1	neg	32.5	4.63	ABC	60.7	3.16	ABCDE
14	1, 2	neg	32.2	4.63	ABC	61.1	3.16	ABCDE
10	0	neg (control)	31.6	4.63	ABC	59.5	3.16	ABCDE
7	nt	pos	31.5	4.63	ABC	60.8	3.16	ABCDE
24	nt	pos	31.4	4.63	ABC	60.2	3.16	ABCDE
6	3	neg	31.4	4.63	ABC	58.1	3.16	ABCDE
3	1	pos	31.1	4.63	ABC	57	3.16	ABCDE
4	3	pos	30.4	4.63	ABC	63.7	3.16	AB
23	3	neg	30.4	4.63	ABC	63.2	3.16	ABC
8	1, 2	neg	30.2	4.63	ABC	61.7	3.16	ABCD
22	4	pos	30.1	4.63	ABC	60.9	3.16	ABCDE
19	2	pos	29.3	4.63	ABC	58.9	3.16	ABCDE
12	nt	pos	28.8	4.63	ABC	55.2	3.16	BCDE
21	nt	pos	28.6	4.63	ABC	60.2	3.16	ABCDE
16	3, 4	pos	26.4	4.63	ABC	59.7	3.16	ABCDE
5	1	pos	25.8	4.63	BC	60.7	3.16	ABCDE
25	0	neg (control)	25.8	4.63	BC	56.3	3.16	ABCDE
26	nt	pos	24.5	5.71	BC	59.3	3.89	ABCDE
2	1	pos	21.9	4.63	C	52.1	3.16	E
17	2, 3	neg	21.7	4.63	C	54.2	3.16	CDE
1	1	pos	20.9	4.63	C	52.7	3.16	DE

[a] Plants ranked according to LS (least squares) mean dry weights.
[b] nt = not tested.
[c] As determined by starch plate assay on heat treated samples.
[d] Values connected by the same letter are not significantly different, P = 0.05.

was less than for the Mn-P field test. This was probably a reflection of the fact that these transplants were smaller than the ones used in the Mn-P field test and their root systems were not as well-developed. This also accounts for the yield data of control plants from this study being considerably less than those seen in the Mn-P study. Plant dry weight and height (Table 3 and Fig. 6) were more uniform for the alpha-amylase set of plants than for those expressing Mn-P. There were significant differences between height and weight for plants at the extremes of the range but the performance of most of the group was not significantly different. The outlying plants probably reflect somaclonal variation commonly seen amongst regenerated plants. The pattern of T grouping for Mn-P and alpha-amylase clearly shows the differences in the two sets of plants. Expression of alpha-amylase has no effect on field performance.

Progeny testing. Only 5 of 120 seeds derived from crosses of transgenic alfalfa producing Mn-P with Blazer-XL failed to germinate. All of the failed seed was derived from crosses of high-expressing plants. Growth and development of the seedlings was normal for the first 17 days, at which time symptoms of foliar yellowing appeared in some of the seedlings. Results

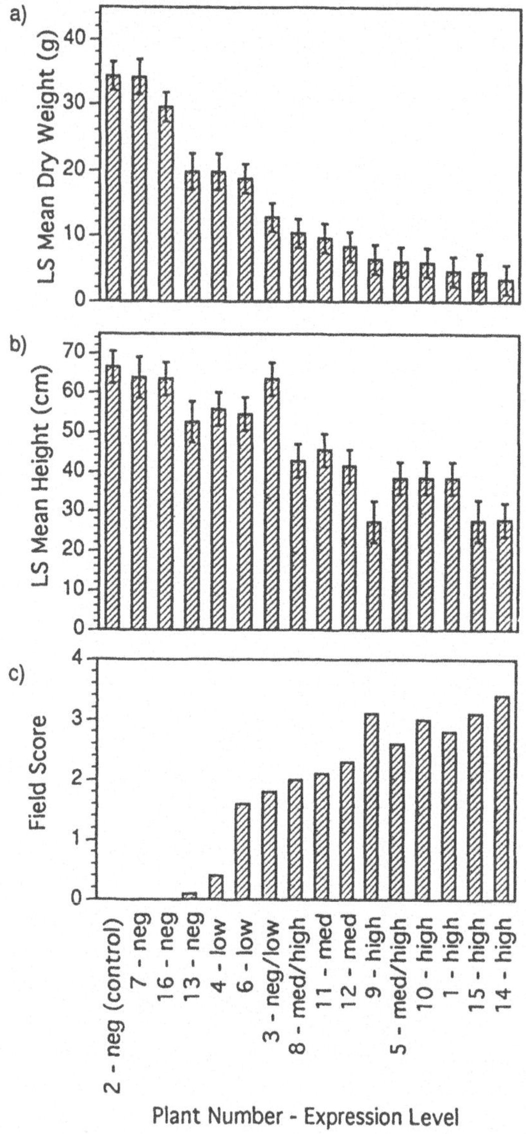

Fig. 5. Dry weight, height and phenotypic score from the first harvest of a field test of transgenic alfalfa expressing *P. chrysosporium* Mn-P.

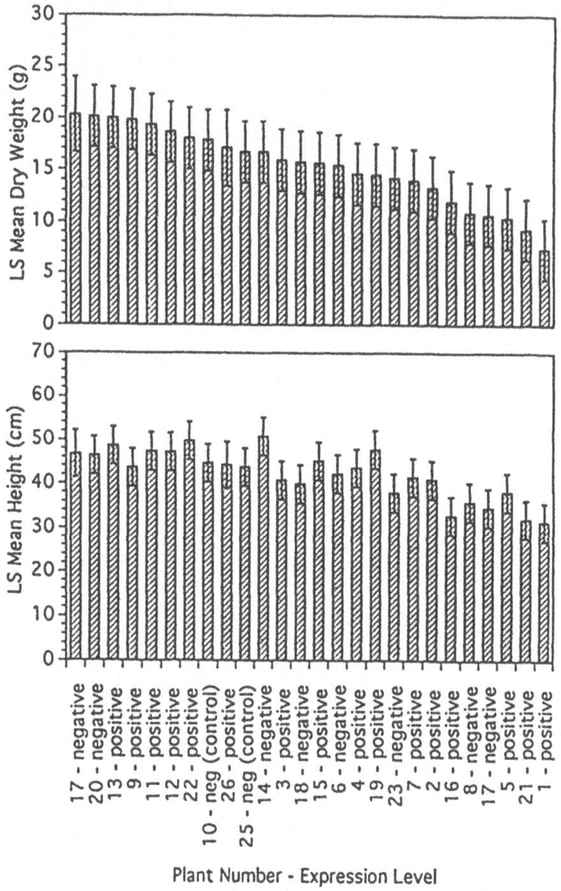

Fig. 6. Dry weight and height data from the first harvest of a field test of transgenic alfalfa expressing *B. licheniformis* alpha-amylase.

of the visual assessment after 4 weeks of growth (Table 4) showed that symptom expression segregated in the progeny. Western analysis of seedlings from 7 of the 12 sets of progeny confirmed that symptom expression was associated with production of the fungal protein (data not shown). Interestingly, parent plant number 13 had no detectable Mn-P protein production but 2 of the progeny did produce Mn-P. Thus, a silent or very low-expressing copy of the gene in the original transgenic became capable of high expression in some of the progeny. Progeny obtained from high-expressing parents tended to be more severely affected than those from low- or medium-expressing plants. The progeny which are hybrids with an elite cultivar tended to have larger trifoliates and a more vigorous growth habit than RSY27. Crude estimates of Mn-P levels in progeny ranged from 0.09–0.35% total soluble protein in extracted samples. As a general observation, the deleterious effects of relatively high amounts of Mn-P protein were not as severe in the progeny plants. Further crosses are currently in progress.

Discussion

We have presented data showing the expression of a bacterial and fungal protein in transgenic alfalfa. The transformation system used is similar to that of other researchers (e.g. Shahin et al., 1986; Chabaud et al., 1988; D'Halluin et al., 1990; Hill et al., 1991; Schroeder et al., 1991; Bagga et al., 1992). The culture

Table 4. Assessment of progeny derived from transgenic alfalfa expressing manganese-dependent lignin peroxidase

Parent	Expression level in parent	Visual score of progeny 4 weeks after germination									
		a	b	c	d	e	f	g	h	i	j
16	neg (control)	–	–	–	–	–	–	–	–	–	–
13	neg	–	–	–	–	–	++	–	–	+	–
6	low	–	–	+	++	–	–	–	–	–	–
12	med	++	+	–	–	–	–	–	–	–	+
11	med	++	++	–	–	++	+	–	++	–	–
8	med/high	–	–	+	–	–	++	–	+	–	++
5	med/high	–	–	+	++	–	+	–	–	–	–
1	high	–	–	++	+++	+++	–	++	–	–	+
15	high	+++	ng	++	–	–	++	ng	+	+	+
9	high	++	–	ng	++	+++	–	+++	–	+	–
10	high	++	+++	+++	–	++	++	–	+	++	++
14	high	–	ng	–	+++	ng	++	+++	++	+++	+

– No significant phenotypic effects.
+ 10–20% of leaves affected; growth slightly stunted.
++ 20–50% of leaves affected; growth slightly stunted.
+++ > 50% of leaves affected; growth stunted.
ng = Seed did not germinate.

conditions were a little different since they were optimized for an easily-regenerable genotype of alfalfa. The major difference in our system is that kanamycin is selective for inhibition of embryoid formation but not for callus formation. This is probably the reason that our system produces plants in a much shorter time (10–12 weeks) than that reported by others (20–30 weeks). In these and in other experiments using this transformation system, 20–50% of treated explants gave rise to transgenic plants. This high transformation frequency and the speed at which plants can be recovered makes this an ideal system for our studies.

Expression of *B. licheniformis* alpha-amylase had no effect on the yield of field-grown transgenic alfalfa. The expression of active alpha-amylase in transgenic alfalfa was lower than that reported by Pen et al. (1992) for tobacco. It is difficult, however, to compare absolute levels of foreign protein production between plant species. Expression is usually based on the amount of soluble protein in the extract prepared for analysis, for example, protein levels can vary enormously between plant species, between different parts of the same plant and on the conditions used to extract the tissue. We are currently analysing transgenic tobacco to determine whether the relatively low levels of expression are a function of our construct. The field plants will be maintained over two more seasons and used to look at large-scale recovery of *B. licheniformis* alpha-amylase from plant juice.

Levels of Mn-P in transgenic alfalfa were relatively high but expression of the protein had a deleterious effect on plant growth and development. We have not been able to demonstrate that the enzyme is active in the plant. The correlation between the effects on growth and development and protein production strongly suggests that at least a portion of the expressed protein is active enzyme. The delay in the onset of foliar symptoms in the transgenic plants is interesting. It is not clear if it is triggered by a development stage in alfalfa or reflects increased production/accumulation of the protein. If protein expression were deleterious in the very early stages of plant development we would not have recovered any transgenic plants. We have no explanation for the deleterious effects of Mn-P in alfalfa, particularly since these were not seen when we expressed the gene in tobacco using the same constructs (Mathews et al., 1993). The enzyme requires manganese for its activity and converts Mn^{2+} to Mn^{3+}. Foliar symptoms could be due to a deficiency of Mn^{2+} or the result of the damaging action of Mn^{3+}. It is not clear what effect, if any, expression of this enzyme has on lignin content of the transgenic alfalfa. Preliminary data suggest that there are qualitative changes in lignin in transgenic plants expressing Mn-P. This is being investigated further since it may have potential for increasing the ruminant digestibility of this widely-used forage legume. Mn-P production may be an ideal system to test the use of an inducible promoter. It may be possible to express the protein a few days before harvest and thus be able to recover the enzyme and avoid most of the deleterious effects on growth and development.

Acknowledgements

The authors thank S. Carrera, W. Coches, D. Drott, M. John and L. Spatola for technical assistance. E. Santos and P. Crump of UW-CALS Computer Services advised on, and assisted with, the statistical analysis of the field data. The work was supported by a USDA grant from the Consortium for Plant Biotechnology Research (formerly the Midwest Plant Biotechnology Consortium) and in its early stages by Rohm and Haas, Dupont and Procter and Gamble.

References

Austin, S , R G Koegel, D Matthews, M Shahan, R J Straub & R Burgess, 1994 Production of industrial enzyme in transgenic alfalfa Annals NY Acad Sci In Press

Bagga, S , D Sutton, J D Kemp & C Sengupta-Gopalan, 1992 Constitutive expression of the β-phaseolin gene in different tissues of transgenic alfalfa does not ensure phaseolin accumulation in non-seed tissue Plant Mol Biol 10 951–958

Bingham, E T, 1991 Registration of alfalfa hybrid Regen SY germplasm for tissue culture and transformation research Crop Sci 31 1098

Bingham, E T, L V Hurley, D M Kaatz & J W Saunders, 1975 Breeding alfalfa which regenerates from callus tissue in culture Crop Sci 15 719–721

Bradford, M M , 1976 A rapid and sensitive method for the quantification of micro organism quantities of protein utilizing the principle of protein-dye binding Anal Biochem 72 248–254

Brown, D C & A Atanassov, 1985 Role of genetic background in somatic embryogenesis in *Medicago* Plant Cell Tissue Organ Culture 4 107–114

Chabaud, M , J E Passiatore, F Cannon & V Buchanan Wollaston, 1988 Parameters affecting the frequency of kanamycin resistant alfalfa obtained by *Agrobacterium tumefaciens* mediated trans formation Plant Cell Rep 7 512–516

Comai, L , P Moran & D Maslyar, 1990 Novel and useful properties of a chimeric plant promoter combining CaMV 35S and MAS elements Plant Mol Biol 15 373–381

D'Hallum, K , J Botterman & W D Greef, 1990 Engineering of herbicide-resistant alfalfa and evaluation under field conditions Crop Sci 30 866–871

DeZoeten, G A , J R Penswick, M A Horisberger, P Ahl, M Schultze & T Hohn, 1989 The expression, localization and effect of a human interferon in plants Virology 172 213–222

During, K , S Hippe, F Kreuzaler & J Schell, 1990 Synthesis and self-assembly of a functional monoclonal antibody in transgenic *Nicotiana tobacum* Plant Mol Biol 15 281–293

Glenn, J K & M H Gold, 1985 Purification and characterization of an extracellular Mn(II)-dependent peroxidase from the lignin-degrading basidiomycete, *Phanerochaete chrysosporium* Arch Biochem & Biophys 242 (2) 329–341

Hiatt, A , R Cafferkey & K Bowdish, 1989 Production of antibodies in transgenic plants Nature 342 76–78

Hill, K K , N Jarvis-Eagan, E Halk, K Krahn, L Liao, R Mathew son, D Merlo, S Nelson, K Rashka & L Loesch-Fries, 1991 The development of virus-resistant alfalfa, *Medicago sativa* L Bio/Technology 9 373–377

Horsch, R B , J E Fry, N L Hoffman, D Eicholtz, S G Rogers & R T Farley, 1985 A simple and general method for transferring genes into plants Science 227 1229–1231

Krebbers, E & J Vandekerckhove, 1990 Production of peptides in plant seeds Trends Biotech 8 1–3

Lee, I , A Bleeker & R M Amasino, 1993 Analysis of naturally occurring late flowering in *Arabidopsis thaliana* Mol Gen Genet 237 171–176

Mason, H S , F D Guerrero, J S Boyer & J E Mullet, 1988 Proteins homologous to leaf glycoproteins are abundant in stems of dark grown soy bean seedlings Analysis of proteins and cDNAs Plant Mol Biol 11 845–856

Mathews, D , S Austin, J Monorama, J Will, M Shahan, R Amasino & R Burgess, 1993 Expression of fungal lignin degrading enzymes in plants Plant Physiol (Supp) 102 960

McBride, K E & K R Summerfeldt, 1990 Improved binary vec tors for *Agrobacterium* mediated plant transformation Plant Mol Biol 14 269–276

Murashige, T & F Skoog, 1962 A revised medium for rapid growth and bioassays with tobacco tissue cultures Physiol Plant 15 473–497

Pease, E A , A Andrawis & M Tien, 1989 Manganese-dependent peroxidase from *Phanerochaete chrysosporium* J Biol Chem 264 13531–13535

Pen, J , L Molendijk, W J Quax, P C Sijmons, A J J van Ooyen, P J M van den Elzen, K Rietveld & A Hoekema, 1992 Production of active *Bacillus licheniformis* alpha-amylase in tobacco and its application in starch liquification Bio/Technology 10 292–296

Reid, R A , M C John & R M Amasino, 1988 Deoxyribonucle ase I sensitivity of the T-DNA *ipt* gene is associated with gene expression Biochemistry 27 5748–5754

Saito, N , 1973 A thermophilic extracellular α amylase from *Bacil lus licheniformis* Arch Biochem Biophys 155 290–298

Schroeder, H E , M R I Khan, W R Knibb, D Spencer & T J V Higgins, 1991 Expression of a chicken ovalbumin gene in three lucerne cultivars Aust J Plant Physiol 18 495–505

Schwardt, E , 1990 Production and use of enzymes degrading starch and some other polysaccharides Food Biotechnol 4 337–351

Shahin, E A , A Spielmann, K Sukhapinda, R G Simpson & M Yashar, 1986 Transformation of cultivated alfalfa using disarmed *Agrobacterium tumefaciens* Crop Sci 26 1235–1239

Sijmons, P C , B M M Dekker, B Schrammeijer, T C Verwoerd, P J M Van den Elzen & A Hoekema, 1990 Production of correctly processed human serum albumin in transgenic plants Bio/Technology 8 217–221

Vandekerckhove, J , J Van Damme, M Van Lijsebettens, J Botterman, M De Block, M Van de Wiele, A De Clercq, J Leemans, M Montagu & E Krebbers, 1989 Enkephalins produced in transgenic plants using modified 2S seed storage proteins Bio/Technology 7 929–932

Yuuki, T., T. Nomura, H. Tezuka, A. Tsuboi, H. Yamagata, N. Tsukagashi & S. Udaka, 1985. Complete nucleotide sequence of a gene coding for heat- and pH-stable alpha-amylase of *Bacillus licheniformis*: comparison of the amino acid sequences of three bacterial liquifying alpha-amylase deduced from the DNA sequences. J. Biochem. 98: 1147–1156.

Euphytica **85:** 395–401, 1995.

The phenotypic characterisation of R_2 generation transgenic rice plants under field and glasshouse conditions

P.T. Lynch[1], J. Jones, N.W. Blackhall, M.R. Davey, J.B. Power, E.C. Cocking, M.R. Nelson[2], D.M. Bigelow[2], T.V. Orum[2], C.E. Orth[3] & W. Schuh[3]
Plant Genetic Manipulation Group, Department of Life Science, University of Nottingham, Nottingham NG7 2RD, U.K.; [1] present address: Plant Biotechnology Group, Division of Biological Sciences, University of Derby, Kedleston Road, Derby DE22 1GB, U.K.; [2] Department of Plant Pathology, College of Agriculture, University of Arizona, Tucson, Arizona, U.S.A.; [3] Department of Plant Pathology, Pennsylvania State University, 210 Buckout Laboratory, University Park, Pennsylvania, U.S.A.

Key words: *Oryza sativa* L. cv. Taipei 309, rice, protoplasts, direct DNA uptake, kanamycin-resistant transgenic plants, field trial, glasshouse trial, neomycin phosphotransferase II (*npt* II) gene, gene expression and inheritance

Summary

The phenotypes of seed progeny (R_2 generation) of *Oryza sativa* L. cv. Taipei 309, which carried the neomycin phosphotransferase II (*npt* II) gene, were compared with those of non-transformed, protoplast-derived plants of the same generation and non-transformed, seed-derived plants under field and glasshouse conditions. Under both conditions the transgenic plants were generally smaller, took longer to flower and had reduced fertility. Significant differences were observed between individuals within the group of transgenic plants. The *npt* II gene was present in most of the transgenic plants, but NPT II activity was only detected in a minority of individuals.

Introduction

Although a range of different approaches has been used to introduce genes into rice, direct DNA uptake into protoplasts remains one of the most widely applicable (Kothari et al., 1993). The production of transgenic rice plants of the Japonica cultivar Taipei 309, reported previously from this laboratory (Zhang et al., 1988), utilised an established protoplast culture procedure (Abdullah et al., 1986) and electroporation-mediated uptake of pCaMVNEO (Yang et al., 1988). A subsequent study (Davey et al., 1991) reported that seeds from one R_0 plant, which contained and expressed the *npt* II gene, germinated on medium supplemented with kanamycin sulphate to give 16 green R_1 generation plants. When grown to maturity under glasshouse conditions at Nottingham, these plants could be divided into 3 phenotypically distinct groups, based primarily on their fertility. Transgenic plants of the R_1 generation were shorter and took longer to flower compared with non-transformed R_1 protoplast-derived plants. Some of the phenotypic characteristics of these transgenic rice plants may have been a result of protoclonal variation. Characteristics, such as reduced fertility, have been previously observed in protoplast-derived plants grown under field conditions (Abdullah et al., 1989). Changes in plant fertility may be particularly important if transgenic rice plants are to be used in conventional plant breeding programmes. A comparative study was made under field and glasshouse conditions, of the agronomic characteristics of R_2 generation, protoplast-derived, transgenic rice plants with non-transformed R_2 protoplast-derived and seed-derived plants of Taipei 309.

Materials and methods

Production of transgenic plants

The conditions for electroporation-mediated uptake of pCaMVNEO into protoplasts isolated from cell suspensions of Japonica rice (*Oryza sativa* L.) cv. Taipei 309, the selection of kanamycin-resistant, protoplast-derived colonies and shoot regeneration from transformed callus, were as described by Zhang et al. (1988). Plants were also regenerated from protoplasts electroporated in the absence of plasmid, to provide non-transformed material for comparison with the transgenic plants. The transfer and maintenance of the R_0 generation plants and the subsequent seed generation (R_1), produced after self-fertilisation, under glasshouse conditions at the University of Nottingham, UK were described previously by Davey et al. (1991). The R_2 generation plants used in this study were from seeds produced after selfing, under glasshouse conditions at Nottingham, 4 R_1 generation plants (previously designated H6A4, H6A5, HF3 and HF4 (Davey et al., 1991) and 1 protoplast-derived, non-transformed R_1 plant. Non-transformed, conventionally produced, seed-derived plants were also grown as controls. The population of transgenic plants was subdivided into 4 groups based on their R_1 parents. Thus, transgenic plant subgroup 1 was derived from plant H6A4, subgroup 2 from H6A5, subgroup 3 from HF4 and subgroup 4 from HF3.

Plant growth and maintenance

Seeds for both the glasshouse and field trials were germinated *in vitro* as described by Schuh et al. (1993). Approximately half the seedlings (25 transgenic protoplast-derived R_2 seedlings; 25 non-transformed, protoplast-derived R_2 seedlings and 25 non-transformed, seed-derived seedlings), randomly selected from each plant group and subgroup, were shipped to Maricopa Agricultural Experimental Station, Arizona, USA. On arrival in Arizona, the seedlings were transferred to compost and, once acclimatised, transplanted to a fibreglass tank (2.5 × 5.0 × 1.0 m deep), where they were maintained under field conditions, as described by Schuh et al. (1993). The plants were arranged in a randomised design, surrounded by 2 rows of seed-derived plants of the Japonica rice cultivar Lemont. The seedlings remaining at Nottingham (28 transgenic protoplast-derived R_2 seedlings; 25 non-transformed, protoplast- derived R_2 seedlings and 41 non-transformed, seed-derived seedlings) were transferred to compost and subsequently grown under glasshouse conditions as described previously for the R_1 generation plants (Davey et al., 1991).

Phenotypic characterisation

The phenotypic characteristics of the plant material in the glasshouse and the field were scored when tillering first occurred and at maturity/harvest. Plant height, number of tillers per plant, the length, width and length/width ratio of the youngest fully-developed leaf were determined when the plants first tillered. At maturity, plant height, number of tillers per plant, flag leaf length, width and length/width ratio and the number of panicles per plant were also measured. For each panicle, measurements were taken of length, number of primary branches and secondary branches, number of spikelets, grains, mean grain weight, length and width and length/width ratio. Subsequently, the mean grain weight was recorded for each plant. The date on which a panicle was 'bagged' was taken as the number of days to flowering.

DNA isolation, PCR and Southern blot analysis

The presence of the *npt* II gene in the transformed protoplast-derived plants was assessed by PCR analysis. DNA was extracted from leaves using the method of Edwards et al. (1991). The reaction mixture (25 μl total volume) consisted of 2.5 μl of DNA extract and 1 unit of thermostable DNA polymerase enzyme (TaqXL, Northumbria Biologicals Ltd., Cramlington, Northumberland, UK) in a buffer solution (10 mM Tris-HCl, pH 8.8, 50 mM KCl, 15 mM $MgCl_2$ and 0.1% v/v non-ionic detergent) supplemented with 200 μM of each dNTP and 0.2 μM of two *npt* II specific primers. The sense primer sequence was 5′-GTCGCTTGGTCGGTCATTTCG-3′ while the antisense primer sequence was 5′-GTCATCTCACCTTGCTCCTGCC-3′. This primer pair gave an actual fragment length of 540 base pairs. Amplification reactions were carried out in a PHC-3 programmable Dry-Block Cycler (Techne (Cambridge) Ltd., Duxford, Cambridge, UK) and consisted of 35 cycles (95° C for 48 sec, 65° C for 30 sec and 73° C for 150 sec) followed by a final extension stage (73° C for 10 min). A 5 μl aliquot was removed from each tube and analysed by electrophoresis (2.8 Vcm^{-1}) in a 3% w/v agarose gel (NuSieve 3:1 agarose, FMC BioProducts, Rockland, ME, USA). The presence of the *npt* II gene was indicated by the production of a 540 base pair product.

For Southern analysis, DNA from young leaf material was isolated and purified using the method of Doyle & Doyle (1990). The conditions for restriction of the isolated rice DNA with *Bam* HI, agarose gel electrophoresis, blotting and probing with the P^{32}-labelled 1.0 Kb fragment of pCaMVNEO carrying the *npt* II gene, were as described by Schuh et al. (1993). The presence of the NPT II enzyme in the plants grown in Arizona was assayed with an NPT II ELISA assay kit (5 Prime → 3 Prime, Inc. West Chester, Pennsylvania, USA). NPT II activity in leaf samples of plants maintained in Nottingham was assessed as described by Davey et al. (1991). The ploidy level of individuals, selected at random from the 3 plant groups was determined using flow cytometric techniques as described by Lynch et al. (1992).

USDA-APHIS permits for importation of the plant material and field testing of transgenic plants were obtained using the NBIAP Permit Application System (National Biological Impact Assessment Program, USDA-CSRS, Washington, D.C. 20250-2200, USA). Plant material was transported under MAFF licence No. PHF 1202/105 (81) and phytosanitary certificate No. EEC/UK/E&W/022529.

Results

For the majority of the phenotypic characteristics examined under field and glasshouse conditions, the mean values for the transformed plants were lower and more variable than those for the non-transformed, protoplast-derived and seed-derived plants (Tables 1 and 2). Analysis showed significant ($p = 0.05$) differences between the plant groups for all variables, except for the number of primary branches per panicle. Under glasshouse conditions, the transformed plants also had significantly ($p = 0.05$) narrower flag leaves at maturity.

At both assessment times, the transgenic plants were shorter, with fewer tillers than plants of the other groups whose height and tiller development were not significantly different. The flag leaf length of transgenic plants was significantly less ($p = 0.05$) than for the protoclonal and seed-derived, non-transformed control plants. In the field, this difference decreased during the growing season but was retained in plants under glasshouse conditions. All plant groups grown in the glasshouse had significantly ($p = 0.05$) longer flag leaves. Although the flag leaf widths of all the plant groups, at both sites, were not significantly different at maturity, the leaf length/width ratio was lower for glasshouse-grown plants.

Transgenic plants had significantly fewer panicles than plants in the other groups, but the numerical difference was small. At maturity, the percentage of tillers producing panicles was similar in the three plant groups. Plants in the glasshouse, however, formed fewer panicles compared with plants under field conditions. Panicles on plants in the transgenic group were significantly ($p = 0.05$) shorter than those of the non-transformed plant groups. Panicle lengths of plants in the latter groups were not significantly different. Although the trend in panicle length was observed at both sites, panicles produced under glasshouse conditions were consistently longer than those of field-grown plants. Except for fewer secondary branches, the panicles produced by plants from each of the groups had comparable branching patterns. The number of secondary branches produced by glasshouse-grown plants was consistently higher than for field-grown material. The number of spikelets produced per panicle was significantly different between the plant groups; protoclonal plants had the greatest number and transgenic plants the lowest. Transgenic plants also produced fewer mature seeds per spikelet, compared with non-transformed plants. Consequently, the number of seeds per plant and grain weight per panicle/plant were significantly lower for the transformed plants compared with the plants in the other groups. The number of seeds and their size were consistently lower in the glasshouse-grown material. Transgenic plants at both sites had significantly longer, but narrower grain, than non-transformed protoclonal and seed-derived plants. Although numerically the difference in grain width was small, seeds from transformed plants had high length/width ratios. Seeds from all the plant groups grown in the glasshouse were shorter and narrower than those produced in the field, but the length/width ratio was constant between sites.

The phenotypic characteristics of the four subgroups within the transgenic plant group grown at both sites varied significantly. Under field conditions, transgenic plants of subgroup 2 were shorter, with fewer tillers than plants in the other subgroups. Significant differences between the subgroups were observed in the characteristics of the flag leaves and the panicles, but the differences were not consistent between plants at the two sites. The numbers of grains per panicle and per plant, together with grain weight per panicle and per plant, were not significantly different between the subgroups or sites.

Flow cytometric analysis showed that randomly-selected individuals from the 3 plant groups were

Table 1. Comparison of the phenotypic characteristics of the transgenic protoplast-derived R_2 plants, non-transformed, protoplast-derived R_2 plants and non-transformed, seed-derived plants grown under field conditions at Maricopa Experimental Station, Arizona, USA

Phenotypic trait	Non-transformed seed-derived plants	Non-transformed protoplast-derived R_2 plants	Transformed protoplast-derived R_2 plants
At first tillering			
Plant height (cm)	29.88 ± 2.01 A	29.20 ± 1.67 A	24.20 ± 4.00 B
Number of tillers per plant	3.20 ± 0.85 A	3.05 ± 0.79 AB	2.12 ± 0.71 B
Leaf length (cm)	19.40 ± 1.71 A	18.40 ± 1.57 A	15.52 ± 2.53 B
Leaf width (mm)	7.00 ± 0.06 A	6.60 ± 0.60 B	6.10 ± 0.60 C
Leaf length/width ratio	27.83 ± 2.77 A	28.25 ± 2.94 A	25.56 ± 3.61 B
Number of days to flowering	68.40 ± 10.25 A	72.15 ± 8.83 A	86.92 ± 19.33 B
At maturity/harvest			
Plant height (cm)	88.64 ± 4.29 A	90.45 ± 3.39 A	70.00 ± 19.97 B
Number of tillers per plant	21.33 ± 2.82 A	22.10 ± 4.62 A	19.00 ± 4.52 B
Flag leaf length (cm)	21.54 ± 4.44 A	22.42 ± 4.66 A	19.52 ± 4.52 B
Flag leaf width (mm)	12.80 ± 1.70 A	12.60 ± 1.65 A	11.31 ± 1.06 B
Flag leaf length/ width ratio	16.86 ± 2.60 A	17.78 ± 2.28 A	16.08 ± 4.33 B
Panicle length (cm)	13.15 ± 3.55 A	13.58 ± 3.66 A	11.18 ± 4.16 B
Number of panicles per plant	17.40 ± 2.75 A	17.80 ± 2.21 A	15.60 ± 6.37 A
Number of primary branches per panicle	7.60 ± 2.90 A	7.40 ± 2.80 A	7.32 ± 3.40 A
Number of secondary branches per panicle	4.75 ± 4.00 A	5.05 ± 7.13 B	2.39 ± 6.49 B
Number of spikelets per panicle	71.99 ± 38.28 A	73.94 ± 39.21 A	55.58 ± 36.69 B
Number of grains per plant	178.88 ± 98.47 A	219.00 ± 73.77 A	8.54 ± 5.26 B
Grain weight per plant	4.22 ± 2.33 A	4.99 ± 1.65 A	0.44 ± 0.44 B
Grain length	6.70 ± 0.20 A	6.60 ± 0.20 A	6.90 ± 0.50 B
Grain width	3.50 ± 0.10 A	3.50 ± 0.06 A	3.34 ± 0.02 B
Grain length/width ratio	1.90 ± 0.06 A	1.90 ± 0.06 A	2.06 ± 0.17 B

Values are the mean of 25 non-transformed, seed-derived plants, 20 non-transformed, protoplast-derived plants and 25 transformed protoplast-derived plants ± standard deviation. Least significant means followed by the same letter within a row are not significantly different at $p = 0.05\%$.

Table 2. Comparison of the phenotypic characteristics of the transgenic protoplast-derived R_2 plants, non-transformed, protoplast derived R_2 plants and non-transformed, seed-derived plants grown under glasshouse conditions at Nottingham University, UK

Phenotypic trait	Non-transformed seed-derived plants	Non-transformed protoplast-derived R_2 plants	Transformed protoplast-derived R_2 plants
At first tillering			
Plant height (cm)	33.30 ± 4.44 A	30.20 ± 3.29 B	27.88 ± 4.04 C
Number of tillers per plant	2.78 ± 0.62 A	2.58 ± 0.49 AB	2.60 ± 0.49 B
Leaf length (cm)	26.46 ± 3.23 A	24.08 ± 2.59 B	14.00 ± 10.43 B
Leaf width (mm)	8.09 ± 1.15 A	7.25 ± 0.81 B	7.00 ± 1.08 B
Leaf length/width ratio	3.34 ± 0.68 A	3.35 ± 0.44 AB	3.10 ± 0.54 B
Number of days to flowering	164.97 ± 9.68 A	171.58 ± 9.43 AB	183.44 ± 28.4 B
At maturity/harvest			
Plant height (cm)	95.47 ± 8.24 A	93.77 ± 7.27 A	67.94 ± 16.88 B
Number of tillers per plant	25.37 ± 4.12 A	24.72 ± 5.60 A	10.20 ± 5.95 B
Flag leaf length (cm)	61.03 ± 5.66 A	61.88 ± 7.78 B	48.41 ± 7.81 C
Flag leaf width (mm)	14.38 ± 1.14 A	13.87 ± 1.62 AB	11.92 ± 1.85 B
Flag leaf length/ width ratio	4.30 ± 0.47 A	4.54 ± 0.68 A	4.06 ± 0.76 B
Panicle length (cm)	15.85 ± 30.08 A	16.88 ± 34.77 B	14.11 ± 16.93 B
Number of panicles per plant	6.90 ± 3.33 A	5.21 ± 2.66 B	4.70 ± 3.21 B
Number of primary branches per panicle	7.37 ± 2.19 A	8.03 ± 2.19 A	7.79 ± 2.50 A
Number of secondary branches per panicle	7.02 ± 4.79 A	6.17 ± 3.93 B	6.43 ± 6.07 B
Number of spikelets per panicle	67.55 ± 26.63 A	66.46 ± 23.62 A	57.01 ± 26.46 B
Number of grains per plant	25.05 ± 12.64 A	15.62 ± 8.75 B	11.22 ± 8.29 C
Grain weight per plant	2.93 ± 0.14 A	2.90 ± 0.14 A	2.04 ± 1.03 B
Grain length	5.82 ± 0.31 A	5.78 ± 0.34 A	4.28 ± 2.15 B
Grain width	2.93 ± 0.86 A	2.90 ± 0.19 A	2.04 ± 1.03 B
Grain length/width ratio	2.00 ± 0.14 A	1.71 ± 0.16 A	1.62 ± 0.72 A

Values are the mean of 41 non-transformed, seed-derived plants, 25 non-transformed, protoplast-derived plants and 28 transformed protoplast-derived plants ± standard deviation. Least significant means followed by the same letter within a row are not significantly different at p = 0.05%.

diploid. The *npt* II gene was detected in all the plants in the transgenic group, with the exception of one individual in the glasshouse trial. However, the NPT II enzyme was only detected in 9 plants grown in the glasshouse and in none of the transgenic plants in the field trial. Individuals from each of the transgenic plant subgroups both contained and expressed the *npt* II gene. Comparison of the glasshouse-grown plants, which contained but did not express, the *npt* II gene, with those which both contained and expressed the gene, showed no significant difference in their vegetative characteristics. However, the plants which showed NPT II activity took significantly longer to flower ($p = 0.05$). Although the panicles produced by these plants were structurally not significantly different from plants that did not express the gene, their form was more variable. Seeds of plants which expressed the *npt* II gene were significantly ($p = 0.05$) shorter and narrower compared with those from transformed plants which did not express the gene. The grain weight, however, was not significantly different.

Discussion

Under field and glasshouse conditions, the agronomic potential of R_2 generation, transgenic rice plants was reduced as compared with that of non-transformed, seed-derived plants. The reduction in fertility of transgenic rice plants, the most significant phenotypic change observed, has also been described by other workers in studies of Japonica and Indica rices (Hayashimoto et al., 1990; Peng et al., 1992). Peng et al. (1992) reported that transgenic R_0 plants of the Indica variety IR58 were all male sterile. R_1 plants were produced only after back-crossing the transgenic plants with non-transformed, seed-derived plants. The pollen from the R_1 generation parents of the plants used in the present study has been shown to be viable (Davey et al., 1991). Despite the occurrence of significant differences between the phenotypes of plants grown at the two experimental sites, probably due to different environmental conditions, the general trends in plant phenotypes observed between plant groups and sub-groups were comparable.

Some of the phenotypic traits of the transgenic plants observed in this study may be related to somaclonal variation (Scowcroft & Larkin, 1988). In a field trial of R_1 generation, protoplast-derived, non-transformed plants of Taipei 309 (Abdullah et al., 1989), changes in panicle morphology, fertility and days to flowering were observed, comparable with those reported here. Although changes in ploidy level may be associated with protoclonal variation (Stafford, 1991), the DNA content of the plants examined was constant and the phenotypic characteristics normally associated with polyploid and aneuploid rice plants were not observed (Hayashimoto et al., 1990). The phenotypic characteristics of the transgenic plants suggest that protoclonal variation was greater in this plant group compared with the non-transformed, protoplast-derived plants and/or that factors specific to the transgenic plants influenced their phenotypes. The similarity in the phenotypes of the seed generations of transgenic plants reported here and by Davey et al. (1991) indicate the heritable nature of these characteristics. Somaclonal variation in rice has been shown to be transferred between seed generations (Xie et al., 1990).

The protoclonal plants were derived from protoplasts which were electroporated under the same conditions as the protoplasts which gave rise to the transformed plants but in the absence of plasmid. It is therefore unlikely that the phenotypic differences observed between these two plant groups were due to the method of plasmid delivery. The transformed cells from which the R_0 generation plants developed were selected on medium containing kanamycin sulphate. Antibiotics have been shown to influence, for example, plant regeneration (Zhang et al., 1989) and DNA synthesis (Santos & Salema, 1989), but as yet the effect of antibiotics on the phenotypes of transgenic rice plants has not been defined. It is also known that the insertion of foreign DNA can disrupt gene activity, with associated alteration in plant morphology (Luehrsen & Walbot, 1990; Van Lijsebetens et al., 1991).

Although the majority of transformed plants contained the *npt* II gene, only a minority expressed the gene. The loss of expression of this gene in transgenic rice has been reported previously (Davey et al., 1991). The reason for the loss of expression is not known, however, it may relate to DNA methylation, which has been shown to alter nopaline synthase and NPT II activity in transgenic *Arabidopsis* (Errampalli et al., 1990). The lack of foreign gene expression in transgenic plants in the field may relate to an environmental effect and/or differential sensitivity of the methods used to determine expression of the *npt* II gene. The phenotypic differences between transgenic plants, which expressed the *npt* II gene compared with those which did not express the gene, may relate to the production of and/or the presence of the NPT II protein in these plants.

Although foreign genes can be stably inherited, their expression is less consistent. This, combined with the reduced vigour, fertility and increased phenotypic variation of transgenic rice compared with non-transformed material, may complicate the use of such plants in conventional breeding programmes. Thus, careful selection of the transgenic plants to be included in a breeding programme is vital, to ensure their successful integration.

Acknowledgements

This work was supported by the Rockefeller Foundation and, in part, by the Overseas Development Administration. The assistance of the Ministry of Agriculture, Fisheries and Food, the United States Department of Agriculture and the US Animal Plant Health Inspection Service, is gratefully acknowledged.

References

Abdullah, R., J.A. Thompson & E.C. Cocking, 1986 Efficient plant regeneration from rice protoplasts through somatic embryogenesis. Bio/Technol. 4 1087–1090

Abdullah, R., J A Thompson, G S Kush, R P Kaushik & E C Cocking, 1989. Protoclonal variation in the seed progeny of plants regenerated from rice protoplasts Plant Sci. 65 97–101

Davey, M.R, S.L. Kothari, H.M. Zhang, E L Rech, E C Cocking & P.T. Lynch, 1991. Transgenic rice Characterisation of protoplast-derived plants and their seed progeny J Exp Bot 42 1159–1169

Doyle, J F & J L Doyle, 1990 Isolation of plant DNA from fresh tissue. Focus 12 13–15.

Edwards, K, C Johnstone & C Thompson, 1991 A simple and rapid method for the preparation of genomic DNA for PCR analysis Nucleic Acids Res. 19 1349

Errampalli, D, D. Patton, L Castle, L Mickelson, K Hanson, J Schnall, K.A Feldmann & D Meinke, 1991 Embryonic lethals and T-DNA insertional mutagenesis in *Arabidopsis* The Plant Cell 3 149–157

Hayashimoto, A., Z. Li & N. Murai, 1990. A polyethylene glycol-mediated protoplast transformation system for the production of fertile transgenic rice plants. Plant Physiol. 93 857–863.

Kothari, S.L, M.R. Davey, P.T. Lynch, R.P. Finch & E.C Cocking, 1993. Transgenic rice. In S-d. Kung & R. Wu (Eds) Transgenic Plants, Volume II, pp 3–20. Butterworths, New York, London.

Luehrsen, K R & V Walbot, 1990. Insertion of *Mu1* elements in the first intron of the *Adh* 1-S gene of maize results in novel RNA processing events. The Plant Cell 2 1225–1238.

Lynch, P.T, J. Jones, H. Zang, E.L. Rech, P.S. Eyles, S.L Kothari, N.W. Blackhall, E.C Cocking & M.R. Davey, 1992. Transgenic rice plants Characterisation of two generations of seed progeny Physiol. Plant. 85 362–366

Peng, J.-Y., H. Kononowicz & T.K. Hodges, 1992. Transgenic Indica rice plants. Theor. Appl. Genet. 83 855–863.

Santos, I. & R. Salema, 1989. Penicillins and activity of nitrogen metabolism enzymes in plant tissue cultures. Plant Sci. 59 119–125.

Scowcroft, W.R. & P.J. Larkin, 1988. Somaclonal variation. In G Block & J Marsh (Eds) Application of Plant Cell and Tissue Culture, Ciba Foundation Symposium 137, pp. 21–35. Wiley, Chichester.

Schuh, W, M.R. Nelson, D.M. Bigelow, T.V Orum, C E. Orth, P T Lynch, P S. Eyles, N W. Blackhall, J. Jones, E C Cocking & M R Davey, 1993 The phenotypic characterisation of R_2 generation transgenic rice plants under field conditions Plant Sci 98 69–79

Stafford, A., 1991. Genetics of cultured plant cells In A. Stafford & G. Warren (Eds). Plant Cell and Tissue Culture, pp. 25–47. Open University Press, Milton Keynes, UK.

Van Lijsebetens, M., R Vanderhaeghen & M Van Montagu, 1991. Insertional mutagenesis in *Arabidopsis thaliana* isolation of a T-DNA-linked mutation that alters leaf morphology Theor Appl Genet. 81 277–284

Yang, H, H M Zhang, M.R Davey, B.J. Mulligan & E C Cocking, 1988. Production of kanamycin-resistant rice tissues following DNA uptake into protoplasts. Plant Cell Rep 7 421–425

Xie, Q.J, M C Rush & J Cao, 1990. Somaclonal variation for disease resistance in rice (*Oryza sativa* L). In B T Grayson, M.B. Green & L G Copping (Eds). Pest Management in Rice, pp 491–509 Elsevier Applied Science, London

Zhang, H.M, H Yang, E L. Rech, T.J Golds, A S. Davis, B J Mulligan, E.C. Cocking & M.R. Davey, 1988 Transgenic rice plants produced by electroporation-mediated plasmid uptake into protoplasts. Plant Cell Reps. 7 379–384.

Euphytica **85:** 403–409, 1995.

Breeding of transgenic orange *Petunia hybrida* varieties

Johan S.N. Oud, Harrie Schneiders, Ad J. Kool & Mart Q.J.M. van Grinsven
S&G Seeds B.V., Research, P.O. Box 26, NL 1600 AA Enkhuizen, The Netherlands

Key words: Petunia hybrida, petunia, flower, colour, DFR, maize dihydroflavonol-4-reductase, anthocyanin biosynthesis, genetic modification

Summary

Colour is the major contributor to the total ornamental value of a flower. The combination of biochemical knowledge and genetic engineering technology has resulted in the addition of a new colour to the existing colour range of *Petunia hybrida*. This has been achieved by expression of the maize dihydroflavonol-4-reductase (*dfr*) gene in a suitable petunia acceptor which leads to the accumulation of pelargonidin-derived pigments in flowers. The resulting flower colour, however, was a pale brick-red, which is commercially unattractive in petunia.

Our objective was to produce a product suitable for commercialisation by introducing the *dfr* gene into our breeding material via normal sexual recombination. Although the initial transformant exhibited many negative characteristics, first analyses indicated that it was feasible to obtain suitable material for creating commercial hybrids. Experimental hybrids based on F_4 lines were obtained with improved phenotypic expression of the orange flower colour in combination with a good general performance.

In order to assess consumer-related characteristics, selected experimental hybrids were tested under field conditions. All transgenic plants had a normal appearance when compared with non-transgenic control plants. No linkage was observed between the transgenic trait and any negative characteristic. From these studies it can be concluded that through a combination of biochemistry, breeding and genetic engineering, it is possible to generate unique flower colours in a cultivars with commercial potential.

Introduction

Within the Solanaceae, the genus *Petunia* consists of about 40 species, mostly native to South America; only one species, *P. parviflora* is native to warmer parts of the United States (Munz, 1968; Willis, 1973). The common garden petunia: *Petunia hybrida* is a popular garden plant, appreciated for its flowers which are highly variable in colour and shape. The species is a hybrid between *P. axillaris* and *P. violaceae*, both originating from Argentina (Everett, 1981). In *Petunia hybrida* two morphologically different flower types can be distinguished based on the gene *Un*, which determines the undulata shape of the corolla (Bianchi, 1959). The dominant *Un* genotype is known as 'grandiflora', whereas varieties with the recessive gene *unun* are known as 'multiflora'. The *Un* genotype predominantly modifies plant habit; besides the undulata corolla, plants carrying this genotype have wide, short petals, thick, short filaments, carry less flowers, and are less vigorous than multiflora individuals.

Flower colour is predominantly due to three types of pigments: flavonoids, carotenoids and betalains. Flavonoids are the most common flower pigments which provide the greatest range of flower colours. The most important class of flavonoids with respect to flower colour are the anthocyanins. In addition to the flavonoids, co-pigmentation with other components and vacuolar pH also influence flower colour (Brouillard, 1983). The pathway for flavonoid biosynthesis (Fig. 1) is well-documented both at the genetic and the biochemical level (Van der Meer et al., 1991).

The available colours in *Petunia hybrida* are predominantly based on derivatives of the flavonoids, cyanidins and delphinidins, giving a colour range from white via salmon, pink and red to violet. Orange

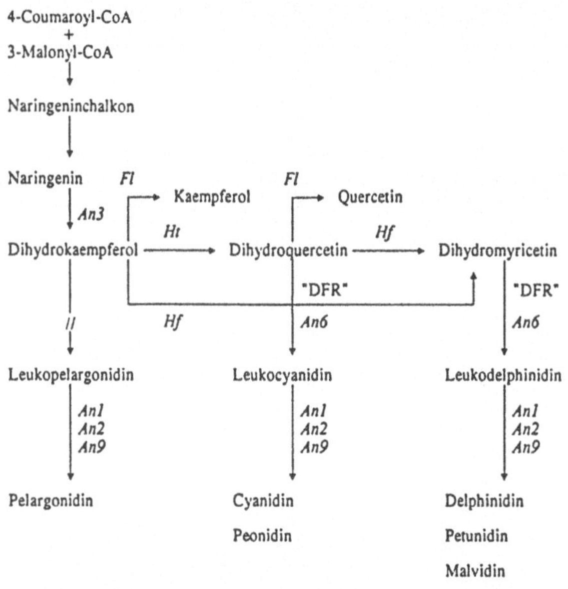

Fig. 1. Biosynthetic pathway of anthocyanins in *Petunia hybrida*, where available gene names are indicated. Nomenclature of *Petunia hybrida* loci is according to Gerats et al., 1982.

shades do not exist in petunia. One of the important enzymes in the formation of coloured anthocyanidins is dihydroflavonol-4-reductase (*dfr*). The enzyme catalyses the reduction of dihydroflavonols to unstable pro-anthocyanidins. Since the petunia *dfr* does not accept dihydrokaempferol as a substrate, pelargonidin-derived pigments, which are responsible for orange shades, are absent in petunia. In contrast, maize *dfr* has a different substrate specificity and is able to convert dihydrokaempferol and dihydroquercetin into leucoanthocyanidins.

Maize *dfr* therefore, when expressed in plants, is able to produce pelargonidin-derivatives provided that dihydrokaempferol is present as a substrate. The availability of the cloned maize *dfr* gene (Schwarz-Sommer et al., 1987) enabled Meyer et al. (1987) to complement the endogeneous petunia *dfr* gene. A number of independent petunia transformants that accumulated pelargonidin were obtained, but the pale brick-red flowers of these plants were not commercially attractive.

Our objective was to investigate the possibility of employing the maize *dfr* gene in a product with commercial potential for the bedding plant market, based on the added value of a unique orange colour. Our primary emphasis was on the introgression of the maize *dfr* gene in the multiflora plant type. A prerequisite for commercial feasibility is that the flower colour should be clearly distinct from the existing colours salmon and red. Moreover, the overall quality of the final product should fit into the profile description of the existing multiflora F_1 hybrids with respect to greenhouse and field performance. Here we present our successful efforts to create a commercially-interesting variety of *Petunia hybrida* that, owing to the expression of a heterologous *dfr* gene, has a bright orange flower colour.

Material and methods

Material and methods

The starting material used in breeding was acceptor line RL01 (Stotz et al., 1985), a genotype not present as such in the S&G Seeds gene pool. The genotype of RL01 for flower colour is *AnAnhththfhfflfl*, in which *An* stands for anthocyanin, *Ht* for flavonoid-3^1 hydroxylase, *Hf* for flavonoid-3^1,5^1 hydroxylase and *Fl* controls the synthesis of kaempferol and quercitin. Two transformants MPI-15 and MPI-17 (Linn et al., 1990), with a single gene copy of the maize *dfr* gene, were used as donors for the transgene.

S&G Seeds breeding material that served as acceptors differed considerably in genetic backgrounds and flower colours from the original RL01 type. The acceptor lines were L2134 (pink), L2124 (rose), F5 (salmon) and Y1 (red). All lines are based on the genotype (*AnAnHtHthfhfflfl*). Recombinations using the transgenic lines as male and female were made with the four selected elite lines. Approximately 5 were selected and self-pollinated per recombinant offspring. For comparisons the Polo series (S&G Seeds) were used, which are representative of the *Petunia hybrida* multiflora market. All characteristics were visually assessed by comparing with these references.

Chromatographic identification of anthocyanin pigments

Extraction of the flower pigments

The corollas (2 for dark colours and 5 for light colours) were incubated overnight in 5 ml methanol with 0.5% HCl at 4° C. The methanol-HCL solution was subsequently lyophilised and the anthocyanin pigments were dissolved in 1 ml 2M HCl and incubated for 40 minutes at 100° C. After cooling for 5 minutes on ice, the solu-

Table 1. Rf values of control pigments in either Forestal or Isopropanol-HCl buffer. Detection is under visible light conditions with the exception of Kaempferol. Colours are indicated

	Forestal	Isopropanol -HCl	Colour
Pelargonidin	0.66	0.56	Orange Red
Cyanidin	0.49	0.38	Red
Peonidin	0.64	0.43	Red
Delphinidin	0.31	0.20	Violet
Petunidin	0.47	0.22	Violet
Malvidin	0.62	0.26	Violet
Quercetin	0.42	}	Yellow fluorescent
Kaempferol	0.56	}	under ultra violet light

tion was centrifuged for 10 minutes at 1000 g and the supernatant was transferred to a small tube. Four drops of iso-amylalcohol were added and the solution was shaken vigorously for 2 minutes. The samples were stored at 4° C until use.

Cellulose plate chromatography of anthocyanins

The iso-amylalcohol was carefully spotted using a capillary (repeated six times on the same spot) 2 cm from the bottom of the cellulose plate (DC-Fertigplatten Cellulose, Merck), with 1.2 cm spacing between samples. The iso-amylalcohol was vaporised and the plates were put in the chromatography tanks with 1.5 cm of running buffer (Forestal: 23% aqua dest-70% concentrated acetic acid - 7% concentrated HCl or ISO : 50% 2-propanol - 50% 2 M HCl). On completion of the run (Forestal after ± 14 hours, ISO after ± 20 hours) the plates were dried. Subsequently the Rf values of the pigments was determined, identified with reference to the standards (Table 1) and the relative amount of the anthocyanins were estimated.

Greenhouse trial

16 experimental hybrid crosses between orange F_4 breeding lines, sown on February 1, 1993 in small seedboxes and transplanted after 3 weeks in small containers (approx. 200 plants per m^2), were tested. The temperature regime was: 18° C (night–20° C (day). Plant material was treated 3 times with a growth regulator (Daminozide) to maintain the plant quality and compactness. Flowering started about 75 days after sowing and 28 plants were analysed per entry. The trial was carried out under natural light conditions

Table 2. Pigment analysis of original breeding material used in the recombination programme, relative contents of anthocyanidins on a scale of 1–9. The flower colour description corresponds with the terminology used in the market

Line	Flower colour	Relative anthocyanidin contents					
		pel.	cya.	peo.	del.	pet.	malv.
RL01	Pale Pink		+		+		
MPI-15	Brick-red	6			1	3	
MPI-17	Brick-red	8	2				
L2124	Rose		1	9			
L2134	Pink			9			1
Y1	Red		8	1	1		
F5	Salmon		9		1		

Note: pel = pelargonidin, cya = cyanidin, peo = peonidin, del = delphinidin, pet = petunidin, malv = malvidin, + = trace. Where no pigment was present in a sample, it has been left blank.

(15.000–25.000 J/cm^2) and day length in the period February–April (9–13 hours).

Experimental hybrids were assessed for many characteristics, such as plant habit, plant vigour, earliness of flowering, flower colour and floriferousness. Colour observations were made by visual assessment of entries under different circumstances such as early morning and midday conditions, in springtime and autumn and by thin layer chromatography.

Field trial

The field trial took place at a breeding station in Naples (Florida, USA). USDA-APHIS approved the application and granted permit # 92-260-01 to conduct this trial. 25 experimental hybrid crosses between orange F_4 and F_5 breeding lines were tested under high light (60,000–70,000 J/cm^3) and high temperature (± 25° C) conditions. 15 plants were planted per experimental hybrid. The main goal of the trial was to evaluate the material under outdoor conditions. Observations were made of plant habit, plant vigour, flowewr traits and standing ability. Data were collected every two weeks during the growing season (April–May) over a period of 8 weeks.

Results

In order to have a significant commercial value, *Petunia hybrida* plants should be compact, basally-branched

Table 3. Colour description and pigment analysis of several F_3 breeding lines and their corresponding F_4 offspring, relative contents of anthocyanidins on a scale of 1–9

Crossing	F_3 parental colour	Relative anthocyanidin content						Colour F_4 offspring
		pel	cya	peo	del	pet	malv	
MPI-15 × L2124 (rose)	P2253-1 orange	8			1	1		orange, light salmon
	P2254-1 rose-orange edge	2		7		1		rose-orange edge
	P3007-2 rose-orange edge	7		2		1		rose-orange glow, rose-orange edge
	P3009-2 orange-grey	8			1	1		orange
MPI-17 × L2134 (pink)	P2255-1 rose-orange glow	9					1	rose-orange glow, pink
	P2255-3 light pink	1			+	+	9	light pink, light salmon
	P2256-1 salmon	8			1	1		orange, salmon
MPI-15 × Y1 (red)	P2260-2 orange	8			1	1		orange-grey, light salmon
	P2260-6 red	1	8	1				orange, red
	P3025-4 light pink				7	3		light pink, light salmon
MPI-15 × F5 (salmon)	P3015-3 orange	8			1	1		orange
	P3016-1 red	2	8					orange, red
	P3015-2 orange-grey	8			1	1		orange

Note: pel = pelargonidin, cya = cyanidin, peo = peonidin, del = delphinidin, pet = petunidin, malv = malvinidin, + = trace. Where no pigment was present in a sample, it has been left blank.

and vigorous with a good standing ability. The erect, non-basally-branched plant habit of the transgenic *dfr* gene donor RL01 is, in this respect, not commercially useful. In order to develop good quality breeding lines, a recombination programme was initiated to introduce the maize *dfr* gene into a number of different genetic backgrounds. Initial crosses were made between transformants MPI-15, MPI-17 and breeding lines L2124, L2134, Y1 and F5. These breeding lines were selected for good quality parameters as indicated above, in combination with a genetic background for flavonoid synthesis comparable with that of line RL01. Pigment characteristics of the relevant lines are indicated in Table 2.

Recombinant progeny between the MPI-15, MPI-17 and the non transgenic breeding lines accumulated low quantities of pelargonidin (relative contents: ≤ 1) in combination with cyanidin and peonidin (data not shown). The flowers of the resulting F_1 progeny were not orange, due to the presence of dominant *Ht* alleles from the S&G Seeds breeding material. Several F_1 plants from each recombinant offspring were self-pollinated and in the F_2 populations, plants with an orange flower colour were selected. These plants contained pelargonidin as the major anthocyanidin. Since different genetic backgrounds were used, new combinations of anthocyanidins were found yielding different colour shades. This is illustrated in Table 3 which summarizes the results of the pigment analysis of a number of F_3 individuals and their F_4 progeny.

Although F_3 progeny contain individual plants which accumulate pelargonidin it should be realised that F_3 individuals still segregate for colour in their offspring. Table 3 also illustrates that pigment compositions of all entries are clearly distinct from the current commercial assortment. Moreover, by combining the RL01 donor background with elite breeding material a higher level of colour expression was achieved. It should be noted that ratios and not absolute amounts of pigments are presented in Table 3. These data therefore do not allow a comparison of the flower colour intensity between the F_3 individuals.

Intense flower colours such as rose, red and violet, appear more intense under high light conditions. This phenomenon is not observed with less intense colours such as pink and salmon. Strikingly, the newly-created orange colour also intensifies under high light conditions.

Besides improving the expression of the orange colour we were also able to improve the plant habit significantly. It was possible to combine the orange colour gene with most beneficial genes without any negative effect. Linkages between colour expression and negative plant performance characteristics were not observed. Under a variety of different greenhouse conditions the transgenic plant material appeared very healthy. Pollen production and seed yield were even above standard, in contrast to the donor RL01, which had somewhat reduced fertility. By introgression of the *dfr* gene via normal sexual recombination into breeding material, different combinations of anthocyanins can be created, resulting in novel colour shades, incorporated into excellent performing varieties.

Expression of the orange flower colour is sometimes covered by a grey shade, i.e. 'grey filter'. Orange flower colours with 'grey filters' could be found in the progeny of all initial F_1 crosses and was not related to the quantity of pelargonidin. 'True' orange flower colour was always a combination of pelargonidin, delphinidin and petunidin. Both shades of orange appear side by side in the F_3 and F_4 offspring. 'Grey filter' orange could be eliminated by selection (Table 3, e.g. P3009-2 and P3015-2).

The red flower colour in plants with the heterologous *dfr* gene, accumulating pelargonidin, appeared brighter than that based on only cyanidin and peonidin in the normal red flowers. Existing rose flower colours are based on the combination of peonidin, petunidin and in some cases cyanidin. With the addition of pelargonidin the new colour may vary from a rose-orange edge on flowers which only contain low amounts of pelargonidin, to a rose-orange glow with a high concentration of pelargonidin (see Table 3, cross MPI-15 * L2124). Pelargonidin is also present in low amounts in flowers with colours such as salmon and pink, where delphinidin, petunidin and malvidin are present. Synthesis of pelargonidin seems to decrease the colour expression of the other compounds (Table 3, P2255-3).

Field trial

The field trial was evaluated every two weeks during the growing season. Data were collected on plant performance aspects such as vigour, plant habit, flower characteristics and standing ability. The quality parameters of the transgenic plants were similar to the control cvs. Polo Salmon and Polo Pink. The colour expression of the maize *dfr*-containing hybrids under high light and high temperature conditions was good and was clearly distinct from the current assortment of *Petunia hybrida* multiflora colours. Moreover, the trait is highly stable in many different genetic backgrounds.

Discussion

In all four elite lines used to cross with the transgenic donor lines, the dominant *Ht* allele was introduced into the progeny. All resulting F_1 progeny accumulated large amounts of cyanidins in addition to pelargonidins. This indicates that the maize *dfr* is not able to convert all available di-hydrokaempferol into leuco-pelargonidin. Apparently the K_m for the reduction reaction is similar, if not lower than the K_m for the hydroxylation by the *Ht*-encoded enzyme. In order to accumulate as much pelargonidin as possible, the *Ht* allele derived from the S&G Seeds elite material should be replaced by its recessive counterpart originating from the RL01 background. In the subsequent F_2 and F_3 progeny, plants were selected which contained the recessive *ht* allele, and therefore have an orange flower colour.

The accumulation of delphinidin and derivatives in F_3 lines of all tested combinations (Table 3) can only be explained by assuming that both the breeding lines and the original RL01 lines have a 'leaky' *hf* allele. This was confirmed by analysis of RL01 which showed that small amounts of delphidin and cyanidin still accumulated. This does not explain, however, the predominant presence of delphinidin and derivatives in some of the F_3 lines (e.g. P2255-3 and P3025-4, Table 3). More elaborate analyses of crosses with well-defined tester lines are needed to elucidate this apparent contradiction.

Introgression of the heterologous maize *dfr* gene through classical breeding procedures led to selection of commercially-interesting lines. The products fitted our profile of quality parameters for varieties with good greenhouse and field performance.

The occurrence of grey-orange shades in the progeny of all initial F_1 crosses and in the consecutive generations is puzzling. Analysis of the segregation data did not lead to a clear conclusion, since both dominant and recessive behaviour of the 'grey-filter' was observed. Wiering (1974) pointed out that grey shades of other colours co-segregated in the presence of relatively high amounts of delphinidin in combination with specific glucosidation patterns. The 3-glucoside and

3-rutinoside derivatives of delphinidin showed grey shades, the acylated 3-rutinoside-5-glucoside forms never gave rise to grey shades. The expression of a 'grey-filter' in our orange flowering plants is probably due to the leakage of the *hf* locus leading to accumulation of delphinidin in MPI-15, Y1 and F_5 (Table 2), in combination with segregating glycosidation genes.

Previous trials performed in Germany (Meyer et al., 1992) using the original transgenic lines (Meyer et al., 1987) showed frequent reversion of the brick-red flower colour to the original RL01 pale pink colour. This reversion occurred either in sectors of the corollas or in complete flowers. When these original lines were recombined with elite material such reversions were not observed. Such reversions were never observed in further introgression up to the F_5 generation of the original maize *dfr* loci derived from lines MPI-15 or MPI-17. The different transgenic petunia generations were cultured under varying conditions year-round, indicating that the stability was not sensitive to changes in culture practices. The field trial performed with hybrids under the extreme conditions at the trial location did not lead to the identification of any revertant. Based on these findings it can be concluded that the expression of the maize *dfr* gene is stable during the introgression process.

Meyer et al. (1993) presented evidence that the revertant colour patterns were based on a reduction or loss of the heterologous *dfr* gene transcription, which was correlated with an increase in DNA-methylation. Another possible explanation for the observed phenomenon could be that the A1 instability observed in the original RL01 background is due to instability of heterochromatin formation in the immediate environment of the insertion site of the transgenes. Our observations suggest that introgression from genotypes containing the maize *dfr* gene into commercial elite material may stabilize transgene expression. It can be speculated that either the genetic backgrounds, used in this breeding programme, contain 'stabilizers' preventing erroneous euchromatin formation at or surrounding the inserted maize *dfr* gene, or that a 'destabiliser' from the original vicinity of the *dfr* insertion has been deleted. One should realize that in breeding programmes, breeders have for commercial reasons consistently selected against genetic backgrounds that give unstable flower colour expression. Further molecular analyses are needed to study the molecular mechanisms underlying genetic instability of flower pigmentation.

It can be concluded from our study that the orange flower colour trait now present in transgenic *dfr* petunia breeding material, behaves normally in all tested aspects and has been successfully used in breeding programmes aimed at developing commercial F_1 varieties with this trait.

Acknowledgements

We thank Ir. Wim van Kester, Dr. André Schram and Dr. Ir. Peter de Haan (S&G Seeds) for critically reading this manuscript and continuous stimulation during the breeding process. The scientific discussions with Dr. Gerrit-Jan van Holst (S&G Seeds, Biotechnology Dept.) and Dr. Peter Meyer (MPI) were greatly appreciated. The collaboration with our colleagues at Rogers Seed Co., Naples Florida is acknowledged.

References

Bianchi, F., 1959. Onderzoek naar erfelijkheid van de bloemvorm bij *Petunia*. Academisch proefschrift, Amsterdam.

Brouillard, R., 1993. The *in vivo* expression of anthocyanin colour in plants. Phytochemistry 22, 6: 1311–1323.

Everett, T.H., 1981. The New York Botanical Garden Encyclopedia of Horticulture. Garland, New York and London: 3596 pp.

Gerats, A.G.H., P. de Vlaming, M. Doodeman, B. Al & A.W. Schram, 1982. Genetic control of the conversion of dihydroflavonols into flavonols and anthocyanins in flowers of *Petunia hybrida*. Planta 155: 364–368.

Linn, F., I. Heidmann, H. Saedler & P. Meyer, 1990. Epigenetic changes in the expression of the maize A1-gene in petunia; Role of numbers of integrated copies and state of methylation. Mol. Gen. Genet. 222: 329–336.

Meer van der, I.M., A.R. Stuitje & J.N.M. Mol, 1991. Regulation of general phenylpropanoid and flavonoid gene expression. In: D.P.S. Verma (Ed.). Control of plant gene expression, Telford Press.

Meyer, P., I. Heidmann, G. Forkmann & H. Saedler, 1987. A new petunia flower colour generated by transformation of a mutant with a maize gene. Nature 330: 677–678.

Meyer, P., L. Felicitas, I. Heidmann, H. Meyer, I. Niedenhof & H. Saedler, 1992. Endogenous and environmental factors influence 35S promoter methylation of a maize A1 gene construct in transgenic petunia and its colour phenotype. Mol. Gen. Genet. 231: 345–352.

Meyer, P., I. Heidmann & L. Niedenhof, 1993. Differences in DNA methylation are associated with a paramutation phenomenon in transgenic petunia. Plant Journal 4: 89–100.

Munz, P.A., 1968. A California Flora. University of California Press, Berkeley and Los Angeles: 1681 pp.

Schwarz-Sommer, Z., L. Leclerq, E. Göbel & H. Saedler, 1987. Cin$_4$, an insert altering the structure of the *A1* gene in *Zea mays*, exhibits properties of nonviral retrotransposons. EMBO J. 2: 287–294.

Stotz, G., P. den Vlaming, H. Wiering, A.W. Schram & G. Forkmann, 1985. Genetic and biochemical studies on flavonoid-3'-

hydroxylation in flowers of *Petunia hybrida*. Theor. Appl. Genet. 70: 300–305.

Wiering, H., 1974. Genetics of flower colour in *Petunia hybrida* Hort. Genen Phaenen 17 (1–2): 117–134.

Willis, J.C., 1973. A Dictionary of the Flowering Plants and Ferns. Eighth Edition. Cambridge University Press. Cambridge et alibi: 1245 pp.

Euphytica **85:** 411–416, 1995.

A study of different (CaMV 35S and *mas*) promoter activities and risk assessment of field use in transgenic rapeseed plants

J. Pauk[1], I. Stefanov[2], S. Fekete[1], L. Bögre[2], I. Karsai[3], A. Fehér[2] & D. Dudits[2]
[1] *Cereal Research Institute, H-6701 Szeged, P.O. Box 391, Hungary;* [2] *Institute of Plant Biology, Biological Research Center, Hungarian Academy of Sciences, H-6701 Szeged, P.O. Box 521, Hungary;* [3] *University of Debrecen, H-4010 Debrecen, P.O. Box 37, Hungary*

Key words: Agrobacterium, Brassica napus, CaMV 35S promoter, *mas* promoter, gene expression, risk assessment

Summary

Gene fusions between the β-glucuronidase (GUS) reporter gene and the promoters of the cauliflower mosaic virus 35S RNA transcript (CaMV 35S) and the mannopine synthase (*mas*) genes were introduced into rapeseed varieties via *Agrobacterium*-mediated transformation. Fluorometric assay of β-glucuronidase activity indicated different expression patterns for the two promoters.

In seedlings, the CaMV 35S promoter had maximum activity in the primary roots, while the *mas* promoter was most active in the cotyledons. Etiolated seedlings cultured in the dark showed reduced activity of the *mas* promoter.

Before vernalization at the rosette stage, both promoters were more active in older plant parts than in younger ones. At this stage the highest activity was recorded in cotyledons.

After the plants had bolted reduced promoter function was detected in the upper parts of the transformed plants. Both promoters were found to be functional in the majority of the studied organs of transgenic rapeseed plants, but the promoter activity varied considerably between the organs at different developmental stages.

The ability of pollen to transfer the introduced genes to other varieties and related species (e.g. *Brassica napus* and *Diplotaxus muralis*) by cross-pollination was studied in greenhouse experiments, and field trials were carried out to estimate the distance for biologically – relevant gene dispersal. In artificial crossing, the introduced marker gene was transferable into other varieties of *Brassica napus*. In field trials, at a distance of 1 metre from the source of transgenic plants, the frequency of an outcrossing event was relatively high (10^{-3}). Resistant individuals were found at 16 and 32 metres from the transgenic pollen donors, but the frequency of an outcrossing event dropped to 10^{-5}.

Introduction

Genetic modifications and the introduction of new characters into crop plants have opened up new possibilities for analysis of the activity of different transcriptional control elements during the life time of the plant. The use of gene fusion constructs carrying a defined promoter sequence from the gene of interest and a reporter gene has contributed significantly to the understanding of the molecular components involved in the regulation of gene expression (Jefferson, 1987; Benfey et al., 1989). Currently one of the most widely used reporter genes is the *uid* A gene from *E. coli*, genetically modified to make it suitable for expression experiments in plants. It encodes for the protein β-glucuronidase. This enzyme catalyses the cleavage of several β-glucuronides, including commercially-available substrates. Fluorometric and colorimetric quantitative assays and procedures for tissue-specific detection by histochemical staining have been developed. It has been demonstrated that the expression pattern of a promoter may vary with the plant species (Harpster et al., 1988; Benfey et al., 1989; Blaich, 1992); thus, a comprehensive analysis of expression pattern is presumed to be based on a broader set of host plants. It has also been found that the use of alternative reporter genes may result in different expression patterns (Jefferson et al., 1987; Schneider et al., 1990).

The promoter of gene IV of the cauliflower mosaic virus (CaMV) encodes the 35S RNA transcript. It drives the synthesis of an RNA serving as a non-reusable template for CaMV DNA synthesis. Although the natural host range of CaMV is limited to species in the Cruciferae, the promoter is active in the genome of the large variety of monocotyledonous and dicotyledonous plants (Harpster et al., 1988; Boulter et al., 1990; Blaich, 1992). This promoter has been considered to be constitutive (Odell et al., 1985; Benfey et al., 1989), but highly-sensitive methods have shown some tissue-specificity (Jefferson et al., 1987; Battraw & Hall, 1990) and cell cycle stage-dependent expression (Nagata et al., 1987). The dual bi-directional promoter TR1′-2′ is derived from the genes for mannopine synthase (*mas*) on the TR-DNA of an octopine-type Ti plasmid (Velten & Schell, 1985). This promoter has been preferably used for studies on the expression of chimeric genes both in transient experiments and in transgenic plants (Velten & Schell, 1985; Debleare et al., 1987; Saito et al., 1991; Stefanov et al., 1991). It has been assumed that its transcriptional activity is constitutive (Debleare et al., 1987; Rogers et al., 1987). Recent results, however, demonstrated the tissue-specific, organ-dependent and auxin-enhanced expression of the *mas* promoter (Langridge et al., 1989; Leung et al., 1991).

The aim of the present work was a detailed analysis of the expression pattern of these nominally constitutive promoters in *Brassica napus*, with GUS being used as the reporter gene. Results on the risk assessment of the field use of transgenic rapeseed are also described.

Materials and methods

Plant material

'Arabella' and 'Santana' rapeseed (*Brassica napus*) cultivars were used in the experiments. Transformation was achieved by *Agrobacterium* infection (Stefanov et al., 1994) using the following plasmids: plasmid pPCV701 GUS (a kind gift from Dr. A. Zilberstain, Tel-Aviv University, Tel-Aviv, Israel, and Dr. Cs. Koncz, Max-Planck-Institut für Züchtungsforschung, Köln, Germany) contains the *uid* A gene coding sequence under the control of the CaMV 35S promoter and linked to the *nos* termination signal (Koncz & Schell, 1986); plasmid pIS 412 was constructed by replacing an *Eco* RI/*Hind* III fragment of pGA 482 (An, 1987) with an *Eco* RI/*Hind* III fragment of the transient expression vector pIDS 411 (Stefanov et al., 1991), which contains the GUS coding sequence under the control of the *mas* promoter and the *nos* termination signal.

As hosts for the binary vectors, *A. tumefaciens* strains LBA 4404 (Ooms et al., 1982) and GV3101 (pMP90RK) (Koncz & Schell, 1986) were used. Before plant transformation, the structures of the binary vectors were checked by restriction endonuclease digestion of plasmid DNA.

Conditions of transformation, selection and plant regeneration were as detailed previously (Stefanov et al., 1994). Transformation was confirmed by Southern hybridization of genomic DNA from primary transformants and PCR analysis of DNA from second generation seedlings.

Analysis of rapeseed transformants and their progeny

The NPT II tests were performed as described by Reiss et al. (1984). Transformation was confirmed by Southern hybridization of genomic DNA from primary transformants and PCR analysis of DNA from offspring (Stefanov et al., 1994).

Assays for GUS activity and histochemical staining of plant tissues were carried out according to Jefferson (1987) with minor modifications. Plant tissues were ground with quartz sand in 200 ml of ice-cold extraction buffer in an Eppendorf tube, using a glass pestle connected to a mixer head. After centrifugation in an Eppendorf centrifuge for 5 min, the cell debris was precipitated and 25 μl samples were taken from the clear phase. One sample was used for protein determination, according to Bradford (1976), while the others were incubated for 0 min, 60 min or 120 min with 75 μl of assay buffer containing 1 mM methyl-umbelliferyl β-glucuronide (MUG) in microtitre plate wells. The reaction was stopped with 25 μl of 1 M Na_2CO_3 and the fluorescence of methylumbelliferone (MU), the product of the reaction catalysed by GUS, was measured with a Perkin-Elmer fluorometer. The specific activity of the reporter enzyme was expressed in nmoles mg^{-1} protein min^{-1}.

Cross-pollination experiments were carried out in the greenhouse in winter to avoid natural pollination by insects. In the field trial, six genetically-modified

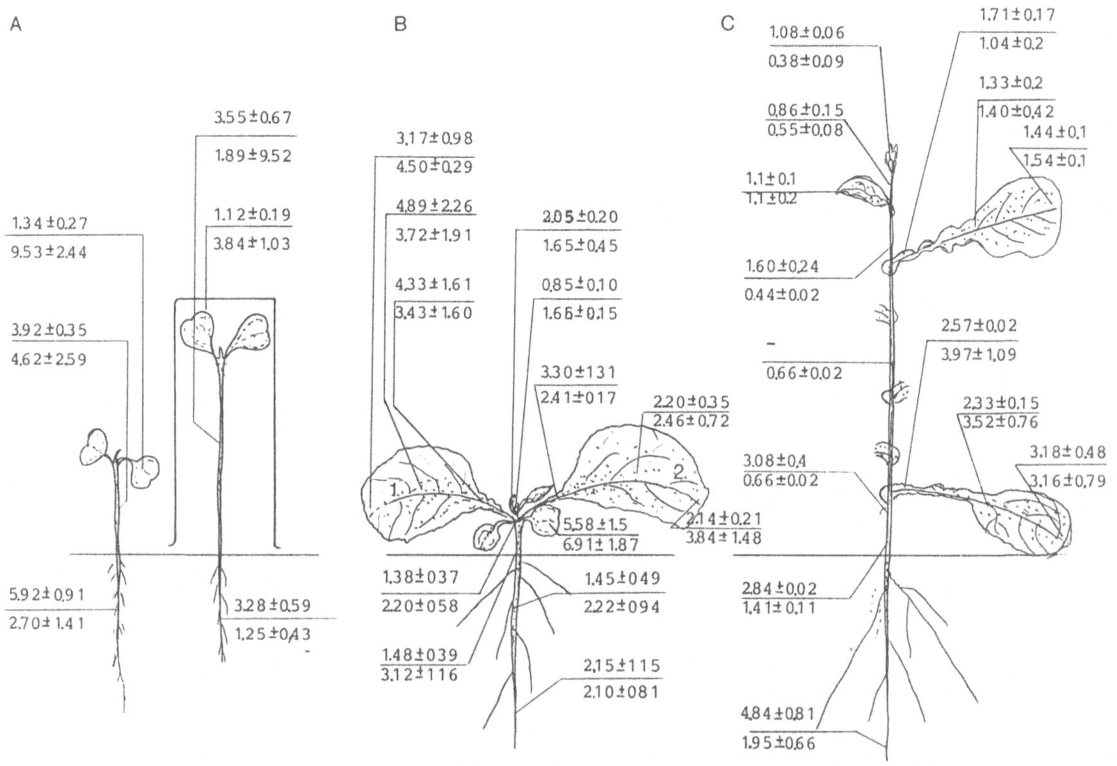

Fig 1 GUS activity in second generation (R_1) transgenic *Brassica napus* plants at different stages of development A Seedlings grown under light (left) or darkness (right) B Young plant at rosette stage C Mature plant before flowering The row indicates the GUS activity (nmoles mg^{-1} protein min^{-1}, ± standard error) detected in transformants with the CaMV 35S (upper), or *mas* (lower) promoter

plants were grown at the centre of the field (1 ha) of non-modified rapeseed plants. The field trial was carried out in isolation, surrounded by forest, meadow and cereal experiments. An attempt was made to minimize the genetic-environmental interaction. In the field trial, the ratio of transformed/non-transformed plants was 1 2×10^5 Seed of transformants was harvested and seed samples were collected from 1-, 2-, 4-, 8-, 16-, 32- and 50-m sectors at the four cardinal points.

Results and discussion

Expression pattern of the chimeric β-glucuronidase gene in transgenic rapeseed plants

The expression of the transgene was tested in the offspring of R_0 transformants. After self-pollination of independent transformants, seedlings were grown on medium containing 100 mg l^{-1} Kanamycin and the resistant segregants were used for analysis of the expression of the *gus* reporter gene.

Transgenic seedlings grown either in the dark or under conditions of 16/8 hours light/dark period exhibited significant differences In seedlings, the CaMV 35S promoter had maximum activity in the primary roots, while the *mas* promoter was most active in the cotyledons Etiolated seedlings cultured in the dark showed a reduced activity of the *mas* promoter (Fig. 1A).

The CaMV 35S promoter activity was affected by illumination only in the root. Roots grown in the dark displayed a reduced growth rate and lower CaMV 35S promoter activity

Activity of the *mas* promoter was strongly influenced by the light conditions. The GUS activity in etiolated seedlings was 2–2.5 times lower than that in non-etiolated ones The differences could be observed in all parts of the seedlings. Down-regulation of the *mas* promoter in etiolated seedlings may be caused by the presence of inhibitory factors.

GUS activity was also analysed in plants at different stages of development. When 6 to 8-week-old plants grown in the greenhouse without vernalization were tested, the activity of both promoters was highest in the cotyledons and both promoters were more active in older than in younger plant parts (Fig. 1B). The fully-expanded leaves (first leaf) exhibited a substantially higher promoter activity than developing ones. At this developmental stage, the activity of the CaMV 35S promoter was found to be weak in the meristematic buds. The leaves surrounding the apical bud had an activity similar to that of the growing leaves.

The plants were vernalized to induce the generative phase. After transfer to the greenhouse, the plants reached a height of 70–90 cm in 3–4 weeks. This process involves an increased production of growth hormones, especially gibberellins (Trewavas, 1981). Accordingly, considerable changes may be expected in the activities of the promoters analysed (Fig. 1C). After the plants had bolted, reduced promoter activity was detected in the upper parts of the vernalized plants. A decrease in activity of both promoters was seen in the upper leaves. An even greater decrease in promoter activity was detected in the leaves beneath the flower. In the stem and the root, the CaMV 35S promoter displayed increased activity in comparison with the *mas* promoter. We detected a sharp decrease in activity of the *mas* promoter especially in the upper part of the shoots. This is a condition similar to that found in etiolated seedlings. Since gibberellic acids are the main hormones involved in etiolated growth and bolting (Trewavas, 1981), it may be speculated that the lower activity of the *mas* promoter in these two cases was caused by the increased level of gibberellic acids.

To monitor changes in promoter function during morphogenesis *in vitro*, callus and shoots were induced from hypocotyl explants of R2 plants. Morphogenic and non-morphogenic calli, and shootlets were sampled. The CaMV 35S promoter had significantly higher activity in the morphogenic callus. The first organized structure to emerge from the callus is often a leaf, and the process of regeneration is similar to the development of a lateral bud. The first leaflet had a lower *mas* promoter activity that of the shoot. The trend to lower activity of this promoter is organized structures has already been seen in tobacco (Langridge et al., 1989).

The transformation and regeneration system of diverse *Brassica napus* cultivars was thus established by induction of callus formation from transformed cells before plant regeneration. Use of the *gus* reporter gene permitted a detailed analysis of the activity of the CaMV 35S and *mas* promoters in the transformed rapeseed plants studied. The results demonstrated consistent differences in the expression of the promoters studied. Both regulatory elements can be considered constitutive in the sense that they are expressed in all plant parts, but the level of expression was strongly dependent on the developmental stage of the plants and on the specific organ. We also detected differences between the two promoters in tissue-specificity. The growth conditions of the individual plants also influenced promoter function. Since actual GUS enzyme activity can be modified by some factors, we propose that the promoters analysed are the key factors generating the observed differences.

Study of transferred gene dispersal

Cross-pollination

In the greenhouse, artificial crosses were made to study the frequency and extent of cross-pollination between transformed crop plants containing the marker gene, and other plants of related species (e.g. weeds). When we used genetically-transformed 'Arabella' and 'Santana' rapeseed plants as pollen donors, incompatibility was not found with commercial, non-transgenic rapeseed varieties. The pollinated flowers (85 samples) produced healthy fruits in 83.5% of the crossed flowers. In the hybrid seeds (4220), the introduced marker gene was expressed in 3:1 and 1:4 ratio for resistant/sensitive plantlets, depending on the pollen donor parent, 'Arabella' and 'Santana', respectively. The 1:4 segregation ratio means that the transgene is not efficiently transmitted. Gene transfer from rapeseed to relatives (*Brassica rapa, Diplotaxus muralis* etc.) – using embryo rescue in *D. muralis* – has not been confirmed so far. The experiments of sexual transfer of the transgene are continuing.

Field trials

Field trials were carried out to examine the interaction of genetically-modified plants with the cultivation environment. Six modified plants were grown at the centre of the plot, surrounded by 1 ha of non-modified rapeseed plants (Fig. 2A). Pollen dispersal was achieved by natural pollination with bees (Fig. 2B). Marker gene dispersal from the transformed plants was monitored by screening progeny from the non-modified plants for the presence of the introduced

Fig. 2. Rapeseed field trial (A) to examine the interaction of genetically-modified plants with non-modified ones, by natural pollination by bees (B).

character (*nptII* gene). Selection for Kan^R was made in each case after a second exposure (selective liquid and agar-solidified shoot culture) to kanamycin and the resistance was tested by *npt*II assay. So far, kanamycin-resistant individuals have been found in the immediate area (within 1 m of the test plants) in 10^{-3} frequency, but resistant individuals were found at 16 and 32 metres from the transgenic pollen donors in very low frequencies (10^{-5}). Screening for very low frequencies still continues. The harvested yield of the 1 ha field surrounding the trial was burnt.

Acknowledgements

The authors express their thanks to OMFB Budapest for supporting this research and B. Dusha for photographic work. The conference participation was supported by Soros Foundation.

References

An, G., 1987. Binary Ti vectors for plant transformation and promoter analysis. Methods in Enzymology 153: 292–301.

Battraw, M.J. & T.C. Hall, 1990. Histochemical analysis of CaMV 35S promoter-β-glucuronidase gene expression in transgenic rice plants. Plant Mol. Biol. 15: 527–538.

Benfey, P.N., L. Ren & N.H. Chua, 1989. The CaMV 35S enhancer contains at least two domains which can confer different developmental and tissue-specific expression patterns. EMBO J. 8: 2195–2202.

Blaich, R., 1992. Function of genetic material: Activity of genes in transgenic plants. Progress in Bot. 53: 207–220.

Boulter, M.E., E. Croy, P. Simpson, R. Shields, R.R.D. Croy & A.H. Shirsat, 1990. Transformation of *Brassica napus* L. (oilseed rape) using *Agrobacterium tumefaciens* and *Agrobacterium rhizogenes* – a comparison. Plant Science 70: 90–91.

Bradford, M.M., 1976. A rapid and sensitive method for the quantitation of microgram quantities of protein utilizing the principle of protein-dye-binding. Anal. Biochem. 72: 248–254.

Deblaere, R., A. Reynaerts, H. Hofte, J.P. Hernalsteens, J. Leemans & M. Van Montagu, 1987. Vectors for cloning in plant cells. Methods in Enzymology 153: 277–304.

Harpster, M.H., J.A. Townsend, J.D.G. Jones, J. Bedbrook & P. Dunsmuir, 1988. Relative strength of the 35S cauliflower mosaic virus, 1′,2′, and nopaline synthase promoters in transformed tobacco sugarbeet and oilseed rape callus tissue. Mol. Gen. Genet. 212: 182–190.

Jefferson, R.A., T.A. Kavanagh & M.W. Bevan, 1987. GUS fusions: β-glucuronidase as a sensitive and versatile gene fusion marker in higher plants. EMBO J. 6: 3901–3907.

Jefferson, R.A., 1987. Assaying chimeric genes in plants: the GUS gene fusion system. Plant Mol. Biol. Rep. 5: 387–405.

Koncz, Cs. & J. Schell, 1986. The promoter of TL-DNA gene 5 controls the tissue-specific expression of chimaeric genes carried by a novel type of *Agrobacterium* binary vector. Mol. Gen. Genet. 204: 393–396.

Langridge, W.H.R., K.J. Fitzgerald, C. Koncz, J. Schell & A.A. Szalay, 1989. Dual promoter of *Agrobacterium tumefaciens* mannopine synthase genes in regulated by plant growth hormones. Proc. Natl. Acad. Sci. USA 86: 3219–3223.

Leung, J., H. Fukuda, D. Wing, J. Schell & R. Masterson, 1991. Functional analysis of cis-elements, auxin response and early developmental profiles of the mannopine synthase bidirectional promoter. Mol. Gen. Genet. 230: 463–474.

Nagata, T., K. Okada, T. Kawazu & I. Takebe, 1987. Cauliflower mosaic virus 35S promoter directs S phase specific expression in plant cells. Mol. Gen. Genet. 207: 242–244.

Odell, J.T., F. Nagy & N.H. Chua, 1985. Identification of sequences required for activity of the cauliflower mosaic virus 35S promoter. Nature 313: 810–812.

Ooms, G., P.J.J. Hooykaas, R.J.M. Van Veen, P. Van Beelen, T.J.G. Regensburg-Tüink & R.A. Schilperoort, 1982. Octopine Ti-plasmid deletion mutants of *Agrobacterium tumefaciens* with emphasis on the right side of the T-region. Plasmid 7: 15–29.

Reiss, B., R. Sprengel, M. Willi & H. Schaller, 1984. A new, sensitive method for qualitative and quantitative assay of the neomycin phosphotransferase in crude cell extracts. Gene 30: 211–218.

Rogers, S.G., H.J. Klee, R.B. Horsch & R.T. Fraley, 1987. Improved vectors for plant transformation: expression cassette vectors and new selectable markers. Methods in Enzymology 153: 253–277.

Saito, K.M., H. Yamazaki, I. Kaneko, Y. Murakoshi, I. Fukuda & M. Van Montagu, 1991. Tissue-specific and stress-enhancing expression of the TR promoter for mannopine synthase in transgenic medicinal plants. Planta 184: 40–46.

Schneider, M., D.W. Ow & S.H. Howell, 1990. The *in vitro* pattern of fire-fly luciferase expression in transgenic plants. Plant Mol. Biol. 14: 935–947.

Stefanov, I., S. Fekete, L. Bögre, J. Pauk, A. Fehér & D. Dudits, 1994. Differential activity of the manopine synthase and the CaMV 35S promoters during development of transgenic rapeseed plants. Plant Science 95: 175–186.

Stefanov, I., S. Iljubaev, A. Fehér, K. Margóczi & D. Dudits, 1991. Promoter and genotype dependent expression of a reporter gene in plant protoplasts. Acta Biologica Hungarica 42: 323–331.

Trewavas, A., 1981. How do plant growth substances work. Plant Cell and Environment 4: 203–228.

Velten, J. & J. Schell, 1985. Selection-expression plasmid vectors for use in genetic transformation of higher plants. Nucleic Acids Research 13: 6981–6998.

Euphytica **85**: 417–423, 1995.

Assessing the risks of wind pollination from fields of genetically modified *Brassica napus* ssp. *oleifera*

A.M. Timmons, E.T. O'Brien, Y.M. Charters, S.J. Dubbels & M.J. Wilkinson
Scottish Crop Research Institute, Invergowrie, Dundee DD2 5DA, Scotland, U.K.

Key words: gene flow, transgene, pollen movement, genetically modified oilseed rape, wind pollination, risk assessment, feral populations

Summary

Intensive research over the past 10 years has produced many genetically-modified lines of oilseed rape with market potential. Assessment of these lines in statutory trials prior to their release as cultivars is necessary, owing to concern over the likelihood of transgene escape from such crops. Here, we examine the movement of airborne pollen grains from oilseed rape fields and assess their capacity for long-range geneflow.

Pollen dispersal from isolated rape fields was monitored over two seasons and related to the distribution of fields and 'feral' (domesticated plants growing outside cultivations) populations of the crop in Tayside and North East Fife regions of Scotland. Airborne pollen density declined with distance and at 360 m was 10% of that at the field margin. Pollen counts of 0–22 pollen grains m^3 were observed 1.5 km from source fields and apparently were sufficient in number to allow seed set on emasculated bait plants. Oilseed rape pollen has greater capacity for long-range dispersal than had been suggested by small-scale field trials. Mean separation of oilseed rape fields in the survey area was 410 m and the mean distance from 'feral' populations to commercial fields was 700 m. Sixty percent of 'feral' populations with more than 10 plants occurred downwind and within 2 km of an oilseed rape field. Provided that the flowering biology of genetically-modified oilseed rape does not differ from the conventional crop, these data suggest that transgene movement to non genetically-modified fields or 'feral' populations is likely following commercial release.

Introduction

Annual production of oilseed rape (*Brassica napus* ssp. *oleifera*) has grown dramatically in recent years and 232,190 million tonnes were produced worldwide in 1992 (Anon, 1993). Development of a transformation system for the crop (Ooms et al., 1985) has enabled the production of modified lines containing transgenes for herbicide resistance (Miki et al., 1990; Mariani et al., 1991), increased methionine in seed meal (Altenbach, 1992), male sterility and restorer genes (Mariani et al., 1990, 1992, respectively), heavy metal tolerance (Misra & Gedamu, 1989) and antibiotic resistance (Arnoldo et al., 1992).

The performance of genetically-modified oilseed rape under agricultural regimes has been assessed in numerous small-scale experimental field trials (e.g. Gressel & Ben-Sinai, 1985; De Greef et al., 1989; Arnoldo et al., 1992). Eighty-two such field releases were approved worldwide by 1992 (Chasseray & Duesing, 1992; Dale et al., 1993), and the number submitted for approval is rising annually. Attention is now focused on the prospects for commercial release of genetically-modified oilseed rape. Crawley et al. (1993) identified three areas of concern associated with the release of genetically-modified crops:

i. genetically-modified plants themselves may become weeds and/or invade natural habitats

ii. the release of genetically-modified crops may enable the sexual transfer of the inserted genes to neighbouring commercial or natural populations whose offspring may then become more weedy or invasive

iii. genetically-modified plants may be a direct health hazard to humans or animals.

Brassica napus is predominantly self-pollinating (Olsson, 1960) and most groups employ breeding strategies designed for self-pollinating crops (Becker et al., 1990). Outcrossing rates of 5–55% have been reported, however, under field conditions (Olsson, 1952, 1960; Huhn & Rakow, 1979; Rakow & Woods, 1987). Outcrossing is mediated by wind and insects. Oilseed rape flowers contain nectaries which make them very attractive to honey bees (Free, 1970, 1993), although good yields have been reported without insect pollination (Olsson & Persson, 1958; Free & Nuttall, 1968; Langridge & Goodman, 1982). Furthermore, several authors have reported yield reductions of 33–50% when oilseed rape plants are grown in the still air of a glasshouse (Harle, 1948; Jenkinson & Glynne-Jones, 1953; Williams, 1978), and this has led to the widespread belief that wind plays an important role in pollination of the crop.

Geneflow may occur from a field of genetically-modified oilseed rape to wild relatives, 'feral' populations or to other (non genetically-modified) fields of the crop. *B. napus* has been shown to hybridize with a range of species including wild populations of *B. rapa* (Kapteijins, 1993), *B. adpressa* (Lefol et al., 1991), *Raphanus raphanistrum* (Chevre et al., 1992) and *Hirschfeldia incana* (Darmency et al., 1992). Nevertheless, Raybould & Gray (1993) indicated that oilseed rape displays a low probability of geneflow between the crop and its wild relatives. The extent of geneflow between genetically-modified and non genetically-modified fields, or between the former and 'feral' populations, is largely dependent on the scale of pollen emission and dispersal, and on the distance between source and recipient populations.

Several groups have monitored oilseed rape pollen emission and dispersal (McCartney & Lacey, 1991; Mesquida & Renard, 1982), and geneflow (Manasse, 1992) from small field trials, using models to assess the potential for pollen movement over greater distances. Differences in plot size and design (Levin & Kerster, 1974) and genotypes and environmental factors, have contributed to considerable variation between these models. McCartney & Lacey (1991) reported a 77% decrease in pollen concentration 6 m downwind of a 20 m × 20 m plot of oilseed rape and a 90% decrease at distances over 10 m. A reduction in pollen concentration of up to 96% and 98% at 50 m and 100 m, respectively, was extrapolated from these results. Manasse & Kareiva (1991) reported outcrossing rates of 0.022% at 50 m and 0.011% at 100 m, whereas Scheffler et al. (1993) reported outcrossing rates of 0.00033% at 47 m. Such trials have done little to remove uncertainty over the scale of pollen emission and dispersal likely to emanate from genetically-modified fields of the crop. In this study, the aerobiology of commercial oilseed rape fields was monitored over two years (1992, 1993). Results are related to an extensive survey of the distribution of commercial oilseed rape fields and feral populations encompassing a 70 km^2 area in the Tayside and North East Fife regions of Scotland.

Materials and methods

Mapping of pollen movement

Two commercial fields of oilseed rape (*Brassica napus* cv. Falcon) in the Tayside region of Scotland (grid reference: NO 295/275 in 1992 and NO 321/285 in 1993) were selected to monitor pollen movement in 1992 and 1993 respectively. The field used in 1992 measured approximately 375 m × 500 m (≈ 10 hectares) and the one used in 1993 measured 250 m × 250 m × 375 m (≈ 3 hectares). Both fields were isolated by at least 1 km from other rape fields in the direction of the prevailing wind (S/SW), and by > 360 m in any other direction. Seven-day volumetric spore traps (Hirst, 1952; Burkard Manufacturing Co., Rickmansworth, UK) were positioned on a linear transect in the direction of the prevailing wind at distances of 0 m, 100 m and 360 m from the field.

Long-range pollen movement was assessed using a volumetric spore trap positioned 10 m above the ground at the Scottish Crop Research Institute. The nearest field of oilseed rape to this trap was 2.5 km in 1992 and 1.5 km in 1993. In both instances the trap was sited in the direction of the prevailing wind from the fields. Traps were free to rotate through 360° and were fitted with a vane that ensured the aperture faced into the wind. The mean number of pollen grains m^{-3} air was determined for each 24 h period, according to the manufacturer's instructions. Each 24 h recording period commenced and ceased at 09.00 British Summer Time. In 1992 and 1993, 24 h average counts (pollen grains m^{-3}) were obtained throughout the flowering period of the crop and were presented as 11-day running means (McCartney et al., 1986; McCartney & Lacey, 1990, 1991). Hourly averages were recorded for selected days in 1992.

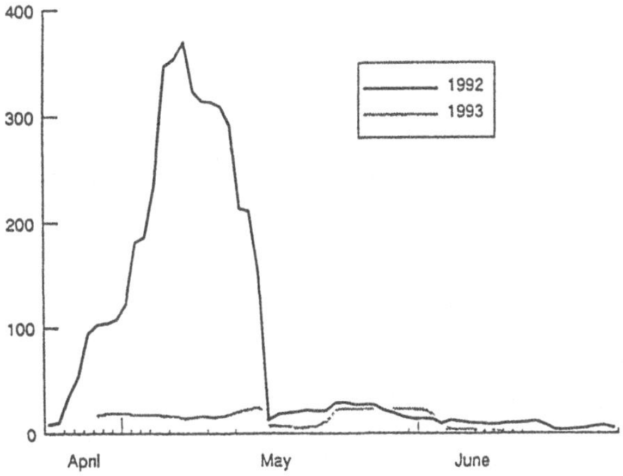

Fig. 1. 24-hour average pollen concentrations (grains m^{-3} air) at 0 m from the field margin (trap 1) for 54 days in 1992 and 47 days in 1993. Note: values are presented as 11-day running means.

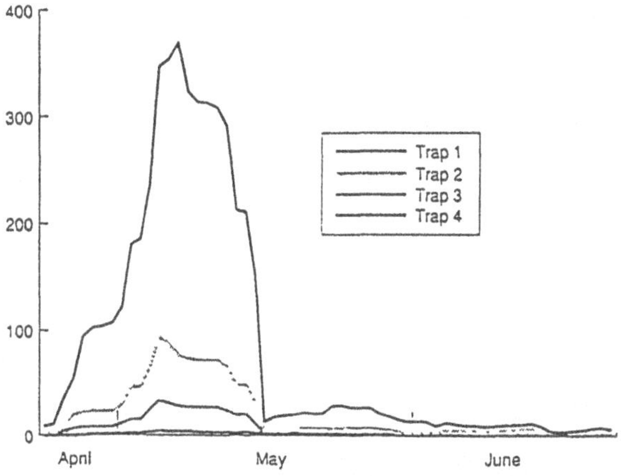

Fig. 2. 24-hour average pollen concentrations (grains m^{-3} air) at 0, 100, 360 and 2500 m from the field margin for 54 days in 1992. Note: values are presented as 11-day running means.

Pollen viability

The possibility of long-range fertilization by airborne pollen was assessed using emasculated oilseed rape 'bait' plants. The petals were removed from emasculated flowers to discourage insect pollination, and immature buds that had not been emasculated were detached prior to opening. In 1992, six plants (two of the winter oilseed rape 'Cobra' and four of 'Tapidor') were placed beside the elevated trap at SCRI. In 1993, three 'Tapidor' plants were sited in the same position. Approximately 100 flowers were emasculated on each plant. The emasculated flowers were exposed for two weeks at the end of the winter oilseed rape flowering period (29/5–12/6 in 1992 and 21/5–4/6 in 1993) to allow pollination. The inflorescences were then bagged to prevent further pollination and allowed to set seed. Where seed was set it was extracted and germinated on moist filter paper at 21° C for three days and transplanted into 7.5 cm pots to encourage active root growth for chromosome counts.

Chromosome counts were made from root tips of seedlings secured from bait plants. Actively growing roots were pre-treated in ice water for 24 h and fixed overnight in 3:1 ethanol:glacial acetic acid. Fixed roots were washed three times in distilled water (5 min each) and hydrolysed in 5N HCl at 21° C for 90 min. After rinsing in distilled water, they were stained in Schiff's reagent for 1 h at room temperature, washed briefly in distilled water and mounted in 45% acetic acid.

Distribution of commercial oilseed rape and suspected 'feral' populations

An aerial and ground survey was conducted in the spring of 1993 in North East Fife (28 km^{-2}) and Tayside (42 km^{-2}) regions of Scotland. The survey coincided with the flowering period of winter oilseed rape cultivars grown in the region. The distributions of fields and 'feral' populations of oilseed rape were recorded on a 1:25,000 scale Ordnance Survey map. Distances separating 'feral' populations and agricultural fields were measured from the centre of the 'feral' site to the edge of the nearest flowering oilseed rape field, along one of 16 compass points. Distances between neighbouring fields were measured as the shortest distance between field margins.

Results

Pollen emission

Total pollen emission detected at 0 m was significantly higher (t-test, $p < 0.05$) in 1992 than in 1993 (Figs 1 and 3). Differences were noted in the highest daily counts (1,247 in 1992, 120 in 1993), the number of days exceeding 50 grains m^{-3} (12 in 1992 and 3 in 1993) and in the mean pollen density over the entire season (77.27 in 1992 and 12.51 in 1993).

Pollen density declined with distance (Fig. 2). The rate of decline was more pronounced in 1992 than in 1993 (Fig. 3), although, during both years, pollen

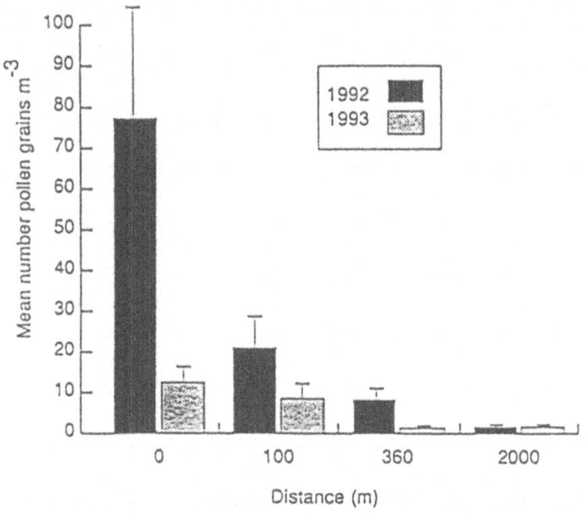

Fig. 3. The mean daily pollen count for 1992 and 1993 (54 days in 1992 and 47 days in 1993) recorded at four distances from the field margin (0 m, 100 m, 360 m and 2.5 km in 1992 and 2.0 km in 1993).

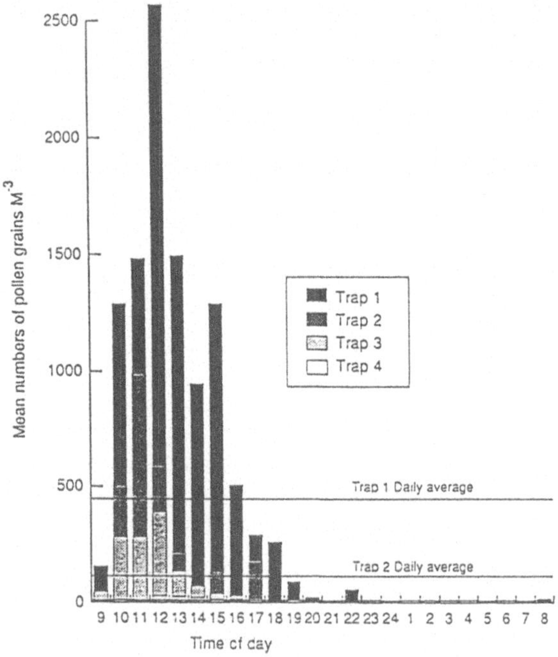

Fig. 4. Mean pollen counts for 3 days in 1992 (4, 13 and 15 May) divided into hourly intervals. The mean daily average for the same days is also shown.

concentrations at 360 m were 10% of those observed at the field margin. Mean pollen concentrations at the most distant trap were similar in 1992 and 1993 (1.58 grains m^{-3} and 1.57 grains m^{-3}, respectively).

Hourly changes in pollen density were calculated for each trap over three days in 1992 in which wind was blowing from the prevailing direction for more than 12 h (i.e. 4, 13 and 15 May). Pollen concentrations varied throughout the 24 h period (Fig. 4). Maximum pollen concentrations were observed as a single peak between 11.00 and 13.00 hr at all distances from the field margin, although pollen density declined from 2,571 grains m^{-3} at the field margin to 27 grains m^{-3} at the most distant trap. Low pollen counts were consistently recorded between 01.00 and 07.00 hours. The hourly average pollen concentrations were up to 3.5 times higher than the 24 h average value for 5–6 hours at 0 m, 100 m and 360 m, and for 10 h at distances over 2 km.

Long-range fertilization

The viability of pollen at the most distant sampling site was assessed using emasculated bait plants. In 1992, one of these plants (*B. napus* cv. Tapidor) produced two seeds in a single pod. In 1993, a single *B. napus* cv. Tapidor plant produced 1 pod with 14 seeds. This represents 0.08% and 1.2% of the potential for seed set for the flowers emasculated in 1992 and 1993, respectively. The 14 seeds produced in 1993 were germinated and chromosome counts made from root tip squashes. All plants contained 38 chromosomes and appeared phenotypically normal for *B. napus*.

Distribution

Three hundred and twenty-one fields of winter-sown oilseed rape were found in the survey area. Many occurred in close proximity to one or more of the 100 'feral' populations identified. 'Feral' populations ranged in size from isolated individuals to stands containing more than 1,000 plants. Fifty populations consisted of 10 or more individuals. The most commonly encountered habitat types were roadside verges, margins of agricultural fields and sites with disturbed soil (often associated with road or building construction). A number of more unusual sightings included individuals growing in domestic gardens, drainpipes and guttering.

The mean distance between cultivated fields in the survey area was measured as 0.41 ± 0.35 km for winter-sown oilseed rape cultivars. Ninety-five fields occurred within a 2 km radius of mapped 'feral' populations containing in excess of 10 plants. The mean distance from cultivated fields to 'feral' populations of oilseed rape was 0.70 ± 0.45 km. Among the mapped 'feral' populations with 10 plants or more, 10% and 60% occurred downwind and within 360 m and 2 km, respectively, of an oilseed rape field. Considering only those fields situated S/SW of a 'feral' population, the average distance from cultivated fields to 'feral' sites increased to 0.89 ± 0.55 km.

Discussion

The difference between pollen emissions from fields of the same cultivar in 1992 and 1993 emphasises the importance of environmental factors in wind-mediated pollen dispersal. The single peak observed between 11.00 and 13.00 hr at all distances suggests oilseed rape pollen moves rapidly from the source and does not remain airborne for significant periods of time. This is further supported by the virtual disappearance of airborne pollen at all traps between 01:00 and 07:00.

Wide day-to-day fluctuations in airborne pollen levels in both flowering seasons are in accordance with previous reports using smaller scale field plots of oilseed rape (e.g. McCartney & Lacey, 1991). The most striking feature of the present study, however, is the large disparity between the amount of airborne pollen we detected at all distances and those reported previously from trial plots. McCartney & Lacey (1991) report a reduction to 2–11% of source emissions at 100 m compared with 27–69% reported here. Moreover, geneflow levels of 0.022% (Manasse & Kareiva, 1991) and 0.00033% (Schefler et al., 1993) at 47 m and 50 m, respectively, compare with 0.08 and 1.2% reported here at distances of at least 2.5 and 1.5 km, respectively. Differences in plot size represent the most probable explanation of these discrepancies (Levin & Kerster, 1974). If this is so, great care should be exercised before extrapolating information obtained from small-scale release experiments to predict the likely performance of genetically-modified crops under standard agricultural conditions.

Low pollen counts were observed during both years at the greatest sampling distance. Comparable levels of oilseed rape pollen were recorded at British Aerobiology Federation metropolitan sampling sites in Derby (J. Gordon, personal communication) and Leicester (J. Wentworth, personal communication), suggesting such densities may represent 'background' levels of airborne oilseed rape pollen. There are problems in relating airborne pollen density to geneflow, however. Pollen detected on volumetric spore traps may not be viable or indeed may be from another *Brassica* species with morphologically identical pollen (e.g. *B. oleracea* or *B. rapa*). Seed set on emasculated bait plants, suggests there are sufficient viable *B. napus* pollen grains at distances of at least 1.5–2.5 km to affect geneflow, provided they do not arise from parthenogenesis or interspecific hybridization (G.R. Mackay, pers. comm.). In *B. napus*, haploids are known to arise spontaneously (Morinaga & Fukushima, 1933; Thompson, 1969; Stringham & Downey, 1973), after interspecific hybridization or, in some cases, they may be parthenogenic in origin (see Prakash & Hinata, 1980 for review). None of the seedlings obtained in 1993 was found to be haploid. Unreduced 'maternal' seeds have been reported in various Brassica species including *B. napus*. These arise parthenogenetically following stimulation by foreign pollen or when induced by prickle pollination (Tokumasu, 1970; Mackay & Low, 1974; Eenink, 1974). In the absence of pollination or prickle pollination, however, no seed set occurs (U, 1935; Mackay, 1972). It seems improbable, therefore, that the seeds obtained here are entirely of maternal origin, although this cannot be firmly discounted. There are two related species growing in the Tayside region with which *B. napus* (2n = 38) hybridizes: *B. rapa* (2n = 20), with which it crosses freely and *B. oleracea* (2n = 18) with which it forms occasional hybrids. Hybrids between these species possess intermediate chromosome numbers (Stace, 1975). All of the 14 seeds produced in a single pod in 1993 contained 38 chromosomes and so are unlikely to be hybrids between the above species. Thus, oilseed rape pollen is capable of affecting geneflow over distances of at least 1.5 km. The mean distance between oilseed rape fields was 0.41 ± 0.35 km and between 'feral' populations and fields was 0.70 ± 0.45 km in the survey area. These results suggest, therefore, that wind-mediated pollen dispersal of oilseed rape could enable widespread geneflow between agricultural fields and 'feral' populations of the species under current farming practice. Provided pollen viability and dispersal characteristics of genetically-modified oilseed rape do not differ from the conventional crop, transgene movement to other non genetically-modified fields and/or to 'feral' populations following commercial release seems likely. Such releases should be reviewed, therefore, on a case by case basis.

Acknowledgements

We thank George Mackay and Derrick Lydiate for their helpful advice, Mike Hardy and Marshall Muirhead for the use of their fields, and Ian Pitkethly and Sheena Forsyth for helping to prepare the manuscript.

Thanks are also due to the Scottish Office Agriculture and Fisheries Department (Contract No. FF 340) and to the Department of the Environment Research Programme (Contract No. DOE 5133) for financial support of the work.

References

Anonymous, 1993. Food and Agriculture organization of the United Nations, Rome. F.A.O. statistics, series no. 112.

Arnoldo, M., C.L. Baszczynski, G. Bellemare, G. Brown, J. Carlson, B. Gillespie, B. Huang, N. MacLean, W.D. MacRae, G. Rayner, S. Rozakis, M. Westecott & R.J. Kemble, 1992. Evaluation of transgenic canola plants under field conditions. Genome 35: 58–63.

Altenbach, S.B., C.C. Kuo, L.C. Staraci, K.W. Pearson, C. Wainright, A. Georgescu & J. Townsend, 1992. Accumulation of a Brazil nut albumin in seeds of transgenic canola results in enhanced levels of seed protein methionine. Plant Molec. Biol. 18: 235–245.

Becker, H.C., B. Karlsson & C. Damgaard, 1990. Genotype and environmental variation for outcrossing rate in rapeseed. In: Rapeseed in a changing world. (Proc. 5) GCIRC 8th International rapeseed conference.

Chasseray, E. & J. Duesling, 1992. Field trials of transgenic plants: an overview. AGRO Food Industry hi-tech July/Aug. 1992: 5–10.

Chevre, A., M. Renard, F. Eber, P. Vallee, M. Deschamps & M.C. Kerlan, 1992. Study of spontaneous hybridization between male-sterile rapeseed and weeds. In: Reproductive biology and plant breeding. Book of poster abstracts, XIIIth Eucarpia Congress, 67–68.

Crawley, M.J., R.S. Haus, M. Rees, D. Kolin & J. Buxton, 1993. Ecology of transgenic oilseed rape in natural habitats. Nature 363: 620–623.

Dale, P.J., J.A. Irwin & J.A. Scheffler, 1993. The experimental and commercial release of transgenic crop plants. Plant Breeding 111: 1–22.

Darmency, H., E. Lefol & R. Chadoeuf, 1992. Risk assessment of the release of herbicide resistant transgenic crops: two plant models. In: ANPP annales 1992. IXth International Symposium on the Biology of Weeds, 513–523.

De Greef, W., R. Delon, M. De Block, J. Leemans & J. Botterman, 1989. Evaluation of herbicide resistance in transgenic crops under field conditions. Biotechnology 7: 61–64.

Eenink, A.N., 1974. Matromorphy in *Brassica oleracea* L. I. Terminology, parthenogenesis in Cruciferae and the formation of matromorphic plants. Euphytica 23: 429–433.

Free, J.B., 1970. Insect pollination of crops. Academic Press, London.

Free, J.B., 1993. Insect pollination of crops (2nd Edition). Academic Press, London.

Free, J.B. & P.M. Nuttall, 1968. The pollination of oilseed rape (*Brassica napus*) and the behaviour of bees on the crop. J. Agric. Sci., Camb. 71: 91–94.

Gressel, J. & G. Ben-Sinai, 1985. Low intraspecific competitive fitness in a triazine-resistant, nearly isogenic line of *Brassica napus*. Plant Sci. 38: 29–32.

Harle, A., 1948. Ist der Rapsglanzkafer (*Meligethes aeneus* Fabr.) nur ein Schadling? Nachrichtenblass für den Deutschen. Pflanzenschutzdienst Berlin 2: 40–42.

Hirst, J.M., 1952. An automatic volumetric spore trap. Ann. Appl. Biol. 39: 257–265.

Huhn, M. & G. Rakow, 1979. Einige experimentelle Ergebnisse zur Fremdfruchtungsrate bei Winterraps (*Brassica napus oleifera*) in Abhängigkeit von Sorte und Abastand. Z. Pflanzenzüchtg. 83: 289–307.

Jenkinson, J.D. & G.D. Glynne-Jones, 1953. Observations on the pollination of oil rape and broccoli. Bee World 34: 173–177.

Kapteijns, A.J.A.M., 1993. Risk assessment of genetically modified crops. Potential of four arable crops to hybridize with the wild flora. Euphytica 66: 145–149.

Langridge, D.F. & R.D. Goodman, 1982. Honeybee pollination of oilseed rape, cultivar Midas. Aust. J. Exp. Agric. Anim. Husbrandry 22: 124–126.

Lefol, E., V. Danielou, H. Darmency, M.-C. Kerlan, P. Vallee, A.-M. Chevre, M. Renard & X. Reboud, 1991. Escape of engineered genes from rapeseed to wild Brassiceae. In: Proceedings of the British Crop Protection Conference on Weeds, 1049–1056.

Levin, D.A. & H.W. Kerster, 1974. Gene flow in seed plants. Evolutionary Biology 7: 139–220.

Mackay, G.R., 1972. On the genetic status of maternals induced by pollination of *Brassica oleracea* L. with *Brassica campestris* L. Euphytica 21: 71–77.

Mackay, G. & R. Low, 1974. Spontaneous triploids in forage kale, *Brassica oleracea* var. *acephala*. Euphytica 24: 525–529.

McCartney, H.A. & M.E. Lacey, 1990. The production and release of ascospores of *Pyrenopeziza brassicae* Sutton et Rawlinson on oilseed rape. Plant Pathology 39: 17–32.

McCartney, H.A. & M.E. Lacey, 1991. Wind dispersal of pollen from crops of oilseed rape (*Brassica napus* L.). J. Aerosol. Sci. 22 (4): 467–477.

McCartney, H.A., M.E. Lacey & C.J. Rawlinson, 1986. Dispersal of *Pyrenopeziza brassicae* spores from an oilseed rape crop. J. Agric. Sci. Camb. 107: 299–305.

Manasse, R., 1992. Ecological risks of transgenic plants: effects of spatial dispersion on gene flow. Ecological Applications 2 (4): 431–438.

Manasse, R. & P. Kareiva, 1991. Quantifying the spread of recombinant genes and organisms. In: L. Ginzburgh (Ed.). Assessing Ecological Risks of Biotechnology, pp. 215–231. Butterworth-Heinemann, Boston.

Mariana, C., M. De Beuckheleer, J. Truettner, J. Leemans & R. Goldberg, 1990. Induction of male sterility in plants by a chimaeric ribonuclease gene. nature 347: 737–741.

Mariani, C., W. De Greef, M. De Block, M. De Beuckeleer, V. Gossele & J. Leemans, 1991. Genetic analysis of engineered sterility in oilseed rape. In: D.I. McGregor (Ed.). Proc. of the GCIRC Eighth International Rapeseed Congress, pp. 352–357. Saskatchewan.

Mariani, C., V. Gossele, M. De Beuckeleer, M. De Block, R.B. Goldberg, W. De Greef & J. Leemans, 1992. A chimaeric ribonuclease-inhibitor gene restores fertility to male sterile plants. Nature 357: 384–387.

Mesquida, J. & M. Reynard, 1982. Étude de la dispersion du pollen par le vent et de l'importance de la pollinisation anémophile chez le colza (*Brassic napus* L., var. *oleifera* Metzger). Apidologie 13: 353–366.

Miki, B.L., H. Labbe, J. Hattori, T. Ouellet, J. Gabard, G. Sunohara & P.J. Charest, 1990. Transformation of *Brassica napus* canola cultivars with *Arabidopsis thaliana* acetohydroxyacid synthase genes and analysis of herbicide resistance. Theor. Appl. Genet. 80: 449–458.

Misra, S. & L. Gedamu, 1989. Heavy metal tolerant transgenic *Brassica napus* L. and *Nicotania tabacum* L. plants. Theor. Appl. Genet. 78: 161–168.

Morinaga, T. & E. Fukushima, 1933. Karyological studies on a spontaneous haploid mutant of *Brassica napella*. Cytologia 4: 457–460.

Olsson, G., 1952. Undersökning av graden av korsbefruktning hos vitesnap och raps. Sveriges Utsädesförenings Tidskrift 62: 311–322.

Olsson, G., 1960. Self-incompatability and outcrossing in rape and white mustard. Hereditas 46: 241–252.

Olsson, G. & B. Persson, 1958. Inkorsningsgrad och sjalvstorilitet hos raps. Sveriges Utsädesförenings, Tidskrift 68: 74–78.

Ooms, G., A. Bains, M. Burrell, A. Karp, D. Twell & E. Wilcox, 1985. Genetic Manipulation in cultivars of oilseed rape (*Brassica napus*) using agrobacterium. Theor. Appl. Genet. 71: 325–329.

Prakash, S. & K. Hinata, 1980. Taxonomy, cytogenetics and origin of crop brassicas, a review. Opera Botanica 55: 25–26.

Rakow, G. & D.L. Woods, 1987. Outcrossing in rape and mustard under Saskatchewan prairie conditions. Can. J. Plant Sci. 67: 147–151.

Raybould, A.F. & A.J. Gray, 1993. Genetically modified crops and hybridization with wild relatives: a UK perspective. J. Appl. Ecol. 30: 199–219.

Scheffler, J.A., R. Parkinson & P.J. Dale, 1993. Frequency and distance of pollen dispersal from transgenic oilseed rape (*Brassica napus*). Transgenic Research 2: 356–364.

Stace, C.A., 1975. Hybridization and the flora of the British Isles. Academic Press, London.

Stringham, G.R. & R.K. Downey, 1973. Haploid frequencies in *Brassica napus*. Can. J. Plant Sci. 53: 229–231.

Thompson, K.F., 1969. Frequencies of haploids in spring oilseed rape (*Brassica napus*). Heredity 24: 318–319.

Tokumasu, S., 1970. Studies on pseudogamy in the case of tetraploid plants in Raphanus. Jap. J. Breed. 20 (1): 15–21.

U, N., 1935. Genome analysis in Brassica with special reference to experimental formation of *B. napus* and peculiar mode of fertilisation. Jap. J. Bot. 7: 389–452.

Williams, I.H., 1978. The pollination requirements of swede rape (*Brassica napus* L.) and of turnip rape (*Brassica campestris* L.). J. Agric. Sci. Camb. 91: 343–348.

Euphytica **85:** 425–431, 1995.

The release of transgenic plants from containment, and the move towards their widespread use in agriculture

Philip J. Dale & Judith A. Irwin
John Innes Centre, Colney Lane, Norwich NR4 7UJ, U.K.

Key words: consumer, food, genetic modification, transgenic plants, risk assessment

Summary

The use of transgenic plants in breeding makes it possible to utilise a wide variety of novel genes from unrelated plants, microbes and animals. Because of the diversity of genes that have now become available for modifying crop plants, it is agreed internationally that there should be a risk assessment before transgenic plants are grown outside the laboratory or glasshouse. Various aspects are considered in a risk assessment including any non-target effects of the transgene, changes in plant persistence and invasiveness, and the possibility of movement of the transgenes to wild populations by cross pollination. It is generally argued that the need for risk assessment and regulation should be determined by an analysis of certain products of transformation, rather than a risk assessment being required for all plants modified by the process of transformation. A possible consequence of considering the product only, however, could be that some of the products of conventional breeding may need to be assessed by the risk assessment procedures developed for transgenic plants. There are discussions with interest groups on the use of transgenic plants in the environment and in food products. It is likely that some form of labelling will be required for certain foods containing ethically-sensitive genes. There is little doubt that transgenic plants will make a significant contribution to agriculture in the coming decades. Developments in the patenting of genes, release regulations, food labelling, consumer reaction etc., will influence the rate of progress and should be considered in the strategic planning of plant breeding programmes.

Introduction

Worldwide, there have now been more than 1000 field experiments with transgenic plants (OECD, 1993), and the first transgenic varieties will go into commercial production in the near future. Over the coming decades, transgenic varieties will be produced that have modifications in a wide range of characters and have genetic changes not achievable through the use of conventional plant breeding methods (Dale, 1993). It is agreed internationally that, before transgenic plants are grown in field plots outside containment, there should be a risk assessment. This is not because transgenic plants are innately hazardous, but rather that the choice of genes available for introducing into plants is large and varied and no longer limited by sexual incompatibility. Genes can now be transferred into plants from a wide range of organisms, including unrelated plant species, microbes, animals (including humans) and from DNA synthesised in the laboratory.

Many of the plant phenotypes modified by transformation, such as disease resistance, pest resistance, herbicide tolerance and modified plant products, are familiar in traditionally-bred varieties. Some very different phenotypes, or the genetic mechanisms used to produce familiar phenotypes, fall outside the experience of conventional plant breeding, for example viral genes are being used to confer resistance to viral diseases, bacterial genes to confer resistance to pests and tolerance to herbicides, and animal genes are being evaluated for the production of pharmaceutical substances (see Table 1). Undesirable phenotypes are occasionally produced in conventional plant breeding, for example high glycoalkaloids in potatoes or high glucosinolates in brassicas, and procedures have been developed in breeding programmes to eliminate vari-

Table 1. Examples of genes being inserted into crop plants[a]

Plant character modified	Transgene product	Source of transgene	Examples of crop plants modified
Pest resistance			
Various insect pests	Bt insecticidal protein	*Bacillus thuringiensis*	cotton tomato corn
Various insect pests	Trypsin inhibitor protein	*Vigna unguiculata*	tobacco
Viral disease resistance			
Papaya ringspot viuus	Viral coat protein	Papaya ringspot virus	papaya
Cucumber mosaic virus	Viral coat protein	Cucumber mosaic	cucumber virus
Tomato mosaic virus	Viral coat protein	Tobacco mosaic virus Tomato mosaic virus	tomato
Potato leafroll luteovirus (PLRV)	Viral coat protein	Potato leafroll luteovirus	potato
Potato virus X and virus Y	Viral coat protein	Potato virus X or Y	potato
Fungal disease resistance			
Alternaria longipes (Brown spot fungus)	Chitinase	*Serratia marcescens*	tobacco
Rhizoctonia solani (damping off, seedling blight etc.)	Bean endochitinase	*Phaseolus vulgaris*	tobacco rapeseed
Rhizoctonia solani	Ribosome-inactivating protein	*Hordeum vulgare*	tobacco
Environmental stress tolerance			
Cadmium tolerance	Metallothionein-binding protein	mouse	tobacco
Frost protection	Antifreeze protein	*Pseudopleuronectes americanus* (fish)	tomato tobacco
Cold tolerance	Glycerol-3-phosphate acyltransferase	*Arabidopsis thaliana*	tobacco
Herbicide tolerance			
glyphosate	Analogue of EPSP synthase	Various plant and microbial genes	soybean cotton flax
sulfonylurea	Acetolactate synthase	*Arabidopsis thaliana*	maize rapeseed flax
glufosinate	Phosphinothricin acetyltransferase	*Streptomyces hygroscopicus*	sugarbeet rapeseed cabbage tall fescue tomato alfalfa potato wheat

Table 1 continued

Plant character modified	Transgene product	Source of transgene	Examples of crop plants modified
bromoxynil	Bromoxynil-specific nitrilase	*Klebsiella ozaenae*	cotton
2,4-dichlorophen-oxyacetic acid	2,4-D mono-oxygenase	*Alcaligenes eutrophus*	cotton tobacco
Food processing/quality			
Improved storage	Antisense polygalacturonase	*Lycopersicon esculentum*	tomato
Increased stearic acid	Antisense stearoyl-ACP desaturase	*Brassica rapa*	rapeseed
Increased mannitol	Mannitol dehdrogenasen	*Escherichia coli*	tobacco
Increased methioine	Seed storage protein	*Bertholletia excelsa*	soybean rapeseed
Improve protein quality	chicken ovalbumin	chicken	alfalfa
Increased starch content	ADP-glucose pyrophosphorylase	*Escherichia coli*	potato
Flavour enhancer	synthesized monellin	Artificially synthesized	tomato
Speciality chemicals, breeding system & flower colour			
Increased lauric acid	Lauroyl-ACP thioesterase	*Umbellularia californica*	rapeseed
Serum albumin	Human serum albumin	*Homo sapiens*	potato
Enkephalins	Leu-enkephalin	chimeric gene, part from *Homo sapiens* & *Arabidopsis thaliana*	rapeseed
Cyclodextrins	Cyclodextrin glycosyltransferase	*Klebsiella pneumoniae*	potato
Male sterility system	Ribonuclease and ribonuclease inhibitor	*Bacillus amyloliquefaciens*	rapeseed
Biodegradable thermoplastic	Polyhydroxy butyrate (PHB)	*Alcaligenes eutrophus*	*Arabidopsis*
Flower colour	Dihydroflavonol 4-reductase (FR)	*Zea mays*	petunia
		Gerbera ssp	petunia
Experimental studies			
Gene expression	Chloramphenicol acetyl transferase	*Escherichia coli*	tobacco rapeseed
	Neomycin phosphotransferase	*Escherichia coli*	potato rapeseed
Gene regulation	ADP glucose pyrophosphorylase	*Escherichia coli*	potato tomato
	Sucrose phosphate synthase	*Zea mays*	tomato

Table 1 continued

Plant character modified	Transgene product	Source of transgene	Examples of crop plants modified
Pollen dispersal	Neomycin phosphotransferase	*Escherichia coli*	cotton
	Acetolactate synthase	*Arabidopsis thaliana*	potato
	Phosphinothricin acetyltransferase	*Streptomyces hygroscopicus*	rapeseed
Plant persistence	Phosphinothricin acetyltransferase and neomycin phosphotransferase	*Streptomyces hydroscopicus*, *Escherichia coli*	rapeseed potato sugarbeet

[a] For references see Dale et al , 1993

ants of this kind. The wide diversity of genes that can be inserted by transformation, however, means that special internationally-agreed assessment procedures are being applied to the experimental release and to the commercial use of transgenic varieties.

Small scale experimental releases

When transgenic plants are first produced they are evaluated in the laboratory, growthroom and containment glasshouse to determine the nature of expression of the transgenes, and whether the plant phenotype is modified in the way intended. This is accompanied by molecular analysis to determine the number of copies of the transgenes present, and the integrity of the inserted DNA. Plants with a single copy of the transgene construct are usually selected because the inheritance and expression patterns of the gene in subsequent generations are simpler and more predictable. This also avoids the possibility of interaction between different transgenes present in the same plant (Matzke et al., 1993).

Preparing for a field release

Preparing for a field release of transgenic plants requires the preparation of a proposal which includes the basic information necessary for an assessment of any risks the transgenic plant may pose to the environment or to human health. The proposal usually goes first to a multidisciplinary institutional (or company) biosafety committee. Members of this local biosafety committee will frequently have close knowledge of the particular crop species, the environment, the release conditions and probably of the transgenes inserted. Once approved by the institute biosafety committee, the release proposal is presented to the national regulatory authorities who consider the risk implications of a release and give their approval or ask for further information. The field evaluation of novel transgenic plants will then be carried out under the terms agreed with the regulatory authorities. Special procedures may be needed, such as physical isolation from sexually-compatible species, deflowering, the use of non-transgenic buffer crops, special disposal procedures or plot monitoring after harvest.

Risk assessment

The details required in a typical release proposal made to the national regulatory authorities have been reviewed elsewhere (Dale et al., 1993a). A selection of the questions considered in a risk assessment is summarised here.

Non-target effects

Does the transgene have an effect on non-target organisms? In the case of insecticidal proteins, for instance, is there an effect on species of insects that are not the main target of the protein? Do any residues remain in the environment after the crop has been harvested, and is fo, is there an effect on soil micro-organisms?

Persistence and invasiveness

Does the transgene influence the persistence of a plant in an agricultural environment or its invasiveness in natural habitats? Does the transgenic plant have the potential to become a weed or to be damaging to natural populations? It may be necessary to compare the transgenic plant with its non-transgenic counterpart to determine whether its persistence or invasiveness characteristics have been modified. While the incorporation of a herbicide tolerance gene is often cited as an example of a modification that may increase the invasiveness of plants, this should not be the case in natural habitats where herbicides are not applied. The potential for persistence in the agricultural environment will depend on the availability of alternative methods of chemical or mechanical weed control (Dale, 1994).

Transgene movement into natural populations

The likelihood and consequences of the movement of transgenes into natural populations also need to be considered. Are transgenes for herbicide tolerance, disease resistance or pest resistance, transferred to wild relatives by cross-pollination (Dale, 1992)? Some crop species do have sexually-compatible relatives that are found as arable weeds. Sugar beet can be accompanied by related wild beet, and there is well-documented evidence of gene transfer between the two (Boudry et al., 1993). Oilseed rape is able to cross pollinate under field conditions with the crop species *Brassica rapa* and *B. juncea*, and recent studies have shown that *B. adpressa* and *Raphanus raphanistrum* will hybridize with male-sterile oilseed rape under field conditions (Scheffler & Dale, 1994).

The consequences of hybridization between crop and wild species will depend on the nature of the transgene and the fertility of the hybrid and any of its progeny (Dale et al., 1993b; Scheffler et al., 1993; Dale, 1994; McPartlan & Dale, 1994). For a transgene to make a significant change to the persistence or invasiveness of a species, it will be necessary for it to confer a selective advantage. The potential for greater invasiveness or persistence under field or natural conditions may need to be examined more closely when genes for improved tolerance to environmental stress (e.g. drought, salt, high temperature, heavy metal) become available.

Other issues

The potential allergenic effects of introducing novel proteins into plants also need to be considered. These could include hay fever-type allergies where transgenes are expressed in pollen, and the effect of incorporating novel proteins into food products. The use of viral genes to confer virus resistance is also being assessed because of the possibility of a heterologous virus becoming encapsidated in the coat protein synthesised by the transgenic plant. This may result in the virus being transmitted to new host plants. There is also the possibility of aberrant recombination between the viral RNA genome and transgene RNA transcripts made by the plant. Host range and vector specificity have been attributed to the capsid proteins of several plant viruses. As transgene mRNA is available to recombine with replicating RNA viruses, the possibility that RNA recombination could generate a virus with an altered host range cannot be excluded. A more detailed description of these and other issues will be found elsewhere (Tepfer, 1993; Dale, 1994; Falk & Bruening, 1994; Greene & Allinson, 1994).

Product vs. process debate (or horizontal vs. vertical assessment)

It is frequently argued that the need for regulation should be determined by a detailed analysis of the product of genetic modification, rather than having to regulate all the plants modified by the process of transformation. The main point of the argument is to emphasise that it is the effect of introducing certain traits known to be related to risk that should be the focus of the assessment, rather than testing and retesting the hypothesis that recombinant DNA techniques *per se* alter risk-related characters.

While this is a reasonable argument, it is desirable that the products of conventional breeding methods remain outside the regulations developed for transgenic plants. The merit of considering the process by which a modified plant is produced is that it defines a class of modified plants about which we need to ask certain questions. The challenge for the regulators is, in the light of assessment and experience, to define particular crops and transgenes that can be given a fast-track through the regulatory system. The development of fast-tracking is already in progress in the UK and USA (Lehrman, 1993; DOE, 1994).

Public perception and the release of transgenic plants

Several surveys of opinion on the release of transgenic organisms into the environment have been carried out (Durant, 1992). Some groups have expressed the view that modern methods of genetic modification interfere with nature to an extent that is unacceptable. It has also been argued that in parts of the world where there is a food surplus, there is little need for the development of transgenic plants. Some groups are concerned about the transfer of animal genes to plants. Does eating a gene (and its protein product) from a pig or cow present an ethical problem for some people in our society? Does the fact that the animal gene is propagated in a bacterium before it is inserted into a plant, or that the corresponding DNA sequence can be synthesised artificially in a laboratory, make a difference to the arguments. These issues are already being discussed by interest groups (Aldridge, 1994).

Many media articles over the past few years have presented the application of modern methods of genetic modification in an emotive manner. It is important that the discussions with the general public are well-informed, and based on a knowledge of the principles of genetic modification and the opportunities provided by transgenic plants. To do this effectively, we have a responsibility to educate at all levels of society from school age to adulthood.

Labelling the products of genetic modification

Issues of public preference and choice, raise questions of how much information should be given to the public about foods produced from transgenic plants. In most cases there will be no greater risk from eating a particular transgenic plant than from its non-transgenic counterpart. Should people who have ethical objections to the use of transgenic organisms, or to the incorporation of animal genes into plants, though, have the right of access to the information they need to make a choice. A UK committee, set up to study the Ethics of Genetic Modification and Food Use, has recently recommended the labelling of foods containing certain ethically-sensitive genes which, in the case of plants, include the use of animal or human genes that might be unacceptable to certain vegetarian or religious groups (Aldridge, 1994). This issue could be important for the commercial future of transgenic plants.

In a different but related tissue, the Food and Drug Administration (FDA) in the USA has decided that milk from cows injected with the hormone bovine somatotropin (BST) does not have to be labelled as such, but milk produced without the use of BST can in certain circumstances be labelled, and there is currently a debate about what kind of labelling is acceptable (Gershon, 1994; Lehrman, 1994).

Cheese made with recombinant rennet is an example where labelling is used to indicate a positive advantage for certain consumers, because it does not use rennet obtained from the linings of calves' stomachs (Dughan, 1994). One food retailer in the UK (CWS Retail), labels such cheese as 'produced using gene technology and so free from animal rennet'. CWS has also recently made a voluntary statement that it will always label food if gene technology is used in its production, whether or not the food itself contains recombinant genes. The company has also announced that its branded products will not contain fruit or vegetables modified with animal genes, that all products containing genes from unrelated species will be labelled and that it will not sell any products containing human genes (Dughan, 1994).

Calgene has chosen to emphasise the benefits to the consumer of its transgenic tomato by labelling it 'Flavr Savr' to highlight the claimed improvement in flavour over conventional tomato varieties (Miller, 1994).

The potential of public perception to influence the application of a technology is well illustrated by the debate surrounding the use of irradiated food. The experience of food irradiation and its contribution to reducing food losses and preventing foodborne diseases, is based on 30 years of scientific research. Irradiation of food has been approved in 33 countries for about 30 food products. Proposals not to label irradiated food have generally been rejected in favour of providing full information, on the grounds that consumers have the right to be informed about the food products they buy and use. One of the main reasons why food irradiation is not yet in general use is that governments are uncertain about consumer acceptance. Without public endorsement, food irradiation will remain largely neglected in both developed and developing countries (WHO, 1988).

Strategic planning in plant breeding

In discussions on the development and use of transgenic plants in agriculture, there is often greater emphasis on risk than benefit. Genes that are already available and genes that will become available in the future, present many new opportunities for modifying crop plants. Global food production needs to increase as the world population is projected to reach 10,000 million early in the next century. Progress on the production of transgenic plants for developing counties has been modest, but there is active research aimed at developing transgenic lines for specific crops and environments in some of these countries.

It is difficult to predict the precise pattern of events that will lead to the widespread use of transgenic plants in agriculture. Progress will undoubtedly be influenced by developments in a number of areas including: intellectual property, patenting of genes, plant variety rights, release regulations, labelling policies, and consumer reaction. Modern methods of genetic modification will undoubtedly provide important opportunities for plant breeding in the future because, as with conventional breeding, the production of transgenic varieties aims to provide choice. In the final analysis it will be the consumer, distributer and grower who will make that choice.

Acknowledgements

We acknowledge the financial support of the EEC BRIDGE programme and the UK Agricultural and Food Research Council. We thank Eddie Arthur and Colin Morgan for reading the manuscript.

References

Aldridge, S., 1994. Ethically sensitive genes and the consumer. Trends Biotech. 12: 71–72.

Boudry, P., M. Morchen, P. Saumitou-Laprade, Ph. Vernet & H. Van Dijk, 1993. The origin and evolution of weed beets: consequences for the breeding and release of herbicide-resistant transgenic sugar beets. Theor. Appl. Genet. 87: 471–478.

Dale, P.J., 1992. Spread of engineered genes to wild relatives. Plant Physiol. 100: 13–15.

Dale, P.J., 1993. The release of transgenic plants into agriculture. J. Agric. Sci. Cambridge 120: 1–5.

Dale, P.J., 1994. The impact of transgenes in hybrids between genetically modified crop plants and their related species – general considerations. Mol. Ecol. 3: 31–36.

Dale, P.J., J.A. Irwin & J.A. Scheffler, 1993a. The experimental and commercial release of transgenic crop plants. Plant Breeding 111: 1–22.

Dale, P.J., P. Parkinson & J.A. Scheffler, 1993b. Dispersal of genes by pollen – The Prosamo Project. In: D.J. Beadle, D.H.L. Bishop, L.G. Copping, G.K. Dixon & D.W. Hollomon (Eds). Opportunities for Molecular Biology in Crop Production. 55th ed. Farnham, Surrey, UK: British Crop Protection Council, pp. 133–142.

D.O.E., 1994. Fast tracking procedures for certain GMO releases. UK Department of the Environment/ACRE Guidance note 2.

Dughan, L., 1994. UK supermarket labels recombinant foods. Biotechnology 12: 351.

Durant, J., 1992. Biotechnology in public: A review of recent research. Science Museum, London: pp. 1–201.

Falk, B.W. & G. Bruening, 1994. Will transgenic crops generate new viruses and new diseases? Science 263: 1395–1396.

Gershon, D., 1994. Monsanto sues over BST. Nature 368: 384.

Greene, A.E. & R.F. Allinson, 1994. Recombination between viral RNA and transgenic plant transcripts. Science 263: 1423–1425.

Lehrman, S., 1993. Rules eased for US field tests. Nature 362: 483.

Lehrman, S., 1994. BST enters battle for consumers. Nature 367: 585.

Matzke, M.A., F. Neuhuber & A.J.M. Matzke, 1993. A variety of epistatic interactions can occur between partially homologous transgene loci brought together by sexual crossing. Mol. Gen. Genet. 236: 379–386.

McPartlan, H.C. & P.J. Dale, 1994. The transfer of introduced genes from field grown transgenic potatoes to non-transgenic potatoes and related solanaceous species. Trans. Res.: In press.

Miller, S.K., 1994. Genetic first upsets food lobby. New Scientist 1927: 6.

OECD, 1993. Field releases of transgenic plants, 1986–1992, an analysis. Organisation for Economic Co-operation and Development, pp. 1–40, Paris.

Scheffler, J.A., R. Parkinson & P.J. Dale, 1993. Frequency and distance of pollen dispersal from transgenic oilseed rape (*Brassica napus*). Trans. Res. 2: 356–364.

Scheffler, J.A. & P.J. Dale, 1994. Opportunities for gene transfer from transgenic oilseed rape (*Brassica napus*) to related species. Trans. Res.: In press.

Tepfer, M., 1993. Viral genes and transgenic plants. Biotechnology 11: 1125–1132.

WHO, 1988. Food Irradiation – a technique for preserving and improving the safety of food. World Health Organisation/Food and Agriculture Organisation of the United Nations. WHO publications Centre USA, pp. 84.

Euphytica **85:** 433–440, 1995.

The use of conventional and molecular genetics to produce new diversity in seed oil composition for the use of plant breeders – progress, problems and future prospects

Denis J. Murphy
Department of Brassica & Oilseeds Research, John Innes Centre, Norwich, NR4 7UH, U.K.

Key words: designer crops, fatty acids, novel crops, oilseeds, seeds, storage products

Summary

Considerable advances in the manipulation of seed oil yield and quality have been made using conventional breeding methods. Newer methods such as induced mutation and wide crosses have also resulted in significant achievements. These techniques, however, are all limited by the gene pool available in the crop of interest and its near relatives. The prospect of diminishing supplies of non-renewable fossil hydrocarbon reserves and the reality of edible crop surpluses have focused attention on the development of industrial oilseed crops. This will require the production of types of seed fatty acids which are not available in the oilseed crop species that are grown at present.

The introduction of novel oil crops via domestication is contrasted with the insertion of alien genes into existing oilseeds to create designer crops. The use of molecular genetics to alter the yield and quality of rapeseed oil is discussed, and strategies for the engineering of seven new designer rapeseed varieties are presented.

Introduction

Seed storage reserves such as oils, proteins and carbohydrates, are the most important products of global agricultural systems. Within a given crop species, plant breeders have been able to take advantage of considerable variation in the ratios of storage oils: proteins: carbohydrates, in order to select for the desired seed composition. Whereas in most wild brassicas, the amount of oil per seed typically ranges from 20% to 30% total seed weight, in the various brassica oilseed crop species, plant breeders have been able to increase this from 45% to 50% total seed weight. This has been achieved by reducing the levels of fibre in the seeds, and by changing the oil: protein: carbohydrate ratios in favour of oil. Nevertheless, further increases in oil content using classical breeding approaches are becoming progressively more difficult. This is one area where molecular genetics may lead to the manipulation of the metabolic pathways leading to storage reserve synthesis, in order to achieve a particular oil: protein: carbohydrate ratio that lies beyond the normal range of genetic variability within a particular crop species.

For each crop species, the quality or type of each category of storage product tends to be relatively constant. The fatty acid composition of the seed oil, or the amino acid composition of the seed storage proteins both tend to be characteristic of a particular species, genus, or even family. Brassicaceae seed oils, for example, tend to contain fairly high levels (20%–50%) of erucic acid, while seed oils from the Umbelliferae have high levels (60%–85%) of petroselinic acid (Gunstone et al., 1986). In both cases, these fatty acids are rarely found in other plant families. Plant breeders have sometimes found intra-specific diversity in seed fatty acid quality in natural plant populations. Often this diversity is due to mutations which block various parts of the fatty acid biosynthetic pathway, leading to the accumulation of an intermediate, rather than the end-product which is normally characteristic of the species. An example of this is rapeseed, a member of the Brassicaceae, whose seed oil formerly had the normal high erucic phenotypic characteristic of this family. During the past 30 years, new varieties of rapeseed and some of its brassica relatives have been bred, which contain mutations in the fatty acid elongation system which would normally produce erucic acid (Scarisbrick & Daniels, 1986). The result is a number of new varieties of brassica oilseeds containing high levels of oleic acid, an edible oil, rather than erucic acid, which is predominantly used as an industrial oil. These mutations were naturally-occurring variants selected by normal plant breeding methods.

A more recent technique has used induced mutation to create additional diversity in seed quality. Perhaps the most dramatic example of this technique is the development of low α-linolenic or 'linola' varieties of linseed by researchers in Australia and Canada (Green & Marshall, 1984; Green, 1986; Rowland, 1991). Normal varieties of linseed contain high proportions of α-linolenic acid (50%–60%), which forms an inedible seed oil with excellent characteristics as a drying agent in the manufacture of paints, varnishes, putties and other coverings. By treating normal linseed plants with chemical mutagens, researchers were able to select two independent lines, each with a single mutation in the linoleate desaturase system which normally gives rise to α-linolenic acid accumulation in the seed oil. By crossing these two mutant varieties, a double mutation was obtained which contained α-linolenic acid levels as low as 2% (Green, 1986). These new 'linola' varieties of linseed contained high levels of the edible polyunsaturate, linoleic acid, and were crossed into a superior agronomic background with no significant yield penalties. 'Linola' linseed is currently undergoing field trials in Australia, North America and Europe, and now has the potential to be a new edible oil crop with a seed fatty acid composition resembling that of sunflower.

The remarkable achievements of plant breeders both ancient and modern are, however, limited by the genome of the crop species in question. The use of natural variation or induced mutation, allows for the selection of variants with either complete or incomplete sets of functional genes responsible for storage product formation. It is not possible to add novel genes to the crop species by using these breeding approaches. The only caveat here is that, under some circumstances, it is possible to use wide crosses to introduce genes from related species into the crop plant of interest. An example of this is the recent successful introduction of disease resistance genes from wild populations of *Brassica oleracea* from the Mediterranean into a resynthesised *Brassica napus*, which was then backcrossed to an agricultural variety of rapeseed in order to produce a new rapeseed variety with enhanced disease resistance characteristics (Mithen & Herron, 1992). This rapeseed variety is now undergoing field trials in the UK. The use of such wide crosses is, however, limited by the range of interspecific crosses possible, and would not normally include the transfer of genes from unrelated plant species, or indeed animal or microbial species. Gene transfer between distantly related species requires the intervention of the cell biologist or molecular geneticist, using asymmetric cell fusion or DNA transformation methods such as particle bombardment or *Agrobacterium*-mediated gene transfer. It is the application of gene transfer to crop species which promises to revolutionise agriculture in the coming decades.

Designer crops *versus* Novel crops

The concept of 'designer crops' involves the introduction of alien genes into a crop species in order to enhance its agronomic performance. This is particularly relevant for oil crops at the present time, due to the prevalence of large surpluses of edible crops in the developed world. Over the past few years, many countries in Europe and North America have seen as much as 15%–25% of agricultural land taken out of production, i.e. 'set aside'. The result is a rekindling of interest in the growth of crops, and particularly oil crops, for industrial rather than edible use. Most oilseeds crops have been bred to produce high quality edible oils. By the addition of alien genes from other organisms, however, molecular geneticists can produce 'designer' varieties of oilseeds producing industrial-grade oils. In principle it is now possible to clone genes responsible for the formation of a particular fatty acid from virtually any organism into the oilseed crop of interest. This technology is particularly advanced in the case of rapeseed, which at the time of writing is the only major oilseed crop generally amenable to routine transformation at high efficiency. In addition to the two naturally–occuring varieties of rapeseed, i.e. high oleic and high erucic, at least three new designer varieties have recently been produced which, respectively, contain elevated levels of stearic, lauric and γ-linolenic acids in their seed oils (Murphy, 1994). Within the next few years it is likely that as many as ten varieties of rapeseed will be available, each with a totally different fatty acid composition, designed either for edible use or to serve as feedstocks for the manufacture of a wide range of non-edible products.

Over the next 5–10 years, transformation technology in the other major, annual oil crops such as sunflower, linseed and soybean, should allow for the production of new designer varieties of these crops. Although these developments will result in large numbers of new varieties of current crop species and a huge diversification of their possible end uses, they will not result in increases in crop species diversity at the farm level. In the UK, for example, by far the most dominant oil crop is rapeseed. The production of designer varieties of sunflower or soybean will not alter the climatic performance of these crops, and therefore it is likely that UK farmers would only select designer rapeseed varieties, rather than diversifying into other designer oilseed crops.

In this regard, it is germane to note that the European Union (EU) has made it a matter of policy to encourage the adoption of new crops into European agriculture, in order to diversify both the range of crop species and the types of crop products available to farmers. This policy was recently outlined as follows:

'At present the majority of Europe's agriculture is based on less than ten crop species; cereals, sugar crops, potatoes, oilseed crops and grasses. The increasing trend towards 'monoculture-based production', primarily due to subsidies and guaranteed markets, increases the dominance of these species and intensive production. With the reform of the CAP (Common Agricultural Policy) and inevitable reduction in subsidies, increased diversification and a decrease in intensification are stated Commission objectives. A more diversified production will give a broader spectrum of raw materials and bring advantages to farmers, industry and the consumer' (Rexen, 1992)

The development of novel oil crops, and indeed novel crops in general, is obviously a desirable objective, both from an agricultural and environmental point of view. There is an enormous number of uncultivated plant species which have the potential to be useful oil crops. In a survey carried out by the US Department of Agriculture from the late 1950s to the 1970s, more than 7000 plant species were screened and over 75 new fatty acids discovered (Princen, 1983, 1989). Many species were found that produced seeds containing industrially-useful fatty acids, several of which are now under development as novel oil crops. Some examples include crambe and meadowfoam to produce very long chain fatty acids, coriander to produce petroselinic acid, and *Cuphea* spp. to provide short to medium chain fatty acids. The problems with introducing novel crops include the very long timescales involved in their domestication and the frequent requirement for the retooling of agricultural machinery and downstream processing equipment. These factors add to the cost of the novel crop and hence reduce its competitiveness *vis a vis* existing crops. Most existing major crops have been domesticated for several hundred years at least and, in some cases, over tens of thousands of years. For new crops to compete with existing crops in regard to their yield, harvestability, disease resistance, etc., a large investment in germplasm acquisition and plant breeding is required. To a great extent, modern plant biotechnology can assist in this process, particularly through techniques such as marker-based gene selection and *in vitro* cell and tissue culture. This may allow for the cultivation on an economic scale of several new oil crops, such as *Cuphea*, coriander and meadowfoam, within the next few decades. Nevertheless, the imperative to use currently-available crops for the production of industrial raw materials, makes it necessary to proceed with the development of designer oil crops, at least as a medium-term strategy for the diversification of crop products.

Ideally, the development of novel oil crops should proceed in parallel with that of designer oil crops. It is likely that each will eventually find specific market niches, as well as allowing the agricultural industry to respond to the ever-changing demands for renewable feedstocks in both the edible and non-edible sectors.

Development of designer oil crops

Changing seed oil yields

The manipulation of storage product profiles in oilseeds is being carried out at both the quantitative and qualitative levels. Quantitative manipulation by molecular genetic methods is aimed at altering the oil: protein: starch ratios in oilseeds, particularly rapeseed. During seed development, photosynthetically-fixed carbon is imported into the seed in the form of simple soluble compounds such as sucrose. Once in the seed, sucrose can potentially serve as a substrate for the accumulation of storage oils, proteins and/or carbohydrates. The mechanism of carbon partitioning from sucrose towards the various seed storage products will determine the eventual composition of a given seed. In most plant species, the seeds contain all the enzymes required for the conversion of sucrose into all three classes of storage product. It is the rate of sucrose uptake by the various competing biosynthetic pathways, which eventually gives rise to a seed containing a particular storage product composition (Murphy et al., 1993).

The use of new molecular techniques, such as antisense technology, will allow for the selective modulation of key enzyme activities in the developing seed, while keeping the rest of the genetic background of the plant absolutely constant (Knutzon et al., 1992). This approach has the potential to elucidate the mechanisms involved in carbon partitioning to oil, protein and carbohydrates during seed development and thereby provide the basis for a non-empirical approach to its manipulation. This should in the future provide a powerful new tool for plant breeders to use to increase or decrease seed oil yield. Indeed, it may enable breeders to design a specific ratio of seed storage products in order to suit the final use of the seed, whether this is for animal feed or as a chemical feedstock, or for human nutrition.

The biochemical pathways involved in the conversion of imported sucrose into storage products are extremely complex. The formation of storage oil from sucrose alone involves dozens of enzymes, any one of which may be a potential candidate for genetic manipulation (Murphy, 1993a; Murphy et al., 1993). As a first step to investigating this process, several groups have selected the first and last enzymes of this pathway, in order to ascertain their contribution to oil yield in a developing seed.

The first committed step in oil formation is catalyzed by the enzyme, acetyl-CoAcarboxylase, which is responsible for the conversion of acetyl-CoA to malonyl-CoA. Genes encoding various isoforms and subunits of this enzyme have recently been cloned from at least four important crop species including rapeseed (Schulte et al., 1993; Deka & Rawsthorne, personal communication), alfalfa (Shorrosh et al., 1994), wheat (Gornicki et al., 1994), maize (Egli et al., 1993) and pea (Sasaki et al., 1993). In several laboratories, including our own, part of this gene has been inserted in the antisense orientation into rapeseed plants in order to test whether this enzyme is indeed a key regulatory step in the accumulation of seed oil (Kang et al., 1994). The results of these experiments should permit us to determine the metabolic flux control coefficient of acetyl-CoA carboxylase for the biosynthesis of storage oils. This information will be crucial in determining whether acetyl-CoA carboxylase is an appropriate target for genetic engineers wishing to manipulate the relative oil yield of a seed. The use of antisense constructs for its down-regulation, or the addition of extra sense copies of the complete acetyl-CoA carboxylase gene(s) for its upregulation, may allow us to generate far wider levels of variation in seed oil content than were possible hitherto. This variation could then provide useful material for plant breeders attempting to produce novel high-oil or low-oil varieties of a given crop plant. In the case of most industrial oilseeds, the seed oil is the highest value component of the crop. The objective for breeders of industrial oilseeds would, therefore, be to maximise oil content at the expense of protein, carbohydrate or fibre. For other applications, however, it may be desirable to produce oilseeds with a low to medium oil content and a high protein content. An example of this would be the production of animal feed, where the oil component of a seed would give a high calorific value, while the protein would give superior nutritional qualities.

The final enzyme involved in storage oil formation is diacylglycerol acyltransferase. Indeed, this is the only enzyme which is unique to storage oil biosynthesis, since all other enzymes in the pathway can also potentially contribute to membrane lipid formation. This enzyme is of great interest, therefore in our attempts to understand the regulation of storage oil versus membrane lipid biosynthesis in oil crops. Diacylglycerol acyltransferase, like many enzymes of lipid metabolism, is an insoluble trans-bilayer membrane protein which is much more difficult to purify than most soluble proteins. For this reason, attempts to clone its gene, both in our laboratories and elsewhere, are as yet at a relatively early stage (Weslake et al., 1992; Wilson et al., 1992; Hobbs & Hills, personal communication). Nevertheless, the potential importance of this enzyme will make the elucidation of its structure, function and regulation, of enormous interest, both academically, and for the manipulation of seed oil content in crop plants.

Changing seed fatty acid composition

The qualitative manipulation of seed oils involves the modification of their fatty acid compositions. It is the fatty acid profile which determines the chemical properties responsible for the particular end use of the seed oil. Oil-bearing plants produce many hundreds of fatty acids that can serve as actual or potential industrial raw materials. In attempting to demonstrate the effectiveness of biotechnology for producing new industrial oil crops, it is therefore important to decide initially upon a relatively limited number of target fatty acids to be produced. It is equally important to select an appropriate host species, such as rapeseed, which is both an established, high-yielding oil crop and is amenable to genetic transformation and regeneration via tissue culture. While routine transformation of the other major annual oil crops will probably be developed within the next 5–10 years, the transformation and regeneration to maturity of the important perennial oil crops such as palm, coconut and olive, is unlikely to be developed for several decades (Murphy, 1994). For much of the present decade, therefore, rapeseed will continue to be the best candidate for molecular manipulation to produce industrial oils (Murphy, 1993b).

Target seed fatty acids

The selection of target fatty acids for designer rapeseed varieties has been largely determined by the availability or the ease of isolation of the relevant gene clones. It has proved relatively straightforward to purify many of the soluble enzymes associated with fatty acid formation *de novo* and therefore to clone their genes. In contrast, the vast majority of the membrane-bound enzymes of oil synthesis have proved to be singularly recalcitrant to purification, despite efforts spanning as much as 30 years. Luckily, for the state of progress in this field (not to mention the sanity of protein chemists!) it has recently been possible to use molecular genetic techniques such as T-DNA tagging and chromosome walking, to clone genes encoding enzymes which have proved impossible to purify by conventional biochemical approaches (Murphy, 1994). This has now opened up the prospect of being able to clone genes encoding all of the enzymes involved in storage oil formation in seeds. At the time of writing, there are reports of at least seven target fatty acids which molecular geneticists are attempting to produce in transgenic rapeseed plants.

Stearic acid
This is a C_{18} saturated fatty acid that serves as a major ingredient in margarines and cocoa-butter substitutes. The Δ_9 stearoyl-ACP desaturase gene which normally converts stearic to oleic acid was partially inactivated in rapeseed using antisense methods, resulting in the accumulation of a seed oil containing up to 40% stearic acid, in contrast to levels of < 5% stearic acid in conventional rapeseed varieties (Knutzon et al., 1992). Transgenic rapeseed plants containing a stable level of 25%–30% stearic acid in their seed oil are currently in their second season of field release trials in Scotland and in their fourth season of trials in the USA. Preliminary reports indicate no other phenotypic or pleiotropic effects due to the presence of the transgene. The transgenic plants show normal establishment, vigour, yield and pest/disease-resistant characteristics (Kridl, 1994).

Lauric acid
This is a C_{12} saturated fatty acid that is an important industrial feedstock for the manufacture of detergents, soaps and surfactants. Rapeseed oil does not contain lauric acid, but it is found at high levels in the seed oil of the California Bay plant, *Umbelluria californica*, due to the presence in the latter species of a lauroyl-ACP thioesterase. This gene was duly cloned from the Bay plant and inserted into rapeseed, resulting in a novel variety with a seed oil containing almost 25% lauric acid (Voelker et al., 1992). Transgenic, lauric-containing rapeseed plants are currently undergoing field release trials similar to those described above for the high-stearic plants and likewise show no discernable pleiotropic effects of the transgene.

Petroselinic acid
This is a C_{18} mono-unsaturated fatty acid isomer of oleic acid with its double bond in the Δ_6, rather than the Δ_9 position. Oxidative splitting of petroselinic acid into lauric and adipic acids generates potential feedstocks for the detergents and plastics industries. While rapeseed oil contains no petroselinic acid, it accumulates at very high levels (> 80%) in most species of the Umbelliferae. A Δ_4 palmitoyl-ACP desaturase gene involved in petroselinic acid formation has been cloned from the spice plant coriander, *Coriandrum sativum*, and inserted into transgenic rapeseed plants (Cahoon et al., 1992; Fairbairn et al., 1992). Insertion of this gene alone may allow for production of a new petroselinic acid-containing rapeseed variety, although recent reports indicate that additional coriander genes may be required for the production of high levels of petroselinic acid in rapeseed oil (Cahoon & Ohlrogge, 1994a, 1994b; Ohlrogge, 1994).

Ricinoleic acid
This is a C_{18} hydroxylated mono-unsaturated fatty acid found in very high quantities (90%) in castorbean, *Ricinus communis*, seed oil and absent from rapeseed oil. Castor oil is a valuable industrial raw material for the production of polymers, lubricants, cosmetics, perfumes and pharmaceuticals. In our laboratories we are

using a variety of differential screening approaches for the isolation of the oleate hydroxylase gene from castorbean (which cannot be grown as a crop in the UK) for insertion into rapeseed (Murphy, 1993a, 1993b; Murphy et al., 1994).

Erucic acid

This is a C_{22} mono-unsaturated fatty acid that is found in most of the Brassicaceae, but it is normally only present on the first and third carbons of the glycerol backbone of the triacylglycerol molecules that make up most the seed oil. Hence the maximum potential erucic content in rapeseed oil is only 66% and the normal amount in presently available high-erucic cultivars is only 45%–55%. Erucic acid is a valuable feedstock for the manufacture of plastic coatings, polymers, lubricants, etc., and the greater its yield in a seed, the more favourable are the economics of its commercial utilisation. The meadowfoam plant, *Limnanthes alba*, contains a 1-acylglycerol 3-phosphate acyltransferase capable of inserting erucic acid onto the second carbon of the triacylglycerol backbone, and it is hoped that the insertion of its gene into rapeseed will allow for a very high erucic (> 90%) seed oil. Efforts to achieve this are now underway in our own laboratories (Hills, Chamberlin & Dann, personal communication) and elsewhere (Taylor et al., 1992; Peterek et al., 1992; Morand & German, 1993; Brown et al., 1994).

Jojoba wax

This is an ester of mainly C_{20}–C_{22} mono-enoic fatty acids and fatty alcohols found in the seed oil of the North American desert shrub, jojoba, *Simmondsia chinensis*. It is a liquid wax and is a high-value source of cosmetic and pharmaceutical products and high-grade lubricating oils. Of the two key genes involved in its synthesis, the cloning of one gene has already been announced and progress on the cloning of the other gene is reportedly well-advanced (Lassner et al., 1994). The intention is then to insert both genes into rapeseed in order to product a novel jojoba-wax variety of this crop.

γ-linolenic acid (GLA)

This is a C_{18} trienoic acid isomer of α-linolenic acid with its double bonds in the $\Delta_{6,9,12}$ positions. γ-linolenic (also called GLA) is the active ingredient of evening primrose or borage oils and is widely used as a therapeutic agent to alleviate symptoms of disorders ranging from skin complaints to pre-menstrual tension. The Δ_6-desaturase gene has been cloned from several species of cyanobacteria (Murphy, 1993a, 1993b) and its insertion into rapeseed has led to the accumulation of γ-linolenic acid in the seed oil. Field assessment of GLA-containing, transgenic rapeseed plants is now underway but no further details on their phenotype or performance characteristics are available at the time of writing.

Future prospects

From this brief survey, it can be seen that designer rapeseed crops will shortly become a reality. While progress over the last few years has been most impressive, the production of designer rapeseed oils as economically-viable, industrial feedstocks will require the accumulation in the seed of very high levels of specific fatty acids. It will be necessary for the novel genes to be under the control of suitable promoters to achieve the appropriate high levels of expression and the correct spatial and temporal regulation. The problems to be overcome in achieving this are far from simple and our approach has often been hampered by ignorance of the fundamental mechanisms of both the biochemical regulation of the metabolic pathways responsible for seed oil accumulation and the developmental regulation of gene expression in seeds. Nevertheless, the results of the initial attempts to engineer new varieties of rapeseed are most encouraging. All the indications from four seasons of field trials of transgenic rapeseed varieties are, that apart from an altered seed oil composition, there are no other phenotypic effects of the transgene. Most importantly, there are no adverse effects on such agronomically-important characteristics as seed germination, seedling establishment and vigour and overall seed yield. It is probably only a matter of time before transformation techniques are refined enough to allow for the engineering of virtually any desired oil composition in the annual oilseed crop of choice. This should then give farmers a realistic alternative to the overproduction of edible crops that currently leads to unpopular policies, such as the removal from production, i.e. set-aside, of so much valuable farmland in the developed world. It will additionally allow for the production of renewable, environmentally-friendly and CO_2-neutral raw materials as alternatives to the rapidly depleting, non-renewable fossil hydrocarbons that currently supply much of these raw materials to industry.

Acknowledgements

This author is grateful to the members of the Brassica & Oilseeds Research Department for their assistance and comments. Particular thanks are due to Drs. M.J. Hills and S. Rawsthorne. Financial support for the John Innes Centre oilseeds programme comes from BBSRC core funds, the Ministry of Agriculture, Fisheries and Food, the British Council and a variety of private contractors whose support is gratefully acknowledged.

References

Brown, A P, A M Tommey, M D Watson & A R Slabas, 1994 Complementation of an E coli 1 acyl-sn-glycerol-3-phosphate acyltransferase mutant with a *Zea mays* cDNA clone In Proceedings of the 4th International Congress of Plant Molecular Biology Abstract 1465, ISPMB, Amsterdam

Cahoon, E B & J B Ohlrogge, 1994a Metabolic evidence for the involvement of a Δ4-palmityol acyl carrier protein desaturase in petroselinic acid synthesis in coriander endosperm and transgenic tobacco cells Plant Physiol 104 827–837

Cahoon, E B & J B Ohlrogge, 1994b Apparent role of phosphatidylcholine in the metabolism of petroselinic acid in developing Umbelliferae endosperm Plant Physiol 104 845–855

Cahoon, E C, J Shanklin & J B Ohlrogge, 1992 Expression of a coriander desaturase results in petroselinic acid production in transgenic tobacco Proc Nat Acad Sci USA 89 11184–11188

Egli, M A, S M Lutz & B G Gengenbach, 1993 Cloning and sequence analysis of a maize acetyl-CoA carboxylase gene In Proceedings of the National Plant Lipid Cooperative, poster A2 Plant Lipid Symposium, Minneapolis, USA

Fairbairn, D, J H E Ross, S Bowra, S & D J Murphy, 1992 Molecular & biochemical studies of the biosynthesis of storage oil & petroselinic acid in embryos & cell cultures of Umbelliferae In A Cherif (Ed) pp 67–70 Metabolism, Structure and Utilisation of Plant Lipids 91 6860–6864

Green, A G, 1986 Genetic control of polyunsaturated fatty acid biosynthesis in flax (*Linum usitatissimum*) seed oil TAG 72 654–661

Green, A G & D R Marshall, 1984 Isolation of induced mutants in linseed (*Linum usitatissimum*) having reduced linolenic acid content Euphytica 33 321–328

Gunstone, F D, J L Harwood & F B Padley, 1986 Occurrence and characteristics of oils and fats In F D Gunstone, J L Harwood & F B Padley (Eds) The Lipid Handbook, pp 49–170 Chapman & Hall, London, U K

Kang, F, C R Ridout, C L Morgan & S Rawsthorne, 1994 The activity of acetyl CoA carboxylase is not correlated with the rate of lipid synthesis during development of oilseed rape (*Brassica napus* L) embryos Planta 193 320–325

Knutzon, D S, G A Thompson, S E Radke, W B Johnson, V C Knauf & J C Kridl, 1992 Modification of Brassica seed oil by antisense expression of a stearoyl-acyl carrier protein desaturase gene Proc Natl Acad Sci USA 89 2624–2628

Kridl, J C, 1994 Modified oils in transgenic rapeseed In Proceedings of the 4th International Congress of Plant Molecular Biology, Abstract 193 ISPMB, Amsterdam

Lassner, M W, K Lardizabal & J G Metz, 1994 Fatty acyl-CoA elongation in developing embryos of jojoba (*Simmondsia chinensis*) In J C Kader & P Mazliak (Eds) Proceedings of the 11th International Conference on Plant Lipids, L53 26 June–1st July, Paris

Mithen, R F & C Herron, 1992 Transfer of disease resistance to oilseed rape from wild *Brassica* species Proceedings of the 8th International Rapeseed Congress pp 244–249 Saskatoon, Canada

Morand, L Z & J B German, 1993 Rescue of the phenotype of an *E coli* mutant defective in 1-acyl-sn-glycerol 3-phosphate acyltransferase by complementation with an *Arabidopsis* expression library In Proceedings of the XV International Botanical Congress, pp 404 Yokohama, Japan

Murphy, D J, 1994 Biotechnology of Oil Crops In D J Murphy (Ed) Designer Oil Crops, pp 219–251 Weinham, Germany

Murphy, D J, 1993a Structure, function and biogenesis of storage lipid bodies and oleosins in plants Prog Lipid Res 32 247–280

Murphy, D J, 1993b Designer Oilseeds for Industry The Agronomist 2 6–9

Murphy, D J, D Richards, R D Taylor, J Capdevielle, J -C Guillemot, R Grison, D Fairbairn & S Bowra, 1994 Manipulation of seed oil content to produce industrial crops Ind Crops & Prod in press

Murphy, D J, S Rawsthorne & M J Hills, 1993 Storage lipid formation in seeds Seed Sci Res 3 79–95

Ohlrogge, J B, 1994 Design of new plant products engineering of fatty acid metabolism Plant Physiol 104 821–826

Peterek, G, V Schmidt, F P Wolter & M Frentzen, 1992 Approaches for cloning 1-acylglycerol acyltransferase from oilseeds In A Cherif (Ed) Metabolism, Structure and Utilization of Plant Lipids, pp 401–404 CNP Press, Tunisia

Princen, L H, 1983 New oilseed crops on the horizon Econ Bot 37 478–492

Princen, L H, 1989 In R S Cambrie (Ed) Fats for the Future Horwood, Chichester, U K

Rexen, F, 1992 Crops for Industrial Use In Towards sustainable crop production systems 11 pp 79–80 Monograph series, Royal Agric Soc England

Rowland, G G, 1991 An EMS-induced low-linolenic acid mutant in McGregor flax (*Linum usitatissimum* L) Can J Plant Sci 71 393–396

Sasaki, Y, K Hakamada, Y Suama, Y Nagano, I Furusawa & R Matsuno, 1993 Chloroplast-encoded protein as a subunit of acetyl-CoA carboxylase in pea plant J Biol Chem 268 25118–25123

Scarisbrick, D H & R W Daniels, 1986 Oilseed rape Collins, London

Schulte, W, J Schell & R Topfer, 1993 Molecular cloning of a gene encoding acetyl-CoA carboxylase from *Brassica napus* In Proceedings of the National Plant Lipid Cooperative, poster A6 Plant Lipid Symposium, Minneapolis, USA

Shorrosh, B S, R A Dixon & J B Ohlrogge, 1994 Molecular cloning, characterization and elucidation of acetyl-CoA carboxylase from alfalfa Proc Natl Acad Sci USA 91 4323–4327

Taylor, D C, J R Magus, R Bhella, J Zon, S L MacKenzie, E M Giblin, E W Pass & W L Crosby, 1992 Biosynthesis of triacylglycerols in *Brassica napus* L cv Reston, Target, Trierucin In S L MacKenzie & D C Taylor (Eds) Seeds Oils for the Future, pp 77–102

Voelker, T A, A C Worrell, L Anderson, J Bleibaum, C Fan & D J Hawkins, 1992 Fatty acid biosynthesis redirected to medium chains in transgenic oilseed plants Science 257 72–73

Weselake, R.J., M.K. Pomeroy, T.L. Furukawa & R.L. Oishi, 1992. Partial purification and characterization of diacylglycerol acyltransferase from microspore-derived embryos of oilseed rape. In: S.L. MacKenzie & D.C. Taylor (Eds). Seed Oils for the Future, pp. 103–115.

Wilson, R.F., P. Kwanyuen, R.E. Dewey & S.B. Settlage, 1992. Recent developments in the molecular biochemistry and genetics of diacylglycerol acyltransferase from soybean. In: S.L. MacKenzie & D.C. Taylor (Eds). Seeds Oils for the Future, pp. 116–135.

Euphytica **85:** 441–449, 1995.

Genetic engineering of Indica rice in support of sustained production of affordable and high quality food in developing countries

I. Potrykus, P.K. Burkhardt, S.K. Datta, J. Fütterer, G.C. Ghosh-Biswas, A. Klöti, G. Spangenberg & J. Wünn
Institute of Plant Sciences, Swiss Federal Institute of Technology (ETH), ETH-Centre, CH-8092 Zurich, Switzerland

Key words: Oryza sativa, Indica-type rice, genetic engineering, vitamin A endosperm, insect resistance, virus resistance, fungus resistance, essential amino acids

Summary

Indica-type rice provides the staple food for two billion people in Third World countries. Several problems involved in the stable and sustained production of high quality food cannot be solved by traditional breeding. Methods have been established for gene transfer to Indica rice breeding lines to study possible contributions from genetic engineering. Experiments are in progress on the development of transgenic resistance towards Yellow Stem Borer, resistance towards Rice Tungro Virus, accumulation of provitamin A in the endosperm, increase of essential amino acids in the endosperm such as lysine, cysteine and methionine and resistance towards fungal pests such as Rice Blast and Sheath Blight. Transgenic clones from Indica rice breeding lines have been recovered from several of the approaches mentioned, some of which have been regenerated to plants.

Introduction

Indica-type rice (*Oryza sativa*) feeds more than two billion people, predominantly in developing countries. In humid and semihumid Asia, where rice is the basic food, the population is expected to increase by 58 percent over the next 35 years. In thirty years the world will need 70 percent more rice than it requires today. These ca. 800 million tons of rice will have to be grown with considerable reduction in the input of agrochemicals under sustainable conditions (IRRI, 1993). This immense task requires that traditional plant breeding is supported by every possible contribution from novel technical developments. Genetic engineering, applied appropriately and with care, has the potential to contribute to the sustainable production of affordable food for the increasing population in developing countries. To achieve maximum benefit for these countries, the application of gene technology should focus on important problems for which solutions by conventional approaches are not available. For Indica-type rice, such problems have been identified and described in a joint study of the International Rice Research Institute (IRRI), Manila, and the Rockefeller Rice Biotechnology Program, New York (Khush & Toenniessen, 1991). Among the high priority problems to be solved are a) resistance to fungal diseases, b) resistance to Tungro virus, c) resistance to Yellow Stem Borer (*Scirpophaga incertulas*), d) stable supply of provitamin A, and e) improvement of nutritional quality. The problems mentioned are especially severe for people depending on Indica-type rice. It is, to date, relatively easy to genetically engineer Japonica-type rice, but relatively difficult to do the same with Indica-type rice. The tasks mentioned can be solved only by focusing work on Indica-type rice. The IRRI has a world mandate to develop breeding lines for the benefit of small rice farmers, and has already released numerous successful varieties to the rice-growing countries (IRRI, 1992). The following research projects were, therefore, performed in collaboration with IRRI and using IRRI breeding lines, to assure direct transfer of successful results to the target population. The best possible solutions can be found only with the close involvement of the international scientific community. This is being achieved by collaboration with the Rockefeller Rice Biotechnology Program (Toenniessen et al., 1989).

Gene transfer for integrative transformation

Breeding with transgenes requires independent populations of fertile, transgenic plants to chose the most stable and best-expressing lines for subsequent traditional breeding. This in turn requires routine and efficient gene transfer protocols leading to fertile transgenic plants. Although recovery of transgenic Indica-type rice plants and their offspring has been described from our laboratory (Datta et al., 1990, 1992) and from others (Christou et al., 1991), gene transfer to IRRI breeding lines is not yet routine and efficient, and recovery of fertile, transgenic plants is still also rather inefficient and requires further optimisation. Gene transfer to Indica-type rice is, to date, possible via direct gene transfer to protoplasts (Datta et al., 1990; Ghosh-Biswas et al., 1994) and our group is using this technique routinely with IR43, IR72 and other Indica rice varieties. With IR72, however, the most advanced breeding line, we still face severe fertility problems. With IR43 we see better chances of raising sufficient numbers of independent and fertile, transgenic plants in the near future. Transgenic Indica rice plants have also been recovered from biolistic treatment of immature embryos (Christou et al., 1991) and from electroporation of split embryos from mature seeds (Xhu & Li, 1994). We have a series of transgenic clones and plants from IR43 and IR72 under investigation, and hope to present data permitting identification of the more appropriate of the two techniques in the near future. We have also applied electroporation to cells of immature embryos for transformation of cereal cells. Although gene transfer to scutellum cells of wheat is routine and efficient (Klöti et al., 1993), it has not been possible to repeat these results with rice. *Agrobacterium*-mediated transformation of rice has been attempted by other laboratories (e.g. Rainieri et al., 1990) but *Agrobacterium*-mediated transgenic rice plants have not been recovered from any laboratory.

Gene transfer for transient expression

Integrative transformation in totipotent cells is the basic prerequisite for adding valuable new characters to a breeding programme. Transient expression of transgenes in differentiated tissues is experimentally far less demanding. This technique provides an experimental tool which facilitates and speeds up vector development for integrative transformation, and allows us to study the functionality of transgenes and expression signals prior to integrative transformation. We use transient expression of marker genes in biolistic treatments and direct gene transfer to protoplasts and electroporation, not only to study expression signals and function of structural genes, but also to optimise gene transfer protocols for integrative transformation.

Analysis of expression signals

The expression of novel, beneficial genes in transgenic plants requires expression signals. So far, in most cases, the transcription control signals of cauliflower mosaic virus have been used to obtain strong and uniform expression in almost all tissues of monocotyledons and dicotyledons. Recently, different promoters allowing strong constitutive or regulated expression have been isolated from many different plants and can now be tested for their usefulness in rice transformation. The availability of different constitutive promoters will be useful in avoiding problems of recombination or gene silencing (Matzke & Matzke, 1993) in plants expressing a variety of different transgenes, while the use of tissue-specific or inducible promoters will allow regulated expression of transgenes at the desired site or time. For constitutive expression, in addition to the CaMV 35S promoter, we use a maize ubiquitin promoter (Christensen et al., 1992), the rice actin 1 promoter (McElroy et al., 1991), and the 1′2′ double-headed promoter from the *Agrobacterium* T-DNA (Uchimiya et al., 1986).

For expression in green tissues, promoters derived from the genes for wheat chlorophyll a/b binding protein (Lamppa et al., 1985), wheat ribulose-1,5-bisphosphate carboxylase small subunit (Broglie et al., 1983), and maize phosphoenol pyruvate carboxylase (Yanagisawa & Izui, 1989) are under investigation. For expression of genes in particular tissues, the phloem-specific promoter of rice tungro bacilliform virus (RTBV) (Bhattacharyya-Pakrasi et al., 1993), promoters from root and shoot-specific rice genes (dePater et al., 1990; dePater & Schilperoort, 1992), from rice glutelin genes specific for the endosperm (Kim et al., 1993) and from pistil and anther-specific Brassica genes are being analysed. In addition, a search for rice tapetum-specific promoters has been initiated. Pathogen-inducible promoters from a maize seed PR protein (Casacuberta et al., 1991) and from a rice chitinase gene (Zhu et al., 1993) are being used for the expression of antifungal proteins.

Approach towards Yellow Stem Borer resistance

Insect damage is one of the major causes of yield loss in rice farming all over the world. In Southeast Asia, the value of lost production caused by insect damage reaches more than US $ 600 million per year. More than two thirds of these losses are caused by two insect species, the rice Brown Plant Hopper (BPH; *Delphax oryzae*, Homoptera) and the Yellow Stem Borer (YSB; *Scirpophaga incertulas*, Lepidoptera: Herdth, 1991). In contrast to BPH for which restance genes are known in the gene pool, no such genes are available for resistance against YSB.

The entomocidal spore-forming, soil bacterium *Bacillus thuringiensis* offers a promising range of genes which encode specific endotoxins. To date, more than 40 different nucleotide sequences of such genes have been determined. They are clearly related to each other, and have been classified into 17 distinctly-different crystal protein genes, the so-called *cry*-genes (Peferoen, 1991). These genes encode proteins of some 130–140 kDa or some 70 kDa which are first dissolved and then proteolytically cleaved in the midgut of the larvae of lepidopteran insects, to toxic fragments of approximately 60 kDa (Faust & Bulla, 1982). After cleavage, these toxins bind to specific proteins on the brush-border membranes of the insect gut (Hofmann et al., 1988), resulting in disruption of the epithelium, disturbance of the ionic balance and thereby paralysis of the gut.

Different *Bt* formulations have been used as biological insecticides for many years. Disadvantages like poor persistence and short duration of effect under tropical conditions especially during the rainy season, could be overcome by a transgenic approach, the expression of *Bt*-genes in the rice plant itself.

Success has been reported from tobacco (Vaeck et al., 1987), potato (Peferoen et al., 1991), tomato (Fischhoff et al., 1987) cotton (Perlak et al., 1990) maize (Koziel et al., 1993) and Japonica rice (Fujimoto et al., 1993). These plants showed a high level of insect resistance and it is expected that some of these crops will be released on the market in the next few years (Peferoen, 1991).

The aim of our project is the transformation of advanced Indica rice breeding lines with genes conferring resistance to YSB. We are using the lepidopteran-specific *cry* I A(b)-gene, which has been shown to be effective against YSB and rice leaf folder (D. Bottrell, personal communication).

Since it is known that the highly-conserved, C-terminal half of the *Bt*-toxin is not necessary for the toxicity of the protein, truncated forms of the *cry*-genes have been cloned (Fischhoff et al., 1987). These truncated genes encode only the N-terminal half of the crystal proteins and show significantly enhanced expression levels of the *Bt* toxins in transgenic plants (Fischhoff et al., 1987; Delannay et al., 1989; Perlak et al., 1990; Fujimoto et al., Koziel et al., 1993).

As plants in general show a codon usage different to that found in bacteria, a synthetic *cry* I A(b)-gene was constructed (Koziel et al., 1993) which shows a high G-C content in the coding region. As the maize codon usage resembles the codon usage pattern of monocotyledons in general (Murray et al., 1989), this should also lead to a sufficient expression of the *cry* I A(b)-gene in rice.

As a first step towards YSB-resistant rice plants, we have transformed a truncated version (645 codons) instead of a wild type and a synthetic *cry* I A(b)-gene to elite Indica rice breeding lines. Resistant clones were analysed by Southern Blotting and showed a clear integration of the *cry* I A(b)-gene in the rice genome. Plants were regenerated, transferred to soil and are growing under greenhouse conditions. For further analysis, R_0 and R_1 plants will be checked by Southern, Northern and Western analyses and by insect-feeding studies.

To minimise the possible development of resistance in insects, we put this gene under the control of a tissue-specific promoter which directs the expression of the *cry* I A(b)-gene only to the leaf sheath, the primary target site of the YSB. Furthermore we are planning to use a second *Bt*-gene, the *cry* II A-gene which is known to bind to a different receptor site in the brush-border membrane of the insect gut, as resistance to *Bt*-toxins could be due to a change in such a receptor site (Ferré et al., 1991).

It has to be emphasised, however, that such transgenic rice plants should be planted under field conditions only, according to the concept of Integrated Pest Management (IPM), to keep the *Bt*-toxins as an effective, sustainable and durable tool (McGaughey et al., 1992).

Approach towards tungro virus resistance

The rice tungro disease is caused by a complex of two viruses, rice tungro spherical virus (RTSV) and rice tungro bacilliform virus (RTBV, Hibino et al., 1978). Severe symptoms are caused by RTBV (Dasgupta et al., 1991). RTBV is a member of the newly-assigned group of badnaviruses (Hay et al., 1991; Qu et al., 1991) and is related to the better-studied caulimoviruses in its life cycle and genome organisation (Hohn & Fütterer, 1992). While engineered virus resistance has been achieved for a number of RNA plant viruses, for the DNA-containing badna and caulimoviruses, no successful strategy has been reported (Wilson, 1993).

Since these viruses have very different replication cycles, it is not clear whether approaches that worked with RNA viruses (i.e. constitutive expression of viral coat proteins or wild-type or mutated replicases) will also work for RTBV. It is, however, to be expected that for RTBV, expression of functional viral proteins at the onset of virus infection could interfere with an ordered progression through the viral life cycle, and that expression of mutated viral proteins might interfere with the function of normal viral proteins by competition. We have therefore introduced into Indica rice breeding lines (IR43, IR72) constructs designed to express RTBV proteins 1, 3 and 4. Protein 3 is a precursor from which the viral coat protein, reverse transcriptase and probably a variety of other proteins are generated by proteolytic processing. The locations of these proteins within the precursor can at present be deduced only from sequence homologies to related viruses. A variety of constructs for direct expression of the coat protein or the reverse transcriptase have been prepared on the basis of such estimations. Transgenic plants will be screened for expression of the proteins with antisera obtained from R. Hull (John Innes Institute, Norwich UK) or with antisera against a protein tag that has been incorporated into the expression constructs.

To create mutant proteins that might act as competitive inhibitors for critical viral functions, mutations have been introduced into the coat protein and the reverse transcriptase. In the coat protein region, a sequence motif containing three invariable cysteines and one histidine is conserved between almost all viruses using reverse transcriptase in their replication cycle (retro and plant pararetroviruses; Covey, 1986). This motif, which is involved in several RNA-binding steps, has been mutated to glycine in the RTBV coat protein, since such a mutation has been shown to preserve the structure (and thus part of the function) of the remaining part of the coat protein but to abolish infectivity of a retrovirus (De Rocquigny et al., 1992; Morellet et al., 1992). In the reverse transcriptase region, a mutation was introduced into a highly-conserved sequence motif containing two aspartates (Argos, 1988). In addition, subfragments of the polyfunctional reverse transcriptase have been cloned in an attempt to repeat successes obtained with some RNA viruses where subfragments of the polymerase gene produced high levels of protection (Wilson, 1993).

Since functions for the remaining RTBV proteins are unknown, changes are difficult to design. In protein 4 we localised a leucine zipper similar to those that are involved in protein-protein interactions (Gruissem, 1990). In a yeast system (Fields & Song, 1989) we found that protein 4 indeed has the capacity to dimerise but the dimerisation domain resides outside the leucine zipper which may, however, interact with another protein. We have cloned subfragments of the protein 4 coding region containing either the leucine zipper or the dimerisation domain to express proteins that lack one of the interaction domains. They will probably not have a complete function but will still interact with one of the original partners, therefore acting as a competitive inhibitor.

In addition to these approaches involving expression of a protein, we have also introduced a construct expressing antisense RNA against the leader sequence of the RTBV pregenomic RNA. Antisense RNAs had little effect on RNA viruses (Wilson, 1993) but viruses like RTBV with a nuclear phase might be more susceptible.

Most of these strategies are based on the work performed with RNA viruses. Approaches that would more specifically exploit the particular molecular biology of RTBV require a detailed knowledge of the viral cycle. We are also studying viral gene expression mechanisms and the potential function of viral gene products in transient expression systems (Fütterer et al., 1994).

Approach towards provitamin A accumulation in rice endosperm

According to UNICEF statistics world-wide, over 124 million children are estimated to be vitamin A-deficient (Humphrey et al., 1992). Improved vitamin A nutrition could prevent approximately 1–2 million deaths annually among children aged 1–4 years. An additional 0.25–0.5 million deaths may be avoided if improved vitamin A nutrition can be achieved during later childhood. Improved vitamin A nutrition alone, therefore, could prevent 1.3–2.5 million out of nearly 8 million late infancy and preschool-age child deaths that occur each year in the highest-risk countries (West et al., 1989).

Rice in its milled form, as it is consumed by most people in South East Asia, is characterised by the complete absence of provitamin A. The milled rice kernel consists exclusively of the endosperm. The embryo and the aleurone layer have been removed during processing of the rice grain. The aim of this project is to initiate carotenoid biosynthesis in the rice endosperm tissue to increase the daily vitamin A-uptake of people predominantly living on rice. In maize and sorghum cereal endosperm cells can produce and accumulate carotenoids (Buckner et al., 1991). Furthermore the starch storage tissues of potato and cassava (Pentedao & Almeida, 1988) accumulate carotenoids in considerable amounts. To provide the minimum requirements of relevant carotenoids to young infants, and assuming rice as the sole dietary source, 1–2 μg β-carotene per gram uncooked rice would be needed in rice endosperm (The Rockefeller Foundation, 1993). This is roughly 25–50 per cent of the amount produced in maize endosperm, and enough to turn the rice noticeably, but not dark, yellow.

The carotenoid pathway is a branch of the central isoprenoid pathway which is characterised by four key enzymes necessary for carotenoid biosynthesis. These are phytoene synthase, phytoene desaturase, ζ-carotene desaturase and lycopene cyclase. The genes encoding for these enzymes are available from both higher plants (Fray & Grierson, 1993; Ray et al., 1987; Linden et al., 1993; Bartley et al., 1991) and bacteria (Armstrong et al., 1989).

Our strategy is to produce transgenic Indica rice varieties which contain either single genes or several genes in combination. Experiments to produce plants that contain a phytoene synthase cDNA (Ray et al., 1987) under the control of endosperm-specific promoters (Okita et al., 1989) to ensure exclusive expression in the endosperm are underway. The analysis of these plants will be done in close collaboration with Dr. P. Beyer, Freiburg (Germany).

Furthermore we want to develop a transient expression system which allows us to assess the function and activity of genes and promoters which we will use for transformation. The aim is to bombard immature endosperm of maize, wheat and rice with a particle inflow gun (Finer et al., 1991). Via visualisation of the gene products or by HPLC, we will test the constructs used for stable transformation. The first construct to be tested will be phytoene synthase under the control of a CaMV 35S promoter and two different rice glutelin promoters (Okita et al., 1989). We will complement a phytoene-deficient maize mutant (*y1*), described by Buckner et al. (1991), to assess the proper function of our constructs. This work will be continued with the other necessary key enzymes tested in appropriate maize endosperm mutants before being used to transform Indica rice varieties.

Approach towards improvement of nutritional quality

Milled rice not only lacks vitamins, it is also deficient in essential amino acids such as cysteine, methionine and lysine. In higher plants, lysine is synthesised from aspartate via the aspartate biosynthetic pathway (Bryan, 1980) which also leads to the synthesis of methionine and threonine. The biosynthesis of lysine in plants is regulated by several feed-back loops. Dihydrodipicolinate synthase (DHPS) from *E. coli* is far less sensitive to feed-back inhibition than is DHPS from plants. This key enzyme in the lysine biosynthetic pathway has been expressed in chloroplasts of tobacco leaves (Shaul & Galili, 1992), and expression of the bacterial enzyme was accompanied by a significant increase in the level of free lysine. To study whether expression of the feed-back-insensitive enzyme in rice endosperm might lead to an increase in free lysine, we have fused endosperm-specific promoters from rice (Okita et al., 1989) to the *E. coli dap A* gene coding for DHPS (in collaboration with G. Galili, Rehovot, Israel). Putative transgenic clones are under selection.

In addition, an approach aimed at increasing the content of the essential sulphur-containing amino acids, cysteine and methionine, in milled rice will be undertaken. It will be based on the expression of the sunflower seed methionine-rich 2S albumin SFA8 (Kortt et al., 1991) under control of rice endosperm-specific promoters (Okita et al., 1989). The sunflower protein SFA8 is particularly suited for transfer and expression into food and forage plants which are deficient in sulphur-containing amino acids, since it is a single polypeptide chain with 8% of its residues as cysteine and 16% methionine (Kortt et al., 1991).

In our ongoing biotechnology research programme with forage grasses, we are following a similar approach to improve the supply of sulphur-containing amino acids limiting for wool production. This strategy is based on the expression, in vegetative parts of the forage plant, of the sunflower albumin SFA8, a rumen-resistant (and digestible in the abomasum) edible seed protein containing unusually high amounts of methionine (Kortt et al., 1991; Wandelt et al., 1992; T.J. Higgins, pers. comm.). Plasmids bearing a sunflower albumin SFA8 cDNA, including the ER-retention signal KDEL under the control of 'light-regulated' promoters have been generated and are currently being transferred into tall fescue and perennial ryegrass (in collaboration with T.J. Higgins, Canberra, Australia).

Approach towards fungal disease resistance

Rice blast (*Magnophorthe grisea*) and sheath blight (*Rhizoctonia solani*) are fungal diseases of rice that cause significant yield losses (Reissig et al., 1986; Toenniessen, 1991). Important productivity gains would be possible if these problems – in the case of rice blast particularly in association with upland drought – were overcome (Herdth, 1991). In addition, conventional approaches to improve these traits in rice have been ineffective even after substantial research (Herdth, 1991).

Multiple natural host response mechanisms, including the accumulation of defensive enzymes (e.g. chitinases, β-1,3-glucanases, etc.), are involved in plant resistance to phytopathogenic fungi (Boller, 1988). Chitinase preparations, especially in combination with β-1,3-glucanases, inhibit fungal growth *in vitro* (Mauch et al., 1988; Arlorio et al., 1992; Sela-Buurlage et al., 1993). Chitinases have also been shown to accumulate around invading hyphae *in planta* (Benhamou et al., 1990; Collinge et al., 1993). Transgenic approaches based on the constitutive expression of a bean endochitinase gene in tobacco (Broglie et al., 1991) and canola (Benhamou et al., 1993), or based on the wound-inducible expression of a barley seed ribosome-inactivating protein (RIP) in tobacco (Logemann et al., 1992) have been reported to lead to increased protection against *Rhizoctonia solani*. In addition, osmotin-like proteins inducible by osmotic stress, such as tobacco AP24 (Melchers et al., 1993), have been shown to be pathogen-induced proteins with inhibitory activity toward fungal pathogens (Woloshuk et al., 1991).

The development of gene transfer systems for Indica and Japonica rice opens up possibilities for testing the effects of expression of these candidate antifungal (and stress-response) genes as strategies to overcome sheath blight and upland drought/rice blast constraints. Putative transgenic rice clones are under selection for: a) barley RIP cDNA under control of the inducible rice chitinase promoter RCH10 (Zhu et al., 1993), and b) β-1,3-glucanase and chitinase driven by the Ti-1′2′ dual promoter (in collaboration with J. Mundy, Copenhagen, Denmark). In addition, research aimed at expressing in transgenic rice plants: a) tobacco osmotin-like AP24, b) bean endochitinase, and c) tobacco $\times$-1,3-glucanase, individually and in concert, has recently been initiated (in collaboration with G. Selman-Housein, Havana, Cuba).

Concluding remarks

The goal of our scientific work is to contribute to future sustained production of affordable and high quality food in developing countries. In this applied project, the aim is to work on problems which are a heavy burden on a great number of poor people and to apply biotechnology in such a way that it complements traditional plant breeding. We can reach our goal only if the novel characters we introduce into Indica rice are used in breeding programmes; this is guaranteed through our collaboration with IRRI. The novel characters will be successful in breeding only if they are stable and effective. This requires that we can provide breeders with a collection of many transgenic plants for each novel character to permit selection of the most appropriate individuals for the breeding programme. This in turn requires more efficient gene transfer protocols than those available to date.

Success or failure of our goal will, however, not only depends on success or failure of our experiments and subsequent breeding programmes. It will also depend on political and social circumstances in those countries in which the novel, genetically-engineered varieties are supposed to help to solve problems. Risk assessment will be an integral part of the projects. The judgement of scientists and national biosafety committees on the security of transgenic plants or of food produced from transgenic plants, however, will not necessarily lead to an acceptance of these plants or food by the local population. If we are, for example, able to recover a transgenic rice which accumulates sufficient provitamin A to prevent vitamin A deficiency, there is no guarantee that people would be willing to eat this rice. There is much educational and political work ahead of us in addition to what we are trying to achieve scientifically.

Acknowledgements

This work was supported by a grant from the Rockefeller Foundation Rice Biotechnology Program.

References

Argos, P, 1988 A sequence motif in many polymerases Nucleic Acids Res 16 9909–9916

Arlorio, M, A Ludwig, T Boller & P Bonfante, 1992 Inhibition of fungal growth by plant chitinases and β-1,3-glucanases A morphological study Protoplasma 171 34–43

Armstrong, G A, M Alberti, F Leach & J E Hearst, 1989 Nucleotide sequence, organization, and nature of the protein products of the carotenoid gene cluster of *Rhodobacter capsulatus* Mol Gen Genet 216 254–268

Bartley, G E, P V Vitanen, I Pecker, D Chamovitz, J Hirschberg & P A Scolnik, 1991 Molecular cloning and expression in photosynthetic bacteria of soybean cDNA coding for phytoene desat urase, an enzyme of the carotenoid biosynthesis pathway Proc Natl Acad Sci USA 88 6532–6536

Battacharyya-Pakrasi, M, J Peng, J S Elmer, G Laco, P Shen, M B Kaniewska, H Kononowicz, F Wen, T K Hodges & R N Beachy, 1993 Specificity of a promoter from the rice tungro bacilliform virus for expression in phloem tissues The Plant J 4 71–79

Benhamou, N, M H A J Joosten & P J G M de Wit, 1990 Subcellular localization of chitinase and of its potential substrate in tomato root tissues infected by *Fusarium oxysporium* f sp *radicis-lycopersici* Plant Physiol 92 1108–1120

Benhamou, N, K Broglie, I Chet & R Broglie, 1993 Cytology of infection of 35S-bean chitinase transgenic canola plants by *Rhizoctonia solani* cytochemical aspects of chitin breakdown *in vivo* The Plant J 4 295–305

Boller, T, 1988 Ethylene and the regulation of antifungal hydrolases in plants Oxf Surv Plan Mol Cell Biol 5 145–174

Bryan, J K, 1980 Synthesis of the aspartate family and branched chain amino acids In B J Miflin (Ed) The Biochemistry of Plants Volume 5, pp 403–452 New York Academic Press, New York

Broglie, R, G Coruzzi, G Lamppa, B Keith & N H Chua, 1983 Structural analysis of nuclear genes coding for the precursor to the small subunit of wheat ribulose-1,5-bisphosphate carboxylase Bio/Technology 1 55–61

Broglie, K, I Chet, M Holliday, R Cressman, P Biddle, S Knowlton, C J Mauvais & R Broglie, 1991 Transgenic plants with enhanced resistance to the fungal pathogen *Rhizoctonia solani* Science 254 1194–1196

Buckner, B, T L Kelson & D S Robertson, 1991 Cloning of the *y1* locus of maize, a gene involved in the biosynthesis of carotenoids The Plant Cell 2 867–876

Casacuberta, J M, P Puigdomenech & B San Segundo, 1991 A gene coding for a basic pathogenesis related (PR-like) protein from *Zea mays* Molecular cloning and induction by a fungus (*Fusarium moniliforme*) in germinating maize seeds Plant Mol Biol 16 527–536

Christensen, A H, R A Sharrock & P H Quail, 1992 Maize polyubiquitin genes structure, thermal perturbation of expression and transcript splicing, and promoter activity following transfer to protoplasts by electroporation Plant Mol Biol 18 675–689

Christou, P, L F Tameria & M Kofron, 1991 Production of transgenic rice (*Oryza sativa*) plants from agronomically important Indica and Japonica varieties via electric discharge particle acceleration of exogenous DNA into immature zygotic embryos Bio/Technology 9 957–962

Collinge, D B, K M Kragh, J D Mikkelsen, K K Nielsen, U Rasmussen & K Vad, 1993 Plant chitinases The Plant J 3 31–40

Covey, S N, 1986 Amino acid sequence homology in the gag region of reverse transcribing elements and the coat protein gene of cauliflower mosaic virus Nucleic Acids Res 14 623–633

Dasgupta, I, R Hull, S Eastop, C Poggi-Pollini, M Blakebrough, M I Boulton & J W Davies, 1991 Rice tungro bacilliform virus DNA independently infects rice after agrobacterium-mediated transfer J Gen Virol 72 1215–1221

Datta, S K, K Datta & I Potrykus, 1990 Fertile Indica rice plants regenerated from protoplasts isolated from microspore derived cell suspension Plant Cell Reports 9 253–256

Datta, S K, A Peterhans, K Datta & I Potrykus, 1990 Genetically engineered fertile Indica-rice recovered from protoplasts Bio/Technology 8 736–740

Datta, S K, K Datta, N Soltanifar, G Donn & I Potrykus, 1992 Herbicide resistant Indica rice plants from IRRI breeding line IR72 after PEG mediated transformation of protoplasts Plant Mol Biol 20 619–629

Datta, K, I Potrykus & S K Datta, 1992 Efficient fertile plant regeneration from protoplasts of the Indica rice breeding line IR72 (*Oryza sativa* L) Plant Cell Reports 11 229–233

De Rocquigny, H, C Gabus, A Vincent, M C Fournie-Zaluski, B Roques & J L Darlix, 1992 Viral RNA annealing activities of HIV 1 nucleocapsid protein require only peptide domains outside the zinc fingers Proc Natl Acad Sci USA 89 6472–6476

Delannay, X, B J La Valle, R K Proksch, R L Fuchs, S R Sims, J T Greenplate, P G Marrone, R B Dodson, J J Augustine, J G Layton & D A Fischhoff, 1989 Field performance of transgenic tomato plants expressing the *Bacillus thuringiensis* var *kurstaki* insect control protein Bio/Technology 7 1265–1269

dePater, S, L A M Hensgens & R A Schilperoort, 1990 Structure and expression of a light-inducible shoot-specific rice gene Plant Mol Biol 15 399–406

dePater, S B & R A Schilperoort, 1992 Structure and expression of a root specific rice gene Plant Mol Biol 18 161–164

Faust, R M & L A Bulla jr, 1982 Bacteria and their toxins as insecticides In E Kurstak (Ed) Microbial and Viral Pesticides Dekker, New York, pp 75–208

Ferre, J, M D Real, J Van Rie, S Jansens & M Peferoen, 1991 Resistance to the *Bacillus thuringiensis* bioinsecticide in a field population of *Plutella xylostella* is due to a change in a midgut membrane receptor Proc Natl Acad Sci USA 88 5119–5123

Fields, S & O-K Song, 1989 A novel genetic system to detect protein-protein interactions Nature 340 245–246

Finer, J J, P Vein, M W Jones & M D McMullen, 1992 Development of the particle inflow gun for DNA delivery to plant cells Plant Cell Reports 11 323–328

Fischhoff, D A, K S Bowdish, F J Perlak, P G Marrone, S M McCormick, J G Niedermeyer, D A Dean, K Kusano-Kretzmer, E J Mayer, D E Rochester, S G Rogers & R T Fraley, 1987

Insect tolerant transgenic tomato plants Bio/Technology 5 807–813
Fray, R G & D Grierson, 1993 Identification and genetic analysis of normal and mutant phytoene synthase of tomato by sequencing, complementation and co-suppression Plant Mol Biol 22 589–602
Fujimoto, H, K Itoh, M Yamamoto, J Kyozuka & K Shimamoto, 1993 Insect resistant rice generated by introduction of a modified δ-endotoxin gene of *Bacillus thuringiensis* Bio/Technology 11 1151–1155
Futterer, J, I Potrykus, M P Valles Brau, I Dasgupta, R Hull & T Hohn, 1994 Splicing in a plant pararetrovirus Virology (in press)
Ghosh Biswas, G C, V A Iglesias, S K Datta & I Potrykus, 1994 Transgenic indica rice (*Oryza sativa* L) plants obtained by direct gene transfer to protoplasts J Biotechnology (in press)
Gruissem, W, 1990 Of fingers, zippers and boxes Plant Cell 2 827–828
Hay, J M, M C Jones, M L Blakebrough, I Dasgupta, J W Davies & R Hull, 1991 An analysis of the sequence of an infectious clone of rice tungro bacilliform virus, a plant pararetrovirus Nucleic Acids Res 19 2615–2621
Herdt, R W, 1991 Research priorities for rice biotechnology In G S Khush & G H Toenniessen (Eds) Rice Biotechnology, pp 19–54 C A B International & IRRI, Wallingford/Manila
Hibino, H, M Roechan & S Sudarisman, 1978 Association of two types of virus particles with penyakit habang (tungro disease) of rice in Indonesia Phytopath 68 1412–1416
Hoffmann, C, H Vanderbruggen, J Van Rie, S Jansens & H Van Mellaert, 1988 Specificity of *Bacillus thuringiensis* δ-endotoxins is correlated with the presence of high-affinity binding sites in the brush border membrane of target insect midguts Proc Natl Acad Sci USA 85 7844–7848
Hohn, T & J Futterer, 1992 Pararetroviruses and retroviruses a comparison of expression strategies Seminars in Virology 2 55–69
Humphrey, J H, K P West Jr & A Sommer, 1992 Vitamin A deficiency and attributable mortality among under-5 year-olds Bulletin of the World Organization 70 (2) 225–232
IRRI Annual Report of the International Rice Research Institute, 1992 IRRI, Los Banos, The Philippines
Khush, G S & G H Toenniessen (Eds), 1991 Rice Biotechnology C A B International & IRRI, Wallingford/Manila
Kim, W T, X Li & T W Okita, 1993 Expression of storage protein multigene families in developing rice endosperm Plant Cell Physiol 34 595–603
Kortt, A A, J B Caldwell, G G Lilley & T J Higgins, 1991 Amino acid and cDNA sequences of a methionine -rich 2S protein from sunflower seed (*Helianthus annus* L) Eur J Biochem 195 329–344
Koziel, M G, G L Beland, C Bowman, N B Carozzi, R Crenshaw, L Crossland, J Dawson, N Desai, M Hill, S Kadwell, K Launis, K Lewis, D Maddox, K McPherson, M R Meghji, E Merlin, R Rhodes, G W Warren, M Wright & S T Evola, 1993 Field performance of elite transgenic maize plants expressing an insecticidal protein derived from *Bacillus thuringiensis* Bio/Technology 11 194–200
Lamppa, G, G Morelli & N H Chua, 1985 Structure and developmental regulation of a wheat gene encoding the major chlorophyll a/b binding protein Mol Cell Biol 5 1370–1376
Lee, L, R Schroll, H D Grimes & T K Hodges, 1989 Plant regeneration from indica rice (*Oryza sativa*) protoplasts Planta 178 325–333
Linden, H, A Vioque & G Sandmann, 1993 Isolation of a carotenoid biosynthesis gene coding for ζ-carotene desaturase from *Anabena* PCC 7120 by heterologous complementation FEMS Microbiol Lett 106 99–104
Logemann, J, G Jach, H Tommerup, J Mundy & J Schell, 1992 Expression of a barley ribosome-inactivating protein leads to increased fungal protection in transgenic tobacco plants Bio/Technology 10 305–308
Matzke, M & A J M Matzke, 1993 Genomic imprinting in plants Parental effects and trans-inactivation phenomena Annu Rev Plant Physiol Plant Mol Biol 44 53–76
Mauch, F, B Mauch-Mani & T Boller, 1988 Antifungal hydrolases in pea tissue II Inhibition of fungal growth by combinations of chitinase and β-1,3 glucanase Plant Physiol 88 936–942
McElroy, D, A D Blowers, B Jenes & R Wu, 1991 Construction of expression vectors based on the rice actin 1 (Act1) 5′ region for use in monocot transformation Mol Gen Genet 231 150–160
McGaughey, W H & M E Whalon, 1992 Managing Insect resistance to *Bacillus thuringiensis* toxins Science 258 1451–1455
Melchers, L S, M B Sela-Buurlage, S A Vloemans, C P Woloshuk, J S C Van Roekel, J Pen, P J M van den Elzen & B J C Cornelissen, 1993 Extracellular targeting of the vacuolar tobacco proteins AP24, chitinase and β-1,3-glucanase in transgenic plants Plant Mol Biol 21 583–593
Morellet, N, N Jullian, H De Rocquigny, B Maigret, J-L Darlix & B P Roques, 1992 Determination of the structure of the nucleocapsid protein NCp7 from the human immunodeficiency virus type 1 by ^{1}H NMR EMBO J 11 3059–3065
Murray, E E, J Lotzer & M Eberle, 1989 Codon usage in plant genes Nucleic Acids Research 17 477–493
Okita, T W, Y S Hwang, J Hnilo, W T Kim, A P Aryan, R Larson & H B Krishnan, 1989 Structure and expression of the rice glutelin multigene family J Biol Chem 264 12573–12581
Peferoen, M, 1991 Engineering of insect-resistant plants with *Bacillus thuringiensis* crystal protein genes In A M R Gatehouse, V A Hilder & D Boulter (Eds) Biotechnology in Agriculture 7, C A B International, pp 135–153
Penteado, M V C & L B Almeida, 1988 Occurrence of carotenoids in roots of five cultivars of cassava (*Manihot esculenta*, Crantz) from São Paulo Rev Farm Bioquim Univ S Paulo 24 39–49
Perlak, F J, R W Deaton, T A Armstrong, R L Fuchs, S R Sims, J T Greenplate & D A Fischhoff, 1990 Insect resistant cotton plants Bio/Technology 8 939–943
Potrykus, I, 1990 Gene transfer to plants assessment of published approaches and results Bio/Technology 8 535–542
Potrykus, I, 1993 Gene transfer to plants approaches and available techniques In M D Hayward, N O Bosemark & I Romagosa (Eds) Plant Breeding Principles and Prospects Chapmann & Hall, London, pp 126–137
Qu, R, M Bhattacharyya, G S Laco, A DeKochko, B L Subba Rao, M B Kaniewska, J S Elmer, D E Rochester, C E Smith & R N Beachy, 1991 Characterization of the genome of rice tun gro bacilliform virus comparison with commelina yellow mottle virus and caulimoviruses Virology 185 354–364
Raineri, D M, P Bottino, M P Gordon & E W Nester, 1990 Agrobacterium-mediated transformation & rice (*Oryza sativa* L) Bio/Technology 8 33–38
Ray, J, C R Bird, M J Maunders, D Grierson & W Schuch, 1987 Sequence of pTOM5, a ripening related cDNA from tomato Nucl Acids Res 15 1057
Reissig, W H, E A Heinrichs, J A Litsinger, K Moody, L Fiedler, T W Mew & A T Barrion, 1986 Illustrated Guide to Integrated Pest Management in Rice in Tropical Asia IRRI, The Philippines

Sela-Buurlage, M B , A S Ponstein, S A Bres-Vloemans, L S Melchers, P J M van den Elzen & B J C Cornelissen, 1993 Only specific tobacco (*Nicotiana tabacum*) chitinases and β-1,3 glucanases exhibit antifungal activity Plant Physiol 101 857–863

Shaul, O , G Gaalili, 1992 Increased lysine synthesis in tobacco plants that express high levels of bacterial dihydrodipicolinate synthase in their chloroplasts The Plant J 2 203–209

Toenniessen, G H , R W Herdt & L A Sitch, 1989 Rockefeller Foundations international network on rice biotechnology In A Mujeed-Kai & L A Sitch (Eds) Genetic Manipulation in Crops, pp 265–274 CIMMYT, Mexico, IRRI, Philippines

Toennissen, G H , 1991 Potentially useful genes for rice genetic engineering In G S Khush & G H Toenniessen (Eds) Rice Biotechnology, pp 253–280 C A B International & IRRI, Wallingford/Manila

Uchimiya, H , T Fushimi, H Hasimoto, H Harada, K Syono & Y Sugawara, 1986 Expression of a foreign gene in callus from DNA-treated protoplasts of rice (*Oryza sativa* L) Mol Gen Genet 204 204–207

Vaeck, M , A Reynarts, H Hofte, S Jansens, M De Beuckeleer, C Dean, M Zabeau, M Van Montague & J Leemans, 1987 Transgenic plants protected from insect attack Nature 327 33–37

Wandelt, C I , M R I Khan, S Craig, H E Schroeder, D Spencer & T J V Higgins, 1992 Vicilin with carboxy-terminal KDEL is retained in the endoplasmic reticulum and accumulates to high levels in the leaves of transgenic plants The Plant J 2 181–192

West, K P jr , G R Howard & A Sommer, 1989 Vitamin A and infection public health implications Annu Rev Nutr 9 63–86

Wilson, T M A , 1993 Strategies to protect crop plants against virus es Pathogen derived resistance blossoms Proc Natl Acad Sci USA 90 3134–3141

Woloshuk, C P, J S Meulenhoff, M Sela-Buurlage, P J M van den Elzen & B J C Cornelissen, 1991 Pathogen-induced proteins with inhibitory activity toward *Phytophthora infestans* The Plant Cell 3 619–628

Xhu, X & B Li, 1994 Fertile transgenic Indica rice plants obtained by electroporation Plant Cell Reports (in press)

Yanagisawa, S & K Izui, 1989 Maize phosphoenolpyruvate carboxylase involved in C4 photosynthesis nucleotide sequence and analysis in the 5′ flanking region of the gene J Biochem 106 982–987

Zhu, Q , P W Doerner & C J Lamb, 1993 Stress induction and developmental regulation of a rice chitinase promoter in transgenic tobacco The Plant J 3 203–212

Euphytica **85:** 451–455, 1995.

The integration of protoplast fusion-derived material into a potato breeding programme – a review of progress and problems

S. Millam, L.A. Payne & G.R. Mackay
Crop Genetics Department, Scottish Crop Research Institute, Invergowrie, Dundee DD2 5DA, U.K.

Key words: Solanum tuberosum, somatic hybridization, regeneration, asymmetric fusion

Summary

This paper reviews investigations into the application of protoplast fusion to the genetic and agronomic improvement of potato. Fusion studies involving *Solanum tuberosum* are reviewed under the categories of: fusion with wild relatives, dihaploid fusion and asymmetric strategies. The selection and characterisation of putative somatic hybrid material is identified as a critical stage in the process and certain specific aspects of this technology are identified. Future prospects for the wider uptake and integration of these techniques into breeding programmes are also discussed.

Introduction

Despite many significant scientific developments in plant biotechnology in recent years, the anticipated application of such technology to advancing agricultural production and biodiversity remains largely unfulfilled. Undoubtedly, some of the tools of molecular biology, such as marker-based technology, will provide greater precision in crop breeding programmes (Waugh & Powell, 1992), but one of the key areas in the realization of biotechnological promises to agriculture is in the application of plant cell and tissue culture techniques to the resolution of specific problems. Plant tissue culture technology offers a number of well-documented advantages over conventional approaches (Mantell et al., 1985). One of the most promising is the capability to circumvent practical breeding limitations by overcoming problems of sexual incompatibility between lines of interest using somatic hybridization techniques. Protoplast fusion also allows the transfer of both mono and polygenic traits between lines or species. Protoplasts have been isolated from a range of species and regeneration protocols have been developed for a large number of lines (Binding, 1985; Millam & Burns, 1993). The first report of protoplast regeneration in potato was by Shepard & Totten (1977) and the technology was expanded notably by Nelson et al. (1983) and Haberlach et al. (1985), amongst others. One of the most important uses of protoplasts is in somatic hybridization, where protoplasts of different cultivars or species can be induced to fuse and the resultant hybrid product, under certain defined conditions, regenerated into an intact plant. Protoplast fusion was first reported by Carlson et al. (1972) using *Nicotiana* species and, since then, many fusion studies have been reported, notably in the Cruciferae and Solanaceae. Indeed, the first application of electrofusion involving plant protoplasts of a food crop was the somatic hybridization of *Solanum tuberosum* with *S. phureja* (Puite et al., 1986).

Fusion studies involving *Solanum tuberosum*

Fusion with wild relatives

Many experiments using *Solanum tuberosum* for somatic hybridization have been designed to incorporate the resistance characteristics of wild diploid *Solanum* species into the *tuberosum* genome (e.g. Austin et al., 1985). Wild species that have been used as potential fusion partners include: *S. circaeifolium* – exhibiting resistance to *Phytophthora infestans* and *Globodera pallida* (Mattheij et al., 1992a), and *S. brev-*

idens showing potato leaf roll virus resistance (Preiszner et al., 1991). *S. brevidens* (a non-tuberizing species that will not readily hybridize with *S. tuberosum*) has been extensively used as a fusion partner and hybrids expressing late blight and potato leaf roll virus resistance genes have been reported (Helgeson et al., 1986). Frost tolerance and cold-hardening properties have also been transferred from *S. commersonii* to *S. tuberosum* by somatic hybridization (Cardi et al., 1993). All the wild species cited here are primitive diploid 1EBN species and thus incapable of crossing with tuberosum dihaploids, which are 2EBN (Hawkes & Jackson, 1992).

Dihaploid fusion

Protoplast fusion, in addition to presenting a technique for overcoming sexual incompatibility, facilitates the resynthesis of a tetraploid genotype from diploid or dihaploid parental lines. For heterozygous, clonally-propagated crops such as potato, the resynthesis of tetraploid genotypes from pre-selected donor dihaploids would enable the fixation of heterozygous gene combinations in a single step (Ross, 1986). Approaches include the fusion of *S. tuberosum* dihaploids (Waara et al., 1992; Mollers & Wenzel, 1992) to produce tetraploid potato hybrids, or the fusion of dihaploids of *S. tuberosum* and the cultivated diploid *S. phureja* (Mattheij & Puite, 1992). The latter paper also describes the results of field studies of such somatic hybrid material. A study involving combinations of tetraploid, transformed tetraploid, triploid and dihaploid lines has been reported (Deimling et al., 1988), claiming low levels of somaclonal variation under the conditions described, which would, if applicable to lines and conditions other than those used, allow a more rapid uptake of material into a conventional breeding programme.

Asymmetric strategies

Whilst these former symmetric fusion approaches may allow a rapid incorporation of certain traits from, for example, a wild species into an adapted, cultivated form compared with conventional breeding techniques, they are unlikely to lead to an immediate product i.e. a new cultivar. Such systems are more likely to provide access to genetic variation for the starting point of a 'conventional' breeding programme involving backcrossing and selection to the adapted form. Asymmetric hybrids offer an alternative approach, whereby limited gene transfer from one species (the donor) to another (the recipient) may allow the improvement of existing, adapted cultivars and produce a new, improved cultivar in one step – particularly in vegetatively-reproduced crops such as potato, tolerant of one or two additional chromosomes (Wilkinson, 1992). Asymmetric hybrids of tetraploid *S. tuberosum* and diploid *S. brevidens* have been obtained, for example, with the wild species acting as a donor after irradiation of its protoplasts by X-rays (Feher et al., 1992), and irradiated protoplasts of *S. pinnatisectum* were used in genetic reconstruction studies (Sidorov et al., 1987). In the extreme asymmetric fusion model, the nuclear genome of one of the genotypes can be entirely eliminated and the resultant products are termed cytoplasmic hybrids (cybrids) consisting of the nuclear genome of one genotype in the cytoplasm of another (Aviv et al., 1984). This technique has been used to introduce cytoplasmic male sterility into *Brassica* (Chuong et al., 1988) and is potentially of use in the production of potato cultivars from true botanic seed.

Selection and characterisation strategies for hybrid material

Most investigations have been performed using regenerated plants rather than cell colonies or microcalli due to technical limitations with the uncertainty of any regeneration response from the protoplasts being achieved, in many instances. In normal lines of agronomic interest, morphological or cytological markers are generally absent (Deimling et al., 1988). Methods of heterokaryon identification, however, were reported by Puite et al. (1988) and Waara et al. (1991) who, using micro-manipulator techniques coupled with, in the former report, light-bleached shoot culture protoplasts and, in the latter, a dual fluorescent system, selected hybrids after three to seven days culture. The selection and characterisation of regenerated (rather than at earlier stages of culture), presumptive somatic hybrid material has previously been undertaken by such methodology as intermediate morphology (Austin et al., 1986), isozyme markers (Gleba et al., 1984) and restriction fragment length polymorphisms (Pehu et al., 1989). RFLP approaches were further refined, with ribosomal DNA probes used for somatic hybrid identification (Pental et al., 1988; Rosen et al., 1988) and species-specific sequences of DNA used to analyze somatic fusions of *S. brevidens* and *S. tuberosum* (Pehu et al., 1990). All these approaches rely on the use of relatively large amounts of plant material, however. The possibility of early selection of hybrids of certain combinations based on the more vigorous growth of hybrid colonies (Debnath & Wenzel, 1987) has been substantiated by molecular evidence (Polgar et al., 1993). Recently, the use of PCR-based methodologies such as RAPD markers, has enabled identification of fusion products at a much earlier stage (Baird et al., 1992; Xu et al., 1993). These methods could be applied at the early microcalli stage, or, as DNA isolation and analysis methods are further refined, possibly at even earlier stages. The disadvantages remain in that the assay is destructive, and that the certainty of regeneration of an intact plant from such an early regeneration stage cannot be guaranteed. There also have been problems with the repeatability of PCR-based results.

Confirmation of the hybrid nature and the genome dosage of dihaploid fusions by Giesma-C banding were reported by Waara et al. (1992). The dosage response effects were also observed from a morphological assessment of the material. There have been several reports of morphological assessment analyzed by multivariate analysis (Preiszner et al., 1991) of parental and somatic hybrids, and morphological, cytological and molecular assessments of regenerants (Pehu et al., 1989; Mattheij et al., 1992). It seems, however, the rapid, reliable discrimination between hybrid and non-hybrid fusion products, at the earliest possible stage, remains a critical point in the routine use of somatic fusion in germplasm enhancement studies.

Incorporation of material into breeding programmes

The fertility of interspecific somatic hybrids was dependent on the phylogenetic distance between the parental species, and that sterility problems often arose (Harms, 1983; Feher et al., 1992). The use of 'intermediate hybrid material' for the practical incorporation of PLRV resistance from *S. brevidens* into *S. tuberosum*, via the crossing of somatic hybrid material with conventional breeding lines, was the subject of a report by Ehlenfeldt & Helgeson (1987), where it was found that lower fertility and abnormal chromosome pairing was a problem. Beyond the initial glasshouse assessments reported, an evaluation of sexual progeny derived from the somatic hybrids of *S. brevidens* and *S. tuberosum* (Helgeson et al., 1993) indicated that some of the sexual progeny were improved compared with the somatic hybrids, but still retained the disease resistance characteristic. Accordingly, it can be suggested that the methodologies for the production and selection of somatic hybrid material are applicable to generating material for use in a conventional breeding programme. Pehu et al. (1990a) detected resistance expression to PLRV and PVX in a range between the parents *S. brevidens* and *S. tuberosum*. They found that genotypes with morphology resembling *S. brevidens* were generally more resistant to viruses. They found that the hexaploid hybrids were divided into high and low resistance groups with the resistance being associated with two doses of the *S. brevidens* nuclear genome. Once again, however, those resembling *S. brevidens* in morphology exhibited higher resistance. The authors suggested from these findings that resistance/susceptibility to the different viruses was associated with the loss of individual *S. brevidens* or *S. tuberosum* chromosomes. If this is the case, then some means of incorporating the resistance genes from the wild species into the *S. tuberosum* genome by encouraging homoeologous chromosome exchange will be needed, and recent evidence (Wilkinson, personal communication) supports this possibility.

Discussion

Certain recurrent problems associated with the application of protoplast fusion technology, and other tissue culture-related problems directly applicable to potato programmes are highlighted. In general terms, the costs involved in developing appropriate skills, technologies and facilities are a feature of most plant cell and tissue culture work; certainly, to develop a programme of somatic hybridization without existing facilities and protocols would be an expensive and time-consuming undertaking. Specifically, protoplast isolation, purification, culture and regeneration protocols require continuous development for the specific genotypes under investigation. As yet, no truly genotype-independent regeneration systems exist. The inherent randomness of the protoplast fusion event is a further problem. Techniques for circumventing the inherent randomness of fusion are diverse and include the careful selection of specific parental lines for desired traits, and specific regeneration protocols favouring one parental genome.

Another difficulty is in the screening of somatic hybrid material, and identifying the products sought quickly and efficiently. This represents a major bottle-neck in the efficient use of somatic hybrid technology. In most of the cited identification techniques, the regenerated material has to be maintained for a substantial period before enough is available for the prescribed analysis. Molecular marker-based strategies may assist in reducing the time required for identification of hybrids, and early selection would also reduce the input of labour and resources required to maintain substantial amounts of material. An important but somewhat neglected area of research of great relevance to somatic hybridization technology is in detecting and monitoring chromosome loss from hybrid material though findings have been reported on potato protoclones (Fish & Karp, 1986). The advanced molecular techniques used for RAPD marker detection may assist in this area if markers could be assigned to individual chromosomes rather than the use of RFLP markers for such studies (William et al., 1990). The development of PCR-based protocols applicable to the specific set of circumstances required in a targeted, somatic hybridization programme, however, may take time, effort and facilities without the scope of such a programme. Similarly, advances in methodology for the early identification of somatic hybrids by the use of flow cytometry or refinements such as Flow Activated Cell Sorting (Fahleson et al., 1988; Millam & Burns, 1993), which may have an important role, are not widely available.

The outcome of a somatic hybridization programme can be the generation of a range of fairly diverse material containing a mix of the two parental genomes, plus or minus any changes attributable to somaclonal/protoclonal variation. Depending on the ploidy levels and chromosome complements, a series of backcrossing stages to introgress the required characteristic into the cultivar may be needed. This may provide a substantial time saving over conventional practices if, for example, the number of backcross generations can be reduced to two or three as opposed to sexual hybridization and backcrossing for six or seven generations to achieve similar ends. Another important potential advantage, particularly in the asymmetric approach, is that the recipient genome is retained intact whilst sexual hybridization will invariably result in a massive reassortment of genetic information via the process of meiosis and recombination, resulting in a loss of valuable and unique combinations of genes. The bridge between laboratory-based technology and the work of the plant breeder requires strategic input to enable protoplast fusion to contribute to the production of novel material. Finally, traits that serve as potential targets for incorporation into the cultivated potato are often of a polygenic nature and, as such, do not lend themselves readily to gene transfer by current *Agrobacterium*-mediated transformation technology methods. Thus the fundamental rationale of developing effective somatic fusion systems in potato has justification. Moreover, the products of cell fusion between related species of plants (which may also be currently achieved by 'traditional' methods, albeit with difficulty, perhaps via bridging crosses and techniques such as embryo rescue) are not currently subject to the restrictions imposed on the release of genetically-manipulated organisms.

Acknowledgements

We thank John Bradshaw for his helpful comments. SCRI is supported by the Scottish Office Agriculture and Fisheries Department.

References

Austin, S., M.A. Baer & J.P. Helgeson, 1985. Transfer of resistance to potato leaf roll virus from *Solanum brevidens* into *Solanum tuberosum* by somatic fusion. Plant Science 29: 75–82.

Austin, S., M.K. Ehlenfeldt, M.A. Baer & J.P. Helgeson, 1986. Somatic hybrids produced by protoplast fusion between *S. tuberosum* and *S. brevidens*: phenotypic variation under field conditions. Theor. Appl. Genet. 71: 628.

Aviv, D., P. Arzee-Gonen, S. Bleichman & E. Galun, 1984. Novel cytoplasmic *Nicotiana* plants by donor-recipient protoplast fusion: Cybrids having *N. tabacum* or *N. sylvestris* nuclear genomes and either or both plastomes and chondriomes from alien species. Mol. Gen. Genet. 196: 244–253.

Baird, E., S. Cooper-Bland, R. Waugh, M. De Maine & W. Powell, 1992. Molecular characterisation of inter- and intra-specific somatic hybrids of potato using randomly amplified polymorphic DNA (RAPD) markers. Mol. Gen. Genet. 233: 469–475.

Binding, H., 1985. Regeneration of plants. In: L.C. Fowke & F. Constabel (Eds). Plant Protoplasts, pp. 21–38. CRC Press, Boca Raton.

Cardi, T., F. D'Ambrosio, D. Consoli, K.J. Puite & K.S. Ramulu, 1993. Production of somatic hybrids between frost-tolerant *Solanum commersonii* and *S. tuberosum*: characterization of hybrid plants. Theor. Appl. Genet. 87: 193–200.

Carlson, P.S., H.H. Smith & R.D. Dearing, 1972. Parasexual interspecific plant hybridisation. PNAS USA 69: 2292.

Chuong, P.V., W.D. Beversdorf, A.D. Powell & K.P. Pauls, 1988. The use of haploid protoplast fusion to combine cytoplasmic atrazine resistance and cytoplasmic male sterility in *Brassica napus*. Pl. Cell Tiss. Org. Cult. 12: 181–184.

Debnath, S.C. & G. Wenzel, 1987. Selection of somatic fusion products in potato by hybrid vigour. Potato Research 30: 371–373.

Deimling, S., J. Zitzlsperger & G. Wenzel, 1988. Somatic fusion for breeding of tetraploid potatoes. Plant Breeding 101: 181–189.

Ehlenfeldt, M K & J P Helgeson, 1987 Fertility of somatic hybrids from protoplast fusions of *Solanum brevidens* and *Solanum tuberosum* Theor Appl Genet 73 395–402

Fahleson, J , J Dixelius, E Sunberg & K Glimelius, 1988 Correlation between flow cytometric determination of nuclear DNA content and chromosome number in somatic hybrids within Brassicaceae Pl Cell Rep 7 74–77

Feher, A , J Preiszner, Z Litkey, Gy Csanadi & D Dudits, 1992 Characterization of chromosome instability in interspecific somatic hybrids obtained from X-ray fusion between potato (*Solanum tuberosum* L) and *S brevidens* Phil Theor Appl Genet 84 880–890

Fish, N & A Karp, 1986 Improvements in regeneration from protoplasts of potato and studies on chromosome stability 1 The effect of initial culture medium Theor Appl Genet 72 405–412

Gleba, Y , N N Kolesnik, I V Meshkene, N N Cherep & A S Parokonny, 1984 Transmission genetics of the somatic hybridisation process in *Nicotiana* Theor Appl Genet 69 121–128

Haberlach, G T , B A Cohen, N A Reichert, M A Baer, L E Towill & J P Helgeson, 1985 Isolation, culture and regeneration of protoplasts from potato and several related *Solanum* species Plant Science 39 67–74

Harms, C T , 1983 Somatic incompatibility in the development of higher plant somatic hybrids Q Rev Biol 58 325–353

Hawkes, J G & M T Jackson, 1992 Taxonomic and evolutionary implications of the Endosperm Balance Number hypothesis in potatoes Theor Appl Genet 84 180–185

Helgeson, J P , G J Hunt, G T Haberlach & S Austin, 1986 Somatic hybrids between *Solanum brevidens* and *Solanum tuberosum* Expression of a late blight resistance gene and potato leaf roll resistance Plant Cell Rep 3 212–214

Helgeson, J P , G T Haberlach, M K Ehlenfeldt, G Hunt, J D Pohlman & S Austin, 1993 Sexual progeny of somatic hybrids between potato and *Solanum brevidens* potential for use in breeding programs Am Pot Journal 70 438–452

Mantell, S H , J A Mathews & R A McKee, 1985 Principles of Plant Biotechnology Blackwell Scientific Publishing (Oxford)

Mattheij, W M & K J Puite, 1992 Tetraploid potato hybrids through protoplast fusions and analysis of their performance in the field Theor Appl Genet 83 807–812

Mattheij, W M , R Eijlander, J R A de Koning & K M Louwes, 1992a Interspecific hybridization between the cultivated potato *Solanum tuberosum* subspecies *tuberosum* L and the wild species *S circaeifolium* subsp *circaeifolium* Bitter exhibiting resistance to *Phytophthora infestans* (Montd) de Bary and *Globodera pallida* (Stone) Behrens Theor Appl Genet 83 466

Millam, S & A T H Burns, 1993 Plant protoplast technology In A T H Burns (Ed) Strategies for Engineering Organisms, BIOTOL Series, Butterworth-Heinemann (London) 93–124

Mollers, C & G Wenzel, 1992 Somatic hybridization of dihaploid potato protoplasts as a tool for potato breeding Bot Acta 105 133–139

Nelson, R S , G P Creissen & S W J Bright, 1983 Plant regeneration from protoplasts of *Solanum brevidens* Plant Sci Letts 30 355–367

Pehu, E , A Karp, K Moore, S Steel, R Dunckley & M G K Jones, 1989 Molecular, cytogenetic and morphological characterization of somatic hybrids of dihaploid *Solanum tuberosum* and diploid *S brevidens* Theor Appl Genet 78 696–704

Pehu, E , M Thomas, T Poutala, A Karp & M G K Jones, 1990 Species-specific sequences in the genus *Solanum* identification, characterisation and application to study somatic hybrids of *S brevidens* and *S tuberosum* Theor Appl Genet 80 696–698

Pehu, E , R W Gibson, M G K Jones & A Karp, 1990a Studies on the genetic basis of resistance to potato leaf roll virus, potato virus Y and potato virus X in *Solanum brevidens* using somatic hybrids of *Solanum brevidens* and *S tuberosum* Plant Science 69 95–101

Pental, D , A Mukhopadhyay, A Grover & A K Pradham, 1988 A selection method for the synthesis of triploid hybrids by fusion of microspore protoplasts (n) with somatic cell protoplasts (2n) Theor Appl Genet 76 237–243

Polgar, Zs, J Preiszner, D Dudits & A Fehrer, 1993 Vigorous growth of fusion products allows highly efficient selection of interspecific potato somatic hybrids molecular proofs Pl Cell Rep 12 399–402

Preiszner, J , A Feher, O Veisz, J Sutka & D Dudits, 1991 Characterization of morphological variation and cold resistance in interspecific somatic hybrids between potato (*Solanum tuberosum* L) and *S brevidens* Phil Euphytica 57 37–49

Puite, K J , S Roest & L P Pijnacker, 1986 Somatic hybrid potato plants after electrofusion of diploid *Solanum tuberosum* and *S phureja* Plant Cell Rep 5 262–265

Puite, K J , W T Broeke & J Schaart, 1988 Inhibition of cell wall synthesis improves flow cytometric sorting of potato heterofusions resulting in hybrid plants Plant Sci 56 61–68

Rosen, B , C Hallden & W K Heneen, 1988 Diploid *Brassica napus* somatic hybrids characterisation of nuclear and organellar DNA Theor Appl Genet 76 197–203

Ross, H , 1986 Potato breeding – problems and perspectives – Advances in plant breeding 13 (Suppl J Plant Breeding) Verlag Paul Parey, Berlin and Hamburg

Shepard, J F & R E Totten, 1977 Mesophyll cell protoplasts of potato Isolation, proliferation and plant regeneration Plant Physiol 60 313–316

Sidorov, V A , M K Zubko, A A Kuchko, I K Komarnitsky & Y Y Gleba, 1987 Somatic hybridisation in potato use of irradiated protoplasts of *Solanum pinnatisectum* in genetic reconstruction Theor Appl Genet 74 364–368

Waara, S , A Wallin & T Eriksson, 1991 Production and analysis of intraspecific somatic hybrids of potato (*Solanum tuberosum* L) Plant Science 75 107–115

Waara, S , L Pijnacker, M A Ferwada, A Wallin & T Eriksonn, 1992 A cytogenetic and phenotypic characterization of somatic hybrid plants obtained after fusion of two different dihaploid clones of potato (*Solanum tuberosum* L) Theor Appl Genet 85 470–479

Waugh, R & W Powell, 1992 Using RAPD markers for crop improvement TIBTECH 10 186–191

Wilkinson, M J , 1992 The partial stability of additional chromosomes in *Solanum tuberosum* cv Torridon Euphytica 60 115–122

Williams, C E , G J Hunt & J P Helgeson, 1990 Fertile somatic hybrids of *Solanum* species RFLP analysis of a hybrid and its sexual progeny from crosses with potato Theor Appl Genet 80 545–551

Xu, Y , M S Clark & E Pehu, 1993 Use of RAPD markers to screen somatic hybrids between *Solanum tuberosum* and *S brevidens* Plant Cell Reports 12 107–109

Euphytica **85:** 457–464, 1995.

Potato germplasm enhancement for resistance to biotic stresses at CIP. Conventional and biotechnology-assisted approaches using a wide range of *Solanum* species

Kazuo N. Watanabe[1,2], Matilde Orrillo[1] & Ali M. Golmirzaie[1]
[1] *The International Potato Center, Apartado 1558, Lima, Peru;* [2] *Department of Plant Breeding and Biometry, Cornell University, Ithaca, NY 14853-1902, U.S.A.*

Key words: genetic engineering, introgression, molecular markers, potatoes, resistances, *Solanum*, technology transfer

Summary

Potato genetic improvement has been facilitated using new knowledge of potato reproductive biology and new techniques. Many wild diploid species as well as landrace cultivars have been used in breeding at the diploid level, a strategy which is supported by 1) 2n gametes and 2) haploids from tetraploid cultivars. Different categories of wild species which have been under-utilized are now being exploited further in more systematic enhancement programmes using semi-conventional and biotechnological methods. Molecular maps of the potato genome are used actively to achieve marker-assisted introgression and improved selection among the germplasm collections to facilitate the use of valuable wild genetic resources. As an alternative method to incorporate a high level of resistance, genetic engineering has been employed to facilitate the initial breeding process using various gene constructs for controlling major biotic stresses in the world.

Introduction

Relatively narrow genetic diversity from landraces and a limited extent of wild germplasm were employed in conventional potato breeding for resistance up to the mid 1970s (Ross, 1986). While a high level of resistance to e.g. potato virus X (PVX) and Y (PVY), which are single dominant inherited traits, was incorporated from wild species, only a small proportion of the existing huge germplasm collections in genebanks was used (Ross, 1986; Watanabe, 1991, 1994a). Most wild species were under-exploited for their potential in providing highly important traits, such as resistance to biotic stress because of the lack of systematic enhancement methods. Furthermore, a high level of resistance to cucumber mosaic virus (CMV), one the most devastating diseases of vegetable crop species worldwide (Valkonen et al., 1994a), was found in potato germplasm. Thus, exotic potato genetic resources in a usable form may be very important in crops other than tuber-bearing *Solanum* taxa for crop genetic improvement (Watanabe et al., 1995).

Introgressing such valuable resistance characters into potato varieties is particularly important for growers in developing countries as they have relatively limited resources to control pests in their Integrated Pest Management (IPM) systems. A resistant variety could alleviate such limitations and simplify the cropping system for farmers with restricted access to crop protection and chemicals and/or following 'modern' concepts. Even in the industrialized regions of the world, concerns regarding environmental and sustainability issues are increasing and the use of pesticides is gradually being restricted by regulatory agencies as well as by the perception of the general public. Thus, avoiding the bottlenecks in potato breeding by exploiting the wild genetic resources which provide resistance, is of major importance, and one of the essential activities at CIP.

Constraints in conventional potato breeding

The major constraints in potato breeding have been discussed widely and the following are aspects to be alleviated or improved (Ross, 1986; Watanabe, 1991, 1994a). 1) The genetic base of the present potato cultivars is narrow, and could be very susceptible to newly-arising pests as exemplified by potato late blight in Ireland which resulted in the Great Irish Famine. 2) Tetraploid genetics is far more complicated than the diploid situation, exacerbated by the outcrossing nature of the tetraploid cultivars which causes further complications during segregation. 3) The breeding objectives often involve quantitative traits with large environmental effects. 4) Interlocus interaction (epistasis) and heterozygosity are important factors, while additive components also contribute to quantitative traits. 5) No major information on genetic markers was available to breeders until DNA markers were developed. 6) Identification of individual chromosomes is very cumbersome by conventional cytogenetic methods, while monitoring recombination and introgression is very difficult by phenotypic evaluation.

Modified conventional germplasm enhancement at the diploid level using haploids and 2n gametes

Landmark achievements in potato genetic improvement have occurred in the past two decades by establishing systematic germplasm enhancement methods and subsequent breeding approaches. Germplasm enhancement with diploid tuber-bearing *Solanum* species, including some diploid cultivated taxa has been conducted widely in many potato breeding programmes (Hermsen, 1987; Ortiz et al., 1994; Peloquin et al., 1989; Ross,1986; Watanabe, 1991, 1994a). These methods alleviate some bottlenecks in conventional breeding at the tetraploid level. This involves two major tools; 1) haploids from 4× cultivars and 2) 2n gametes. Haploids from tetraploids can be obtained easily by pseudogametic parthenogenesis using a haploid-inducing pollinator (Hermsen & Verdenius, 1973), and are used to introduce into cultivated potato valuable genes from wild or exotic germplasm. Breeding at the diploid level with disomic inheritance facilitates simultaneous selection for target resistance traits as well as some agronomic characters (Peloquin et al., 1989; Ortiz et al., 1994). Gametes with a sporophytic chromosome number, i.e. 2n gametes, occur spontaneously and the inheritance of traits is simple (Watanabe & Peloquin, 1989, 1993; Werner & Peloquin, 1990). Transmission of important traits from diploid breeding lines is successfully achieved using 2n gametes, especially first division restitution 2n pollen which occurs widely and frequently in the diploid genera of tuber-bearing *Solanum* species (Watanabe & Peloquin, 1989, 1993). Furthermore, the concept of the Endosperm Balance Number (EBN) has greatly assisted prediction of success in interspecific and/or interploidy crosses (Johnston et al., 1980).

Major achievements were made on quantitative disease and insect pest resistance using the above strategy. These traits were acquired from wild species and transmitted to tetraploid breeding lines with in 5–8 years via germplasm enhancement methods (Watanabe et al., 1991; Ortiz et al., 1994; Watanabe, 1994a). Major achievements using the scheme were made on introgression of quantitative resistance to: bacterial wilt (Watanabe et al., 1992b), early blight (Ortiz et al., 1994), potato tuber moth (Ortiz et al., 1990; Watanabe et al., 1993a, b), and root-knot nematode (Iwanaga et al., 1989; Watanabe et al., 1993a, b).

Diverse diploid germplasm can be used for improvement of tetraploid breeding stocks and cultivars through the use of disomic inheritance, 2n gametes and haploids from tetraploids (Watanabe et al., 1994b). Thus, knowledge of the reproductive biology of potato and its wild relatives has provided more effective ways to introduce valuable genetic characters from wild species into the cultivated potato genepool over the past two decades at CIP (Ortiz et al., 1994; Watanabe, 1991, 1994a). For further details see Peloquin et al. (1989), Watanabe (1991, 1994a) and Ortiz et al. (1994).

Interspecific sexual hybridization between non tuber-bearing and tuber-bearing species

Accumulated knowledge of the reproductive biology and genetics of *Solanum* species has made it possible to use disomic tetraploid species (2 EBN) (Iwanaga et al., 1991; Watanabe et al., 1992a) and diploid wild species (1 EBN) including some non tuber-bearing species (Watanabe et al., 1992a, 1994a), which could not be crossed directly with cultivated potato lines. Protoplast fusion is another approach to the utilization of distantly-related taxa (Watanabe et al., 1995; Wenzel, 1994). Since many discussions on somatic hybrids have been reported in other papers by CIP collaborators (Dodds, 1991; Wenzel, 1994), we would like to describe the use of sexual filial progeny in interspecific hybridization here. Previous approaches including somatic hybridization required complicated ploidy manipulation and/or bridging to introgress the valuable genes from these species; using now new strategies interspecific crosses can be made by simpler methods.

Low profile biotechnology such as embryo rescue combined with rescue pollination (second compatible pollination) will break down major cross-incompatibility barriers in many wild species which contain potentially valuable resistance genes not present in the gene pool of cultivated potatoes (Iwanaga et al., 1991; Watanabe et al., 1991, 1992a, b, 1994a, 1995).

Recent development of the enhancement methods will make the introgression genetically and logistically more efficient than previous conventional breeding schemes with tetraploid wild species such as *S. acaule* and *S. stoloniferum* (Watanabe et al., 1992, 1992a, 1994a) and the diploid non tuber-bearing species such as *S. brevidens* (Watanabe et al., 1995); generalization of these methods should be exploited to achieve genotype-independent success. Alleviating problems in obtaining successful filial generations using different wild species also permitted an opportunity for an intriguing comparison with somatic fusion-derived hybrids with respect to genetic efficiency, using such species as *S. brevidens* (Watanabe et al., 1995). The method may have additional value, namely to exploit additional genes derived from the rescue pollinator such as IvP 35 (Hermsen & Verdenius, 1973; Iwanaga et al., 1991; Watanabe et al., 1992a, 1994a, 1995) by enhancing intergenomic translocation (Clulow et al., 1993; Valkonen et al., 1994c) between the tuber-bearing and non tuber-bearing species in sexual hybrids.

Application of flow cytometry

Introgression of filial generations from distant interspecific crosses often induces aneuploids (Ross, 1986; Watanabe, 1994a; Watanabe et al., 1994a). Identification of such aneuploids can be conducted by regular light microscopy using simple chromosome staining, but the work is very cumbersome unless the operator is sufficiently skilled. Employment of flow cytometry may accelerate the identification of aneuploids compared with the experience and skill-dependent microscopic method. A recent study on flow cytometry measurement of aneuploid potato lines of somatic fusion-derived backcross progeny, indicated that these aneuploids could be identified with ease by this method using thousands of cell samples in contrast to 10–20 cells per sample in light microscopy to survey such aneuploids (Valkonen et al., 1994b). Further required the procedures to make a chromosome count using root tips involve time-consuming pre-treatment, fixation and hydration, which would take up to 2 days before making observations (Watanabe & Orrillo, 1993). On the other hand, flow cytometry merely requires chopping leaf tissues, and it should be possible to measure up to 30 clonal samples per hour, while a standard chromosome count by an experienced person would provide data on 30 clones per day (Valkonen et al., 1994b).

Molecular marker-assisted introgression and selection systems in potato genetic improvement

DNA markers such as RFLP, and PCR-based tools such as RAPD, SCAR and DAF, are useful in understanding the structure of the plant genome (Kesseli et al., 1992; Tanksley et al., 1992; Watanabe, 1994b). These tools have provided an extensive opportunity to comprehend the complicated genetic nature of potatoes by monitoring introgression, especially with RFLP and RAPD, to facilitate the introgression of genes from exotic germplasm and to aid selection in potato breeding in general. CIP has received strong collaborative support from several leading research groups on the use of marker technologies (Table 1).

The present molecular map of the potato has been constructed mainly using RFLP markers in a joint project with several major research groups (reviews in Tanksley et al., 1992; Watanabe, 1994b). Map positions of these markers are well-conserved, compared with alignment of the markers on the tomato molecular map, except for five major inversions on five chromosomes (Tanksley et al., 1992). Thus, the tomato map information can be applied directly to the ongoing potato-mapping activities. These potato maps are being integrated in a comprehensive analysis of the potato genome and in applications to potato germplasm enhancement and breeding.

An employment of molecular markers provided a clearer picture and accelerated selection even, for single, dominant inherited traits such as resistance to PVX (Ritter et al., 1991) which is actually controlled by

Table 1. Status of representative potato molecular marker research in the collaborative network supported by CIP

Subject	Target traits	Status	Future plan	Collaborator
Basic marker technology	QTLs in resistance to insects and markers for virus resistances	Significant markers identified for glandular trichomes and Colorado potato beetles Preliminary chromosome(s) identified for PVY loci	Testing the present significant markers for breeding populations and obtaining of more significant markers for PCR assays Central Database supported through Solgene of USDA	Cornell University Univ. Helsinki, Finland
Basic marker technology	QTLs in resistance to late blight and/or cyst nematodes	Significant chromosome regions identified Other mapping populations from different resistance sources developed	Further characterizations of the regions Comparison of chromosome regions among different mapping populations	Max Planck, Germany SCRI, UK CPRO, the Netherlands
Biosystematics/Genetic resources management	Genetic diversity	Phylogenetic information is confirmed with previous taxonomic information	Core collection establishment and determination of population size for true seed preservation	Univ. of Wisconsin, USA Kobe Univ., Japan
Basic marker technology	Monitoring introgression of genes from wild species and mapping of pest resistance genes	Introgression of chrom. segments from species are monitored. Preliminary identification of resistance genes from wild species *S. brevidens*; PLRV *S. bulbocastanum*, RKN	Further characterization of introgression lines	Univ. Helsinki, Finland (*S. brevidens*) Univ. of Washington and USDA, USA (*S. bulbocastanum*)
Application of technology	Marker assisted selection of insect resistance	Breeding populations have been selected for marker testing	Applying markers to different populations	UPLB, Philippines INIA, Chile INTA, Argentina KARI, Kenya
Application of technology	Marker assisted selection on late blight resistance	Breeding populations have been distributed for phenotypic resistance testing	Testing markers provided from the above collaborative groups	UPLB, Philippines INIA, Chile INTA, Argentina KARI, Kenya

two independent loci that are derived from different species. Furthermore, a recent preliminary study by Valkonen et al. (1994c, 1994d) indicated that at least three independent loci from different species could be involved in extreme resistance to PVY. This can be further verified by the use of molecular markers and this finding will be used to facilitate the selection of PVY-resistant genotypes at an early seedling stage without tedious virus inoculation for phenotypic selection. The concept and strategy may be applied generally to other important traits to facilitate breeding processes.

Many quantitatively segregating resistance traits have been under-exploited due to: 1) inadequate screening methods, 2) the complicated nature of the inheritance, and 3) difficulties in monitoring introgression from wild species into the cultivated potato gene pool. Molecular marker information on QTLs is being used to facilitate genotypic selection in potato breeding instead of relying on cumbersome phenotypic evaluation. Joint efforts are in progress to select simultaneously for quantitative traits such as the glandular trichomes, tuberization traits, tuber dormancy etc. in a diploid breeding population at Cornell University and CIP through a collaborative research project (Bonierbale et al., 1994; Watanabe, 1994b). While more intensive molecular genetic information is essential for the accurate detection and employment of such quantitative traits, advances have been made on the marker-assisted selection of such traits (Watanabe, 1994b).

The technologies available are to be applied to marker-assisted selection in potato breeding. For this purpose and to achieve the mission of CIP to help client programmes, especially in developing countries, an informal working group using already-existing potato regional networks has been formed with some representative programmes in developing countries such as Kenya, The Philippines and Chile as well as involving some advanced institutions such as Cornell University, Scottish Crop Research Institute, and The Centre for Plant Breeding and Reproductive Biology in The Netherlands on the use of this technology. This is supported by the UNDP biotechnology programme for research and development collaboration by various groups advanced research institutions generate basic technologies on molecular markers; CIP acts as a catalyst for marker technologies and as a facilitator in searching for essential resources, and the client developing countries apply the technologies for alleviating problems in potato breeding. Emphasis is placed on a multilateral flow of information and technology requiring extensive feedback from the users in developing countries. Focus is on two breeding targets for quantitative resistance: one on general insect resistance provided by glandular trichomes (Bonierbale et al., 1994) and another on the global disease constraints, including late blight (Leonard-Schippers et al., 1994). By sharing the same genetic resources and information for testing the use of markers on the target traits in multi-location experiments in different countries, international information is exchanged to improve the marker-assisted selection scheme.

The glandular trichome project employs the map information from the backcross one (BC1) generation to test the significant markers in the breeding populations with an *S. berthaultii* background. Furthermore, using graphical genotyping (Young & Tanksley, 1988), BC1 genotypes were selected to carry out further experimental introgression to recover cultivated traits, while retaining the chromosome regions responsible for the trichome characters (Yencho et al., Cornell Univ., unpublished data; Watanabe, 1994b).

Genetic engineering on potatoes; global partnerships with various groups

Since 1985, CIP has formed an international network to enhance activities in genetic engineering (Dodds, 1991; Dodds & Jaynes, 1987). The major targets and status of the collaborative research network are listed in Table 2. In-house research has resulted in the development of transgenic potato genotypes, while several technical aspects require improvement. Antibacterial lytic peptide genes for example (Dodds & Jaynes, 1987; Beltran-Destefano et al., 1992), were introduced into tetraploid potato cultivars by *Agrobacterium*-mediated transformation (Watanabe et al., 1993c). No prominent phenotypic resistance was obtained, however, by inoculating the transgenic plants with *Pseudomonas solanacearum* which is the causal agent of bacterial wilt, the most devastating bacterial disease in tropical and subtropical regions. This problem may be attributed to various factors such as genotype dependence, the type and method of making promoters, vector, constructs, etc.

CIP now focuses on making a representation on biotechnology transfer from the private sector to developing countries by collaborating, for example, with Plant Genetics System (PGS) in developing transgenic potato varieties with *Bacillus thuringiensis* (*Bt*) endotoxin genes. All technology involving *Bt* genes (Peferoen, 1992) is donated by PGS, while CIP catalyzes the technology transfer to the client national programmes.

Biosafety is one of the concerns in the use of transgenic crops (Persley et al., 1991; Lesser & Maloney, 1993). Local communities in each country may have a different perception of transgenic crops (Macer, 1994), so that social and political aspects as well as biological safety are very important issues to consider. With assistance from various international organizations and the understanding of client countries as well as the autonomous establishment of regulatory guidelines by local governments in developing countries, CIP has made some progress in clearing the way for field testing of transgenic potatoes. Trials on *Bt* transgenic potatoes

Table 2. Status of representative potato genetic engineering work in the collaborative research network supported by CIP

Target constraint	Type of gene(s)	Status	Future plan	Collaborator(s)
Bacterial disease (Bacterial wilt and *Erwinia* spp)	Three groups of antibacterial lytic peptide genes	Transgenic plants made and phenotypic resistance tested: Low level of resistance expressed	Improve gene expression and phenotypic resistance	Louisiana State Univ., USA Univ. Pacifico, Chile
Fungal disease (Late blight)	The same as above	The same as above	The same as the above	Univ. of Tuscia, Italy Louisiana State Univ., USA
PLRV	Coat protein	Transgenic plants made and low level of phenotypic resistance observed	Different strategies to use alternative types of viral genes such as replicase genes	ILTAB (Scripp Inst.), USA Cornell Univ., USA
PVY and PVX	Coat protein	Transgenic plants made and modest phenotypic resistance observed	Field testing of transgenic crops	ILTAB (Scripp Inst.), USA
Insects	*Bt* endotoxin genes	Transgenic plants made and phenotypic resistance observed in quarantine conditions	Field testing in several countries	PGS, Belgium Tunisia, Egypt, Peru

will be initiated in 1995 in North African countries and Peru by a interdisciplinary team which consists of scientific and regulatory representatives from client national programmes, CIP staff and external international consultants.

Conclusion

CIP has been generating breeding lines and true botanic seed populations towards cultivar development in client countries using precious wild and landrace genetic resources. With the employment of advanced technology such as molecular markers and flow cytometry, introgression can be easily monitored to accelerate breeding processes. While safety issues are of concern for transgenic potatoes, genetically-engineered crops may offer an alternative strategy to improve potatoes without using conventional genetic resources.

Acknowledgements

The authors would like to thank Dr. R.L. Plaisted, Dr. R. Tenney, Dr. K.V. Raman, Dr. T. Suzuki and Ms. J.A. Watanabe at Cornell University for their careful reading of the manuscript.

References

Bonierbale, M.W., R.L. Plaisted, O. Pineda, S.D. Tanksley, 1993. QTL analysis of trichome-mediated insect resistance in potato. Theor. Appl. Genet. 87: 973–987.

Clulow, S.A., M.J. Wilkinson & L.R. Burch, 1993. *Solanum phureja* genes are expressed in the leaves and tubers of aneusomatic potato dihaploids. Euphytica 69: 1–9.

Destefano-Beltran, L., P. Nagpala, S. Cetiner, J.H. Dodds & J.M. Jaynes, 1990. Enhancing bacterial and fungal disease resistance in plants. In: M. Vayda & W. Park (Eds). The Molecular and Cellular Biology of the Potato. C.A.B. International, Wallingford, UK, pp. 205–221.

Dodds, J.H., 1991. Molecular Methods for Potato Improvement. Rept. Plan. Conf. 'Application of Molecular Techniques to Potato Germplasm Enhancement'. CIP, Lima, 181 p.

Dodds, J.H. & J.M. Jaynes, 1987. Crop plant genetic engineering: Science Fiction to science fact. Outlook on Agriculture 16 (3): 111–115.

Hermsen, J.G.Th., 1987. Efficient utilization of wild and primitive species in potato breeding. In: G.J. Jellis & D.E. Richardson (Eds). The Production of New Potato Varieties. Technological Advances, Cambridge University Press, London, pp. 172–185.

Hermsen, J.G.Th. & J. Verdenius, 1973. Selection from *Solanum tuberosum* group Phureja of genotypes combining high frequency haploid induction with homozygosity for embryo spot. Euphytica 22: 244–259.

Iwanaga, M., R. Freyre & K. Watanabe, 1991. Breaking the cross-ability barriers between disomic tetraploid *Solanum acaule* and tetrasomic tetraploid *S. tuberosum*. Euphytica 52: 183–191.

Iwanaga, M., P. Jatala, R. Ortiz & E. Guevara, 1989. Use of FDR 2n pollen to transfer resistance to root-knot nematodes into cultivated 4× potatoes. J. Amer. Soc. Hort. Sci. 114: 1008–1013.

Johnston, S.A., T.P.M. den Nijs, S.J. Peloquin & R.E. Hanemann Jr., 1980. The significance of genic balance to endosperm development in interspecific crosses. Theor. Appl. Genet. 57: 5–9.

Kesseli, R.V., I. Paran & R.W. Michelmore, 1992. Efficient mapping of specifically targeted genomic regions and the tagging of these regions with reliable PCR-based genetic markers. Proc. Symp. Applications of RAPD Technology to Plant Breeding. CSSA-ASHS-AGA, Minneapolis, Minnesota, Nov. 1, 1992, p. 31–36.

Leonards-Schippers, C.W. Gieffers, R. Schäfer-Pregl, E. Ritter, S.J. Knapp, F. Salamini & C. Gebhardt, 1994. Quantitative resistance to *Phytophthora infestans* in potato: A case study for QTL mapping in an allogamous plant species. Genetics 137: 67–77.

Lesser, W. & A.P. Maloney, 1993. Biosafety: A report on regulatory approaches for the deliberate release of genetically-engineered organisms. Cornell International Institute for Food, Agriculture and Development, Ithaca, USA, 68 pp.

Macer, D.R.J., 1994. Attitude to genetic Engineering Eubios Ethics Institute, Christchurch, New Zealand, 164 pp.

Ortiz, R., M. Iwanaga, K.V. Raman & M. Palacios, 1990. Breeding for resistance to potato tuber moth, *Phthorimaea operculela* (Zeller), in diploid potatoes. Euphytica 50: 119–126.

Ortiz, R., M. Iwanaga & S.J. Peloquin, 1994. Breeding potatoes for developing countries using wild tuber bearing *Solanum* spp. and ploidy manipulations. J. Genet. & Breed. 48: 89–98.

Peferoen, M., 1992. Engineering of insect-resistant plants with *Bacillus thuringiensis* crystal protein genes. In: A.M.R. Gatehouse, V.A. Hilder & D. Boulter (Eds). Plant Genetic manipulation for Crop Protection. C.A.B. International, Wallingford, UK, pp. 135–154.

Peloquin, S.J., G.L. Yerk, J.E. Werner & E. Darmo, 1989. Potato ploidy manipulation with haploid and 2n gametes. Genome 31: 1000–1004.

Persley, G.L.V. Giddings & C. Juma, 1992. Biosafety – The safe application of biotechnology in agriculture and the environmnent. International Service for national Agricultural research, The Hague, The Netherlands, 39 pp.

Ritter, E., T. Debener, A. Barone, F. Salamini & C. Gebhardt, 1991. RFLP mapping on potato chromosomes of two genes controlling extreme resistance to potato virus X (PVX). Mol. Gen. Genet. 227: 81–85.

Ross, H., 1986. Potato Breeding – Problem and Perspectives. Verlag Paul Parey, Berlin, 132 pp.

Tanksley, S.D., M.W. Ganal, J.P. Prince, M.C. de Vicente, M.W. Bonierbale, P. Broun, T.M. Fulton, J.J. Giovannoni, S. Grandillo, G.B. Martin, R. Messeguer, J.C. Miller, L. Miller, A.H. Paterson, O. Pineda, M.S. Röder, R.A. Wing, W. Wu & N.D. Young, 1992. High density molecular maps of the tomato and potato genomes. Genetics 132: 1141–1160.

Valkonen, J.P.T., S.A. Slack & K.N. Watanabe, 1994a. Resistance to potato cucumber mosaic virus. Ann. Appl. Biol. Accepted.

Valkonen, J.P.T., K. Watanabe & E. Pehu, 1994b. Analysis of correlation between nuclear DNA content, chromosome number, and flowering capacity of asymmetric somatic hybrids of diploid *S. brevidens* and (d) haploid *S. tuberosum*. Jpn. J. Genet. 69: 525–536.

Valkonen, J.P.T., S.A. Slack, M. Orrillo & K.N. Watanabe, 1994c. Testing virus resistances in sexual hybrids between a diploid potato breeding line and non-tuber bearing *Solanum brevidens*. Am. Potato J. 71: (Abstr.) 707.

Valkonen, J.P.T., S.A. Slack, R.L. Plaisted & K.N. Watanabe, 1994d. Extreme resistance is epistatic to hypersensitive resistance to potato virus Y^o in a *S. tuberosum* ssp. *andigena*-derived potato genotype. Plant Disease. 78: 1177-1180.

Watanabe, K., 1991. Bottlenecks in germplasm enhancement and application of the biotechnology. Proc. Plan. Conf. on the Application of Biotechnology to Germplasm Enhancement of Potatoes. International Potato Center (CIP), Lima, Peru, pp. 135–140.

Watanabe, K., 1994a. Genome and Chromosome Manipulation in Potatoes. Advance in Plant Breeding Vol. 35, Yokendo, Tokyo, Japan, pp. 104–109 (in Japanese).

Watanabe, K., 1994b. Potato Molecular Genetics. In: J. Bradshaw & G. MacKay (Eds). Potato Genetics. C.A.B. International, Wallingford, UK, pp. 213–235.

Watanabe, K. & M. Orrillo, 1993. An alternative pretreatment method for mitotic chromosome observation. Amer. Potato J. 70: 543–548.

Watanabe, K. & S.J. Peloquin, 1989. Occurrence of 2n pollen and *ps* gene frequencies in cultivated groups and their related wild species in tuber-bearing *Solanums*. Theor. Appl. Genet. 78: 329–336.

Watanabe, K. & S.J. Peloquin, 1993. Cytological basis of 2n pollen formation in a wide range of 2×, 4×, and 6× taxa from tuber-bearing *Solanum* species. Genome 36: 8–13.

Watanabe, K., M. Iwanaga, H. El-Nashaar & P. Jatala, 1991. Transmission of resistances on root-knot nematodes and bacterial wilt by 2n pollen in potatoes. Proc. 2nd. Intern. Symp. Chrom. Eng. Plant, Univ. Missouri, Columbia, MO, USA, pp. 128–132.

Watanabe, K., C. Arbizu & P. Schmiediche, 1992a. Potato germplasm enhancement with disomic tetraploid *S. acaule*. I. Efficiency in introgression. Genome 35: 53–57.

Watanabe, K., H. El-Nashaar & M. Iwanaga, 1992b. Transmission of bacterial wilt resistance by FDR 2n pollen via 4× × 2× crosses in potatoes. Euphytica 60: 21–26.

Watanabe, K., M. Orrillo, S. Vega, R. Ortiz, M. Iwanaga, R. Freyre, H. El-Nashaar, K. Raman, M. Palacios, P. Jatala, E. Guevara, J. Tenorio, A. Golmirzaie, G. Yerk & S.J. Peloquin, 1993a. Germplasm enhancement with diploid potato genetic resources. I. Generation and use of haploid-wild species F1 hybrids. Am. Potato J. 70: 854 (Abstr.).

Watanabe, K., M. Orrillo, H. El-Nashaar, K. Raman, P. Jatala, M. Palacios, E. Guevara, S. Vega, A. Hurtado, J. Tenorio & A. Golmirzaie, 1993. Germplasm enhancement with diploid potato genetic resources. II. Use of a wide range of breeding stocks. Am. Potato J. 70: 854–855 (Abstr.).

Watanabe, K., J.H. Dodds, H. El-Nashaar, V. Zambrano, J. Benavides, F. Buitron, C. Segueñas, A. Panta, R. Salinas, S. Vega & A. Golmirzaie, 1993c. Agrobacterium mediated transformation using antibacterial genes for improving resistance to bacterial wilt in tetraploid potatoes. Am. Potato J. 70: 853–854 (Abstr.).

Watanabe, K., M. Orrillo, S. Vega, R. Masuelli & K. Ishiki, 1994a. Potato Germplasm enhancement with disomic tetraploid *Solanum acaule*. II. Assessment to breeding value of tetraploid F_1 hybrids between *S. acaule* and tetrasomic tetraploid potatoes. Theor. Appl. Genet. 88: 135–140.

Watanabe, K., M. Orrillo, M. Iwanaga, R. Ortiz, R. Freyre & S. Perez, 1994b. Diploid germplasm derived from wild and landrace genetic resources. Am. Potato J. 71: 599–604.

Watanabe, K., M. Orrillo, S. Vega, A. Hurtado, J.P.T. Valkonen, E. Pehu & S.D. Tanksley, 1995. Sexual hybrids between non tuber-bearing and tuber-bearing *Solanum* species. Genome 38: 27–35.

Wenzel, G., 1994. Tissue culture. In: J. Bradshaw & G. Mackay (Eds). Potato Genetics. C.A.B. International, Wallingford, pp. 173–195.

Werner, J.E. & S.J. Peloquin, 1990. Inheritance and two mechanisms of 2n egg formation in 2× potatoes. J. Hered. 81: 371–374.

Young, N.D. & S.D. Tanksley, 1988. Restriction fragment length polymorphism maps and the concept of graphical genotypes. Theor. Appl. Genet. 77: 95–101.

Euphytica **85**: 465–476, 1995.

Criteria for decision making in crop improvement programmes - Technical considerations

Peter W. Jones & Alan C. Cassells
Department of Plant Science, University College, Cork, Ireland

Introduction

Plant breeding involves three stages: namely, creation of variation, selection of desired genotypes and trialling to confirm characteristics. Since the 1900s, plant geneticists and plant breeders have worked together to develop methods to increase the options and efficiencies of the different stages in plant improvement. During the intervening years, inputs from plant pathologists and plant physiologists have been assimilated. Recently, new techniques for plant improvement have been developed by molecular biologists and plant tissue culturists. The challenge is to incorporate these and other non-conventional methods efficiently into the development of strategies for plant improvement.

Large demonstration projects, particularly in the area of genetic transformation, have tended to convey the impression that these methods offer a one-stage approach to plant improvement. The precision with which a gene construct can be made cannot be extrapolated to the prediction of the consequences of the incorporation of the gene into the target genome. Pleiotropy and other consequences of transformation with sense or antisense constructs are possible, even likely, and strategies for improved genotype selection are needed.

The aim of this overview is to evaluate the strengths and weaknesses of non-conventional methods for plant genetic manipulation and selection, and to consider the criteria which can be used to identify techniques appropriate for specific breeding programmes.

Methods in plant genetic manipulation

Somatic hybridisation

Hybridisation with distantly-related species is an important strategy for introducing new traits into crop varieties, but there are problems, such as sexual incompatibility and the need to eliminate large amounts of unwanted donor plant genetic material.

Somatic hybridisation is a method for the potential achievement of novel wide crosses, especially between sexually-incompatible species, based on techniques for protoplast isolation, culture and plant regeneration developed by plant physiologists. Among the major crops, continuing problems of regeneration from protoplasts of many species have restricted somatic hybridisation to mainly solanaceous and cruciferous taxa (Warra & Glimelius, this volume). Even within these two families, exploitation of the advantages of protoplast fusion over sexual hybridisation has been limited. Few commercially-important examples of cybrids and other novel nuclear-cytoplasmic combinations, such as the transfer of cytoplasmic male sterility from radish into *Brassica* crops (Pelletier et al., 1983), have been reported. Where somatic hybridisation has resulted in the production of hybrids between sexually-incompatible species, problems have generally arisen with the development of agronomically-acceptable crop varieties. Sexual hybridisation with wild species, for example, has made a major contribution to potato breeding but, to date, the PLRV-resistant somatic hybrids between the sexually-incompatible *Solanum tuberosum* and *Solanum brevidens* (e.g. Austin et al., 1985) have not led to the development of clones with adequate field performance. Strategies for reducing the quantity of alien genetic variation transferred to the initial hybrid, such as asymmetric protoplast fusion (Sjödin & Glimelius, 1989) and microprotoplasts (Ramulu et al., this volume), may reduce this problem.

Improved understanding of the factors determining inter-specific incompatibility, such as Endosperm Balance Number (Watanabe et al., this volume) and the role of late-replicating heterochromatin, has increased the range of inter-specific hybrids possible using the more widely-applicable *in vivo* hybridisation techniques. Long-established procedures such as bridging crosses, stigma treatment (irradiation, excision or hormone treatment), *in vitro* fertilisation and embryo rescue have proved effective in certain genera (Pickersgill, 1993); more research is needed to determine how broadly relevant they are.

In many wide crosses, introgression is blocked because of lack of recombination events. Although manipulation of genes controlling homoeologous chromosome pairing is possible in several amphidiploid species, such as wheat and oats (Thomas et al., 1980), more generally applicable, empirical methods are also available, such as induced translocations following mutagenesis (Sears, 1972) or adventitious regeneration (Dore et al., 1988), or non-homologous chromosome pairing during anther culture (Wernsman, 1992).

In addition to asymmetric protoplast fusion, partial genome transfer is also possible by sexual hybridisation, by use of irradiated pollen (Powell et al., 1983). This method has been shown to be effective in the transfer of small genome fragments between sexually-compatible species (Pandey, 1975), and a re-evaluation of its potential for accelerating introgression (and its extension to gene transfer between incompatible combinations) is overdue.

Genetic transformation

Genetic transformation permits the use of the most exotic germplasm, as well as the modification of the expression of extant genes within the plant's genome. In principle, at least, transformation permits the directed introduction of characters not possible by other techniques, resulting from the transfer of genes from unrelated species, modified regulation of specific gene expression, and down-regulation of the expression of existing genes, particularly those in dispersed multigene families. A major difference between transformation and other sources of genetic variation lies in the level at which the target gene is selected, that of the mechanism of action (e.g. chitinase production) rather than that of the phenotype (disease resistance). This targeted approach is very exciting and undeniably attractive; the role of plant breeders and molecular biologists is to identify the most appropriate areas of use.

Technical problems are being addressed and transformation techniques are now available for all major crops, principally via *Agrobacterium* and microballistics (microprojectile bombardment). The development of microballistics and subsequently of micro-targetting for use with plant meristems (Potrykus et al., this volume), have provided genotype-independent systems for transformation of a wide range of species. To recover solid (non-chimeral) transformants, reference to the body of literature on this topic published by plant cell and tissue culture researchers is strongly advised (George, 1993). Sectoral chimeras are likely to result from meristem bombardment, while adventitious regeneration from *Agrobacterium*-transformed tissue makes chimeras less likely. Adventitious regeneration from transformed cells may result in background somaclonal variation, however, depending on the host genotype. This can be eliminated, in fertile species, by backcrossing the transformants to the parental variety; or, in vegetatively-propagated crops, by clonal selection.

Pleiotropy is likely in all single-gene manipulation techniques, including genetic transformation, mutagenesis and back-crossing; it occurs in all breeding programmes, but it is the overt effect of pleiotropy in single-gene manipulations which has attracted most attention. Problems of pleiotropic effects in selected, fertile transgenic plants can be minimised by following the strategies used by mutation breeders. The principal procedure for reducing/eliminating pleiotropic effects in mutagenesis programmes is to transfer the gene into different genetic backgrounds. For undirected pleiotropic effects this is most simply achieved by hybridising the mutant with a randomly-selected range of other elite varieties. Oud et al. (this volume) demonstrated the effectiveness of this simple approach (using the transformant as the female parent) to eliminate the detrimental floral and habit traits associated with the maize dihydroflavonol-4-reductase transgene conferring orange flower colour in petunia, by the simple expediency of selecting segregants combining the novel flower colour with acceptable field performance. Where the pleiotropic character is directed, such as delayed flowering, appropriate early-flowering genes can be stacked into the genetic background by hybridising the transformant with a parent selected for this characteristic.

Transgenics also suffer from additional problems such as position effects (determined by lack of control over the site of insertion) and gene silencing (determined by lack of control of gene copy number), affecting endogenous genes and/or the transgene (Matzke & Matzke, 1993). In terms of crop improvement, the problem lies with the instability of the mechanisms (e.g. methylation, co-suppression) causing these changes in gene expression. Co-suppression, for example, can result in high frequencies of environment-specific altered expression. In the hemizygous state, the petunia flower colour transgene was stable in the glasshouse but not in the field (Meyer, this volume).

Such instability would cause problems for the breeder at the DUS stage of registration of transgenic plant varieties, or for growers (and subsequently the breeder) if not detected prior to the release of the variety. Co-suppression can be stabilised by transfer of the transgene into a new genetic background, while transformation techniques to achieve better control of transgene integration (site and copy number) are being developed. Of the principal transformation methods, microballistics operates no control over copy number or integration site, while *Agrobacterium* offers some control. The use of gene constructs with flanking regions homologous to appropriate segments of the host plant genome (for example, corresponding to the so-called matrix-associated regions of animal chromosomes; Phi-Van et al., 1990) may reduce problems associated with inappropriate sites of transgene integration. Until this is achieved, careful pre-screening of transformants should always be conducted to eliminate non-transformants and plants with multiple copy numbers of the transgene, as well as polyploids and aneuploids (Meyer, this volume). Individuals should then be selected from large transformant populations for stable expression of the character following testing at different sites and over several generations, i.e. standard trialling practice for putative potential varieties in conventional breeding programmes. Field assessment of transgenic plants in the early generations of a transformation programme is also important because expression of the transgene under growthroom or glasshouse conditions (low light, steady nutrient and water supply, regularised environmental conditions) may well differ from that under field conditions.

Somaclonal variation and induced mutagenesis

Although widely used in developing countries (Maluszynski et al., this volume), induced mutagenesis in breeding programmes in Europe is currently confined largely to aspects such as the improvement of clonal species and ornamental crops. This is due to the widely-held belief that characteristics of mutation breeding (the demand for extremely large ($n = 10^5$–10^6) M_2 populations, problems of detrimental associated characters, the existence in existing germplasm of the so-called 'novel' alleles induced by mutagenesis) mitigate against the incorporation of this technique into conventional breeding programmes. Nevertheless, mutation breeding has resulted (largely, after cross-breeding) in the development of important crop varieties in Europe, such as the *GPert* barley variety Golden Promise which dominated the Scottish malting barley acreage from the 1970s to the mid 1980s. Mutation breeding certainly has value in European breeding programmes, and is currently under-represented.

For a procedure often condemned for its 'hit and miss' nature, mutation breeding is probably the 'unconventional' breeding method which has been subjected to most analysis to maximise its efficiency (IAEA, 1977): selection of mutagen, mutagen dosage, target tissue, target genotype (homozygous or heterozygous), population size, collection of M_2 seed from apical or axillary meristems, for example. Yet, the expectation that specific mutants will be isolated at a very low frequency often results in the design of mutation breeding programmes which fulfil that expectation. In a large M_2 population, only mutations with large phenotypic effects (major gene mutations) are detectable; such mutations (e.g. conferring race-specific disease resistance) do occur at a very low frequency, especially where the mutant allele is dominant (as usually occurs with disease resistance). Hence, low mutant frequencies in mutation breeding programmes represent a self-fulfilling prophecy. On the other hand, intensive screening of small ($n \leq 1{,}000$) M_2 (or M_3) populations, or preferably M_3 progeny rows, permits identification of mutants with altered quantitative traits, such as partial disease resistance, an important trait in many crop improvement programmes. These occur at relatively high (even percentage) frequencies, making this a highly cost-effective approach, easily incorporated into conventional programmes (e.g. Konzak, 1956). The high mutation frequency means that the occurrence of multiple mutations is common, but backcrossing to the parent variety eliminates undesirable mutations (Bagnara et al. 1971).

Although the occurrence of induced mutations is largely random, changes in expression of a particular quantitative trait (e.g. partial disease resistance) in mutants will generally (and logically) be in the direction away from the selection history of the parent variety (Brock, 1967), i.e. the altered height mutants induced from a tall variety will be predominantly shorter than the parent.

The phenomenon of mutant heterosis (Maluszynski et al., this volume) has altered our perception of the role of major gene mutations in crop improvement. Following hybridisation of two mutants in the same parent line, the binary recombinants can exhibit yields far in excess of that of the parent and of both mutant lines in the field. This heterosis occurs even when one or both mutants are clearly inferior to the parental variety. For a complex, empirically-selected trait such as yield, it is perhaps presumptuous to believe that manipulation of a single major gene in a highly-bred crop can markedly improve performance of the original variety; it is unlikely that such a simple improvement would not have occurred (and been selected) in the segregating populations involved in the production of the original variety. Introduction of a second mutant gene changes the genetic background in which the first mutant allele is expressed, and superior genotypes can then be selected.

Somaclonal variation, a phenomenon discovered by plant tissue culturists in the 1980s, can also result in a high mutation frequency, such that, as with mutagenesis, screening of only small populations of adventitious regenerants may be necessary to isolate variants exhibiting alterations in quantitatively-inherited traits. Exploitation of somaclonal variation in plant improvement has been limited by problems previously encountered (and largely solved) by mutation breeders: chimeras, pleiotropy, multiple mutations and poor correlation between glasshouse and field performance of selected variants.

Novel problems also arise with somaclonal variation, principally the high frequency of unstable variation, resulting from methylation and sequence amplification among other effects. As yet there is little knowledge of the factors which cause the various classes (desired and detrimental) of somaclonal variation, which could then be manipulated to minimise the detrimental effects. Reduced culture phase duration, avoidance of mixoploid explants (such as cotyledons), and delay of *in vivo* screening to progeny rows from individual R_1 plants (to eliminate epigenetic effects and certain classes of chromosome mutation and unstable gene modifications) are all practical steps which can be undertaken, but the highly genotype-dependent nature of somaclonal variation makes optimisation work (as conducted for mutation breeding) less feasible.

On the other hand, the spectrum of variants induced following adventitious regeneration (e.g. transposable element activation) is frequently different to that obtainable from mutagens (Gavazzi et al. 1987), and somaclonal variation has already resulted, directly or indirectly, in the development of several commercial cultivars in both field and ornamental crops (Karp, this volume).

Selection

The development of more efficient selection strategies is probably the most cost-effective use of plant breeding research efforts. In conventional programmes, selection is primarily empirical and phenotype-based, with family selection used to increase the selection response for low-heritability traits, such as yield. Further increases in selection efficiency can be achieved by the use of selection indices (Baker, 1986), based on phenotypic traits which are components of, or correlated with, the complex target character. Unfortunately, difficulties in weighting the value of individual phenotypic traits (such as high nitrate reductase activity as a possible component of the high grain protein content trait in cereals; Hageman, 1979) have limited the use of these indices in breeding programmes. The use of molecular markers (RFLP,RAPD, etc.) linked to QTLs provides an indirect but objective basis for selecting for complex traits. Marker-assisted selection (MAS) has the potential to revolutionise selection protocols within conventional breeding programmes, using smaller families but with increased efficiency.

MAS is particularly appropriate for traits which are highly variable in expression, such as yield or partial disease resistance. QTL-based MAS, for example, could prevent the Vertifolia effect (van der Plank, 1963) in disease resistance breeding programmes, maintaining resistance polygenes by MAS in the presence of major resistance genes against the same pathogen, thus increasing the durability of the latter. It applies a more intensive selection pressure than phenotype-based selection, and could be used in the early generations of breeding programmes where inefficient phenotypic selection currently predominates. Problems remain, however. At present, the cost of MAS is prohibitive for use in conventional breeding programmes; costs will need to be reduced by several orders of magnitude to make MAS cost-effective in these circumstances. On the other hand, MAS is appropriate for use in small-population situations; in backcrossing, for example, MAS could reduce generation number (Tanksley et al., 1989), permit the simultaneous selection of several genes and facilitate selection for recessive genes, reducing several serious problems associated with introgression. The reduction in backcross generations would be particularly valuable for crops with long juvenile periods, such as forest trees. It should also facilitate the selection of parents with complementary QTL genotypes, (e.g. in F_1 hybrid breeding) and the evaluation of accessions in germplasm collections.

One point that must be remembered is that the field selection component of modern breeding programmes is valuable in that the putatively superior lines are trialled in different years, under various environmental conditions, in addition to the standard two to three-year National Trials. Any improved selection procedure which shortens the length of the field trialling programme, runs the risk of missing a possibly deleterious [genotype × environment] interaction such as the fertility problems associated with winter wheat cv. Moulin in the UK in 1986.

There has been considerable interest over recent years in *in vitro* selection as a rapid and efficient means of isolating individuals with specific phenotypes generated by *in vitro* techniques such as transformation, protoplast fusion and adventitious regeneration (Dix, 1990). Knowing the mode of action of the gene concerned, e.g. in transformation or somatic hybridisation programmes, appropriate *in vitro* selection conditions can be developed. In an asymmetric protoplast fusion project to transfer resistance to *Phoma lingam* (expressed as resistance to the pathogen-derived toxin, sirodesmin PL) from *Brassica juncea* into *Brassica napus*, incorporation of lethal (for *B. napus* protoplasts) concentrations of the toxin into the regeneration medium selected effectively for protocalli carrying the targeted *B. juncea* genome fragment (Sjödin & Glimelius, 1989).

For selection of novel uncharacterised traits (e.g. following adventitious regeneration), however, *in vitro* selection is, at best, a selection index using a component of the target character as the indirect selection criterion (e.g. toxin resistance as a measure of *Phoma*-resistance). Use of sirodesmin PL would select only one class (toxin resistance) of *Phoma*-resistance among somaclones. Such indirect selection may not be wholly effective: The *Ve* gene in the natural tomato gene pool confers resistance to the phytotoxin produced by *Verticillium dahliae* race 1 but not to the pathogen itself, which multiplies freely within the largely asymptomatic plant (Nachmias et al., 1987). There can also be problems with differences in expression of the gene in question *in vitro* (in a protoplast, callus or shoot, at low light and high nutrient and humidity levels) and *in vivo* (as whole plants in the field). Following *in vitro* selection, for example, Duncan (this volume) isolated no sorghum somaclones which were aluminium tolerant in the field but three tolerant lines were identified when previously-unselected somaclones were subjected to growth under high-aluminium conditions in the field. The value of *in vitro* selection is inversely correlated with the frequency of occurrence of the desired variant/mutant/transformant. With respect to somaclonal variants exhibiting quantitative changes in a trait, where high variant frequencies have been reported, direct *in vivo* screening of several hundred somaclones (or somaclone progeny rows) would be more efficient, permitting identification of all classes of, say, disease resistance, unlike *in vitro* selection where, for example, only toxin resistance could be detected. *In vivo* screening of the R_1 generation (the selfed progeny of R_0 adventitious regenerants) would also permit identification of recessive mutants in the homozygous state, whereas, during *in vitro* selection at the R_0 level, such genes would, generally, not be expressed.

When attempting to isolate induced variants (resulting from mutagenesis or adventitious regeneration) by either *in vivo* or *in vitro* screening/selection, it should be recognised that false positives may arise as the result of pleiotropic effects. Plants with altered rates of development, for example, may appear to be tolerant of certain biotic or abiotic stresses, whereas the true mechanism would be 'stress escape', by not being at the sensitive stage when the stress is applied. As many more genes affect development rate than resistance to a specific stress, the frequency of altered-development mutants is higher than that of the true stress-resistant mutants. The 'salt-tolerant' somaclone of flax cv. MacGregor (marketed as 'Andro') is early developing (McHughen et al., 1987), while several potato somaclone programmes have reported that 'late blight resistant' somaclones were predominantly late-developing lines (e.g. Cassells et al., 1991), which had relatively immature (and hence resistant) foliage when the inoculum arrived. Millam and Craig (this volume) reported that high frequencies of *in vitro* regeneration in oilseed rape is associated with the undesirable trait of high erucic acid in the rapeseed oil.

Criteria for decision making

Identification of the sources of genetic variation appropriate for a particular breeding programme should take into account features of the programme such as:

(a) *the crop*

- whether sexually-propagated (self or cross-pollinated) or clonally-propagated;
- whether for food, industrial, ornamental or pharmaceutical use;
- whether recently or anciently-domesticated;

- whether a major or minor crop.

(b) *the character(s)*
- whether major gene and/or polygene;
- whether available in sexually-compatible germplasm;

(c) *the infrastructure*
- what technical expertise/facilities is/are available;
- whether DUS requirements are required for cultivar registration;
- whether breeding is funded by public or private bodies.

(a) The crop

The breeding system of the plant has implications for the genetic manipulation techniques which can be applied. For certain crops, sexual reproduction is either unavailable as a breeding technique because of sterility (e.g. banana) or apomixis, or must be eschewed, as in crops such as coffee, vines and chrysanthemum, where sexual reproduction would break up important, co-adapted gene complexes, causing outbreeding depression. For these crops, single-gene manipulation strategies (transformation, mutation breeding, somaclonal variation) are the only alternatives to clonal selection. The absence of sexual reproduction reduces the chances of achieving homozygosity (and hence expression) of recessive mutations, and the frequently-polyploid nature of sterile/apomictic plants exacerbates this problem. Mutagenesis is the principal source of genetic variation in chrysanthemum and banana breeding programmes, while somaclonal variation is being widely adopted for coffee improvement (Karp, this volume).

Asexual, single-gene manipulation strategies would accelerate improvement of perennial crops with long juvenile periods, such as forest trees, although in practice, our lack of knowledge of major genes controlling agronomically-important traits such as growth rate and quality in these species, makes clonal selection more viable in all but those species with narrow genetic bases.

The availability of sexual reproduction for crop improvement work opens up a wider range of breeding techniques (including conventional recombination breeding), while providing valuable strategies for eliminating certain sources of unwanted variation. As mentioned earlier, meiosis acts as a sieve, reducing or eliminating the frequency (in subsequent generations) of certain chromosome mutations and epigenetic variants including unstable mutants resulting from methylation and sequence amplification. In addition, sexual reproduction permits the elimination of deleterious phenotypic effects resulting from pleiotropy, co-suppression, multiple mutations and undesirable alien genetic variation (linked or unlinked).

The capacity of a plant to undergo self-fertilisation (natural, or inducible, e.g. by bud-pollination) is useful to fix novel genes present in heterozygous (especially for recessive genes) or hemizygous plants isolated from introgression, mutagenesis or transformation programmes. For species lacking this capacity (obligately allogamous, sterile or apomictic), the use of heterozygous breeding material may help offset this problem. Alternatively, recessive mutations have been recovered in the homozygous state at high frequencies in the R_0 generation following adventitious regeneration of diploid and amphidiploid species, apparently as the result of mitotic recombination (George & Rao, 1982; Evans et al., 1984), suggesting that adventitious regeneration could be the induced variation 'method-of-choice' for crops which cannot be self-pollinated.

Applications of *in vitro* techniques are restricted to those crops amenable to *in vitro* cell and organ culture and regeneration. Although adventitious regeneration has been achieved for a wide range of dicotyledons and an increasing number of monocotyledons, culture of and regeneration from plant protoplasts is still the exception rather than the rule. This is illustrated by the restriction of somatic hybridisation largely to members of the Solanaceae and the Brassicaceae, as described earlier, a limitation which has been a factor in the increasing disenchantment with this technique. Microballistics has helped prevent similar marginalisation of genetic transformation, although four of the seven crops most commonly represented in transgenic field trials up to 1992 belonged to the two families mentioned above (OECD, 1993).

Large-scale commercial breeding programmes concentrate on the major crops with established markets, but there is an increasing number of minor crops, many of which, like evening primrose, sesame and borage, have only recently entered domestication. In these crops, recombination breeding with mass selection can meet with rapid success, with mutagenesis or wide crosses introduced to extend the gene pool if necessary. Reductions in public and seed company-funded breeding research on these crops has led increasingly to the end-users of these speciality crops developing breeding programmes, vertically integrated with production, as illustrated by the evening primrose breeding programme initiated by Scotia Pharmaceuticals in the UK. In such in-house programmes, with specific goals, high-technology genetic manipulation techniques could, where appropriate, be introduced cost-effectively.

The end-use of a crop may also affect the choice of breeding methods. Consumer antagonism towards transgenic plants appears to be restricted to their perceived impact on food. Proposed legislation on the labelling, at point-of-sale, of food material from transformed plants will, if implemented, undoubtedly adversely affect the sales of these products; whether these regulations will distinguish between sense and antisense constructs, or between transgenes expressed in economic and non-economic organs is unclear. If widely adopted, such legislation could see the end of commercial transgenic plant improvement of food crops. Instead, plant transformation would focus or ornamental crops and on field crops grown for industrial, pharmaceutical or fibre uses; antisense ACC synthase would be acceptable in chrysanthenum (to extend vase-life), but not in tomato (to increase shelf-life).

There is a widely-held misconception that, in the development of new varieties of ornamental crops, novelty *per se* is adequate, unlike field crops where any improvements must be made in the context of acceptable agronomic performance. In fact, breeders of ornamental plants work within very tight quality control parameters (imposed by growers, retailers and consumers), such as production costs, floriferousness, habit and stability. This is illustrated by the work on the orange-flowered petunia carrying the maize transgene, where the novelty of the new flower colour in the original transformants was outweighed by problems of germination, flower size and plant habit (Oud et al., this volume). Nevertheless, novelty possibly carries more weight in ornamentals than in field crops. In addition, the trend in ornamental breeding towards producing a 'family' of similar varieties differing with respect to a single character makes single-gene manipulation methods (such as mutagenesis, to create the 'Westland' and 'Miros' chrysanthemum 'colour families', or transformation) particular appropriate.

(b) The character(s)

The principal selection criteria in breeding programmes will probably stay the same in the short and medium term (yield, disease and pest resistance, quality), albeit with changes in emphasis. Instead of maximum economic yield, there may be an increasing demand in the industrialised countries for varieties making more efficient use of resources, with less dependence on high levels of agrochemicals. Excess food production capacity will also lead to an increased acreage sown to crops capable of generating products for non-food purposes such as industrial oils, pharmaceuticals and chemical feedstocks.

Decisions as to which breeding strategy is appropriate for a particular character need to take into account features of the different techniques. For characters determined largely by additive polygene effects, recombination breeding is, at present, the only viable option. For major gene traits, however, more options exist.

Where the gene in question exists in the gene pool of the targeted crop, hybridisation and introgression is the logical strategy but a decrease in the number of (usually) publicly-funded germplasm enhancement programmes (introducing valuable genes from wild species into adapted genotypes of the commercial crops) has led to a fall in the use of exotic germplasm, especially in highly-bred crops; only 6% of European barley breeding programmes use alien parents, even when the principal target character is disease resistance (Vellvé, 1992).

Genetic transformation, on the other hand, can introduce novel characters by modifying existing traits, or by introducing entirely new genes (as in herbicide resistance). For genes absent from the appropriate gene pool for the crop in question, induced genetic variation and genetic transformation would appear to be the appropriate strategies. For traits where the gene or gene product is known, such as the *tfdA* gene from *Alcalagines eutrophus* encoding a degradative 2,4-D dichlorophenoxyacetate monoxygenase enzyme to confer 2,4-D resistance (Streber & Willmitzer, 1989), or human albumin, transformation is the only route. For other traits (especially those which are readily identifiable), induced genetic variation would be effective, as in the introduction of resistance to *Verticillium* wilt into peppermint cv. Mitcham (Murray, 1969). With respect to silencing of endogenous genes, both transformation (with antisense constructs) and mutagenesis have proved effective, although clearly the former permits greater control.

Pleiotropy in the form of a yield penalty is frequently a consequence of the introduction of a gene encoding a novel product, although this can be balanced by the positive benefits of the introduced gene. Further reduction of the yield penalty could be achieved in genetic transformation by using tissue- or stimulus-specific promoters, as Murphy (this volume) discussed with respect to modifications of fatty acid composition of rapeseed oil. Buiatti and Bogani (this volume) indicated that pleiotropy is minimised when the transgene is additional or peripheral to existing pathways in the recipient plant. The most agronomically-successful transformants certainly fall into this 'dead-end product' category: the *Bacillus thuringiensis* endotoxin genes (conferring insect resistance), the *bar* gene for phosphinothricin acetyltransferase from *Streptomyces hygroscopicus* (resistance to the herbicide glucophosinate) and the antisense polygalacturonase construct (non-cracking of tomato fruits). Plants in which the direction of existing metabolic pathways is modified tend to suffer from pleiotropic disturbances of this and related pathways, with consequent effects on the phenotype of the transformed plant. Antisense modification of the expression of ACC synthase (to reduce ethylene biosynthesis), for example, would be expected to generate a syndrome of ethylene-mediated phenotypic effects. Picton et al. (this volume) reported pleiotropic effects of this construct in tomatoes such as delayed leaf production, and reduced fruit coloration.

On the other hand, transgene modification of endogenous pathways may be restricted by compensation by regulatory mechanisms within the plant (Buiatti and Bogani's 'homeostasis'; this volume). In most antisense studies, for example, suppression of gene expression has been less than 100%, and in many cases, subsequent compensation has interfered with expression of the desired phenotype.

(i) Economic yield

Our knowledge of the physiological and molecular mechanisms underlying 'economic yield' is still so paltry that, in the foreseeable future, the predominant breeding strategy will be empirical selection within recombination breeding programmes, aided possibly by marker-assisted selection for yield QTLs at certain stages, such as the early generations of pedigree selection programmes, if MAS costs can be reduced significantly. The phenomenon of mutant heterosis also opens up the prospect of a major gene basis of hybrid vigour, with fixation of heterotic genotypes.

Breeding for high yield under intensive conditions involved a journey into the unknown, in which crop physiology (the 'retrospective science'; Evans, 1977) was able to explain but not predict how plant design resulted in increased yield. The availability in germplasm collections of older varieties selected for high performance under less intensive agricultural conditions may permit the more efficient use of morphological and physiological characters in selection indices for crop varieties for sustainable agriculture.

(ii) Disease and pest resistance

Stable yields under lower-input conditions will place a high premium on durable disease and pest resistance. Whereas control of pests and diseases by the development of resistant cultivars has been one of the major successes of plant breeding, the deployment of resistance genes for short term benefits has tied us into the inexorable boom-and-bust cycle. With our knowledge of the causes and consequences of this phenomenon, breeders and state agencies should now strive to develop and release resistant varieties (produced by conventional or non-conventional means) with the aim of maximising the longevity of resistance. The acclamation surrounding the development of transgenic *Bt* crop varieties drowned out the voices urging caution for fear (now shown to be with foundation) of the development of resistant pest strains (Holmes, 1993). Breeder-operated gene deployment strategies have proved effective in some areas in extending the durability of major disease and pest resistance genes, e.g. the sequential release programme of IRRI rice varieties with different resistance genes against the brown planthopper (Khush, 1977).

Generally, monogenic resistance has proved to be relatively durable against certain classes of pest and pathogen: soil-borne fungi and pests, and, to some extent, viruses (with obvious exceptions such as CMV resistance in cucumbers), bacteria, toxigenic fungi and single-generation fungi. For air-borne, polycyclic pathogens, such as the mildews and rusts, this approach is, at best, short-term, and alternative approaches are needed.

Certain gene constructs could result in novel types of disease resistance, such as durable monogenic resistance, even broad-spectrum monogenic resistance. Incorporation into a cultivar (already carrying a hypersensitive resistance gene against a particular pathogen) of a 'resistance cassette' comprising the corresponding avirulence gene from the same pathogen regulated by a broad-spectrum, pathogen-induced promoter, should result in resistance (by hypersensitivity) to attack by any fungus which triggers the promoter. The binary nature of this sensor system should result in high durability of the resistance (de Wit & van Kan, 1993).

Relatively durable oligogenic disease resistance can be achieved by stacking genes conferring complete and/or partial resistance to the same pathogen, using MAS to aid selection of the desired genotype. Jongedijk et al. (this volume) successfully used the transgenic approach to this problem, stacking two genes with different mechanisms of action against *Fusarium oxysporum* f.sp. *lycopersici* into a tomato line. Similar strategies can be achieved using other sources of genetic variation; with mutagenesis or adventitious regeneration, for example, stacking can be achieved by either crossing mutant lines carrying non-allelic partial resistance genes with complementary modes of action and selecting binary recombinants, or by successive cycles of mutagenesis and selection (Shahin & Spivey, 1987).

There are still many diseases against which no adequate source of resistance is available within the host crop gene pool. In 1986, Fraser reported that there were at least 25 virus diseases of vegetable crops in Britain that were not adequately controlled by host resistance; these could be amenable to tested transgenic methods such as coat protein or satellite-mediated resistance. Different genetic manipulation strategies can be used to achieve a certain breeding objective. Resistance genes for use in oilseed rape against *Sclerotinia sclerotiorum* have been identified from barley (encoding oxalate oxidase to metabolise the fungal toxin, oxalic acid; Thompson et al., this volume) from B genome *Brassica species* and from induced mutants of *Brassica napus* (Mullins et al., 1995).

Tolerance of abiotic stresses e.g. drought, extreme temperatures, salinity, is a trait amenable to improvement by utilisation of alien genetic material (e.g. salt-tolerance from *Lopophyrum elongatum* into wheat; Dvorak et al., 1988) or by induced genetic variation if a selection pressure can be applied uniformly over a suitably large trial area (Duncan, this volume); Unlike biotic stresses, no single gene product has been associated unequivocally with tolerance of an abiotic stress, restricting the use of genetic transformation in this area.

Although ostensibly causing abiotic stress, modern herbicides tend to affect plants at a specific recognisable metabolic point, resembling biotic stresses such as those caused by pathogens more closely than natural abiotic stresses. Consequently, genetic transformation is a highly appropriate strategy. By 1992, herbicide-resistant lines accounted for more than 70% of the approved transgenic release programmes in Europe (OECD, 1993). Although herbicides represent the major sector of pesticide usage in Europe, herbicide tolerance is not regarded by farmers as a major target character; the demand for herbicide-tolerant crop varieties is primarily industry-led (i.e. by seed-cum-agrochemical conglomerates), the result of so-called 'pull-through' breeding.

(iii) Quality

Of all the breeding objectives, quality is probably the best understood in terms of physiology, proteins and genes. Knowledge of the molecular bases of these traits permits a more targeted genotype-based approach to improvement. Identification, based on SDS-PAGE analysis, of the high quality alleles of the HMW glutenin genes of wheat, for example, has improved selection for bread-making quality in breeding programmes (Payne et al., 1984). To achieve genetic manipulation of quality traits such as the fatty acid composition of rapeseed oil, recombination breeding, mutagenesis and genetic transformation have been used (Murphy, this volume), but such traits show the greatest potential of all for transgenic manipulation. Quality traits are frequently organ-specific (e.g. fruit ripening in tomato) and controlled by multigene families (e.g. seed storage proteins). Suitable constructs can be produced to achieve tissue-specific quantitative changes (e.g. antisense down-regulation of the tomato polygalacturonase gene) or qualitative changes (e.g. introduction of the gene encoding Δ_{12} oleate hydroxylase from castor bean, to produce ricinoleic acid; Murphy, this volume).

(c) Infrastructure

In countries where production of new crop varieties is predominantly in the hands of private breeding companies, constraints on the use of breeding programmes arise for commercial as well as technical reasons. With the yields of each successive year's new varieties resulting in annual yield increases of approximately 1% for the major field crops, variety production by direct use of non-transgenic, single-gene manipulation strategies (mutagenesis, somaclonal variation, backcrossing) is rare. The duration (including trialling) of a mutation breeding or backcrossing programme on an elite variety would require that the introduced gene results in a yield increase of at least 7%. An increase of this magnitude by introduction/manipulation of a single gene is unlikely in a highly-bred crop. In addition, problems could be encountered at the DUS stage of registration, depending on the distinctness of the new variety from the original elite parental variety. On the other hand, mutation breeding is a major component of field crop breeding programmes in countries such as China and India. In these countries yield is often limited by single traits (usually pest or disease susceptibility); single-gene manipulation of elite lines is a rapid and cost-effective strategy, and the DUS problems do not arise.

Genetic transformation, another single-gene manipulation strategy, presents a different balance of technical and commercial constraints. The large beneficial effects which can be exhibited by tranformants, e.g. bollworm resistance in *Bt* cotton, especially in countries with low-input agriculture (characterised by yield instability from year to year), can more than compensate for the relatively low yield (by the time of the release of the transformant) of the by-then out-dated elite parental variety. The high recurrent seed sales of transformants, following protection by utility patents, increases the attractiveness of this crop improvement strategy to private seed companies, although this phenomenon would be restricted to countries where patents on transgenic plant varieties (and the transgene) can be implemented.

Potrykus et al. (this volume) discussed the role of transformants in developing countries. As other single-gene manipulation strategies have been warmly embraced in these countries, the prospects for transformants would appear to be very promising. The cost of production of varieties by this route (by primarily public breeding interests) would be high, except where existing gene constructs could be used. The impact of suitable transformants on the agriculture (e.g. leafhopper-resistant rice) and public health (e.g. high vitamin A rice) of the region could be dramatic, fully justifying the investment. Any discussion on the value of transformants is predicated on the assumption that regulatory bodies in the countries concerned sanction the release of genetically-transformed crop varieties.

In some cases, the observation that a particular breeding technique is widely and successfully used in one country but not in others, can be traced back to the actions of one or a small group of scientists acting as mentor for the technique. One example is the use of mutation breeding in Sweden, promoted by successive teams at Svalöf.

Concluding remarks

In the medium and probably even the long-term, crop improvement will be based principally on recombination breeding programmes, supplemented by non-conventional techniques for generating genetic variation and for selecting desirable genotypes. The main objective of the non-conventional strategies is to replace empirical phenotype-based aspects of recombination breeding with more directed and predictable genotype-based approaches. To approach this situation involves intensive analysis of agronomic traits, including the cloning of key genes, and the development of phenotype and marker-based selection indices for traits where such genes have not been identified. This requirement has led to a significant increase in plant science research (both public and private-funded) over the past decade, and a closing of the gap between the achievements of the plant genetic engineers and the demands of the plant breeders.

Nevertheless, the plant biotechnology revolution has, in the words of Robert Hay, 'created an illusion of exactitude'. With the possible exception of the 'dead-end' genes mentioned above, an introduced gene must operate in the context of probably hundreds and possibly thousands of other genes, and over a wide range of environments. Even when the problems of transgene instability have been solved, there will still be the question of how to decide on the appropriate genetic background. The immediate solution is the empirical one: introduce the gene into a range of genetic backgrounds and select the one giving the most appropriate expression of the transgene. Understandably, there is reluctance to follow such an approach: a specific gene is cloned, ideally targetted (as a single-copy insert) into a specific locus on a host chromosome, and then a random 'hit-and-miss' strategy to optimise expression is employed. The reality is, however, that we do not have and are unlikely to acquire in the near future, sufficient knowledge to achieve the same result in a more directed way. Pragmatism is required. Empirical selection, properly used, is a powerful and highly effective tool, not simply a last resort if all else fails.

Serendipity is a form of empiricism. The initial aim of the anti-polygalacturonase construct was to achieve increased shelf-life of the transformed tomato fruit; the transgene caused other, commercially-desirable effects, such as improved fruit flavour and greater serum viscosity. In the foreseeable future, then, the most effective use of transformants is in concert with empirical selection.

The reductionist and integrationist approaches to plant improvement are complementary, and it is important to tailor the methods employed in a crop improvement programme to suit the specific breeding objectives. To improve certain traits, phenotype-based selection may prove more effective than the more specific genotype-based selection. Attempts to improve plant productivity by increasing the ratio of photosynthesis to photorespiration in C_3 plants has long been a goal. Genetic engineering studies, such as site-directed mutagenesis of the *rbc*L gene, have been unsuccessful to date (Spreitzer, 1993), while several mutant tobacco seedlings selected for growth at CO_2 levels close to the CO_2 compensation point, exhibited increased total dry matter productivity associated with higher rates of photosynthesis (Delgado et al., 1993).

The infrastructure of plant breeding has seen major changes in the past quarter of a century. The trend in developed countries towards privatisation of plant breeding, and the pressure from governments (especially in the USA and the UK) on publicly-funded research bodies (e.g. the Land Grant universities) to move away from near-market plant breeding research, has had a marked impact on the research carried out in the universities. The involvement of plant biotechnology in plant improvement, starting in the 1970s, helped to fill the void left by the loss of more conventional breeding research. In Europe and North America in the mid-1990s, university research on the development of improved crop plants is predominantly based on molecular biology and genetic transformation.

As research is the life-blood of teaching, an unfortunate consequence of this shift in emphasis has been a steady decline in the number of universities in the developed world which offer undergraduate or postgraduate training in classical plant genetics and plant breeding. Although a knowledge of plant molecular biology and genetic engineering should be an important part of the training of a breeder, its value is undermined if not presented in the context of conventional plant breeding programmes. If steps are not taken to broaden the plant breeding research (and hence teaching) of universities, the subsequent generations of plant breeders may suffer from a narrow base of experience, with serious consequences for society.

The objective of this conference has been to assess the contributions to plant improvement (potential and realised) of the various techniques available for plant genetic manipulation and selection. A critical evaluation of the strengths and weaknesses of the different methods, identifying appropriate techniques for particular breeding objectives, will support plant breeders in their efforts to maintain sustainable supplies of food and other products into the next century.

The focus in this overview on genetic transformation and, in particular, on its practical limitations, has been deliberate and, it is hoped, constructive. The enthusiasm for this technique has been overwhelming. The developments in plant molecular biology have revolutionised our ability to study how plants function while underlining how little we know and understand, and have stimulated public and private plant science research. Uncritical enthusiasm, however, has lead to unrealistic expectations. There is already evidence of disillusionment from certain major seed companies. To halt this slide, a pragmatic approach must be taken in deciding when genetic transformation (or other genetic manipulation techniques) should be used.

References

Austin, S., M. Baer & J.P. Helgeson, 1985. Transfer of resistance to potato leaf roll virus from *Solanum brevidens* into *Solanum tuberosum* by somatic fusion. Plant Science 39: 75–82.

Bagnara, D., L. Rossi & G. Porreca, 1971. Use of radiation induced mutants in durum wheat breeding. Genetica Agraria 25: 189–230.

Baker, R.J., 1986. Selection indices in plant breeding. CRC Press, Boca Raton, 257 pp.

Brock, R.D., 1967. Quantitative variation in *Arabidopsis thaliana* induced by ionising radiation. Radiation Botany 7: 193–203.

Cassells, A.C., M.L. Deadman, C.A. Brown & E. Griffin, 1991. Field resistance to late blight (*Phytophthora infestans* (Mont) de Bary) in potato (*Solanum tuberosum* L.) somaclones associated with instability and pleiotropic effects. Euphytica 56: 75–80.

De Wit, P.J.G.M. & J.A.L. van Kan, 1993. Is durable resistance against fungi attainable through biotechnological procedures? In: Th. Jacobs & J.E. Parlevliet (Eds) Durability of disease resistance. Kluwer Academic Publishers, Dordrecht, pp. 57–70.

Delgado, E., M.A.J. Parry, A.J. Keys, D.W. Lawlor, J. Azcon-Bieto & H. Medrano, 1993. Photosynthesis and leaf characteristics of tobacco lines selected for survival at low CO_2. Aspects of Applied Biology 34: 147–154.

Dix, P.J., 1990. Plant cell line selection: Procedures and applications. VCH Verlagsgesellschaft, Weinheim, Germany, 354 pp.

Dore, C., Y. Cauderon & M.C. Chueca, 1988. Inflorescence culture of *Triticum aestivum* × *Secale* cereale hybrids: Production, characterisation and cytogenetic analysis of regenerated plants. Genome 30: 511–518.

Dvorák, J., M. Edge & K. Ross, 1988. On the evolution and adaptation of *Lopophyrum elongatum* to growth in saline environments. Proceedings of the National Academy of Sciences, U.S.A. 85: 3805–3809.

Evans, D.A., W.R. Sharp & H.P. Mediña-Filho, 1984. Somaclonal and gametoclonal variation. American Journal of Botany 71: 759–774.

Evans, L.T., 1977. The plant physiologist as midwife. Search 8: 262–268.

Fraser, R S S , 1986 Genes for resistance to plant viruses Critical Reviews in Plant Science 3 257–297

Gavazzi, G , C Tonelli, G Todesco, R Arreghini, F Raffaldi, F Vecchio, G Barbuzzi, M G Biasini & F Sala, 1987 Somaclonal variation versus chemically induced mutagenesis in tomato (*Lycopersicon esculentum* L) Theoretical and Applied Genetics 74 733–738

George, E F, 1993 Plant propagation by tissue culture Exegetics, Basingstoke, UK

George, L & PS Rao, 1982 Yellow-seeded variants in *in vitro* regenerants of mustard (*Brassica juncea* Cross variety Rai-5) Plant Science Letters 30 327–330

Hageman, R H , 1979 Integration of nitrogen assimilation in relation to yield In E J Hewitt & C V Cutting (Eds) Nitrogen assimilation of plants Academic Press, London, pp 591–611

Holmes, B , 1993 The perils of planting pesticides New Scientist, 28 August 34–37

IAEA, 1977 Manual on mutation breeding, Second edition International Atomic Energy Agency, Vienna

Khush, G S , 1977 Disease and insect resistance in rice Advances in Agronomy 29 265–341

Konzak, C F, 1956 Induction of mutations for disease resistance in cereals Brookhaven Symposium of Biology 9 141–156

Matzke, M & A J M Matzke, 1993 Genome imprinting in plants parental effects and *trans*-inactivation phenomena Annual Review of Plant Physiology and Plant Molecular Biology 44 53–76

McHughen, A , 1987 Salt tolerance through increased vigour in a flax line (STS-II) selected for salt tolerance *in vitro* Theoretical and Applied Genetics 74 727–732

Mullins, E , C Quinlan & P Jones, 1995 Analysis of mechanisms of partial physiological resistance to *Sclerotinia sclerotiorum* using induced mutants of *Brassica napus* Aspects of Applied Biology 43 (in press)

Murray, M J , 1969 Successful use of irradiation breeding to obtain *Verticillium*-resistant strains of peppermint, *Mentha piperita* L In Induced mutations in plants IAEA, Vienna, pp 345–370

Nachmias, A , V Buchner, L Tsror, Y Burstein & N Keen, 1987 Differential phytotoxicity of peptides from culture fluids of *Verticillium dahliae* races 1 and 2, and their relationship to pathogenicity of the fungus on tomato Phytopathology 77 506–510

OECD, 1993 Field releases of transgenic plants, 1986–1992 An analysis Organisation Economic Co-operation and Development, Paris, 39 pp

Pandey, K K , 1975 Sexual transfer of specific genes without gametic fusion Nature 256 310–313

Payne, PI, L M Holt, E A Jackson & C N Law, 1984 Wheat storage proteins their genetics and their potential for manipulation by plant breeding Philosophical Transactions of the Royal Society, London 304B 359–371

Pelletier, G , C Primard, F Vedel, P Chetrit, R Remy, P Rousselle & M Renard, 1983 Intergeneric cytoplasmic hybridisation in Cruciferae by protoplast fusion Molecular and General Genetics 191 244–250

Phi-Van, L , J P von Kries, W Ostertag & W H Stratling, 1990 The chicken lysosome 5′ matrix attachment region increases transcription from a heterologous promoter in heterologous cells and dampens position effects on the expression of transfected genes Molecular and Cellular Biology 10 2302–2307

Pickersgill, B , 1993 Interspecific hybridisation by sexual means In M D Hayward, N O Bosemark & I Romagosa (Eds) Plant Breeding Principles and Prospects Chapman and Hall, London, pp 63–78

Powell, W , PD S Caligari & A M Hayter, 1983 The use of pollen irradiation in barley breeding Theoretical and Applied Genetics 65 73–76

Sears, E R , 1977 Analysis of wheat-*Agropyron* recombinant chromosomes Proceedings of the 8th EUCARPIA Congress, Madrid, Spain, pp 63–72

Shahin, E A & R Spivey, 1986 *In vitro* breeding for disease resistance in tomato In D Nevins & R A Jones (Eds) Tomato biotechnology Alan R Liss Inc , New York, pp 89–97

Sjodin, C & K Glimelius, 1989 Transfer of resistance against *Phoma lingam* to *Brassica napus* by asymmetric somatic hybridisation combined with toxin selection Theoretical and Applied Genetics 78 513–520

Spreitzer, R J , 1993 Genetic dissection of Rubisco structure and function Annual Review of Plant Physiology and Plant Molecular Biology 44 411–434

Streber, WR & L Willmitzer, 1989 Transgenic tobacco plants expressing a bacterial detoxifying enzyme are resistant to 2,4-D Bio/Technology 7 811–816

Tanksley, S D , N D Young, A H Paterson & M W Bonierbale, 1989 RFLP mapping in plant breeding new tools for an old science Bio/Technology 7 257–264

Thomas, H , W Powell & T Aung, 1980 Interfering with regular meiotic behaviour in *Avena sativa* as a method of incorporating the gene for mildew resistance from *A barbata* Euphytica 29 635–640

Van der Plank, J E , 1963 Plant disease Epidemics and control Academic Press, New York

Vellve, R , 1992 Saving the seed Genetic diversity and European agriculture Earthscan Publications Ltd , London

Wernsman, E A , 1992 Varied roles for the haploid sporophyte in plant improvement In H T Stalker & J P Murphy (Eds) Plant breeding in the 1990's CAB International, Wallingford, pp 461–484

Developments in Plant Breeding

1. H. Schmidt and M. Kellerhals (eds.): *Progress in Temperate Fruit Breeding*. 1994 ISBN 0-7923-2947-3
2. O.A. Rognli, E. Solberg and I. Schjelderup (eds.): *Breeding Fodder Crops for Marginal Conditions*. Proceedings of the 18th Eucarpia Fodder Crops Section Meeting, Loen, Norway (August 1993). 1994 ISBN 0-7923-2948-1
3. A.C. Cassells and P.W. Jones (eds.): *The Methodology of Plant Genetic Manipulation: Criteria for Decision Making*. Proceedings of the Eucarpia Plant Genetic Manipulation Section Meeting held at Cork, Ireland from September 11 to September 14, 1994. 1995 ISBN 0-7923-3687-9